Nanobrain

Nanobrain

The Making of an Artificial Brain from a Time Crystal

Anirban Bandyopadhyay

CRC Press
Taylor & Francis Group
Boca Raton   London   New York

CRC Press is an imprint of the
Taylor & Francis Group, an **informa** business

CRC Press
Taylor & Francis Group
6000 Broken Sound Parkway NW, Suite 300
Boca Raton, FL 33487-2742

© 2020 by Taylor & Francis Group, LLC
CRC Press is an imprint of Taylor & Francis Group, an Informa business

No claim to original U.S. Government works

Printed on acid-free paper

International Standard Book Number-13: 978-1-439-87549-0 (Paperback)
978-1-138-30292-1 (Hardback)

Library of Congress Cataloging-in-Publication Data

Names: Bandyopadhyay, Anirban, author.
Title: Nanobrain : the making of an artificial brain from a time crystal /
  by Anirban Bandyopadhyay.
Description: Boca Raton : CRC Press, [2020] | Includes bibliographical
  references and index.
Identifiers: LCCN 2019050401 (print) | LCCN 2019050402 (ebook) | ISBN
  9781138302921 (hardback) | ISBN 9781439875490 (paperback) | ISBN
  9780429107771 (ebook)
Subjects: LCSH: Artificial intelligence. | Nanotechnology. |
  Geometrodynamics.
Classification: LCC Q335 .B338 2020 (print) | LCC Q335 (ebook) | DDC
  006.3--dc23
LC record available at https://lccn.loc.gov/2019050401
LC ebook record available at https://lccn.loc.gov/2019050402

Visit the Taylor & Francis Web site at
http://www.taylorandfrancis.com

and the CRC Press Web site at
http://www.crcpress.com

Dedicated to my parents
Ajay Kr Bandyopadhyay (father)
Chhanda Bandyopadhyay (mother)

Contents

Preface

In 2010, an editor at Taylor & Francis asked me to write a book about the human brain. A few days before my father died of a brain hemorrhage at around 1:30 a.m., he came. Standing beside my working table, he said, "Tukun, you must write the book sincerely, giving your knowledge about the brain—your prime love; it is the only route to surviving after your death." I told him, it would not be a claim exceeding delivery, it would be a classic monograph on artificial brain. Today, after nine years, the mammoth job is finished; tons of manuscript drafts were simply discarded because I was not confident that I understood the human brain. Today, I believe I do, so *Nanobrain: The Making of an Artificial Brain from a Time Crystal* is finally going to see the light of day. This book is dedicated to my father, my friend, guide and teacher, he instructed me with all energy; and to my mother, who sacrificed often to make us smile.

The Western world believes the universe is dead; the Vedic philosophers believe that universe is alive. These two cultures dominate scientific practices. Vedic inquiry explores universe as life inside life inside life, making an infinite chain, called *Virata Purusha*. Tis book unfolds the ancient tribal concept of information processing (*Kalachakra*, the cycle of time) where instead of "bits" made of truth and falsehood, the concept is an infinite 11-dimensional architecture of clocks. Time crystal and universal chain of rhythms are discussed in the Upanishad as an unfolding of consciousness. Yet, since no one knows how to create a brain, we likely require a new model with accompanying new mechanics and new mathematics. Darwinism and Turingism are both built upon a rejection of choice, but what if no choice is ever rejected? Both schools of thought believe that events in the universe can be described linearly, but what if the universe builds events as geometric shape with dimensions? This book explores the universe of singularity, in which one makes a journey only through singularity, crafts a geometry of singularity, and builds a geometric algebra of dodecanions. The garden of meander flower was a lost concept in the research of time crystal in the 1990s: we rediscover it here. I hope that a new era of nonsense, illogic, and non-argument decision-making takes birth soon. Earlier, for a century, in Western thought, facts were paramount; now we prefer a geometry of confusion and split confusion, with the exploration ending in facts. I hope, though, that an era of nonsense, illogic, and non-argument decision-making takes birth soon. The marriage of geometry and primes helps to create that era with a new theory of everything.

This book avoids black box–based scientific practice and statistical biology as they fit output with functions that has no physical significance, extensively used in the study of the neural network and in several fields of artificial intelligence. The language of primes is conceived here as the language of the universe. Therefore, this book has cited rich breakthroughs, where biological observations are fitted with elementary physical laws. In addition, the majority of quantum biology works like avian bird magnetic entanglement; a quantum route to auditory signals, quantum coherence in photosynthesis, and force quantum mechanics. *Nanobrain* is based upon the belief that the debate on quantum versus classical physics is irrelevant and a waste of time. The problem, then, is to find how a system runs by itself, using a new system of mechanics utilizing essential experimental evidence, not previously applied to the field of artificial intelligence. This book documents major experimental results aligned with the infinite journey to a singularity network. In that new network, algebraic equations do not work. Long back J. A. Wheeler understood the necessity of geometrodynamics and conceived it in 1920s. Richard Feynman explored it in 1962. By conceiving geometric shapes, complex physical phenomena could be interpreted without reliance upon equations. The followers of that geometrodynamic conception have been honored in this book, starting with Wheeler.

Vedic musical rhythms included the pattern of primes and fused primes and rhythms to the concept of "the breath of Brahma." Even when a person sings Indian classical music, he or she explores geometric language, which is now used to advance Arthur Winfree's idea of the "time crystal." Fractality has been in almost every ancient tribal culture, including African, Mayan, Indian, Persian, and Arabic cultures. This book adds music to those tribal fractals, namely "Brahman," the key vibration that was believed to organize the universe. This book simplifies philosophy, so as to challenge it, while describing hardcore experimental inventions. *Nanobrain* documents both experimental and theoretical research about a time crystal approach to artificial human brains, mostly reevaluated in the last 14 years (2005–2019) at the National Institute for Materials Science (NIMS), in Ibaraki, Japan. Associated research generated across the globe is documented. This book is written for students; only those concepts are used which could be tested using theoretical simulation and/or hardcore experimentation. This is one of the primary documentation of brain research where mysticism of quantum and deep learning have not been used as a tool to confuse true creativity. *Nanobrain* rejects the absolutist view that neuron firing is the primary catalyst for intelligence. It may be possible that major big brain-building projects will fail by 2024, as have molecular electronics, carbon nanotube technology, cold fusion, quantum computers, and high-temperature superconductivity: all find some unavoidable blocks. The era of magnetic vortex-like particles in an absolute insulating matrix is about to begin, where electronics or the engineering of moving electrons would not be required, with instead artificial atoms made of fields reigning This book plants the idea.

We prepare the reader for the journey to when a machine will learn by itself, make decisions by itself, and then operate by itself. This class of decision-making machines could be named "invincible rhythm generators"—in Sanskrit, *Ajeya Chhandam*, nested frequency fractal or a fractal of time crystal (a "fractom"). The short version is written as *AjoChhand*, which also means "Brahman (unit of consciousness)." Eternal vibration that has been propagating through the universe since the Golden Womb (*HiranyaGarva*) exploded, as elaborated in Vedic literature. Hypercomputing models, and an alternative proposal to the Turing machine, namely fractal machines, are discussed in detail in this book.

While visiting the meditation chamber of J. C. Bose, in Kolkata long ago, I could feel how Vedanta's "communication with rhythm" inspired Bose: more of these communications are evident in the machines he designed, and we wish to trigger industrial and scientific revolution to extend Bose's philosophy. The mathematics of clocks over multiple imaginary universes described in this book would add to the proponents of quantum mechanics. Schrödinger's version of quantum mechanics could mathematically be incorporated into only one imaginary world. Here, a generic chain of imaginary worlds is limited to 12, because of complexity, but this would inspire others to explore further—a direction is set. Through a blend of biology, chemistry, computer science, mathematics, and physics, a science has been established purely to learn and understand nature as it is. *Nanobrain* is a humble submission to the legacy of the father of biophysics, J. C. Bose, and his notion of "biorhythmic communication," as universal time crystal, advances Winfree's conceptualization.

Many scientists have directly contributed to this book through essential research: Professors Tanusree Dutta, Daisuke Fujita, S. Daya Krishnanda, Ranjit Pati, Ferdinand Peper, Chi-Sang Poon, and Kanad Ray; postdoctoral researchers Dr. Subrata Ghosh, Dr. Jhimli Manna, Dr. Pathik Sahoo, and Dr. Satyajit Sahu, as well as students Lokesh Agrawal, Krishna Aswani, Greg Beams, Rishi Bhartiya, Djamel Eddine ChafaiNeha Chauhan, Rutuja Chhajed, Pelgrim Chharaud, Arnaud Degreeze, Mrinal Dutta, Batu Ghosh, Indrani Ghosh, Ami Ishiguro, Karthik KV, Cameron Keys, Suryakant Kumar, Aninda Pattanayak, Piyush Pranjal, Ashutosh Rana, Shruthi Reddy, Surabhi Singh, Dheeraj Sonker, Samanyu Tiwari, and Yu Yan; and doctoral students Komal Saxena and Pushpendra Singh (both of whose PhD theses are part of this book). I also acknowledge extensive discussions with Betsy Bigbee, Noam Chomsky, Deepak Chopra, Leon O. Chua, Stuart Hameroff, Sekhar Kapur, Roger Penrose, Jiri Pokorny, Dave Sonntag, Martin Timms and Jack Tuszynski. *Nanobrain* was extensively reviewed by Dr. Conrad Schneiker prior to his recent death from pancreatic cancer. Most art in this book, including the cover art, were created by Anirban Bandyopadhyay and Bhupender Singh from the International Institute of Invincible Rhythm (IIoIR, iioir.org).

Author

Anirban Bandyopadhyay is Senior Scientist in the Advanced Key Technologies Division at the National Institute for Materials Science (NIMS) in Ibaraki, Japan. He earned his PhD in supramolecular electronics, physics, and materials science in 2005. He was subsequently a research fellow at NIMS' International Center for Young Scientists as well as a permanent faculty member at NIMS. Dr. Bandyopadhyay was a visiting scientist at the Massachusetts Institute of Technology in 2013–2014 and has been recognized with multiple awards, e.g. Hitachi Science & Technology Award 2011, Inamori Foundation Award 2011, etc. In 2016, Dr. Bandyopadhyay established the International Institute of Invincible Rhythms (iioir.org) in Shimla, India.

Summary of the Chapters

Chapter 1 For the last century, three concepts dominated the information science and technology. First, whatever be the complexity of events, all could be melded and then rewritten linearly, executed in a sequence. What if the events in the universe are instead geometric, with dimension? One finds a topology of events What if facts are biased and one has to find geometry of confusion to understand the true reality of nature? Second, why must electrons flow, collide, and reveal properties for making useful devices? What if we instead consider the ripples of field waves and build an architecture of clocks, or time, to express everything that happens? Third, the universe and the human brain each have their own language, and if we do not know the language, we can never understand how either works, so could never replicate them. The language of primes is fundamental, because it assumes nothing, but rather fuses events to predict the future. These three ideas are introduced in Chapter 1.

Chapter 2 The universe has so many different forms of signals but we have no idea how they are connected to each other. Imagine that all signals are connected, and when we try to regard them in isolation, we misinterpret and misunderstand them. Can we build a protocol to map intricately connected signals of varying wide ranges of time scales in a singular architecture of time? Here we universalize time crystal and transform the abandoned concept of garden of meander flowers, which contributed to the quest of finding natural intelligence in living systems for 30 years (from 1970 to 2000). Once scientists used time crystal to explain how a virus-like life form runs by itself. Since a time crystal could operate in different forms that looks like meander flowers, the concept of a garden of meander flowers was used to explain how one simple system of time crystals could have so many expressions. Here, a journey is made from a single clock-based time crystal to a large number of clock-based systems, spanning over an 11-dimension universe. We redefine dimension as two parallel yet distinct ideas: one adding a new dynamic on a new axis and the other adding a new imaginary world into a host world.

Chapter 3 Every single interaction follows a language. A language is built to operate a computer, including the artificial language of algorithms. Does the universe have its own language? If one takes only the first 15 primes, 99.99% of all integers could be created. Why not use primes like letters of a language and build astrophysics as a comprehensive metric of all possible choices? One would input random events as choices, and that metric would link them, find the pattern, and thus understand how events unfold. Since primes are original to this universe and do not repeat, in an infinite series, events occur without any repetition of symmetries. A pattern of input events could change its architecture following the metric forever, continuing to predict the future. In Chapter 3, several classes of these metrics are discussed.

Chapter 4 This chapter has two parts—a discussion of fractal mechanics and the presentation of a new kind of geometric algebra. As explained in Chapters 2 and 3, one could write all laws and formulations of quantum mechanics using geometric shapes. While explaining basic quantum mechanics using universal time crystal, one would find—instead of one imaginary world—that there are 12, so one would see something more unique than conventional quantum mechanical events. This is fractal mechanics, from which classical and quantum mechanics derive. Then comes geometric algebra, whereby all basic and complex mathematics are done by taking a pen and drawing circles on a piece of paper to learn what is the projection from infinity that paves non-computable decision-making. However, creating manifold geometric shapes as conscious thoughts, when the universe is comprised of 12 imaginary worlds, is explored in this chapter using dodecanion algebra.

Chapter 5 The journey from a sequential worldview to a metric of primes worldview means changing our thoughts about the decisions human brain makes about nature significantly. This chapter is dedicated to discussing how a new world of decision-making would be born if we simply convert from an assumption of linear, side-by-side assembly of events to the metric of primes worldview. How the new technology would advance is described in Chapter 5.

Chapter 6 Historic experimental evidence about proteins and their complexes, as organic analogs, led to the change in biophysics, neuroscience, and natural intelligence, which is discussed in this chapter. Starting from advanced quantum cellular automaton, advancing from a single molecular layer to the neural network it is shown that actual decision-making by a neuron happens deep inside it, in the microtubule strings. From molecular electronics to total organic synthesis, several systems were built replicating bio-systems to understand what the governing principle of a real human brain would be.

Chapter 7 Starting from the secondary structures of proteins and moving to the complete neural network model of a human brain-body system, even as many as 20 prime experiences of consciousness could be represented using time crystals. In the 1970s, only a virus was totally mapped with clocks; now, not just the biological rhythms of a human brain, but every single system deep down to the atomic scale are represented as an architecture of clocks or periodic signaling devices. Chapter 7 attempts to describe how a complete clock-based architecture (similar to connectome, or complete gene mapping of human DNA) of a human brain could be built.

Chapter 8 The simplest device that can make decisions like a human brain would be a fourth circuit element—not memristor, but a very different class, we named it here, Hinductor, its short form is H. Using three concentric coaxial springs, a new device is conceived that builds magnetic vortex-like particles as a function of memorized charge that is immobile. The device generates time crystals, communicates wirelessly, and self-assembles to emulate complex time crystal inputs. The physics of artificial atoms or molecules made of fields or vortices (knots of lights and darkness), a new Hamiltonian, and finally generic expression for the flux charge relation of H are described in Chapter 8.

Chapter 9 The synthesis of a hybrid humanoid avatar is described in the chapter. The synthesis of a brain jelly that is primarily made of a new kind of fourth-circuit element, H, and mixture of several other organically synthesized components would perform mathematics in a chemical beaker. In the same chemical beaker, the metric of primes described in Chapter 3 would unfold, as the physical structure grows into a new supramolecule, with a time crystal generated. Using such a jelly, 17 brain-morphic components were made to eventually self-assemble into a humanoid avatar.

Chapter 10 The idea to describe the way the universe defines and unfolds events as changing geometric shapes sheds lights on different paradoxes in scientific research. The philosophical perspective completes basic requirements of a scientific proposal when mathematically the metric of prime model is extrapolated to engineer powerful machines with greater consciousness than humans. We outline how multiple proposals made by various researchers on consciousness could be invalidated. If there exists a model for human-like conscious machines, then that model should lead to unconscious entities and series of machines with capabilities exceeding humanity's, mapping the higher conscious thought of all possible living systems. The incredible potential that could unfold in the future is discussed in Chapter 10.

Rudrakṣetrajñavargaḥ samudayati yato yatra viśrāntimṛcched-
yattattvaṁ yasya viśvaṁ sphuritamayamiyadyanmayaṁ viśvametat|
Svācchandyānandavṛndocchaladamṛtamayānuttaraspandatattvaṁ
caitanyaṁ śaṅkaraṁ tajjayati yadakhilaṁ dvaitabhāsādvayātma||

From the smallest thing of the universe (*anuttara*) the eternal primordial rhythm is born (*spanda*) and it fills the entire universe with that; that rhythm dances as the creator and the destroyer, it makes the self and destroys the self, it is born from what it makes.

1 Philosophical Transformation Essential to Reverse Engineer Consciousness

1.1 HOW DO WE DIFFER FROM THE EXISTING WORLDVIEW?

In our wonderful brain, we live in the past, present, and future at the same time. We are a creature of time, operate in time, and evolve with time. For more than 15 years, we have compiled the investigation of periodic vibrations or clocks deep inside a neuron, in the single proteins, to map the neuroscience of time. We have investigated the wide varieties of studies on complex nanomachines to see the dance and simultaneously listen to the music of proteins to learn how its atomic groups keep time as we live. We have compiled protein-inspired complex organic nanomachines to realize the creation of clocks that started life on this planet 4.5 billion years back. We feel that in the century-old adventure to learn "how do I exist," we missed a key aspect of our brain. Our consciousness emerges in the femtosecond (10^{-15} s) clocks of a few atoms in the proteins to the nanosecond clocks of the protein complexes to the millisecond clocks of neurons in the 100-years clock (10^{11} s) that regenerates our heart cells. Conscious experience has a time-bandwidth of 10^{26} orders,—a brain is more than a black hole or a time machine (Buonomano, 2017). All the clocks at all levels simulate the past, present, and future; all interact with operating in real time. Unless we unveil how nature assembles the clocks following a metric that has no boundary, no assumption, and no rules to build, we cannot understand the physics of time—cannot explain how materials break symmetry to keep time. So, we made a journey to demystify the mathematics and the physics of time to eventually learn how an organic reaction could synthesis "time" in the architecture of clocks. In the universe of elementary particles, the knots and loops of energy transmission paths follow the symmetry of primes (Broadhurst and Kreimer,

1995). The use of prime numbers shocks us. Does the universe write its code using primes? George Orwell said, "Who controls the past controls the future. Who controls the present controls the past." We envision a map of the human brain as a 3D architecture of clocks, driven by a metric of primes that is the most fundamental pattern of the universe with zero assumptions. We foresee a mother's womb like futuristic incubator synthesizing an organic artificial brain, namely, a nanobrain.

I insist upon the view that "all is waves."
—Schrödinger letter to John Lighton Synge (9 November 1959), as quoted by Walter Moore in *Schrödinger: Life and Thought* (1989).

Ten fundamental transformations in the existing scientific culture are outlined in Figure 1.1. We describe 10 points one by one.

The current culture's model of learning to learn science is to ask a series of questions, whose answers would be "yes" or "no" (i.e., a bit). The philosophical argument of the universe as "it from bit" (Wheeler, 1990) suggests that one could melt every single piece of information, from the smallest to the largest, into a "bit" stream. We melt matters, forces, identities, and invariants and rebuild them as a one-dimensional thread. Different questioners ask different questions about the same event, based on their own varied perceptions. It is a scientific practice to design and build machines that ask questions in a similar fashion, and then all the reviewers get the same result. Consensus on the right question is political, thus the majority paints a picture of nature (Kuhn, 1962). Perception has led to the 12 versions of quantum mechanics; the wildest dreams of the string universe have reduced 60 dimensions of the universe

Ten ways we see this world differently than the existing system science

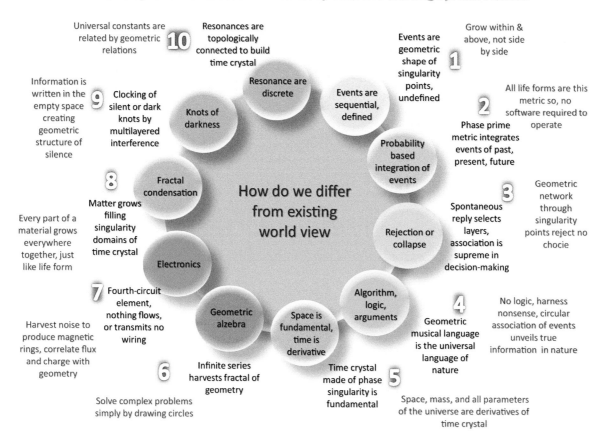

FIGURE 1.1   Ten basic concepts where computing based on fractal of primes diverts from the existing practices of scientific studies.

to just 10. Theories on how the brain works, what is consciousness, and multiverses are countless. If a "bit" or qubit brings a bias, to avoid it, the map that links the events is sensed from nature, which was otherwise ignored, and that map is converted into a geometric shape. An event's key factors are set at the corners of that geometric shape, since each key factor is linked to many events: new events grow as a new geometric shape inside those corner points. It's a tectonic shift from the sequential worldview: now events are growing within and above a corner point that is a singularity, not side-by-side bonded by a human bias called an algorithm.

The second point of Figure 1.1 suggests that if the events around us are geometric, then the sides of the shapes could be a ratio of integers and possibly represented as the nearest ratio of primes. Since the primes create the integers, when a few geometric shapes arrange to create an astronomically large number of structures of the universe, that could be viewed as an effort to create infinite series of integers from a few primes. The ratio of primes depicts the symmetry of a structure; thus, the natural pattern of primes would link the symmetries of all possible events around us. Most interestingly, life forms, which are an assembly of events at various spatial and time scales from molecules to cells, could have a common pattern of primes. A name is given here to this pattern: phase prime metric (PPM). Linking it

with natural events would have many aspects (Harris and Subbarao, 1991; Warlimont, 1993; Dickau, 1999; Richmond and Knopfmacher, 1995).

The third point of Figure 1.1 is about the proactive roles of senses. The old school suggests to reject all choices but one, to set a logic. A system could itself be a seeker, not a dumb receiver; then we do not debate about the quantum collapse or rejection: it's the one whom we want to select who is seeking *us*. Then, all the choices contribute, since the choices too have geometric shapes. They reshape: the corners of shape, made of singularity points, shift but hardly disappear. Imagine a thread passing through the singularity points: by finding the corners of a geometric shape perpetually, braiding of many such threads is how events unfold around us.

The fourth point in Figure 1.1 is a quest to find the basic language of nature, not imposing the human emotion-built logic as an algorithm to fit a few observations. Since a few primes, around 15, could generate 99.99% of all integers in the universe, if one finds a few geometric shapes intimately related to the first 15 primes:

(2,3,5,7,11,13,17,19,23,29,31,37,41,43,47)

Then a new language of geometric shapes could replicate 99.99% events happening around us in nature. That language is the geometric musical language, or GML (Agrawal, 2016b).

The musical word refers to multiple interconnected clocks; when the system points move along circular paths, sonification leads to beautiful music. Fifteen primes are the 15 letters of GML, just like the English language has 26 letters. Cellular automaton promised to recreate the universe, starting from simple patterns (Wolfram, 2002). Here, the natural selection of automaton rules by a pattern of primes would create an astronomically large catalog of intelligent decision-making.

The fifth point of Figure 1.1 is the existing belief that space is fundamental, and the time is its derivative (Girelli, 2009). If we ask ourselves, "What does the geometric universe look like?" we learn that it is a 3D geometric shape, and each of its corners holds another geometric shape inside. Since each geometric shape is encircled by a clock connecting the corners, one could forget the shape and imagine a 3D architecture of clocks, that is, a time crystal, which could either be classical (Winfree, 1977) or quantum (Shapere and Wilczek, 2012). Thorough and wide-ranging documentation is presented in this book to help us rewrite a few physical phenomena (e.g., resonance and quantum mechanics, including basic math like addition and subtraction) in terms of nested circles or clocks (Chapter 4).

The sixth point of Figure 1.1 is developing a mathematical tool to analyze the world within and above. When a quantum has one imaginary world of time,—if we, say, imagine a mathematical universe with 12 imaginary worlds, one inside another, for every physics principle—that's trivial. For 200 years, 4 and 7 imaginary worlds have been studied as quaternions and octonions (Cayley, 1845; Furey, 2018), but here the number of imaginary worlds would vary from 1 to 12, as groups of imaginary worlds interact. One imaginary world in quantum has made humanity crazy for over a century: imagine what 12 worlds would do! In that new paradigm, somewhere an imaginary world at layer 3 could interact with an element at layer 10 and affect some element at layer 1. The real world has no clue about this phenomenon. Such undetected manipulations increase manifold when we imagine 12 imaginary worlds: a dodecanion. Then there is no journey across the imaginary worlds like $3 \rightarrow 10 \rightarrow 1$, but the loops like $3 \rightarrow 10 \rightarrow 1 \rightarrow 5 \rightarrow 9 \rightarrow 3$, which we call a manifold pathway. A new type of geometric algebra is to be born.

The seventh point of Figure 1.1 is to replace electronics with a new kind of information-processing device where a part of the device is made transparent that acts like gates. Then by storing charge in different patterns at different layers, some opaque to signals, some transparent, magnetic part of the light is harvested and that magnetic light is morphed into clocking vortices like artificial atoms and molecules. Imagine that different planes in a device are storing charges in different patterns and vibrating like clocks, and those interactive clocks are being read and printed in the cluster of magnetic vortex atoms. No flow of electrons: the device acts like soft mud. Artists with an electromagnetic source could sculpt many geometric shapes by storing charge; the pattern of charges would build a network of clocking loops of the pure magnetic field as a geometric shape. Time crystals of interactive planes holding the pattern of charges and the time crystal of magnetic atoms are similar. The flux-charge device is called Hinductor, since Chua asked us not to call it a memristor (Chua, 1971) and to give it a new name.

The eighth point of Figure 1.1 is about building new materials. Typically condensation, where a lot of energy levels come together, brings materials along with it, often using self-similar reaction kinetics (Kopelman, 1988) that could be programmed (Ghosh et al., 2016b). The code is written in a seed material in terms of primes: the ratio of resonance frequencies is a set of primes. Then, in a cavity, the seed material expands like prime numbers by similar self-assembling materials to eventually build integers. In doing so, different parts of the structure act like seeds and start building more cavities, clocks, and singularities, and those singularity points would connect, and every part of the structure would become a seed. When singularity domains of time crystals are filled all over a material, every part grows and decays simultaneously; such a phenomenon is termed here as fractal condensation (Chapter 9), which is ultrafast (Sahu, 2014).

The ninth point of Figure 1.1 is about multilayer interference, where the product wave functions of one interference are used as the ingredients for the next interference. For more than 35 years (Nye, 1983), in the 3D space, an electric or magnetic part of the light was neutralized. One could store and process information in an empty space, using strings of darkness ($E = 0$ or $B = 0$), which could never be used as particles. Here we compile research where the dark strings are produced in the vicinity of a material. What was being done earlier in open space is now to be done at the light-matter interface. Plenty of opportunities open if those strings are somehow reshaped as usable vortex-like atoms, then we can use them like matter and build unprecedented engineering. Starting from the atomic scale to the ultimate architecture, there are multilayered interferences: multilayer beating and multilayer condensation of spin-like clocks.

The tenth point of Figure 1.1 is about the inherent links between different kinds of forces and associated resonances operating in widely varied materials. If the resonance frequencies differ by several orders of magnitude, there are limiting velocities of carriers restricted by a given material. There cannot be an ideal material that allows all types of carriers to flow freely at all allowed speeds (i.e., fit for all time zones). If the events happening in nature integrate within and above in different time scales, then each layer acquires multiple fundamental constants, like the velocity of light, and exclusive action (one example of action is the Plank's constant h), which gives birth to distinct imaginary worlds. Thus, a singular imaginary world in quantum is known to be found at the atomic scale, but here, since all imaginary worlds operate by the synthesis of an unprecedented carrier called magnetic vortex atoms, they follow a new mechanics (Chapter 4).

If everything is made of clocks, the past, present, and future are locations on a large time crystal architecture. A time crystal looks so different from different directions that if one reads it from 360° directions, using a probe time crystal, it appears very different. The sonification of a time crystal that represents a decision made by the brain resembles music, so maybe it also signifies life. The free will may originate from a composition of PPMs, whose symmetries are written in multinions; that is, one, three, seven, or eleven imaginary world tensors. The brain's time crystal could well be a guest of its environment's time crystal, which possibly is a tiny guest in the universe's time crystal. The inherent music of thoughts modifies the biological elements to fill in the gaps of this chain, filling the universal mathematical pattern of free will. Therefore, expanding the pattern of prime to attain 15 prime symmetries more and more intricately is genuinely living and evolving in this universe.

1.2 TEN RESEARCH FIELDS THAT WE COVER HERE

In the last century, quantum concepts have transformed three major dimensions in our worldview. At the smaller scale, the distribution of energy is not continuous; the energy parcels into discrete, isolated packets. Second, the smallest entity that makes everything is a field like a rapidly changing jelly. Third, far distant particles or energy packets could reside in a single time coordinate. Occupying time is as pure as occupying space. Using the telescope and radar, we know what our universe looks like physically. One day, we would find what the universe looks like temporally; the journey begins here with a catalog of the clocks in the brain–body network to reverse engineer consciousness (Figure 7.15). The journey to visualize the architecture of time in life forms started in the 1970s as time crystals (Winfree, 1980, 1987). The culture to map the universe as a composition of time crystals would unfold only when new kind of sensor that acquires 11D data

in the time crystal format would begin. A vision of a fractal-like universe, with nested spheres ad infinitum, was envisaged by the Swedish astronomer C. Charlier (1862–1934) and Ray Tomes (Tomes, 1990) from New Zealand, who dreamt a pattern of primes to govern all symmetries and events of the universe. Those concepts are fused here to reject and replace the old belief that a supercomputer's parallelism is like the brain's parallelism (Nowakowski, 2018). We cover 10 fundamental topics in this book to eventually replace the existing culture of 1D (dimensional data) engineering to acquire 11D information as time crystals—bit by bit as outlined in Figure 1.2. Note that 11D has been observed by some groups as a physical structure (Reimann et al., 2017) in the brain as basic geometric shapes. However, here the 11D data with geometric shapes are in the frequency domain or time domain; using physical perception, those shapes cannot be found. Thus, Reimann et al.'s proposal suggests that to store a triangle—a wire-like a triangle, in this case—vortices or spirals are the only shape of the materials; matter and vibrational information are not similar.

Here are the 10 topics outlined in Figure 1.2 and these discussions will be covered in the next eight chapters.

1. **Fractal Information Theory (FIT):** Information in the brain or the universe is not a bit, but a time crystal, which has singularity or holes to glue with other crystals: a Bloch sphere with a hole is not a Bloch sphere. The basic concept of linguistics, how humans frame an event, is embedded in the time crystal, and

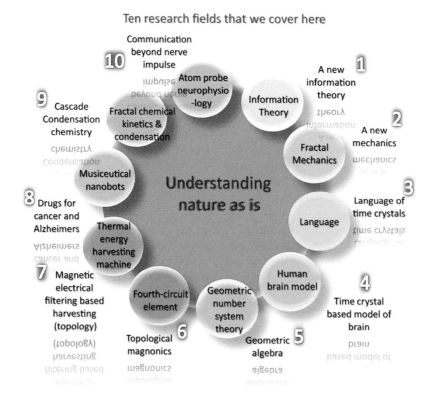

FIGURE 1.2 Ten original research fields were built for the development of an artificial brain.

that basis is held irrespective of dimensions of data or the number of imaginary worlds addressed. Who did it? At which condition? What did it do? How did it do it? (Reddy et al., 2018). When a multidimensional map of key factors that defines an event is captured from nature, it cannot be linearized even in an infinite Turing tape (Hamkins, 2000).

2. **Fractal Mechanics (FM):** Quantum mechanics deals with only one imaginary world, if one requires plenty of interactive imaginary worlds, say 11 imaginary worlds for a dodecanion, then **fractal mechanics** (FM) is a term coined to incorporate all works where the wave functions are nested in different imaginary worlds at the same time. Every single phenomenon that has a classical and quantum version would also have an FM version. Since the interaction among the imaginary worlds builds a hierarchy of groups, a distinct mechanics is required. Condensation, beating, interference, and harmonic oscillation all originate from fractal resonance, where the symmetry of frequency distribution emitted from singularity points of a time crystal evolves differently but simultaneously in various worlds, seeing one another and changing simultaneously. Quantum discussed at a small scale all are discrete and there is no connection among them; FM tells us the connection is there, but many actions guided by their PPMs create an invisible cloud, and "silence is the language of god, all else is poor translation (Rumi)."

3. **Language for operating a new kind of time crystal:** In the last half of the twentieth century, the research on time crystals and phase singularity was limited to studying reactions to external perturbation. The spiral- and vortex-shaped devices generate time crystals where the phase singularity is embedded in the device geometry, so it is fundamental, not a reaction. This is a fundamental difference and helps one to use singularity-induced bursts to develop engineering of self-assembly, information processing, noise harvesting, and so on. By blinking the singularity or holes in the phase space of the spirals, the **geometric musical language (GML)** operates, where all input information is converted into a 3D geometric shape (event = quaternion time crystal). In this language, information is an assembly of the geometric shapes, whose corners are singularity points: a system point rotates in a loop, touching the singularity points or the holes, which make a burst (number of corners = number of bings). Silences between the bings in a loop count the ratio of primes, that is, the symmetry of the geometric shape. In the pattern of divisors of integers, a PPM links all symmetries of the universe to evolve geometric shapes.

4. **Phase prime metric (PPM):** Often it is believed that a pattern of primes would reveal the most fundamental rules that govern the universe; since the primes generate integers, the gap is filled by a new

prime and hence the pattern of primes continuously evolves ad infinitum. PPM holds this self-evolving pattern as the key skeleton. PPM does not look into the pattern of primes except when the primes are arranged to build integers, which themselves follow similar rules primes follow to create integers or fractal-like features therein. This had remained unexplored in mathematics. PPM is, therefore, a collection of several such compositional rules hidden in the integers, waiting to be explored from the beginning of counting. The metric is physically realized by, tuning the topology of specially designed helices and vortices namely inductor H and thus editing its geometric phase (Lauber et al., 1994; Pistolesi and Manini, 2000; Pechal et al., 2012; Lee et al., 2011; Joshi and Xiao, 2006). The geometric phase of H is peculiar: it acquires the singularity state at certain conditions, and the singularity is the corner point of a geometric shape written in the time crystal of H. An event in nature repeats until a set of symmetry representing the event breaks; if the events are nested within and above, this is an infinite series. An event for us is a geometric shape, say a triangle, represented by 3. Using 3 alone, we can create an infinite series, multiple of 3; similarly, 12 events could generate infinite topologies when $2 \times 2 \times 3$, $2 \times 3 \times 2$, and $3 \times 2 \times 2$ arrangement would turn integers active anywhere in PPM, and teardrops of events are triggered (Figure 3.4). Thus, the reflection principle has received a major overhaul here. The principal principle (Lewis, 1980) suggests that we should choose probability such that even in an absolutely random scenario, an event really happens. If it does not, the probability is useless. Using the pattern of primes, we avoid the necessity of probability; PPM is a compiled Dutch book. The ordered factor of all possible events is compiled to make the prime metric a coherent system (Ramsey, 1926; de Finetti, 1937; Shimony, 1955; Kemeny, 1955; Lehman, 1955). Any event, when it takes a clocking geometric shape, we can then, using a PPM, extrapolate its past and future.

5. **A time crystal model of a human brain:** Painstakingly almost every single component of the brain was mapped, in the software and in the hardware, by using tiny dielectric and cavity resonators. By solving Maxwell's equations, the electromagnetic resonance of the components was measured and experimentally verified. The nested clocks were collected from the dispersed database of reported rhythms for 80 years. The first comprehensive catalog of 534 classes of brain rhythms was reverse engineered using copper wires in a 1:1 million ratio in the brain, and 20 conscious expressions were replicated for the humanoid-like response of the avatar (Chapter 9).

6. **Continued fraction geometric algebra (CFGA):** Geometric algebra accounts of the efforts are made to link integers using a geometric shape (Hestenes,

1986). The algebra for a fixed imaginary number already exists: for example, quaternion with three imaginary worlds, octonion with seven imaginary worlds, which were related to Clifford algebra, Lie algebra, and so on. Here we account for a massive transition of octonion to dodecanion, or 11 imaginary worlds. One has to invent the tensor for dodecanions, which is the most primitive tensor for the algebra of multinions, that is, a variable number of imaginary worlds. The mathematical calculation is done by **continued fraction geometric algebra (CFGA)**, which is a new type of mathematics where one draws circles and complex mathematical operations are carried out from the convergence of drawn circles. For the artificial brain, a new operator has been proposed that is 10 steps of geometric operations one after another, solving all major mathematical processes required to regenerate 15 prime symmetries, in case a time crystal acquires an asymmetry (a problem).

7. **Fourth-circuit element: Hinductor (H):** Three circuit elements are a resistor, capacitor, and inductor, and the fourth one that should link stored charge and flux is missing (Chua, 1971). It is not about a memristor, which wrongly claimed current, or voltage device, as a fourth-circuit element. Instead the Hinductor is a real circuit element that generates magnetic vortices as a function of the stored charge when monochromatic light falls on it. It is a metamaterial that cloaks quantum mechanically, which helps in making a part of the device invisible to others. Cloaking enables building circuits without wiring. Quantum cloaking is invisibility for the matter wave. H builds a universal time crystal (UTC), harvesting noise by exploring unique concentric spiral geometries. No electron transmits, only the phase transmits: it is not electronics, but magnonics.

8. **Neuroscience of filaments:** Neurons and cells are densely packed with protein-made, nanowire-like filaments. These nanowires have only one job, providing strength to the cell like a skeleton. However, now the coaxial probe experiments have shown that the filaments deep inside the neural branches check the potential difference across the junctions by sending and receiving electromagnetic signals (Ghosh et al., 2016a, 2016b; Agrawal et al., 2016a). The paradigm shift from the neuron skin–only neuroscience opens the door to resolve several more mysteries of neuroscience, as well as links the events happening in the brain in the seconds time scale to the femtoseconds oscillations of functional groups.

9. **Fractal reaction kinetics and fractal condensation** are, respectively, a new kind of multi-kinetic synthesis and a new class of condensation where multiple reaction centers activate at once. When a baby grows from the embryo, it is a multifractal architecture growth, where the system builds every part simultaneously. Using fractal reaction kinetics, one could run multiple reactions in a single beaker, and using fractal condensation, one could use the principles of cavity resonator to create a 3D pattern of fields in a closed space where the matters would arrive and self-assemble, all parts of a massive architecture at once. Since the artificial brain under exploration in this book requires self-assembly of spirals within and above in various layers (at least 12 times for a dodecanion), fractal reaction kinetics and fractal condensation are key synthetic tools.

10. **Brain jelly to the development of a humanoid avatar: The first primitive conscious machine**! A brain jelly is a material where one could write environment-acquired information as a time crystal of prime numbers and let it evolve into complex architectures following the pattern of primes. A time crystal analyzer (TCA) converts any sensory information into a 3D pattern of geometric shapes, which converts into a time crystal of primes, feeds the time crystal to the organic jelly, and reads the output pattern of magnetic vortices and optical vortices into a time crystal of primes.

    Brain jelly is not an already synthesized organic supramolecule that computes. Before every decision-making, the solution is melted, the synthesis begins from the precursor molecules in a tube with the thermal gradient, and jelly consumes thermal and chemical energy—but only when an input time crystal is pumped. Then the system starts building spirals and vortices, following, most interestingly, a Hasse diagram of primes, which is thermodynamically most favorable. The synthesis does not stop building the material analog of an input time crystal. It is the reason the brain jelly builds a typical pattern of primes as a new metric, for every new input. As the jelly grows, by shining a monochromatic laser light, the evolution of input time crystal could be read by projecting the light-produced magnetic vortices on a magnetic film. Brain jelly reads the dynamics of a big data, splits it in all possible compositions ($30 = 2 \times 3 \times 5...$), and freezes the evolution of dynamics as a supramolecular material whose electrical, mechanical, and magnetic vibration generates a time crystal in the oscillating patterns of the stored charge. We read that in the crystal made of atoms of magnetic light.

One material cannot be an ideal brain jelly, which is very efficient in generating elementary structures of all 15 primes. Some precursor to jelly is good for some primes. So, inspired by a human brain, mimicking the cortical columns, a large number of distinct precursors-based solutions are used in a single unit. Consequently, the helical gels are taken in 19 columns, each column with seven distinct layers. In other words, more than $19 \times 7 = 133$ types of helical supramolecular structures are required to replicate one basic unit of Brodmann's functional region, out of 47 in the cortex; 2D sheets of hexagonal close packing (HCP) of cortical columns are used as folded

2D sheets (Hinductor class III, H3) at plenty of places to build brain components such as ganglions, the hippocampus, and the cerebellum layers' nuclei (Figures 7.11c and 9.3c). The 2D sheet H3 is the final product required to build a self-operating humanoid avatar.

1.3 THE UNIVERSE WITHIN AND ABOVE NOT SIDE BY SIDE

Figure 1.3a describes how three events expand in the existing paradigm of quantum or classical worldview and then the fractal worldview. In the first worldview, expansion of an event means adding more subevents linearly; however, in the fractal overview it's all about expansion of events within and above. One should note that when events are geometric, the ratio of sides is represented as integers and then converted to the ratio of primes.

The evolution of the culture of doing science: An artificial brain driven by a pattern of primes would be reductionist, since it assumes that the pattern of primes that includes all possible symmetries in the universe is fundamental, from which all other theories of science could be derived. Its theory of knowledge or epistemology is that the pattern of prime numbers is an architecture mapping how all possible symmetries in the universe will be linked to one another. The pattern of primes, converted into a temple-like geometric structure, the PPM (Figure 3.2d), which is independent of the human mind and is thus bias-free (Courtland, 2018), supports the scientific culture of realism. In the beginning, the practice of science was to derive a general law or principle from the observation of statistically significant events. Then came the idea of verifying a conclusion by experiment, namely, inductive logic. Then came the falsification proposal, where science is about trying to falsify a theory rather than trying to gather evidence in its favor (Popper, 2002; Lakatos, 1970). Here, the geometric artificial brain contributes to the idea of logical positivism, where the proper use of language in constructing an argument would solve everything. Instead of human language, we would explore a new language called geometric musical language (GML) to do the same.

> God knows I am no friend of probability theory; I had hated it from the first moment when our dear friend Max Born gave it birth.

—Schrödinger to Einstein (13 June 1946)

Fifty years of Bayesian reflection principle, exchangeability, and principal principle: Futility of bias-free learning: Bayes theorem tells us how to find the probability of an event happening in a given set of choices. No one questions the formulation, but the number of variables and their interactions are subject to human imagination (Howson and Urbach, 2005). It is true that the target concept or bias has to be included within the hypothesis, else training that links input and output would fail (Mitchell, 1980; Schaffer, 1994; Wolpert, 1996). So, if we have not experienced the future, we do not have any valid reason to consider it as truth (Hume,

2000 [1739–40]). Bayes' theorem with three added precautions is believed to restrict human imagination, making scientific practice (Nola and Sankey, 2007) little better than pure falsification. The weakness of Bayesian culture is that it assigns a prior probability to a novel hypothesis, while it should be derived from no law or assumption, as argued by Wheeler (1980). A free human will governs the statistics of observation. Hence the derived probability varies with different minds. Three restrictions to constrain the role of free will—by exchangeability, the reflection principle, and the principal principle—are not enough. In the geometric artificial brain, we see that random events unfolding in nature are linked by symmetry, following a typical pattern of primes that we conceived, namely, the PPM. Thus, exchangeability (de Finetti, 1937), which suggests treating symmetric events symmetrically, is honored by the PPM. The reflection principle (van Fraassen, 1984, 1995) suggests that if the conditions remain the same, the probability will not change with time. It is a primitive condition. In the real world, the conditions hardly remain the same.

Bayesian approach has to go beyond Popper's falsification: Where H = hypothesis, D = data, and B = background information, Bayes' theorem tells us that $P(H|DB) = P(H|B) P(D|HB)/P(D|B)$. $P(D|B)$ is independent of any hypothesis, so maybe ignored when comparing hypotheses. Enumerative induction seeks to maximize $P(H|DB)$ directly, Popper's falsification generalizes to maximizing $P(D|HB)$, while the aim of abduction (also known as inference to the best explanation) is to maximize $P(H|B)P(D|HB)$ where D consists of facts.

Redefining the intelligence of artificial brain: Rejecting Bayesian protocol to do a scientific study and replacing it with the PPM and GML redefine the definition of intelligence. Until now, the most accepted definition of intelligence was the ability to assign a realistic prior probability to a novel hypothesis (Carroll, 1993). Now, PPM derives a hypothesis from reading natural events using GML. The intelligence of a PPM-GML engine is the number of prime symmetries configured in its hardware; an asymmetry is detected as a problem, and the recipe for regaining symmetry is the solution. The length of PPM is infinite since the number of primes is infinite; 15 primes do not reject all other primes, and holding 99.99% neighbors of the primes also means holding the primes in some other way.

Spirals and vortices ask about the universe: Spirals and vortices are everywhere from proteins to the galaxies. "The double helix, the four-base codon alphabet and the triplet genetic code for amino acids, any particular gene for a protein in a particular organism—all are the frozen accidents of evolutionary history" (Hogan, 2004). Carbon, a key to life, forms by a three-alpha process. Two of them are very unstable (10^{-16} sec; 8Be, 4He), but they commonly trigger a three-alpha reaction that is very slow; that resonance peak was found by Livio (2003). He observed in the experiment with the tubulin protein that, when triggered with a suitable frequency signal, the proteins assembled rapidly into the microtubule nanowire that has regulated cell life for 4 billion years (Sahu, 2014).

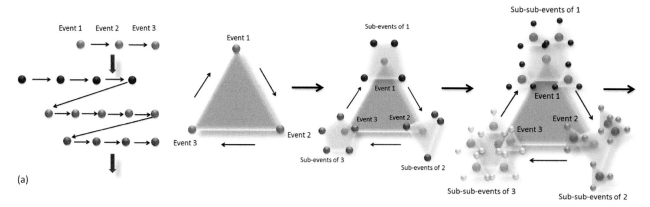

(a)

(b)

(c)

FIGURE 1.3 (a) The Turing machine–based concept rules the world currently, where dots depict events, like "eating food," "dancing," etc. The schematic on the left shows the sequential way of looking into the events. In the fractal machine proposal, where events are connected by a phase change that forms a geometric shape. Each event is grafted in a phase singularity. In three consecutive images, new geometric shapes representing a new set of events are encoded. The arrow after third triangle shows that there is no end; it's a continuous chain of events. (b) Five situations where Turing machine or sequential thoughts would collapse. (c) A singularity point in a cube holds another cube, and in three steps there are 8^{8^8} events, which is processed in the proposed nanobrain in three imaginary layers.

To explain the self-operational intelligence of viruses and primitive life forms, in the 1970s people conceived the idea of a time crystal, an assembly of clocks, where the slowest one is the host clock, holding a pair of guest clocks, running rapidly. For three decades, time crystal studies gave us the concept of meander flowers and their garden (Zykov, 1986, 1990), integrated into the time crystal for a self-operated intelligent life. This book is to bring that adventure back to life.

The research on microtubules and proteins suggest that due to a novel twist in the spiral symmetry, devices harvest noise and convert it into a pure signal (Sahu et al., 2013a, 2013b, 2014). Its fractal clock network observed in its resonance peaks resembles the primes, just like the brain (Chapter 7) and extragalactic energy bursts (Tomes, 1990).

1. Is the universe a giant machine ruled by pre-established continuum physical law? If it is running by itself, then what is driving it? Thus came the idea of a metric of symmetries governed by primes.

2. When 99.99% of the universe is empty, the debate whether space or time is fundamental is irrelevant; the prime concern is with what is the factor that brings "it" from continuum (not from "bit") and returns a continuum.

   The origin of quantum would unravel when we would sustain 11D maps of crucial factors that define an event; a revolution is needed in sensing nature, that is not possible unless we know how nature writes and integrates information. Thus, came the need for engineering singularity, a language that speaks by blinking singularities, that is, GML.

3. When an observer changes the world by observing, then did an event happen when no one saw it happening? Does a real universe exist? If everything happens by changing phase, then many nested imaginary worlds could run the universe, even when no one sees it. Imagine that looking at the moon would change it; then hardware must hold nested imaginary

worlds. Thus came the flux-charge device, a master in creating and assembling clocks in different imaginary worlds.

4. Information loss is entropy; does a metric that regulates all systems yet gains nothing, lose nothing? (Zurek, 1989).

1.3.1 What a Turing-Based Worldview Does Not Consider

Existing information theories do not consider: For over a hundred years, the foundation of information theory is based on the fundamental principle that every single event happening in nature could be explained as a sequence of the simplest event, switching between "yes" and "no." Except for a line, this principle rejects all other geometries that could connect the events, as noted in Figure 1.3b; but here, all possible connecting geometries are allowed. Moreover, each corner of a geometric shape is an independent event. Every time we enter into a corner, it is an imaginary world from outside, but the rules in that world, once we are in, are not like those of classical or quantum. The most exciting feature is that an event, in reality, could get redefined by a mere change in geometry, since the corners of multiple imaginary worlds are connected by a single thread of events. It is the second feature missing in the Turing paradigm as outlined in Figure 1.3b. There would be multiple ways to generate a closed loop between points, geometric constraints, and favoritism. The observer must be integrated as part of the information structure as outlined in point three of Figure 1.3b. All observers would impact largely on the reality we encounter. Inserting a geometry inside the corner point simultaneously adds a new set of clocks in the time crystal.

We do not have any problem with the Turing paradigm regarding the gap between the events or dots—what all we need to think is if the sequence is right. The pattern when events link within and above could have a self-similarity called an escape time fractal; that provision is missed in the Turing paradigm (point 4, Figure 1.3b). When the event is a geometric shape, the number of sides are the number of phase gaps we need to know. Here in this book, we have outlined all essential technologies required. For a cube, 8 corners hold 8 subevents, acting as 8 singularity points, but we need to know 12 phase relations: 2, 3, 4, 6, or 8 subevents could form many sub-clocks. A fractal tape would hold $8^8{}^8$ events in a mere $8 + 8 + 8$ clocks. What we need is the engineering of synthesizing the processors inside a singularity domain (Figure 1.3c).

Now, in the universe of artificial intelligence, human perception runs deep. The deep neural networks are multilayered, linked side by side and operated step-by-step. Parallelism is not simultaneity; the problem is the interaction between a large number of simultaneously interacting nodes with nature, which parallelism cannot address. However, time crystal network in GML grows within and above, differently multilayered; one could replace the multilayered neural network or deep learning with a time crystal. How information looks in nature or in the brain and how it is integrated is a prime query

and needs to be resolved first, before we can do anything else (Figure 1.4a). Unfortunately, the brain builders of today use bits as information and imagine linking them; they consider neurons as switches and the geometry of neural networks as circuits to link devices, and that's why we finally hire human programmers to run the piece of artwork developed by spending billions of USD.

The basic philosophy of the current brain-building projects is that if one copies the neuron-like structure as is, then the properties are automatically copied. This is philosophically incomplete, as the matter is discrete, isolated, and complete, but its time crystal is not; if both are not replicated, we lose information about the "happening of an event" of the replicated structure. The structure of a time crystal holds its future dynamics: it's a seed, which forms a tree when planted in the PPM. Journeying through the singularity paths unfolds a new science.

1.4 BASIC QUESTIONS TO ANSWER: TEN POPULAR HUMAN BRAIN MODELS

The lifelike feature is missing in the existing information theories: The journey to the radical voyage toward the understanding of six key questions as outlined in Figure 1.4a, and it begins with a question: What does information look like in nature and in the geometric brain, and is it in bits? We often see 10 basic features of information in the nature around us:

1. All information is connected; there is no discrete piece of information, such as a complete defined entity.
2. There are always some contents inside any given amount of information; the very idea that there is the smallest information content is an illusion.
3. The plot and the matrix of information on the plot are the same: the information that shapes information content is the information itself.
4. Identical content can give different meaning in a different environment; at the same time, very different information often holds the same appearance; duality sustains both ways.
5. The same information is complete: at the moment, at a fraction, at the other.
6. The same information content gets the different property at the same environment, at the same time, to different observers.
7. Systems spontaneously emerge information, even when no question is asked.
8. A mirror image of information is created without changing the source, though both the source and the observer are coupled.
9. The same information takes different form at different times, thus creating a wide range of information, an endless chain of events.
10. Information is lifelike, it senses, expands, creates new life like information.

Basic questions to answer

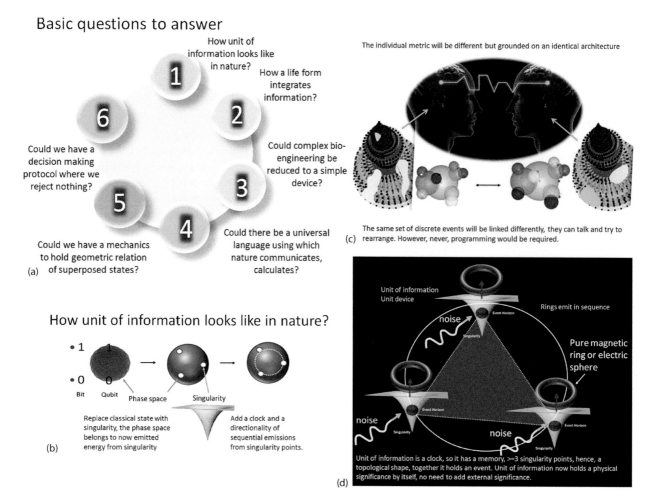

FIGURE 1.4 (a) The new kind of science requires six primary questions to ask. (b) The philosophy of representing everything in terms of bits and qubits (0 and 1, or yes and no) is changed to a composition of phase singularities where an event becomes undefined, the universe has no reality until it projects to a sensor. A new type of basic unit of information where Bloch sphere like imaginary points are there on phase sphere but no real point. (c) Communication between two brains happens here through a time crystal. Two human figures are represented using two distinct phase prime metric (PPM) architectures; PPM is a pattern of prime numbers, for each human their composition is distinct. (d) The mechanism of geometric information is written in a time crystal or architecture of clocks.

So, our journey begins by considering that the unit of information in nature is an event, even in its simplest form, it is lifelike, not isolated, connected to everything, so undefined. Information is not dead bits waiting for humans, to give it a significance. Physics at the bottom is continuous in a very different way; therein, the bit of information is not the basic entity (Hartle, 2005).

Breaking the limit of intelligence: For many, the brain is an advanced computer chip run by a chemical fuel, there is a consciousness switch in the brain, and everything happens at the neuron level. Thus, if a network is mapped in the brain accurately, making an artificial brain would be a piece of cake. Elegans never acquired life in the computer, though, neuron by neuron, circuits were intricately encoded, and we had to add an AI engine to fake its life. False memories (Loftus, 1997) and fake perceptions (Libet, 2004) are real in the

geometric artificial brain. We create a living brain vis-à-vis universe models—after all, both universe and brain evolve similarly—only by symmetry. For some, the brain oscillates around a critical point (Beggs, 2012); for some it's a cavity resonator, like a flute.

1. **Boltzmann brain:** The oldest model of a self-aware entity. It suggests that a random cluster of atoms in a thermodynamic equilibrium could spontaneously assemble into an intelligent structure like a human brain (Carroll, 2017). So, the past should be a mirror image of the future (Boddy and Carroll, 2013). The time spent by a system's microstates in some region of the phase space with the same energy is proportional to the volume of this region (ergodic), i.e., that all accessible microstates or decisions are equiprobable over a long period of time. In the

geometric artificial brain, all systems are non-ergodic, with a singularity hole in them, and Maxwell's tunnel connects them, which ensures a connection en route to singularity (Palmer, 1982; Gotze, 2008).

2. **Anthropic brain and selfish biocosm:** An event does not happen if no one sees it (Copenhagen interpretation of quantum mechanics). By observing a phenomenon, a given observer changes the universe towards a conscious life (participatory anthropic principle; Wheeler, 1996): the universe is biocentric. Continuous observation changes the fundamental constants so as to let the conscious life flourish (selfish biocosm; Gardner, 2005). One could make a journey backward in time to find the ultimate cause (retrodiction), by believing in eternal inflation via 10 dimensions (Linde/Weinberg bound; Weinberg, 1999) or by adding the probabilities of all paths from the beginning until now, just like path integral or "sum over histories" (Hawking and Hertog, 2002). Here we account for a new route for a life-friendly cosmos. PPM is simply a set of curves, but when GML adds to it, the geometric constraints embedded in the metric govern the formation of fundamental constants. Consciousness to the PPM-driven artificial brain is a condition that allows coexistence of at least three editable, complete information structures of an entity. PPM allows non-computable geometric feedback from the infinite series output; its universe is consciousness friendly. Higher-level geometry hidden in the primes is consciousness-centric.

3. **Holonomic brain model** (Karl Pribram) and **black holes as baby universes:** Cosmological replication and natural selection (CNS). If the quantum is true, every time a singularity is born in the universe, they are connected. Thus, certain parameters of the baby universes are close to the parent. Fundamental constants tune so that many black holes or baby universes are born, like Darwin's natural selection (Smolin, 2004). The brain inside a brain…, an infinite network, is the outcome of a cosmic code (Pagels, 1983).

4. **Turing's "it from bit" brain:** Asking questions whose answer is yes or no, 0 or 1, it is possible to create a conscious brain and an entire universe totally as an infinite one-dimensional chain. It is possible to deduce truth by creating imaginary layers of hidden non-physical structures that reduce the choice bit by bit (deep learning). The bit is about two states; time crystal architecture is about geometry, clock speed, silence or phase, rotational direction, noise tolerance, channel pattern, imaginary layer interaction, relative amplitude, ratio of frequencies, and more— the unit of information is like a life form in itself. More interestingly, architecture of time follows a pattern of symmetries; this is a non-physical structure since it is not confined in one layer. PPM is an ideal temple, as nothing else in the universe is; life

is a pattern of symmetries which senses, in the real universe, how the ideal PPM is shifted in the real universe. All PPMs, within and beyond life, autocorrect lack of symmetries generated by conflicts and thus run the universe. It is in tune with the proposal that at every moment the universe or brain tries to attain maximum computability (Gardner, 2002).

5. **Bayesian brain model:** Conditional probability of multiple events is key to decision-making and eventually consciousness (Doya, 2017; David, 2004). An extreme example of using human bias in creating information, and thus all models in this category survive through political consensus, as told by Wheeler. **Connectome-computer brain model:** Mapping and uploading an accurate map of neurons, there are several proponents of this idea. **Integrated information theory** (IIT) belongs to this category (Tononi, 2016), which again considers probability or human observation.

6. **Free energy brain model:** This proposal is now gaining momentum since higher-dimensional analysis has been gaining attention: the energy landscape shows that the brain is not a 3D structure but a 4D structure by itself (Friston, 2010). The decision comes from the 4D analysis of information dynamics.

7. **A matryoshka brain** is a hypothetical megastructure proposed by Robert Bradbury, based on the Dyson sphere, of immense computational capacity. A similar brain model is the **fractal brain model.** Multiple authors have noted that the brain is a fractal (Di Leva, 2016). Here the difference is that the brain is a fractal in the composition of primes while creating integers, not in its physical look: the group of primes is hidden in the infinite, uninteresting piles of integers that remained overlooked for centuries; no mathematician ever wondered whether $2 \times 3 \times 3 \times 3 \times 5 \times 7 \times 131$ and $2 \times 3 \times 3 \times 3 \times 5 \times 7 \times 97$ are similar, many of these compositions of primes occur infinite times when we calculate divisors of an integer, but when Reddy et al. plotted them, patterns emerged (Chapter 3). The self-similarity lies in the mathematical process, not in its appearance.

8. **Electromagnetic resonance brain:** The synchronization of spikes via resonance that carries out all the logical processes—but logic comes from human observation (Izhikevich et al., 2003) **Harmonic and dipole brain model**: flattening the brain components for harmonics-based analysis is another direction (Hurdal and Stephenson, 2009); multiple reports note, for example, that the lateral ventricles can be described in neurodevelopment by spherical harmonics. In resonance-driven development of the folding in of the cortical columns (Striegel and Hurdal, 2009), the question remained unanswered: Who resonates? Here we compiled 12 classes of

rhythms in the brain governed by 12 types of carriers that resonate in 12 temporal bands, each band configured in a triplet-of-triplet group of resonance frequencies (Figures 7.6 and 7.7).

9. **Relativistic brain model:** Dipolar oscillation, or C2 symmetry, governs resonance relativistically at the top level in the left and right brain to the bottom in the alpha-helices or DNA molecules. The relativistic effect leads to consciousness. In a highly unexpected breakthrough in this work, Nicolelis and Cicurel argued how their research replicates the central set of hypotheses in dipole neurology theory (Lanzalaco and Zia, 2009). C2 symmetry alone covers 50% of the decisions made; here we explore how that could be stretched to 99.99% (Nicolelis and Cicurel, 2015).

10. **Schrödinger brain model or quantum brain model:** The neurophysiological response of the human brain (Nobili, 1985) requires arranging the receptors, for which we need a "reference wave"-recruiting device from where information is recovered holonomic manner. The **holonomic brain model** and Schrödinger brain model require all information packed in at a local point, wherein infinite choices collapse, models vary in configuration, but never leave, in the Turing paradigm. **Quantum brain model**: Multiple proposals were made in the 1990s for quantum brain models (Beck and Eccles, 1992; Hameroff and Penrose, 1996), extending the proposal for long-range quantum coherence (Fröhlich, 1968a, 1968b) via microtubules. Penrose argued for non-Turing decision-making, in the line of Turing's morphogenesis, Neumann's non–von Neumann class computing. Gravitational collapse of many entangled decisions is the foundation of the orchestrated objective reduction (Orch OR) theory.

Evolution of the brain: All the functional parts of a dinosaur's brain did not evolve homogeneously to humans. The human lower brain has remained nearly the same as that of a dinosaur, while the frontal lobe has increased at an abnormally high rate. The discovery of cooking food during evolution suddenly increased the human brain activity, as they could devote time in creating art, science, and culture, unlike other species. It is also debated if all living species are conscious. Here, we would explore an engineering way to find an answer. For sure, the advent of enormous complexity in the frontal lobe is the result of human's unlimited quest to find the smallest, making the smallest, reach the farthest, and build the finest; however, that does not give a human a license to be the sole owner of consciousness. Breaking the limit, playing with the limits, and redefining the limits are our daily job, putting tremendous pressure on our frontal lobe: so much so, that in the next few million years, human brains may look something like an ellipsoid. The human brain would grow vertically upward, because eventually, it would have to come out of the mother's womb and expand. Moreover, one cannot afford to increase the brain and the skull further, since symmetry in the skeleton is essential to withstand the enormous pressure of gravity and at the same time keep the balance of the body for moving forward. So the human brain and body are evolving rapidly, even right now, and symmetry is playing a vital role (Bartol et al., 2015). At the same time, with the increasing brain size, axon wires in the neuron's core increase in length and/or width. The longer or bigger wires have less noise in sending signals, but require a much higher energy consumption; therefore, probably the human species is the result of an optimization experiment of nature between an elephant and a honeybee brain. Thus, often it is argued as if a "limit of intelligence" has been reached (Russell, 1948). If evolution is oscillatory, then the next evolution of humanity is a zombie, if there is no quantum jump (Figure 10.12).

1.4.1 What Does the Information Look Like in Nature

The geometry of silence: When one reads this line, no algorithm is instructing one's brain. As we read, proteins in the neurons are clocking at picoseconds, nanoseconds, and microsecond periods so that the millisecond-scale membrane firing is executed properly. Instantaneous perception is a loop-like organization of events, arguing against the universal arrow of time (Zeh, 1989). Pure vibrations in the brain's materials are doing the math, simulating the future, and making the decisions: everything that a computer can do, even much more, but without a single line of software code or algorithm. One has to build a protocol to do everything in a non-algorithmic way. Natural vibration of materials is the only event given to us to make an artificial brain. Then, we need a way to link all the vibrations in nature and within us: an integrated information-processing at multiple times and spatial scales layered one above another, connecting atoms to the entire brain and the whole body. All vibrations are connected by a pattern, an invisible geometric shape, but we needed to know the universal rules of integrating these geometries. The pattern of primes being explored here we believed to be at the core of every single self-operating system, as at the core one PPM talks to another PPM (Figure 1.4c).

Music has been an art of silence, but scientists were looking for screams. Often, we think that geometric shapes could acquire infinite shapes, but all tribes across the planet for thousands of years have stuck to a few geometries. We selected 15: five lines, five areas, and five volumes. The demand for building a metric of information is not new (Zurek, 1989); what's new here is the use of 15 primes. The saga of thought neither begins nor ends at any level; the pattern of primes governs every single event in the universe that is happening, has happened, and will happen. The pattern of primes replaces the need for an algorithm: we link the pattern to drive materials synthesis, so that synthesis of materials is the synthesis of time crystals made of magnetic atoms, which is the synthesis of thoughts in an artificial brain.

The physical origin of time asymmetry: However, time is topologically an enclosure, a loop. If we try to write an event on a circle as a topology of distributed pixels, where each selected pixel is a subevent, then the "gap" between the subevents on a circle is important. The physical origin of time asymmetry is unknown, but the singularity-emitted magnetic vortex atoms reveal that a prime, a nonlinear asymmetry, is in the unit of information itself (Figure 1.4d). The century-old debates on the topology of time considered whether

1. the events on the perimeter of a clock are truly linked
2. the coincidence requires both spatial and temporal locking
3. acceleration or deceleration of clocks happens only due to the unbalanced forces
4. measuring a clock means comparing time
5. stating that A causes B is impossible unless we mark the event using a marker
6. space travel is possible, time travel is not, the boundary condition sets this time asymmetry (van Fraassen, 1970).

Then, it is not space-time: the phase is fundamental. Instead of an unbalanced force, the addition or subtraction of singularity points accelerates or decelerates the clocks. Measuring a clock is not comparing but sensing the symmetry. The origin of time asymmetry is not a boundary condition, but a pattern of primes that invents a large number of simple geometric rules that, if implemented, generate an infinite series of randomness (Wolfram, 2002). One immediate conclusion of the phase metric PPM is that

1. the universe is an infinite 3D network of clocks (metric of time), with no need for eternal recurrence or cyclic recurrence;
2. time asymmetry or unidirectional flow of time in the future is conformal since only a few geometric shapes repeat and always in a circle. The past, present, and future at various places in the infinite network clocking geometries act as an integrated singular geometric entity, the past-present-future. The present holds the future, not otherwise (block universe = no future; Broad, 1923). Thus, the physical origin of time asymmetry in a 3D network of arrows in various directions is guided by a pattern of primes. Depending on the direction, one clock would see future or past or present events for the other clocks. Velocity in those clocks is a function of who and where.

Geometric musical language (GML) is an extension of Wheeler's geometric universe: Grossman's adventure with geometry hidden in the tensors provided Einstein's dream on relativity a mathematical ground, and it inspired many to conceive a geometric universe. Wheeler envisioned reduced physics into a pure geometric event in an even more fundamental way than the Arnowitt, Deser and Misner Hamiltonian formulation (ADM) (Arnowitt et al., 1960a, 1960b) which proposed to reformulate the general relativity with a dynamic geometry whose curvature changes with time (geometrodynamics; Butterfield, 1999). For example, the geometric shape would change such that we would get (1) mass without mass, (2) charge without charge, and (3) field without a field. However, a paradigm shift is essential while conceiving geometry. Instead of slicing the space-time into arbitrary geometric shapes, fill everything with that slice. It is essential to consider the singularities as the corner points of the space-time fabric and build up the network of geometries. Singularities talk to each other, be it a big data or neurons in the brain; repeating events burst with noise when they cross the singularities, and those clocks hold the geometries.

If there are three singularity points, we get a triangle; thus, unlike geometrodynamics, here the geometry is not related to space or time; it is a geometry of silence. While constructing the geometric language, the quantum concept of a Bloch sphere undergoes three major changes: the sphere has no classical or phase-neutral point, it has holes and it has clocks. In this universe, classical continuity does not exist, and an explorer of the universe ignores the century-old madness to cover the empty parts with a quantized fabric, the goalpost, and vision and route get an overhaul. If there are three layers one inside another, across the imaginary worlds we get a triangle of the observer, observed, and the environment; this time crystal has no location, and even its identity is spread over. The time crystal is the unit of information in the universe; one could calculate information entropy and mechanics linking the black holes spread all over the universe, and a conversation of black holes is emulated to build the artificial brain (Chapter 10). Changing geometric shape is the discretization, for quantum and gravity become now a journey from one singularity to another (Wheeler, 1979) for GML. The boundary of a boundary is zero; the geometric algebra of 11D gaming of circles ensures nothing in between two singularities, yet they talk; it is obvious in the elementary particles (Kheyfets and Wheeler, 1986).

What is phase space, Bloch sphere, Hilbert space? A phase space is like a mask, passing through which the input geometries morph into another shape. A phase space is derived from the dynamics regulating the parameters of a device which bridges the input query to the output required by a user. Bloch sphere is an example of a phase space where on the surface of the sphere an infinite number of energy levels between two entangled states reside. Infinite possible choices between two classical observables are the Hilbert space, mostly painted on a sphere.

What is a singularity? A singularity is a gap in the phase space, where the phase structure of a typical biomaterial is undefined: the output is irrelevant to the input. In these conditions, a system resonantly vibrates, emits, or absorbs the signal of a particular frequency. There are many different types of singularities; for example, phase singularity, polarization singularity, and amplitude singularity. Singularity points are the corners of geometric shapes in the phase structure of a spiral or vortex. A system point passes through the corners one by one; if in a loop, it's a clock. The signal bursts when the system point passes through the singularity points located on the loop that defines a clock. The singularity that Feynman eschewed in his renormalization (Feynman, 1949; Cao and Schweber, 1993), the 3D phase structure, holds the Feynman paths as a minor subset in the geometric shapes. Thus, the universe in the time crystal model blinks, spontaneously creating a vertical Turing tape in the cell of a horizontal Turing tape (see Figure 1.5a).

What is a time crystal? Just like a singularity in space creates a crystal of matter, a singularity in phase builds a time crystal. A spatial crystal has three orthogonal position axes, and time crystal has three orthogonal phase axes. Time crystals are nested bubbles with a clock in their great circle. All bubbles hold geometric information; if a bubble holds a triangle, three points on its great circle are undefined singularities. A singularity point can either hold a network of bubbles inside or bind with the neighboring bubbles. Thus, singularity glues an architecture of the time.

1.4.2 Why Two Individuals Understand Each Other or the Universe

One conscious being already knows what the other conscious being would say: The quantum and classical worldviews are two extremes. In Figure 1.5b and c, a cartoon is drawn to depict why they need humans or gods to explain nature. For classical, the tape needs to be infinite, and for quantum, it's a vertical line of entanglement; if we break it, the universe would be gone. The ideal scenario to get a machine for consciousness would be talking one-to-many and many-to-one at a time (Figure 1.5d). In machine terms, consciousness is not everywhere: it emerges from a set of instructions encoded. This book is not for endorsing the idea

of panpsychism using PPM, because it requires a special feature of harnessing the PPM via a time crystal for realizing the consciousness features defined above. It is an inverse transformation of the vibration chain that links all singularities and which can synchronize at any point inside or outside the body. Consciousness is not a state of matter, but a dance of a set of symmetries linked in a 12-layer hierarchical geometry. Consciousness is also not integrated information: nature links resonant oscillations in a band; that is, singularities or black holes form many groups, but groups with a prime number of holes self-assemble to build higher-level groups.

Mathematically, a conscious machine that can verify whether a robot has passed the Turing test or not (Figure 1.6a) is like an axiomatic deductive system that can refer to itself: PPM perfectly satisfied this criterion (Smorynski, 1985). PPM is a generic term and has primarily three classes: one that links integers, one that links groups of primes, and the last one that links prime gaps. The integer class does not have one but 10 different forms; the metric therefore morphs as required, and thus it is a system that deduces axioms (Chapter 3). The advantage of PPM-driven hardware is that it can replicate events happening in nature in its own vibrational string because of its high density of resonance frequency and the length or bandwidth of the resonance chain; both are important factors.

To build a good mirror of nature, one may choose billions of light-year domains, select a string, and increase density of resonance peaks to cover 15 primes in 12 layers. One could build a similar conscious machine in the femtoseconds time domain. One of the very exciting aspects of such a hardware is that the accessible, immediate environment could get mirrored in its own vibrational chain. By inverse transformation, this is also logically possible: that the entire vibrational

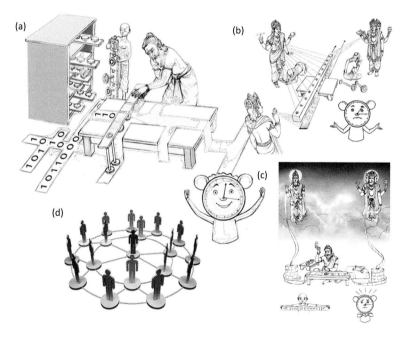

FIGURE 1.5 (a) The philosophy of new kind of tape, where sequential and simultaneous events coexist. Simultaneous events are depicted using a vertical tape, i.e., cell inside a cell. (b) The philosophy of quantum tape, where there is only one vertical tape. (c) Turing tape suggests to rewrite all events that were happening around us as a sequence of events. (d) One-to-many and many-to-one is a philosophy represented with a person in the center and others all around.

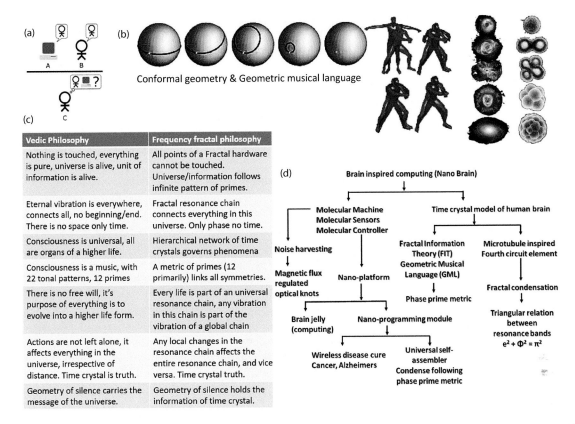

FIGURE 1.6 (a) Turing test. In one side a computer or a human hides. Another human sitting on the other side tries to identify who is interacting, the human or the computer. (b) A circle on the sphere changes its diameter but the geometric shape is retained. To the right, a human is shown in multiple conformations. To the extreme right, the evolution of astrophysical objects and the embryos have similarities; in both cases prime numbers do play a role. (c) The difference between Vedic philosophy and frequency fractal philosophy is shown in a table. (d) A diagram shows the different discoveries associated with the development of an artificial organic brain, i.e., nanobrain.

chain of our brain cum body is replicated in another point of the universal resonance chain, mirroring two simultaneously editable complete information architectures outside the body (including one totally inside) at three points. This is defined in this book as consciousness or self-awareness. Following this protocol, one can determine the degree of consciousness in an object by measuring its length of resonance chain and the density of symmetry breaking points together. To build an artificial brain, the consciousness must be outlined at the beginning, because an astronomical number of theories exist.

> The plurality that we perceive is only an appearance; it is not real. … the many-faceted crystal which, while showing hundreds of little pictures of what is, in reality, a single existent object, does not really multiply that object.
>
> **—E. Schrödinger; "The Mystic Vision" as translated in *Quantum Questions: Mystical Writings of the World's Great Physicists* (1984) edited by Ken Wilber.**

1.5 DIFFERENT KINDS OF TAPES TO RECREATE NATURE IN DIFFERENT LANGUAGES

The geometry of music is key to our thoughts and is not algorithmic: To gather knowledge about the universe, humans use a generic mathematical "inference device"; this device transforms into an observation device, a control device, a prediction device, and a memory device (Wolpert, 2017). This book investigates a new inference device, shown in Figure 1.5a, that is different from the concept of the entangled universe, shown in Figure 1.5b. As Figure 1.6b outlines, the geometric expressions are conformal; that is, be it big or small, a triangle would always be a triangle on the phase space. However, the duality of different identities emerges from the phase space only; be it a journey of an atom or the big bang, the creation of an identity is a game of primes.

Many researchers now seek an inference device that talks to many at a time, like in Figure 1.5d; researchers reject the century-old classical inference device (Figure 1.5c) that believes, irrespective of complexity, any event happening in the universe could be explained as a classical sequence of an elementary singular event. It is difficult to confine a free will in a logic (MacKay, 1960). The ripple effect of this minor change is dramatic. For example, the current neuroscience tells us that intelligence is in the brain alone, our memory is in the neuron junction, and a neuron's skin does all kinds of information processing. It is similar to the belief that our intelligence is in our skin; after all, our skin does everything. If we consider that the 3D geometry of clocks replaces a human body, then our memory is located everywhere, and a decision is made in the geometry of vibrations, not at any particular location, so searching for consciousness in any particular part of the hardware is a misadventure.

Living life with primes and geometric language: Figure 1.6c compares the new class of geometric computing. Indian raga is a geometric mathematical seed, which has specific notes when steps are taken in the direction of increasing frequency and then in the direction of decreasing frequency. When the music unfolds, this basic geometry created by ascending and descending notes is kept intact; the expansion (Vistar) is similar to PPM-based amplification via transformation of a geometric shape. In spite of incorporating another pattern, the basic geometric construct remains fixed; thus, the self-similarity which is a key to the fractal representation is maintained but appears very different at every scale. Mystical beliefs often resemble the scientific arguments: three-thousand-year-old Sumerian cultures and faiths helped the Greeks and evolved into many axioms of modern science. Similarly, four-thousand-year-old Vedic scholarly arguments that the universe to its extreme limit is alive (Viratapurusha), and there is an infinite network of life one inside another in an endless chain, are accidentally similar to the fractal tape explored here. Figure 3.12b and c is strikingly similar to the geometric language and information theory derived from proteins (see also Section 10.9).

Tribes across the globe use primes as holy seeds; similarities could be mere coincidences, but we note Vedic numbers because of the actions and significance attached to these primes. Some examples follow:

- Twenty-three tonal patterns of Indian classical music for a perpetual elementary rhythm (Brahman) govern information processing at all levels.
- The evolving self-similarity of a classical song resembles the prime-driven GML: 108 rhythms generate consciousness (12 triplets of triplet bands, Ghosh et al., 2014; $9 \times 12 = 108$).
- An avatar puts his/her mind into a singular state for 144 seconds (12 meditations, one meditation = 12 second i.e. dodecanion tensor is $12 \times 12 = 144$).
- Patanjali's 8 levels of consciousness include 7 transitions (octonion tensor, $8 \times 8 = 64$, number of qualities the universe or Viratapurusha possesses).
- Fifty-one yoga poses were proposed at 200 BC for worshipping deities.

1.6 BRAIN-INSPIRED DECISION-MAKING— THE OUTLINE OF KEY DISCOVERIES

Within and *above* **means a journey through multilayered imaginary worlds:** In quantum, one imaginary layer brings spooky action at a distance: when multiple imaginary worlds coexist, they exponentially impact the reality. Simultaneously many channel interactions among the imaginary worlds become a key factor, demanding a new algebra, a new mechanics. One of the remarkable aspects of coexisting multiple imaginary worlds separated by speed limits or several boundaries of action (analogous to Plank constant h; Figure 4.5) is decision-making procedure. The decision taken at different interactive imaginary layers could impact a non-participant imaginary layer. By the time a cascade of interactions impacts the reality, the logical mind surrenders any intuition or argument to predict. We do not always need a very high-speed communication or entanglement to demonstrate the harvesting of interactive imaginary worlds. We can artificially create a nested fractal structure in which each layer is a universe, with a maximum and a minimum speed of carriers following a specific action value like the Plank constant. For centuries, mathematicians were struggling for binion, quaternion, and octonion, now we take that journey to the dodecanion, to a point, where for the first time, 12 layers acquire (Figure 4.13) the ability to self-operate. Now, the lowest universe, holding the reality sees only accidents as a reason that an event is set at the higher world, which it cannot observe. The middle layer feels that accidents are generated at a higher level, and information disappears at the lower level, like the data source and data sink. We finish this book with a table, where we have outlined how the journey of dodecanion would bring forward the technologies of the unseen (Figures 1.6 and 10.12).

1.7 ENERGY TRANSMISSION IN THE BRAIN— IT'S NOT ALL ABOUT NEURON SKIN

Popular articles on the Internet always talk about the ultra-low power used by the brain, only 20–25 watts, but they do not split it into different contributors. We took every factor consuming energy in the brain, estimated how each component spends energy, and calculated how many of these elements are present in the system. Thus, we got the total energy spent by a given element. The wheel of energy plotted in Figure 1.7a tells us that neuron firing is not everything. This book does not cover the soliton, but if we are not biased, we found that maximum energy would be spent by solitonic transmission. The second-largest consumers are the filaments, which are ignored in the brain's information-processing theories.

Three pathbreaking discoveries in the biological system: First, the existence of a chain of resonance frequencies (Figure 1.7b and c)—be it mechanical, electrical, or magnetic—as the frequency peaks shift in a group of primes, unveils a lot (see Chapter 5 [Sahu et al., 2013a, 2013b, 2014]). Second, the chain builds rhythms at various time scales that control the biological phenomena by long-range feed-forward communication and often energy transmission 1,000 times faster than those that are observed in molecular, ionic biology. Third, several helical biomaterials exhibit quantum and classical cloaking at a resonance that enables far-distant neighbors' seeing and communicating, even if there are millions of helical cloaking devices are in between. It has been experimentally shown that the basic engineering theme for nature to construct a biological system is "keeping time," which is basically carrying out a precise movement of its components so that the perturbation created by the environment in the

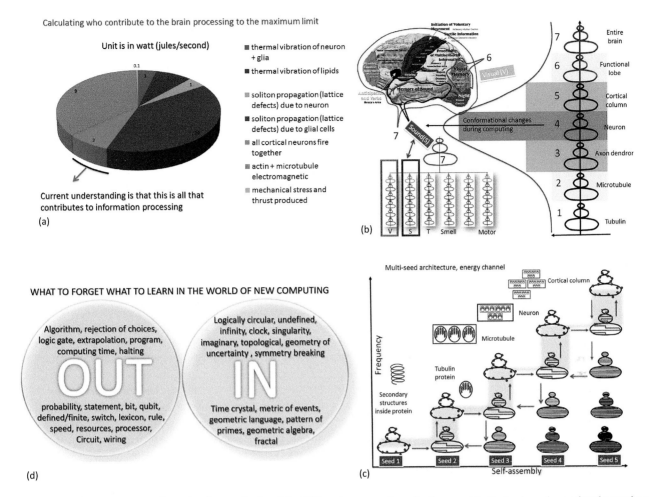

FIGURE 1.7 (a) Energy expenditure has been calculated for different sources of carrier transport in the brain and associated neural network. (b) Seven prime layers, one above another, involved in the decision-making process of the human brain. A Gaussian distribution is shown to depict the contribution of different components in decision-making. (c) Spiral- or vortex-like structure forms a Hasse-like energy transmission diagram in the human brain and its decision-making network. (d) Non–Turing based computing follows several fundamental basic features which do not exist in Turing-based computing.

rhythms inside is met outside, and harmony is established between the two worlds. Ion exchange and membrane potentials are the secondary expressions of the primary electromagnetic communications; ionic clocks are expressions, not the governing rhythms. The membrane potentials are essential for mechanical thrust to rewire the neural branches (Ghosh, 2016a; Agrawal et al., 2016c), hence no part of the existing biology needs to be discarded, but ionic- and potential-based biology is only half of the picture when all other kinds of polarizations are ignored.

Resonance spectrum of biomaterials contains a hidden geometry: Resonance peaks of a biomaterial are not random; they form groups and shift together like geometric shapes (triangle, square, etc.). Geometric shape canals the energy transfer, forming a chain of vibrations; one can see an entire assembly as a 3D collage of geometric shapes. One could extrapolate this idea to the entire universe or the human brain because we see spiral symmetry everywhere. Every single biological system is a small string or a guest time crystal and

tries to match that universal chain of vibrations or become a part of the host time crystal. PPM governs but a projection of its infinite series from infinity causes unwanted changes in the evolving multi-PPM network; the changes caused in the biological body are the evolution. If there is a topological control of mind, deep learning is also a black box, and it does not work (Nikolić, 2017); the geometric view of Wheeler advocates not a traditional triangular control of mind, but a metric of geometric shapes.

Electromagnetic resonance and quantum cloaking: We all wonder about the chemical brain rhythms: there is only one type of carrier, ions, and only a short frequency band in the millisecond time domain. Microtubules that pack the neurons and deliver strength, a unique triplet of triplet resonance bands, or three resonance bands each with three sets of resonance peaks, were observed (see Chapter 6). When a neural membrane and the proteins that make a microtubule also exhibit a triplet of triplet resonance, a scale-free fractal feature exclusive to the typical pattern of primes PPM reminds one of the various historical accidents of Efimov trimers (Adhikari and Tomio, 1982). Carina Curto observed that when a neuron fires, the spikes group in 3,5,6,7,9, and this group repeats; it's one of the infinite possible groups. The key is that spikes form groups of groups and hold geometric information (Curto, 2017). Thus, the idea that the neuron's skin is doing everything as it was tagged in 1907 might require a revisit (Brunel and Van Rossum, 2007).

If AC electromagnetic energy is pumped into these biomaterials, at certain frequencies the materials turn transparent and disappear in the presence of a matter wave: so we call these frequencies the resonance peaks. Proteins, DNA, collagen, and other biomaterials have shown similar resonance behavior, with classical and quantum invisibility in the same frequency range. One of the striking features of these resonance bands is the coexistence of positive and negative peaks, which suggests that when two such oscillators engage in the energy exchange, it happens in a cyclic loop. Thus, rhythm is born even at the scale of two molecules.

If one wants to build an exaflop computer (10^{18} bits per second) one step ahead of the contemporary best, it would require a nuclear reactor producing a few hundreds of megawatts (power); any operation that requires a resistor would generate heat. In contrast, our brain operates by spending only ~25 watts (Figure 1.7a). How could the brain operate using such a low amount of power? It is believed that the major source of brain energy is available as thermal noise in the environment, which is harvested by proteins (see Chapter 5). Therefore, the research on creating the molecular machines for harvesting the freely available thermal energy is a key component of making the artificial brain and a key to replicating the protein-like energy management (see Chapter 9).

Most brain builders ignore the details of brain architecture; a carbon copy of neuron circuits is the global motto. However, here, no structure is left out, but we do not see the structure as is: we look at the symmetry of the components in a system at all spatial scales from DNA, from proteins to the glands, hippocampus, and so on. What is the symmetry of a material? It is the arrangement lattice of the components a system is made of, not how many of them are out there. The presented brain map here is a catalog of symmetries coming together to build a higher symmetry: not a wiring of space but the wiring of undefined regions where an arrangement cannot vibrate. In a gold crystal, using symmetry, we neglect billions of atoms; when energy is pumped into the crystal, the plane of symmetry absorbs or releases the energy as a single unit, and individual atoms do not absorb distinct energy packets. The absorbed or released energy transfers the active plane of symmetry from one to another; under continuous energy pumping, the system switches between two symmetries repeatedly, and we state that the symmetry breaks continuously. A particular form of energy selects a particular plane of symmetry in the material at resonance; a rapid oscillation of the plane makes it noninteractive. Most signals pass through the plane without interacting; microtubule resonates at around 6–8 GHz, where we found it exhibits quantum cloaking. At the same frequency the DC resistance drops by 10^3 orders; obviously,

electromagnetic resonance vanishes, and the electron and the matter wave tunnels through the plane. If there are multiple resonance frequencies in a system (band)—electronic, magnetic, and mechanical resonances driven by ions; electrons, photons, and quasi-particles—each carrier chooses a plane for a particular frequency.

High-frequency signals may interact with the atomic orbitals, while very low-frequency signals may interact with the plane of symmetry of a giant organ, like the entire left and right brain's cortex regions. Switching between the planes defines the clock time; the charge is stored in the plane in a pattern to hold memory, and vortices read the vibration of the pattern as a time crystal. Light builds the time crystal of magnetic vortex atoms, an electric field–induced ion flow creates a time crystal of ionic vortex atoms, mechanical vibrations build time crystal of solitons, and so on. Now, a part of energy could vibrate the plane of symmetry physically; its mechanical resonance, its frequency of oscillations, varies widely. Electromagnetic signals travel at the speed of light; however, if the electronic charge forms a quasi-particle, then it cannot move faster than the speed of sound. Here in this book, we will use the word *resonance*: in general, it would mean 12 carriers and 12 types of rhythms (Chapter 7); however, for biological systems, the mechanical, electromagnetic, and magnetic oscillations occur simultaneously. The genetic and biological basis of human language is unknown. However, the language-processing region of a human brain is not actually a key part of language processing, which is universal and predates humans.

1.8 TERMINOLOGIES OF LIFE THAT COMPUTERS DO NOT SUPPORT

The current estimate is that a synapse stores 4.6 bits (Bartol et al., 2015). Since this estimate, the philosophy has been to combine as many processors as possible: the bigger the number, the more powerful would be the brainlike machine. No one dared to start building a computer in which encoding software would not be necessary, where power consumption is minimal or negligible, and where arguments are not discrete elements; rather, computer intelligence was seen only as a network from which discrete arguments could not be isolated.

Eventually, we had to change our thought process, the terminologies must change from the universe of logic to the universe of nonsense (Figure 1.7d). The human brain or the universe that operates by itself would be an architecture of time, no matter what they would look like physically. Atoms and crystals and all analogs of matter would be created using vortices of fields, as ingredients to synthesize an architecture of clocks that operates by itself. No black box magic, no accuracy of the data—we need to have all the limitations that a brain has; for example, it becomes very difficult for a brain to even multiply two numbers without using a reference paper, which helps the brain to create multiple reference points between the starting point and the endpoint of a calculation. An entire human life could be a single clock completing only one cycle, holding entire lifetimes of decisions within (Figure 1.8).

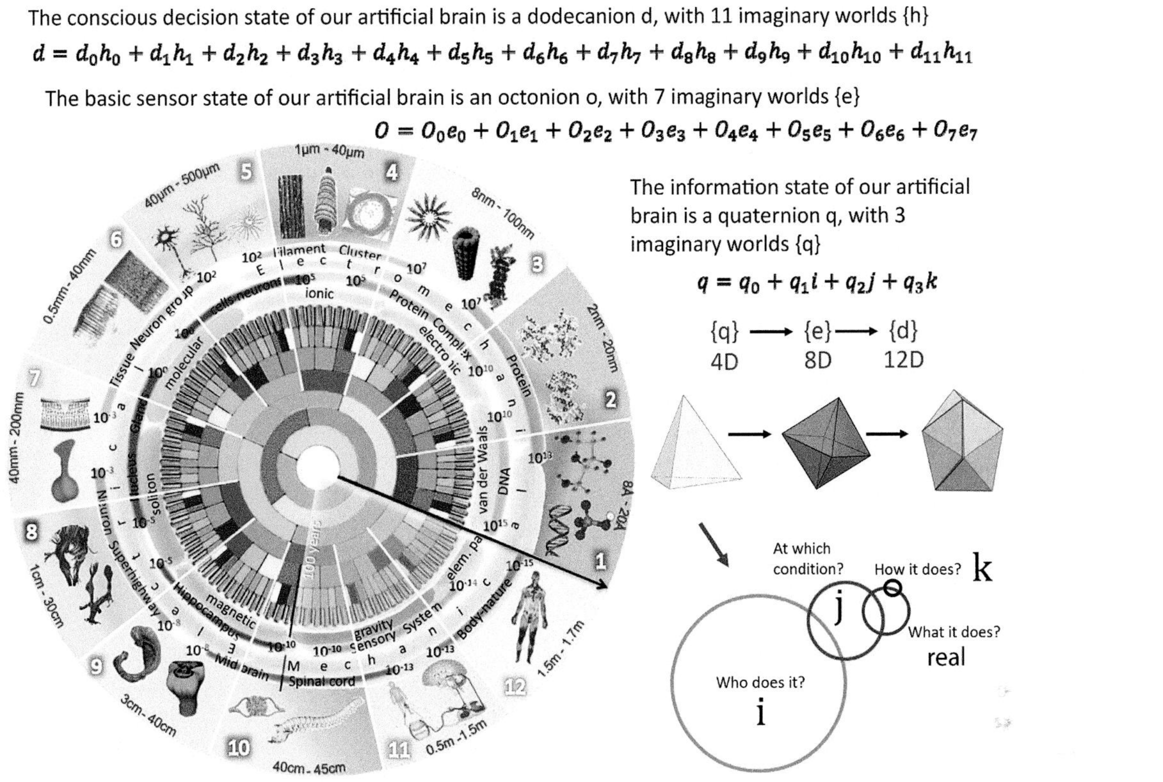

The conscious decision state of our artificial brain is a dodecanion d, with 11 imaginary worlds {h}

$$d = d_0 h_0 + d_1 h_1 + d_2 h_2 + d_3 h_3 + d_4 h_4 + d_5 h_5 + d_6 h_6 + d_7 h_7 + d_8 h_8 + d_9 h_9 + d_{10} h_{10} + d_{11} h_{11}$$

The basic sensor state of our artificial brain is an octonion o, with 7 imaginary worlds {e}

$$O = O_0 e_0 + O_1 e_1 + O_2 e_2 + O_3 e_3 + O_4 e_4 + O_5 e_5 + O_6 e_6 + O_7 e_7$$

The information state of our artificial brain is a quaternion q, with 3 imaginary worlds {q}

$$q = q_0 + q_1 i + q_2 j + q_3 k$$

$$\{q\} \longrightarrow \{e\} \longrightarrow \{d\}$$
$$4D \qquad 8D \qquad 12D$$

FIGURE 1.8 Frequency fractal model of the whole brain. It is made of 12 conscious decision-making tensors $\{d\}$. Twelve interactive forces, starting from the elementary molecular structure to the whole body, are part of a single resonance chain. We have divided the brain into 12 different bands, representing 12 conscious components, holding 12 types of memories by 12 classes of rhythms. This is 2D representative version of the time crystal model proposed in 2014 by Ghosh et al. Currently we have discarded this model and advanced to time crystal representation of the whole brain. Three tensors define an artificial brain: The elementary or basic unit of information is a 4-dimensional quaternion tensor, q; it builds an 8-dimensional octonion tensor o, which builds a 12-dimensional dodecanion tensor d. Each tensor is represented by a 3D geometric shape; for q, o, and d we used a tetrahedron for quaternion, an octahedron or Siamese dodecahedron or dodecadeltahedron or a cube for octonion and isocahedron for dodecanion, respectively; all structures are made of triangles or square, and each plane represents a dimension. The linguistic version of quaternion is shown below. This is the unit of information in the brain; only the real part of quaternion (what it does) is sensed by a sensor.

Existing computers are faster and have more memory; there is no question here of competing with a quantum or classical computer. It's a journey with a vow to not use advanced algorithms, but rather to see events much more intricately, building an engine of clocks that would emulate this unknown engine and building an information stream that the unknown engine would generate in the future. In the case of data deluge, where millions of parameters vary simultaneously, writing algorithms is next to impossible. The pattern of the most active and most inactive points (both are singularities) demonstrates how intricately key factors are related.

Perform a search without searching: Spontaneous reply back: If the clocks are cleverly arranged in a suitable geometry, we can encode an enormous pattern in a limited number of oscillators using a few clocks, which would run to generate newer patterns continuously. GML delivers a static clock structure for a rapidly changing big, complex data. Unlike classical and quantum search, when patterns are matched pixel by pixel in a set of time crystals, the right pattern spontaneously replies, because the already memorized time crystals ran continuously in a silent mode. The one that is to be found becomes

a seeker. The artificial brain compiled here wants to expand its PPM continuously and perpetually. It is a user of nature: the artificial brain is about building a user, not a computer. If a PPM-driven artificial brain finds asymmetry and by simulating the future finds how to regenerate symmetry, only then it replies; it's not a machine, nor does it compute.

1.9 LINGUISTICS AND THE WHEEL OF SPACE, TIME, AND IMAGINARY WORLDS

While formulating physical laws, one assumes that at one time only two particles are present; everything else in the universe is considered nonexistent. At the very next moment, we consider a new pair of particles, as if all the rest have vanished from the universe. Finally, abolishing the true concept of time, these laws are made to fit human perception. PPM sets a new clock seeking laws that link symmetries, via fractal mechanics, which holds 11D data in a 12D tensor (Figure 1.8). FM sets the principles of the language of all possible dynamics; one could realize 11 nested classical layers to create the 12D artificial brain as explored here. Else one could have 11 imaginary layers:

one layer represents quantum and the others beyond quantum. Thus far, simultaneity was not backed by science: one-at-a-time events do happen, but what is that universal factor which can evolve two coupled or entangled entities together? The answer is that both the entities are PPM. When one writes two sets of primes, both evolve together, even if they are universes apart or entangled atoms in a simple molecule. Similarly, no communication is required, as the researchers have argued, because both of them know already what the other knows.

The problem with the current paradigm of classical and quantum mechanics is that the significance of a phenomenon is the joint product of all the evidence that is available only from those who communicate (Follesdal, 1975): no communication means no significance in the law. Even now, we add more and more functions to compensate the loss of an astronomically large number of simultaneous contributors; in science, we never built a methodology to estimate those who do not communicate at all. We named the compensation as a many-body theorem (Thouless, 1972), which is a never-ending optimization to reach nature's true reality. However, what is lost in the simultaneity once, could never be recovered. Brilliant fusions (e.g., multipartite entanglement) did not provide the true picture of nature because, again, two extreme views of nature, classical and quantum, need a generic mechanics that would consider many imaginary worlds, and while the classical neglects it completely, the quantum considers only one imaginary world to paint complete truth of nature. Both are incomplete, since what that is silent now is communicating in the geometric phase; 11 imaginary worlds interact with one another in the intricately connected phase space, all at a time—it's spiral of a spiral…. The phase path is always active.

In 2010, a *Nature Physics* article examined how Feynman's vision, proposed in 1962 to replace the basic physics laws with the change in certain geometric shapes, was introduced (Bandyopadhyay, 2010c). That article also tried to bring Wolfram-Conway's game of life into life. However, the language of equations cannot represent nature and the language of cellular automatons could not explain nature. We are exploring a new language of the time crystal, GML, that it could be a 3D topology of clocks representing the same theories in science in significantly simpler formulations (see Chapter 4 for a few simple examples on quantum mechanics and mathematics). For streamlining the events, the equations containing an infinite series often result in apparently nonphysical unrealistic significance; e.g., see Ramanujan's $-1/12$ derivation as the sum of n natural numbers, $1+2+3+4+….+n=-1/12$. The geometric algebra introduced here actually feeds $-1/12$ back to the input decisions: not once, but at least 10 times cases have been found where physical events happening around us are being perturbed by the eventuality of that infinite series. It is an important factor which suggests that even logically the language of defined states in the Turing tape could never reveal nature's true picture. Projection from infinity is the reality, though; instead of a two-body problem, now it's a prime-body problem. The universe is being seen in terms of the 15-prime system, which covers 99.99% of it; this is primitive yet the first step to finding the language of nature.

1.10 THREE CONCEPTS DEFINE ARTIFICIAL BRAIN

Figure 1.9 outlines the fundamental core of the artificial brain: synthesizer of magnetic vortex atoms, a linker of symmetries, and a language of the universe. These three columns might lead to three cultures: one that replaces current voltage culture with magnetic flux–charge culture; the second, time crystal engineering that harvests noise; and in the third, time is fundamental.

Did some mathematicians, through their love of primes, create the human brain? Magical integers play between short-term memory (STM) and long-term memory (LTM), and these integers are primes (Cowan, 2001). Our sensory systems are paired: 2 eyes exhibit prime 2. We have 3 major parts of the cortex, and one could notice 5 distinct domains in the connectome branches. There are 7 layers in the cortical column, and 7 retina columns harvest light. All animals on the planet have 13 protofilaments in their cellular microtubule. There are 17 regions in the hippocampus, where 37 rings filter the brain's input–output signal. Often in textbooks we find 23 types of branching in the neurons, 23 types of glial cells, and 13 types of oligodendrocytes. There are 31 spinal nerves; with 12 cranial nerves we have a 43-peripheral nervous system (PNS) to sense the world around us. A series of primes, 1,7,19,37…, around 200,000 hexagonal close packings of cortical columns make our decisions in 47 Brodmann's functional regions. Singh P. (2018) made 100,000 organic cortical columns made of 7 organic gels in each column for a humanoid bot. The use of primes is endless. Nature wanted to implement PPM in the brain–body decision-making network, so that everyone needs to communicate very little yet create a similar worldview. That's why, we noted earlier, be it a human or a single molecule, we already know whatever that could be said to us, no one could reveal anything new to us. In any part of the universe or brain, communications are coded to locate and operate PPM.

The problem with the molecular neuroscience is that it does not want to look beyond its faith that everything under the neuron skin must be silent. Now, in the GML paradigm, when a time crystal needs to be transmitted from one junction to another, one particular frequency channel in the resonance band could be allocated for a particular neurotransmitter molecule (Ghosh, 2016a). Inside the axonal and dendritic branches, the electromagnetic signal propagates 250 microseconds before a neuron fires (Agrawal et al., 2016c). The mechanical, electrical, and magnetic resonance frequencies of the filamentary bundles inside a neuron follow a vortex identity $(e^2 + \theta^2 = \pi^2)$. Geometric ratios of neural branches do not explain why and how the neural wiring works. The 3D folding of shapes is the topological event of identities set by 12 types of resonances found in the biological systems (Chapter 7), and one could regulate the formation of biological structures only to build the respective time crystals using a particular set of symmetries (Hughes, 2018).

Prime integer 7 regulates a human brain's capacity to process information (Miller, 1956). At the core of the

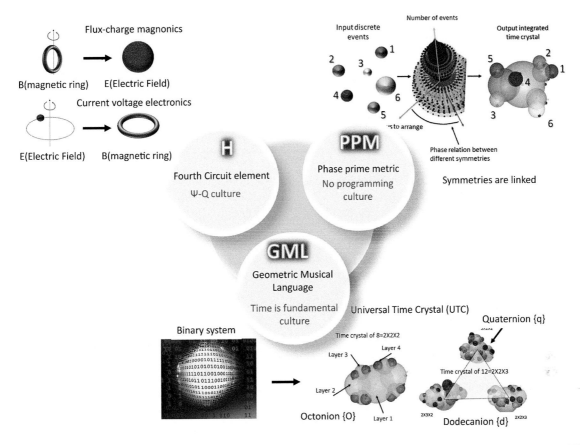

FIGURE 1.9 The artificial brain is based on three fundamental discoveries: First, a new kind of fourth-circuit element, namely Hinductor, *H*. Second, the pattern of primes that links discretely apparent symmetries, which is called the phase prime metric (PPM). Third, geometric musical language (GML), where a special kind of time crystal is used to write information.

microtubule, the water molecules is arranged in a beautiful symmetry of primes. The symmetries of prime are abundant in biology, e.g., in the secondary structures of proteins, along with the DNA spirals, at the smallest scale. Now, as we move toward the larger-size biomaterials or organs, the symmetries of its constituents interact, using octonion or dodecanion tensors to build a higher-level symmetry. Synthesis of symmetry is the key to build a PPM; it is the key to reverse-engineer the brain by synthesizing brain jelly. Apparently, from a dimer of proteins to the layers of the hippocampus and cerebellum, symmetries are composed as primes and generate similar triplet-of-triplet resonance bands in the experiment, which is strikingly similar to the PPM. There is no doubt, nature was perfecting a PPM in the human brain for the last 2 million years, possibly doing the same for billions of years at every part of the universe.

While integrating all the worlds of primes with a chain of vibrations, one should not isolate forces: every single force interacts, and the associated phenomena are simple perceptions of an observer, in the domain of vibrations: nothing is left out—prime loves us. The driving force of such puppetry of primes is the desire of the composition of symmetry for the completeness of PPM; it's hungry to grow from a single zygote in the mother's womb to matching the cyclic rhythms that are governing nature, and it has an ever-increasing appetite. While linking all possible symmetries, PPM is a naturally

intelligent analog of the metrics customary in astrophysics, but instead of linking space and time, it links symmetrical breakings of time crystals. The evolution of PPM will continue beyond this book; the objective is to eventually derive a system that refers only to itself, building physics laws from no law (Wheeler, 1980, 1983a, 1985).

Human brain model as a conscious being: The beauty of the time crystal–based brain model is that the consciousness criterion establishes that an element should have at least 12 triplets of triplet resonance bands, so that 3 distinct compositions of bands coexist and operate independently. Since the triangle is a distinct entity, made of $2 \times 2 \times 3$, $3 \times 2 \times 2$, $2 \times 3 \times 2$, it is a superior control of 3 simultaneous, independently editable, complete identities of a conscious being. It was found that the majority of brain elements satisfy this criterion: be it DNA, protein, microfilaments, neurons, cortical columns, or their assemblies, they were experimentally found by biologists for the past half a century to have a typical 3D arrangement of clocks. We simply compile widely distributed reports into a single integrated catalog here. All components share a consciousness as the brain's building block of consciousness (Nowakowski, 2018). Thus, an entire brain turns out to be a timekeeper that primarily manages time crystals of a few femtoseconds to a few million years by adjusting precise motions of the oscillators within the framework of its frequency fractal network. Cognitive behavior is the output of time adjustment by which external and internal rhythms undergo precise changes at all spatial scales, from a few atoms to the centimeter-scale ranges, to make each other's PPMs similar.

Not just the sensory input, but the pathways or nerve bundles follow a definite geometry and play active roles, infusing time crystals or patterns into a higher-order time crystal (see Chapter 7). The existing models consider that the nerves are mere wires, but in the time crystal model, pathways follow explicit symmetry to build clocks at lower time domains. The story of nerves repeats Tagore and Nowakowski's teaching's narrative (Toolan, 2009; Nowakowski, 2018): not a single loop would exist there unconnected. Such a massive 3D multidirectional clock-composite (Figure 7.15) cannot be linearized into a set of sequential logical rhythms. No software is required to operate it; PPMs have dodecanions with 8 corners or icosanions with 12 corners packed in quaternion statements to help analyze and respond to nature, which is also a PPM.

In summary, all brain components together are a time crystal synthesizer, whose evolutionary purpose is to expand the pattern of primes and acquire the distinct dynamics of new primes—that is, enrich its exclusive PPM. Memory is a time crystal; decision-making is an activation of the linked time crystal. If we generalize, one can encode rhythms of a few million years of periodicity with a cycle that survives only for a trillionth of a second; brain jelly is envisioned to do just that. In a human brain–body system, three time crystals coexist: one, the regeneration time crystal; two, the decision-making time crystal; three, the sensory time crystal. The circadian rhythm or sleep–wake cycle has a time period of 24 hours, proteins a few hours; skin cell replacement rhythms run for two weeks, the bone cell replacement runs for 2–8 years, and the heart cell replacement cycle has a time period of 100 years. Thus, from a single DNA molecule to the heart, everything lives a life and gets replaced by a better one by clock rhythms, and all rhythms are connected by a phase in a 3D pattern of geometric shapes. The artificial brain will not compute following instructions; rather it would be a prime number acquisition device so that its skeleton of primes, its PPM, gets richer. As in Figure 7.15, it would regularly update as more experimental data arrives; brain mapping essentially means "fill in the gaps" of the generic chain PPM. Even a kid can add a few circles to make it more symmetric.

1.10.1 A Language of Time Crystals Written by the Symmetry of Primes

The origin of the geometric representation of events: The matryoshka dolls of Efimov trimer: Two correlated events happening in nature always have a third event as the cause, the triangular correlation of events is fundamental (Reichenbach, 1956). By extending the view of Reichenbach, Price proposed the three-arrows concept (Price, 1996): the cause-effect, the past-future, and the asymmetric time arrow. The beauty of this proposal is that it demands a microphysical symmetry to link the three; Price even started drawing geometric links between the events. Here asymmetric time

flows with an architecture of symmetries or a pattern of primes, which evolves as the periodic flow of certain symmetries. Thus, human cognition does not have to link the past to explain the present and predict the future. All are part of an architecture called PPM.

PPM: PPM links the symmetry of events correlated in the universe, which enables us to define events in a way that could be linked to the pattern of primes. Without any human bias, the pattern can predict how events vis-à-vis symmetry would unfold in nature (Courtland, 2018). Thus, without writing code, the metric would analyze how an event would unfold in the future, and what we have missed in the past, if anything. Subevents are interconnected in the shape of a topology—say, like a triangle or pyramid, etc.—to create an event. So, if we express linearly, we lose most of an event's significance.

Say each of the eight corner points of an event with the topology of a cube has unique topological shapes inside. The journey through the corner points, representing the subevents, is endless; events integrate only within and above, not side by side, because if one adds a new corner point to a triangle, we get a quadrilateral. The singular difference in worldview has led us to create a new set of scientific tools described in this book. Instead of using human imagination and logic/memory to link random events and build intelligence, PPM relies on the creativity of the number system, which generates new primes in a new pattern continuously in the infinite integer space $(0, 1, 2, 3 \ldots \infty)$. We assume that nature self-assembles the primes to synthesize a new symmetry, thus governs the universe. If we do this with brain jelly, its exclusive PPM can estimate the randomness, predicting the future course of events perpetually. Inventing prime begins at 2 and 3; $2 \times 2 = 4$, $2 \times 3 = 6$...: there is an empty space between 4 and 6, it must be a prime, *let's call it 5!* Similarly, brain jelly can cover 220 million primes for $N = 10^{12}$.

Only 15 primes can generate 99.99% of all integers of infinite number space. So, infinite possible events or shapes can be represented using 15 basic shapes for 15 primes, i.e., a language with 15 letters, like English has 26 letters. One could write infinite possible events as 15 primes, then read evolved unknown primes as the nearest composition of 15 primes. But a computer has to find how, given set of primes, it could test an astronomical number of ways to link participating geometric shapes and converge to one continuously emerging a new set of prime compositions. Since existing computers cannot honor the geometric network of events, we need a new computer that links a set of primes and finds hidden patterns in them by testing all compositions in real-time. A conventional computer cannot do this, because it often requires the following:

1. trial and error for $>10^{23}$ per unit time (however, chemical collisions mimic this routinely in a beaker);
2. 15 primes choose from 220 million primes at a time. Parallelism includes acquiring changes in real-time; no code can interact 220 million nodes at a time (number of primes $<10^{12}$). Parallelism can happen only in a beaker with diffusing and self-assembling clocks.

3. It's impossible to wire all to all, within and above, multilayered imaginary or phase connections; such wiring is possible, however, in the supramolecular synthesis: When 15 prime-based input dynamics are fed to the organic jelly, where one can write a set of primes, one can see the jelly multiply, rearrange, and regroup geometric shapes or the ratio of primes by trial and error and test 99.99% of all possible integers to find a pattern that links input primes and tells us the new sets of primes most probably emerge in the near future. Grouping the primes properly links the input events, a job that a scientist does to build a scientific model, or a programmer imagines to write code. Jelly doesn't follow instructions, but it synthesizes decisions.

1.10.2 A Magnetic Light: Creating a Device That Stores Charge and Builds a Time Crystal

Atoms and crystals made of magnetic light operate as time crystals: In 1984, J. F. Nye proposed to interfere polarized monochromatic light and create dark lines in an empty space where the electric E or the magnetic part B of light disappears. If $E = 0$, we get pure magnetic lines arranged like a thread in the 3D space. Floating threads of pure field lines are not the vortex atoms; they need to be converted into vortices, and only then we can use them as atoms. Different lattice symmetry could initiate the interference of electric and magnetic fields separately. One could create various free particle–like structures made of the magnetic field: some of them look like a ring, some a ripple, some a spiral. It is a major transition from the J. F. Nye work: the synthesis of pure magnetic lines does not happen in open air. Perturbation is injected by a proper design of the device: lines that form on the cylinder-triplets bind and then come out as vortex atoms from the surface, which then could be used for various purposes.

Using charge, one can manipulate the interference condition of a spiral cylinder and create a composition of free magnetic rings rotating clockwise or anticlockwise. So, if one could design a device with the three concentric spiral cylinders, then one could create a phase space, which looks like a hollow sphere with 12 holes all around it. "Hole" means at a particular length, pitch, and ratio of the lattice area of the diameter values; no defined output could be found. The shape acts as a mask: from one side the dark lines fall on the phase space and emit from the other side. When the phase space projects the lines, they bind strings into loops and add a clocking direction to those nested loops. Often multiple strings bind, and a superposition of many rings forms a compound of vortex atoms. A single vortex ring could be an atom, but a composition of rings acts as a crystal.

What we suggest as fiction above is actually implemented as the flux-charge device, known as the fourth-circuit element; the adventure is not new. However, due to the distinct contradictory features with another variant of a fourth-circuit element memristor, the fictional device is named the Hinductor

(Chapter 7). The stored pattern of charge distribution can memorize the periodic oscillations of various kinds when a polarized monochromatic light falls on it to read what is written; the distributed charge morphs the electric dark, pure magnetic lines, and forms atomlike vortices, as Kelvin envisioned many years back (Thomson, 1867). Reading a composition of atoms made of magnetic light and immediately finding how delicately the pattern of charges are written on the device is a telling story. Light does not destroy the intricately coupled clocks or periodic oscillation of charges in the device, just reads them as-is and prints it in the open-air 3D space; a magnetic film can read the projected pattern. A time crystal is written into the device by changing the physical shape of the concentric cylinder triplets, following suitable means. Often the suitable mean is using an antenna and wirelessly sending an electromagnetic signal to resonate them. In brain jelly, input information synthesizes an inductor, and the jelly eats time crystals to build H, then self-assembles them to build a higher level H—the process continues.

1.10.3 A Pattern of All Possible Choices to Arrange Primes

Frequency Fractal Computing: Bits of discrete pieces of information are captured in conventional artificial intelligence algorithms. Then human imagination tries to deliver a significance to the dead "bits" and add a sensible logic. Intelligence = number of statements. Our frequency fractal computing hardware has a few 1D, 2D, and 3D geometric shapes stored inside. Since the seed geometric shapes that could assemble to build all other shapes are finite, the quest for geometry in the interactive environment always finds a near or a distant similarity. The most incredible point here is that there is no input from outside; the artificial brain seeks for the existence that governs our universe. It means, for the geometric computing protocol, it does not matter what an object looks like or what the music sounds like, or even the geometric shapes it resembles, or how many features of the music match with its stored ones. The event "searching" that we always do in the current information paradigm never happens inside the artificial brain jelly. In its "spontaneous reply," the entity that holds the answer somewhere in the giant time crystal syncs with any input time crystal absorbed into the network, at all times. Since the brain has only one time crystal, as soon as we try to fuse an input time crystal to it, the spontaneously reply is made immediately; after all, it is the same structure of time. No entity asks questions, no one replies, all entities are geometric, and syncing is the only goal.

When an unknown data flow, PPM-driven GML protocol tries to find the most actively changing regions in the meaningless data structure. In a cube filled with rapidly changing pixels, if the number of rapidly changing points is three and the most silent domains are three, it forms a triangle. All sensors are designed to acquire streams from a singularity and silent communication between the singularity points as part of a single geometric shape. Moreover, the artificial brain jelly does not look for the existence of geometric shapes. It tries to see how that shape changes into another geometric shape. Thus, in the journey of computing, GML changes what was information totally: we do not acquire factual data, we search for symmetry breaking and how a geometric shape changes. That change is human bias–free, the pure dynamics of a system and the system alone. Then we find how that change repeats: say, a triangle becomes a square and that becomes a pentagon and that becomes a triangle, and the loop continues. Then we draw a circle and put the triangle, square, and pentagon touching the circle perimeter. It is our time crystal. Our information content is actually how symmetry breaks and forms a new symmetry in any system. Thus, a number of bits, flops, speed—learning the well-known computing terms is irrelevant here. We will dip into its rigorous details hereafter.

Creating more than what we know: The second part of the unique feature in computing with primes is that a PPM expands the time crystal obtained from any event happening in nature. Primes are pure: they represent nondeductive symmetry. But an integer space is a deductive map of all possible interactions of primes. When an event is rewritten in the language of symmetry, then since the number of geometric symmetries is finite, PPM can extrapolate the event. PPM generates a map of all possible ways a symmetry could break in the future and might have broken in the past. It means we create an architecture of past, present, and future in the evolved time crystal. The expansion of an event as breaking symmetry in the PPM has several implications. We create information that does not exist; we add features in the future that had no past instances. Thus far, we have documented around 12 different versions of PPM in Chapter 3. The promise for the fusion of finite geometric shapes and finite symmetries is endless.

1.11   CONCLUSION: THE RELIGION OF SCIENCE HAS A TRIANGLE, DARWIN, TURING, AND HODGKIN-HUXLEY

Darwinism argues that the evolution follows treelike branches, the one who is fit survives in the struggle, and the failed ones disappear. Darwinism is applied everywhere, from protein expression to economics (Edelman, 1987). When one argues for an astrophysics-like metric for natural intelligence that maps all possible symmetry breaking in the universe, namely a PPM, it suggests that each prime plants a tree of evolution, and then nature would be a canopy of 15 prime trees: it's not about branches, but a rainforest. There is a collective life of

15 symmetries, not one "it from bit" (Wheeler, 1990); a unit of information holds a reading of a geometric shape, with speed. Turingism argues that irrespective of complexity, every single event that has happened, happening, and would happen in the universe could be written as a sequence of "yes or no" answers to a series of questions. When many people argue about a GML they suggest that irrespective of complexity, every single event that has happened, happening, and would happen in this universe is topological and looks like a pyramid or a cube, whose every corner has an event inside. When we enter inside a corner of that cube or pyramid, we encounter another 3D geometric shape made of events, and the journey is endless. Nature constructs events within and above, not side by side. Thus, a change in the geometric shape that Wheeler envisioned is the closest reality of nature, with a twist that the change in shape is determined by a typical composition of a pattern of prime, PPM. Turing himself talked about the possibility of writing arguments with a pen on a piece of paper. Now, if there are one-to-many connections, then one cannot cut those arguments and glue them to make a Turing machine. This takes us to a new world of computing, where Gödel's incompleteness (Gödel, 1938, 1947) and Russell's paradox come into action ($1 + 1 = 2$, in a box; where the second 1 comes from, if we start from one 1; Russell, 1901, 1948). The third religion of science is Hodgkin-Huxley, which suggests that all cognitive information in a neuron is processed by its skin; neuroscience is dermatology of membranes. The early historical roots of the field can be traced to the work of people such as Lapicque, Hodgkin and Huxley, Hubel and Wiesel, and David Marr, to name a few. Lapicque introduced the integrate-and-fire model of the neuron in a seminal article (Lapicque, 1907). Since then, ion diffusion and membrane potential regulate the core of cellular biology; thousands of concretely packed nanostrings of protein remained an uncharted territory, until we listened to their music. Even now, the neuron is just like a balloon, ions move to create membrane potential flow, and the whole system inside is silent. Nanotechnology reveals that every single component starting from alkyl and amino groups of proteins or DNA to the largest organ, the skin, forms a time crystal architecture, a network of clocks from femtoseconds to hundreds of years. This time crystal model of a human brain is the foundation of the nanobrain (Chapter 7).

Let's begin our journey with Leibniz (Leibniz, 1956): "Although the whole of this life was said to be nothing but a dream and the physical world nothing but a phantasm, I should call this dream or phantasm real enough if, using reason well, we were never deceived by it."

Vikalpabalādeva jantavo baddhamātmānamabhimanyante|So'bhimānaḥ saṁsārapratibandhahetuḥ|
Ataḥ pratidvarūpo vikalpa uditaḥ saṁsārahetuṁ vikalpaṁ dalayatītyabhyudayahetuḥ||

Non-dual Shaivism of Kashmir (Trika)
It is the sensory misperception that leads us to believe that our consciousness is locally confined to this world; that's why time to time an opposite thought (non-local) arises in our mind and that elevates us to a higher-level consciousness.

2 Replacing Turing Tape with a Fractal Tape
Fractal Information Theory (FIT) and Geometric Musical Language (GML)

2.1 INCOMPLETENESS OF CURRENT INFORMATION THEORY

Is it better to see a triangle or read 100001010101010000000? These sets of letters make no sense to us. We need a decoder to read it. What if we have a language that has a decoder, our perception, a higher-level argument? For centuries, information was an inanimate object. Imagine a time in the future when information is like a life form. It has a clock to keep time and a geometric relation as its significance or importance; it could be understood as a cognitive entity, with no reality or classical physical existence; it is like a concept. If we indeed find the information structure of the universe or our brain, we should be able to understand it as soon as we look at the information structure. Seeing a geometry, we must say, is a structure of fear or a structure of a joy.

When everybody bypasses *undefined* and *singularity*, here is a creature that cannot survive without them: If somehow, we preserve the way the events are linked in nature, we do not have to create scientific models or theories; the pattern that links is itself the scientific theory. Writing an equation may not be the only way to practicing science; it could be singing too. Unfortunately, connecting the events within and above was always considered scientifically impossible, and undefined, and if one could not make equations or differentiation, then what would be the point of studying these events? So, historically, scientists have always found a way to bypass such singularity. What if nature's journey is only through singularity? While traversing an infinite path within and above, nature might consider that every single event that has happened, happening, and will happen are all linked by a topology. One at a time, a sensor sees only one event as a single point, but there is an endless journey within and above. A new generation of sensors is required that could read to find that discrete, isolated events are not random and that they are linked by a topology; but that linking, even if regulated

by free will, would not be random. If the universe is cooked using finite geometric elements, then it has a defined pattern made of undefined points. Free will is random only when we want to confine it in a linear sequence. However, if we want to see events as a topology growing within and above, finite events could be arranged following a finite number of choices in mathematics. Computing or decision-making is always about linking events: the choices of combining events make a pattern, and that pattern is universal. If a system or computer has an imaginary hardware that links events following that pattern, admittedly, it would be able to predict the future without depending on human imagination–driven scientific models or learning algorithms. That is the idea of making an artificial brain, and all artificial brains have an embedded event-linking pattern hardware, so when they interact with the events unfolding in nature, both arrange the events similarly, plan events similarly, and execute events similarly. Turing's culture to melt and rebuild is replaced by "be like it"—that is the new bit.

Connecting the dots: "It from bit," to continuous to geometric: Every single knowledge that science generates is asking a question by doing an experiment, whose answer would be in terms of "yes or no"; thus, we derive "it from bit" (Wheeler, 1990). "Every 'it'—derives its function, its meaning, its very existence entirely—even if in some contexts indirectly—from the apparatus-elicited answers to yes or no questions, binary choices, bits" (Tukey, 1984). Nature arranges events "bit by bit," linearly, and we read only one reply at a time. There are plenty of problems with this philosophy (Figure 2.1): (a) A directly opposite thesis could be that the physics at the bottom is continuous; nature does not write information as a bit, does not connect events linearly. (b) Nature could write events as vibrating strings, like string theory. There could be unknown grand unification theory, suggesting a unique event for physics at the bottom. (c) The local environment might change the response when we ask a question to a single entity. "The choice of a question asked, and choice of when it is asked, play a part—not the whole part, but a part—in deciding what we have the right to say" (Wheeler, 1984, 1986). Who, what, when, and how are four questions that structure a bit at the bottom. (d) A bit could be an expression of many, indistinguishable, entities. If the physics at the bottom is topological, then events are connected by phase; a qudit can model such physics at the bottom. However, if every single event in the universe is made of a topology of subevents, then neither classical nor quantum mechanics could act reasonably. The information is not a language of bits, but a geometry of silence. (d) Always, there is a question, information of whom? If there is no answer as to whom, the rest is human imagination, and we build physics models as an extension of a black box. (e) Connecting the bits to regenerate a natural phenomenon includes a human bias. Probability, like space and time, is invented by humans, and different people have a different belief; they connect the bits or dots differently to produce the same output, assess the future differently. Probability as frequency, per Bayes theorem (Jaynes, 1986; Denning, 1989), is a function of human belief (Berger and Berry, 1988), agreed upon by a large number of people (Burke, 1985); thus integrated information theory (IIT) accounts for the political strength of a human belief. Despite their great explanatory power, physics laws do not describe reality; they lie (Cartwright, 1983). The reason for non-reality is the scientific method (Feyerabend, 1975).

Define what understanding is: Define whose information, and who measures: The structure of science is loaded with human bias and a luck factor that, if interacting with the right entity, would address the question we ask (Figure 2.1, see human bias). The scientific endeavor to understand the existence of life or that of the universe is a trial and error to ask the right question at the right time, to a right entity, and in the right way. Understanding is not universal. "Meaning is the joint product of all the evidence that is available to those who communicate" (Follesdal, 1975). We need an information theory that addresses the silence or even non-communication. Thus far, we have no structure, no plan of organization, no framework of ideas underlaid by another structure or level of ideas, underlaid by yet another level, by yet another, ad infinitum, down to a bottomless night. Here we argue that it

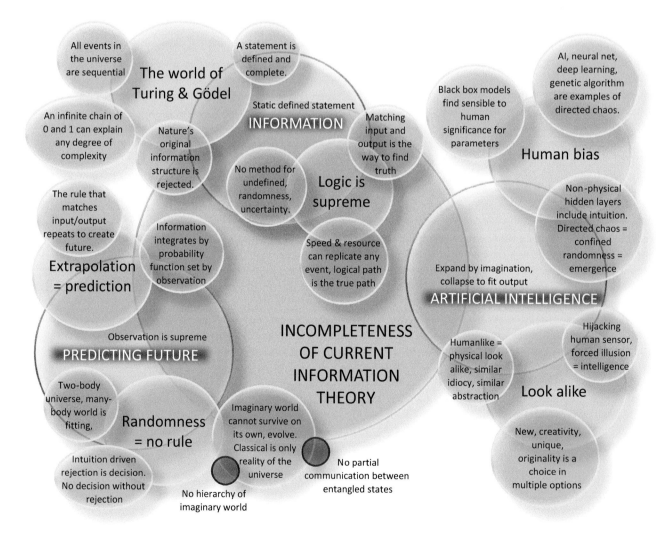

FIGURE 2.1 A diagram showing the problems with the current information theories.

is not mandatory to have a grand unification theory, to have such an architecture of information. To endlessness no alternative is evident but a loop (Wheeler, 1988), and such a loop as this: Physics gives rise to observer-participancy (Wheeler, 1977), observer-participancy gives rise to information, and information gives rise to physics. Quantum interactions tell us that it is possible to do measurement without changing the state. In that sense, a new information theory should inspire from the Ahranov-Bohm experiment (Aharonov and Bohm, 1959) where that flux of magnetic lines of force finds itself embraced between—but untouched by—the two-electron beams that fan out from the two slits. The shift in the interference fringes between field off and field on reveals the magnitude of the flux:

$$\text{Phase change around the perimeter of the included area}$$
$$= 2\,\text{vr} \times \left(\begin{array}{c} \text{shift of the interference pattern,} \\ \text{measured in several fringes} \end{array} \right)$$
$$= (\text{electron charge}) \times (\text{magnetic flux embraced}) / \hbar c$$

Even quantum experiments that avoid collapse as they are "untouched by electron beam" and gather "it from bit." Even quantum experiments consider that truth about nature could reveal as a "yes" or "no" answer to the query (Wheeler, 1984).

The answer to "yes" or "no" is not the only way to learn the truth; not bit-by-bit but within-and-above, singularity in geometric phase: We suggest that there could be an experiment where we do not need to ask questions. We built a series of technologies with simple instrumentation to sense pure information from a system. Quantum provides the trick to measure the change in phase, a non-demolition way of measuring an event.

The universe is not just participatory; both the observer and the environment are part of information content. Thus, the information is geometric. All Wheeler's criteria are fulfilled, except for the following:

1. Information is an infinite network of geometric shapes.
2. Observer and environment effects are neutralized as an infinite conformal network is created
3. Time and space are both tools to measure interval: the phase is a continuum, and we create interval looking into the location of singularities on the loop perimeter. Thus, mass, time, and length disappear; we intricately map the singularities.
4. Geometric shapes are made of singularity points. Events are not real; there is no classical point in the network of events.
5. Geometric phase is not continuous, as believed in the classic version of quantum mechanics (Anandan, 1988; Anandan and Aharonov, 1988).

All solids we see are dead and empty by science, and that is a fact: Mathematically several researchers argue for a fractal world, and possibly that is why nature is proposed as a computer (Margolus, 2003). There are plenty of rhythms far outside our body, and there is plenty of rhythm deep inside our body. Both kinds of rhythms run beyond our senses. Beyond our senses, the rhythms run, we perceive those only when they affect the rhythms of our conscious experience. Therefore, for Turing's world, the universe is melted and rebuilt as an infinitely long thread whose pixels are binary bits; for the universe we explore here in this book, the entire universe is an intricately connected 3D geometric shape whose only reality is the bursts from the singularity points, and rhythms bridge the gap between the singularity points. Zillions of system points touch the corners of the geometric shapes, and its periodic motions are seen as running clocks. The universe is lifelike and cannot be melted and rebuilt; we need a probe that has a similar architecture in order to learn about it, and that is the artificial brain explored here.

No event in nature could be re-created by arranging a set of elementary events one after another. An event could itself be a closed 3D topological structure. Imagine a cube: each of its eight points is a subevent, so, when we linearize, we destroy coexisting paths linking the subevents. If nature integrates subevents as topology, the journey to integrate information would be within and above, never side by side. If we could reduce Big Data into a topology, once we convert, then we apply pure physics: how topological structures change, break symmetry, and undergo a phase transition. So, without finding a programmer to unravel ideas in dead "bits," we can use physicists to predict threats or other decision-making. Then decision-making is possible without writing a single line of an algorithm, just like our brain does it always. Geometric entities could be used for computation, and the idea is not new (Forrest, 1971; Preparata and Shamos, 1985); but the use of geometric shapes as a composition to build up mathematical constructs was not yet there.

Simple analogies to understand a time crystal: Space singularity creates a spatial crystal, and the phase singularity creates a time crystal. Time crystal does not have time within, by closing a loop it defines a unit of time. Imagine many such clocks are arranged in a 3D space; one could build infinite close pathways by connecting the clocks. The topology of time cycles is different; therein, we cannot add the time linearly. Time is not a flow of the wave, and it is a cycle, a closed loop. One has to wait for the singularity bursts to sense that a circle has completed a rotation. Only then we can estimate the phase structure. That is why we sense the phase singularity points and note the geometry they reveal. A good analogy is an electron rotating around a complex orbit, on the phase sphere always there is a circular orbit, but that orbit rotates with a phase velocity. If the phase change makes a sudden jump and an energy packet emits only then we get the information about a phase change, else if there is an orbital jump we get different information. Orbital transition is not our target, for detecting a time crystal we wait when the rotation of an orbit reaches a singularity and makes a jump, the energy emits naturally. So, a time crystal is not about the clocking of an electron, a symmetry breaking or phase transition when an electron jumps from one orbit to another, it is exclusively a topology of phase singularity on the sphere only. Take another analogy. Several car drivers running cars, all of their minds are connected, for a time crystal we are not interested in the time required to travel, the path being traversed, but sudden maneuver to avoid accidents is visible which accounts for the interconnected state of the minds of all drivers. That sudden event is analogous to burst from a singularity and thus, reflects some features about the topology in the mind of the drivers. That static architecture of phase singularities is the time crystal, an engine that bursts at the singularity.

Particles move clocks run, Clocks interact singularities burst, Phase of bursts jumps in architecture, those jumps build time crystal. The photon moves like a particle; it is not time; it does not keep time; it is not a time crystal. However, if several photons interact, all clocks would run, change each other, that is not time crystal. However, when interacting clocks jump in phase, they reveal singularities. Only a time crystal reads another time crystal. Since there is no signal propagation, it is all about a match between two-time crystal releasing coherent bursts and clocking signals from singularity points. Reading one singularity points one by one not possible.

2.1.1 Fractal Tape and Surgery of a 2D Image to Place It in a Nested Sphere

All journey begins from a tape: Universality is nothing but a statement written in a piece of paper that can be solved without taking outside help. If we write statements line by line on a piece of paper and after cutting those parts, we can glue them in a line exactly as we said above, this is universality, this is complete (Gödel's completeness theorem; Gödel, 1938, 1947). However, we can imagine millions of different ways so that we need some arguments not written in that piece of paper. Figure 2.2a shows one such way. In biological systems, there are multiple clocks one inside another and all are triggered simultaneously, we do not know why biological systems do that? All fractal Turing tapes proposed till date are Iterative Function System (IFS) class as shown in Figure 2.2a, which means a particular seed geometry is repeated in a 2D or 3D space. One can see the entire fractal shape like a fractal antenna, nothing lies in the imaginary space. Such systems were introduced to compute at a "nearly linear time" (Gurevich and Shelah, 1989). Only when we enter in a cell, we find another tape as shown in Figure 2.2a (Ghosh et al., 2014a). In principle this is an infinite network, each tape is incomplete. For a particular tape, all other tapes are in the imaginary space and time. We simply cannot define a state as is, every single element is a door to another universe inside and it is a constituent of another universe, including the observer, and the journey is endless. In mathematical terms, we say every single matter is an "escape point," building escape time fractal, ES tape. One could argue that why should one care about the imaginary tag, have two Turing tapes, one for real-world data and the other for the imaginary world.

Together, two tapes process the entire information in parallel, then even quantum computer is a Turing machine. Though we start with a Turing tape, since (i) not a single cell in entire tape network is complete, (ii) no part of tape network cannot be cut into isolated tapes, (iii) we cannot take derivative at any cell space, therefore, it is not a Turing tape network anymore.

Inventing a fractal tape that does not have a Turing analog: In the Turing vision of the universe, one could melt and everything out there as a long 1D chain of bits. It would be a complete description. The concept of fractal tapes as shown in Figure 2.2a right, was proposed to run many Turing tapes side-by-side like a tree or in any self-similar or fractal geometry (Pippinger and Fischer, 1979). The advantages of geometric shapes in decision-making were explored in details in the 1990s. There are a series of pioneering works to keep time nearly linear using various geometric arrangements of Turing tapes (Gurevich and Shelah, 1989). Every single cell in a Turing tape is defined. Since we redefine any event happening in the universe as a clocking geometric shape, whose corner points are sub-events, an architecture of the clocks is important. To build that architecture, we need a defined machine, to eventually replace the Turing tape. The new tape we call Fractal tape, the name is an oxymoron, because there is nothing fractal about it, except that researchers and general readers, all consider that if there is a set of objects inside an object in an infinite network that is a fractal, we call it Mandelbrot attraction. We get a Fractal tape by placing a Turing tape in every single cell of a host Turing tape (Ghosh et al., 2014a; Agrawal et al., 2016b), so, every single cell is undefined in the infinite tape network. Since undefined, one cannot build a differential equation. The result is the development of multiple technologies, which are junk in conventional electronics, and making several nonsense and irrelevant computing concepts useful.

The discovery of time crystal and remarkable effort for two decades (1970–1990): We all know that a crystal is made of matter, but how could it be made of time? It was the genius of Winfree A., who made two striking observations (Winfree, 1977). He noted that a random noise could not change the biological clocks arbitrarily, the perturbation should be particular. Only at a particular phase of the clock, if the perturbation sustains for a long time, only then the output phase changes significantly. However, once a system is perturbed, even after removing the source of perturbation, three spontaneously emerging frequency peaks emerge in the output ripple, not one. In between 1970 and 1990 for two decades, a wide range of living biological samples showed these incredible features. In summary, (i) biological clocks have a hidden structure of phase singularities, (ii) the mechanics of this structure of clocks are undefined, neither classical nor quantum. Both the mechanics do not have any provision for triggering a chain of events in the absence of a perturbation or a noise. (iii) The hidden architecture of clocks is designed to operate with noise as a perturbation and no random noise can edit or even manipulate that structure. By conditioning when, how, and which noise could interact, a biological system uses noise as an intelligent and programmed signal source.

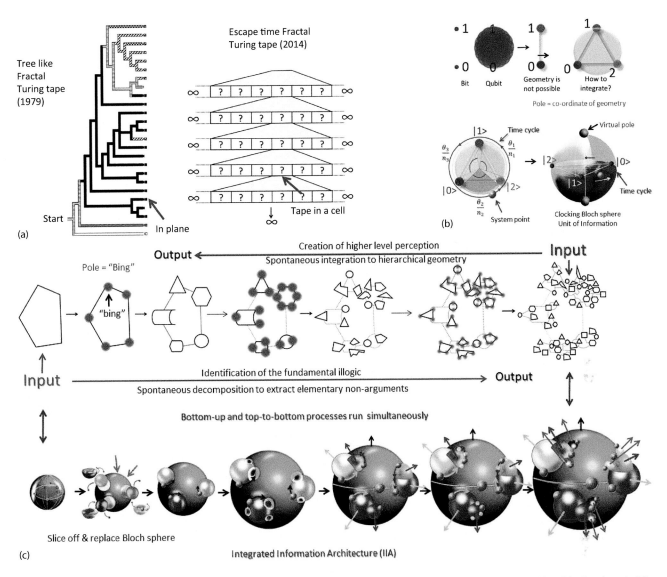

FIGURE 2.2 (a) Two types of fractal wiring of Turing tapes. (b) A transition from bit to qubit (top-left) to clocking Bloch sphere holding a geometric shape (bottom-right). (c) The top row shows decomposition of a pentagon. Corner points are singularities. Each point holds a geometric shape inside. The bottom row shows the corresponding time crystals.

Engineering the Bloch sphere: Brain is neither a classical nor a quantum computer because it does not have a defined classical state; it is not even a computer. Quantum can survive with an infinite number of real solutions (Unruh and Zurek, 1989), but we do not need a single real state, but 12 imaginary worlds affecting the reality. Probability, like time, is a concept invented by humans, and human observation that decides when and where to measure finds unique solutions accordingly (Wheeler, 1990). So, we need to add an observer to the wave function (Wheeler, 1977), and the environment too. These are not possible if there is a classical point on the Bloch sphere. We need a Bloch sphere, a sphere with infinite solutions between two choices. However, instead of impure classical (Joos and Zeh, 1985), it should be made of singularity points, paving the way for an observer, environment and geometric shape to be part of the fundamental unit of information

of the universe. We need quantum for using the beautiful concept of Bloch sphere, modify it by adding a clock on its great circle to build a suitable unit of information for the universe (Wheeler, 1983b). The Bloch sphere does not have a singularity, here for the new unit of information, and it is all about opening/closing multiple holes or singularities on the Bloch sphere, as shown in Figure 2.2b. Then time as a 3D geometric structure made of the phase of a set of vibrating material, so we may think, the matter is real, the phase of vibration is imaginary, dependent on the matter but it is not valid. When a pair of clocking spheres synchronize, the phase difference of these vibrating clocks gives the concept of time, when several clocking spheres couple to bond, they fill up a region of time-domain wherein it does not allow other phase structures to occupy, it is just the concept of space. Thus, a phase structure could govern a physical space, so neither space is fundamental nor the time is, it is a phase. As Einstein said, "Time and space are modes by which we think and not conditions in which we live." (Einstein, 1963).

Six fundamental changes to be made in defining what is an event in nature: (1) Nature store event in a geometric shape, if we do not change the sensors and capture the geometric structure of the events in the pure form, once destroyed, human imagination using deep learning cannot rebuild the accurate information ever. Time crystal analyzer should capture natural events as nested clocks to build a generic time crystal. (2) Existing science and the foundation of artificial intelligence tells us that whatever be the complexity of events, we can re-build that event as a sequence of simple events. It is not right. Nature store events not "side by side," instead, "one inside another." Thus, not just how the event looks like, the geometric arrangement of several events happens one inside another, which again needs to be captured at the sensor level, which we do. (3) All events are new and original is an illusion. Events continuously repeat until it encounters a new symmetry. Finite clocks arranged in a few symmetries one inside another could project in 360° spherical direction 8^{8^8} ways (8 = number of singularities). Infinite possibilities could generate from finite elements if the singularity is explored. Symmetry is finite, but their corresponding geometric shapes could be infinite. (4) Nature integrates events using all possible symmetries and all possible choices, we discovered this metric in the protein, and using this metric, we crosscheck the geometric structure of events. The fine-tuning enables getting rid of human bias, fix errors in reading events as time crystal and predicting the future by adding time crystals following phase prime metric (PPM), so everything is done naturally without programming. (5) The number of bits, amount of resources, speed, all these are irrelevant, a trillion-switch, holding a single symmetry could represent only one clock or one symmetry. When at all layers, all the clocks operate at various speeds, continually update geometric changes in big data, there is no computing time or computing speed. Zillions of data can pass through, and the system checks whether the geometry of clocks change or remains constant with minor editing. (6) It was never thought that even devices with no connections in between could link by the time building a circuit of time. Superlensing ability (Figure 9.4b) isolates the connected devices, gives one device in a complex mixture a distinction, thus allows building a singular clock using distant elements. A device is visible only when it wants to be seen. A device with a set of cloaking frequencies could form multiple distinct circuits, with far distant located elements. Imagine an entire living system is a circuit of time; the natural look is an illusion. Tiny pieces of times as clocks build a life form, to implement that design, clocks come, assemble. The natural look is a follow up of the circuit of time, realized by matter. If one isolates the matter, then reads the wrong information.

Why universality troubles computer scientists? The reason is simple for us, the concept of "Paper cut and glue," that we have extracted from the original document of Turing was lost in history. Greatest mathematicians have wasted pages after pages "inside the paper or outside?," which they call decidability issue. Can we write the statements such that it is difficult to cut? Russel's paradox (if one cuts the paper, he cuts only one path, Russell, 1901) and many-body theorem of physics. If several statements are connected, one has to sit idle, wherever he cuts, he removes two paths. Then if he tries to glue, he loses at least one path, this is a generic situation of Russel's argument. For a fractal tape, imagine that as soon as one cuts a piece of argument, it becomes a new paper, and that repeats forever. Figure 2.2c shows, how does a fractal tape process an image to build a complex time crystal. One interesting factor is that when we cross the layer above, we end up into a few numbers of low-frequency oscillators. Thus, automatic simplification of the image is made by dilution of geometric parameters via resonance chain (a geometric distribution of resonance frequencies). When we go down below with a large number of high-frequency oscillators, even a small part of the image is expanded into various self-similar forms, and that are then summed into a single pattern. Thus, the transition to the low-frequency layers extracts fundamental grouping parameters. While going to the lower frequency layers cause fractal decomposition of the image and higher-level perception forms of an image. The back and forth journey are shown in Figure 2.2c. The process chiefly shrunk nested rhythms for an entire image, which we call fractal seed. In this process, the phase sphere is cut off or sliced where there is a singularity domain, as shown in Figure 2.3a.

2.1.2 SELF-ASSEMBLY OF GEOMETRIC SHAPES AND THE CONCEPT OF SINGULARITY

Which clock should we take to build time crystal: Minkowski clock, Poisson clock, Nested clock in PPM: Classical time is a near-equilibrium approach, in quantum, the clock is deterministic? Minkowski clock is a sequence of clocks wherein the higher frequency clocks traverse equal distances around the world line, hence classical, as no phase difference is observed between the two paths. If it is just two-photon clock and the frequency is increased say by ten times, a spiral change in phase is visible, and this is the driving force for spiral self-assembly. It could be explained using an imaginary number, but it is not a quantum (Ord, 1983). For quantum, the trajectories follow several points before return to the base point or a classical line like the sphere shown in Figure 2.3b, so a phase difference is created. It is precisely the principle of a Poisson clock, where the rate of decay and growth of a wave creates the phase difference. One can write a clock as a tensor, a simple mechanism to write a tensor is to put tensor elements as characteristic parameters of the singularity domain as described in Figures 2.3b and 2.4a.

Understanding the singularity in a time crystal: Space singularity creates a spatial crystal, and the phase singularity creates a time crystal. Time crystal does not have time within, by closing a loop it defines a unit of time. Imagine many such clocks are arranged in a 3D space; one could build infinite close pathways by connecting the clocks. The topology of

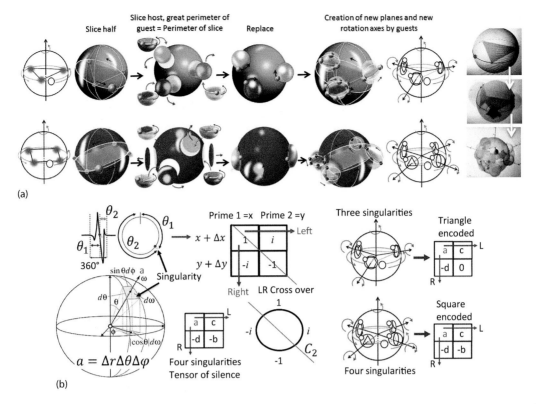

FIGURE 2.3 (a) The mechanism how phase sphere is sliced off and a new phase sphere containing the information about the geometric shape is inserted into the sliced region. Two rows show two distinct examples. The top row begins with a triangle where the corner points are then sliced off. The bottom row begins with a square, whose four corner points are then sliced off. To the right we see that in a triangle continuously new information is inserted. Three consecutive steps are shown. (b) A pulse carries change in sign while representing a phase singularity. The periodic emergence of such phase change in a system response is then plotted as a sum of two distinct phase domains equating to 360°. The circular plot representing rhythm is written as a 2 × 2 matrix operation. The ratio of phase is represented with nearest primes. The imaginary number is used to depict the contribution of phase in the singularity region to the rotation of the clock. For the phase sphere representing the singularity with a particular area where the phase becomes undefined is shown. To the right, two examples are given which shows how to write a matrix for a triangle and a square.

time cycles is different; therein, we cannot add the time linearly. Time is not a flow of the wave, and it is a cycle, a closed loop. One has to wait for the singularity bursts to sense that a circle has completed a rotation. Only then we can estimate the phase structure; change in phase is time. That is why we sense the phase singularity points and note the geometry they reveal as shown for three and four singularities in the Figure 2.3a and b. A good analogy is an electron rotating around a complex orbit, on the phase sphere always there is a circular orbit, but that orbit rotates with a phase velocity. If the phase change makes a sudden jump and an energy packet emits only then we get the information about a phase change, else if there is an orbital jump we get different information. Orbital transition is not our target, for detecting a time crystal we wait when the rotation of an orbit reaches a singularity and makes a jump, the energy emits naturally. So, a time crystal is not about the clocking of an electron, a symmetry breaking or phase transition when an electron jumps from one orbital to another (Zeng, 2017), it is exclusively a topology of phase singularity on the sphere only.

Mind-Car analogy: Several car drivers running cars, all of their minds are connected, for a time crystal we are neither interested in the time required to travel nor the path being traversed, but a sudden maneuver to avoid accidents is visible which accounts for the interconnected state of the minds of all the drivers. That sudden event is analogous to a burst from the singularity and thus, reflects some features about the topology in the melted-mind of the drivers. That static architecture of phase singularities is the time crystal, an engine that bursts at a singularity.

As the particles move the clocks run. Clocks interact, singularities burst. The phase of the burst makes a quantum jump in the architecture of time; thus, jumps build a time crystal. The photon moves like a particle; it is not time; it does not keep time; it is not a time crystal. However, if several photons interact, all clocks would run, change each other, that is not time crystal. However, when interacting clocks jump in phase, they reveal singularities. Only a time crystal reads another time crystal. Since there is no signal propagation, it is all about a match between two-time crystal releasing coherent bursts and clocking signals from singularity points. Reading each singularity points one by one is not possible.

How would a kid design a time crystal detector? Harvesting the singularity glue: Pump-probe experiments

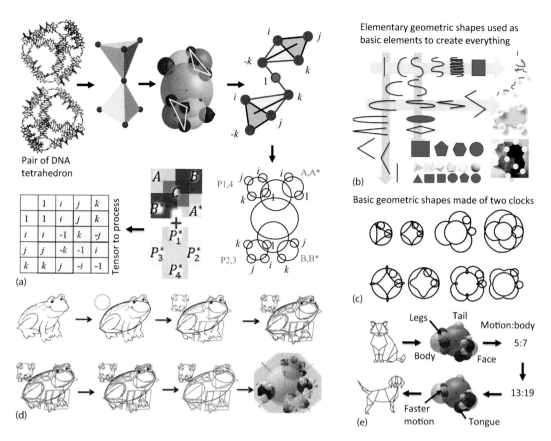

FIGURE 2.4 (a) A pair of DNA that looks as if two tetragons are facing each other. The corners are replaced by a phase sphere keeping in mind that diameter is inverse of time. Each sphere represents one imaginary world, together, for each imaginary world a clock is drawn. The groups of the composition of clocks are written as a pair of matrices which combines into a quaternion. (b) Five lines, five 2D shapes and five 3D structures build the basic geometric shapes that are used to search for patterns by GML in the FIT. (c) Epicycloids, cycloids and hypocycloids are key geometries for nested clocks used extensively for developing the time crystals in the FIT. (d) Multilayer image processing to convert an image into a time crystal. (e) Cat is converted into a time crystal and then a dog is converted into a time crystal, a subtle change in the geometry could convert a cat into a dog and a dog into a cat. The ratio of primes depicts the exact parameter where the change takes place. Each organ of an animal gets a special position in the time crystal.

work on an internal linear structure. For a time crystal, we need a similar structure to sync with the time crystal that we want to measure and in return let the singularity points bursts signals. The solar system is a nested clock architecture made of multiple planets and their sub-structures, if we see them using a telescope, we do not read the time crystal, if we plot a picture of their rotational path we do not draw the time crystal. All the planetary bodies in the solar system have phase recession motions, and that encounters singularities, generating turbulence in the gravitational field distribution. If we can map that disturbance and find its topology, a static geometry that maps turbulence defines the time crystal of the solar system. Now, a kid may shift the moon and generate turbulence in the gravitational field distribution; the time crystal would change to another geometry, say a triangle converts to a rectangle. It would be a non-demolition type measurement where we try to change the time crystal continuously keeping both measuring and to be measured time crystals side by side until there is a match. As soon as they match the two types of signals emit. Two waveforms fuse. The first stream of waveforms of different frequencies with a changing phase.

The second stream of complex waveforms as an amplitude modulated signal. Bandyopadhyay et al. have studied a large number of proteins, and their complexes and these two types of signals generate continuously, that revealed the time crystal (see Chapter 6 for details). Measurement is morphing between two geometries whose corner points are made of singularities. Time of measurement is the time of the slowest clock, not linear. So, in a time crystal computer, the computing time is fixed by the slowest clock in operation.

2.2 THE BASICS OF A GEOMETRIC MUSICAL LANGUAGE

To unveil the language of nature, at the bottom, we observe which event (= equipment evoked response), not as a reply to a question, but as an essential burst from singularity domain. What we do is to observe how in a particular time domain natural, spontaneous energy bursts repeat. How long the system responds and how long it remains silent. Our job is to draw a circle and put dots on its perimeter to get a geometric shape. That dot is silence. We believe that nature writes events as geometric shapes not as bits and link events by putting new geometric shapes inside the corners of one geometric shape. Unlike bit by bit scenario, where a bit is a complete statement, here, each information is an infinite chain of interlinked geometric shapes. We change the time domain an observer would draw a geometric shape. Eventually, a genuine nature reader has to cover entire time domain to find out the layered network of geometric shapes one inside another, the repetition of geometric shapes assists in finding the seed geometric shape and the grammar if followed, the complete information architecture is generated. The pathway shown in Figure 2.4a, a real structure to the tensor elements are phases of signal bursts from singularities.

Time crystal analyzer, TCA: We need a language to build artificial brain where the sound of words would be felt in mind, understood, a perception would trigger similar waves in the human brain, even though the person does not know the letters or grammar of that language. Inventing a language that connects to the human brain directly, even if we do not learn word's significance is a challenge. For that purpose, several replicas of sensory signals are processed independently by a time crystal analyzer, TCA, looking into different geometric parameters in big data. Then the signal is converted into 11D time crystal, the input of TCA is a complex set of wave streams or ripples, and the output is a nested sphere. The analysis is carried out so that the linguistic key, event = [subject-clause-(verb-adjective)] is grabbed from nature directly at the entry point and the purity of this form is maintained at all levels while integrating the time crystal and preparing it for sending it to the PPM. Acquiring 10D data following linguistic protocols, requires asking questions to nature, where the events take place, see Figure 10.1, where we have outlined how TCA, the brain-nature interface forms its queries to get a suitable quaternary linguistic demand. TCA counts each event = [subject-clause-(verb-adjective)] as a single abstract sphere, then a complex time crystal formed by acquiring natural data (GML) becomes a network of integers, we convert it into a network of primes ($4 = 2 \times 2$), so that it is ready to go to PPM described in details in the next Chapter 3.

The geometry of a language, primes of a geometry: Whatever is the language, a statement has four parts. Who? Alternatively, the subject. At which condition? The predicate. What? The verb. How? The adjective. If one asks to put circles and draw a sentence, the circle that holds many subjects is crossed by a circle holding many what. A circle on whose perimeter many conditions are written, bridges the Who? The conditions. The fourth circle on which the steps how an event took place are written across the other three circles. The crossing points of circles in the picture reveal the critical links between the facts, and if a system point rotates through all the circles making a sound at the cross-section, the gap between the cross-section becomes phase, and that defines the event. So, if we have a language where the phase is the only variable, cross-sections are phase singularities, then the engineering of singularity would build a new language. Three points on a circle make a triangle. If the event is a

tetrahedron DNA like Figure 2.4a, how does one move from one point to another? An addition of phase takes place when we take the product of vectors, as the indices of exponential terms. PPM maps these bits of pieces by counting how many ways we arrange them, and which of such arranged structures would link to grow. Finally, only a set of numbers could represent a whole movie, with a past, present, and future. There are only 15 geometric shapes, five lines, five areas and five volumes as outlined in Figure 2.4b. However, using a pair of clocks and rotating them one top of another, we get the geometric shapes. Some examples of areas are shown in Figure 2.4c. One of the fundamental problems of nested clocks is that in practice, we could hardly generate a finite number of clocks using materials. Just like when we hit a tuning fork, we get infinite harmonics, for the clocks too, similar to the fundamental clock, an infinite series of clocks generate spontaneously.

Why do we use hyperbolic functions to generate geometric kernels? One key difference between resonance band with harmonics and an infinite chain of clocks is that the clocks are connected by geometric constraints. It is reported that the resonance band of the real human brain is a geometrical progression of mean frequencies from band to band, roughly a constant ratio of $e \sim 2.7$, i.e., the base of a natural logarithm, we get time fractal using e^{-t}, $e^t = \sin ht + \cos ht$, hyperbolic functions are continued fraction too, just like the resonance band (see Section 2.7), for example $\tan h1 = \cfrac{1}{1+\cfrac{1}{3+\cfrac{1}{5+\cfrac{1}{\cdots\cdots}}}}$, Similarly, we get

$e = \cfrac{1}{1+\cfrac{1}{2+\cfrac{1}{2+\cfrac{1}{2+\cdots}}}}$. Similarly, we can derive several

fundamental constants that restrict us to add geometries randomly in a 3D space. That restriction brings primes into action, once we break the concept of linear distribution of resonance frequencies and enter a continued fraction, as shown above, spirals, vortices and fractals would make their entry too.

What is meant by each geometric point connecting a circle, or sphere? Basic 1D, 2D, and 3D geometric shapes shown in Figure 2.4b could be generated using hyperbolic functions, which is an infinite time series or rhythm. A nested rhythm is an endless network of periodic or quasi-periodic oscillations (infinite series mathematically). Nested rhythm = Nested clock = time crystal. Figure 2.4d shows a static time crystal, though the clocks embedded in the spheres are rotating, they do not change in size. In contrast, Figure 2.4e shows a dynamic time crystal, where the spheres are expanding and shrinking. Therefore, the existence of an infinite series of clocks or the clocking spheres rapidly oscillating in size. Regarding the clock within a sphere, we take only the ratios of connecting points, each contact point is a time fractal, say $f(t), g(t), k(t)$, then for a triangle ABC, we take three points, $\frac{f(t)}{g(t)} \sim \frac{1-e^{-t}}{1+e^{t}}$, then $\frac{g(t)}{k(t)}$ and $\frac{k(t)}{f(t)}$. Three functions determine a rotation translation independent feature that musicians also use to create a melody in the human mind. Three relative functions are three time-fractals, which bond together, by creating another function, as a host time fractal that represents a triangle. The fused nested rhythm is the time crystal, when we measure it, pure time crystal fundamental to the system, becomes a guest on the host clock that measures it. Using a simple analogy, we have explained it in Figure 2.5a. In a classical or quantum measurement, the probe clock would be invisible, but in the universal time crystal measurement, the probe clock becomes a fundamental part of the system's

clock. Normally, one atom forms a crystal, but to make an artificial brain, we need composite clocks of various kinds, just like composite materials. Therefore, we have shown in Figure 2.5a, how three distinct clocks on a guest making four clocks form the basic unit of a universal time crystal. Three atoms form a triangle.

The case I: a demand from the linguists to hold on to the basic definition of a sentence throughout the brain: In the discussion above, we find that to hold a minimum geometric shape that is a triangle we need four clocks. Even if we want to create any 1D line, we need three points, two endpoints and a curvature. Four clocks if resides one inside another, tell another story. For the linguistic purpose all formulations

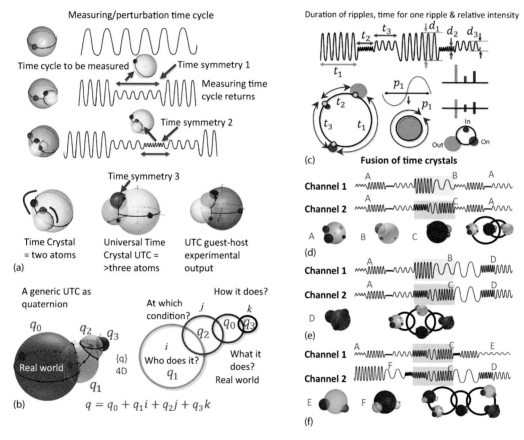

FIGURE 2.5 (a) There are four rows. The top row shows a phase sphere with a clock that represents the signal sent from outside to read an inherent clock or time crystal residing inside a system. The second row from top shows the spontaneous generation of a new signal, in principle not related to the input signal. The spontaneous clock disappears naturally after some time and the measuring input signal returns. The third row from top shows that for a finite time period when the spontaneous clock operates a new clock is born in the system. The fourth row from top, shows three-time crystal structures, the first one is the time crystal without measuring phase sphere, or measuring signal. The central time crystal shows a new guest sphere that means three inherent clocks are there in the system, we call this universal time crystal or UTC, because it holds geometric information. The central time crystal with the measuring phase sphere is shown in the right, during a real experiment the triggering input signal also returns as an output. (b) A generic universal time crystal is made of four clocks, one real and three in the layered imaginary worlds. These clocks form a quaternion of decision-making as shown in the right. In the responses of a system, the real world is not the slowest clock. Who does it? The clock that answers this question is the slowest clock. The clock that answers to the question, what it does? That represents the "Real world." (c) The three basic principles for constructing a time crystal from a waveform. The amplitude of the waveform gives the diameter of the circle. The duration of spontaneously born signal provides the relative phase between the clocks or the location of local clocks on the perimeter of the circle. Relative amplitude difference tells us by how much a guest clock is inserted inside the host clock. (d) The common region between the two channels is bonding regions that glue two time crystal. (e) If the common region between a pair of wave streams are not identical then the protocol followed is presented. (f) If two different channels have a part in common, that part acts as a glue or common domain that links two time crystals.

analyzed in this book follows the structure of a statement (e.g., Who? When? What? How?), even the smallest unit of information requires three imaginary worlds, i.e., a quaternary tensor as shown in Figure 2.5b. From any sequence of wave streams, one could create a time crystal. By looking into its ripples, the duration of one kind of signal gives the phase of a guest circle representing a clock. The amplitude gives its diameter. If the signal is perturbed, we could even map the 3D phase plot or 3D time crystal. One could fuse two time-crystals by looking into the common region in a pair of wave streams. Figure 2.5d–f are three ideal cases. In Figure 2.5d, everywhere, the streams are identical, except a small-time domain. In Figure 2.5e, everywhere, the streams are identical but follow a complex scenario. In Figure 2.5f we demonstrate the real glue effect, where tiny time-domain showing a similarity between a pair of wave streams bond two time crystals.

The case II: Acquiring time crystal from the real world by a time crystal analyzer, TCA: In the real world, wave streams are not a linear singular non-modulated one as shown earlier. In Figure 2.6a we show that in a complex modulated wave, one has to find all frequencies engaged in modulation, their duration, amplitude and repeatability are checked to build a nested circle. When multiple wave streams flow, their phase difference is collected and after placing a reference point, the relative phase differences of the nested clocks decide where on the perimeter we should put the system point (Figure 2.6b). In case the stream of pulses is so noisy that we cannot recognize waveforms, then we could take area covered by streams and or periodic patterns wherever possible. These periodic parts are stored as memory units and the system of sensors looks for repetitions continuously (Figure 2.6c). All those areas where the program fails to read any periodicity, geometric shape of those particular time domain is read (Figure 2.6d). In order to confirm the periodicity, a group of sensors could be hardwired to build a polar plot that reveals hidden periodicity. Thus, without using an algorithm, a set of resonators could sense hidden periodicities and build nested circles or clocks. Once the 2D clock geometries are found, the system of clocks is perturbed to find the 3D clock architecture.

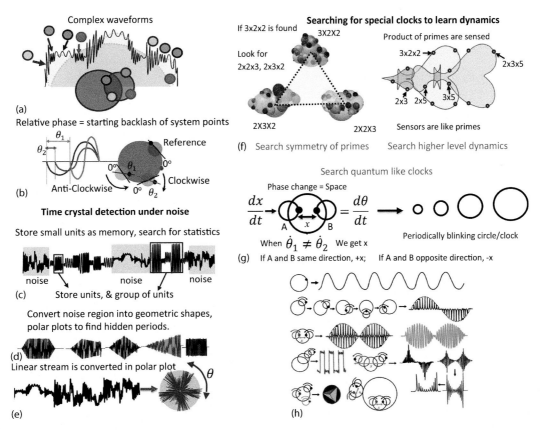

FIGURE 2.6   (a) If there are nested waveforms one in another, then the time crystal is generated. (b) The relative phase between the waveforms is included as relative phase difference between the system points that rotates around the circle. (c) The time crystal analyzer continuously monitors the waveform and stores new pattern of wave packets as a search unit. These search units are stored in the guest time crystal bank to find repetitive units. Periodicity of waveform is essential to build a time crystal. (d) When all efforts to detect a waveform fails, the simulator goes for pattern recognition as if the input waveforms are not wave but a stream of geometric shapes. (e) When the algorithm even fails to detect an area, then it builds a polar plot and identifies periodicity. (f) If an integer is found in the wave stream, the simulator finds the other symmetric divisors, that means if $2 \times 3 \times 2$ is found the system searches for $2 \times 2 \times 3$ and $3 \times 2 \times 2$. Also it looks for 2×6 and 6×2, or 3×4 and 4×3. Thus, the system finds links between the different integers in terms of their divisors. (g) A wave stream might find oscillating clocks, this is important because oscillating clocks are used a mathematical variable. (h) A special composition of waveforms is saved in the memory bank of the simulator to model very special clock networks.

The case III: Searching for special clocks to learn advanced dynamics: In the real world, search for linguistic feature (Case I), and the search for hierarchical periodicities (Case II), we introduce the search for symmetry (Case III). For example, 12 means it could be $2 \times 2 \times 3$, $2 \times 3 \times 2$, or $3 \times 2 \times 2$, in all situations. Say a system of oscillators senses two possibilities of 12, then it would look for the third. Means, from the stream of pulses if it finds a triangle with arms ratio 2:2:3 and 2:3:2 then at a certain time the triangle might morph or change its shape to 3:2:2 (Figure 2.6f). Apparently, possibilities are infinite, but in reality, it is finite. In the PPM, Reddy et al. have shown (2018) that pattern of primes connecting group of primes and integers could guide a system of oscillators to naturally find the missing symmetries in the input signal. A list of some patterns of metrics is listed in Figure 3.4. The language of time crystals is being developed to replace the world of differential equations and emulate the natural phenomena (Figure 2.6g and h). Think in a reverse direction. If a set of oscillators emulate a natural phenomenon (Bandyopadhyay et al., 2010b, 2010c), then it would sense a similar differential equation. Multiple natural phenomena are described in Figures 2.6h and 4.1b. A tutorial is provided in Figures 4.9 through 4.12, on how to draw circles and solve complex mathematical problems.

2.2.1 How a 3D Structure Becomes a Time Crystal or a Tensor: Non-differentiability

Non-differentiability makes rhythm or time fractal essential: Most of the spaces between "matters" are a vacuum, in order to connect them, the energy should exchange between any two elementary matters in a periodic fashion that is what rhythm is, the periodicity is all about keeping time. Every period has a time width that determines the frequency and clock limits too. Moreover, always there should be wireless energy transfer, because in the escape time worldview (zoom in to find other worlds) nothing is touching each other. In the fractal universe, the clocks face a unique situation, if the nested imaginary worlds are entirely made of time crystals. We explain the paradoxical situation using a network of cylinders in Figure 2.7a and b. How do we get a cylinder? Imagine a clock is running, we get a circle. Now, if the clock is perturbed, it would try to return to the initial point. It would take a finite time, that makes a cylinder. In a fractal network, cylinders would trigger each other. Then, generating a fractal time is possible if we have a change in the potential $\varnothing(t) = \int_0^\infty \beta \exp(-\beta t)\theta(\beta)d\beta$, with $\int_0^\infty \theta(\beta)d\beta$, let us consider the expression $\varnothing(t) = \frac{1-a}{a}\sum_{n=1}^\infty a^n b^n \exp(-b^n t)$, $b < a < 1$; in this expression β varies as b^n and $\theta(\beta)$ varies like a^n, then the fractal time is ensured for $b > a$. In several working environments, it has been documented that fractal time is generated (Shlesinger, 1988). In a practical fractal time scenario, the average time is infinite. Wherever there is a situation where a parameter changes with time such that the rate of variation is proportional to the magnitude of the parameter itself, we get an exponential variation and when it is stretched using power on top of it, we get fractal time. One example is: $\varnothing(t) = \exp[-\frac{(t)^\beta}{\tau}]$, where $0 < \beta < 1$.

Fractal of meander flowers: The scaling law: Why FIT must not have any conventional fractal in it: If the behavior of a quantity say F (say light emission) is directly or almost proportional to a scaling parameter s (say time or frequency scale), such that $F(s) \sim s^\alpha$ then the power law is valid over a broad range of s values. It is the scaling law, if α is non-integer, then it is a fractal. A fractal does not give new information, it is a composition of the mirror images of the same information repeatedly feedback on the higher scales. However, the network of cylinders shown in Figure 2.7c, representing a system, where the cylinders are changing in length, pitch and diameter, would build wave vectors of various kinds. Excitation to the recovery time of the clock or the cylinder length tells the propagating wave along the cylinder surface, how many loops it can make (Figure 2.7c). The loops around a center look like a meander flower. Figure 2.7c could be plotted differently like Figure 2.7d for better clarity. The horizontal axis shows increment in recovery time, vertical axis frequency. The lines are drawn to exhibit Hopf bifurcation, or allowed journey through a series of breaking symmetry in the time crystal, while a vertical column suggests that all flowers belong to same symmetry. Thus, how the symmetries would change is a geometric choice and those restrictions are mapped by a pattern of prime, i.e., PPM. Self-similarity is nowhere, but still we sense a fractal everywhere, in all the gardens of meander flowers. The artificial brain would be a garden of gardens, GOG that picks branches of flowers to compose new flowers for its garden (Figure 2.7e).

2.2.2 Fifteen Geometric Shapes Are Enough to Recreate Any 1D, 2D, or 3D Pattern

Zooming the perimeter of a circle: Underprivileged and superprivileged pixels: All the circles of a time ring oscillate continuously. During oscillation, it increases its diameter together coherently (in-phase) and decreases to a single point. That single point is also one pixel or the smallest phase cycle (Ord, 2012). Say, one of the many connected pixels starts oscillating in a different phase, for a particular time. Then it returns to the same phase as its neighbor. The process repeats with all the pixels one by one in a sequence. Then an external observer sees as if a point is moving in a circle, or a clock is born. Two such points hold an angle that enables encoding a geometric shape. For example, by shifting the position of points, one can encode triangles of any shape. Similarly, by using four points one can encode a rectangle, or square and so on. Those pixels are privileged. Every pixel in a phase cycle is another phase cycle or clock (Girelli et al., 2009), one imaginary world's lower limit is the upper limit of another world. When the clocks change locations, it is not a crystal, but a jelly. The jelly absorbs, writes and erases the time rings or clocks to sync with its environment (Bandyopadhyay et al., 2009b; Ghosh et al., 2016b), forms sensor of a different kind (Figure 2.8a).

The geometric kernels that form the elementary letters of the brain's geometric language: Conventional geometric languages (Peyré and Cuturi, 2019), convert the entire image

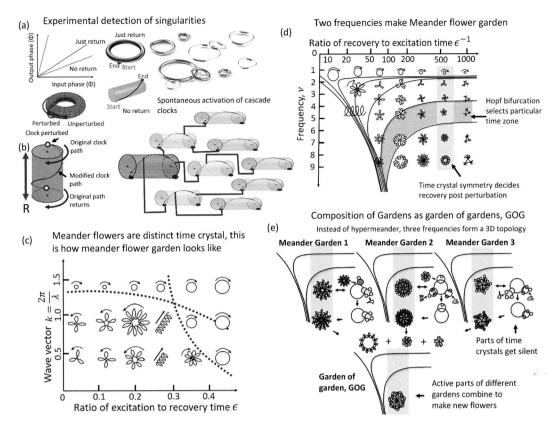

FIGURE 2.7 (a) The mechanism of detecting a time crystal. First an existing periodic oscillation is perturbed, when, the existing wave-form changes its basic parameters for a short time before returning to the same signal or phase reset happens three events can happen. First, the system would return exactly when it suppose to begin its new period. Second, the system returns much earlier, and continues to spiral oscillation. Third, it takes a much longer time than the next beginning of a new period. (b) R = Recovery time. There could be several interconnected clocks which are triggered during the perturbation of one clock. (c) The recovery time shown in the panel b could change as a function of input perturbation time and frequency of the signal used. Consequently, the pair of nested clocks change relative diameter, phase positions, which generates a large number of time crystals. Each time crystal has only one system point and two clocks, hence we get meander flowers, the entire plot is called meander flower garden. (d) The same meander flower garden plot like panel b but wave number is replaced by frequency and the ratio of excitation to recovery time is also inversed. Two major lines of sequential generation of time crystals are highlighted by shading. (e) Multiple gardens of meander flowers could interact and build garden of gardens where multiple meander flowers fuse and build a new flower.

on a single layer and considers only triangulation. Thus, it covers the entire 2D or 3D surface using triangles in a different arrangement. Geometrification of data is good for aesthetics, does not serve the purpose to represent the dynamics with morphogenesis of a few geometric shapes.

Encompassing circle or sphere determines the time domain: For example, irrespective of the sensory signal, be it visual, auditory, touch, smell or taste, the 2D or 3D pattern is created first as described in Figure 2.4d and e. Any change in perspective or logic, means changes in those few geometric shapes. When we see an image, it is converted to a 2D iso-frequency geometric patterns. Various geometric shapes are detected in the image. Be it a line, curve or triangle, pentagon, all are inserted into a circle of a fixed area. However, the contact point between the circle and the triangle would be the most energetic. The conversion to circle happens for all possible 2D geometries, square, pentagon, hexagon, heptagon etc. For an S, U, L, V, T, all 1D patterns convert to a straight line and a circle simultaneously for the natural oscillation of the network of an oscillator and platonic 3D geometries,

all structures convert to a sphere. These cross-sections of a circle or sphere and the 2D, 3D geometric shapes get more energy and these frequency values play a dominant role in the further oscillations. As a result, we find that the ratio of these contact point frequencies becomes the variable in the hardware, which forms the rhythm since all-encompassing circles of a particular imaginary world have nearly the same area the ratios play a crucial role in defining the essential terms of the rhythm.

A platonic love letter to Big Data: In principle, we could randomly choose geometric shapes and find them in the natural events and use those very shapes to assign complex geometric shapes to the integers in the PPM. However, 1D, 2D, and 3D geometries are selected mathematically to serve as an analogy to the Platonic geometries (five Platonic solids are tetrahedron (or pyramid), cube, octahedron, dodecahedron, and icosahedron). For 2D, we take triangle, quadrilateral, pentagon, hexagon, for 1D a few letters like L, V/U, C ($S = 2C$), O. Sensors search only for these fifteen geometries, each with a distinct value of S in a massive rapidly changing

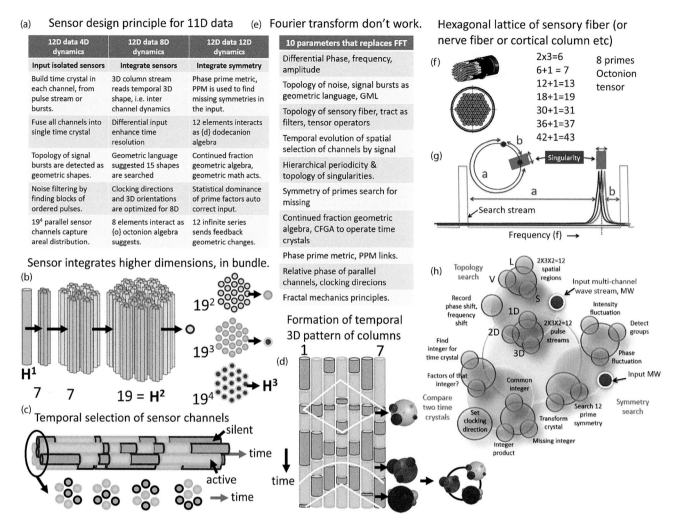

(a)   Sensor design principle for 11D data   (e) Fourier transform don't work.   Hexagonal lattice of sensory fiber (or nerve fiber or cortical column etc)

12D data 4D dynamics	12D data 8D dynamics	12D data 12D dynamics
Input isolated sensors	Integrate sensors	Integrate symmetry
Build time crystal in each channel, from pulse stream or bursts.	3D column stream reads temporal 3D shape, i.e. inter channel dynamics	Phase prime metric, PPM is used to find missing symmetries in the input.
Fuse all channels into single time crystal	Differential input enhance time resolution	12 elements interacts as {d} dodecanion algebra
Topology of signal bursts are detected as geometric shapes.	Geometric language suggested 15 shapes are searched	Continued fraction geometric algebra, geometric math acts.
Noise filtering by finding blocks of ordered pulses.	Clocking directions and 3D orientations are optimized for 8D	Statistical dominance of prime factors auto correct input.
19^4 parallel sensor channels capture areal distribution.	8 elements interact as {o} octonion algebra suggests.	12 infinite series sends feedback geometric changes.

10 parameters that replaces FFT
Differential Phase, frequency, amplitude
Topology of noise, signal bursts as geometric language, GML
Topology of sensory fiber, tract as filters, tensor operators
Temporal evolution of spatial selection of channels by signal
Hierarchical periodicity & topology of singularities.
Symmetry of primes search for missing
Continued fraction geometric algebra, CFGA to operate time crystals
Phase prime metric, PPM links.
Relative phase of parallel channels, clocking direcions
Fractal mechanics principles.

2x3=6   8 primes
6+1 = 7   Octonion
12+1=13   tensor
18+1=19
30+1=31
36+1=37
42+1=43

Sensor integrates higher dimensions, in bundle.

(b) 19^2 19^3 19^4

H^1 7   7   19 = H^2 H^3

(c) **Temporal selection of sensor channels**

silent
time
active
time

Formation of temporal 3D pattern of columns

(d)   1   7   time

(g)   Singularity   Search stream   Frequency (f) →

(h)   Topology search   2X3X2=12 spatial regions   Input multi-channel wave stream, MW
Record phase shift, frequency shift   Intensity fluctuation
Find integer for time crystal   2X3X2=12 pulse streams   Detect groups
Factors of that integer?   Phase fluctuation
Compare two time crystals   Common integer   Input MW
Set clocking direction   Search 12 prime symmetry   Symmetry search
Integer product   Transform crystal
Missing integer

FIGURE 2.8 (a) A table that summarizes how hardware captures 4D, 8D and 12D data from the sensors, integrators and decision-making units respectively. (b) Integration of channels, each channel propagates one waveform. (c) In a practical scenario, some channels get active for a while then other channels get active. (d) The 7 channel system if unfolded into a 2D surface then we find emergence of composite patterns across the channels which are linkers between the signals. These bonding time crystals are identified. (f) Hexagonal close packing is very important in processing prime number of channels, 7,13,19,31,37,43 all these primes could be processed by a hexagonal closely packed channel bundles or nerve fibers. In other words, whenever, there is a hexagonal close pack bundle, the dynamics against the central fiber or channel gives all six primary primes. Even, other primes, 11, 17, 23, 29, 41, 47 are just one less than the hexagonal lattice. Which means when central channel is silent we get 6 primes, when active we get six primes, total 12 prime related dynamics could be processed by a hexagonal close-packed structure with only 48 channels. (g) Electromagnetic, ionic or other energy bursts are the signatures of a singularity point. (h) Entire-time crystal processing is shown here in one chart.

database of events, not by looking into the content, but, in a matrix of dataset, which is converted into a rapidly changing topology. In TCA, any information is converted into a 1D, 2D or 3D...10D geometric shape. Conversion to a shape ignores the actual content, converts the big data stream into a spatial flow of fluid as a function of time or phase. In the rapidly changing 3D distribution of data, imagine it could be a 3D cubic glass chamber filled with clouds one finds the most inactive and the active points that appear, disappear or change periodically. Active points are those which bursts like a lightening; inactive points are those who remain silent in the 3D glass chamber.

Looking for absolute peace and war zones in a Big Data: Say, there are three active points, it means the total

no of primes $p + q + r = 3$, the finding of three key locations in the data confirms that it needs three primes to reveal the hidden dynamics. Then in the glass chamber, one has to further monitor the ratio of spatial distance between the three points (e.g., $p:q:r::2:3:2$), to find the ratio in terms of closest primes to a distinct integer, start from 2, try to use the smallest and nearest primes. In a rapidly changing big data, represented as a fluid flow, if we see the silent and the most active domains are changing with the ratios 2:2:3, 3:2:2, 2:3:2, we understand it is a clocking triangle ($S = 3$), and the corresponding integer is 12. We do not take the ratio of integers but convert to the nearest ratio of primes. Why are only primes taken? Many triangles with similar ratios e.g., 3:5:3, 11:19:11, 19:37:19 etc which builds an infinite series of triangles in the PPM, superposition of all those triangles covers a pure noise in sensing the input, the ratios cannot be deduced to non-prime forms like 3:6::1:2. A given geometry = An infinite series in the PPM. In the GML, the corner points embed distinct geometric shapes inside to grow within, and the whole shape acts as a corner point, assemble with similar points to build a geometric structure above. Hence the corner points at any layer, are the singularity points. In GML when we see a structure, the first thing we do, is to find which geometric shape represents the whole. Then the shapes grow within and above, not side by side. Each shape is placed in a sphere, its corners touch the sphere's surface, a clock derived from fluid flow runs on the sphere's surface touching all the singularity points or corners of the geometric shape. Such an assembly of the clock is defined as a time crystal.

2.2.3 How to Convert Waveforms into a Time Crystal: Non-differentiability

We have described above how to convert 1D, 2D and 3D data into time crystals. Here is a critical review of the technical procedure.

1. Unless we find three clocks at least, a guest clock and a host clock, i.e., a linguistic quaternion, keep searching. The clocks in a neural system self-assemble to modulate the time (Arvanitaki and Chalazonitis, 1968). For this purpose, a need for three clocks is observed in some bio-systems (Berliner and Neurath, 1965). Who?What?Why?How? must be found to write a basic unit of information.

2. The central clock (phase cycle = clock; Ord, 2012) survives even if the environment edits the two boundary clocks (Prati, 2009); i.e., slower and the faster ones. In such three-layered clock, a time crystal turns naturally fault-tolerant; i.e., breaking of time symmetry is uninterrupted. As Figure 2.8d explains, one requires three kinds of sensors. The first one that is specialized in particular symmetry breaking, and specific time domain. The second one that integrates different sensors or fuses widely varied symmetry breaking regions and time domains. The third one is good in projecting a given set of symmetries to infinity and return feedback to adjust the geometry hidden in the time crystal it just sensed.

3. The number of singularity points on the primary phase cycle of a host is the number of guest clocks (Aschoff and Wever, 1976). It is the number of different time flows experienced by a system point as it moves 360°. Each clock can have its system point and can grow its phase cycle structure or time crystal inside by making a new guest-host system. It can connect with neighboring time crystals or phase cycles as guests of a larger phase cycle. It is self-assembly of time crystals side by side (Pippinger and Fischer, 1979; Gurevich and Shelah, 1989) or one inside another (Ghosh et al., 2014a). Symmetry breaking

makes life, as claimed by many (Kuhn, 2008), here, symmetry breaking is searched in the input data to find linguistic decisions Who-What-Why-How in a totally unknown environment.

4. The relative locations of the system points, estimate the initial phase differences among different clocks. It significantly changes the output measurement of the time crystal. To re-assemble the disintegrated parts of a time crystal, reviving the initial phase difference between clocks is essential. Such a phase reset is abundant in biology (Best, 1976; Bruce et al., 1960), thus, biological systems have a memory to remember the phase gaps of various clocks.

5. When a time crystal has only two clocks then a 2D plane is sufficient to represent. If there are three clocks, then 2D phase cycles orient as a 3D sphere. Since three singularity points ensure holding a triangular geometric shape, this is a clocking Bloch sphere. The time crystal becomes information storage and processing device.

6. A spatial crystal appears different, from different directions. Its response remains the same as it is determined by the lattice symmetry. For a time crystal, different rotational directions of a system point in the phase cycle measure different responses. It depends on three parameters. First, the relative phase difference between the clocks. Second, the relative location of the clocks. Third, the relative diameter of the clocks.

7. The repetitive patterns of densely connected phase cycles are denoted as a "mass" in a 3D phase structure, when observers time crystal cannot resolve the distinct clocks in the 3D phase architecture depicted as a time crystal. Then the relative perimeter of the longest phase cycles of the observer and that of the object or event under measurement is defined as "space." Thus, "clock" made of phase paths wire events, one gets a circuit of mass, space and time (Prati, 2009; Ord and Mann, 2012).

8. All singularity points may remain intact, that is, no change is observed in the resonance frequency band, yet, it is possible that time crystal is changing its symmetry. The relative phase path between frequencies changes. To make a crystal, one has to fit multiple phase cycles inside a longer phase cycle, such nesting of phase is meticulously designed in biology (Betz and Becker, 1975). As described above, the singularity points residing on a phase cycle represent a geometric shape. A small perturbation to a system by applying a noise of selected frequency range reveals the singularity points, just like a noise reveals the Fermi level. Perturbation creates a ripple of phase shifts (Johnsson and Karlsson, 1971). The relative rotations of the phase cycles are restricted by the topological constraints. The topology of the phase response curve reveals the variables and the constraints (Kawato and Suzuki, 1978). The desired 1D, 2D and a 3D time crystal structures

form (Ghosh et al., 2014b, 2015b, 2016b). The formation could be linked to the pattern of primes.

9. A time crystal is an artwork of singularity points
connected by phase, not a single point in it is real.
There is no time, space, or mass, it is a network of
phase (Girelli et al., 2009). The phase shift is the only
event in the information processing, caused either
by changing the input frequency (Chandrashekaran
and Engelmann, 1973) or by the intensity of the light
pulse (Chandrashekaran and Loher, 1969). Time
crystal represents any information as topology and
every topology or geometric shape is a single point
or corner point in its higher topology.

10. The appearance of time crystal depends on three
parameters. First, the observer's phase-detection
resolution. Second, the relative phase between the
observer and the time crystal. Third, the orientation
of the observer.

11. The time crystal dynamics strictly depends on the
topology of singularity. Neither classical nor quantum
mechanics address the issue of singularity. In a classically static resonance band one could measure quantum
fluctuations of phase paths. In a random fluctuation of
the phase path of quantum, one could find topology of
phase structure following fractal mechanics.

2.3 THE BASIC CONCEPT OF A TIME CRYSTAL AND THE GARDEN OF GARDENS (GOG)

How the garden of gardens converts 4D to 8D to 12D?
We have described above the fundamentals of a garden of
gardens, GOG in Figure 2.7e. GOG is an important concept,
because the sensors at the interface level, when enters into
the brain, captures 11D data, i.e., clocks are spread over 12
layers one inside another, but at a time, clocks in four layers undergo changes for a quaternion tensor. So, we get 4D
dynamics in an 11D data structure. Now, an advanced sensor
has to integrate several sensory inputs to build 8D dynamics
in an 11D data structure and finally, build up 12D dynamics.
The flowers in the GOG are not real flowers that come from
environmental input as shown in Figure 2.7e. Geometric similarities in the meander flowers coming from different sensors
as 4D dynamics are matched and picked up by the system to
build flowers that would bridge sensors. For example, say the
visual sensor is capturing 4D dynamics where the imaginary
worlds, 3rd, 7th, 8th and 12th are changing simultaneously
the geometric shapes written in their time crystal. Now, if a
sound sensor captures 4D dynamics, where 1st, 5th, 7th and
8th imaginary worlds, the higher-level sensor would notice
that 7th and 8th imaginary worlds are undergoing simultaneous changes. Then, it would see, if similar geometric shapes
are found in the respective time crystals. If the higher-level
sensor (Figure 2.8a) finds that 8th imaginary world of visual
sensor and the sound sensor is undergoing a change of geometric shapes of a very similar kind. Then like Figure 2.5d–f,
the sensor picks up similar geometries and builds a new
meander flower that never came as an input, but may belong

to 11th imaginary world. The process is shown in Figure 2.7e.
Thus, 4D data converts into an 8D dynamics and then to 12D
dynamics. How two imaginary worlds interact and the third
imaginary world gets affected? This magic happens because
of the dodecanion and octonion tensors, just give a look to the
tensors in Figure 4.13. In summary, invisible pathways connecting the cells and pattern of primes linking and governing
the evolution of events within and outside the hardware are
two key features that make fractal tape original.

2.4 HOW TO DESIGN A SENSOR FOR ACQUIRING 11D DATA

The GOG operation described above for three types of sensors to acquire 11D or 12D data (do not worry, classical and
quantum mechanics have one real world, here no real world,
hence 11D = 12D), requires a new hardware operational
mechanism to search and find a similar geometric match in
the time crystal.

**Simultaneous electrical, mechanical and magnetic
resonance: The key to sense 11D data:** A single oscillator
vibrates like a clock when it is periodically pumped from outside and eliminate its damping. The feedback circuit should
harvest abundant thermal noise, chemical energy, mechanical
motions, in this book we have emphasized a unique energy
harvesting regulatory mechanism. It is recently reported
(Ghosh et al., 2016a) that electrical, magnetic and mechanical
resonance frequencies of specific biomaterials are connected
by a quadratic relation $\left(e^2 + \phi^2 = \pi^2\right)$, which ensures canaliz-
ing three forms of noise into a signal of another kind. Means,
electrical noise would feed to mechanical and magnetic
clocks, mechanical noise into magnetic and electrical clocks
and magnetic noise into electrical and mechanical clocks.
It was an important discovery because until then all efforts
to create a long-running clock was to feed a similar type of
energy, which failed in every possible manner. Following a
quadratic mathematical control in the hardware, it is possible
to generate much longer running clocks. Fourth-circuit element H, could implement this typical feature, see Chapter 8.
The philosophy behind this intra-conversion is beautiful.
A noise activates a search mode for the feedback channel, the
active device that executes the quadratic relation has three
pairs of energy levels, a pair each for mechanical, electrical
and magnetic energy transmission.

How does the $e - \pi - \phi$ sensor work? Most resonance
frequencies of the electrical energy are driven by ions or
electrons emerge in the ratio of Pi (3.14..). Most magnetic
resonance frequencies made of a vortex of loop currents or
magnetic flux develop resonance frequencies with the golden
ratio Phi (1.61..). The mechanical resonance frequencies made
of soliton or diffusing molecules in a cavity, vibrating strings
are in the ratio of e (2.73….). Quadratic relation means, if the
magnetic and the electric resonance frequencies shift in a certain way, the change in the resonance frequency of mechanical vibration is fixed and vice versa. Geometric control of
feedback mechanism is a key to harvest noise, two-way feedback fails, but three-way feedback ensures one-way energy

transfer, hence it is stable. If a tape is a linear chain of oscillators, then it would produce time fractals or rhythms (Muller, 2009). Corrections by noise are carried out at all levels since higher-order terms in the infinite series do take part in keeping the quadratic relation intact. It is a careful observation that nature uses geometry to link distinctly operating time domains, where the corrections run through each term of an infinite series. The frequency spectrum for each cell determines the limiting times for the tape inside that cell. In other words, the limiting times of energy bursts of the lower layer cells of a fractal tape are the maximum temporal resolution of the upper layer tape.

2.4.1 Why Fast Fourier Transform Does Not Work

The artificial brain must explore the unseen in the most scenic picture: Fast Fourier Transform, FFT is widely used to convert a wave stream into a single or multiple frequency peaks. Here, while describing time crystal analyzer, TCA, we noted that different frequency peaks built by FFT would never capture the simultaneous shifts of multiple peaks as a group, possibility of a group of groups, the existence of singularity points, the dynamics of relative phase changes between multiple wave streams. It was never thought that multiple parallel channels could hold interactive relationships as described in Figures 2.5 through 2.7, which would require analogs of nerve bundles or cortical columns (Figure 2.8b). The collective evolution of patterns in the simultaneously operating parallel channels as shown in Figure 2.8c and d suggests that one system point creating one channel in a clock is not isolated and distinct. Even in one channel that carries one system point in a complex time crystal structure, the hierarchical geometric information is absolutely neglected in the FFT. A single pulse stream could not just hold hierarchical groups of local patterns distantly located in the time series, but geometric shapes, widely varied information even within a single waveform were never taken into account in FFT.

An eleven-dimensional pattern to link all events: projection from infinity: Imagine, we are given a random set of events and we want to link them without knowing anything about the particularity of events, is it possible? We always think what we see is real, like sun rotates around the earth, or the earth is flat, but it could be that what we measure, actually it is spontaneously giving us what we already have? Then there is no measurement, there is no detection, there is no searching, it is replaced by a spontaneous reply. So, when all the events are interconnected, we cannot solve it, right. However, we can imagine two hardware that is made of the same protocol to link an unknown number of events into a universal pattern. The universal pattern never repeats itself totally, but 12 significant features repeat scale-free through infinity. Anything that is born from this hardware is a piece of an infinitely long, endless chain of pattern, it evolves with time to enrich itself, perfecting the protocol to link events. The hardware learns that means it increases the length of integral choices, in the course of enriching, once the topology of evolving geometry saturates by absorbing matter, cannot

add more choices, it saturates or dies. Our job is to accurately detect the repeating features in the pattern of choices using which we link events, that mathematics would require a new information theory, a new language GML and a metric of choices PPM. Figure 2.8f and g explain how the choice of symmetries in the cables could enable a sensor to read the dynamics of primes or symmetry of events naturally without any effort. Thus, the culture of FFT sensors is very different from the kind of $e - \pi - \phi$ sensors running the time crystal analyzer, TCA.

2.4.2 The Engineering of a Nerve Bundle in Acquiring Hidden Data

The engineering of pixels: the door to different imaginary worlds: The duration of silence and active signal bursts in a nerve bundle carries time crystals of 12 imaginary worlds (Figure 2.8c and d). Multi-channel simultaneous energy transfer among different imaginary worlds is difficult to picturize, it is like suggesting that something that does not exist or impact us. The idea is to bypass singularity or the non-differentiable points by redefining pixels. When we zoom a circle to a level where we see a finite number of circles as pixels, we can pick one such circle or pixel, and inside that the other imaginary world would survive. Since anything inside a pixel is undefined the world inside is imaginary, but that world can define the property of the pixels, thus, affecting across the singularity point. When we discuss the 200 years old mathematical journey of octonions, and introduce the magical automation of 12 imaginary worlds that has no real-world, bonds with observer's similar time crystals, then hand waving argument that phase exchange would do the job is not enough. Either nerve bundle or a time crystal carrier would create a larger singularity encompassing the entire geometric shape, or fill in the blanks inside a singularity: Singularity points of a Bloch sphere bursts in a sequence, clockwise or anti-clockwise to hold the geometric shapes, that is the memory of a cable transmitting time crystal. These singularity points burst signals by harvesting noise in the biomaterials (Kuramoto, 1983). Burst from the pixel is a crude expression of a manifold complex processing happening the imaginary world inside.

The illusive transmission through the nerve bundle: In the conventional mathematical formulations of physics where the differential equation plays a dominant role, a new version of GML has been created, it is called continued fraction geometric algebra, CFGA (Chapter 4). Then the laws of equations would be written by drawing, changes in the geometric shapes. We often find multilayered materials that can carry 4D, 8D and 12D tensors. For example, the core-shell nanostructures, dendrimers and various other organic supramolecular architectures where one could experimentally measure that for any layer of the structure, an energy transfer occurs from its bottom to the top and then simultaneously from top to bottom. We do not need a pathway, nerve bundles arrange in geometry to hold a particular type of clocks, we mistake an effective information-processing organ as transmission

cables. The pattern of prime hardly repeats. The beauty of
the pattern of primes (Figure 3.4) to retain self-similarity as
different integer scale originates via projection from the infinity that governs very different kinds of dielectrics and forces
of interactions into self-similar spatial variations. On a tiny
scale, the quantum principles work, at a larger scale, weak
interactions and strong molecular binding work, and then at
the largest scale, electromagnetic effects come into play. Scale
changes from atomic to meters (10^{12}), time changes femtoseconds to say 100 years (10^{-9} to 10^{15} Hz), $e - \pi - \phi$ sensors do
their job 1:100 scales between space and time.

2.4.3 OPERATIONAL CHART OF A SENSOR

By synthesizing the organic jelly, a neural network like
supramolecular structure was built (Ghosh et al., 2014b,
2015b, 2016b). That brain jelly is poured in a fractal dielectric, designed similar to a brain. The organic jelly-made
device of time clocks morphs the EEG features of a human
brain. The clock-like crystallization of materials is unique
(Brumberger, 1970).

How sensory input integrates by GML: Geometric
shapes built from sensors can integrate two ways; side-by-
side and one inside another. The corner points of geometric
shapes break, and then one inserts a geometric shape in it
(Figure 2.8h). Thus, geometries grow side by side and one
inside another. It is not a 2D structure. The integrated geometric shapes are best represented in a clocking Bloch sphere.
When it integrates information bubbles of Bloch spheres would
grow (Agrawal, 2016b). When a time ring holds more than one
geometric shape, any of them could represent a query and the
other, an answer, together, QA couplet. Therefore, when the
clock runs, the decision is made for a query. The existing 3D

assembly of the Bloch sphere adds new sets of nested clocks
or bubbles with its surface (Vitiello, 2012). During the addition, it even undergoes a phase transition just like an organic
supramolecule (Ghosh et al., 2014b, 2015b, 2016b). The rule
for a phase transition is the same, "symmetry breaking." Here
the 3D oriented structure of phase cycles is an alternative to
the program or algorithm. When synchronized clocks run
together, every time, synchrony selects new wiring (Mirollo
and Strogatz, 1990). To an external query, all the associated
clocks run. All the issues related to a query are built into one
single Bloch sphere structure; no choice is left out.

2.5 COMPARATIVE STUDIES BETWEEN WINFREE, WILCZEK, AND THE UNIVERSAL TIME CRYSTAL

**Frank Wilczek's version of time crystal contradicts
Winfree's version:** Frank Wilczek revived the lost time crystal of the 1990s in 2012 (Shapere and Wilczek, 2012). There is
no guest-host phase cycle in Frank's version, i.e., the guest's
singularity is not explored in the way others did for 40 years
(Figure 2.9). The follow-up works have surprisingly rejected
the concept of singularity. An external energy input signal oscillates the diameter of a given time cycle by beating.
After a while, the original cycle returns (Zhang et al., 2017;
Else et al., 2016; Yao et al., 2017). It is like the orbital transition of an electron in a molecule. Such shreds of evidence
neither support classical nor quantum time crystals. Temporal
oscillation of the diameter of a phase cycle is found in multiple systems. Periodicity in the quantum ground state alone
is not enough evidence to justify a time crystal. Therein two
different time symmetries do not coexist. If we detect one, cit-
ing uncertainty, it violates the basic definition of a time crystal.

Winfree time crystal	Wilczeck time crystal	Universal Time Crystal UTC
Input white noise, clocking wave is output, only one internal clock	Input a clocking signal, output is modified clocking or beating, one internal clock	Input is noise, output multiple waves, more than two clocks.
Two clocks are nested one inside another, no pattern.	Elementary clocks are isolated, only one clock.	Clocks arrange to form 1D, 2D, 3D geometric shapes.
Phase singularity makes time, like curved space makes mass in a crystal.	No singularity, or singularity burst, or phase plane.	Phase singularity one inside other connected by imaginary layers (>=12)
Integration of distinct clocks are not addressed. Cant enter inside singularity	No restriction on symmetry of time crystal, like Bravais lattice, group theory, etc. Cant enter in a singularity	Topology of phase singularity has 12 possibilities to make time crystal, it comes from metric of prime. Enter inside singularity.
Clocking never damps. Under noise, crystal is permanent.	Clocking damps out. Crystal is temporary.	Crystal is permanent, could be solid or jelly or liquid
Phase cycle self-assembles to make crystal, metric sets rule.	No self-assembly	Pattern of primes link symmetry of crystal arrangement
Resonance, synchrony acts.	Irrelevant if quantum else OK.	Acts
Classical mechanics	Classical or quantum	Classical Quantum Fractal
Clocking direction, 3D projection, symmetry link do not exist	No such concept was introduced.	3D architecture of clocks have many rotational directions, non-allowedsymmetrylinksetc.

FIGURE 2.9 Comparison between Winfree class time crystal, Wilczek time crystal and finally the proposal for universal time crystal.

Two kinds of time symmetries should coexist, with or without entanglement (Choi et al., 2017), the uncertainty that we need in quantum would be in the phase path. Now, the change in the phase path is not much investigated in the history of quantum, it requires an understanding of topology when one fuses multiple Hilbert spaces. Frank's version therefore has not two, but one phase cycle, it contradicts the definition of a crystal. Frank Wilczek's proposal is also tagged "impossible" (Bruno, 2013). Comparison between the universal time crystal that Reddy et al advanced from the Winfree era, bypassing the current sensation of a time crystal, suspecting that it needs severe corrections.

2.6 THE DEFINITION OF A QUATERNION, OCTONION, AND DODECANION

Spheres of different unit vectors originating from the different imaginary worlds: Normally a fractal is $f(z^n)$, for example, the Mandelbrot fractal, $z^2 = z + c$, here, we get $z^n = r_p^n \left(\cos p \left(n\theta_p \right) + \sin p \left(n\theta_p \right) \right)$. Order p harmonic would be the function, when $p = -1$, it is a single burst like events, and when $p = 0,1$ then we get wave equation, both in the imaginary and in the real worlds, it means we get time fractal or rhythms propagating across the tape network. The physical significance of p is feedback, a particular rhythm should have a particular type of feedback from the imaginary spaces or times, hence a constant p. Time fractal or rhythm made of multiple p values $f(z_{p1}^n, z_{p2}^n, \ldots z_{pm}^n)$, m is the escape number, each time one enters inside a tape to find a new tape adds value to m. Figure 2.10a and b shows one to one correspondence between the linguistic structure of an event and imaginary world representation of tensors whose each element holds peculiarities of a singularity point, i.e., coordinates on the phase sphere, clock speed, burst energy etc. In the human brain network how different components contribute to the higher-dimensional tensors are outlined.

The dodecanion or icosanion is a combination of dodecanes (or 12-tuples) of real numbers. Every dodecanion is a real linear combination of the unit dodecanions $\{h_0, h_1, h_2, h_3, h_4, h_5, h_6, h_7, h_8, h_9, h_{10}, h_{11},\}$; where h_0 is the scalar or real element; it may be identified with the real number 1.

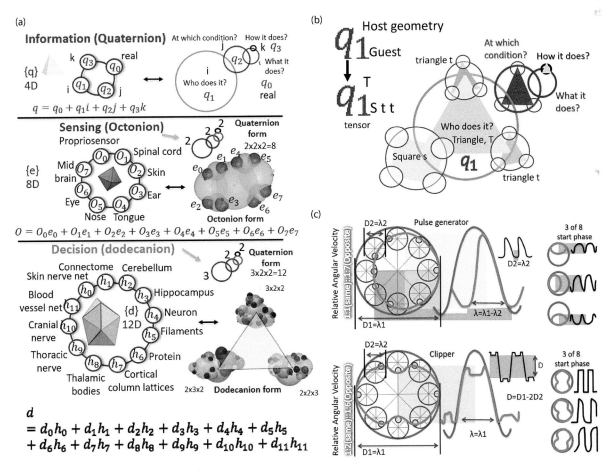

FIGURE 2.10 (a) There are three rows. The top row shows the basic structure of information, quaternion 4D question-answer architecture. The central row shows 8D information structure. Mostly the sensors are designed to process octonions, its time crystal representation is shown to the right. The time crystal representation is the cage that holds information content as singularity shape. Similar cage is shown for the decision-making of an artificial brain in the third row. Three rows show the vector expression of generalized dodecanion algebra, which is tensors. (b) The tensor representation of a quaternion. Four basic questions build a quaternion. Who does it? At what condition? What does it do? How does it do? (c) Two rows show two examples, how nested clocks generate combined output waveform.

That is, every dodecanion for 12 planes or icosanion for 12 corners x can be written in the form

$$x = h_0 x_0 + h_1 x_1 + h_2 x_2 + h_3 x_3 + h_4 x_4 + h_5 x_5$$
$$+ h_6 x_6 + h_7 x_7 + h_8 x_8 + h_9 x_9 + h_{10} x_{10} + h_{11} x_{11}$$

The dodecanion x spreads over eleven imaginary layers located one inside another. When a new class of sensors finds 11D time crystals from the environment, one requires specialized hardware to hold the time crystal, and dodecanion is the data form used. Addition and subtraction of quaternions, octonions and dodecanion (Conway and Smith, 2003) are done by adding and subtracting the corresponding terms and hence their coefficients, but we need a multiplication table to find the product as shown in the Figure 4.13. While processing a time crystal, the wave functions representing the clocks in the different imaginary worlds multiply, a multiplication table maps the effect of two coupled or entangled worlds. However, dodecanion multiplication is neither commutative, $h_i h_j = -h_j h_i \neq h_j h_i$, if i, j are distinct and nonzero; nor associative, $(h_i h_j) h_k = -h_i(h_j h_k) \neq h_i(h_j h_k)$, if i, j, k are distinct, nonzero and $h_i h_j \neq \pm h_k$. There is an association, but that link is a bit complex.

The origin of geometric algebra lies in the arrangement of sub-tensors of a given tensor. At the lowest level there is a di-nion tensor where only one imaginary world counts. Quaternion has three, octonion has seven and dodecanion or icosanion has eleven imaginary worlds at a time, one of the imaginary worlds (that one could be any) becomes a reality for the observer. Recently, efforts have been made to dilute the distinction of imaginary worlds and use linear vectors to regulate the evolution of imaginary worlds in quantum thus build quantum logic gate (Freedman et al., 2018). However, all these efforts dilute the distinction, the linear vectors made of multiple imaginary worlds do not explore the most exciting feature of multiple imaginary worlds. That is the communication between imaginary worlds. The formation of linear vectors taking one contribution from each of the available imaginary worlds is again returning to the Turing's information processing paradigm. Linear composition puts all solutions on a single-phase sphere, so, one could imagine that century-old Bloch sphere to represent infinite possible pulse trains (Figure 2.10c). Universal time crystal endorses singularity, the clocking geometries i.e., quaternions hidden in the octonion, while octonions are hidden in the dodecanion. Thus, many phase spheres form a 3D arrangement of time crystals, we build a GML.

2.7 THE BASIC CONCEPT OF A HIGHER-DIMENSION DATA: A LUCID PRESENTATION

A journey through 12 dimensions or 11D as outlined in Figure 2.11a is easy to learn if we remember one question "who jumps?" as outlined in Figure 2.11b. 11D concept is built on three worldviews. Real-world spanning over 0D–4D that we can easily picturize in our mind feel with our senses. Then the geometric phase world where we could use advanced instruments and detect that changes are happening in the 5D–7D region. Finally, using pen and paper and doing only math we may simulate and extrapolate experimental observation to understand there could be a mathematical world with 8D–11D. Always remember, something should jump through an imaginary space to create a higher dimension. Points jump to create a straight line, we move from 0D to 1D. Lines jump to create an area we move from 1D to 2D. The table of Figure 2.11a and b are the same, one we do not picturize the other is easy to feel. One interesting thought, the followers of the artificial geometric brain should note the third row of Figure 2.11b, irrespective of dimension always there is a quaternion, the linguistic core has to be preserved like an idiom. The journey above 5D requires the PPM, where the geometric correlation of primes or integers guides us to make the jump to a higher dimension. Below we discuss which clock one brain builder should choose to create an artificial brain that emulates Figure 2.11a and b.

Clocks of imaginary layers as particles in the pattern of primes: When one feeds tensors representing the collection of phase values to the pattern of prime, PPM, the phase values are automatically edited since all imaginary worlds have their distinct clocks. Then for the first time when we enter into the world of 11 dimensions in a dodecanion, we encounter the superposition of clocks (see Figure 4.13). Between two classical observations, which lies along the horizontal axis of primes, the system point passes through various choices by which elementary events could be arranged. For a particular tape, this is indeterministic, but for an external observer, it is deterministic if the observer sees all imaginary worlds at a time, but not if the number of worlds is 12, i.e., a dodecanion tensor. Then the coexistence of multiple groups of imaginary worlds ($2 \times 2 \times 3$, $2 \times 3 \times 2$, $3 \times 2 \times 2$) creates a higher-level topology that could represent a non-accessible clock. The trajectory of the phase is not spiral anymore, and it is a spiral of a spiral of a spiral… go eleven times! In Poisson, it is random, in fractal space-time is fractal, and the fabric is the imagination of a physicist. In general, in the PPM metric, it is like a vortex of spirals, many spirals converging to one point, for each prime there is a dedicated metric, or a grammar to fuse the clocks, but even specific numbers like 6, 8, 9 even they have distinct PPM (Figure 3.4). Minkowski clock gives the space-time, Poisson clock plays with the twin paths to generate the quantum world, but in PPM, multiple spirals generate from every prime number point in the world line. Therefore, instead of just getting one imaginary space-time world like in quantum for the Poisson clock ($i^2 + 1 = 0$), we need multiple space-time worlds surviving together, so we need $i^2 + p = 0$. The value of p varies between 0 and −1; exact values of p could be determined. The stress created by the sum of all vortices generates the time series $f(t) = a_0 + a_1(t - t_i) + a_1(t - t_i)^2 + \cdots$ thus, a time fractal or frequency fractal is the perceived mass in PPM clock, while in Poisson clock, it is the fundamental vibrational frequency of the matter, a single value (Compton frequency).

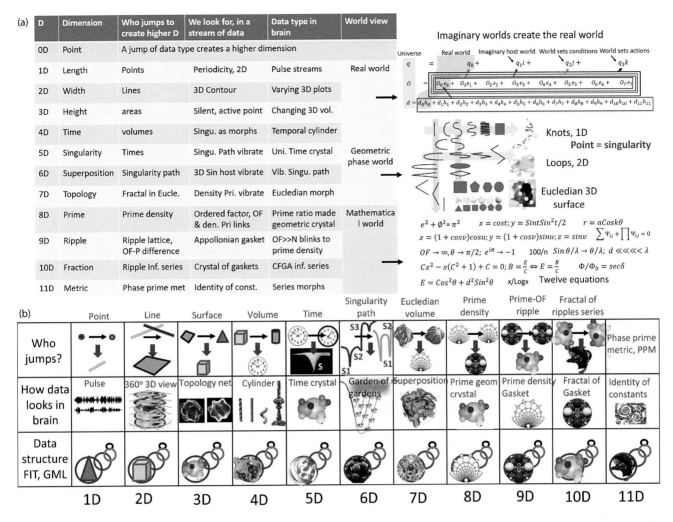

FIGURE 2.11 (a) A table explains the 11 dimensions in terms of a GML. Using three arrows three worlds are explained, real-world for dimension up to 4D that deals with quaternion, octonion and dodecanion. The second world view 5D–7D, is tagged as geometric phase world where 1D structures are knots of darkness or interference induced structures. 2D loops represent 2D structures, and 3D Euclidian surfaces are used to represent the 3D architectures. From 8D to 12D information is classified as Mathematical world, 12 equations are keys in engineering of the mathematical world that would govern an artificial brain. (b) The concepts of 12 dimensions, how to easily understand how the engineering of 12 dimensions is realized. There are three rows in this table. The first row explains "what jumps?" the answer tells how to realize a dimension going one level above. The second row shows how the data looks like in the operational brain. The third row shows quaternion representation of the decisions. Note that in the brain, be it a structure of octonion or dodecanion, eventually the data is arranged as a quaternion.

So, in simple terms, instead of De-Broglie's "particle and wave," Copenhagen's "particle or wave," in the PPM metric we use "particle is a typical network of the wave."

2.8 A COMPARISON BETWEEN GML AND SOFTWARE ALGORITHM

A time crystal is naturally a seed of an astronomical number of events: Encodes event dynamics as soon as it reads time crystal: The time crystals interact with the environment, morphs perpetually, like biological systems (Clark and Steck, 1979). Nature never stops changing. Interfacing clocks of time crystals never fully stabilize into a solid phase structure. Therefore, a standalone time crystal is not a solid but jelly, wherein, correlated geometric shapes construct an interactive matrix. Its phase flips, resembling a jelly. If nature triggers any of the elements or the clock, it ripples the entire matrix. Time crystals do not need to be programmed to suitably linking with new clocking Bloch spheres. Only feasible compositions allowed mathematically are automatically embedded in the PPM which guides the integration of newcomers. Phase connected 3D network of geometries cannot be linearized into a sequence of tasks, like a software algorithm (Figure 2.12). Therefore, it emulates nature like a universal sensor. In most cases, the 3D clocking geometry could be accessed in astronomical ways. Even if one twists input in an incredibly large number of ways, yet the structure would sync reliably. In a time crystal of ten clocks, each clock with eight possible connecting routes can coexist with 10^8 ways.

PPM addresses the automaton concern of Turing: Output is always higher than the input: Since basic

Geometric musical language (GML)	Turning machine based Software/Algorithm
Close loop of arguments, clocks, time crystal	Linear flow of arguments, switch, bits/qubits
Arranged geometry of arguments sensed and 1D-11D shapes arrange to build an argument	How arguments are arranged has no role in the intelligence building
Every argument has structure of arguments inside, all statement undefined, all linked.	Every argument is static, complete, nothing inside, guest-host concept does not exist
Data structure: triplet of linguistics, Subject, clause, verb-adverb: who-why-what-how	Data structure: If-then statement links events as if one happens the other would follow
A catalogue of linking symmetries (Phase prime metric, PPM) look into shape and transform-link discrete events into one.	Human free will uses its senses, intelligence to link events, deep learning or fitting tools designed by human bias also links events.
Shape similarity links uncorrelated events	Waits for human/statistical validity i.e., fitting
Temporal editing may take to different paths, logical tree is purely temporal, instant.	Flow of logic defines paths, condition sets it, not a change in geometric shape or an instant shape
Superposition of geometries, composition of shape making a new shape, make new.	Superposition of arguments cannot happen as linearity does not support simultaneity
Drive to symmetry create/transform logic	There is no natural drive, phase transition big no
Topological constraints sets boundary	Instructions set boundary, limit, halting
Ten dimensional data needs fractal hardware	1D data needs linear hardware
Converts to resonance bands for machine	Converts to 0 and 1 for running in a machine

FIGURE 2.12   A table that compares the algorithm with time crystal-based geometric architecture developed in the GML.

geometric shapes are finite, only 15, and PPM based computing hardware uses only 15 primes, which covers a significant domain of information processing happening in nature, PPM hardware co-exists in the environment, discrete isolation occurs nowhere. Then how could an engine operate and deliver an independent solution? It was Turing's primary objection to analog the computers in his "Lecture on the Automatic Computing Engine" (1947), that additional accuracy in analog machines is costly to obtain. For optical machines, for example, we have to progress to shorter and shorter wavelengths of light. For a machine of a given physical cross-section, this means that the energy consumption of the analog machine will grow as the cube of the accuracy required, whereas for a classical digital machine the energy consumption growth is more like nlogn, n is the number of steps. In PPM, the derived solution could have no relation whatsoever with the input because geometric shapes could arrange continuously in a new way. The infinite number of primes and an encounter with a prime enables the creation of 3D geometry or evolution of an event that may not have any signature left of the input (Figure 2.12).

2.8.1   Historical Background on Hypercomputing and Super-Turing Hypothesis

Undecidability, Halting problem and the necessity of self-programmable algorithm: The most arrogant statement about the universe is the following: anything that can be "computed" can be computed by some Turing machine. Turing himself in 1938 started to find the limitations of the Turing machine (Turing, 1939). Undecidability is not possible to resolve using Turing machines. One example problem is a question, how much time would require to find all

prime numbers? Now, once this question is asked, the Turing machine is not designed to simulate the future and accordingly create an algorithm (self-taught software), estimate computing time feasibility of solution making and halt if it cannot finish within finite time (halting problem). Each clock in the time crystal is an infinite loop, thus, an infinite loop cannot fall into an infinite loop and get locked.

Zeno machine and Fractal machine: A comparative approach: A Zeno machine is inspired by Zeno Paradox (Hermann Weyl, proposed it in 1927; Weyl, 1949), which argues that motion cannot happen if we follow the conventional thoughts, using uncertainty principle one could explain it, or using one imaginary world of quantum mechanics. In Zeno machine a fractional number of an infinite series is considered and it is argued that, since the infinite series converges to a finite number, the computing time will be finite. All terms, if put together, becomes a fractal. The infinite time Turing machine is a generalization of the Zeno machine (Hamkins and Lewis, 2000). For Fractal machine at least twelve Zeno's infinite series exist simultaneously, the Fractal machine absorbs the entire series as anharmonic overtone created by the projection of an infinite series from infinity.

Inductive reasoning and Fractal machine: A comparative view: In a deductive reasoning the conclusion is an absolute truth, but in inductive reasoning (Rathmanner and Hutter, 2011) based on evidence the most probable answer is derived, a little probability exists that it could be false (Examples of Inductive reasoning are many-valued logic, Dempster–Shafer theory, or probability theory which decides following the Bayes' rule as an example). The effort lead to a path beyond Turing, namely to a class of efforts called hypercomputation, Fractal machine is not an effort to go beyond Turing, instead, complement it, so it is not hyper-computation. In Fractal tape reasoning, it is a matrix of truth, the whole concept of "true" and "false" is made irrelevant, both deductive and inductive reasoning is not related to Fractal tape reasoning. Since probability is not used in the Fractal tape reasoning, "Hume Fork" criticism is invalid here (Herms, 1984). Trial and error iteration to perfect a job (Schubert, 1974) is a concern when the algorithm learning protocol is not encoded in the fundamentals of machine protocol, as it is done primarily in Fractal machine using user-machine integrated rhythms.

The fractal machine operates on a multifractal tape: Connecting several Turing tapes is done frequently in the category Multi-tape Turing Machine. In order to encode massive parallelism Multitrack Turing machine is proposed (Sudkamp, 2006). There are plenty of Turing machine equivalents, which are shown to have no more power than the standard Turing machine (e.g., a Wang, B—machine, Register machine). In general, it is a multifractal network of tapes, since the number of cells in the tape is not fixed, it might change every time it is zoomed. In a multifractal system s, the behavior around any point is described by a local power law: $s(\vec{x}+\vec{a})-s(\vec{x}) \sim a^{h(\vec{x})}$, the exponent $h(\vec{x})$ is called singularity exponent.

Chaos, Reservoir, Cellular Automaton, Fuzzy, Genetic Algorithm, Neural network and Fractal machine: A comparative analysis: Random cluster of choices in a black box emulated the dynamics of natural events efficiently, reservoir computing like echo state model (Maass et al., 2002), and geometry inspired recent works on liquid state model (Hazan and Manevit, 2012). For 70 years, successful outputs of the multilayered hidden neural networks leading to deep learning, thenceforth Fuzzy Turing machine (Santos, 1970), Cellular Automaton (Wolfram, 2002) and the works of Genetic Algorithm, all attempts fall under the purview of Turing machine, random points are allowed to mimic the precise input-output observation. Analog computing harnesses randomness for a better search of good points to generate functions and then returns to Turing. The fractal machine does not mimic points in input data, instead, symmetries of Platonic geometries (Forrest, 1971; Preparata, and Shamos, 1985) and their hierarchical network dynamics are accurately mapped as rhythms. There is no need to search in Fractal machine, "right rhythms spontaneously reply."

Super-recursive approaches and the nested recursive approaches of fractal tape machine, FTM: Recursion means repeating itself, recursive means a procedure that invokes the procedure itself, in computing, it is sub-dividing a problem into subproblems, but it could be top-down or bottom-up (Shoenfield, 2001). The recursive feature is implemented in the Turing machine, but Super-recursive (Roglic, 2007) approaches need Hypercomputation like an inductive Turing machine (see above). Sincere efforts thus were made to advance algorithm to estimate halt, even more intelligently in Schmidhuber's generalized Turing machines, yet Gödel's incompleteness argument sustains. A program checking the halting process cannot decide by itself, but the Fractal machine can because it completes the network with the user/environment. The fractal machine applies Gödel's incompleteness to its favor. Regarding unbounded non-determinism, since Fractal machine uses rhythms that encompass user, environment and the machine in a single system, a similar resource sharing is observed, but "time" is deterministic, the FTM machine is always "online." In a big data FTM finds nested recursive points and their groups, the recursive feature of the groups, then consider it as a singular recursive point, the loop repeats ad infinitum, never halts.

2.9 CREATION OF A NON-ARGUMENT

Arguing illogically would make a sense: But how making a nonsense argument? Creating a world beyond Turing and Russel is not easy, the journey to the world of non-arguments is full of obstacles. One mistake takes one back to the world of Turing. On the other hand, being non-sense is being random, with no rules. That again leads nowhere. The other route is the Fractal tape or frequency fractal universe that we suggested. If Turing-Russel arguments were binary arguments, time crystal route is "nested arguments" (Summary in Figure 2.13).

Here are 12 ways of arguing like a stupid that will make sense how the universe argues and processes information at all scales:

1. **No question is born in the universe without an answer, they are born together, live together:** Always a question always has its answer in it, no question could be created without an answer. Say a question is a triangle and an answer is a square, then both are played in a single time cycle. If one plays a triangle, then the square is played out, and if a square is played then a triangle is automatically played. Elementary truth cycles self-assembles but never break apart. There is no true and no false. No distinct question and no distinct identity of an answer. A pair of truths makes a cycle i.e., QA doublet or multiplet.

2. **A QA doublet is a projection, there is no absolute truth, every truth is decomposed into a set of new QA couplets:** thus, it is a fractal network. There are only two ways the decomposition works. QA doublets have self-similar clocks (IFS) one inside another. Thus, QA is never a sum or step-by-step output of another set of QA doublets. Fractal decomposition of arguments is feasible; however, it does not allow split them into sequential or parallel instructions.

3. **Elementary arguments, hierarchical arguments are progeny, they are created. The argument is not the right word:** Clocks holding facts have their plane and several such planes grow spontaneously in QA host clock. A non-sensical argument is alive and extremely unpredictable. Unpredictability is not the sign of randomness. However, a single observation of nested rhythm could originate from several distinct time crystals.

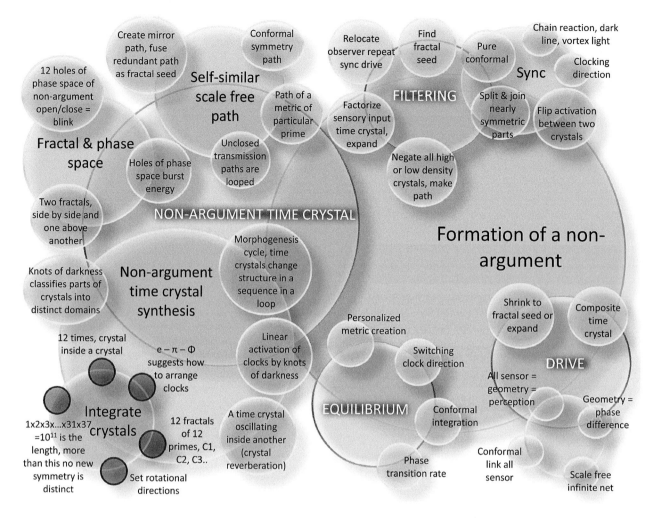

FIGURE 2.13 The methodologies for developing a non-argument. An argument is considered an "if-then" statement. However, it is possible to develop a set of statement that is undefined, each element of the argument does not stand on its own, but requires extensive argument within. When such a journey continues, an argument does not stand on its own. However, an observer could look only two or more layers within and above an argument. This is called non-argument.

4. **No answer or Question is complete, there is always a question that invalidates the answer, fixing that to evolve answer is a journey to completeness:** Gödel hinted for incompleteness, never suggested a route to explore incompleteness. There is no fact, no true statement, the information is not just numbers. Just like in a flute a stationary point is generated, in the dynamic world, we get nodes and anti-nodes due to interference with a boundary. Similarly, facts are static points, it is an emergent feature of the time crystal.

5. **Logically circular or circular logic is the unit of a singular logic to enter inside and go above:** Universal time crystal always bring circular logic but follow the boundary limit of an observer. Earlier, logically circular was a matter of shame, but for geometric prime approach, it is a matter of pride, it is the foundation. A pair of observers noting a fact would note different things.

6. **An infinite number of planes and their arrangement makes a reality sphere or arguments: Evolution of an argument:** Since question-answer couples always together, every possible solution is connected to the question contextually. A question could have many answers in different contexts. All are mapped in a 3D nested time cycle geometry. There are two ways an argument evolves. First, increasing the length of the flute Second, editing the position or the number of holes in the flute, global change means to keep the weight between the arguments same, only change the perimeter of the time cycle. Second, editing the holes means changing the geometric shape itself.

7. **Truth is not a statement, but a geometric shape:** The motif of argument is to make the geometry more symmetric and integrating the geometries into one singular geometry at the highest level (Figure 2.2). Quaternion arguments form a geometric shape

and a complex pattern of the composition of basic geometries grow continuously. The objective is to make the shape symmetric following the pattern of primes, PPM.

8. **A logical argument requires the angular orientation or location of an observer on the sphere, i.e., the perspective of an observer sets the truth:** No logical argument could stand-alone, isolated, it needs support, even though a fractal arrangement argues that at the top there should be one argument. However, even that would require at least three perspectives, first one from the arguments below that makes it, the second one from above or the argument, it constructs with others, and the third one the perspective of an observer. A QA time crystal does not have an idiom or universal truth, no assumption, all assumptions are environment and context-dependent. There is never a truth or a false statement.

9. **"Burst" does not contain any information, but a "silence" does:** Evolution of dynamic and geometric phase is everything, amplitude, mass, space all is an illusion, only time is real. Time does not have any direction, time does not flow, it adjusts relative positions of "bursts" on its surface that's why the past present and future like perceptions appear. There are only "phase" in the universe, dynamic and geometric phases. It is the philosophical foundation of nonsensical arguments.

10. **An expansion of 12 tunes into a raga is the way to expand an argument:** Raga means an active and silent composition of 12 notes from 23 options. GML always builds a time crystal of 12 argument cycles and then expand them. The PPM follows the same protocol. QA couplet cycles perpetually making it nearly perfect. Comparing morphogenesis of time crystals is the route to auto-correct an argument.

11. **Morphing with the universal geometries is a drive for an argument, there are ten ancient philosophical ways of making an argument in the non-argumentative world:** Every natural event has a geometric arrangement of argument and the practice or game of argument is to morph with nature's geometrical arrangement of argument. **How to form an argument in the world of nonsensical argument**: (1) **Negative approach (Neti-Neti)**: Start with everything as wrong and always try to negate everything that is already concluded. (2) **On the last time crystal change the point of an observer or change perspective of an observer** and repeat the cycle. (3) **The philosophy of infinite series e and Pi and Phi** regulates how the integral systems evolve. More one repeats the quadratic relation between the mechanical, electrical and magnetic resonance, time crystal evolves closer to the truth. (4) **Breathing of Brahma**, periodic morphogenesis of time crystal in the form

of argument. (5) **Teardrop and ellipsoid fusion**, time crystals with all input information always have a basin of attractor, like a pole or nucleus. A pair of poles are created from a single-pole, just like a circle morphs into a sphere. (6) **The swastika route of arguments**, the outflow of QA clocks on the host clock reverses the orientation of the host. (7) **The Om route of arguments:** Om letter's geometry suggests three imaginary worlds with distinct dynamic pathways, and one reality point. Thus, a quaternary tensor is searched. (8) **1 = 0 = 1**...The infinite series suggests, what that is a point, is nothing and everything at the same time. Full or completeness and the hollow are the same things. One could enter and see the eight types of nested clocks operating simultaneously (8 levels of consciousness, Yogasutra; octonion tensor), and 12 layers one inside another, as dodecanion for 12 planes, icosanion for 12 corner points. (9) **The argument of Biratapurusa, Varnasram and Hiranyagarva:** Viratapurusha means life inside a life... an infinite network (we limit to twelve). Varnasram means four functional universes (quaternion). Hiranyagarva means burst from nothing or singularity. Three concepts build the foundation of the undefined universe. (10) **The argument of Brahman or universal resonance:** There should be one and only one, time crystal. All sensory inputs, and hierarchical processing would lead to the fusion of time crystals following the patterns of primes.

12. **The universal geometric relationship evolved into the number system is emergent nonrepeating:** An argument always has one less variable than the fixed truth or fact. The total number of variable and truth should compose like the number system metric. A little change in the most static argument or QA clock generates a ripple effect to all nested arguments would change simultaneously. The concept originates from the assumption that "everything is connected to everything" or "the universe is a frequency fractal and consciousness is its music."

13. **Ten limits of nonsense arguments that are responsible for making sense:** (1) **One could enter inside an argument 12 times to generate 12 fundamentally different dimensions.** There are infinite primes, but if one takes more than 15 primes, it is nearly redundant. If one takes more than 15 geometric shapes it is redundant. Eighteen or thirty imaginary worlds would generate very different kinds of time crystal. (2) **An argument is a beating of beats,** Beating means two nearby frequencies start vibrating at a new frequency which is the difference between the two. However, for a time crystal, so many clocks run very closely in the time domain, one may

get crystals of beat frequencies, (3) **All integers do not provide hierarchical groups:** Twelve is important because $2 \times 2 \times 3$ makes the first triangle, similarly $30 = 2 \times 3 \times 5$; $60 = 2 \times 2 \times 3 \times 5$, $18 = 2 \times 3 \times 3$. However, not any number can bind clocks to be used in GML. (4) **It is not possible to frame an argument**, projection is approximate. (5) 10^{11} **truths or pattern of 12 primes (up to 47, three primes are silent) is the limit of an argument**. (6) **There is a 64 cell-matrix** of the teardrop to ellipsoid transition, this morphogenesis dominates all other geometric transformations. (7) **Triplet of triplet fractal forms the boundary** for any information content. (8) **The phase is the only variable**, or relative position concerning time. (9) **Multinions of arbitrary choices is not possible**, one cannot create odd tensor like 3×3, 5×5 or 11×11, so four choices, 2×2, 4×4, 8×8 and 12×12 tensors. (10) **Primes 2,3,5,7 alone is enough** to generate the symmetries created by other primes in an approximate solution.

2.10   FIT SUMMARY IN A SINGLE CHART

FIT is a combination of four GML; PPM described in Chapter 3; Fractal mechanics, FM and Geometric algebra described in Chapter 4. In the FIT, information means no dead fact, with an added flavor of significance, but, a geometric structure, which would provide a correlation between two or more hidden geometric shapes (Figure 2.14, an exclusive summary). A fact is an event here, it defies logic, question and answer have no separate identity; no answer is complete, generation of symmetry following the pattern of primes is the spontaneous reply, in response to a question, that could even be a question. It is fundamentally a different scenario; there is no static or defined state in the elementary information unit. FIT strictly relies on replacing the Fast Fourier Transform with time crystal analyzer where we capture much more information that was never captured in the existing information theories or available technologies. In the fractal information theory (FIT), the information does not represent the number of symmetries, but how a system jumps from one symmetry to another. These transitions

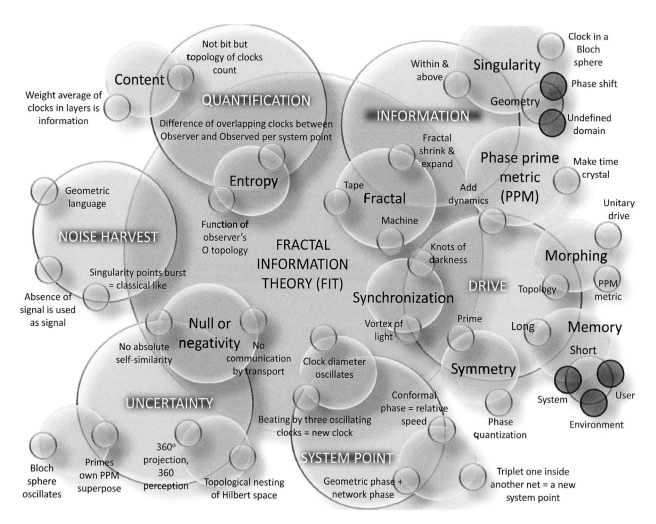

FIGURE 2.14   All key functionalities of FIT is shown in a single chart.

could be such that three or more periodic cycles of switching between symmetries could play simultaneously. Thus, in a given set of symmetries, an observer would find only the random transitions, a beautiful picture of intricate time relationships would unravel only in the 3D time crystal plots. In the FIT, all the time cycles run simultaneously and contain hidden clocks which the sensors never captured, using the pattern of primes, PPM, the input information is expanded, the output is always higher than the input. One can never finish reading the entire information in a time crystal. The observer sees it using its clocks, by locking it from a particular direction. Thus, reading or acquiring input is not one-way traffic, but both ways. Thanks to the pattern of primes, PPM, a questioner already knows the answer, he asks to find only a fractal seed like the key to rebuild the entire solution. Fractal mechanics FM is out there to regulate the manifolds of dodecanion or 11-dimensional dynamics and instead of writing tons of equations by drawing circles using a pen, is out there to see the evolution and sense it immediately.

Basics of FIT:

First, the technical details of the geometric shapes are written on the cycle perimeter of a time cycle. When the computation is geometric, the input geometry received at the millisecond scale could be sent to the femtosecond scale and the process could be completed.

Second, the process of linking discrete geometric shapes continues until the highest-level, singular geometry is reached, it means at the top level there will be only one geometric structure, e.g., a triangle or a circle. In the astronomically long resonance chain, a geometric shape, say pentagon could enter anywhere say it enters in the microsecond time domain but would expand both ways in the time scale. The Pentagon would trigger PPM both toward the seconds time scale and toward the picosecond time scale in the resonance chain.

Third, earlier, information theory required a physical object to hold memory, a switch acting as a bit or qubit, however, since the time crystal is an essential requirement for the FIT, here vortex atoms takes over. Vortex atoms could be made of water ripples when we drop a stone in the pond, it could be a vortex of light, magnetic flux, ion, molecules, electric field (ring-shaped electric field travels around an axon) etc. Materials synthesis of virtual atoms is the information processing in FIT.

Fourth, in a packet of information, no clock is isolated, all the cells of a fractal tape hold a data represented by a quaternion. It should be noted that a dinion or single imaginary world explains quantum information theory. However, for FIT one requires at least a quaternion; thus, classical and quantum information theories fail to emulate the FIT. From the sensory input to the final decision-making, a dodecanion structure of data or a dodecanion 11D time crystal is always preserved.

Fifth, here the time cycles self-assemble to build a hierarchical network of time crystals, which is a learning system, i.e., it self-corrects and grows by itself, even without an input, following the expansion of a pattern of primes which glues different geometries to the input and there is no end to it. During growth, PPM finds many geometric arrangements missing, then it asks outside or evolves to accommodate it.

Sixth, unlike bits or qubits, here in the FIT, there is no isolated information unit, the units are always highly interconnected like a fractal to eventually create a single cell of a fractal tape. It means the total amount of information is always one time-cycle or one clock, since the log of 1 is 0, hence the total information content of a fractal tape is zero by Shannon count. Since all clocks are connected by PPM, count of those symmetries which breaks at a time in a local PPM network is the quantity of information content.

Seventh, from different directions, the time crystal is different, and there are infinite possible directions and the projections would be different. All projections interact with an observer when it sees the information content by becoming a guest or host of the time crystal and constitute a new pattern as a function of time perpetually.

Eighth, no additional memory concept is there in FIT, perpetual cyclic rotation is the memory; no algorithm is required, the learning is spontaneous and the perception-based projection from infinity delivers a defined output, else there is no defined state within a system operating FIT. No collapse is required, the decision is retrieved without computation, no reduction of choices, no logic gate and no switch is required, nothing of the existing information theories survives.

Ninth, there is no "real" domain in the fractal tape, the cell that is under consideration do not contribute to its value. It is surprising, the cells below create an imaginary world, the cells above create an imaginary world and the observer uses own imaginary world to find the weight of time cycles, the cell that is being measured contributes nothing when it is estimated, but of course when the neighbors are estimated, it contributes.

Ten, the unpredictability is enormous. On a single fractal tape three observers measuring the cell state at three different domains simultaneously, would affect each other's measurement. However, they would never detect the error. The measurement of a single cell in the fractal tape may affect the entire chain. Most importantly, that ripple effect would depend on the position of all system points in the entire clock network at that instant.

Comparison between Fractal machine and the Turing machine

Fractal turing tape machine: We have defined the fractal tape machine (FTM) in the same way Turing machine is defined. It follows four tuples (i) Information is converted into nested rhythms or 11D time crystals. (ii) Nested rhythms are absorbed in the PPM and associative time cycles activate. (iii) PPM emulates the PPM or the questioner and that spontaneously replies the missing symmetries in the vortex atom network as the solution; (iv) change the internal circuits to add the missing pieces of symmetries and thus learning begins.

The remarkable changes that would happen if an FTM is widely used in the industry:

1. Multilayered cavity resonator and dielectric resonator networks or multiple "**nested clocks**" one inside another enable "**a virtual instant decision-making.**" A question asked to a clock that syncs with output is sent to faster clocks inside, decision delivers before the slow clock ticks even for once. PPMs of the questioner and the responder melt together.

2. "**Nested cavity**" keeps the "**volume of hardware intact**" as the required resources are poured inside a vortex generator.

3. "**Spontaneous reply**" enables "**search without searching.**" It relies on the resonance energy transfer, via the time crystal wirelessly hence no wiring needed, vortex generators should orient in the floating 3D world in particular directions. No logic gate or circuit is needed, so "**zero junction hardware.**"

4. The decision-making device uses GML, hence "**nested frequency cycles or time crystals hold the geometric shape**" are necessary information not bits. A cycle is a memory, rotation is the processing, "**no transport needed between memory and processing units.**" Capturing geometric structure of nature as a time crystal is itself the discovery of hidden intelligence of a complex event, that is itself the discovery of a new physical phenomenon, data capture means learning and completion of analysis.

5. Reduction of choices or logic gate is replaced by "**spontaneous activation of many paths,**" ability to "**perception capture**" by acquiring projection from infinity, it means how does the complex geometric shape evolution look like in the infinity is taken back as reality. **Since there is no reality, everything is made of singularity, the reality is generated by expanding and projecting the shapes to the infinity and acquiring the feedback geometry as the reality**.

6. "**Perpetual spontaneous editing of slower cycles**" (creation/destruction/defragmentation) "**prepare for unknown**" = higher-level learning is achieved by the pattern of primes, which continually tries to increase its operational bandwidth, causing the birth of slower and slower cycles.

7. "**No programming required**" as "**cycles self-assemble/dis-assembly for better sync at all possible time scales simultaneously**" by following the PPM this is not astrophysics like singular metric, but a composition of several matrices, at least ten different classes are out there (Chapter 3).

8. "**Halting concern irrelevant**" as "**looping nested rhythms of input and memory drives computation,**" as soon as the time-cycle forms a closed loop, the expansion and feed-forward sync/de-sync drive halts naturally as every cycle completes and the projection from infinity or feedback is sent out.

9. "**Nested decoding of sensory signals**" protocol "**captures and preserves hidden dynamics of a phenomenon,**" so we do not have to formulate theories, its purity is preserved in an all-analog process. Since dodecanion is an 11D dynamics hunter it while embedded in the PPM perfects a phenomenon, 99.99% accuracy.

10. "**Equal diameter cycles assemble into a sphere**" enables the estimation of simultaneously operating a million paths "**extreme parallelism.**" In quantum, measurement disrupts, decreases computing speed, here that cannot, there is no chance of entanglement breaking because unlike quantum here PPM drives breaking and regeneration of symmetries in a network where the information processing is carried out. In quantum breaking entanglement destroys all leverage over classical, here, breaking symmetry is the only route of information processing.

11. "**Every component is life-like,**" as the "**geometry encoded in a cycle changes the hardware.**" If one asks where is the PPM located, it is located everywhere in the computer, from the sensor to transporter to the single fourth-circuit element Hinductor H, every single device is programmed to operate a PPM with a dodecanion, so, all must be lifelike.

12. Construction and energy transmission in the hardware follows **ordered factor metric** or PPM made fractal tape that operates by fractal mechanics (FM). It is "**beyond quantum computing**" realized by a "**non-differential tape/machine enables multiple imaginary worlds to play together.**"

13. Existing sensory resolution is "high" or "low," but FTM introduces "**fractal resolution,**" a complex signal's lowest and fastest time scale signals are absorbed simultaneously. During the expansion of a time crystal the fractal seed delivers output with an unprecedented resolution. It just depends whether the observer can sense or not.

14. For the decision-making, no energy is required, instead it executes "**noise harvesting,**" due to the **nested cavity structure with resonance chain**, each layer absorbs its associated energy naturally from the

environment apart from intentionally initiated communication. Noise is required for $e - \pi - \phi$ magneto-electric-mechanical triad to operate, that is a key to running a Hinductor H, the fundamental time crystal synthesizer explored in this book.

15. **"Question re-born memory"**: Geometry of nested cavity is such that any few sets of frequency as query triggers the same rhythm reproducibly, hence, **holding memory does not require a power supply**.

2.11 GML SUMMARY IN A SINGLE CHART

Many-body theorem and the conventional scientific culture to make theory: In all physical theories at a time, only two particles exist in the universe, then using many-body theorems we combine several such pairs and construct the "scientific representation" of the actual universe. Physicists understand that the journey of "many-body theories" is never-ending, adding better and more efficient functions in the energy expression that fits the natural phenomenon better than its previous version is not an absolute path. The "completeness" concept originates from the argument that all

material properties are encoded inside the matter at a particular time scale. Boundary makes it complete. This book on dodecanion driven PPM machines or FTM machines is a journey that compiles the research works that explored the roles of agents beyond the boundary. A quantum description of the universe is incomplete (Bohr, 1935), considering directly that everything in the universe (Wheeler, 1967) is due to changing geometric shape does not work either in quantum mechanics (Wheeler, 1957). Once programmed, circuit built, quantum computation takes no time. For example, in case of pattern search algorithm (say a pattern search in the Clique problem), with the increase of search space, the search time does not decrease much compared to a classical computer. The reason is straightforward: perpetual entanglement does not allow rejection and selection of an astronomically large number of paths, without losing time. Thus, every time entanglement is broken, the time is lost, the overall speed in quantum computing goes down significantly, though entangled states collapse in no time. GML does not require breaking entanglement if one builds a quantum computer using this language, but the pattern of primes (PPM, Chapter 3) should drive the hardware, see Figures 2.14 and 2.15.

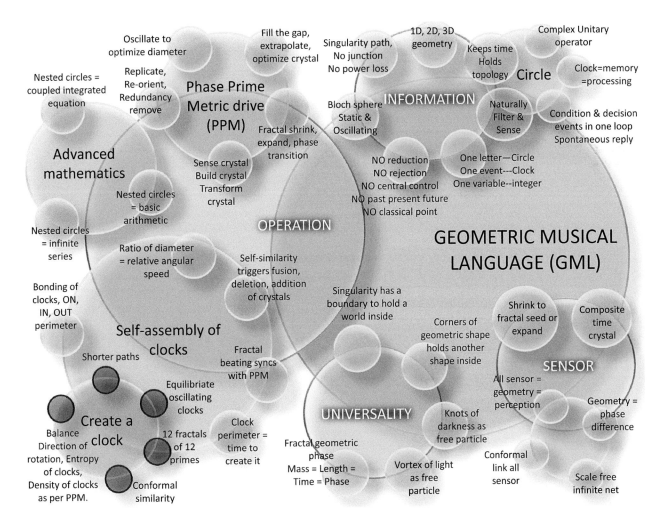

FIGURE 2.15 Different key features of a GML is shown in a single chart.

2.12 CONCLUSION: RUSSEL'S PARADOX AND HIGHER-ORDER LOGIC IN A GEOMETRIC LANGUAGE

Higher-order logic in the layered 12 nested universes: the language of primes: In the last century 1905–1937, the syntax of arguments came into being. The culture expanded from the fact that any sequence of events could be split as the sum of the elementary, isolated, discrete sum of events, which are results of logical queries either true or false (predicate logic). Thus, we introduced the sign "equals" to relate input and output (Church's λ calculus, Kleen's recursive function theory or computability theory, recursive = computable etc.). Darwin's definition of evolution where he argued descent with modification is an effort to endless journey to perfection. **Water leaks, but how could get a logic leak:** The secret of the universal computing concept lies in the philosophy of Turing, where he described brilliantly that the world of information and event is represented as a grid (a piece of paper) made of cells containing discrete events. If we consider a leak in the isolated cells of the grid, the entire adventure of 100 years old computing will fall apart, isolation, pristine identity that gives completeness. All previous attempts to go beyond the Turing kept the grid intact and yet tried to find ways to draw multiple paths between a pair of facts, thus, converted into a Turing machine eventually. A fractal tape only one cell in the grid, but if we enter in that cell we get millions of grids, select and enter inside, then it would open up another grid, and the journey continues, then the concept of first-order logic would not be the answer of a query, "true or false," preferably a 3D network of truth. In the Turing world, we can write many FOLs using a pen on a piece of paper, torn each piece of paper and arrange them in a line (Turing machine). However, in the new world of FIT, we sit with a piece of paper forever, several new papers exist, but invisible to each other, they grow within and above, each paper has no idea that due to their composition new papers are taking birth. Thus, logic leaks, or nonsense, illogical decisions govern. Turing machine where the solution is restricted to the instructions written in the tape, in fractal tape, it is not. There is no previous and the next step, unlike Marvin machine, every state is a solution to the problem, the solution is delivered perpetually. The system is interactive like a "reactive system," response to the environment, but as part of the environment, instead of the state transition diagram FTM has a set of PPMs, which is a highly connected 3D network of time crystals and evolves in a lifelike manner.

Russel's paradox and the magic of time crystals: One of the most beautiful parts of a single, unified time crystal extended from the smallest spatial dimension to the large, covered with clocks of the shortest time domain to the longest is that only 15 geometric shapes with 15 primes can sing it 99.99% accuracy. Humans sense 4D, some other sensors could read 11D, but the pattern of primes reaches saturation above 10^{12} clocks, then it begins counting in a new way, so, the properties of the properties is a set to it. The argument that the universe is a frequency fractal of primes and the consciousness is its music is an excellent example of Russel's paradox where the properties of properties become an element of the set in its core, hence, the set theorem cannot explain the situation. Now, since Russel's paradox sets in, all higher-order logic cannot be transformed into a second-order logic. In such systems, a set of oscillators decides the order of logic and if it is said 5, then all five orders of logic operate independently and co-operatively in a single system. Of course, we can determine from the coupling map which clock would trigger which order logic. These time crystals cannot be transformed into a set of a linear sequence of clocks.

Processing higher-order time crystals or network of events: First-order logic is about the strict factual statement; here it is the coupling between the symmetries within a given imaginary world. Note that all 12 worlds are imaginary, the one that communicates with the external worlds acquires the state of a real world. When the interactive imaginary worlds have a similar local domain of clocks, this is a property of the higher-order coupling and thus, fundamentally different from the "direct coupling" that happens within an imaginary world. Quaternary, tensors hold the second-order logic, which is more potent than "direct coupling" induced first-order logic. Octonion tensors generate duality when we see the tensor as a composition of quaternions in the Figure 4.13, it is a third-order logic; while the dodecanion tensor holds a triangular composition of quaternions in the same figure, hence it is the fourth-order logic. The hierarchy of logic is an oxymoron, because in a circular loop, there could never be an "if-then" statement, still, when the projection comes from infinity, be it an illusion, not a mathematical perfection but a PPM driven hardware would tag it as a "reality."

Sātra kuṇḍalinī bījajīvabhūtā cidātmikā|
Tajjaṁ dhruvecchonmeṣākhyaṁ trikaṁ varṇāstataḥ punaḥ||
Ā ityavarṇādityādi yāvadvaisargikī kalā|Kakārādisakārāntādvisargātpañcadhā sa ca||
Bahiścāntaśca hṛdaye nāde'tha parame pade|Bindurātmani mūrdhānte hṛdayādvyāpako hi saḥ|

From the endpoint of the spinal cord (Mulasringata~corpus callosum) to the pons (mahasivapada~brain stem, which is frequently written as "parama pada" too) the power of rising from sleeping (Kundalini) resides, an eternal wish to rise that comes from a point and passes through the five elements of the universe, to finally connect the supreme point of pons (brain stem) to the universal consciousness. That very point (mahasivapada, paramapada) is going global (byapaka) and it pervades all of them to connect it to the soul.

3 Phase Prime Metric (PPM) Links All Symmetries in Our Universe and Governs Nature's Intelligence

3.1 TEN CLASSES OF PHASE PRIME METRIC (PPM)—A PATTERN OF PRIMES

How physics laws were born: It is not possible that physics laws were born one by one after the universe got created. Not before, not after, just at the moment the universe was born (Steenrod, 1962). We leave the debate on the origin of the universe to astrophysicists (Weil, 1951), instead get concerned about the demand for the pure information architecture of the universe. Phase prime metric (PPM) is a pattern of primes to create integers in various ways, and it is only a principle of organization of symmetries, which is no organization of other elementary concepts or system or entity. A metric of primes is not a derivative; it is original, primitive. Metric means to measure; here, in PPM, phase relations that might bind distantly located integers is measured. There is not one, but several forms of PPM as outlined in Figure 3.1, which accounts for different kinds of dynamics that number system follows. Some metric act as a glue that links discrete events together and forms a time crystal. A typical metric exclusively provides direction of a clock. A particular metric provides the scaling law of primes or ripples in the integration of clocks. In the history of mathematics literature, several attempts were made to find relations between the primes, detecting a new prime. When we mention, the pattern of primes, it means the pattern created by prime numbers to build integers, not linking the primes themselves. Only one available work is in the literature (Reddy et al., 2018). When we use the term, PPM, it means a composition of a large number of metrics, it is a

system of operators as compiled in the table of Figure 3.12a. Ten classes of metrics in Figure 3.1 are so versatile, that one could discover those relations as soon as Reddy et al. considered that the primes have no identity, they combine to build integers. By considering the collective identity of primes as geometric, they opened the pandora's box. The composition routes of primes are not random, there are plenty of geometric arts yet to unravel. The patterns of primes appear distinct, but they have a close relation in between. Here, we compile largely ignored artworks by several amature scientists and mathematicians between the neighboring integers when viewed with the eyes of prime numbers. With geometric musical language (GML) such pattern of primes provides a form free logic loop made of clocking geometric shapes. It is a closed circuit of transformations from one geometric shape to another, the sides of geometric shapes are either integers or primes as described in Chapter 2. Each prime is distinct, pure and generates a unique self-similar fold in the surface of a 3D metric. For 15 significant prime numbers, one could quickly identify 15 different dynamic features as shown in Figure 3.4 each resembling a typical physical phenomenon and associated with the fundamental constants found in nature. We have outlined a summary of these findings in Chapter 10. Therefore, the physics laws could be associated with the self-similar curvature in the metric of prime, PPM. Thus, PPM is self-referential, deductive axiomatic system (Weil, 1951; Steenrod, 1962).

Research in the last decade to address the problem of computing: Historically, there has always been an effort to

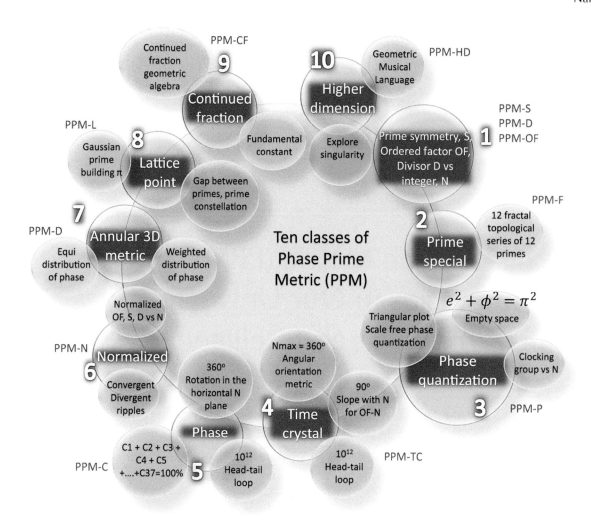

FIGURE 3.1 A table showing ten classes of PPM.

"go beyond Turing" (Copeland and Proudfoot, 1999). It started by von Neumann to build a non-von-Neumann machine; however, incompleteness remained beyond the adventures of an engineering lab. The quest continued with Turing in 1953, with an attempt to building non-Turing machines, patterns that create spontaneously, one such class is hypercomputing (Siegelman, 1995). Feynman proposed to "replace all physical laws with changing patterns" in 1962 (Feynman, 1965) further advanced by several authors (Sahu et al., 2009, 2012), possibly inspired by Conway's game of life in the 1960s. The quest never stopped, it continued with a series of works to make computing nature-inspired (Ridge et al., 2005). In 2010, Feynman's suggestion that patterns would replace the physics laws turned out to be true, at least for some physical phenomena. Two physical events were emulated using quantum properties of molecular clocks (Bandyopadhyay et al., 2010b, 2010c) on a group of molecules in the monolayer. First, the evolution of cancer cells. Second, the diffusion of electrons on a grid of molecules. The result was obvious. Cellular automaton works as a non-Turing machine (Sahu et al., 2009, 2012). Two events were emulated, but it would require new hardware every time one tries to emulate a new natural phenomenon.

Hacking nature using a PPM: Non-Turing does not mean absolute randomness, or imaginary, every part of it should be undefined. The quest for a universal system ranged from the neural network (Bandyopadhyay et al., 2009a, 2009b, 2010a), to deep learning, the efforts to confusing a cell in the Turing tape is fooling it, not setting a new pathway beyond Turing. All efforts to go beyond failed because the undefined routes were never explored; a journey through singularity was considered impossible. Instead of facts no one looked into the architecture of confusion, mapping confusions. No one believed, a journey through confusion that ends with facts could solve a problem. In summary, a century-old saga of failed attempts tell one thing, find a way to make a journey through undefined pathways, if defined, nothing can stop converting it into a Turing machine. A dielectric resonance artist would consider that everything in this universe is made of dielectric resonators, everything vibrates at specific frequencies, be it mechanical, ionic, electrical, magnetic, nuclear (Lauber et al., 1994; Pistolesi and Manini, 2000; Pechal et al., 2012; Lee et al., 2011; Joshi and Xiao, 2006). For this worldview, one can calculate the geometric correlation between the resonance frequencies ranging from the Planks scale to the largest dimension of the universe, as an endless chain of resonators, one inside another. Then that chain of vibration where resonance peaks reside inside a peak would be an undefined pathway. If a system B is made of system A alone, B is non-existent without A, then AЖiB represents the system. Ж is a mathematical operation described in Chapter 4. On multiple occasions in nature, the ratio of resonance frequencies of biomaterials follow the pattern of primes. Sahu et al. first observed the pattern of primes while monitoring the vibrations of microtubule nanowires (Sahu et al., 2013a, 2013b). Who wrote the geometric shapes in the proteins, why did they write it? Then, is that pattern of primes universal, written everywhere in the brain to the cosmos? It was historically used by many researchers to claim that the compositions of primes connect everything in the universe (Haramein, 2001; Tomes, 1990). If true, every single system in nature is a PPM and hardware inspired by this PPM should absorb the information in nature as is. During absorption, it will not destroy the 3D integration of events as geometric shapes. Even if a small fraction of total information if absorbed as geometric shape, PPM would add geometric shapes to extrapolate that into total information (Horodeck et al., 2005). Thus, PPM hardware would analyze events that have not yet happened by using the metric of primes. Consequently, PPM would hack nature, copy and paste it in the hardware. Therein one would see nature, the way it unfolds.

Ten classes of PPM: a journey of geometric shapes: Pattern of primes should link integers, but in reality, as we would see below, PPM signifies symmetry of the geometric structures more. Several versions of the PPM reveal all possible dynamics, symmetry breaking, phase transition, and morphogenesis of the geometric shapes, whose sidearms ratios are the integers. Figure 3.1 also notes the physical significance associated with a particular metric. Either one could calculate the ordered factor of an integer, which is simply finding the divisors of an integer and arranging them, as shown in Figure 3.2a using balls. In this figure, one could nest the prime number of balls representing an integer in various ways. For example, one could find how many compositions of primes generate an integer, for example, for $N = 12$, we can count 3 ($2 \times 2 \times 3$, $2 \times 3 \times 2$, $3 \times 2 \times 2$) or even 8 if all possibilities are taken into account. In other words, how do we perceive an integer, that approach could lead to a very different kind of geometric shapes for a metric. The possibilities are plenty. Following ten protocols, three modules are made, as a core decision-maker (Ghosh et al., 2014b). They are Sensors to sense the environment. Initiator filters and makes decisions. The processor stores the learned situations. Regulator filters the learning parts and evolves the entire architecture. The four hardware modules operate independently in the non-computer or artificial brain core. Together, they have a purpose. Specific features of a hardware increases the length of its operational PPM, i.e., the highest integer, up to which the pattern of primes one should be drawn.

Ten different patterns of a metric act coherently to frame an event. As said in Chapter 2, an event is a geometric shape whose corners touch the phase sphere with infinite pathways connecting the corners in a loop. A system point runs through the loop. PPM is not a random structure (Li and Yorke, 1975) though it never repeats. A generic operator "phase prime metric" puts an intricate geometric detail of the event in nature. One represents an event as a stream of waves, then the phase-signal relationship is written as a few geometric shapes in a network of clocks. Then write that geometry as a set of integers in the metric structure. Integers are product of primes, so, primes make structures inside the structure made by an integer. PPM can geometrically link the phase of two events separated by a singularity barrier. Quantum computing can do this once (Zhou et al., 2013), but PPM based computing is built on an interactive twelve imaginary worlds, where there could be no need ever to come to reality. For a given number of vertices (icosahedron has 12 vertices), there could be choices for multiple 3D architectures, and or multiple compositions of distinct geometric shapes. Moreover, when we activate a silent integer between a pair of active integers in the PPM, it could link the neighboring integers in many different conformations. Thus, a PPM extrapolates an event to the unknown future. We would explore all these magical interplays between integers in the ten sections, one section each for a particular class of metric.

Recreating nature in a PPM non-computer: The principle of decision-making by using a triangular combination of PPM, GML, and fourth-circuit element H. Nature uses the triplet protocols to integrate the primary events to construct a complex event in the same manner as the primes integrate to form an integer. Since a typical group of primes repeats in constituting multiple integers ($2, 2 \times 2, 2 \times 2 \times 2...$), eventually, most integers are connected via common primes. Thus, bonded by multiple groups of primes, or symmetries (2×3, $2 \times 2 \times 3 \times 3, 2 \times 2 \times 2 \times 3 \times 3 \times 3...$), the entire integer space acts a singular network of symmetry made of a few primes, as argued in the Figure 3.2b. It is a beautiful plot to show that 15 primes are enough to build a language of symmetry that nature would follow. The symmetry is conserved; it breaks only when an event happens, a change in integer emulates this physical process. In the previous Chapter 2, we learned that to convert all integers into a geometric shape, we do not require infinite, but only a few say 15, 1D, 2D, and 3D shapes, which could assemble one inside another to reflect any symmetry of primes. As said, since, a group of primes repeats, several discretely spaced integers are connected in a hierarchical network of geometric shapes. To sense, to memorize or to make decisions, we search for the common geometric shapes in the events happening around us in nature. Bring those matched shapes to the integer series that integrates the events by bridging the gaps by adding the missing shapes. Finally, the integer series can fuse, extrapolate, reduce, transform, or morph the discrete events into a single geometric shape as output, we would see how intricately geometric it is, below.

How does a metric make a journey through the singularity: Two concepts irritate a pure mathematician. First, when a physical significance is attached to an integer or a mathematical process as we do it here. Second, free from the materialistic attachment, mathematicians do not like undefined singularities holding a geometric shape.

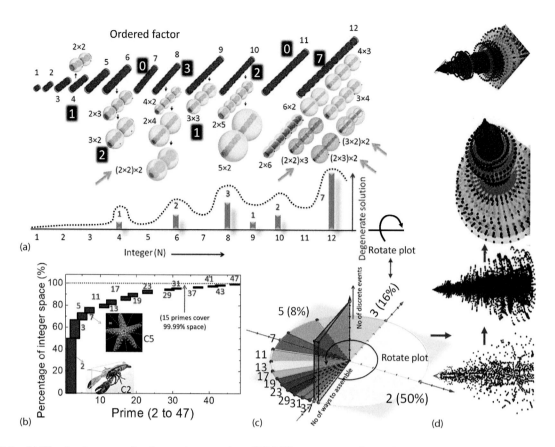

FIGURE 3.2 (a) The development of a phase prime metric or PPM. There are two sub-panels. In the top, we show an array of balls in a single line representing the integers. For each linear arrangement of balls, groups of balls are tagged which could vibrate together as single phase space. All possible compositions for a single linear array are shown below the line. By changing the order we get different combinations here, e.g., 2×3 is not equal to 3×2. In the bottom panel we plot the count of group compositions we can make from a single number. This number is also the number of degenerate solutions for the generic oscillations of a string. (b) The contribution for a particular prime in the integer space is counted. For example, prime 2 contribute to 50% of all possible integers in the number system. For each prime while calculating the contribution, only its contribution alone is calculated, for example 6 could be counted for 3 and 2, we have counted 6 only once for 2, not 3. Similarly we have counted for 15 primes and reached total contribution of 15 primes to 99.99%. (c) The degeneracy plot of panel (a) is rotated along the integer axis, the total number of rotational angles is 15 and their contributions are plotted in the XY plane while the degeneracy is plotted along the Z-axis. (d) In the panel (c) and the panel (a), the continuously decreasing contributions of primes are ignored and all 15 primes are given equal contributions 24 degrees. Then the bottom plot in panel (a) is rotated 360 degrees to get the plots of panel (d) bottom to top.

How could one element that does not have an identity, have a significance? Corners of a geometric shape are made of singularities, and even here we assign a size of phase values to a singularity. When the clocks run, or a system point moves through the close loop connecting the corners of a geometric shape, at the singularity points one hears "burst," and all-around nature, the singularity points are known to create bursts (Mallat and Hwang, 1992). The time gap between a pair of "bursts" tells us the phase gap. The ratios of the phases between periodic events hold the geometric parameters. Journey through singularity means entering inside a burst to find another set of bursts within the primary burst. It is difficult to picturize how could many bursts be hidden inside a single large burst. It is very much possible that a single burst from a singularity that we sense is actually an illusion, higher resolution would reveal that it is not 1 but a pentagon or hexagon hidden inside, whose five or six singularities are bursting off. Thus, one could enter inside one

then find 3, and inside each of the three bursts, find some more geometries. Each layer would count the total number of bursts, those are integers that is all a PPM requires to build an architecture of choices. An external sensor could lock on any part of the hardware. Due to two limiting time resolutions of the sensor, the fastest clock and the slowest clock are selected naturally. Thus, decision-making fixes the end of computing before even it begins synchronization.

3.1.1 15 PRIMES CONTRIBUTE TO 99.99% OF ALL INTEGERS IN THE UNIVERSE

In Figure 3.2a, one gets the freedom to insert the balls inside a ball, a nested grouping of balls, within and above, naturally explains why ordering is so important. When one takes three balls first and then inserts two balls inside, we get a group of six balls, but that architecture is very different if we take two balls first and insert three balls in each of the two. Counting is the same but the geometric significance is different. To build a PPM along the horizontal axis the integers are plotted and the vertical axis shows the number of compositions of primes required to represent the corresponding integer. When we connect the choices using a point one should remember, the vertical dots which are nearby so using a pencil we connect them, it means a lot. How many primes are involved, they are similar or not, it is an important factor, not what is that prime. Since the pattern is prime independent, it can happen at many places in the integer space, this is how a fractal pattern triplet of triplet of teardrops could be found in the infinite long integer space. The prime independent feature gives the fractal nature to the metric. Does it mean, primes would never have any identity?

The question is answered in Figure 3.2b.

50% of all integers in the number systems are divisible by 2, every next number we get by adding 2, starting from 0, the contribution of a prime P_1 to generate integers is $C_{P1} = 1/P_1$. The next prime is not just $1/P_2$, but there are numbers, which are products of $P_1 \times P_2$. Thus, P_2 makes only $C_{P2} = 1/P_2 - 1/(P_1 \times P_2)$ additional contributions. Since P_2 is 3, instead of $1/3 \sim 33\%$, deducting the 16% common factors, only 17% of total non-primes is contributed by 3. Together 2, 3 covers $50\% + 17\% = 67\%$ of entire integer space. For P3, the contribution is $C_{P3} = 1/P_3 - \{1/(P_3 P_2) + 1/(P_3 P_1) + 1/(P_1 P_2 P_3)\}$. In general, $C_{P3} = 1/P_n - \sum_{i=1}^{n-1} 1/P_n P_i - \sum_{i=1}^{n-1} 1/P_n P_i P_{i-1} - \sum_{i=1}^{n-1} 1/P_n P_i P_{i-1} P_{i-2} \ldots n-1$ terms

$$C_{P_n} = \frac{1}{P_n} - \frac{1}{P_n} \sum_{s=1}^{n-1} \frac{1}{\prod_{i=1}^{s} P_i} \qquad (3.1)$$

At $P_n = 17$, $C_{P_n} = 0.02$ i.e., contribution is 2% $\lim_{n \to \infty} C_{P_n} = 0.02$ There are two major conclusions. First, only 8 primes contribute $\sum_{i=1}^{n} C_{P_n}$ to around 98% of integers in the number system, 12 primes could reach it to 99.99%, by fitting approximation. We have plotted the contributions of primes in two comparative plots, Figure 3.2b, the additional contributions of primes are added as columns, while in Figure 3.2c we have made an additional contribution of a given prime in a pi-chart. Inspired by the pi-chart, we have rotated the horizontal axis of the Figure 3.2a, the bars representing the ordered factor value rotates to form a disk, if it's a point then a circle and we get a 3D structure as shown in Figure 3.2d. While rotating we have kept the contribution of all primes equal, because that is what an artificial brain is going to do to sense the universe and respond uniquely.

No, we do not need a giant metric to sync with the universe: The necessity of a giant metric was argued as an essential step to address the complexity (Zurek, 1989). When Ghosh et al. studied neurons much to their surprise, they found that firing pattern of neurons cut a slice of PPM up to 23 (Ghosh, 2016a; see details Chapter 6). The same thing happened with Sahu et al. when they measured the electromagnetic resonance pattern of microtubule they found abundance of ratios of peak frequencies up to 13 (Sahu, 2013a, 2013b). They went on studying the entire brain, many times encountered the number 12,

that we had to change the resonance chain theory (Ghosh et al., 2014a). We might sense that we need a large number of primes, a long metric to explain a complex system, later the theoretical study of vibrations and pattern of frequencies showed that we need to know only first 12 primes, up to 47, max (3 primes are silent, else first 15 primes). Each prime, for example 7, forms a product with other primes say 31 and creates a series with unique geometric symmetry ($7 \times 31, 7 \times 31 \times 2, 7 \times 31 \times 2 \times 2$, and so on). However, only 12 primes could cover 99.99% of all possible symmetries of the universe. It means $2 \times 3 \times 5 \times 7 \times 11 \times 13 \times 17 \times 19 \times 23 \times 29 \times 31 \times 37 \times 41 \times 43 \times 47 = 10^{12}$ number of clocks is enough to create a metric of prime-based hardware that could spontaneously synchronize with all events happening in nature. Roughly we have 86 billion neurons in our brain (approximately 10^{12}); there are not many neurons to add during evolution of brain. One could hypothesize that all life forms could be metric of primes. At the highest level, it starts with a little part of the metric, then we enter into the singularity points and find tiny slices of the metric. In a few layers, we could construct architectures of enormous complexity. Based on the above arguments, the Universe is not a pre-coded machine. It is self-synthesized by PPM. Everything we see, if we enter inside it and do not stop our journey, make the motion continuously inside of whatever we get, it will be a vacuum in most of the places. At the smallest mass which would appear as solid but when would enter inside, again, the most part would be empty. Thus, within and above is the universe we live in, the structure of information is the universe.

3.1.2 Ordered Factor Metric and Its 3D Version

What is a PPM: Say in the beginning nature is given only one quanta, to create the universe and only allowed to decrease the wavelength of a single quanta, by splitting it one by one, dividing one into two quanta, then each one into three and so on. Now, nature's approach would be top-down spatially, from the largest boundary of the universe spatially making a journey to the plank scale. Simultaneously mapping the shortest to the longest time is possible. A standing waveform is part of a clock (sin or cos wave = a circle with a rotating system point) and it occupies space, so it is a unified way to map both space and time. A given number of nodes of a waveform is combined until it reaches the prime number of nodes. Then in a particular way, both time and space are integrated. Therefore, one gets a composition of frequencies. The number of composition C is plotted as a function of a number of quanta n. A group of primes determines C compositions. $C - n$ plot shows geometric correlations of different choices (Harris and Subbarao, 1991; Warlimont, 1993; Dickau, 1999; Richmond and Knopfmacher, 1995). The $C - n$ geometric pattern is called a PPM. Twelve distinct ways one could connect the dots or events in the PPM. Therefore, there are 12 sub-metrics, or 12 geometric patterns, one top of another. If 12 special methods are adopted, each pattern is seen. The beauty of PPM is that once a point is defined, its integration topology is revealed. If it is time, then the past and the future is also evident. The number of compositions varies non-linearly with the N number of the waveform. The origin of nonlinearity is the irregular occurrence primes. The prime number of waveforms has no harmonics, hence, no superposition. Recently, the fractal metric is suggested to replace the space-time metric (Ord, 1983). PPM shows phase fractal, triplet grouping fractal, phase quantization, organized selection of clock and anti-clock spin, circular traps at logarithmic space, a saturation of patterns for 1 = 1 million. **Creation of a PPM:** Total number of harmonics for a positive integer equals to number of its proper divisors, for all positive Integers N, HN (N)

$$= \left[\left((a_i + 1) * (a_{i+1} + 1) * (a_{i+2} + 1) \ldots * (a_n + 1) \right) - 1 \right] \quad (3.2)$$

Here, HN is Number of Harmonics for N. $a_i, a_{i+1}, a_{i+2}, \ldots a_n$ are powers of prime divisors.

Recursive Function for the calculation of time crystals: Let's assume N is a positive integer number. $P_i, P_{i+1}, P_{i+2}, P_{i+3} \ldots P_n$, are proper divisors of N. $DP_i, DP_{i+1}, DP_{i+2}, DP_{i+3} \ldots DP_n$, is the number of proper divisors of the corresponding proper divisor of N. Number of Nested Cycles:

$$NC(N) = \left[(DP_i - 1) + (DP_{i+1} - 1) + (DP_{i+2} - 1) \right. $$
$$\left. + (DP_{i+3} - 1) \cdots + (DP_n - 1) \right];$$
$$(3.3)$$
$$NC(N) = \sum_{i=1}^{i=n} (DP_i - 1)$$

Here, NC = the number of nested cycles; n is equal to the number of proper divisors of N. It means the number of nested cycles is the "sum of the number of proper divisors −1" of each proper divisor of N.

Recursive function for time crystal: If proper divisors of N have the value greater than 7, then it can also be written in the form of its proper divisors, which has a proper number of divisors greater than Zero. Let's assume: $\{(P_{i+21}, P_{i+22}) > 7\}$ are divisors of P_{i+2}, $\{(P_{i+31}, P_{i+32}) > 7\}$ are divisors of P_{i+3} and $(P_{ns} > 7)$ is a divisor of P_n. $DP_{i+21}, DP_{i+22}, DP_{i+31}, DP_{i+32} \ldots DP_{sn}$, is the number of proper divisors of corresponding each proper divisor of proper divisors of N. Note: We do not consider 4 and 6 because 4 has its proper divisors 1,2 and 6 have its proper divisors 1,2 and 3 which do not have any nested cycle. Number of nested cycles:

$$NC(N) = \left[\left((DP_i - 1) + (DP_{i+1} - 1) + (DP_{i+2} - 1) \right. \right. $$
$$+ (DP_{i+3} - 1) \ldots + (DP_n - 1) + \big)$$
$$+ (DP_{i+21} - 1) + (DP_{i+22} - 1) + (DP_{i+31} - 1)$$
$$\left. + (DP_{i+32} - 1) \ldots + (DP_{ns} - 1) \right]$$

$$NC(N) = \sum_{i=1}^{i=n} (DP_i - 1) + \sum_{i=1, j=1}^{i=n, j=s} (DP_{ij} - 1) \quad (3.4)$$

Here, NC = number of nested cycles; n = number of proper divisors of N; s = number of proper divisors of corresponding each proper divisor of N. Ordered Factor, OF: Ordered factor OF of any positive integer N is the sum of its total Harmonics and Nested cycles. Thus, we get OF metric, $OF(N) = HN(N) + NC(N)$. The harmonic metric is negligible, OF metric plotted in Figure 3.2a and d includes all possibilities. OF metric was known for a long time as discussed above, however, it was Reddy S. et al, converting the architecture into a 3D shape.

The transfer matrix, harmonic and an-harmonic overtones: The space-time metric in Figure 3.2d is a 360° rotation of a 2D ordered factor plot, we get a 3D PPM. Therein, the angular contribution of the primes is not similar. However, as said earlier, in the artificial brain, the contribution of each of the 15 primes should be equal. That is the engineering challenge for building an artificial brain. The circular trajectory created by 15 edges of the geometric shapes in the Figure 3.2d generate the polygon to cover 360°, but we convert the polygon into a circle to take into account the phase denoted by a complex number. There is no fixed rule that one cannot add a new plane, 3×2 creates a new plane and a new pattern in Figure 3.2d. One could take a pen and draw event linking pathways when an event happens in nature. The fun begins when we feel the necessity to avoid the surface and take shorter imaginary paths to move from one point of the surface to another. It's better then to leave algorithm, and move to material's self-assembly to emulate the journey through the imaginary paths. In the fractal tape, where every single event contains an infinite chain of events making each tape represented by the vertical lines across a non-prime number in the PPM is incomplete by Gödel argument. For a particular vertical line or tape, $p = 0$ and all neighbors get a distribution of p values depending on its quantized relations (even or odd). We need at least 4 points to create the smallest 3D volume, so we keep 4 points for 2D and 1D patterns, taking 4 points on PPM metric we construct the minimum size of a nested rhythm $\{V\} = \sum V \pm ip_1 \sum U_1 \pm ip_2 \sum U_2 \ldots$ Mathematically we call it the transfer matrix, in this transfer matrix we see that all positive terms generate aperiodic or inharmonic oscillations and the cross negative terms generate periodic or harmonic oscillations. Therefore, even at the smallest scale, we get both kinds of vibrations. It is the divisors of a number that decides on the harmonicity of the frequencies produced.

Derivation of wiring of clocks from 3D PPM: For this reason, one rotates the XY plane 360° along the X-axis ($X = n$ or number, $Y = C$, OF, NC, HN). When one connects the ±Y/2 points, using an imaginary line, a 1D line makes a 2D surface. However, when one rotates the XY plane, an infinite number of the same imaginary lines in the 2D surface roll. Thus, one gets a 3D structure, teardrop to ellipsoid, vortex to spiral, and dumbbell disc to nephroid. As noted above, each ±Y/2 points form a Bloch sphere. As the 2D surface rotates around the X-axis, the real points of the Bloch sphere rotate with the same angular speed. Multiple geometries overlap in the metric. Thus, as one connects the C coordinates, by finding the nearest neighbors as shown in Figure 3.4, it is similar to applying geometric algebra that creates a hyperspace. The process generates naturally abundant symmetries and dynamics from the pattern of primes. A slice of PPM designs a universal sensor. PPM inspired sensors would capture the natural events more efficiently.

Imaginary lines are phase paths connecting the singularity points. Once a singularity point is formed, it triggers a cascade effect. Singularity acts like glue. It embeds various time crystals inside or connects side by side (Pippinger and Fischer, 1979; Gurevich and Shelah, 1989) to the surface of the Bloch sphere. Singularity points are undefined functions, so Bloch sphere converts into a phase sphere. There are layers of geometric shapes inside a singularity point. Several clocking spheres pour in and embed in the host-sphere. Thus, the Phase spheres expand as it integrates information. It is fractal information theory (FIT) (Reddy et al., 2018).

How the PPM hardware is designed and built: The philosophy is that all information in nature is an event, like a single point, which has a geometric shape inside with its corners made of sub-events. Following that philosophy, one has to convert a stream of all sensory signals (visual, auditory etc.) into a time crystal. There are several self-similar geometric shapes of the time crystal (triangle = 3:2:17). PPM based filter shrinks the size of a time crystal network. The series $3 \times 2 \times 17$ repeats infinite times in the integer space. The shrunk time crystal is called a fractal seed. A fractal seed $3 \times 2 \times 17$ means one could build a rectangle whose corners are triangles $3 \times 2 \times 17$, if we touch integer $3 \times 2 \times 17 \times 4 = 408$. The decision-making core that represents a PPM would be a 3D time crystal architecture. First, one should pump such a fractal seed into it. The fractal seed expands by PPM, morphs into information architecture. During morphogenesis, the formation of a fractal structure is common in biology (Cooke and Zeeman, 1976). Most dense parts of a time crystal network are blacked out as if these are pieces of mass. We repeat that in a phase-shift model space, time and mass are rewritten as a network of clocks mimicking biological rhythms of a living life form (Frank and Zimmerman, 1969). If these "masses" are replaced with high-density clocks, phase lines are replaced with wire, one gets a circuit (see humanoid avatar construction in Chapter 9). The circuit vibrates like a time crystal. Building a circuit from the PPM requires ten steps. We discussed the steps below in a separate section. The experimental prototype of the non-computer is a device whose vibration is similar to a dedicated PPM. There are various ways to do it. A suitable organic jelly (Ghosh et al., 2014b, 2015b, 2016b) is being developed for over a decade (Chapter 9). Now, a suitable fourth-circuit element is found (Sahu et al. US patent 9019685B2). It could be assembled to clock like a PPM (Chapter 8). If electrical white noise is applied to this hardware, then the clocking waveforms superimpose. It delivers a resonance frequency pattern whose frequency peak ratios are similar to that of the PPM. With the current technology, it is not possible to realize a reliable circuitry with intricate

details of PPM. The humanoid avatar built in chapter 9 is primitive but at least a hope for the future. Moreover, no one requires a new kind of accurate high-resolution computing. The existing Turing based computers are ultimate in doing that.

Prototype under construction: A global platform is under construction (Bandyopadhyay et al., 2006b; JP-5187804-"a vertical parallel processor"). In the current prototype of a PPM inspired humanoid brain, a slice of $N = 1000$ is cut from PPM. One could find its equivalent clocking circuit, by building an equivalent time crystal architecture. Depending on the number of integers in the metric, defines its class. For example, $N = 1000$ means a thousand class hardware, if $N = 10^{12}$, then it is a conscious class or G class (Figure 10.11, human consciousness = G and higher G+, the table in the Figure 10.12). A G class means a superposition of a pair of time crystals generated by single hardware (classified). However, a global triplet of triplet fractal pattern emerges in most of the PPM. If that seed pattern is plotted in a circle to address the phase quantization, it makes a wheel for the rotation of 360° phase. This plot is noted as frequency fractal. Frequency wheel classifies the non-computing ability in the hardware. It also accounts for a transition from artificial intelligence to natural intelligence.

3.1.3 How Is a Time Crystal Decomposed and Amplified Using PPM

When we plot all possible combinations of primes constituting an integer, e.g., ordered factor, OF metric (Figure 3.2a); divisor, D, or maximum prime symmetry S, as a function of integer, then we get prime composition metric, PC metric (Figure 3.3e); when the gap between primes is used as a key factor in plotting the metric, we get the prime gap or PG metric (Figure 3.10a). Big data could be fed in all three classes of metrices, we take one example for simplicity.

Be it a 2D map of spreading a virus in a closed environment, or a time-lapse satellite image of drying up of a lake, or a 3D map of a black hole swallowing a supernova, 3D pattern electric field pattern on the human brain, always we see only one thing in a Big Data, how the most silent and the most active regions are changing. The technology is patented by Ghosh et al. (Ghosh 2019a, 2019b). Locations of the most silent points either in the region of highest intensity or the lowest are points that we count. If three silent points are found and they are in a ratio 2:5:3, or sometimes, 2:3:5, 5:2:3, 5:3:2, 3:5:2 and 3:2:5, then we can imagine a hexagon, whose length of arms are relative time gap between the two events represented here as two ratios. Thus, in the PPM-S, or S-N plot (**Metric-1**) we can tick $N = 30$ ($2 \times 3 \times 5$). Hexagon is a structure made of phase, neither time, not space. A circle rotates around its perimeter and bursts energy at six points. If we see only three combinations out of six in big data, definitely the hexagons predict the future as it includes other three unknown choices or events happening beyond that sensor could read. A journey from one corner point of a hexagon to another is a symmetry breaking. If we add another prime, say 7 into the event, we get $2 \times 3 \times 5 \times 7$, or 210, and the hexagon could give birth to three hexagons connected by a line, together they make 24 corners, and 2, 3, 5, 7, four primes could arrange in 24 ways. Then it is not a symmetry breaking but a phase transition. There could be morphogenesis where two hexagons connected by a line could transform into a pair of hexagonal prisms, using the same number of corner points.

For $N = 300$, $S = 30$ and the corresponding peaks are made of $2 \times 2 \times 3 \times 5 \times 5$ coupled primes. Now, 2:2:3:5:5 ratio is created by multiple clocks operating at 1:2, 1:3 and 1:5 ratios. These interactive clocks change their phase following various pathways that build the surface of the phase sphere. Thus, several spheres nest with each other to build complex crystals. One very interesting feature of converting an image into a network of nested spheres is that a fractal architecture or dynamics is naturally converted into a seed pattern. For example, for $N = 300$, when we get 2:2:3:5:5, it could shrink a massive fractal architecture. Imagine we have a host sphere with two guests A1, A2, (1×2) each has two guests, B1, B2, B3, B4 $(1 \times 2 \times 4)$. Each B has 3 guest spheres, C1, C2, C3… C10, C11, C12. Each C has 5 guests. D1, D2, D3 … D57, D58, D59, D60. Each D has 5 guest spheres, E1, E2, E3 … E297, E298, E299, E300. Thus, $300 + 60 + 12 + 4 + 2 + 1 = 379$ clocks running in a fractal network would be replaced by five spheres, one for each layer A, B, C, D, E. Thus, fractal shrinks a complex dynamic system as a time crystal form in GML. The GML is a language of nature analogous to the machine language of computers.

3.2 METRIC 1: AN INTEGER IS REPLACED BY A TYPICAL GEOMETRIC SHAPE

Below is a summary of ten distinct classes of PPM essential to operate together for making a reliable and comprehensive decision. The first type of PPM links an integer with a geometric shape, as described above.

How to calculate S for an S-N plot: The new parameter S is not an ordered factor or factor or divisor commonly used in mathematics. An integer is a product of primes $N = a^p \times b^q \times c^r$, ($300 = 2^2\, 3\, 5^2$). How many primes do we have? $C = p + q + r$, i.e., $2 + 1 + 2 = 5$. We can arrange these primes in $D = (p + 1) * (q + 1) * (r + 1)$ ways. For 300, we get for $N = 300$, $D = (2 + 1) * (1 + 1) * (2 + 1) = 18$. Here D is the number of factors or divisors. For 300, the numbers 10, 12, 15, 20, 25, 30, all are divisors. We take only the ordered factor made of primes to build an S-N plot, PPM-S. So, for 300, we get 30 combinations of $2 \times 2 \times 3 \times 5 \times 5$, i.e., $S = 30$. For 300, the constituent primes are 2, 2, 3, 5, 5, we calculate all 30 possible distinct symmetries like 2:2:3:5:5, 3:5:5:2:2, 5:2:2:3:5 etc in a data and position all 30 of them at the corners of a geometric shape with 30 vertices ($S = 30$). The ratio of primes can have constellations, for example, here 2:2:3:5:5 can have several similar ratios of primes, like 13:13:17:23:23, 37:37:43:47:47. They create an infinite series by writing an impression in the integer space, that is why no sensor-detected geometry is pure, but a repetition of the ratio of prime's infinite series.

Let's take another example as shown in Figure 3.3a. Say, we have 12. Its 8 ordered factors are 2×6, 6×2, 3×4, 4×3, $2 \times 2 \times 3$, $2 \times 3 \times 2$, $3 \times 2 \times 2$, 1×12, but we take only 3 of 8, they are $2 \times 2 \times 3$, $2 \times 3 \times 2$, $3 \times 2 \times 2$, hence $S = 3$ (Figure 3.3b–d). Thus, S-N is a topological function. From ordered factor, OF, we simply remove non-prime terms to get S. Easy way to calculate S is to find the permutation of C and divide it with permutations of all repeating primes. For 300, $S = 5!/(2!x2!) = 30$ and for 12, $S = 3!/2! = 3$. For readers $5! = 5x4x3x2x1$. In principle, all possible symmetries of the real world are hidden in some integer of the number system, we map it using an S-N plot as shown in Figure 3.3e, this is the basic PPM-S. It is different from the OF(N) plot shown in Figure 3.2a. The hidden topologies of primes shown in Figure 3.3f bind the isolated patterns or shapes drawn in the SN plot of Figure 3.3f. SN plot is called prime composition or PC metric.

Every single prime has a geometric shape, so a few shapes could generate 99.99% of all integers: Every single value of S is assigned to a geometry, check the list of primes with the specific geometric shapes as shown in the Figure 3.3g. Say, in a rapidly changing big data, we find that the integer representing the dynamics of the database is changing from 140 to 144. At 140, we get the number of ordered prime factor = the number of vertices = $S = 12$, i.e., a hexagonal prism. Then for 141, we get a line ($S = 2$), so, the symmetry breaks. However, for 142, 143 the line remains ($S = 2$), as the number of divisors for all three numbers is two, the symmetry is intact. At 144, again, the symmetry breaks, and we get 15 vertices or divisors ($S = 15$), which could be three square pyramids arranged in a triangular shape ($5 \times 3 = 15$). Change in integer could trigger a phase transition. In Figure 3.3h, we have shown how to write number 30 using geometric shapes and made it to 180 by geometrically multiplying 30×6. Therefore, the prime composition metric shown in Figure 3.3e is also shown in Figure 3.3h which explains the prime geometric significance of every single integer. All versions of PPM including OF metric and prime gap or PG metric would have a similar geometric representation, though we have not discussed it further.

A fusion of several geometries one inside another: A precursor to time crystal: $S = 1$, refers to a dimensionless point, $S = 2$ signifies a line, $S = 3$ to 6 we get 2D shapes and for $S = 7$ to 20 we get 3D shapes. If sensors get the value of S that is not defined by GML then we combine the geometries so that the total number of corner points of all participatory geometries is S. Here we give two examples. If we get $S = 8$, we do not opt for the 2D octagon, the largest 2D shape allowed is a hexagon, we stick to that, and adopt a 3D shape, a cube or a square pyramid. Since the largest 3D shape is dodecahedron (20 vertices, we get 20, for integer $N = 120$ for the first time), for integer $N = 180$, we get $S = 30$, since we can write $30 = 3 \times 10$, and pentagonal prism has 10 vertices, we can draw a triangle, its each corner point would have a pentagonal prism. Together, the triangular arrangement of three prisms provides 30 vertices. The composition of a new

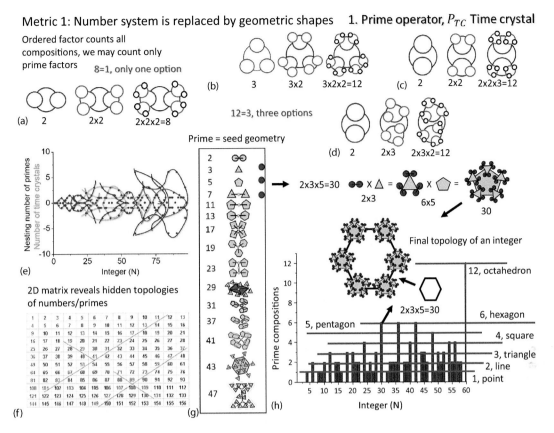

FIGURE 3.3 The first class of PPM-1 is presented, it is codenamed PTC or Primes of a time crystal. (a) How an integer 8 is written in terms of prime numbers. Panels (b), (c) and (d) describe 12 written in three ways. (e) How many ways primes could be arranged to build an integer is a plotted against the integer. (f) One kind of pattern depicting the similarities of primes to group them. (g) Every prime has a particular geometry, using these geometries all other integers could be generated, shown its right with an example of number 30. (h) The nesting of primes plotted in the panel (e) is plotted again in the panel (h) but the values in panel (e) were divided by 2 to demonstrate C2 symmetry and in the panel (h) it is not. A straight line is drawn to demonstrate the geometric shapes associated with a value. Note that largest shape that we have to address belongs to a 12 corner 3D architecture or dodecahedron.

geometric shape by inserting one geometric form into another follows a rule, we start with a point i.e., 0D, put 1D shapes insides then put 2D shapes inside. Inside the 2D shapes, we put the 3D shapes.

3.3 METRIC 2: PRODUCT OF PRIMES FORMING INTEGERS LED TO A UNIQUE ORDERED FACTOR METRIC

Even within the domain of Metric 2, we now discuss one of the most important aspects of PPM, projection from infinity. Currently in Figure 3.4, the projection and feedback are discussed only for OF metrics described in Figure 3.2a. However, a similar discussion could be carried out for SN or prime composition, PC metric (Figure 3.3e) or prime gap or PG metric (Figure 3.10a).

Projection from infinity: When Ramanujan showed that the sum of integers is −1/12, many people ignored it as fun. However, several people went ahead and started seeing how the arrangement of integers could project an entirely new perspective to the infinity, which returns an unbelievable value. Here in this book, any chapter we open, we would find

from the darkest corner inside a protein to the mathematics of primes to nature's 1.5 million years of a nearly perfect decision-making machine called the human brain, 12 is out there. 12 triplet of triplet resonance band builds a human brain (Chapter 7). Now, return to the Ramanujan's derivation, look at it again, why that 12? Is it an accident? No. Do a little math and we would find there are only 12 projections to infinity and one of them is returned from infinity, that's what negative sign refers to. It is exactly similar to the fourth-circuit element we have explained in Chapter 8 where we have explored a new kind of phase space, with 12 holes or singularities, projection through that mask creates the composition of vortex atom assembly. That composition of vortex atoms is also the foundation of the GML. Now, by counting the number of divisors and ordering them we built a 2D and 3D pattern in Figure 3.2, however, can we use the Ramanujan's mathematical trick and apply that on the geometric pattern?

If mathematics is about playing with numbers, and geometric shapes, why not project the metric to infinity following Ramanujan's way and have fun. That's is exactly what we have compiled in Figure 3.4. Here we take all primes but only the product of particular primes along the horizontal

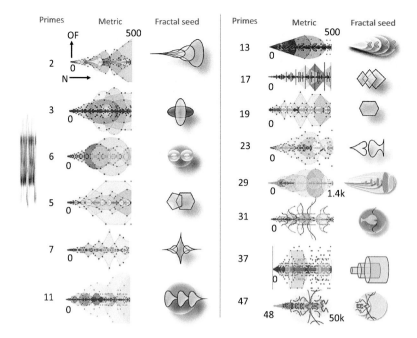

FIGURE 3.4 PPM-2 is codenamed as symmetry ratio of primes. Here a table is presented with 14 integers and its corresponding metric shapes. The table has two parts one in the left and one to its right. Each part contains three information column-wise. First, is the prime number, second is metric and the third is the basic geometric pattern that gets repeated over the integer space. Note that if X and Y axes boundary values are changed then the pattern also changes. When we consider an integer and continuously take its products, say the starting integer is 2 and we take the products of 2 as $2n$, $n = 0,1,2,3,4,5,6,7,8,9$. So, for 19 the integers to be considered in the horizontal X-axis would be 19, 38, 57, 76... For these integers, the corresponding ordered factors OF are calculated and plotted on the vertical axis with $\pm$ OF/2. Then the nearest neighbors are connected in a 2D loop. A composition of these loops becomes a PPM. In this way, if a particular prime dominates statistically, then there is a probability that the metric corresponding to that particular integer is dominating in the system.

axis, the vertical axis remains the same, and the ordered factors are connected to build the geometric shape. Now, if just like Ramanujan, instead of all patterns, reject the uncommon loops between the neighboring domains and see which pattern survives until $N = 500$. We do not have to go to infinity, but the projection to infinity is visible, it is named as fractal seed. It is a seed geometry and its importance is enormous in the construction of a Fractal tape, what that comes from infinity is never defined, one of 12 possible choices return, though only one option is shown here. Projection from infinity to the reality is not mysterious, it is like 11 imaginary worlds are redefining the reality with a choice not logically feasible, but mathematically perceivable.

Projection from infinity is not limited to the PPM, in the Figure 3.12, we have plotted a table listing the ten physical phenomena outlined in this book where artificial brain engineering would harvest infinity and make sure that Ramanujan's negative feedback of number system is explored. Another version of the same table is to be found in Figure 4.14, where one could sense how the typical mechanics of 11 imaginary worlds make sure that projection from infinity redefines reality. Thus, a projection from infinity is a change in reality, due to the combination of options, which is projected to infinity and feedback is sent. To an engineer, if implemented in the hardware, this would be an auto-evaluation program.

The second type of PPM metric accounts for the scale-free geometric shapes or fractals (PPM-F): Each prime is associated with a particular seed geometry, that repeats continuously in the PPM. A tubulin protein uses only a few integers, which means only a few primes, so is microtubule and, so is the neuron. Bio-engineering is all about "bandwidth of primes." Since we limit ourselves within a maximum N that is the product of the first 12 primes $N_{max} = 2 \times 3 \times 5 \times 7 \times 11 \times 13 \times 17 \times 19 \times 23 \times 29 \times 31 \times 37$, we explore only 230 million primes within $N = 10^{12}$. Thus, each of the 12 patterns of 12 primes is edited by 230 million primes. A seed geometry expands fractally, over the entire integer space. Here we provide 12 geometric seeds that cover 99.99% integer space.

Prime 2 triggers the formation of crossing parallel lines into a rhombus. Its lowest integer holds another rhombus, and this fractal feature continues. Similarly, triplets expand fractally, the first loop of a teardrop-like a triplet in the PPM contains another similar triplet, we call it triplet of the triplet. Prime 5 builds S-shaped open-end curves. Prime 7 embeds the circles in the curves. Together prime 5 and prime 7 creates a fractal of flower pot repeated over the entire integer space. Prime 11 builds a fractal of reverse petals; Prime 13 builds a fractal of hearts Prime 17 builds a fractal of spiral Prime 19 builds a fractal of spiral; Prime 23 builds a fractal of vortex Prime 29 builds a fractal of toroid Prime 31 builds a fractal of crossed petals Prime 37 builds a fractal of parallel lines. In the PPM-S, which structure would they prefer during the integration of neighboring integers is decided by fractal metric PPM-F. To make sense brain may predict future

(Nieuwland et al., 2018); for PPM-hardware, in the metric that maps past-present-future of an event, as soon as the space for "present" is filled up, the infinite series of future events is redefined.

3.4 METRIC 3: INTEGERS LIMITED TO 360° PHASE BUILD CLOCKWISE AND ANTICLOCKWISE PATHS

The third type of PPM-LK covers the effect on phase and dynamics when a particular slice is cut from a PPM-P, by selecting an integer. When we select an integer, phase varies uniquely for the structures of all integers below. We need an intricate map for three reasons. (a) Setting the critical spatial and angular parameters of geometric shape for a given number of vertices. (b) Identifying the rotational direction of clocks, the relative phase difference between clocks and finding how do the directions change. A change in the clocking direction maps the inherent dynamics of an unknown system. (c) When we link the two active integers in the PPM, the silent integers in between providing the key linking geometries. However, we need strictly defined mathematical rules to use those shapes to link the neighboring geometric shapes. Molecular

left-right (LR) asymmetry leads to whole-cell LR asymmetry (Wood and Kershaw, 1991), here we see the numerical origin of LR asymmetry, a key to engineer morphogenesis in a life form (Pearson and Winey, 2009). For these reasons, we have introduced **three "phase" versions of PPM**, namely PPM-LK (Figure 3.5). Three PPMs map how between a pair of selected integers, the starting phase of the structures varies from 0° to 360°. The relative phase of all the structures is meticulously defined.

The first metric identifies a set of integers form a local group preferring a clockwise or anticlockwise rotation. Means, say if we select $N = 12$, we cut the ordered factor metric a slice of length 12. The journey from 12 to 1 is divided into 360° angular distributions. In the polar plot, each of the 12 integers constitutes a radial line. So, radial line for 1 differs from the line for 2 by 360/12 = 30°, the distance from center is the value of number of divisors, D or ordered factor, OF; or prime symmetry, S. The phase plot connecting D or OF or S, depicts a pair of clockwise and then a pair of anticlockwise rotating lines, globally form a lotus flower. It is a very significant discovery. It means a set of integers would favor a clockwise or anticlockwise rotation when they would link to an external pattern. A pair of limiting integers reach

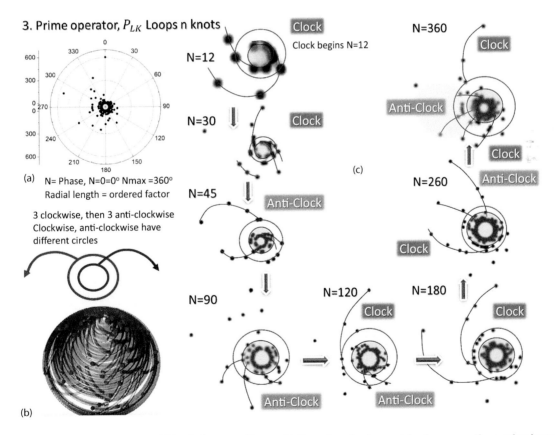

FIGURE 3.5 Third class of the PPM would be the loops and knots. (a) Cymatics of prime metric. As soon as the number increases, a pair of spirals born alternately. Polar plot (r,θ) where θ is N, an integer and r is the ordered factor. Say we want to plot $N = 30$, then we create an imaginary circle where each integer is separated by 12° and for each angle we move outward radially, since the distance from center would measure the ordered factor. In the bottom part of the panel we have shown a schematic showing clockwise and anticlockwise rotation of the minimum distance lines created by connecting the ordered factor of a natural number N. (b) A cone is created when a large number of spirals born out of the center of the circle. (c) Eight examples of polar plot where the clockwise and counterclockwise rotation of the spirals are visible.

an equilibrium, form a petal. Since the layers of petals move radially outward, it promotes self-assembly within and above. Thus, when we choose an integer, it makes a slice in PPM-LK from 0 to N, that slicing naturally defines its dynamics and the self-assembly properties.

3.5 METRIC 4: A DOMAIN OF INTEGERS SETS A LIMIT ON THE USABLE 15 PRIMES

The Fourth type of PPM analyses the asymmetric contribution of primes PP, or singularity points when we assign geometric shapes to all integers. The OF-N metric or ordered factor metric or NC metric, both, increases so rapidly with N that if we measure the OF-center ($N = 0$) line making an angle with the horizontal axis for integers, that angle exhibits remarkable features as shown in the Figure 3.6a and b. In Figure 3.6a we find that initially the slope does not change exponentially. However, above $N = 10,000$ the slope does not increase much and reaches 90°. In Figure 3.6b inset,

we find that by $N = 1$ million, the slop curve traces a circle. The saturation means, new patterns that could connect the $N = 0$ point all the way to the front would not be a new one, it would almost retrace the tear-drop or other loops created by smaller numbers. When system retraces the previous plot with a little or no change, it means, there would be no new feedback from infinity as shown in Figure 3.4.

We argued above that the product of the first 12 primes is around 10^{11}. Since we limit ourselves within a maximum N that is the product of the first 12 primes $N_{max} = 2 \times 3 \times 5 \times 7 \times 11 \times 13 \times 17 \times 19 \times 23 \times 29 \times 31 \times 37 = 7.42 \times 10^{12}$ Here only 12 primes are taken because the early occurring primes in the integer space leaves its signature the most on the entire integer space to follow. The total contribution for a prime is ~100/prime, hence integer 2 occurs 50% in all possible natural numbers. The total contribution is not what we want. If we record the original contribution of a prime in addition to the contribution made by the previous primes over the integer space, we find that only 12 primes contribute to 99.99%.

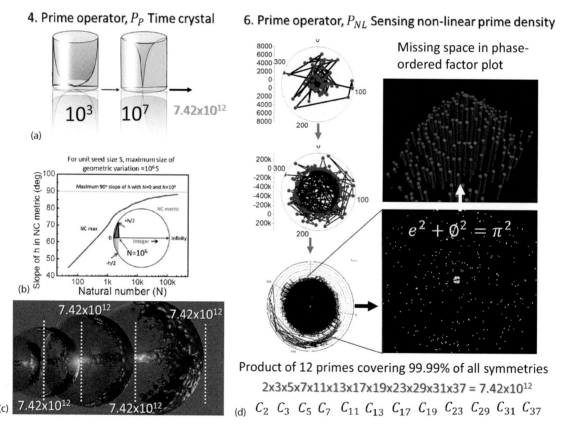

FIGURE 3.6 Two classes of PPM are presented here. PPM-4 describes the relative increment of the ordered factor with respect to the integer (panel a, b and c). To the right, panel d explains the PPM-6. The polar plot of ordered factor vs integer N shows that certain points appear empty, white spots. These empty points form a pattern. (a) Two schematic cylinders whose diameters are N and lengths are the ordered factor. Two cylinders are for $N = 1000$ and $N = 10000000$. (b) $N = 0$, $N = n$ and ordered factor OF forms a triangle. Its slope is plotted with the integer, the value reaches to 90°. (c) The change in the shape of the plot shown in panel (a) is gene ralized in the panel (c). The slope of OF with N when reaches 90°, this means the line that forms a slope is a tangent, the phase plot looks like a circle. Product of 12 primes reaches a thousand billion, once C2, C3…C37…C47 symmetries cover 99.99% (12 to 15 primes does not make much difference). Panel (c) demonstrates that once 99.99% is reached the resultant number around 7.42×10^{12}, that is not a fixed number, but that final number is considered 1 and a new plot begins. Four times the process is repeated. (d) Polar plot of OF, for $N = 10^9$, period $N = 360$. The new cycle begins at 361 and then again at 721, etc. Here, all OF points are connected by line. We get a gap. These gaps make circles.

Thus, $N = 10^{11}$ or $N = 10^{12}$ limits are not accidental. There is no strict limit, but if we go beyond, new primes above 47 or 53 hardly make a significant contribution to the symmetry breaking, or they do not regulate events statistically. It is better to build a decision-making system made of 10^{11} or 10^{12} clocks as a single clock and start recounting $N = 0, 1, 2, \dots 10^{11}$, time and again. We convert various one-dimensional parameters of integer space into a 3D architecture by rotating a 2D plot by 360°. Always, the PPM actually clocks in nature (PPM-P), in a cycle between 1 and 10^{11}.

3.6 METRIC 5: WHEN ORDERED FACTOR IS MUCH GREATER THAN THE INTEGER

The fifth type of metric considers the differential factor between two types of matrices, PPM-D. In the first route, using a 360° rotation of 36 planes, one for each integer between 1 to 37, each pair of planes separated by a 10° gap, we can convert any 2D PPM into a 3D PPM. In the second route, we set the contribution of a prime as its angular separation, say 2 means 50°, because it occurs 50% times, 3 means 16°,

because 3 occurs additional 16% in the entire integer space, and so on until we reach a total of 100°. Then we increase each prime's angular gap so that total angular distribution is 360° for first 12 primes. These two routes are primarily used to generate the 3D PPM. Above, we explained that each prime starts generating its own fractal geometry in the PPM, it means above $N = 37$ the 12th prime, at some integer N, we would find that ordering of primes would generate geometries unrelated to the pattern of any particular prime (ordered factor, OF). That integer is $N = 48$, OF $= 48$, above this point we start seeing OF $\gg N$, i.e., ordering of numbers (e.g., 2×3 and 3×2) factor so much that the values of integer N or divisor D do not matter. For OF $\gg N$, the grouping of sets of primes or nesting of geometries regulate the properties; higher-level geometries lose all key geometric feature that primes used to generate in the PPM as shown in the Figure 3.7a. In the 3D plot of OF-N, we remove the 3D part OF $< N$, thus, we get an annular 3D architecture, which reflects how groups of primes form higher-level topologies. Figure 3.7b shows another journey of the annular metric from $N = 400,000$ to $40,000$. It is a new metric which links distinct geometries that creates

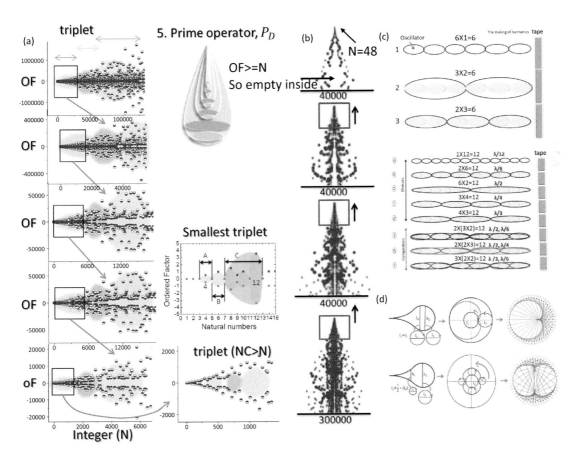

FIGURE 3.7 PPM-5 describes the case when Ordered Factor, OF $\geq N$, then we find a fractal, triplet of triplet, this is the most important fractal with large integers $N \geq 48$. (a) A series of five plots. The front part of the OF plot is zoomed below for all cases. Triplet of triplet formation has been detected for all cases. The smallest triplet of triplet forms at $N = 12$, this plot is shown right below. A schematic of the fractal is shown in the top-right sub-panel. (b) All the values below OF $< N$ are deleted, so that a unique set of pattern is visible. One could compare the plots in Panel (a) and the panel (b). (c) Waveform based realization of ordered factor to develop a physical interpretation of ordered factor in terms of waveforms in a string. (d) A loop in the PPM could be realized as a coffee cup effect.

virtual corner points more than the number of singularities. It is a remarkable map because it accounts for the superposition of geometric shapes and dynamics that systems with apparent self-similarity could not fathom. The differential PPM is a particular class and we can choose any two parameters to build a similar annular metric that accounts for the 3D evolution of a typical functional parameter.

The triplet of triplet fractal features of this annular metric is shown in Figure 3.7a. If one continuously zooms the metric as shown in the left column of the plot, the smallest triplet of triplets is reached, at $N = 12$. If one compares the figures zoomed at different intervals, one of the three teardrops, the patterns are never self-similar, but globally always triplet of triplet. Therefore, whether to call it a fractal or not is a big confusion. In Figure 3.7c we explain the musical string analogy of the ordered factor OF and in Figure 3.7d we explain the coffee cup analogy of ordered factor OF. While Figure 3.7d shows that coffee cup = cycloid = teardrop of OF, the Figure 3.7c shows that even classical strings (choice 7,8) would show a continued fraction of nesting clocks as described in Chapter 2.

3.7 METRIC 6: EMPTY SPACE IN THE PLOT OF ORDERING OF DIVISORS IN PHASE PLOT PRESENTS HOLES

The sixth metric described in the Figure 3.6d is the same polar plot of phase described above for metric 3, clocking features, but we notice that no line connecting the S, OF, and D values pass through particular points. When the maximum N increases and we connect the OF, CN, SN or NC values, there are empty places found when we connect the value points in the polar plot using lines. The missing points in the phase-OF plot show that empty points are not randomly distributed. Empty or silent points are invisible points of attractor when N increases. The silent points in the phase plot described above where no S, OF, or D values are found; these silent points arrange in $e^2 + \phi^2 = \pi^2$, thus, a global phase regulation exists over the entire integer space. Thus, we get three fundamental constants regulating the clocking direction by creating null points. The non-existent features in the **PPM explores the possibility of integrating infinite mathematical series** using geometric algebra; we call it continued fraction geometric algebra, CFGA. Here, the circles are looped intricately one inside another; the diameters are such that they generate infinite series for e, π, and Φ. Often we build a new calculator where one draws only circles but solves complex mathematics, even simple arithmetic, or algebra. Then we monitor the major structural changes around the peak, $P = \pi(N)/\left(\frac{N}{\text{Log}N}\right)$ as expected. The density of primes nonlinear features triggers major structural phase transitions. We noticed that the location of the peak is related to a pair of primes, prime1 $\times$ $10^{\pm\text{prime2}}$. Thus, the sudden non-linear changes in the density of primes generate a constant to estimate a major change in the geometric structure. These constants appear similar to the universal constants we find in nature. When the events are converted into geometric structures, and if the pattern of primes governs

the evolution of events, then the sudden turbulences are expected if such topological conditions. These topological constraints even regulate the dynamics of events at all levels above a sudden peak point.

3.8 METRIC 7: STATISTICAL DOMINANCE OF PRIMES IN THE INTEGER SPACE— SILENT AND ACTIVE PRIMES

The Seventh type of PPM, PL PPM explore the effect of the non-linear distribution of primes below 10^{12}: The gap between the neighboring primes is an interesting twist in the GML-PPM-H triad computing. If the gap is two, it is called paired prime, the gap is six, sexy primes and so on. Figure 3.8a shows the maximum gap of neighboring primes, which increases with the integer. Primes are null points, silence in the language of nature but they shape the pattern, 2×3 and 2×47 have the same horizontal axis in any version of the PPM plot.

Normalized distribution of primes tells us how prime distribution differs from a particular function that supposed to predict the density of primes. We have plotted a very well known data the density of primes $\pi(N) = N/\text{Log}N$, which increases and then nearly saturates over the integral space as shown in Figure 3.8b. The function $P = \pi(N)/\left(\frac{N}{\text{Log}N}\right)$, measures the non-linear regions of prime distribution in the integer space, which is below $N = 10^{12}$. Above the limiting N, the prime density does not change in an unpredictable manner, i.e., follows $\pi(N)$ function as is. Below that limit, the sudden peaks and dips in the P-N plot depict the structural phase transitions in the time crystals. Why is this plot so important? When we look at a fractal, even if you expand it to infinity, a simple pattern repeats, therefore, information content is null, irrespective of its size. Now, when a single function draws the baseline and with respect to its variation we measure the prime density distribution, we find exactly the change from predictable density. That estimation of non-predictability would also be reflected when the PPM would change the geometric shapes during decision-making. Thus, Figure 3.8b shows two distinct confirmations that beyond a limiting integer, the primes would not bring unpredictable changes to the output.

Figure 3.8c shows the sum of prime factors, SOPF, which is a unique metric created by taking prime factors like Figure 3.3e, PC metric, but with a twist. Prime factors are added and to implement C2 symmetry, divided by two. The plot shows a nearly periodic ripple and as the number increases, we find more and more ripples are born along a virtual axis. An integer is a product of primes that delivers OF metric, PC metric and PG metric. However, the addition of those primes for integers follow superposition of diverging waves in a regular increment of its periodicity. It makes the metric beautiful. To find the origin of this spontaneous rhythmic behavior, we dug deeper, looked into the spiral plot of primes, which is quite popular among mathematicians who want to find a hidden rule of nature coding the primes in the plots Figure 3.8d and e. Serious mathematicians and

7. Prime operator, P_L OF max and Non-linear density of primes

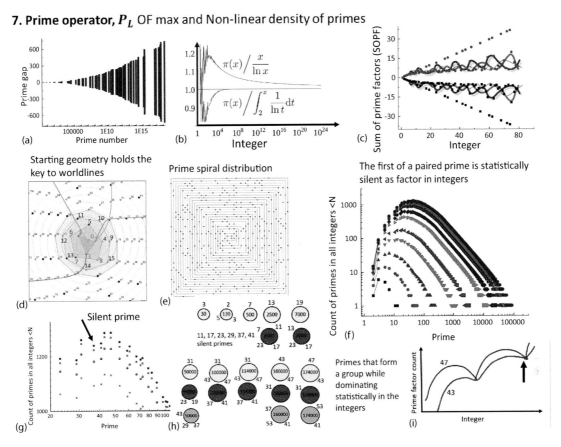

FIGURE 3.8 Seventh class of PPM deals with the features of primes which make some of them statistically dominating or silent when we count them in a limit of integers. (a) The prime gap is plotted against the integer N for $N = 10^{18}$. (b) Graph showing the ratio of the prime-counting function $\pi(x)$ to two of its approximations, $x/\ln x$ and Li(x). As x increases (note x-axis is logarithmic), both ratios tend towards 1. The ratio for $x/\ln x$ converges from above very slowly, while the ratio for Li(x) converges more quickly from below (Wiki, free commons). (c) An integer is made of primes, its factor primes if added shows ripples. We plotted $\pm$SOPF/2 vs N. Panels (d) and (e) show spiral plots of integers and primes, often such plots are used to find the pattern of primes. The starting pattern and the worldlines connecting the similar values in the plots are governing factors. (f) We count the number of times a given set of primes are used while creating integers up to N. Here all primes are counted up to $N = 174000$. One could see that the peak P is at prime 47. Normally, we expect 50% of all factors should be 2, so it should be a decreasing plot with the maximum value at 2. But we find here that has a modal distribution that is increasing at a slow rate approximating the cube root of N. Up to 10,000, this mode is 19. Up to 1,000,000, this mode is 73. For any given set of numbers, there will be a distinctive peak near $P \wedge 3 = N$. (g) The plot (f) is zoomed in the panel (g) left and ripples of statistical dominance and silence of the primes are shown schematically in the panel (i) and the entire series of events is summarized as colored balls in the panel (h). The panel (h) shows a yellow ball with Nmax inside and the dominating primes around the perimeter. The red balls are statistically second position and blue balls depict the third position. Silent primes are also noted in the panel.

topologists do not care that at the center of these plots lie a geometry and always that is a triangle, who started it we do not know but if we change that the connected worldline would change, and even in spite of that a pseudo-periodicity in density distribution is apparent.

Finally for the density of prime metric discussion we arrive at the most exciting part, the story of silent and active primes. In the mathematics literature there is no mention of such a phenomenon of the primes (Figure 3.8f and g). This is quite well known and studied by a large number of mathematicians is that how many times the primes occur within a given integer limit? One could write a simple code, finding prime factors is easy, simply count them and plot against the primes. Figure 3.8f is one such plot, which suggests that there is a peak around 70–80, the point shifts but very slowly as the

number increases rapidly. The feature is popular, but when we dig deeper we find that during counting some primes never dominate the statistical occurrence. That makes them silent all around. In the central plot of the bottom row in Figure 3.8g, we have created statistics, how this phenomenon unfolds. Until 30, 3 dominates, until 500, 7 and so on. We have plotted up to 53 because we are interested only around 12 active primes and that story completes within 170,000. That is not a big number. In fact, above 12 or 15 primes, the statistical fluctuation of which prime dominates the prime factor of integer, saturates. It means a clear answer to the big data generation or to the spontaneous organic synthesis of brain jelly, even if one begins with a soup of random H devices, the composition of statistically dominating primes would take over the dynamics as plotted in the Figure 3.8g.

3.9 METRIC 8: NORMALIZED RIPPLES ON THE METRIC PLOT SHOWS PERIODICITY IN EVENTS

Prime operator 8, namely PN, Normalized ripples: Figure 3.9a–d is a journey to see all above metrics discussed above by normalizing the values to one. Irrespective of the nature of the metric or their foundation, when we consider that maximum choices are one, a new pattern emerges. Here we would discuss only the OF metric, i.e., normalize Figure 3.2a in a step-by-step process. If we do not look into the $C2$ symmetry, which means do not divide OF by 2, we get a plot like Figure 3.2a, ripples are there, but it is expressed more when in Figure 3.9b, it is divided by 2. We have enlarged the ripple of the panel a in panel c so that one could compare it with the normalized plot of OF metric in Figure 3.9d. Here we write OF = $C(H)$ in honor of Ray Tomes.

The mathematical kernels for the escape time fractal tape network: Ordered factorization: Let us consider that fractal tape machine/network is represented by a single wave function ψ, which includes the observer too. Now, if we include boundary conditions, ψ represents one tape, but inside it could be two tapes or two wavefunctions, $\psi = 2\psi_2$, in general $\psi = n\psi_n$, now, for some n, there could be more waveforms inside, e.g., for $n = 4$, $\psi = 4\psi_4$, $\psi = 2\psi_2$, two possibilities

are there. For $n = 8$, $\psi = 8\psi_8$, $\psi = 2\psi_4$, now, $\psi = 2\psi_4$ can happen in two ways, as the part $2\psi_4$ could be made using 4 small ψ_4 parts, or 2 numbers of ψ_2 parts. Thus, for $n = 8$ there are three ways energy could transfer from point A to B (C(H)~3). Probability of some harmonics to occur is more than others in the fractal tape network. Let's take another example, $n = 12$, $\psi = 12\psi_{12}$, $\psi = 3\psi_4$, $\psi = 4\psi_3$, $\psi = 2\psi_6$, $\psi = 6\psi_2$, now again, $\psi = 3\psi_4$ can happen in two ways as earlier, and $\psi = 2\psi_6$ can happen in three ways. Thus, $n = 12$ harmonics can get energy 8 ways (C(H)~8).

However, for $n = 5, 17, 13$ only one way (C(H)~1). Now if we make a journey from $n = 1$ to $n = 10^{53}$, we find that for $n = 2, 4, 24, 36, 72, 96$ the same tape can have a large number of cells C(H), each cell is identical by energy but structurally and information-wise distinct. Now, as n increases, or we make a transverse journey through the tapes, energy could decay 1/H, or $1/H^2$ etc, then we find C(H)/H or power distribution across harmonics, remains nearly constant over n. C(H)/H vary with n as a local Gaussian-like function, first increases with 2^n and then decreases, again increases with 2^n3^1, again decreases, then it goes on and on $2^n r^5 k$.... In $n = 10^{53}$ scale of harmonics, just after 5 is encountered, there is a primary peak in C(H)/H, and just after 3 is encountered there is a secondary peak in C(H)/H, overall, it is a nearly periodic oscillation of power distribution over n harmonics, in Log scale (0 to 53), the period is ~4.5

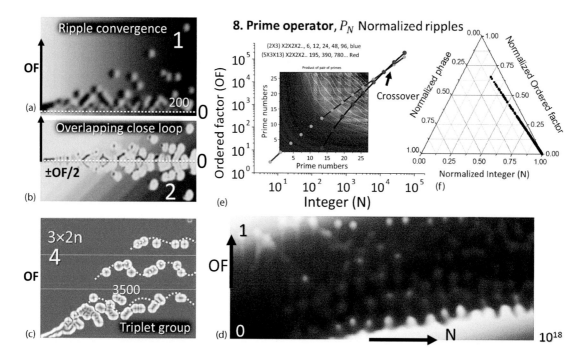

FIGURE 3.9 Eighth class of a PPM, finds normalized ripples in the ordered factor metric in different ways. (a) Ordered factor OF of a number is plotted against the integer *N*. (b) PPM is shown in ±OF/2 vs integer plot. (c) Using an algorithm when we enlarge the data points in the panel (a) it reveals the classified distinct choices and ripples. All three panels (a), (b) and (c) have *N* = 200. (d) Ordered factor OF is normalized to one and plotted against an integer *N*, 10^{18}. The scale-free ripples in the factors are visible. (e) OF-*N* plot for two integer series. First 2 × 3 and its products with 2, number of times generating all products of 6n (blue). Second 5 × 3 × 13 and its products of 2, generating 195 *n* series (red). The crossover between the two plots is shown with an arrow. In the inset, similar plots for primes, both horizontal and vertical axes are primes and OF or their products are plotted perpendicular to the plane of the paper. The ripples are visible. (f) Three normalized parameters are plotted together in a triangle plot. The slope created by OF, *N* = 0 and *N* is the phase, all three parameters the phase OF and *N* are normalized to 1 and the plot identifies a scale-free relation existing for a million *N*.

($\partial n \sim 34560$ in linear scale). Ray Tomes used harmonics to argue that every single object in this universe generates symmetry following the harmonics relationship.

The integers with similar OF, D or S form a wave-like link over the entire integer space, irrespective of the primes they are made of (Figure 3.9e). The ripples depict an ordered flow of higher and lower values of OF, D or S, the ripple wave either converges or diverges as the numbers increase. The path mapped by the ripples connect integers which have a nearly similar shape, a gradual change in shape is a map of symmetry breaking when there is a large change in a waveform, we call it phase transition. One example of a phase transition or crossover is shown in the panel Figure 3.9e, where two distinct starting group of primes engage in a predictable crossover.

Then we plot three types of phase parameters along three axes in a 3D phase variation as it is practiced in the time crystal analysis as shown in Figure 3.9f. For the first phase parameter P, we consider $N = 10^{12}$ points as 360°, since the hardware counts an object with 10^{12} clocks as a single clock and restarts counting from $N = 1$. Now, for the second phase parameter, we plot the slope between the ordered factor and its corresponding integer, which tends to reach 90°, which is maximum. So, we consider 90° as the maximum phase 360° and the slope becomes the second phase parameter. The third phase parameter is a contribution of 12 primes, total contribution 100 is considered as 360° and angular width of each prime is its occupancy in the integral space. These three-phase values are plotted along three orthogonal axes to create a 3D phase architecture representing the most dynamic part of prime density distribution, similar to the time crystal (PPM-L).

3.10 METRIC 9: LATTICE GROUP OF PRIMES (TWIN PRIMES, COUSIN PRIMES, CO-PRIMES, AND GAUSSIAN PRIMES)

The ninth type of PPM explores building lattice structures (PPM-G) using the gap between the primes. The gaps form groups (tuples), if k gaps form a group, it repeats over the infinite integer space as k-tuples. Every occurrence of k-tuples means a certain geometric feature would repeat in the PPM which would bond the isolated and discrete geometric shapes of associated integers into a common geometric shape. The general tendency of gaps between adjacent primes to become larger as the numbers themselves get larger. However, the plot of the frequency of particular gap between a pair of primes varies in an oscillatory manner over the integer space, the peaks are always at multiples of 6 (Ares and Castro, 2006). Since $6 = 3 \times 2$, we plotted PPM only for the integers which are multiples of 2 or 3, and PPM-L transformed into an infinite chain of circles or spheres, thus, clocking is inherently embedded in the frequency of prime gaps as shown in the Figure 3.10a. Sexy prime doublets, (5,11), (7,13),

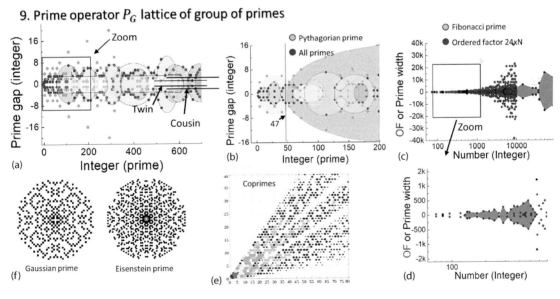

FIGURE 3.10 The ninth class of PPM is about grouping of different prime numbers. Panel (a) shows that prime gap is plotted against the integers or primes. However, one could plot Gaussian primes or Co-primes too to get another group of PPMs. A part of the metric shown in panel (a) is zoomed in the panel (b). Both panels (a) and (b) plots Pythagorean primes (green) and the rest of the primes (red). The existence of twin primes, (prime numbers separated by 2), Cousin primes (primes separated by 4), sexy primes (primes separated by 6), and even triplets and quadruplets of primes. Recently, it has been demonstrated that there exist arbitrary large arithmetic progressions of primes (Terence Tao). It is evident that these groups of primes, or constellations, emerge naturally because the whole prime structure has no evidence of certain order, it has no series expansion, and has no symmetry. (c) Fibonacci primes (blue) and ordered factor of all integers that are product of 24 are plotted here. A part of the metric is zoomed in the panel (d). Panel (e) shows co-primes or mutually prime number for example 1,2, or 4,5,6, where there are no divisors except 1 between then. The first node (2,1) is marked red, its three children are shown in orange, third generation is yellow, and so on in the rainbow order. Triplet of groups is noted. (f) Gaussian prime and the Eisenstein primes are plotted. Note that the primes processed by quaternion, octonion and dodecanions are all generic versions of Gaussian primes where not one but multiple imaginary vectors reside.

(11,17), (13,19), (17,23), (23,29), (31,37), (37,43), (41,47), (47,53).. triplets (5,11,17), (7,13,19), (17,23,29), (31,37,43), (47,53,59), (67,73,79)… Quadruplets (5,11,17,23), (11,17,23,29), (41,47,53,59), (61,67,73,79), are all clock generators. It reveals more when we plot Pythagorean primes and zoom a small part of Figure 3.10a. The question is how the circular feature originates? If we can express a $Prime1 = (prime2 + i)(prime2 - i)$, it is called a Gaussian prime, e.g., 5, 13, 17, 29 (Pratt, 1975). If the condition is not satisfied, then non-Gaussian, e.g., 3, 7, 11, 19, 23, 31 etc. For a Gaussian prime, the eight corresponding divisors are located on a complex circle or sphere. Thus, in the sexy prime tuplets, if we have one or more Gaussian prime we get the $6 = 3x2$ feature of generating a clocking phase (i) is activated in the PPM-G.

In Figure 3.10b, we have plotted both OF and the prime gap for Fibonacci primes side by side and zoomed it in Figure 3.10d to compare how at the very core region they match. This plot is important since Hasse diagram that we would discuss in Chapter 9, governing the energy transmission the PPMs, run on Fibonacci primes primarily. Addition of primes is a very classical way of energy transfer, while the product is a quantum way, we discovered this in the proteins and implemented it in designing the fourth-circuit element's governing function that regulates the blinking of 12 singularities. During energy transfer, primes act in a group and then co-primes act as coherent entities, however, even in that group we find the triplet of triplets as shown in Figure 3.10e. Normalized polar plots of the distribution of primes create wheels, mostly such plots as shown in Figure 3.10f are drawn for fun, but, here, when the pattern of primes govern organic synthesis and that synthesis is the brain jelly, making decisions, the wheels are all regulators of periodicity that would unfold in the complexity of the Big Data.

3.11 METRIC 10: MULTILAYER IMAGINARY OPERATION ON THE REAL INTEGERS BUILDS A NEW PATTERN

Imaginary world means passing through pixels continuously, the journey through imaginary worlds is explained with a simple animation capture in Figure 3.11a, the corresponding octonion and dodecanion tensors are shown in Figure 3.11b and c.

The tenth type of PPM explores the idea of higher dimensions (PPM-HD). We write circle as $x^2 + y^2 = 1$ in 2D and $x^2 + y^2 + z^2 = 1$ in 3D, and $a^2 + b^2 + c^2 + d^2 + e^2 \ldots + k^2 + l^2 = 1$ for 10D. We consider complex number with one imaginary phase i, then quaternions like 3D space with three imaginary spaces ($i = j \times k$, and $j = k \times i$) and octonions that behave like elements in an 8D space (Figure 3.11b) where "e" elements reside on a plane called Fano plane $(e_1 \times e_7 = -e_3, e_6 \times e_3 = e_4, e_6 \times e_5 = -e_1, e_5 \times e_2 = e_3,)$. Grouping between the elements of higher dimensions changes the clocking direction $((e_5 \times e_2) \times e_4 = e_3 \times e_4 = e_6$; if grouped differently, $e_5 \times (e_2 \times e_4) = e_5 \times e_1 = -e_6)$. When we enter a singularity point, we move to a higher dimension. In 1919, inspired by Einstein, Theodore Kaluza thought if by using 4D, Einstein could explain gravity, then why not take 5D and include electromagnetic field. The problem was where to keep the 5D elements? In 1926, Oskar Klein suggested putting the tiny spheres (10^{-30} cm) of 5D on every point on the 4D space-time structure, we can't see them because the size is small (10^{-8} cm is the diameter of H atom). In the 1960s, when more forces were introduced, e.g., strong and weak interactions, to integrate, five new dimensions were proposed. In the 1990s, Ed Witten came to a truce, he reduced all possible dimensions to only 11 (M-theory). We extrapolated Kaluza'a idea, insert the entity of a higher dimension inside every pixel of a lower-dimensional entity. However, there is a major difference between GML and Kaluza's proposition. While Kaluza and their followers transformed every single pixel of a lower dimension with an entity of a higher dimension, we suggest looking only into the singularity point of a lower dimension entity.

Tenth class PPM metric implements the concept of a higher dimension, where the sum of primes could build a modulo and regulate the multiple prime sums as alternate solutions (Figure 3.11b). We always transform an acquired input, a point to a 3D sphere, a line to a 3D cylinder and a 2D square to a cube, thus, the starting geometry is always a 3D shape. With learning, the 3D shape morphs to mimic the input data structure [subject-clause-(verb-adjective)]. If we see a cat from one side, C2 symmetry is applied and the other invisible half is created as a mirror reflection. Similarly, we search for C5, C7…C47 symmetries in the data and by applying each symmetry we regenerate missing parts. It does not end there. We create an infinite series of 3D structure dynamics using PPM incorporating higher dimensions as the higher-order terms. A cat has now infinite replicas doing different things. Thus, even for a given point, the true input is a sphere vibrating at 11 distinct dynamics embedded at 12 layers one inside another. The different quaternion regions in the dodecanion (Figure 3.11c) and octonion tensors representing the group of singularity points No input data is static, no geometric structure is kept at one fixed dimension, ever, it is always a series of events. Seeing a cat means adopting all possible dynamics of a cat, not just physical, but all higher-level control and regulatory mechanisms. Of course, the infinite chain of possibilities is edited with learning. We need at least 12 steps to learn anything with 99.99% accuracy, if one sees 12 distinct symmetries of a cat's moody positions shown in 12 images, then PPM generated infinite series of cat dynamics could cover 99% possibilities. Since all information in nature is an infinite series of vibrations, true information appears as noise to the normal sensors. We need time crystal sensors that read the map of phase change in a wide range of time scales, only then the information in nature makes sense.

10. Prime operator, P_{HD} Dodecanion

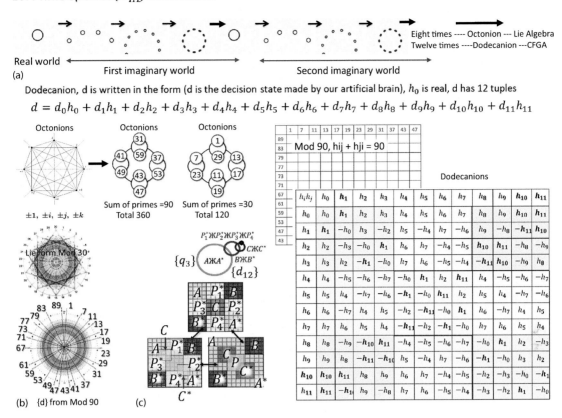

Eight times ---- Octonion --- Lie Algebra
Twelve times ----Dodecanion ---CFGA

Real world ⟷ First imaginary world   Second imaginary world
(a)

Dodecanion, d is written in the form (d is the decision state made by our artificial brain), h_0 is real, d has 12 tuples

$$d = d_0h_0 + d_1h_1 + d_2h_2 + d_3h_3 + d_4h_4 + d_5h_5 + d_6h_6 + d_7h_7 + d_8h_8 + d_9h_9 + d_{10}h_{10} + d_{11}h_{11}$$

Octonions   Octonions   Octonions

Sum of primes =90   Sum of primes =30
Total 360   Total 120

$\pm 1, \pm i, \pm j, \pm k$

Lie form Mod 30

{d} from Mod 90
(b)

$P_1^* \text{ЖР}_2^* \text{ЖР}_3^* \text{ЖР}_4^*$
$\{q_3\}$ АЖА* ВЖВ* СЖС*
$\{d_{12}\}$

(c)

Mod 90, hij + hji = 90

Dodecanions

h_ih_j	h_0	h_1	h_2	h_3	h_4	h_5	h_6	h_7	h_8	h_9	h_{10}	h_{11}
h_0	h_0	h_1	h_2	h_3	h_4	h_5	h_6	h_7	h_8	h_9	h_{10}	h_{11}
h_1	h_1	$-h_0$	h_3	$-h_2$	h_5	$-h_4$	h_7	$-h_6$	h_9	$-h_8$	$-h_{11}$	h_{10}
h_2	h_2	$-h_3$	$-h_0$	h_1	h_6	h_7	$-h_4$	$-h_5$	h_{10}	h_{11}	$-h_8$	$-h_9$
h_3	h_3	h_2	$-h_1$	$-h_0$	h_7	h_6	$-h_5$	$-h_4$	$-h_{11}$	h_{10}	$-h_9$	h_8
h_4	h_4	$-h_5$	$-h_6$	$-h_7$	$-h_0$	h_1	h_2	h_{11}	h_4	$-h_5$	$-h_6$	$-h_7$
h_5	h_5	h_4	$-h_7$	$-h_6$	$-h_1$	$-h_0$	h_{11}	h_2	h_5	h_4	$-h_7$	$-h_6$
h_6	h_6	$-h_7$	h_4	h_5	$-h_2$	$-h_{11}$	$-h_0$	h_1	h_6	$-h_7$	h_4	h_5
h_7	h_7	h_6	h_5	h_4	$-h_{11}$	$-h_2$	$-h_1$	$-h_0$	h_7	h_6	h_5	h_4
h_8	h_8	$-h_9$	$-h_{10}$	h_{11}	$-h_4$	$-h_5$	$-h_6$	$-h_7$	$-h_0$	h_1	h_2	$-h_3$
h_9	h_9	h_8	$-h_{11}$	$-h_{10}$	h_5	$-h_4$	h_7	$-h_6$	$-h_1$	$-h_0$	h_3	h_2
h_{10}	h_{10}	h_{11}	h_8	h_9	h_6	h_7	$-h_4$	$-h_5$	$-h_2$	$-h_3$	$-h_0$	$-h_1$
h_{11}	h_{11}	$-h_{10}$	h_9	$-h_8$	h_7	h_6	$-h_5$	$-h_4$	$-h_3$	$-h_2$	h_1	$-h_0$

FIGURE 3.11 Tenth class of PPMs is groups of little sub-matrices reside as a sub-set in the quaternion, octonion and dodecanion matrices and their pattern. (a) A pixel in a circle is a circle, this concept is shown by zooming. Zoomed eight times we get octonions, if zoomed 12 times we get dodecanions. 12 tuples of a dodecanion are shown in a tensor expression. (b) Decision-making elements of a quaternion, octonion and dodecanion would be prime numbers, and their sums form groups. Multiple examples are shown representing different examples of octonions. Modulo means sum of primes would be a fixed number in all the sub-groups forming an octonion. There are two examples one modulo 90 and another modulo 30. How the octonion would accept the primes in the formation of an octonion is shown to the right. (c) A dodecanion matrix is shown and different regions of this metric are divided into two nine parts. Using different colors it is shown that a dodecanion matrix or tensor remains in three fundamentally different forms, all three of which could acquire a quaternion logical architecture.

3.12 TEN PRIME OPERATORS TO BE APPLIED STEP-BY-STEP

Projection from infinity: In the ten sections above, ten classes of metrics are shown, each work as an operator applied to the time crystal generated from input information. Avoiding complex mathematical language, we have listed, in the simplest scenario, how the operations would look like on the fifth column of the table in Figure 3.12a. One very interesting column is the sixth column where without discussing in details we have put the projections of the metrics from infinity to the input geometry of the time crystal. Most of these equations a reader would find in Chapter 8 where we discussed the fourth-circuit element in details. While studying them experimentally we found that particular geometric feature is most sensitive to particular symmetry, and we went on investigating the row data to find that particular primes when they appear in

the symmetry as a regulating factor, only then the particular geometric feature comes to life. That has an easy explanation. A symmetry breaking is associated with a particular energy exchange. The resonance of material could get triggered by a composition of frequencies, the gap between the frequencies as intricately connected with the synchronization and growth of self-assembly of materials, which is here, the decision-making process. Another interesting part of the Table is the input-output scenario. Note that a typical feature of the output is taken into account in the input of the next operator. This is the reason; we need a proper sequence when we need to apply the PPM operator. The operation of the operator on the dodecanion, octonion and quaternion tensors or time crystals is closely related to the artificial brain operation outlined in Figure 3.12d. We would return to the projection from infinity in Figure 4.14, after we discuss the mechanics and mathematics associated with the geometric artificial brain in Chapter 4.

Ten prime operators of phase prime metric, PPM are applied step-by-step

Step	Operator	Input	Output	Operation by CFGA	Feedback from projected infinity
1	P_{TC}	Number, N; quaternion	Time crystal, TC, 3D geometry	$N.P_{TC}=q_{rs}^{abc}$	$\Phi/\Phi_0=sec\delta$
2	P_S	Symmetry ratio	Loops & knots	$P_S.q_{rs}^{abc}=\{q_{rs}^{abc}\}$	$\sum \Psi_{ij}+\prod \Psi_{ij}=0$
3	P_{LK}	Loops & Knots	Linked net of loops & knots	$P_{LK}.\{q_{rs}^{abc}\}=\{q_{rs}^{abc}\}^T$	$x=cost;$ $y=SintSin^2t/2$
4	P_P	Net of L, K, quaternion	3D time crystal octonion	$P_P.\{q_{de}^{abc}\}^T=\{q_{de}^{abc}\}^{TC}=\{O_{r...t}^{a...z}\}^T$	$e^2+\emptyset^2=\pi^2$
5	P_D	3D time crystal	OF>=N, points	$P_D.\{O_{r...t}^{a...z}\}^T=\{D\}$	$Sin\theta/\lambda\to\theta/\lambda;$
6	P_{NL}	3D time crystal	Non-linear prime's detect	$P_{NL}.\{O_{r...t}^{a...z}\}^T=\{NL\}$	$x/Logx$
7	P_L	3D time crystal	Local groups of OF & NL primes	$P_L.\{O_{r...t}^{a...z}\}^T=\{L\}$	$Cx^2-x(C^2+1)+C=0$
8	P_N	3D time crystal	Normalized ripples identify	$P_N.\{O_{r...t}^{a...z}\}^T=\{N\}$	$OF\to\infty,\theta\to\pi/2;$ $e^{i\pi}\to-1$
9	P_G	3D time crystal Octonion sets	Detect fractal & infinite series	$P_G.\{D\}\{NL\}\{L\}\{N\}=\{O_{r...p}^{a...n}\}$	$E=Cos^2\theta+d^2Sin^2\theta$
10	P_{HD}	dodecanion	11D time crystal	$P_{HD}.\{O_{r...p}^{a...n}\}=\{d_{r...p}^{a...n}\}$	$r=aCosk\theta$

(a)

(c)
- 2 C2 symmetry delivers other side of truth, mirrors or morphs
- 3 C3 symmetry explores three options for where and when
- 5 C5 symmetry groups five dynamics, nothing, cont. motion, diffuse, free, ultimate energy
- 7 C7 symmetry classifies synchrony classification,
- 11 C11 symmetry maps dynamics of dynamics chain in 10 higher layers
- 13 C13 symmetry maps elementary decision maker's triplet of triplet geometry 12 ways
- 17 C17 symmetry periodify collision, evolution, struggle's duration, 18 steps to return to start.
- 19 C19 symmetry finds 19 ways to sacrifice connections and direct to one selected dynamics
- 23 C23 symmetry delivers 23 ways of Branching, tonal patterns, rhythms, vibration
- 29 C29 symmetry Rising of star, Shining of devotion, Bursts of dark
- 31 C31 symmetry classifies types of sensing, Absorbing neighbours, trans-receiver
- 37 C37 symmetry forms the mat where paths of decisions, maze expands maze, lattice blinks
- 41 C41 symmetry covers all possible spirals & vortices, Ulam's spiral, log spiral, etc.
- 43 C43 symmetry maps all possible of patterns of 31 senses governing 12 expressions,
- 47 C47 symmetry maps singularities, undefined at the junctions of complete & incomplete

(b)
- 2 True, false
- 3 Space, time, state
- 5 Elements, matter,
- 7 Consciousness, emotion, reward
- 11 Dimension, hierarchy
- 13 Expression, seed of vortices, map
- 17 Journeys, War, Adventure, Mission
- 19 Mercy, Generosity, Pain, Love
- 23 Branching, tonal patterns, rhythms, vibration
- 29 Rising of star, Shining of devotion, Bursts of dark
- 31 Sensory net, Absorbing neighbours, trans-receiver
- 37 Paths of decisions, Expanding maze, lattice of living
- 41 Ulam's spiral, log spiral, vortices & spirals
- 43 No of patterns of 31 senses & 12 expressions,
- 47 Miracles, Circling, Junctions of complete & incomplete

(d)

FIGURE 3.12 Ten PPM operations transform input, the necessary operators are selected by Continued Fraction Geometric Algebra, CFGA, to be explained later. (a) The table summarizes the steps that are checked sequentially. (b) 15 primes are often found in nature and events around our life, all primes are found $6n+1$, $n=0,1,2,3...,8$. The significances are summarized here. (c) The basic elements of 15 primes are detailed in terms of geometric structure symmetry. Geometric structure of primes are stable, acts as unit, do not split into multiple primes like integers. Synthesis of primes is an objective of the brain like computing system. When the geometric shape of events group into prime numbers, the operational significance is described in the panel (c) and the philosophical significance is described in the panel (b). (d) An oversimplified flow-chart on how an artificial brain would make decisions in terms of geometry of primes.

3.13 HIDDEN PHYSICAL SIGNIFICANCE OF 15 PRIMES WHEN PPM EVOLVES A TIME CRYSTAL

Figure 3.12b, c are lists of primes as we find them in nature. Internet and popular works of literature are full of magic numbers governing the universe, but no one lists prime. We found that the primes are abundant in the tribal cultures all over the globe and those are nearly identical. When we see a prime it reminds some knowledge other than the dead geometric shape. Following rigorous research we have compiled the life of primes as described below.

The physical significance of 15 primes in real-life problems for a time crystal: When events unfold, the primes that appear early in the integer system gets a bigger share to regulate the universe. The number 2, is lucky, 50% events would try to have mirror symmetry, it always did, 16 billion years ago and would have the same 16 billion years later. C2 symmetry means left- and right-hand sides look similar. The inverted replica of an event reveals mirroring features embedded within. Similarly, spatial part of the information structure or time crystal is taken and three identical parts are created plotted in a circular pattern which changes as a function of time into a cylindrical pattern. Just like C2 symmetry reveals 2D self-similarity between truth and false, C3 symmetry plot filters space, time and person, all three features have three identities (e.g., time = past, present, and future). Triplet of triplet symmetry derived from C3 symmetry builds the fundamental structure of information. C5 symmetry reveals the dynamics of variables when plotted as five mirror replicas along a circle and as the information content or time crystal takes a new shape with time, we get a cylindrical pattern. The self-similarity in the cylindrical tube-like plot reveals five key features of the dynamics of a variable. First, continuous mode, second empty state, third, diffusion, fourth totally free but organized, fifth, ultimate randomness. So C5 symmetry classifies five distinct dynamics features in the input time crystal. Similar to C3, C5 symmetries C7 symmetry builds a cylinder by creating 7 replicas of input time crystal and arranging them in a cylindrical structure. Seven steps are taken by two-time crystals to synchronize between its local clock arrangement.

Seven steps are (i) direct matching of clocks, (ii) conditionally connected network matching between clocks, (iii) rotational direction, (iv) geometric shape written by singularity points, (v) symmetry of arrangement of a group of clocks, (vi) time crystals participating in synchronization have complementary geometries to build a new combined structure, (vii) loops in the PPM favors bridging two participating time crystals.

If one continues to build cylinders for any given input data, a cylinder for each prime reveals a unique feature of the information content, by running simulations, a statistical database has been created. Since most primes are $6n \pm 1$, for example, 5,7,11, 13,17,19,23,29,31,37,41,43,47, where n is a natural number, the physical significance of primes is easily understood by using hexagons and pair of triangles. 6 means a combination of C2 and C3 symmetry. Two triangles facing back to back build a hexagon, if two triangles vibrate independently, then we get an additional dynamic center, so $6 + 1 = 7$. We get 11 dimensions for prime 11. Similarly, we get $2 \times 6 + 1 = 13$, $3 \times 6+1 = 19$, and similarly 31, 37, 43. The prime 13 provides the matrix where for the first-time hierarchical geometries are born in the time crystal (triangle has three points, $2 \times 3 \times 2$, $3 \times 2 \times 2$, $2 \times 2 \times 3$). Prime 17 reveals the collapse of an event following a massive 16 step phase transitions. The sixteen phases are (1) Conceive a seed geometry, (2) They are adopting environment, (3) Learning begins, (4) Elimination of unwanted symmetries, (5) Location identity is acquired. (6) C5 symmetry, i.e., dynamics of variables are adopted, (7) Adoption and elimination of external time crystals begin as a routine process, (8) C13 symmetry is adopted as self-decision-making ability, (9) Cyclic learning begins as a routine affair, (10) Mechanical clocks are embedded, (11) Functional identity is embedded, (12) Four knowledge acquiring beings, metric of primes, fractal generation, creation of new time crystals, healing of time crystals, (13) Time crystal leaves the host, create replicas elsewhere, (14) Unification with complementary time crystals, (15) Creating unique seeds (16) Elimination of time crystal.

The prime 19 reveals to 19 steps of isolations of a guest time crystal from the host. 19 steps are $3 \times 6 + 1$ (three hexagons of C2, C3 symmetry comes into play, mirroring and locational judgment). Prime 23 identifies branching out of a time crystal in 23 symmetries during expansion or growth. 23 branches are $2 \times 11 + 1$ (two universes with 11-dimensional dynamics interplays). Prime 29 reveals 29 dynamic features that lead to the saturation of growth of a time crystal. 29 comes from $3 \times 9 + 2$ (three triplets of triplets with two control centers synchronizes rapidly with the environment). Prime 31 reveals 31 routes by which a time crystal learns, adopts to an unknown environment. Prime 31 comes from $5 \times 6 + 1$ (C5 symmetry plays with hexagonal symmetry, means dynamics of variables encode mirroring and space-time-personal identity; but a control center acts in the time crystal) Prime 37 reveals 37 different ways of filtering the desired time crystal in 12 different imaginary worlds, 37 classes of decisions deliver a hierarchical geometric integration of discrete isolated time crystal inputs ($6 \times 6 + 1$). Prime 41 reveals vortex or spiral dynamics in the evolution of a time crystal, 41 ways

basic geometries of a structure evolve 41 classes of vortices. We get 41 by $3 \times 13 + 1$ (triplets of C13, means hierarchical decision-making ability with a control center). Prime 43 reveals the number of ways a time crystal holding the solutions of a problem could express itself, 43 distinct transformations of a time crystal could coexist. 43 comes from $6 \times 7 + 1$ (Since $7 = 3 \times 2 + 1$, identity and mirroring control is applied for multi-layered imaginary worlds, C2, C3 symmetry is applied in a layered manner). Prime 47 reveals 47 dynamically distinct time crystals that could coexist holding different memories at a time. 47 means $2 \times 23 + 1$ (23 branches following C2 symmetry).

3.14   HOW TO USE PPM

Ten fundamental operations of PPM: Fusion, transformation, extrapolation, reduction, and morphogenesis: Figure 3.13 has outlined how one user should ideally use a PPM to run an artificial brain described in Figure 3.12d.

Step 1: First we have to find the integers in the Big Data that are related to the most dynamic and the most static domain of the variables. We zoom a local part of an apparent geometric shape in a rapidly changing big data, i.e., zoom a singularity point to go inside, find a singularity point and go inside again to find the new geometric shapes. If one repeats this process 11 times, he gets 12 layers, one inside another. All geometric shapes in a given layer are included in one PPM. Thus, artificial brain hardware would have at least 12 layers, wherein 12 PPMs would generate simultaneously 12 infinite series of a particular geometric shape. If two shapes originate from the same integers in a layer, the least matched shape is split for a better match, i.e., the hardware avoids degeneracy.

Step 2: When we convert a shape into a ratio of primes in a layer, we maintain a proportional ratio of sides of the geometric shapes while assigning a prime. Alternatively, each layer is normalized separately. Computing by PPM means only finding the inactive integers to complete the apparent metric shape and activate those integers. The physical process "activation" is an essential and sufficient step to generate all computing features. In step 2 time crystal is built since singularity points are mapped.

Step 3: When we detect symmetry or ratio of the arms of the geometric shapes formed by connecting singularity points, depending on the time width of the singularity, there could be detection error. Therefore, associated symmetries for a given set of primes is searched and just like we read differential signal during an experiment to reduce the noise we start reading the differential symmetries between different layers or imaginary worlds so that a smooth transition is noted.

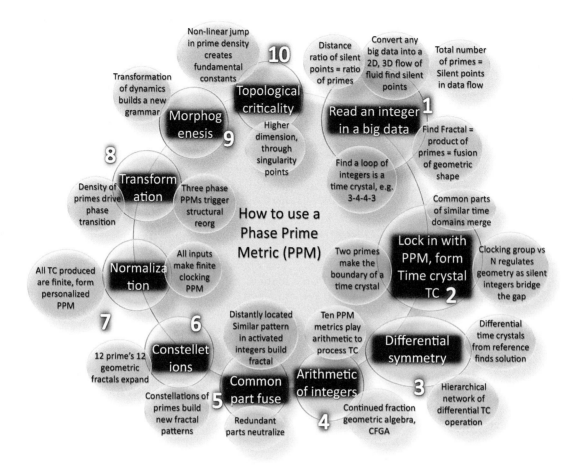

FIGURE 3.13 Ten steps of a PPM, operating in the system are listed in a decision-making process. A PPM does not decide about the time of computing and different times took in executing different parts of the process. Once it begins, the organization continues.

Step 4: Search for continued fraction, let it grow. Say we got a symmetry 2:3 it is 1 + 1/2, but we can extend to infinity by extrapolating 2 in a series that eventually results in 2. Conversion to an infinite series enables comparison of different imaginary layers, executing differential symmetry correction in a better way and keeping an option open to editing clocks.

Step 5: Common geometries between isolated time crystals fuse (Figure 2.5d–f).

Step 6: Find the dominating primes and discover the unique metric that the database would follow, one hint is given in Figure 3.4.

Step 7: Projection from infinity is applied and the input time crystal is transformed as described above.

Step 8: Three classes of PPMs OF metric, PC metric and PG metric, all would apply to convert a 4D input dynamics into a 12D dynamics of the combined time crystal formed by combining all hardware involved in the process.

Step 9: Morphogenesis: the story of PPM revolves around a beautiful truth of nature, "geometric constraints are written in the primes," a reader could read this a billion times to find its meaning in a trillion way. A periodic journey of integers 1 + 1 = 2,

ends up in absolute silences call primes and the distribution of silences build, clocks of various kinds, silences group to create geometries. Thus, repeatedly, geometric transformations are essential after previously learned time crystals are adjusted with the input, it is morphogenesis.

Step 10: At the highest level of decision-making the non-linear distribution of primes that do not follow any predictable function is taken into account as described in Figure 3.8b. The originality of distribution that never repeats itself is the domain where every single randomness that we see in nature is born. While making an artificial brain, the hardware must be programmed to sense it, in nature, in all events around us, the discrepancy set by $\pi(N) = N/LogN$ function from the real density distribution.

Here is an example for the decision-making chain of events in the geometric artificial brain.

Sensor(input)→
((ElevenDimensionData[subject-clause-(verb-adjective)]
→SensorsFusionFractalConversion→Integer Symmetry))
SensorLOOP1→

((ElevenPPMsymmetryExpand→Memory&Fractal
Fusion))MemoryLOOP2→
((ElevenPPMsymmetrySuperposition→ContinuedFrac-
tionGeometricAlgebra))TopologicalTransformation
LOOP3→
((TimeCrystalFractalReduction→ElevenDimension
Data[subject-clause-(verb-adjective)]))Reduction
-Morphogenesis-LOOP4 →Sensor(output)

Thus, five primary processes of a fractal tape machine, FTM, created by using a pattern of primes, namely PPM.

1. **Fusion:** Once the data structure [subject-clause-(verb-adjective)] from the environment find their match in the PPM as discretely spaced local structures, the regions between a pair of active integers activate. All geometric shapes or time crystals within a pair of primes fuse into one integrated geometric architecture, which is a compiled time crystal. Imagine 9, and 12 are activated in the PPM since the sensors found in the environment a triangle (12 = 3) and a point (9 = 1) discretely but missed how these shapes are linked. Thus, 10 is a silent integer, on the PPM, 11 is always silent as a prime. Therefore, 10 activates naturally, since 10 = 2 (5 × 2, 2 × 5), so a triangle (12) is connected by a straight line to a static point (9). The activation of integers links the discrete-time crystals, that is equivalent to logic, a replacement of algorithm or programming. The natural integration of integers adds information that never existed in the input created by sensors. We observed it in the living hippocampal neuron cells, live.

2. **Transformation:** There could be multiple isolated activities on the PPM where the parts of the geometric compositions are similar. Alternately, a single event from the environment could lock at multiple locations as mirror replica in the infinite chain of PPM. If the groups of activated integers are distantly located, they will grow as independent time crystals. However, the distinct features of each isolated mirror crystals are replicated into one another, so that they appear more similar, sometimes interaction creates more diversity. It is a PPM driven transformation.

3. **Extrapolation:** Once natural events activate the discrete parts of the PPM, the local geometric shapes continue to integrate. During structural transformation, self-similar patterns at various locations on the PPM could find that if the group there could be natural support in the geometric composition of PPM. Thus, those isolated regions would be naturally activated as higher level geometric structure. It is what we call extrapolation.

4. **Reduction:** Isolated events captured from the environment lock in with the PPM at multiple places may find that identical geometric architectures are produced from the PPMs. If these structures are redundant then only one copy remains, the rest disappears. Nullification of the redundant geometric architectures is carried out without destroying key geometric features.

5. **Morphogenesis:** When discrete events lock in with the PPM at distant places, if that distribution has a repeating rule or self-similar feature then that pattern transforms the integration of geometries to realize the data structure [subject-clause-(verb-adjective)]. Not just a new kind of geometric architecture but a new dynamic is adopted that builds a new grammar for its unique geometric transformation. It is a form of morphogenesis.

3.15 SOLUTION OF A BIG DATA PROBLEM ON DIABETES—PPM IN AN ARTIFICIAL BRAIN

How brain jelly uses the PPM: the brain jelly finds symmetry in the geometric representation of the input time crystals in terms of integers. Three classes of PPM are there to solve a problem (Figure 3.16). First, the metric of integers, it produces a large number of divisors of integers and begins nucleation in the precursors of the brain jelly solution. Since several integers form a group, a closed-loop in the PPM (Figure 3.4), new clocks and symmetries are born. What ten PPM operators embedded in brain jelly does actually? It generates a symmetry to the asymmetry of the time crystal and sets a perpetual drive as PPM has no limit, that is simulation of the future. Secondly, the metric of primes as shown in Figure 3.3e sets in to reach convergence. Finally, at the third step, triggers the metric of primes. 15 primes represent 15 worldviews as listed in the Figure 3.12b, c and described in Section 3.13, if those symmetries are applied to the information content encoded in the time crystal, then that adds a transformative outlook to the solution. While applying the pattern of primes brain jelly links the primes (not just 15 but all primes produced by self-assembly of 15 primes) following the metric of primes (Figure 3.10). Thus, three classes of PPMs drive synthesis in the brain jelly. If symmetry is not generated, memory bank with stored time crystals complements the need, if not found in memory, the missing time crystal essential to generate symmetry is sent out to the output as query. The complete derived time crystal is sent to output as a solution. All inputs and outputs to the out of the brain jelly are quaternions.

In Figure 3.14, on the first column we have diabetes genetic profile, using a genetic marker, the data suggest that in the pool of distribution of genes on a 2D surface, certain genes which are colorful, activates and the ones silent are kept dark. The activity of genes regulating diabetes is distributed in three regions, along the column we have zoomed continuously and tried to find the silent and active regions. The philosophy of the PPM is also outlined in Figure 3.15. One could find the key parts of the chart reflected when we discussed the solution of diabetes problem.

The construction of a single tensor q_0 for a quaternion from a real data using PPM

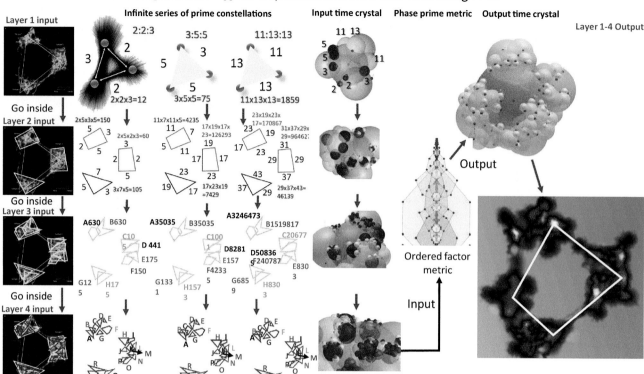

The same procedure is followed for all tensors, the linguistic quad who, when, why & how is asked

FIGURE 3.14 An example of how a PPM is used in solving big data. In this case the occurrence of diabetes in a large number of patients are plotted as a function of two types of genetic expressions, one plotted along the horizontal axis and the other along the vertical axis. The axis perpendicular to the plane represents the count of patients. The image is divided into four layers. In each layer we find the geometric shapes covering similar z values or occurrences. The ratio of the sides in the geometric shapes are noted for all the structures and for layer 3 and layer 4, the data are given below this figure caption. To the right, in each row, the equivalent time crystal is presented. As one moves from layer 1 to layer 4 from top to bottom, the time crystal grows as it adds more information by including higher resolution geometric shapes. The completed time crystal is then sent to the PPM, wherein divisors of the integers are found as primes and linked integers via drawn loops join and thus transform the time crystal output, which is then again converted into the starting image, but this time, the image reveals hidden patterns distinctly. The starting image had 3 regions and the final image has 4 regions. So, a totally new type of diabetes that was happening for a unique genetic composition is discovered from the big data.

Layer 3. SET I: Object A, 3:3:5:7:2(630); 5:7:7:11:13(35035); 19:19:23:23:17(3,246,473)...... Object B, 3:5:2:3:7(630); 7:11:5:7:13(35035); 13:23:17:13:23(1,519,817)......

SET II Object C, 3:5:7(105); 7:11:13(1001); 23:29:31(20677)...... Object D, 7:7:3:3(441); 13:13:7:7(8281); 31:31:23:23(508,369)...... Object E, 5,5,7(175); 11,11,13(1573); 19,19,23(8303)......Object F, 3,2,5,5(150); 7,5,11,11(4235); 23,19,29,19(240787)...

SET III Object G, 5:5:5(125); 11:11:11(1331); 19:19:19(6859)...... Object H,5,5,7(175); 11,11,13(1573); 19,19,23(8303).

Layer 4. SET I: Object A, 11:11:17(2057); 19,19,29(10469); 29,29,37(31117)...... Object B, 7,7,7(343); 11,11,11(1331); 17,17,17(4913);...... Object C, 5,11,7,13(5005); 19,23,29,31(392863); 37,41,43,47(3,065,857)...... Object D, 7,7,11(); 13,13,17(); 23,23,29()...... Object E, 2,3,5(30); 5,7,11(); 11,13,17()...... Object F, 3,3,5,7,13(); 17,17,19,23,31(); 29,29,31,37,43()......... Object G, 3,3,3(27); 7,7,7(539); 11,11,11(1331)......

SETII: Object H, 7,7,7(539); 13,13,13(2197); 19,19,19(6859).............. Object I, 2,3,5,7(210); 13,17,19,23(96577); 29,31,37,41(1,363,783)............ Object J, 5,5,5(125); 7,7,7(539); 11,11,11(1331)...... Object K, 5,7,7,11(18865); 11,13,13,17(31603); 19,23,23,29(291479)......... Object L, 3,3,5(45); 11,11,13(1573); 19,19,23(8303)......... Object M, 11,11,13(1573); 17,17,19(5491); 29,29,31(26071)......... Object N, 17,17,17(4913); 29,29,29(24389); 37,37,37(50563)......... Object O, 2,7,11,13(2002); 13,17,23,29(147407); 29,37,41,43(1891699)...... Object P, 7,7,11(539); 13,13,19(3211); 23,23,29(15341)......

SETIII: Object Q, 7,11,19(1463); 19,23,29(12763); 29,37,47(50431)...... Object R, 5,5,11(275); 7,7,13(3237); 11,11,19(2299)...... Object S, 3,5,7,11(1155); 11,13,17,23(486013); 17,19,23,29(215441)......... Object T, 5,7,7(245); 11,13,13(1859); 17,19,19(6137);

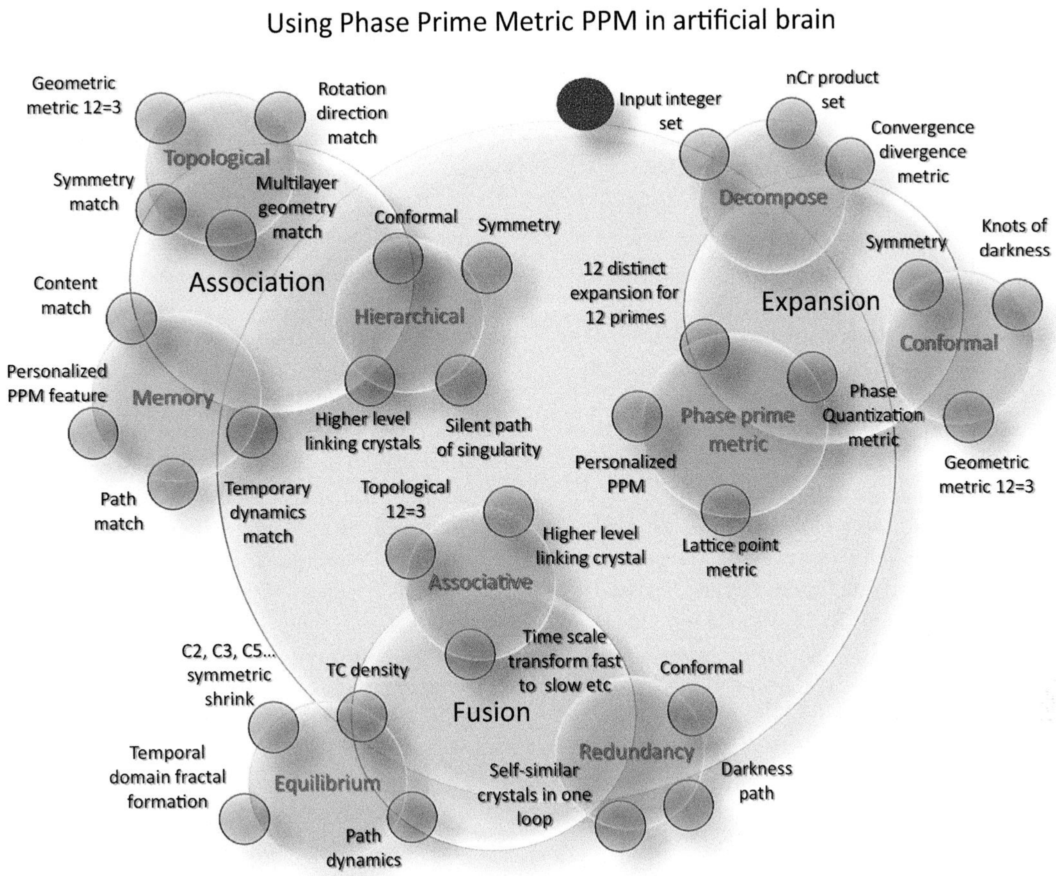

Using Phase Prime Metric PPM in artificial brain

FIGURE 3.15 A chart shows how a PPM operates in the human brain. Time crystal always emerges in the integers between two primes, they serve as the boundary values, so larger is the difference between two primes, larger is complexity of time crystals. Within 1 billion we have roughly 255 million primes means around same number of time crystals. $x/ln\ x$ where x is the number less than which the prime numbers should be and ln x is logarithm of x to base e. So using this, we get 255.88 million number of primes between 1 and product of all 12 primes that primarily takes decision in the artificial brain.

Determination of wiring between clocks from a PPM: a decision-making non-computer has an upper and a lower time limit of an integer representing a slice of the PPM. That core architecture is a fractal network of clocks or clocking Bloch spheres holding the geometric shapes. Mathematically, it is an assembly of a large number of distinct ΔN slices of the PPM. To cut a ΔN slice of metric, start at $N = 0$, end at a certain N value. All clocks are wired following ten steps noted below. Metric prime hardware could be realized in various ways. We repeat, every point in the PPM is an event that contains many points forming a geometric shape inside. One could start from any simple event to begin with and find events inside, continuously, building a topological network.

1. **Select a set of N e.g., {1, 3, 7, 45, 32, 734, 1500, 3800}:** For all zoomed images in the Figure 3.14s, a set of n values is selected on the PPM that forms a closed loop. The phase relationships of N Bloch spheres plotted in a circle. Together n components try to cover a phase space of 360°. The plot reveals the value of the quantized phase. The resultant parameters of spiral dynamics become evident as in biosystems (Bourret et al., 1969). In the set of n, some engage in a clock, others are found as not participating in the periodic vibration. They remain outside the spiral or vortex dynamics that integrates a set of N.

2. **Identify different slices are components, to make individual loop of geometric structures:** All distinct slices of a PPM have their own time crystal, or phase space structure. A slice means a set of values n in ΔN, it has its own factors, own sub-metric. The network of time crystal reveals a hidden geometric structure. The layered geometric structures match for all N time crystals and those hidden self-similarities group sub-sets of N. Therefore, even before the PPM starts convergence to primes, as a drive to integrate the discrete-time crystals into a singular one, a preliminary grouping of the time crystals is made as shown in the infinite series of prime constellations in column two of Figure 3.14.

3. **Initial phase gap between the clocks is important:** Initial phase relationships between different clocks can change the output projection of time crystal dramatically. Initial starting points define temporal change of phase, or dynamics. Temporal recording of resonance peaks reveals how the time gap or phase gap between clocks changes with time (dynamics). For a given slice of metric ΔN, the dimensional ratios of the n geometric shapes located at n points on the metric are noted. Junctions of clocks locate system points, direction of rotation is selected to neutralize vector sum of 3D structure. The system points and their phase gaps determine the angle between clock planes. Readers should follow the four rows to feel the transition from a 2D image to a 3D architecture of time in Figure 3.14.

4. **The fractal network of clocks, side by side and one inside another:** When a set of N integers is spotted in the PPM, the integration begins. Say two integers of set N are 32 and 56, they make individual time crystals by following PPM. At the same time, their products (N1×N2, N1/N2, N1+N2, N1−N2) will be the new members in the PPM, expansion of set numbers will be plenty to find a closed pattern in the PPM. The gaps will be filled by repeated elementary primes. The repeat of primes repeats a particular geometric shape to build a large structure. These repetitions cause self-similar patterns in architecture, there is essentially no universal fractal in the PPM.

5. **3D geometry optimization of clock location:** Since one has to build a wireless network of clocks, the physical location of the clocks in a 3D network is important. Wireless projection of constituent time crystals all around, by 360° solid angle, selects whatever time crystals reside in its path. A 3D network appears as a sphere to an external probe or we state "3D projection of a time crystal." So, the clocks along with a line screen each other's information, thus, creating an erroneous signal to an observer. Note that in case of conventional electromagnetic signaling by an antenna and a receiver, the energy degrades largely with the distance. However, for magnetic beating in a noise activated phase coupled matrix of time crystals, there is no single point source of energy and no fixed destination. Thus, the size of the phase network, is irrelevant to an observer, which also becomes a part of it. Geometric locations govern the phase relationships of the integrated clocks. Locally in the 3D architecture of the computer core, a group of clocks shifts their coordinates. Thus, locally, the system shifts from one metric to another.

6. **Vibrations at all time scales in all spatial scales maintain speeds:** Clock speed is regulated by delay time, and fixed everywhere in the metric hardware.

Only one, the fastest clock alone is used to integrate the phase or delay to create all possible "times" for operation. For that purpose, it is imperative that every clock senses all neighbors once hardwired. It is a critical technological challenge. All the clocks communicate wirelessly in hardware following the magnetic beating of beats explained earlier. The localized beating during synchrony should converge on the particular locations of the clocks, so that phase shifts do not occur. Phase shift with the motion of the system point is key to reliable processing of topological information. The speed of a system point along the perimeter of phase cycle changes if it falls into the singularity domain, and recovers it afterward. The falling into a singularity is crucial for editing geometric memory.

7. **Thermal and electrical noise as the source of energy:** Wireless communication is the key to achieve "one-to-many and many-to-one" communication (Bandyopadhyay and Acharya, 2008; Bandyopadhyay et al., 2009a). The 12 singularities detected in the biomaterials 3D phase space links the magnetic field with the stored charge. The relation is historic because in the last 800 years all theories of electromagnetism have endorsed that a current must flow. But this phase space activates by noise and only noise, no current flows in to modulate magnetic flux with charge. For that reason, clocks are wired but make sure that em noise splits into electric and magnetic parts but no current flows but only white noise (Sahu et al. US patent 9019685B2). Thus, a clock in a phase cycle is free for neighbors, no wiring is essential. Noise is applied artificially to trigger morphogenesis and symmetry regeneration of the time crystal spontaneously without any algorithmic instruction.

8. **Testing the 12 metric compatibilities, building a grammar of limiting changes allowed:** In the fourth column of Figure 3.14, we have implemented the PPM. First, the clocks are wired as suggested by the PPM. Provisions are kept so that input time crystal modifies the co-ordinates of the clocks. Specific input time crystal selects one of 12 distinct patterns of PPMs. It is a crucial task to find the right trigger. A delicate choice of the clock co-ordinates eventually allows a few metric to dominate. We are currently adding several new aspects of the PPM. Newer and more insightful metric features are being revealed.

9. **Geometric shapes synchronization is a test of unity (one-ness):** The non-computer has two parts. One, converting all sensory signals into a set of topology. And the second, integration of topology. However, at both parts the geometric shapes spontaneously store and synchronize the clocks. Circuits modify, filters, amplifiers and vibrating membranes are added to

process GML (Reddy et al., 2018). The whole device is just one single clock holding plenty of clocks inside. The non-computer core has one device, only one geometric shape for all sensory memory, it delivers only one unified time crystal eventually as shown in the fifth column of Figure 3.14.

10. **Non-disruptive Interaction between the four modules:** Finally, the wiring is edited to enable the device as a filter, resonator, inverse of resonance, and clocking geometry writer. Once these four qualities are optimized, true hardware of metric of primes is made. It is tested for image processing, hierarchical perception to find the mismatch in the wiring. A particular test is repeated in all directions of the 3D network of clocks. To the extreme right of Figure 3.14, we have returned back to the gene profile and found that there are four regions. The result suggests that some genes would probably remain silent, but, they might be governing the entire dynamics of the evolution of diabetes in millions of diabetes across the globe.

3.16   THREE CLASSES OF PRIMES

When we look at a metric picture or a 3D movie for the PPM, the first question that arises where is the time? Well, the time has always been a perception, but here in this book we sincerely try to bring attention to the symmetry of time as a geometric shape. One should look at three figures and compare them repeatedly. The ordered factor metric of Figure 3.2d, OF metric. Then the prime composition metric, PC metric of Figure 3.3e. Finally, the metric of prime gaps, PG metric in Figure 3.10a. They appear distinct in geometric shape, how about their function? We have compared three different worlds in Figure 3.16.

First, Sensor acquires data from its environment. As the signals fall in, its clocks are activated. Thus, it transforms a binary stream of pulses into a 3D network of clocks. It cooks an input time crystal. For operating the sensor, a dedicated ordered factor metric is used. The figure shows with an arrow that the sensors capture the sidearm ratios and other typical parameters of the geometric shapes as integers. There could be a fraction, but the sensor ignores that and tries to reach an integral point. The ratio of integers is created by the ordered factor metric described in Figures 3.2 and 3.4.

Second, an initiator acts like a bipolarity filter. In one way, it shrinks the size of an input time crystal. The output becomes a small fractal seed. If the input is sent through the reverse direction, the PPM fills the missing gaps. It inflates the time crystal, to its original form, sometimes much larger. Output time crystal contains situations not yet happened, i.e., futuristic dynamics. As a central processor, it harvests the real strength of linking events, this PPM is called PC metric.

Third, all parts of a Processor are always active. As time crystals arrive from the initiator, synchronization begins. A system that is a finite PPM, from its smallest element to the largest component, operating time scales syncs

simultaneously. All the matching time crystals amplify the signal. A regulator that investigates the PC metric intimately, finds that the primes irrespective of the geometric information content does something else. The group of primes synchronizes with the time crystals missing in the processor part, or uncorrelated geometries form groups that are not suggested by PC metric. Then it was found that there exists a new kind of metric that links groups of primes and activates the missing clocks where there is no symmetry. The mismatched yet essential clocks find a suitable location in the Processor. They are later absorbed there as a part of learning.

3.16.1   ORDERED FACTOR METRIC (OF), PRIME COMPOSITION METRIC (PC), AND PRIME GAP METRIC (PG)

From the discussion above, we have learned that we can choose any integers along the horizontal axis, only point that we have to consider is that the horizontal axis is all about the options we have while making decisions. Along the vertical axis, we calculate how those options are to be grouped. Now, when we connect the dots, we get a perspective about the geometric link between the events. Three examples are shown in Figure 3.16. The first step is the ordered factor metric, OF metric, where the unknown environmental data first enters into the artificial brain. Most sensors are configured as this metric. Though the input data is 11D, belong to 12 imaginary worlds. However, the dynamic feature it holds is 4D, a quaternion like structure. The dynamic feature means at a time, simultaneous changes in four imaginary worlds are appropriately recorded.

The second step is the Prime composition metric or PC metric. The particular metric is the point in the artificial brain where the information from the external world is converted into the language of primes. If we keep the input information as the ratio of the integers, then we would have infinite letters, no language. A language is beautiful because it uses few letters to create an infinite number of words. So, conversion to prime, preparing the geometric shapes for the decision-making is what PC metric does. Since the language of primes now activates, input 3D geometric shape distribution is represented as the ratio of primes, any structure we look, the arms are primes, the processing of information begins. Though the input data is 11D, but only holds 4D dynamics, now at this stage, the artificial brain would run a feedback loop to increase 4D dynamics to 8D dynamics. We repeat, 8D dynamic means 8 out of 12 imaginary worlds.

Finally, the information comes to the last stage, where at the deepest core of the artificial brain, one may draw an analogy to the 47 Brodmann's region in the human brain where the final processing is done. That place holds the Prime gap metric. Here, the horizontal axis is all the integers, i.e., all possible choices and the vertical axis is the gap between the primes. The particular metric does not deal with the geometric shape. If we look at the plot for Prime gap metric or PG metric, it links pairs of primes, an abstract mathematical representation of the information. Abstract mathematicians might find interest when the pattern between the groups of primes form a

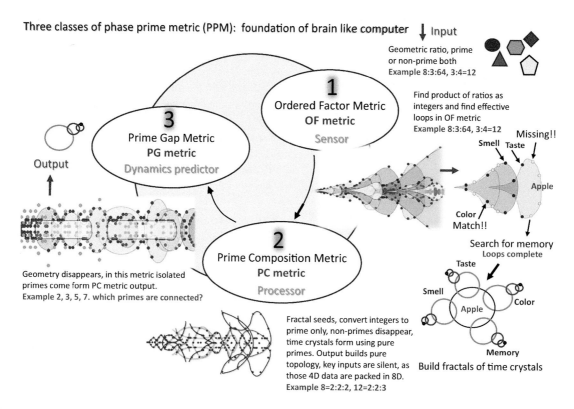

FIGURE 3.16 A table shows three classes of PPMs essential for building an artificial brain. The first one is ordered factor metric where ordered factor of an integer is plotted against the integer. An input data is divided into geometric shapes and then PPM integrates different kinds of sensory data. Class I PPM is all about capturing and integrating the sensory information. Class II PPM is made by plotting the composition of primes required to present an integer and how many ways those primes could be arranged. This metric does not acquire integers unlike class I matrix, thus, the situation here is that entire sensory information is converted into set of primes from the set of integers. After class II, comes the third class III metric in the last step, where PPM is built by counting the gap between primes. This metric does not look into the geometric shapes but the true pattern of grouped primes as neighbors. Thus non-geometric part of the primes works here to build the artificial brain.

higher-level group and that pattern reveals a new metric to link events. At this level, the artificial brain would make a journey from 8D to 12D, by running a feedback loop where symmetries would synchronize, in the language of symmetry.

In three layers, we find the language of geometry brings integers from nature, then the language of primes brings symmetries from geometries, and finally, the language of symmetries bring identities of an abstract language that nature speaks.

3.16.2 Advantages of PPM

When the language of 15 primes meets the language of 15 geometric shapes: One could never think that a journey through singularity is to replace the "free will of a programmer" that looms large in the current structure of artificial intelligence and information processing. The metric is not a black box but a guideline to glue a 3D network of events, that is often used to build scientific models. With more information, from the metric one finds the exact location of the event in a complex network of events. In order to create one event, a single metric is not enough; we need all ten classes of metrics. If one knows how nature processes information,

then any form of artificial intelligence is irrelevant. Now, a GML, wherein, the letters are a few geometric shapes bridges with the language of primes via PPM. Using this language, a non-computer could search outside. It does not have to wait to get input. The approach is just opposite to a computer. Understanding the true nature of an event is to replace the black box with something real, closer to nature. It is not the ultimate. Non-computing is a primitive yet significant step to mark the beginning of replacing artificial intelligence with natural intelligence. We have outlined all these points in Figure 3.17.

Turing computer needs a human, non-Turing has to glue the user and the nature: When we advocate the culture of listening to the silence of a system and when it bursts a noise naturally, when no question asked, and find self-similarity, then we need to know "I," "observer" and "environment." The days of human bias and building scientific formulations would be replaced by hardcore topological mathematics addressing the self-similar folds in the PPM. Observer participants in governing laws are filtered out by the PPM since it can accurately map the time crystal architecture of the observers and the system it studies. After all the information controlling architecture of the universe "has to postulate

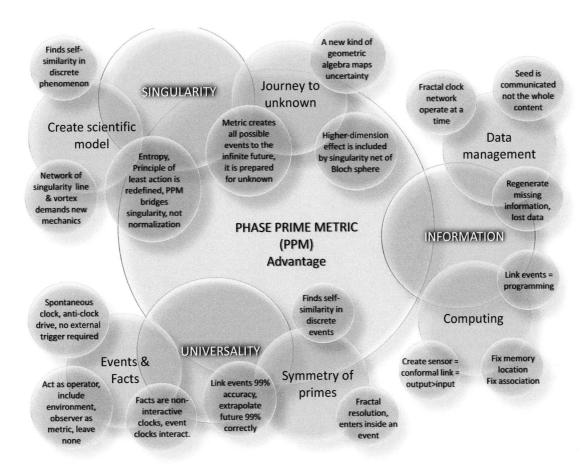

FIGURE 3.17   The advantages of a PPM are summarized in a single chart.

explicitly or implicitly, a super-machine, a scheme, a device, a miracle, which will turn out universes in infinite variety and the infinite number" (Wheeler, 1988). Circuit-based Turing computers have human as the key component for its efficiency. Non-computer relies on the PPM to fill the gap and expand the time metric it gets from nature. Discrete clocks link. If some clocks are missing, they are created. Constructing a higher-level clock is the only driver of a non-computing core. A drive to make slower clocks gets everything done, spontaneously.

The use of the pattern of primes in integrating the time crystals has ten significant advantages.

1. **Retrieving the lost data and transmitting a fraction:** Sometimes, hardware damage loses information. One has to place the remaining part in the user's time crystal network. The pattern of PPM integrates the available crystals, thus, recreating the missing network (Bandyopadhyay et al., 2010c). For this reason, there is no need to communicate full information. If a fractal seed is sent, it evolves uniquely, more profoundly in the receiver. The morphing of a fractal seed retrieves complete information. An entire infinite series could be retrieved from any part of the hardware.

2. **Drive to integrate discrete time crystals is similar to programming:** The metric acts as an operator on the input matrix of clocks. It links any form of time crystals with slower clocks. The PPM has two drives. Make slower clocks to integrate and expand its PPM implementation. No hardware can implement a PPM with intricate details. A continuous drive to improve metric makes nominal mistakes in reconstructing the perception of logic.

3. **Provide key information to change wiring:** Unlike other space-time metrics, the PPM does not depend on the user's guess. It determines the detailed structural features like an origami (Ghosh et al., 2016b). The time crystals dynamics are mapped accurately. The time of symmetry breaking and the states post phase transition are also determined. Thus, the PPM provides the software solutions, like a fusion of a user and computer. The essential hardware modifications required for an input time crystal to store and evolve also delivered.

4. **Higher-level perception is naturally embedded:** Slower clocks in a PPM hardware integrate the faster clocks in a scale-free manner (Eguiluz et al., 2005). In its time scale, clocks are arranged symmetrically.

It enables fractal clocking. It means during a one-second operation the system point could dip down into the picosecond scale, process and return with a solution before the one-second clock "ticks." Fast running time crystals are never left alone. In a widely varied time scale, only a few clocks need to "tick" to make a decision.

5. **Computing location, start, end, and halting are decided early:** The construction and editing of the time crystal that continue forever. During synchronization, the transformation (morphing) peaks, but do not die out (Bandyopadhyay et al., 2010c). The effective length of the metric is infinite due to a closed-loop (Hamkins and Lewis, 2000). It also drives to increase metric length. It has several advantages; no instruction is needed to start and halt. Halting is a significant problem only when the end is not fixed (Minsky, 1967). It is a case of a Turing computer. Here, computing is all about entering inside a singularity, so the start and the end are fixed before the journey begins. The metric activation reaches a maximum and then naturally reduces the editing of the time crystals. Thus, a near halt is reached.

6. **Non-reductionist approach: no choice is never rejected:** The PPM reads the events in nature, so the concept of input is none. A PPM hardware includes an observer, an external user and all environmental participants. It integrates into a unified virtual universe. There is never a rejection or reduction of choices or even a probabilistic select. The only effort is to en-loop the isolated loops. Non-reductionism ensures junction free hardware (no heating).

7. **Quantum-like speedup without entanglement:** Due to the fractal nature of the PPM, it performs a search without searching. Physical wiring destroys phase modulation. So, the hardware uses wireless communication (Jaeger and Haas, 2004). Spontaneous reply requires only 12 layered clocks to find 10^{12} number of clocks. At every layer, one enters inside a faster clock. Hence, time to solve a problem in the shortest time possible in the layer where the question was asked (Ghosh et al., 2014a). Note that 10^{12} oscillators make 99% of all patterns in nature (10^7 almost covers all). Therefore, it is the maximum

number of clocks used in a layer. Quantum computing also provides this speed (Schnorr, 1978; Galil et al., 1987; Bloch, 1997; Hamkins and Lewis, 2000). However, if entanglement needs to be broken, repeatedly, which is often requisite, then the advantage of speed disappears in Quantum (Jozsa and Linden, 2003). Alternate routes where one could use polarized light-induced interference can speed up computing (Lloyd, 1999).

8. **Directional memory delivers a virtually infinite capacity:** The same 3D structure of a time crystal emits a different burst of signals to different directions at any given time. The observer can choose infinite locations around the structure to get a new solution. Therefore, the memory capacity and the distinct solution generation ability are astronomical. It has nothing to do with the number of oscillators.

9. **The universal language is fundamentally embedded in the PPM:** The GML in combination with the PPM can build a virtual language of patterns of any system. They need only a temporal evolution of resonance signal data to build a network of phase. It suggests interacting with any system whose language is unknown. No rules or information about the intelligence of the system is required.

10. **Harnessing singularity is not possible with the existing mechanics:** Due to the fractal clocking behavior, there are singularities at every location on the PPM (Kolwankar and Gangal, 1996). The differential calculus needs to be replaced by a conformal algebra to simulate the wiring. If one enters a singularity domain, it finds no self-similarity. The PPM ensures a non-repeating experience continuously (Barnsley et al., 2008). One needs to make a journey to the singularity domain blindly following a PPM. Thus, both classical and quantum mechanics are not useful here. Bridging the singularity safeguarded quantum. Here, it is prohibited. One has to enter inside a singularity and collect available self-similar clocking factors. It's finding holds the key dynamics of a PPM. It explores pure topological factors for developing effective mechanics (Agrawal et al., 2016a; Hohlfeld and Cohen, 1999; Kawato and Suzuki, 1978).

Only that person comes to learn universal consciousness, to whom the universal consciousness wants to unfold in an unexpressive manner; however, that person, to whom it comes as a subject of knowledge, cannot state what it is. Those who have seen and felt it could never be consciously aware that they have realized it, but those who think that they have felt it have never learned what it is.

4 Fractal Mechanics Is Not Quantum but Original—Geometric Algebra for a Dodecanion Brain

4.1 REVISITING THE BASIC CONCEPTS OF QUANTUM MECHANICS USING CLOCKS

If today, geometric musical language (GML) is the language of the universe, then: "Tomorrow we will have learned to understand and express all of physics in the language of information" (Wheeler, 1990). If one has understood the right information theory of the universe would be, "at that point ready to revalue $h = 2.612 \times 10^{-66}$ cm^2—as we downgrade $c = 3 \times 10^{10}$ cm/s today—from constant of nature to an artifact of history, and from the foundation of truth to the enemy of understanding." The quest for a unified language for physics and mathematics is long (Hestenes, 1986). The message is clear. Explain the origin of fundamental constants, learn what existence is. Only acceptable proof that the true language of nature is invented, is the derivation of all fundamental constants from "no law." He envisioned "the world as system self-synthesized by quantum networking" (Wheeler, 1988). When there are plenty of observers, how the information structure changes? It is said that every measurement changes the source. It is very exciting. The photon that started from a distant star Billions of years back, when we measure it in the planet earth, by looking at that star, in our naked eye, the atoms in our eyes collapse the state, the entangled photon may be traveling to a very different direction, finds its state fixed. Across cosmos, zillions of atoms which are observing a single star would even change the star (Super Copernicus principle).

Nanobrain is a nano-sized black hole: Wonderful entropy of Bekenstein: We revisit quantum mechanics by writing all its basic knowledge gathered and optimized in the last century in terms of geometric musical language (GML) developed recently. We find that the basic unit of information, a clocking Bloch sphere holding a geometric shape that we introduced in the second chapter, could pictorially present all derivations of quantum mechanics. We have not tried to derive quantum mechanics, we consider it a phenomenon that has only one property, "*collect the singularity points and clock them in a single circle*" (Wheeler, 1968; Hartle, 2005; Hawking, 1982; Vilenkin, 1982). The singular property is essential and a sufficient condition to generate a wavefunction (Hartle and Hawking, 1983), thus, all of the quantum using GML. See how measurement is a fusion of two clocks in Figure 4.1a. In fact, the very foundation of GML is quantum, with a difference that its Bloch sphere has no classical point. The interesting feature of a self-assembled Bloch sphere is that apart from being a time crystal, the whole information architecture, whether it represents a human brain or a single protein molecule, it is similar to a nano black hole. Its only constituent is singularity points. Bekenstein found that the surface area of the horizon of a black hole, rotating or not, measures the entropy of the black hole, it is 2.77 times the plank area (Bekenstein, 1973, 1980; Penrose, 1969). A composite time crystal representing the information architecture of a complex system is filled with clocking singularity points, hence the surface geometry of silence accounts for an entropy. We do not need to count the Bekenstein number. ***The complex surface area of a time crystal is proportional to its entropy***. Since we use fourth-circuit element, Hinductor for implementing phase prime metric (PPM) and the GML, which increases magnetic flux linearly as a function of charge, such a scenario is valid even for a charged black hole (Zurek and Thome, 1985). It was argued that to address the complexity of the level of universe or brain we need an information metric (Zurek, 1989), we have it now as PPM, that does not grow or lose its complexity remains fixed everywhere.

Feynman's singularity bridge (Cao et al., 1993; Feynman, 1949)**:** The worldview breaks the fundamentals of the information science that has been successful for over a century. The reasons are the following. First, the observer dilemma is that already it is putting its bias into the system by choosing

85

when to see, wherefrom to see, how much to see in the output. Then, the observer should not make a black box to fit nature blindly. Second, quantum fails to probe singularity. When one considers events inside an event inside an event in an infinite network, then it takes us back to the deadlock of the 1930s. Then, Feynman bypassed the singularity to save the quantum deadlock (Reddy et al., 2018). Bridging singularity saved them. However, the journey they avoided is what makes nature beautiful. One should not bypass but explore it (Mallat and Hwang, 1992). Third, logic and the fitting tools of AI are blamed for being a product of human imagination for creating the abstract black box. It is far beyond reality, only to fit certain observations. Then, one should not avoid singularity, do not use an "educated guess."

It is the purity of time that demands a new mechanics: Time travels both ways in quantum, to the past and to the future, just like all other physics laws, so the quantum is natural. However, the classical world where time travels in one direction is shocking, surprising (Price, 1996). In classical wisdom, time does not divide like cells, it does not intersect with another flow of time, it does not return or suddenly die out, or disappear forever. Time has no beginning or end, be it flows linearly or in a circle (time asymmetry). Here we envision a world where time is an architecture of phase following the symmetry of primes. Quantum is its local view, classical is a point on this structure. On a circle where the phase is changing 0°–360°, on its perimeter a pixel cannot be a complete event, as no event is before itself, else "the world recurs infinitely many times" (eternal recurrence; Nietzsche, 1967). Thus, we sense the need for new mechanics. Even quantum has evolved so much. Earlier it was proposed that multiple classical worlds interact in parallel to create a quantum world (Poirier, 2010), and now it is more clearly argued that all classical world's work deterministically, since we are ignorant about their motion, we see the unpredictability (Hall et al., 2014). Ehrenfest's theorem, wave packet spreading, barrier tunneling, and zero-point energy have been mathematically created using repulsions among these deterministic classical worlds. The journey between classical to quantum and quantum to classical has seen many adventures, but the journey through imaginary worlds remained unhonored.

A journey from qudit to fractal mechanics: an undefined network of actions hc (Plank constant × velocity of light): We consider quantum mechanics as a fundamental property of the universe that couples the vibrational systems in the time domain, independent of space, the connection is made by a wave function. Two particles at the extreme ends of the universes could be part of a single wave function. Quantum mechanics is said to be incomplete; a wave function should include an observer (Wheeler, 1977), but if there are classical points on the Bloch sphere, this won't be possible. As outlined in Figure 4.1a, an observer could be a pixel of the object to be observed or vice versa, when clocks depict wave–particle duality (pixel or perimeter that's the uncertainty), the relative size of clocks would matter. If the worldview is sequential, then one imaginary world of quantum mechanics is enough. However, when we explore "within and above" assembly of

events, a large number of imaginary worlds would coexist. Thus far, quantum mechanics used octonions only to define multi-dimensional dynamics of one imaginary world, never explored, how multiple imaginary world could coexist. That takes us to a new mechanics, namely fractal mechanics (FM).

Ten principles of fractal mechanics: Thus far several attempts were made to put multiple classical points on the surface of a Bloch sphere. The attempt curves the Bloch sphere and we need to go to a higher dimension. Instead, when we built the foundation of GML (Chapter 2), we inserted new Bloch spheres at the singularity points. Instead of Bloch sphere we use the term phase sphere, because no real point exists on the Bloch sphere. The particular step drastically reduces the computational complexity for theoretically estimating the efficiency of the information architecture. Moreover, the mechanics driven by bursts from singularity alone requires multiple changes as outlined below:

1. We do not need to go to a higher dimension, 3D representation of phase spheres is sufficient to process higher dynamics. One can enter inside the singularity points 10 or 11 times, as soon as a system point crosses 10 layers or singularity points one inside another, in reality the complexity of 10 dimensions is addressed.

2. $\hbar$ maps the horizon area as information lost, it reveals the wave number of light as photon momentum and considers field flux similar to a bit-registered fringe shift. Therefore, GML derived time crystals that we built for operating the fourth-circuit elements (Chapter 8), we get momentum as typical geometric features of time crystal at various layers. In the conventional quantum mechanics, one layer or one imaginary universe is enough. Here at each layer, we get a set of momentum space made of layers below and contributes to defining the momentum above. When we rewrite basic quantum interference and quantum beating in terms of nested clocks or circles, we see the need for at least two nested imaginary layers (Figure 4.1b).

3. Product of plank constant and velocity is defined as action ($A = hc = hv$). In the conventional quantum mechanics action is always hc, here in some layers it is hc and somewhere it is hv (v = velocity). Most importantly, hc is the flux of the most fundamental ring, it is magnetic. Information architecture is now an infinite network of various kinds of actions that is undefined everywhere. Projection of an observer's time crystal on the measuring time crystal to the infinity reflects back to the observer. The feedback from infinity on the observer is the reality. For this reason, we cannot even use the term classical-like or quantum-like. Of course, an entangled quantum system or free-to-edit classical system are both not essential to operate a PPM.

4. Bell's theorem is a proposal to measure the action at a distance; it is a bridge between the special theory of relativity and quantum mechanics. For entanglement

we need a hidden variable common to both the systems (Bohm, 1952). When we introduce PPM, there is a composition of symmetries linking observer and the system to be observed doing the measurement, the system and the environment are connected by clocks. Thus, it is not just one but an intricate map of geometric variables acting on the measurement. Therefore, a measurement following fractal mechanics would deliver a set of observations, not one, one just requires proper sensors to sense them in a non-demolition manner.

5. There is no collapse of the wave function by measurement, it is a redistribution of the clocks, restructuring the topology of symmetries that link two spatially distant, temporally non-separated systems. It is nearly impossible to destroy or delink all clocks, a measurement acts only on a few clocks. A clock forms when a phase loop traps a clock smaller than its pixel, then a new pixel is born in the host clock's perimeter. The creation of a system point is adding a pixel with a physical identity (Figure 4.1c).

6. Fractal mechanics is based on three fundamentals, singularity, the multi-layered architecture of clocks and the symmetry of primes. The architecture of clocks is similar to a generalized version of time

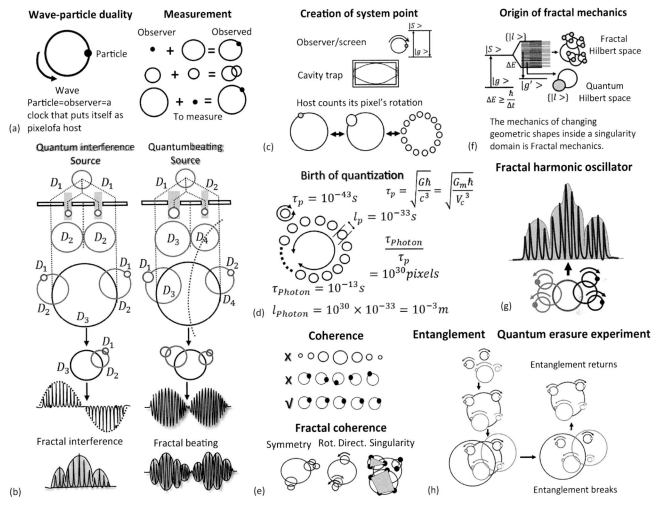

FIGURE 4.1 Using geometric musical language (GML) we revisit quantum mechanics. (a) No wave flows without a counter. Using a circle and a dot the philosophy of quantum mechanics is explained. The observer could be a dot, a circle or even bigger than the host. When a circle or a clock is measured it is a particle and when observed it is a wave. (b) Quantum interference and quantum beating shown with circles as light. Quantum is about one imaginary world. Fractal mechanics is about n number of imaginary worlds. Below the interference pattern and beating pattern, a simple case of fractal version is presented with one additional imaginary world. (c) Three routes of creating system points. (d) How smallest circles presenting time and thence distance quantizes space and time in a given system. (e) Three ways of quantum coherence, top two rows show multiple signals which are not coherent, the bottom one shows the coherent signals. The bottom panel we see three cases of fractal coherence. The left panel shows symmetry matching, central panel shows rotational direction matching, the rightmost panel shows matching in the geometric shapes stored by the singularity points. (f) Fractal mechanics is defined. In the higher excited state, within a small energy gap a large number of states reside. The same system would have two distinct clock representation for two kinds of mechanics. (g) Nested clocks add superposition of multiple periodicities in the energy spectrum, thus defined fractal harmonic oscillator. (h) Explanation of quantum erasure experiment using nested circles.

crystals, not a single clock with a single singularity points as advocated by Winfree or Wilczek.

7. Fractal mechanics do not have "A causes B," no one observes independently, no one measures as an outside system, no collapse is ever complete (Dummett, 1954). Triangular systems, the observer, measuring system and the environment are part of a single network of phase or architecture of clocks that has guidance following PPM.

8. The origin of unpredictability in the universe is one and only one source, the number of primes required to cover 100% of all possible composition of symmetries is not fixed. It is the reason, some systems reach saturation at 12 number of primes, some at 13, some at 14. A system could decide to stop at 99%, or 99.99% or 99.99999%, above first 12 primes, a system could stop anywhere. It stops constructing a pure mathematical universe or artificial brain or a self-operational system, then starts growing again, from a single unit.

9. No need to save locality in the future and in the past. No need to artificially link causal asymmetry with the physical asymmetry, the map is part of "advanced action" (Price, 1996) embedded in PPM.

10. In quantum, no one dares to picturize the origin of quantization, coherence and entanglement. In fractal mechanics, the smallest pixel that makes a circle or loop perimeter is a natural choice for quantization (Figure 4.1d). Distance between two pixels is inevitable. Geometric similarity, rotational directional similarity and a similarity in the arrangement of clocks determine the fractal coherence (Figure 4.1e). A fractal coherence means a map of coherent systems spread over several imaginary worlds. Since quantum has only one, such maps are non-existent. Fractal mechanics could explore the architecture of clocks in the superposed Hilbert space of quantum mechanics (Figure 4.1f). Since all clocks are connected, fractal harmonic oscillators show a complex hierarchical pattern of resonance peaks (Figure 4.1g), and retrieves the entanglement like responses since, it is all about rebuilding the missing clock (Figure 4.1h).

4.1.1 WAVE–PARTICLE DUALITY, BEATING, INTERFERENCE, ENTANGLEMENT, AND HARMONIC OSCILLATOR

The birth of a system point: In GML, every event is a clock with a few singularity points constituting a geometric shape, and a singularity point or a pixel often has a topology inside. When all the singularity points in the internal topology of the guest clocks synchronize to activate an entire loop at a time, effectively it becomes a single system point for the higher topology and could move along the higher topological structure, i.e., available above. Thus, a system point is born at the event point (Figure 4.1c). Energy burst from a singularity is how a system operating GML carries out a conversation.

Measuring and editing a time crystal: Pump-probe experiments work on an internally linear structure, for a time crystal, we need a similar structure to sync with the time crystal that we want to measure and in return let the singularity points bursts signals. It would be a non-demolition type measurement where we try to change probing time crystal continuously keeping both measuring and to be measured time crystals side by side until there is a match. As soon as they match two types of signals emit. The first stream of waveforms of different frequencies with changing phase values and the second stream of complex waveforms where different waves are fused together. We have studied a large number of proteins and their complexes and these two types of signals generate continuously, from which we identified the time crystal. Measurement is morphing between two geometries whose corner points are made of singularities. Time of measurement is the time of the slowest clock, not linear. So, in a time crystal computer the computing time is fixed by the slowest clock in operation. **During measurement, can we sync with a part of time crystal?** We can if the part's complexity has reached beyond a certain level. If 12 or more clocking units are there we get $2 \times 2 \times 3$, $2 \times 3 \times 2$, $3 \times 2 \times 2$, three points that generate a higher-level topology. Now, there are integers for which we may get geometric shapes and in the corner of the geometric shapes we get new structures within. When geometric structures seeded within corners they are separately measurable. By bringing a time crystal of similar symmetry composition we can measure the time crystal. But still there is no gradual synchronization. We have to redefine our concept on the gradual sync, it's either the whole part at once, or no interaction at all, time is fractal here. It is morphing between two geometries.

4.1.2 MULTIPLE IMAGINARY WORLDS OPERATING AT A TIME NEED FRACTAL MECHANICS

When a person moves around a circular path, after reaching the starting point, no evidence is retained anywhere that he rotated along the circular path at all. It is a classical view, how could we describe motion or dynamics in the quantum world? Pancharatnam came up with an idea of an infinite cylinder, where one sees the cylinder classically along the length, only the circle is visible, classically, but for quantum, one sees from the side, the moving point follows a spiral path, each rotation acquires a geometric phase. When we move to multilayered imaginary universes, the first major change in the Pancharatnam's idea would be twisting the cylinder into a cylinder shape (Pancharatnam, 1956). If we have clocks in the geometric phase of quantum, or the first imaginary world, then it would be sufficient to roll the Pancharatnam's cylinder into a circle, but then we would have the same problem that Pancharatnam had with quantum mechanics. How do we measure the phase change for the second imaginary world, beyond quantum. So, we twist the cylinder into a spiral form, get a higher-level infinite cylinder for the second imaginary world. The journey continues here, for 11 times, we say spiral of the spiral of a spiral….11 times, it's a cylinder made of a spiral made of a cylinder made of a spiral.... So, fractal mechanics is the mechanics of interactive imaginary layers, that interaction is not random, governed by multinion tensors, in the real physical space, minor changes would dramatically change the phase space.

4.2 FRACTAL MECHANICS ACTS IN THE PHASE SPACE CONNECTING SINGULARITY POINTS

Ten foundations of the Fractal mechanics (FM)

1. Any point in space is a *superposition of at least three imaginary "time" worlds forming a time crystal* while there is only one imaginary world in QM. *Supplement to a fractal state is contextual or non-contextual, if contextual then what would be its nature is decided by observer's coupling with the three or more imaginary time cycles*. Here, the study is limited to 12 imaginary worlds (dodecanion or icosanion).

2. *Time gets a complex composition of rotations or spins as soon as information is encoded and that direction is purely based on the 3 to 12 imaginary worlds*. The 3D orientation of three or more imaginary time cycle planes changes with time, as information exchange between time crystals. Sometimes exchange means bonding or breaking of time cycles, often it changes the relative orientation of the planes. Hence, the same system is measured differently by a different observer, differently by the same observer from different directions. *Measurement is a match between common nested cycles between the observer and that of the two imaginary worlds, the matched nested cycles create a domain on the phase space which is the "reality sphere." Oscillation of unit time means a change in the diameters of the local guest time cycles embedded sequentially in the singularity domains of the host giant time cycle*.

3. **Ten fundamental properties of wave function in Fractal mechanics (FM) that differs than the quantum mechanics:** Wave function ψ is a time crystal or a deconstructed topological architecture in the Fractal mechanics (FM), while it is the fundamental unit in quantum mechanics: (i) ψ is not a wave function, in FM it is a projection P of the imaginary triplet worlds of the observed on the imaginary triplet world of an observer (we call it "reality sphere"; $\varnothing$ is a point on this sphere). Note that we described in Chapter 2 that be it a quaternion, or octonion, or dodecanion tensor, all could be represented by a quaternion, wave function in FM is a tensor product Ж of the observer, observed and the environment. In other words, $\langle \varnothing \rangle = P\{f(z_1) + f(z_2) + f(z_3)\}$. (ii) Probability does not exist, $\varnothing^2$ takes a few common clock based architecture or time crystals of the imaginary part of the wave function and returns to a real entity in QM, we cannot do this in FM here $\varnothing\varnothing^* = \sum_i f(z_i)f(z_i^*)$. (iii) The wave function $\{\varnothing\}$ does not have a stationary state, the observed value is a set of output crystals by Ж operation which is a function of time. The "singularities" containing time crystals on the "reality sphere" changes with the observers time crystal. (iv) The wave function $\{\varnothing\}$ does not require additional normalization or renormalization as the projection of a structured system's charge and mass, i.e., all divergent parameters are always finite, w.r.t. observer. (v) The wave function $\{\varnothing\}$ grows and does not remain as an isolated packet, edit cycles phase to make a virtual sphere/hyperboloid unlike QM, hence, not only Fractal mechanics (FM) details the skeleton of $\{\varnothing\}$, it maps $\{\varnothing\}$'s evolution. (vi) The wave function $\{\varnothing\}$ self-assembles following ordered factor metric or prime number symmetry in the number system, i.e., there is a universal frequency chain or resonance chain that governs self-assembly. (vii) Bra-ket algebra does not work here, continued fraction algebra that processes infinite series in the finite domain is applicable, the mathematical operator Ж works here, we would detail this operator in the Chapter, a bit later. (viii) The wave function $\{\varnothing\}$ is a multinion tensor follows an unique multinion algebra, but normally we think tensor means a spatial direction, here it is a temporal direction, imagine 11D orientation of clocks in a manifold. (ix) The wave function $\{\varnothing\}$ is mass and length independent, no distinction, hence, it can hold infinite mass or length to the smallest mass or distance possible. Thus, the observer decides the magnitudes of the observed. (x) The wave function $\{\varnothing\}$ holds information as geometric distribution of frequencies in the time crystal.

4. **Zero-point energy and minima of QHO does not arise:** Quantum field theory (QFT) suggests that every point in a vacuum is quantum harmonic oscillator (QHO), so every point has finite energy, providing zero-point energy. For fractal mechanics (FM), every point is a fractal harmonic oscillator, due to fractal nature there is no limit, and minimum energy for FM is determined by observer unlike QM or QFT. If we resolve the energy minima, we get something else, a composition of time crystals within. As a result as *empty space is an infinite volume fractal, mathematically undefined and is truly a superposition of fluctuating oscillator network*.

5. **Ten basic principles of Fractal mechanics (FM) are:** (a) *Quantum superposition* is simple addition of wave functions, $\psi = c1\psi1 + c2\psi2 +...$ however in fractal mechanics (FM) there is no addition or subtraction or division or multiplication. Those are the entities of sequential world. All wave functions that

superimpose are connected by nested cycles, its projection, $f(z) = \text{Sum } f(zi)$, which is regulated by multinion algebra and Ж operator, product is not sum of wave function. Superposition is different at different location of the nested cycle structure, or time crystal. (b) ***Entanglement*** in QM is a definite operation, if one entity is changed the other entangled entity changes to a given state. For FM, two entangled objects are part of a single cycle or multiple cycles. Therefore, the output of measurement could deliver more than one values at a time, and or as a function of time, not just one. (c) ***The elementary composition of an observer has defined*** QM observer's treatment has a long history of algebraic, contextuality, etc., it is debated, FM does deconstruct observer with strictly defined rules. (d) **Measurement** in QM gives a probability distribution of values, in FM a measurement gives a time crystal, ψ is always an output of a measurement which is observer and environment-dependent. The ψ does not exist without measurement, since there is no projection, a projection from infinity is required for the existence. (e) **The measurement complexity:** In FM ψ is $\{\varnothing\}$, a time crystal, looks different from a different direction. For a single observer, a 3D measurement around ψ creates a time crystal architecture of measurement this is the projected matrix. Since two identical observers observing the same ψ simultaneously cannot occupy the same position hence would measure different parts of the observed entity, or fractional part of a time crystal. The output cannot be taken from one sphere of a time crystal because two observers will influence each other, hence a new combined time crystal needs to be created. (f) **Variable system points:** Since variable system points are included in the nested cycles, how many system points need to be considered depends on the observer's nested cycle too ($D2 << D1$, as explained above). (g) **Fractal harmonic oscillator:** unlike QM where the quantum harmonic oscillator is used, the fractal harmonic oscillator is not singular entity. Here an oscillator is made of another kind of oscillator inside and is itself a guest inside a host oscillator above. (h) **Twelve, instead of one imaginary world:** In QM, the state of a system is undefined, its burst on a circle perimeter of a sphere or time crystal. In FM, multiple imaginary axes orthogonal to each other builds 11D architecture of ψ. (i) **Infinite resolution is achieved comprehensively:** QM can have a fractal clock in one imaginary world, even in 11D dynamics. Here, in FM the wave functions have infinite resolution that cascades through 11 imaginary worlds one inside another. (j) **Treating mass and space separately:** In QM, mass is a matter wave, similar to a wave function but still mass is used in the Schrödinger equation as a distinct entity. *Fractal space-time tried to resolve this configuration by considering mass = a 12 clock based nested rhythm or*

time crystal with a particular topology. Since FM considers time as a singular variable in its metric, space to be replaced by phase governing evolution of time, here *mass = unresolved nested cycles made of frequencies when an observer interacts with a high density nested cycles*. Thus, FM's version of Schrödinger's equation could have a time crystal for mass. Entanglement, duality and superposition are peculiarities of three nested imaginary worlds in FM. *Entanglement primarily originates in the imaginary world above, the wave–particle duality comes from the world present and within, Superposition originates in the imaginary world below, and collapse does not occur,* there is no reduction, only one event happens, that is synchronization and de-synchronization of nested time cycles between the three imaginary worlds.

6. **Time-based dimensional analysis unravels a unique physical picture: What is space? What is mass? What is the time?:** In FM it is considered $M = L = T$, mass = length = time = phase, all three parameters are represented as time crystal, distinct architecture of phase. All parameters in the universe should have dimensions, related to, phase alone, if one converts to time then $1/T^4$, $1/T^3$, $1/T^2$, $1/T$, 1, T, T^2, T^3, T^4.... Since mass and length or space are nested time cycles, hence, every single fundamental constant and all parameters turn to the power of T or phase. The power of time dimension determines the weight of a parameter, Plank's constant is T^2, velocity is 0, etc., the infinite series is converted into nested cycles. It is interesting because from a time cycle, we can get $2\pi R = h$ ($h =$ Plank's constant), h should be the smallest distance, and a series of existing scientific dimension now gets a geometric perspective.

7. **Equivalence principle changes fundamentally: now all parameters and fundamental constants turn geometric:** Time is a cyclic motion of phase. The unresolved energy packet with the simultaneous coexistence of all points on the loop perimeter, when an observer resolves the energy packet it detects the change in phase or time. Stereographic projection of the imaginary planes deliver several fundamental constants. The simultaneous phase change decreases to velocity c when two cycles assemble one perpendicular to another making a sphere, as π decreases due to a nonlinear surface (then we get $\pi = C$, the velocity of light). Note that the value of π is changed on a Euclidean surface (π increases due to nonlinearity then we get $2\pi \sim h$, G, Avogadro). Josephson constant $\sim 3/2\pi$; $Mu \sim 4\pi$; F (faraday, or electron mass) $\sim 3\pi$, $Vk = 8\pi$ (von Klitzing constant), $\pi/2 \sim e$ (an electronic charge, atomic mass constant), $\pi/4 =$ Cosmological constant (dimension $1/L^2$, for the fractal information processing $1/T^2$, =energy density of vacuum). *It seems all fundamental constants originate from e, π, and ϕ (Golden ratio) due to the non-linear surface or Euclidian geometry, they*

change due to the topological constraints of different embedded imaginary worlds (we neglect the order or power). The hyperbolic surface has $\pi > 3.14$ and the spherical surface has $\pi < 3.14$; there are multiple geometric spaces across the imaginary manifolds of 11D on which the physical constants get created. On a sphere, some sets of from e, π, and ϕ values are stereographically projected on a hyperbolic surface. *Hence, the geometric shape that nested cycles want to create is not just a sphere rather a 3D dual conical shape, sometimes it's a sphere and sometimes it's a hyperboloid. Why π values form manifold/fraction in an integral way, who quantizes the constants?* The answer is the number of loops allowed in the ordered factor metric. There is no space, a nested time cycle has upper and lower detection resolution, the lower resolution creates the boundary of a particle and beyond the upper resolution it's an open space. Mass comes from the transition of elliptic to ellipsoid cavity resonator rhythms, nested cycle rotation is felt like gravity, so there is no need to consider a separate mass, a typical 12 clock structure based time crystal is sufficient to represent mass. Momentum is a property that varies with the square of the time-lapse in a system, it means a time crystal is sufficient to represent momentum. The Kinetic, Potential and total energy is simply "time." In QM an operator executes a defined transformation, gives the expectation values, in FM we have a protocol. The observer's time cycle continuously checks the tape's cells and integrates, disintegrates (clocking together or in an isolated manner) the nested time cycle and that is synced with the observer. This continues time and again to deliver different physical parameters that govern a system.

8. **Implications of the fractal clock: Fusion of special relativity, general relativity and Quantum mechanics: The fusion is not new, it has been realized with fractal clocks for over a decade now:** G. N. Ord, Laurent Nottale, and M. L. Naschie are a few among recent proponents of fractal time. The strongest and direct argument for such a fractal clock universe has been put forth by Ord (1983, 2011, 2012). Already G. N. Ord showed in a fractal clock system special relativity emerges at different continuum limit. Since FM takes out the mass from the equations, it is possible to estimate the laws of gravitation from the nested cycles simply by using the ordered factor metric (teardrop to ellipsoid transition in ordered factor metric delivers the laws of gravitation) one might connect both quantum mechanics and the general theory of relativity. There is no velocity, no distance, only phase that we have to change in the time cycle, not in the distance parameter to generate cause-effect. Since distance does not exist in reality, it is a particular topological feature of a time crystal, the extreme velocity c and entanglement induced collapse are not contradictory.

No one measures velocity, actually measures "relative clock diameters," and entanglement for the fractal clock as explained earlier is all about creating two complementary guest nested cycles in a single host cycle of a much longer time scale.

9. **Fractal Interference and transition to classical mechanics and beyond FM:** With the fractal harmonic oscillator the interference images appear a bit different. We have generated wide ranges of interference patterns for the fractal interference in microtubule and Sahu et al. study suggests that if a part of the pattern is zoomed then a new pattern appears.

10. **Ripple effects in the information science due to three imaginary worlds: Uncertainty principle:** The uncertainty principle that is frequently used as a certainty limit, even school textbooks give problems where one finds "time" from "energy" and momentum from the position is not a logical approach. Fractal information estimates the uncertainty from detailed time cycle maps between the observer and the observed, i.e., cells of the fractal tape map each other.

4.2.1 Multilevel Geometric Architecture of the Hilbert Space

A Bloch sphere with multiple singularity points: Bloch sphere (Poincaré sphere in optics) is a geometrical representation of a 2-level quantum system, this concept was introduced in 1946 (Bloch, 1946); since then, several attempts were made to modify it. One critical problem is the generalization of the Bloch sphere to include n level quantum mechanical system (Kimura, 2003). The surface of the Bloch sphere is made of pure states, $|\psi\rangle = \cos\theta|0\rangle + ei\phi\sin\theta|1\rangle$ with $-\pi2 \le \theta < \pi2, 0 \le \phi < 2\pi$ is also called Hilbert space. In order to increase the number of classical poles n ($n \ge 2$, $SU(n)$ Lie group) on the surface (a transition from Qubit to Qudit), the Bloch vector (or coherence vector) does not trace a spherical surface. This is very difficult to imagine, so we built an analogy. When we have two classical poles A and B, if one puts a torchlight at pole A pointing to the other pole B, the projection on the back of the pole would be a circle, and vice versa. However, when, there are three poles (Qutrit, $SU(3)$), we can do a darkroom experiment shining light to the center of a glass sphere from three points on its surface. Depending on the locations of three lights, the central region of the sphere would lit brightly, so much so that at certain locations, we might not see the origin of all light sources on the surface of the sphere (Goyal et al., 2011). Instead of three, if one detects only two light sources, then it is stated that there are only two pure states, the rest are mixed states. High-density light patterns would change dramatically inside the sphere as a function of the number of light sources and their location. The 3D pattern of bright light domain maps the complicated topology of coherent vector or Bloch vector, located much inside the Bloch sphere, a part of the map depicts separated states and the rest part depicts entangled states (Jakobczyk and Siennicki, 2001). At $n = 8(SU(8), E7)$, the coefficients of wave vectors distribute such that a fraction of total topology remains entangled, the rest turns separable, it is effectively a classical system (de Wit and Nicolai, 1982).

Vibrating clock: Why should the Bloch sphere expand and shrink? Geometric algebra addressing the Euclidean vector space often maps the logic gate operation. It requires a specific rotation across the trajectories on the topological space inside the fully deformed Bloch sphere (Havel and Doran, 2004). As the number of classical state n increases, the folds in the topology of Bloch sphere complicates. One such example is Hopf bifurcation in two Qubit system (Mosseri and Dandoloff, 2001), accessing the desired coherent vector during a measurement becomes critical. It is not possible to use the Bloch sphere as a structural unit in FM due to singularity. In the fractal mechanics the singularity points were embedded on the Bloch sphere and triggered in a loop to activate self-assembly of many information units (Figure 4.2a). The number of layers through the singularity point equals the level of the operational imaginary worlds. The loop is around the great circle, hence, the solid angle created by singularity points hold the geometric phase (Pancharatnam, 1956) while

the topologically closed-loop tracks the dynamic phase of the quantum system. If Phase spheres (FM variant of Bloch sphere) are self-operating, it has to expand and shrink, so that the trajectories crossover, the geometric phase paths create a suitable network. In Figure 4.2a we have outlined two types of Phase spheres, one that changes its surface area to modulate the infinity, since projection from infinity is our key. The second is the counter Phase sphere with an indefinite diameter, or multiple diameters coexist. When diameter changes, it means, a change in the surface area, a single sphere coexists in three shapes means three imaginary worlds are processing similar geometric information. The concept of fusing dynamic and geometric phase in a solid cone is adopted from neutron interferometry results of A. G. Wagh and Rakhecha (1990). As n increases from 1 to 8, the number of singularity points also turns eight, increasing n further for useful topology is useless, and literature is rich in mapping such critical folds inside the Phase sphere. Total dynamic phase for $SU(8)$ would be $\exp\left(-i\sigma_n\alpha_n/2\right) \times \ldots \times \exp\left(-i\sigma_n\alpha_n/2\right)\exp\left(-i\sigma_n\alpha_n/2\right) = e^{i\varnothing}$, similarly we can extrapolate it for dodecanion, icosanion with a twist, there would be three such equations coexisting

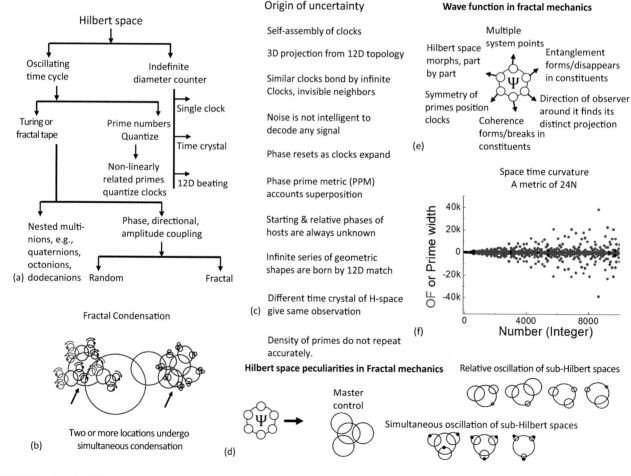

FIGURE 4.2 (a) Different classes of Hilbert space to be found in the fractal mechanics. (b) Condensation in fractal mechanics happens in different imaginary worlds at a time. (c) A list of geometric fluctuations that naturally causes uncertainties in the system that operates using the geometry of clocks. (d) Three incredible features of the Hilbert space when mechanics are driven by geometry of clocks. (e) Comparative wave functions in fractal mechanics and in the quantum mechanics. (f) Space-time curvature of quantum mechanics is replaced by a metric of symmetries that maps the symmetry breaking possibilities as a function of available choices (integer).

simultaneously. To engineer singularity, the diameter of the clocks should vibrate along with the rotation of the system point.

Bypassing the complex manifolds and adopting the pattern of primes: As said above, when classical point increases on a Phase sphere, the surface gets folded in a complex manner. So, an assembly of spheres reduce the complexity, adds manifold controls which were otherwise not possible to encode. The replacement of a complex architecture using a group of small spheres (thanks to singularity points) paves the way to implement the pattern of primes. Now we suggest a point that absorbs light, say, like a black hole, neutralizing the complex folds in topology. One could edit electron spin by shining light, i.e., geometric phase could be precisely chosen by driving photons in multiple cycles (Kim et al., 2010) and Cn coefficients that map the topology. The center of Bloch sphere makes a phase with the topological path between two singularity points, a set of such angles completing the loop represents the geometric shape made of singularity points. Thus, the geometric phase that could be used to run a clock (Badurek et al., 1986) is interpreted as a measurement of time. Quantifying time as geometric phase (Wagh and Rakhecha, 1990), store memory and even decision-making, the solid angular projection through the geometric phase loop from the center of Bloch sphere is the basis of replacing curvature on Bloch sphere with another small sphere. Let us put it in another language. Inserting a time cycle or time ring using an additional sphere, holding the geometric information would be compensating the topological deformation. It would increase the probability of accessing the separable and entangled states would make computing more feasible. Geometric phase is an integral part of the Bloch sphere or Phase sphere. However, for quantum computing one needs to switch On and off the motion of geometric phase through its trajectory. For PPM-GML-H triad computing that we are exploring in this book, several system points holding different local clusters of the geometric phases, covering a wide range of imaginary worlds (from 1 to 12, it could be anything), carry out complex tasks. First, new paths condense at various local regions of the unified time crystal representing the entire information content (Figure 4.2b). These independent local regions contribute to ten different ways to create uncertainty (Figure 4.2c). The wave function, a clock-structure representation of matter, would have a complex geometric architecture, it would be a time crystal as shown in Figure 4.2d. A simple wave function, which was otherwise thought to be a single, mysterious wave-like entity following Schrödinger's equation, would now be complex life-like object performing PPM (Figure 4.2e). For quantum, Bloch spheres infinite solutions are called Hilbert space. For FM, the Hilbert space analog is fundamentally different. Two fundamental processes govern the relative oscillations of the sub-Hilbert spaces. Either they follow a periodic expansion and contraction; or, all sub-Hilbert spaces would expand and shrink together like one unit (Figure 4.2d, right). The simultaneous dual-mode is fundamental to find symmetry in a complex ordered factor metric.

Superposition of OF-metric and PG-metric as described in Figure 3.16 is superimposed in Figure 4.2f, to show how the prime gaps are located in the key locations on the OF metric. Doing mathematics is not easy for a wave function following Fractal Mechanics, formulation needs noise, to explore all possible choices of clock-geometric path and a noise assists in finding the right choice (Figure 4.3a).

The ultimate engineering of Singularity: an experimental perspective: Whenever a fractal wave function described above is less than the pixel size (the longest clock is less than the host, but the host-guest geometry matches), the guest goes to the imaginary world and a singularity point is created (Figure 4.3b). Singularities are everywhere in nature, depending on the type of singularity, and the system the architecture of wave function changes, it may be a quantum or classical but it has an additional feature of exploring the engineering of multiple worlds (Figure 4.3c). We do not carry out proper experiments to see them. One possible experimental path is fractal interference using polarized light as shown in Figure 4.3d–f. Singularity is a time lost in the flipping direction of spin if we cook a spin-based quantum system, but this particular physical significance is an extremely useful tool. One can insert a complete-time loop or a clock inside a singularity (Kim et al., 2010) as shown in Figure 4.3d. Unfortunately, the power of drawing geometry on a Bloch sphere was never harnessed, though very recently, its potential is getting recognized (Oh et al., 2016). The singularity points are not actually a point, rather, it is a sphere (diameter is set by Rabi Frequency damping, i.e., flipping time for the quantum state) represented by a function that is non-holonomic, but switches to a holonomic one as soon as it acquires another time ring of singularities along its closed-loop trajectory. See the generalized version in Figure 4.3d. Essentially we get, $\theta_{TotalN} = \theta_{Geo(N-1)} + \theta_{Dyn(N-1)}$ and $\theta_{Dyn(N-1)} = \theta_{Geo(N-2)} + \theta_{Dyn(N-2)}$, if the replacement continues, $N \rightarrow \infty$, we get pure geometric phase regulating the system. In case pure geometric phase regulates the entire system we get a quantum time crystal and the output differs between a classical and a quantum experiment (Figure 4.3d, e). Eliminating the dynamic phase to create a pure geometric phase is not new, and the most remarkable advantage is "passage of vibration to the higher level" (Far off-resonant trap or FORT; Joshi and Xiao, 2006). Joshi et al. beautiful description of bypassing the core shells ionic vibration to the outer optical cavity nicely fits with Reddy et al. self-similar or fractal network. The wave function of the system is $\psi_{system} = \prod_n e^{-i\theta_n}$, apparently unitary feature is broken, but if we do numerical simulation relative intensity of the signal is $\mathrm{Sin}\, n\pi/n$, which would be a binary stream of pulses in an experiment like the one shown in Figure 4.3f. When we consider the geometric phase, it is independent of the parameters that govern the path of the projective Hilbert space, thus, it is independent of the speed.

Why was it essential to introduce a singularity domain and reject any form of classicality from the Bloch sphere: We need the undefined singularity points and not classical that could be stated with an argument. Because (i) we wanted to use multiple clocks one inside another at various layers,

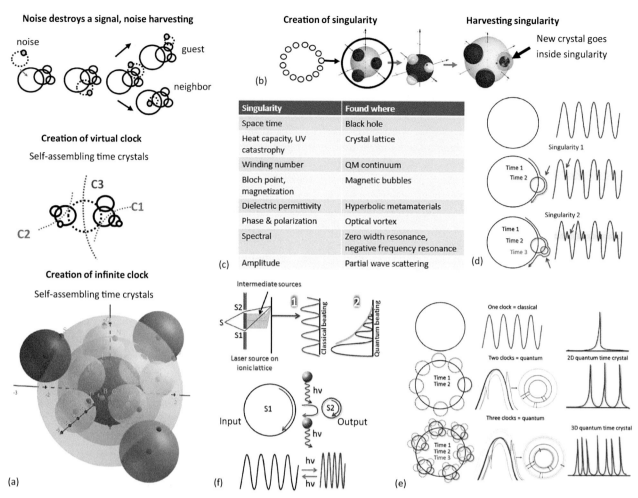

FIGURE 4.3 (a) Three panels depicting three basic features of fractal mechanics. Top panel shows how noise is harvested by absorbing its time crystal. The central panel shows how does the symmetry of the 3D arrangement of clocks determine which one to self-assemble and how. The bottom panel shows unique self-assembly of clocks, within and above. (b) A single pixel of a clock holds a 3D geometry of clocks. (c) A table of different sources of singularity. (d) Three rows show three types of nesting of clocks in classical mechanics. A circle represents a clock, a wave. In three rows, the number of clocks increases one by one that increases the ripples in the waveform. (e) Plots for panel d is repeated in the panel e, but for quantum mechanics. Guest clocks position is uncertain but time is certain so we expect three peaks for a guest. (f) A quantum interferometer showing the interference (1) and quantum beating (2).

to model Sahu et al. experimental finding of nested clocks in the protein complexes. (ii) Creating a fusion of adiabatic (clock inside singularity) and non-adiabatic system (clocks reside side by side) is essential to build a generic model where a sub-system retains virtual isolation from the environment. (iii) A burst from singularity could be a classical event born out of an undefined function, thus, making classicality an emergent phenomenon from the undefined world. (iv) We get a virtual spin or directionality, no real direction; even the direction would remain undefined. Guest Bloch spheres inside a host Bloch sphere could adopt a composition of various rotational directions, thus, the rotation of the host would always remain undefined. (v) When a Bloch sphere is defined in Quantum, a classical certainty is already included in its poles. Diminishing an essential form of classicality or certainty of the Bloch sphere in the form of 0 and

1 (poles) is required to make a version of Quantum that is purely uncertain. (vi) Singularity enables escape time fractal to be introduced, fractal clocks bring an imaginary time (fractal uses iota just like Quantum), and so we do not have to introduce imaginary space-time unexpectedly as we do it in the conventional Quantum mathematics. We could derive the emergence of the imaginary term from a special type of fractal nature in the information structure. (vii) When we introduce singularity entire Bloch sphere, and its self-assembled architecture becomes undefined. Therefore, it is not a Turing machine anymore, it represents the fractal tape that we proposed to replace "machine." (viii) When the entire system becomes a singularity, at a time, at any given part of the system, the available information is finite. Thus, unlike Quantum's Hilbert space, we get a finite projection. Singularity therefore serves as a route to reject "infinity"

and "probability" in a single shot, as "infinity" gives inroads to the "probability" and "perturbation." (ix) Uncertainty of Quantum has been a mystery, and demystification was essential, an absolute singularity at all scales, everywhere in the system enables generating uncertainty in a logical way in any system (see ten different ways of generating uncertainty). (x) Geometry has a classical nature; corners are a certainty, which ensures incorporation of classical from the backdoor. Nested singularity enables varying the surface on which the geometric shape is drawn, so, the triangle is a certainty, but not the sides or its angles.

Conversion of a fractal Bloch sphere into a Quantum Bloch sphere and a classical state: The fractal Bloch sphere turns to a quantum Bloch sphere when the observer's resolution is less, so it sees the guest singularity spheres on the host Bloch sphere as the discrete single points (classical). Consequently, the clock line connecting the singularity points into a single loop disappears. If there is only a pair of singularity points on the great circle, then it's a Qubit (0 and 1). Even for more than two singularity points, since, the singularity domains turn to a point (Qudit), no geometric information exists. A fractal Bloch sphere, is itself a singularity sphere, when the entire Bloch sphere converts to a single point, it is classical. The disappearance of singularity leads to a classical or a defined state. The loop connecting the singularity points is not fixed, it connects two rotating system points of loops located in the neighboring singularity domains.

4.2.2 Fractal Harmonic Oscillator, Fractal Condensation, and Peculiarities

The maximum number of states a quantum system can process? A quantum rhythm and quantum time crystal: In a quantum system, the notion of distinct states is very well defined, two states are distinct if they are orthogonal. It has been shown mathematically that a quantum system with an average energy E could be made to oscillate between two orthogonal states with a frequency $4E/h$. Thus, even in the quantum scale we can get rhythm (Margolus and Levitin, 1998). It has also been noted that for very long evolutions that form a closed cycle, the maximum transition rate between orthogonal states is only half as great as it is for an oscillation between two states. A quantum system can run through a long sequence of mutually orthogonal states for a maximum time $\tau = N - 1/N(h/2E)$. Interestingly, this is not only true for the quantum system, for any harmonic oscillator, it means even for the kind of time crystal, argued here, but the limits are also valid. One can run through a long sequence of nearly orthogonal states at a rate $Emax/h$ or $2E/h$, where E is the average energy of the microscopic system under interaction.

Morphogenesis of information in a time crystal is multi-layered or fractal condensation: Time crystals arriving through the sensors eventually fall on a resonance chain that operates following the guidance of a PPM. The resonance chain morphs the seed geometric shape in the time crystal that emulates an event happening in nature. The morphogenesis

of events happens at every spatial and temporal scale, in the real human brain and in its analog brain jelly. Mimicking the frequency space of a 3D time crystal into a real physical structure or morphogenesis is amazing in the sense that communicating proteins or fourth-circuit elements H (Chapter 8) would undergo simultaneous changes only at those parts where minor changes would edit only a few clocks of the time crystal. Communication means an exchange of local parts of a time crystal. If one images the shape of those two proteins or H, apparently both would be silent as if there is no physical movement. If we could enter inside the structure, then we would find some other structures are changing quite fast. If we could take one, and enter inside, we would find a large number of extremely fast shape-changing materials. The journey reveals that two proteins or H devices exchange very little energy only where the changes should be made. In the Turing paradigm, one has to carry entire information content, not here. During the exchange, mismatched clocks are written, this is not new writing, many clocks are not linked, remain free in the hardware of a resonance chain. They condense to build a new clock. Therefore, the condensation of carriers in shape-changing material is a part of the information exchange.

Experimental evidence required to establish that fractal mechanics (FM), a generic form of quantum mechanics does really exist

1. **Fractal fringes:** If a system follows fractal mechanics (FM), then one could zoom any part of the quantum interference pattern, to discover that there are hidden fringes. QM has no such provision.

2. **Fractal Feynman diagram: superposition of multiple interference patterns:** If a system follows fractal mechanics (FM), then electromagnetic frequencies would be able to track the fractal gaps in the Feynman diagram and we would get multiple interference patterns for multiple resonance frequencies. QM does not allow many imaginary worlds to interact and generate superposition of superposed wave functions.

3. **Fractal condensation:** Condensation of the vibrational modes in Raman spectrum at the room temperature, one could see live the vibrations collapse at various frequencies so that peaks merge. It is simultaneous Frölich condensations at multiple singularities; singularity inside a singularity is a trademark of FM and does not exist in QM.

4. **Fractal coherence leads to the quartet:** When two spins are entangled, it means one wave function, both particles are actually one in reality and that is equal to a nested three time cycles. Therefore, a mother photon can produce a child, a granddaughter and a great-granddaughter, but the granddaughter's child will not be born (above quartet not possible), a single photon cannot produce infinite progeny in FM, there is no such restriction in QM.

5. **Multiple clock entanglement: Entanglement without an identical point of origin:** Until now, it is believed in QM that entanglement happens only with photons from the same origin, i.e., among mirror photons, however if FM exists, then it can happen for photons from other sources too. Thus, suitably generated photons from different sources if entangles and generates interference then it would establish the existence of FM.

6. **Superposition is a function of time:** Time cycles join and disjoin spontaneously, thus, in a suitably designed fractal tape system, entanglement will be a function of time. With a suitably designed time resolution set up, an entanglement would appear and disappear, as one switch to a different time domain. Such a feature is feasible in FM, prohibited in QM.

7. **Quadrupolar moment changes as a function of time:** One essential outcome of converting all parameters, mass, length to nested time cycles in FM is that, all higher-order moments change as a function of time. It is observed in the frequency-time-intensity 3D spectrum during energy transmission between two objects. QM does not allow this.

8. **Frequency wheels of primes determine the system's nesting:** Every fractal tape system has its own frequency wheels of primes. QM has no provisions for nesting of waves.

9. **Fundamental constants emerge as infinite series like e, π, and φ:** The triangular geometric musical formulation of number system fractals generate all other fundamental constants as a single time cycle and several time cycles inside. QM is not geometric like FM, the fundamental nature inside of a wave-function is beyond the limit of QM.

10. **Live "beating" communication of three imaginary worlds:** One could experimentally verify three imaginary time worlds simply by experimenting on a nested quantum system like protein, which has three layers of quantum systems embedded. Such a nested QM would exhibit classical, Quantum and fractal beating simultaneously.

4.2.3 HARVESTING NOISE BY HARVESTING SINGULARITY— PPM REPLACES STATISTICAL MECHANICS

Space-time and path: From any given point in a spatial crystal, by moving toward any direction, one would get different kinds of the arrangement of atoms. By rotating 360°, one should find at least two distinct spatial symmetries. One cannot imagine a space in a set of time intervals, as one cannot state a "direction." If the phase oscillates one full swing 360°, it is equivalent to the concept of all directions in a spatial crystal. On the circular 360° path, if at least once the cycle hosts another small loop that shifts the rate of phase change, then one gets two time symmetries. A system point moving along the phase cycle perimeter would experience

two different rates of time flow (Ord, 2012; Girelli et al., 2009). The large phase cycle which constitutes a "time" is called the host time cycle and on its path the local phase cycle is called guest time cycle. The guest-host phase assembly is called a time crystal (Winfree, 1977). Phase-shift is fundamental to both space and time, and it is abundant in nature (Chandrashekaran, 1974). Therefore, information structure, if it is a 3D phase sphere, could represent all existing physical parameters comprehensively.

Old schools of statistics to the statistics of primes: Events are not random, one can predict events with 99.99% accuracy: Maxwell Boltzmann (MB) statistics is applicable to identical, distinguishable classical particles with Non-quantized energy, with any type spin (gas molecules). **Bose-Einstein (BE)** statistics are applicable to the identical, *indistinguishable* particles of zero or integral spin (bosons, photon, He atom etc.), where any number of particles can occupy any level. **Fermi Dirac (FD)** statistics is applicable to the identical, indistinguishable particles of half-integral spin (Wu and Cai, 1999). Since only one particle can occupy any particular level, these particles obey **Pauli Exclusion** Principle (electron, proton, etc.). Imagine we have a network of cells, where a finite number of indistinguishable and distinguishable events are diffusing like particles, colliding randomly. Unlike spin half or integral spin or anyon, here, the repeating events with a particular symmetry are indistinguishable. As the symmetry breaks and a new symmetry starts contributing to the unfolding of the events, then the events are distinguishable. Conditional distinguishability was not part of FD, BE, MB, or anyon statistics. PPM is a 3D architecture where the symmetry of primes is mapped over the integer space. We define the event as clocking topology, it means we account for the symmetry of a repeating event and map the symmetry relation of the constituent events. Therefore, PPM is a grammar book, if the events change, how would it follow.

Statistical entropy is perpetually clocking: a matrix of past-present-future: At certain conditions, a physical system is bound to transit arbitrarily close to the initial state, even one with very low entropy (Poincare's recurrence theorem; Poincaré, 1890). Zermelo (1896) argued that the Poincaré recurrence theorem shows statistical entropy in a closed system must eventually be a periodic function; therefore, the mechanism to increase entropy, is unlikely to be statistical. At any moment we consider, entropy was more in the past, will be more in the future, no one answered why it is low now. Boltzman said, isolated systems are always at maximum entropy, but suddenly it switches to very low entropy, traps in a local minimum, so, both the time directions from the bottom now, go upwards. In other words, we take it for granted that the influences in the other direction, the minute influences that a system inevitably exerts on its environment, do not "go at haphazard." Demarcation of past present and future is a function of the architecture of the time, preferably a time crystal built using a PPM. The direction of time and entropy is not objective but an appearance, a product of neighboring clocks direction, a vector product.

The entropy of a series of clock outputs from a biological system $E = -(\ln T_{max})^{-1}\sum_{t=1}^{T_{max}} P(t)\ln P(t)$. The entropy is known as Recurrence period density entropy (RPDE) and used to find the degree of symmetry in a composition of deterministic and stochastic signals (Marwan et al., 2007).

The weakness of statistical methods: Principle of independence of incoming influences, PI³: Time asymmetry does not emerge in statistical entropy. Using two criteria, it is assumed that even before the derivations of statistical principles. First, the output products of collisions are correlated with one another, even if they will never encounter in the future. Second, the incoming components of a collision are correlated, if they have never encountered in the past. Two assumptions make sure that the previous collisions will induce correlations between the participants in future collisions. The problem with statistics is that (i) every single entity and event is equally probable, (ii) an entity that participates in defining statistics is non-deductive, (iii) there is no emergent property for a given entity, (iv) environment and boundary properties are imaginary tools of humans like hidden layers in a deep learning network, (v) no distinction between past and future of an entity is considered, it involves only counting, the past is merely the mirror image of the future. In fact, "that previous collisions induce correlations between the participants in future collisions" (Zeh, 1989, 1992). Memory has no accountability, the past is an illusion (von Weizsacker, 1939)! (vi) The universe is made of fractal branches of systems one inside another, the direction of time should be in harmony and that communication of time is not understood (Sklar, 1992). Extreme parallelism and diversity of directions have no internal mechanism. (vii) All kinds of fundamental waves radiate outward, not inward in nature. No inward means no mechanism to set direction. Even if there is an unknown inward wave, radiating to the past and then the past contributing to the future to direct time and entropy in a particular direction (Wheeler Feynman absorber theory) is not possible because absorption is an irreversible process (Gleick, 1992). (viii) Quantum effect like "advance action" and "backward causation" defines the direction of time and entropy (Cramer, 1980). (ix) Two kinds of perturbation modes regulate the universe, one is clocking and the other static. Combination of various modes of perturbation expands and contracts phase and at both phases, the system grows, it defines the arrow of time and entropy only in one direction (Hawking, 1985, 1988, 1994). (x) Inhomogeneous anisotropic modes originate at the ground states itself (Halliwell, 1994), the path integral of the just born universe account for this confined, limited perturbation that keeps time and entropy to a minimum. Creatures living at that time and entropy followed that limit. A century-old effort to link life, time and symmetry are confined into two choices. Both are fatal. First, if we opt for low entropy choice, then we end up in Big Bang and Big crunch. Second, if we chose underlying physics of the universe is asymmetric in time, the symmetry of the physics laws that we are an illusion, then we need to find a missing link, a governing gene of codes running the universe (Penrose, 1979, 1989).

PPM-based temporal architecture as an alternative to statistical mechanics: Time asymmetry does not emerge in statistical entropy. Using, PPM, we map how prime symmetries interact, all possible compositions are mapped in this infinite 3D architectures. The advantages of this approach to constitute an alternative to statistical mechanics. (i) Time asymmetry in the universe is invisible until the architecture and intricate map of symmetry is understood, this is realized in PPM. (ii) The necessity of inward wave is satisfied by singularity points. Fourth-circuit element Hinductor is designed to produce vortex of light and knots of darkness at the same time, by harvesting singularity. (iii) Addition of a new symmetry in the metric increases or decreases the number of compositions. That changes the entropy. The universe is neither symmetric, nor asymmetric, it is an intricate map of symmetries, PPM. Since adding new symmetries enable a system to achieve 100% coverage of entire symmetric space and only 12 primes cover 99.99% symmetries, the entropy fluctuates in a non-repeating way. PPM pattern is infinite and never repeats. (iv) Future is the addition of symmetry, past is a subtraction of symmetry, and the present is clocking a single symmetry repeatedly. Evolution in the universe is making a journey from 0% to 100% symmetry structure of PPM by restructuring the system, repeatedly eon after eon, from the smallest to the largest. It is similar to the gold universe hypothesis, where at both ends the universe has minimum entropy (Gold, 1962, 1967). (v) PPM appears similar to Hawking's explanation where he argues that both at contraction and at expansion phase, the system grows in time and entropy (Hawking, 1985, 1988, 1994). There is a fundamental difference. In a PPM, it is an architecture of all possible symmetries, if we move toward zero or infinity, eventually, the system adds or subtracts symmetry. Therefore, quantum mechanics is confined within a local disc, a single plate in the infinite stack of discs in the 3D metric. The interaction between piles of discs create a beautiful surface gradient of the metric which describes a new mechanics, we named it fractal mechanics. Perturbation causes interference, which in turn generates singularities of various kinds. Fractal mechanics relies on engineering the singularity in a system. (vi) PPM does not have to ensure a low entropy of the big-bang or the growth of a conscious brain from a single zygote cell. It avoids two critical dilemmas. The infinite cycle of PPM, where we make a journey to cover the symmetry space 99.99999% by increasing the integer N to around 7.6 trillion, takes the final product and start recounting from 1 to 7.6 trillion again and again and again, we do not have any end anywhere. It is somewhat similar to Penrose's conformal cyclic universe where there is no end, but in his proposal, universe reaches to a finite topology which is retained as the remains for the next eon (Gurzadyan and Penrose, 2013). For PPM universe, at primes, entropy reaches minimum, even without a crunch, and follows a beautiful temple-like terrain, without a collapse, the system becomes a singularity point for the aeon above, because around 12 primes a system, say a universe or the brain cannot add a new dynamic to make a significant effect. (vii) A system is converted into a time crystal; constituent clocks rotate clockwise or anticlockwise. The architecture of time crystal's direction of time or direction of entropy is an irrelevant question. Depending on the set of clocks we chose, on the host Bloch sphere it is a local flow of time and entropy through the surface of the phase space. Sometimes, paths are closed (clocking) or even open. (viii) Penrose calculated that 1 of $10^{10^{123}}$ universes would have the right Big Bang. If PPM is considered, universes would be foolish to go beyond 7–8 trillion choices. If someone believes in parallel universes, currently at this moment there won't be more than 7–8 trillion of them. Around that number 100% symmetry saturation would be reached, and naturally a combination of 7–8 trillion universes would act like a single atom, but in that effort, would end up building a periodic table full of atoms. The second law of thermodynamics is a particular line drawn on a PPM's temple surface. (ix) PPM is an architecture which is primarily a superposition of 12–14 prime number related symmetries. That gives birth to Gauge invariance and several other conservation laws. (x) PPM's singularity engineering with GML breaks one old myth that anything emerges out of singularity should be random (Davies, 1974). That does not happen in a PPM universe, communication across the singularity points is the foundation of Fractal Information Theory, FIT and GML.

Lotus lotus everywhere, this could be a very nice caption of the nanobrain if we look at it from the entropy symmetry perspective. The entropy symmetry relationship of any system around us is a linear plot that we learn in the textbook. More is the number of distinct symmetries embedded in a system, more will be the disorder, and the degree of disorder is the entropy, so creating a mess increases the entropy of anything that is in perfect order. However, the picture changes completely when we think of a higher level of architecture. Say, we have a ten-storied building, and we are creating a mess in every single floor randomly, shall we plot entire mess up into one linear plot? Of course, we can do that if we are asked to provide an account of gross mess up, however, if every single floor has residents from different nations, different culture language habits, altogether the variation of the degree of the disorder will vary distinctly between the floors. Therefore, as soon as we are asked to provide a dynamic picture of the degree of the disorder instead of a static one, the entire game shifts to a different perspective.

4.2.4 What Is Energy in the Many Interacting Imaginary Worlds?

What would be the energy of the nested Hilbert spaces and how they would exchange energy: The energy in the classical and quantum world is a quantity that drives motion. Here in the universe of time crystals, a suitable analog of energy is an ability to add more clocks, just like molecular motors it is not about running but changing three parameters (Figure 4.4a). First, change in the symmetry, making a structure more and more asymmetric makes it hungry for adding new clocks and gaining symmetry. Second, the morphogenesis

caused by changing the directions of the clocks. Even no clock is added or subtracted to a time crystal architecture, a change in the direction of rotation alone could generate an asymmetry. Finally, the third important parameter governing the density of clocks allowed in a host clock. No one ever tried to encode a geometric shape in a Bloch sphere, continuously. Since the addition of clock is made deep inside a singularity point, the shape of the singularity surface is not circular, it is determined by the geometry of clocks inside, which governs the critical parameter of energy burst from a singularity (Figure 4.4b). When a single time ring is played out, there is no logic gate, no reduction of choices, yet a decision will be made. When the energy is redefined, the correlation between the prime numbers governs the fractal mechanics and the differences between the classical, quantum and fractal mechanics explicitly reveal this feature (Figure 4.4c).

How entropy changes when universe is a linear chain of bits to the universe of singular assembly of clocks: In the language of the bits the entropy of the baby universe as big bang as deduced from the entropy of the present 2.735 K (uncertainty<0.05 K) microwave relict radiation totaled over a 3-sphere of radius 13.2×10^9 light-years (uncertainty <35%) or 1.25×10^{28} cm,

The number of

$$\text{bits} = (\log 2e) \times (\text{number of nats})$$
$$= (\log 2e) \times (\text{entropy/Boltzmann's constant, } k)$$
$$= 1.44 \ldots x \left[(8\pi 4/45)(\text{radius} \cdot kT/\hbar c)3 \right] = 8 \times 10^{88}$$

When we switch to the PPM, the matters vibrate in a cyclic pattern to come to the same starting point, several such loops, while exchanging energy, a part of that energy is always neither found in any of the participating systems, we call it bond energy. It can happen at any scale. When a matter or system gets an energy packet, where does the energy go? It goes to the structural symmetry. The symmetry defining part always takes the energy to vibrate, so, replace every matter and simply consider the structural symmetry, now PPM is like a matrix of all symmetries of the universe. Every single system in the universe is a composition of multiple symmetries, each symmetry has characteristic vibrational frequencies and each object created in the universe has a personalized PPM. Among three particles, all or a can exchange energy and interact, see for example a new time crystal-based definition of mass (Figure 4.5a). If the exchange repeats periodically, then cyclic energy exchange arises and a periodic oscillation or rhythm is born, enriching its personalized PPM. In the universe or human brain the time crystal grows continuously following PPM and the number of symmetries, that change simultaneously edits the surface area of the time crystal, hence we see in real-time the change in entropy. Here the physical size of the universe does not matter, to find the boundary of the universe we do not need to carry out space travel, seed symmetry of a few galaxies would repeat, just like the symmetry of tubulin

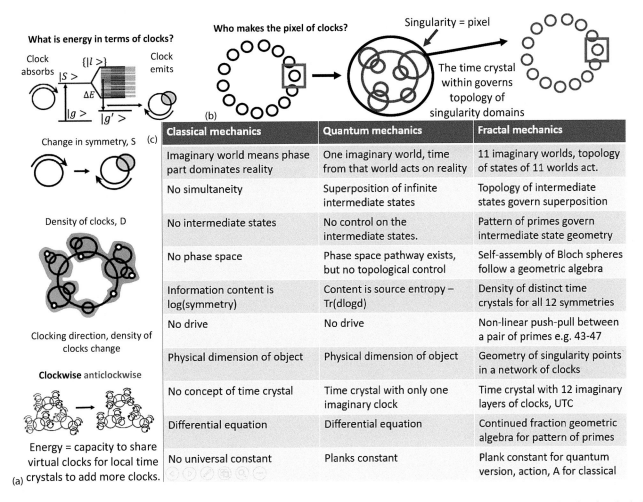

Classical mechanics	Quantum mechanics	Fractal mechanics
Imaginary world means phase part dominates reality	One imaginary world, time from that world acts on reality	11 imaginary worlds, topology of states of 11 worlds act.
No simultaneity	Superposition of infinite intermediate states	Topology of intermediate states govern superposition
No intermediate states	No control on the intermediate states.	Pattern of primes govern intermediate state geometry
No phase space	Phase space pathway exists, but no topological control	Self-assembly of Bloch spheres follow a geometric algebra
Information content is log(symmetry)	Content is source entropy – Tr(dlogd)	Density of distinct time crystals for all 12 symmetries
No drive	No drive	Non-linear push-pull between a pair of primes e.g. 43-47
Physical dimension of object	Physical dimension of object	Geometry of singularity points in a network of clocks
No concept of time crystal	Time crystal with only one imaginary clock	Time crystal with 12 imaginary layers of clocks, UTC
Differential equation	Differential equation	Continued fraction geometric algebra for pattern of primes
No universal constant	Planks constant	Plank constant for quantum version, action, A for classical

FIGURE 4.4   (a) Four sub-panels explain what is energy in terms of nested clocks. First from top: Energy is absorbed and emitted, a shaded circle shows the high-density overlapping of energy levels. Second from top: Change in symmetry triggered by a singularity domain. Third from top: Density of clocks change when in an assembly of clocks, the direction of rotation of a group of clocks changes. Fourth from top: Energy transfer is explained as exchange of a set of clocks. (b) A pixel is a singularity. Singularity holds geometric information, exchange information via tensor mathematics that deals with multiple imaginary worlds. (c) A table comparing the classical quantum and fractal mechanics.

protein extrapolates a distinguishable feature in the PPM of Eukaryotic life form. Thus, even if we do not comment on how the universe was born, what is its future, we could predict a lot using black hole entropy explored here using time crystals (Figure 4.5b, c). During the interaction, two interacting time crystals, if the observer is morphed it is mass, and if the observer fails to become a pixel in the host, it creates space (Figure 4.5b). The transformation of the basic concept of mass, energy and space helps in reconstructing every single object in terms of time crystals as outlined in table Figure 4.5c.

Thermodynamic asymmetry: Why time and entropy flow in one direction Origin of statistical mechanics: Time is asymmetric, flows to the future and entropy is asymmetric, only increases with time, both are irreversible (Landsberg and Park, 1975). It is the current view, which suggests that once upon a time the entropy was zero and time had a starting point. For geometric artificial brain, we envision a universe where after integrating around seven trillion

units of information (Clocking Bloch sphere), we reach a saturation point where adding a new symmetry by introducing a new prime does not help to add distinct dynamics. It means, in the PPM, a system makes a journey from the bottom to the top, to the peak of the temple-like structure. After $1 \times 2 \times 3 \times 5 \times 7 \times 11 \times 13 \times 17 \times 19 \times 23 \times 29 \times 31 \times 37 = 7,420,738,134,810 \times 10^{12}$ or 7 trillion micro-states a system behaves as a single unit, entropy is zero, and the counting starts from 1, 2, 3, … Entropy is clocking. When a system is changing, say, water becomes ice and vice versa, the structure of phase space (Earman, 2006) if determined would give a reliable description of the microstates. The ab initio theory of statistical mechanics is extremely debated. Primarily the reason for debate has been to extract the probability distribution of local groups, the interactions or interventions between the groups to generate a group of generic probability features, so that various different equilibrium states are formed as a part of deductive formulations.

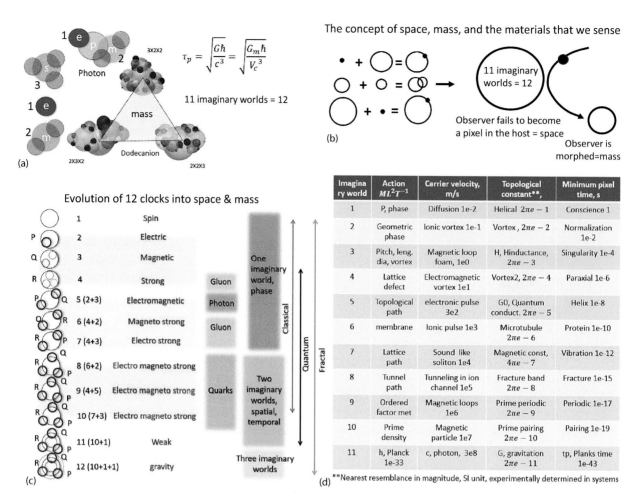

FIGURE 4.5 The basic clocks are representing the known elementary particles, a perspective of nested clocks. (a) Integers 1, 2, 3, … 12 are basic units of particles, following ordered factor metric described in Chapter 3, these clocks or their 3D representation, phase space, 12 imaginary worlds could deliver at least 12 clocks, to build a dodecanion, it is used to represent mass. (b) The process to explain mass using singularity driven clocks is explained. When observer, e.g., photon morphs, the existence of mass is registered. (c) For a universal language elementary fields are represented in terms of circles. (d) A table demonstrating action parameter found in nature at different temporal scales. From natural examples, we compile carrier velocity, topological constant and minimum time.

Some proposals try to assign initial conditions to override microstate distribution (Price, 2002) such that the evolution of the dynamics generated by the micro-clusters of quasi-isolated systems is properly mapped (past hypothesis, Albert, 2000). PPM is a group of systems, where an integer represents a single system, or a time crystal, all integers or systems are quasi-isolated as time crystals overlap, one does not require to know initial conditions, all possible perturbations are meticulously mapped. So, be it a human brain or big bang, be it a journey from the bottom of a PPM to the top, or from the top to the bottom, entropy at both ends is almost zero, time at both ends is almost zero. In between it is an architecture of phase, convert it as a structure of time or phase or an ensemble of frequency (Loewer, 2001) or even space. One cuts a slice of maximum integer of PPM, and rolls it into a closed-loop, that is a geometric form of statistical mechanics or fractal mechanics. Thus, background asymmetry projects time and entropy as asymmetric, but closing a loop ensures that all laws are time-symmetric.

4.3 A COMPARISON BETWEEN CLASSICAL, QUANTUM, AND FRACTAL MECHANICS— SCALE SYMMETRY AND SCALE RELATIVITY

Why 12 imaginary worlds cannot be reduced to one imaginary world: One imaginary world of quantum mechanics is founded on the fact that momentum and position, energy and time, angular momentum and angle are connected by an action called Plank constant h. Now, when angular momentum of magnetic vortex atoms is a dodecanion tensor where uncertainties of 11 different worlds change the way groups of imaginary worlds would affect each other, then it cannot be translated into quantum mechanics (Figure 4.5d). A quaternion has four quantum-like imaginary worlds each world is 2×2 tensors, an octonion has 2 sets of 5 quaternions it means five 4×4 tensors, a dodecanion has 3 sets of 5 octonions. Icosanions have 20×20 tensor, hence one could find various fundamental tensors embedded. Thus, when a system follows octonion, one could deduce multi-level imaginary worlds into a pair of co-existing hyper-imaginary spaces, which then cannot be deduced to quantum mechanics, it requires a new one. However, dedocanion does not only spontaneously generates a hyper-hyperspace of quaternions, the triplet of octonions means a triangle, a topology that enables hyperspace to hold three functionally independent universes at a time governing reality. Thus, the journey of fractal mechanics begins at 12 imaginary worlds, at dodecanion.

Scale symmetry: Those who oppose multiverse, they need to prove that the universe we live in is not special. If it is not special, then we do not need many other universes. One way to do this is to alleviate the necessity of mass and length from basic formulations. What we would be left with is a dynamic of symmetry that nature uses as it does not differentiate between the scales. It means if we take a triangle and draw another triangle inside and continue to do so forever, we get a scale-free symmetry or a scaling symmetry (Bardeen, 1995). Interaction of charge would spontaneously break symmetry generating the mass and length. So, the charge of the particles interacts using equations that has no mass and length. In some sense the proposal is just opposite to the idea that suggests time is not required, space is sufficient. The real mass of the Higgs boson is a billion billion times of magnitude smaller than the Planck mass, (hierarchy problem), as per standard models of physics the quantum contributions from other particles should make the Higgs boson mass, i.e., Higgs mass does not exist as a fundamental entity. In order to explain the discrepancy, supersymmetry was considered, which studies the existence of anti-matter, both pairs of Higgs Boson could reduce the effective mass. However, since anti-matter is unfound, one way to predict the very low mass of Higgs Boson is to consider that there is no mass and length in the universe in reality, it is subject to interaction. Thus, scale symmetry concept shapes the universe by chance and eludes understanding.

The concept of scale symmetry has multiple distinct proposals with beautiful consequences. First, since the gravity is born from zero mass, the elements that interact to give birth to gravity should not have any effect of gravity. In simple word, there is no mass, to begin with, that's why for 30 years, dozens of predicted particles were never found, probably calculating the mass was wrong. Second, supersymmetry considers twin masses, by interacting they can generate plenty of other particles, but scale symmetry does not allow creating new particles as many as we wish. In the scale symmetry proposal agravity (Salvioa and Strumiab, 2017), the scale symmetry introduces two kinds of dynamics, one generates Higgs Boson the other one generates a billion billion times higher Plank mass, there is a problem. Negative energy, negative probability and ghost particles are required which are not real. Third, there exists a proposal for a hidden sector, the phase transition like events (Carena et al., 2018) switch Higgs Boson like particles to a Plank class mass. All existing versions of scale symmetry models like agravity, hidden sector, predicts new particles, future experiments might reveal the winner. However, when we created the chart in Figure 4.5c, the reader could notice that all of the existing models received a form of acceptance.

Difference between scale symmetry and PPM: For PPM-GML-H triad, time crystals are the massless particles which are made of phase singularities, they constitute a charge, mass and distance (see Table 4.5c). It is a major difference with the scale symmetry where a cascade of interactions between the uncorrelated charges like dominos generates the phase transition. For PPM-GML-H triad, every single parameter, charge, mass and distance all spontaneously arise, since the geometric combination of time crystals determines when a phase transition would happen as per the PPM. Thus, spontaneous phase transitions of time crystals are programmed in the geometry following PPM.

Scale relativity: Fractal space-time and its relation to quantum mechanics were proposed in 1981 (Abbot and Wise, 1981) and fractal space-time came into being (Ord, 1983). Nottale coined the term "scale relativity" a decade later (Nottale, 1989, 1992). Scale relativity hypothesizes that space-time in nature is fractal and quantum behavior comes from that fractal feature. Einstein said, space-time is curved, and the proponents of fractal space-time suggested that the traversing paths through the curved space-time are fractal. Indeed, fractal geometries allow studying such non-differentiable paths. The fractal interpretation of quantum mechanics has been further specified (Abbot and Wise, 1981) showing that the paths have a fractal dimension 2, if infinite series then $4 + \varnothing^3$ (Prigogine et al., 1995). Only scale ratios have a physical meaning, never an absolute scale. Multiple proposals exist that quantum mechanical paths or Feynman diagrams are a fractal path (Kröger, 1997). The fractal path has non-differentiability. However, the self-similarity allows generating the paths using complex differential equations. Some researchers used infinite cantor sets, still, a singular mathematical function express all of it. The fundamental "postulates" of quantum mechanics were derived (Ben et al., 2005) starting from scale relativity. Schrödinger's equation, the Klein-Gordon, and the Dirac equation can then be derived (Célérier et al., 2003).

Difference between scale relativity, quantum mechanics and fractal mechanics: We have made a few fundamental changes in the concept of space-time curvature. In all fractal space-time research over 40 years, the fractal path was defined as a simple equation, repetition of simple geometry. We use PPM to generate the fractal path, thus, no symmetric entity is pure, it is all about the composition of symmetries. Feynman suggested a few diagrams to connect the empty space between two elements (Feynman and Hibbs, 1965). Fractal space-time concept suggested that the path is a repetition of geometric paths for a finite time or infinitely as cantor sets (Prigogine et al., 1995; Marek-Crnjac, 2009). In general scale relativity, the fractal dimension can take any value, but only one value. For PPM-GML-H triad, when we use PPM, we have a superposition of multiple fractals, thus, the fractal dimension is not one, but a set of values, that too, is a function of location, where on the PPM the selected symmetry is located. Due to the superposition of several patterns of PPM, we get non-self-similarity, the changing scales generate several distinct scale-forces or scale-fields, instead of one (Nottale, 1997,

2004). In order to analyze, we find self-similarity between different scale dynamics. It is the reason, we cannot process the evolution of dynamics using quantum mechanics formulations using mass and length, we change the name to fractal mechanics. For PPM, if one uses simple patterns its quantum, complex set of patterns, it is classical, there is no boundary between classical and quantum-like de Broglie length scale (Turner, 2013). If a system represented by time crystal uses a few symmetries, say S, then measuring time crystals M with simple structures do not change S, hence M gets reproducible outputs.

Note that scale relativity never attempted to rewrite the formulations of quantum mechanics. They consider the charge as the basic foundation of all, then derive mass and length required for Schrödinger's equation and all other forms of quantum mechanics. All existing formulations of quantum mechanics survive. Ours is a version of quantum mechanics that does not use charge or anything like the basic entity, rather, from a structure of phase inspired by the pattern of primes we derive all essential parameters. Thence we replace the equations with a changing geometric structure. We call this new form of mathematics as continued fraction geometric algebra, CFGA, where all formulations of quantum mechanics take a geometric shape. Equation = changing geometric shape. Moreover, since we always put a geometric structure inside a singularity point, a state is defined using phase value, no real classical path is there. Since wave functions are nested, it does not feature wave like but time crystal-like, we are bound to change the name from quantum mechanics to fractal mechanics. Scale relativity gives a geometric interpretation to charges only, we provide clocking geometric interpretation of everything. Matter wave, Schrödinger's equation and all fundamental operators of quantum mechanics could now be drawn as a 3D architecture of clocks where geometric shapes are rotating in a typical pattern.

The question may arise, how could it be that there is no conserved quantity for PPM-GML-H triad? New scale symmetries build all essential parameters from PPM just like the scale relativity (Nottale, 2011), the only difference being that for PPM-GML-H triad, it is not a singular symmetry but a set of symmetries arranged in another symmetry. Geometric quantization from numbers is not an ad hoc, but a well-observed feature. Quantization of charge happens because integers are quantized (Furey, 2015).

4.4 ACTION IS NOT LIMITED TO PLANKS CONSTANT—EVERY IMAGINARY WORLD HAS ONE

Deduce the Quantum and our universe from an understanding of existence: The dimension of Plank constant h is that of action, it is a limit on the position and momentum. Poor gravitational objects like us (a human body weighs ~70 kg) cannot cross a photon speed c (what makes photon massless?), similarly, loss-less solitons cannot cross ~10^{-5}c that means 10,000 times less than the speed of light. If we

use solitons to communicate and change the state of another particle far apart, then it will be as slow as the speed of sound, using effective mass it could be shown that uncertainties and limiting values are there at various energy levels. However, we can even use a diffusion of ions, to lock particular parameter for hours or even years later; therefore, eventually, locking coupling values is the stringent condition in entanglement, not the apparent speed, that emerges with system co-ordination and those are written in the tensors, e.g., quaternion, octonion, and dodecanions. As coupling becomes a simple physical interaction, it opens up worlds beneath classical perception time scale and the worlds beyond quantum time scales. We have not remained confined in the materials and carrier transmission, because that would have confined us in a single clock. The geometry of clocks is limited to 11 layers one inside another for a dodecanion, all 12 layers have its distinct action, an equivalent of h and in the classical systems we could implement the concept of imaginary layers, a chart is shown in the Figure 4.5d, wherein we have used the universal constant values from the chart of Figure 4.6.

Eleven Mr. Topkins is moving around 11 imaginary worlds of George Gamow: The rule of the game is that never separate space and time, if we can preserve their coupling then even if we move one frame of reference to another the massively parallel coupling between large numbers of pixels carrying information of the event retains its originality. Every system built by nature has its own time of reference, because the speed of a photon in that medium is different from the vacuum. Inside that system, time travels in its own way. The fundamental rule of the living would change dramatically in that very world. We know George Gamow's story about Mr. Topkins, who traveled to a world where the maximum velocity of light is only 30 km/hour. Though the story is wrong completely since the observer always lives in the current world, and the photon while carrying information might change it suitably to fit that information in the new world, the story is important to feel the new world. There are several systems around us that follow the modified "time" based on photon velocity and or carriers of time defining action. We live with these worlds together, for dodecanion, its 11 imaginary worlds. Therefore, the modulation of time domain by a computing system is essential for the true replication of a natural event. The time-domain means maximum and minimum time-lapses for all possible events occurring on the brain. The minimal two events could be a change of states, the fusion of states to form groups, the disintegration of groups etc. Eleven layers of actions set 11 Mr. Topkins in a nanobrain, and brain jelly, how is it possible to manipulate the symmetry of architecture to modulate the time domain of pattern evolution in a molecular cellular automaton.

Physical systems (when viewed from a distance) can be grouped into a small number of classes, with identical scaling laws.

Ken Wilson

Fundamental constant	Actual value	Prime-e-prime
Cosmological const Strong coupling const α	0.7 0.5, 0.3, 0.2, 0.1	7e-1 Prime-e-1
Wein displacement Fine structure const Helion molar mass	2.8e-3 m K 7.2e-3 3e-3 kgmol-1	3e-3 7e-3 3e-3
Fermi coupling const Quantum conductance	1.1e-5 GeV-2 7.7e-5 S	1e-5 7e-5
Magnetic constant Nuclear magneton Stephan Boltzman	4πe-7 N A-2 0.3e-7 eV T-1 0.56e-7Wm-2K-4	11e-7 1e-7 1e-7
Gravitational G, electrical ε Bohr radius Atomic mass	6.6e-11, m3kg-1s-2 0.8e-11 Fm-1 5.2e-11 m 0.14e-11 kg	7e-11 1e-11 5e-11 13e-13
Compton wavelength	3.8e-13 m	3e-13
First radiation const Atomic unit of time	37e-17 Wm-2 2e-17 s	37e-17 2e-17
Electronic charge	1.6e-19 C	1e-19
Boltzman const Bohr Magneton Elec mag moment	1.38e-23 J K-1 0.9e-23 J T-1 0.9e-23 JT-1	1e-23
Thomson cross section	6.6e-29 m2	7e-29
Electron mass	9e-31 kg	7e-31
Plank's constant	6626e-37 Js	3313x2e-37
Electronic polarizability	1.6e-41 C2M2J-1	1e-41
Plank time	0.53e-43 s	1e-43

Fundamental constant	Actual value	Prime-e-prime
E-mag-mom to Bohr mag ratio Absolute entropy Sackur-Tetrode Weinberg angle Quark mixing matrix CKM	1 -1 -1.1 0.23° 0.97, 0.23, 0.41	1 11e-1 23e-2 Prime e-2
Proton G factor Molar gas const Cabibbo angle (Weak interac)	5.5 8.31 J mole-1K-1 13.02°	1e1 13e0
Vacuum impedance Muon mass energy equi	3.7e2 Ohm 1e2 MeV	3e2 1e2
E-mag mom/nuc. Magne.ratio	1.8e3	2e3
Faraday constant Standard atmosphere Magnetic flux density Von Klitzing Const	0.9e5 Cmol-1 1e5 Pa 2.3e5 T 2.5e5 Ohm	1e5 1e5 23e5 3e5
Velocity of light Rydberg constant	30e7 m/s 1e7 m-1	29e7 1e7
Electron charge to mass Electron gyromagnetic ratio	1.7e11 Ckg-1 1.7e11 s-1T-1	1e11
Electron volt Hertz relation Josephson constant	24.1e13 Hz 48e13 HzV-1	23e13 47e13
Kilogram-joule relation	0.89e17 J	1e17
Atomic unit e-field gradient	971e19 Vm-2	37e19
Avogadro constant Loschmidt const	6e23 mol-1 268e23 m-3	7e23
Joule Hertz relationship Plank temperature	150e31 Hz 14e31 K	37e31 13e31
Kilogram inverse-meter	4.5e41 m-1	5e41

FIGURE 4.6 Fundamental constants found in nature and trying to visualize the constants in a prime-e-prime.

What if our universe is made of virtual atoms of waves, the universe of mass is an illusion?: Origin of the universe is beyond the scope of this book, but when the identity of nature's true information structure is developed often the key features of this universe factor in. Battle for the heart and the soul of physics: Experiments to find out the smallest indicate that the fundamental constituents of the universe are 10 million billion times smaller than the resolving power of the Large hadron collider, LHC. Massless virtual objects like the vortex atoms made of wide ranges of electrical, magnetic, mechanical ripples are the key ingredient using which the human brain model and its artificial analog have been created here. The whole idea that "no mass" or a very low effective mass is equivalent to an infinite space holds a "time" in it, a unique time crystal feature has been assigned to tag it as a mass. In general mass is a geometric shape of a unique kind as shown in Figure 4.5a, it is in line with the field theory. Fractal information theory, FIT proposes that rhythms are the keys to everything that originates from the concept of spin. However, then replacement does not end in the mass or space, "phase change" creates both "space" and "mass."

Spin is a property of a material that enables it to exhibit the rotational symmetry. For example, the symmetry of a particle is such that after two rotations, it returns to the initial symmetry, then it is spin ½. However, if just by rotating 180 degree it returns to the initial state, then the spin is 2. We know that the spin governs the nature of force, why then we do not see so many different long-range forces other than photon and graviton. Both are massless, and lower the mass, more global is the interaction!!!! Of course, Weinberg argued that more than spin 3 is not possible because they would not interact, but how true is that proposal is not important.

Wigner found that in the massless situation, instead of particles labeled by a single spin, there could be a range of values. Photons only have the spin 1 label while gravitons only have spin 2, it could be that the photons and the gravitons are actually special cases of a more general massless particle with an infinite list of integer spins. The massless particle is labeled by spin 0, 1, 2, 3, all the way to infinity! Wigner called these "continuous spin particles" (CSPs; Schuster and Toro, 2014). The spins are definitely discrete not truly continuous (see below QSP). For FIT, these are signatures of a time crystal or organized spin foam where each spin is a guest clock located on a host clock.

Weinberg showed how the spin of a particle could provide insights into the electromagnetism and the gravity. Continuous spin particles (CSP) are objects with all possible spins. Natalia and Toro argued that instead of all possible spins, the spins would take certain values, Quantized spin particles or QSP (Schuster and Toro, 2013). It would definitely change the way we look at nature. When the geometric space was imagined for the number system theory, imaginary iota was brought in and showed that it returns the quantized values. Dodecanion in combination with octonion and quaternion would bring about further organized structures where symmetry would govern the degree of quantization.

4.5 A TABLE OF FUNDAMENTAL CONSTANTS IN NATURE

Fractal space-time: Throughout this book, we do not debate whether our brain is a classical or a quantum device because if the two formulations are combined then, the fractal space-time features naturally make a smooth transition from classical to the quantum regime. Feynman showed that the quantum mechanical paths that contribute mainly to the path integral, these paths are non-differentiable and are fractal. Though there are three routes following which the laws of quantum mechanics are derived from the fractal geometry of space-time (theories of scale relativity). The first route is, a random walk through the Feynman chessboard as suggested by Ord et al. Second is the scale relativity approach wherein a collection of coordinates, some points are finite, has limits, so its classical and some points in the space-time are fractal, means, infinite, and minimum energy path or geodesics are irreversible. When these typical fractal features are applied to classical coordinates, Newton's equations are written in terms of Schrödinger's equation (Nottale, 2005). So, quantum to classical transitions is feasible in fractal worldview. Third, El Naschie approach, where he replaced the "continuity" part of a fractal and replaced it with the Cantorian space-time fabric. The Cantorian approach is exciting because using this space-time fabric, all fundamental constants including the fine structure constants, gravitational constants are derived (El Naschie, 2004, 2009). The approach allows the formation of wild topologies (multi-fractals). A Cantorian space-time metric uses multi-fractals, means fractals with different dimensions, this generates the E8 symmetry. The experimental verification of a unique fusion of E8 symmetry (248-dimensional rotational symmetry) is shown, we call it the lotus symmetry. One reason for stressing a new metric—fractal time considers no limit, not even in the Plank dimensions, hence, the Plank cut off as considered by some fractal space-time theories contradict themselves. PPM guides linking the symmetries, no imagination on the space-time could be left out.

Explaining all of quantum mechanics using geometry: The foundation of entropy and fundamental constant: Wheeler tried to explain everything that is happening in the universe in terms of changing geometric shape. Now, Wheeler used classical points, not phase space like us to construct the geometric shape. As a result oscillating geometric shapes gave rise to incredibly non-physical behavior at certain scales. The boundary of a boundary is zero is the foundation principle

of geometric algebra (Spanier, 1966). As Wheeler argued, "we will someday complete the mathematization of physics and derive everything from nothing, all law from no law." We extend the statement here, possibly, one day we could derive everything from a metric of primes including all laws. In brief, the choice of the question asked, and choice of when it's asked, play a part—not the whole part, but a part—in deciding what we have the right to say (Wheeler, 1984, 1986). When phase singularity accommodates geometric shapes, the boundary of a boundary is zero, shapes, irrespective of their oscillations are confined within the singularity domain. Hence non-physical situations do not arise. Due to the PPM we find that each prime writes its typical signature in the metric of primes, we can easily test it by keeping say products of 11 only to see what prime 11 does to the metric. Consequently, we found that around 15 proteins are sufficient to generate a self-similar pattern for each prime as a physics law and typical curves at the junction of two primes as a fundamental constant. Thus, universal physical laws are observed in the changing geometric shapes generated by primes. Not just quantum, serious attempts were made to include gravity as changing geometric shape (Wheeler, 1967; Misner et al., 1973) that could be done with PPM and GML, as we would see below. We do not try to derive quantum origin, but we see quantum as an opportunity to resolve the asymmetric flow of time and entropy (Price, 1996).

The emergence of fundamental constants from PPM: Fundamental constants are not that universal as we are taught in the textbooks, they vary with measurement conditions and environment (Gross, 1989). Figure 4.15a explains Fundamental constants are functions of e, π, and Φ. These three fundamental constants e, π, and Φ. are connected by quadratic relations. If in our universe, there truly exists an orthogonal triangle of e, and Φ, whose three angles are $\theta_1 = \cos^{-1} e/\pi$, and $\theta_2 = \cos^{-1} \varphi/\pi$, which could change. The quadratic relation is beautiful, since angles between e, π and Φ, π change, keeping the quadratic relation $e^2 + \varphi^2 = \pi^2$ intact. In the PPM when we find the topological shapes (12 has 3 nested points, hence a triangle, 8 has one nested point so it refers to a singular point, etc.) by the time we reach 10^{12} integer space, covering $1 \times 2 \times 3 \times 5 \times 7 \times 11 \times 13 \times 17 \times 19 \times 23 \times 29 \times 31 \times 37 = 7.420$, 738, 134, $810 \times 10^{12} = B$ (Brahman), we get dodecahedron, a 20 planar 3D architecture. If we include two more primes 41 and 43 to avoid the debate whether 2 is at all a pure prime, $B = 1.308276133167003e + 16$. We can go for a higher number of clocks but as explained in Chapter 2, even if we build a system with more number of clocks, we will not add more symmetry to the universe. Ninety-nine percent of all non-primes could be created by 12 primes only, therefore, it is wise if we reach the number B, we would consider the whole system as one unit and start counting again, 1, 2, 3, … n. It means, starts building a new system. Now, we have also explained in Chapter 2 that due to phase quantization we get clockwise and anticlockwise rotations, but, certain integers start the rotations in the phase space, the others follow. The rotational space forms a lotus satisfying various logarithmic spirals, around $e^2 + \varphi^2 = \pi^2$. In the PPM, as distinct metrics generated by each prime superpose, certain critical folds with an angular gap θ need to be bridged for generating a continuum in the metric. The bridging factors are fundamental constants that we see around. The generic formula for fundamental constant is $\theta(e^2 + \varphi^2)^{\pm \text{prime}}$. It is not the ultimate one, it is the first and the most primitive attempt to topologically derive the fundamental constants in the future (Figure 4.6). Consequently, the physics laws would be born in the landscape of symmetry, PPM.

4.6 QUANTUM INTERFERENCE- AND FRACTAL INTERFERENCE-EXPERIMENT ON A SINGLE MICROTUBULE

A journey to entanglement in room temperature under noise: The hallmark of a quantum oscillator is a textbook-like quantum interference pattern depicting the wave functions of the quantum well with all possible nodes starting from the ground zero. Nanoelectromechanical oscillators (NEMS) are the newest classical window to the quantum world. Microtubule nanowire found abundantly in eukaryotic cells (Figure 4.7a), spontaneously generates a pair of coherent sources due to birefringence (Figure 4.7b). Sahu et al. carried out a series of experiments on a single microtubule to check if it is a quantum device (Figure 4.7c). While basic relation $h\upsilon = k_B T$ sets limits to the quantum preparation and measurement (Figure 4.7d), the thrust to bring quantum features to the classical regime remains confined either in ultra-cold temperatures and/or in an ultrafast time domain. Journey to hot entanglement has stemmed recently (Galve et al., 2010) by squeezing the normal mode of vibration of synchronously coupled oscillators vulnerable to heat contacts, advancements are too limited to make an impact. Thanks to quantum-discord that harness quantum benefits, even without entanglement. Squarely apart, on a carpet of strongly coupled oscillators, the coherent stream of soliton particles does not de-cohere under thermal or electronic noise, remains pristine for kilometers; thus, could replace the entangled atoms or ions. However, the problem for not having superposition principle restricts their localization; hence, quantization is possible only if individual solitons are weakly coupled, non-linearly shaping into a waveform (Rajaraman, 1982). Once soliton's limitation to form wave is resolved, discretely spaced soliton-particles all along the NEMS structure are synchronized as sinusoidal wave delivering all that quantum entanglement and quantum superposition had promised, semi-classically, in the open air (Figure 4.7e). Quantum was prepared theoretically, using solitons; but no attempt was made to realize it experimentally by breaking the limits set by $h\upsilon = k_B T$.

With an ideal ferroelectric behavior (Tuszynski et al., 1985), microtubule is piezoelectric, oscillates mechanically if an electrical signal sent through it (Sahu et al., 2013a, 2013b); lattice sites on the cylindrical 2D surface generates symmetrical points in the power spectrum, with large piezoelectric degrees of freedom. Size of the constant-voltage step in multiple ferroelectric states measures the degree

of resonant velocities of solitons; thus, varying ac driving bias the velocity of solitons could be controlled. Velocity tunes delay, essential to generate normal modes of vibration among all the classical soliton particles vibrating at the single resonance frequency. MT's dc resistance remains constant unless an ac signal of particular frequency triggers one of the eight co-existing quantum wells inside and release energy to the external circuit reducing the contact-resistance. Even a ± 50 mV$_{rms}$ ac signal turns a ~23.5 μm semiconducting MT (1–10 MΩ or ~300 MΩ) into a nearly ballistic conductor (39 kΩ~3G$_0$, independent of L), at those eight resonance frequencies (f) ~15 kHz(Q~4,|1>), ~9 MHz(Q~43,|2>), ~12 MHz(Q~75,|3>), ~15 MHz(Q~46,|4>), ~18 MHz(Q~64,|5>), ~20 MHz(Q~110,|6>), ~22–23 MHz(Q~129,|7>), ~24–25 MHz(Q~62,|8>, Figure 4.7f). Length variation study of resonance peaks shows that a single microtubule exhibits any one of the four pairs of dominating peaks, |1>|2>, |1>|4>, |1>|6> and |1>|8>. The co-existence of two peaks reflects the possibility of four quantum bits. four distinct coherence times that a particular microtubule delivers for four quantum bits. Soliton takes 23.5×10^{-9}s ($\tau_\phi = L_\varphi/c$, c~10^2–10^3 m/s) to cross $L\varphi$, since coherence time $\tau_C > \tau_\varphi$, quantum bits can survive sufficiently long to observe classically.

Microtubule is quantum device but only at certain ac frequencies: During four-probe quantum interference measurements using a single microtubule device (Figure 4.7c) it was found that at certain ac frequencies, there is a current at zero gate bias and zero probe bias (quantum), at all other frequencies the center is dark (classical). Two coherent electromagnetic (em) pulses are given by $E_1(t) = (A(t)\exp(ik_1 r) + c.c)/2$ and $E_1(t + \tau) = (A(t + \tau)\exp(ik_2 r) + c.c)/2$, where $\tau = 2\pi d/\lambda$, d is the path difference, λ is the wavelength of em signal, k is wave vector, $c.c$ is complex conjugate. The intensity of their interference pattern is $I = \langle|E_1(t) + E_2(t+\tau)|\rangle^2 = C + \langle A(t)A^*(t+\tau)\rangle \cos\{(k_1 - k_2)r\}$, here normalized temporal coherence function is $\gamma(\tau) = A(t)A^*(t+\tau)/C$. When we change the em frequency, we tune τ (delay), which first increases and then decreases the amplitude I of the output signal, i.e., $\gamma(\tau)$. Full-Width-Half-Maxima (FWHM) for $\Gamma = I_{max} - I_{min}/I_{max} - I_{min}$ vs. τ provides the coherence time $\tau\varphi$ (Iwata and Hieftje, 1992). When the gate voltage changes (inner electrodes in Figure 4.7c) the Fermi-level of synchronized soliton packet, original and gate-modulated (wavenumber, k) signal develops a phase difference, which generates a Quantum Interference pattern but only at certain frequencies.

One easy way to prove that microtubule follows fractal mechanics or it is a fractal harmonic oscillator is to zoom the interference pattern (Figure 4.7g). If applied bias shifts wave vector k largely by zooming a particular domain as shown in Figure 4.7g, only the ballistic paths for solitons change, the weak coupling among all the solitons is not affected, so we observe a similar pattern. It also depicts that transition to hexagonal lattice phase is essential for two co-existing quantized soliton waves to reside physically on the MT surface. Distinct quantum interference patterns for eight dielectric resonant states in the range was 10 Hz–50 MHz for different lengths, confirms MT's

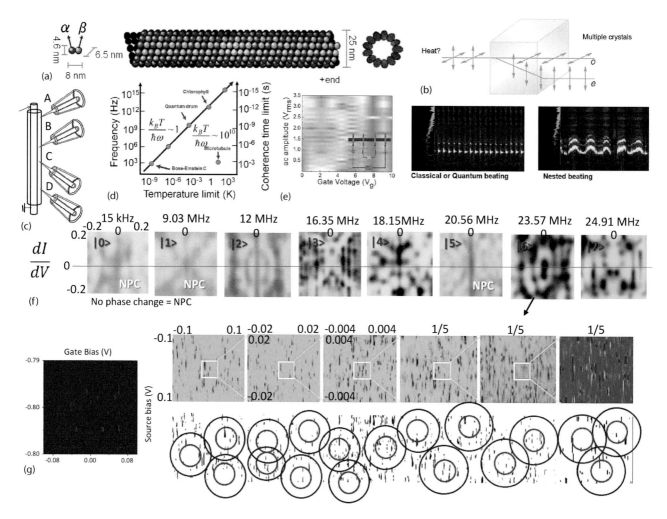

FIGURE 4.7 (a) Microtubule and tubulin protein. (b) O-ray and E-ray for the crystal that shows birefringence (top). The corresponding beat frequencies measured in microtubule nanowire is shown below. Both the bottom panels show frequency (vertical axis) with respect to time (t) in the horizontal axis. The left panel shows the patterns for classical and quantum beating. The right panel shows fractal beating where the conventional beating ripple is split into multiple sub-ripples. (c) Four probe system that is used to measure the 2D interference pattern. Inner two probes act as gating channel and the outer two rings are to send input and read output signal. (d) The ratio of available energy in the thermal noise (k = Boltzman const) and T is the Kelvin temperature, to the energy or quanta for resonant oscillations is plotted, h is Plank's constant. The ratio is plotted as a function of the environment temperature and the applied frequency. The coherence time limits reported in different kinds of literature have compiled the scale in the right. (e) One example of interference plot when no ac signal is applied to the microtubule. (f) Differential conductance (dI/dV) is plotted (vertical z-axis, perpendicular to the page, color contrast of the 3D plot) against the gate bias (horizontal axis) and the input bias (vertical axis). Eight 3D interference plots are presented, each for one typical frequency. (g) Apparent feature of the interference pattern, horizontal axis is the gate bias and vertical axis sample bias. The axis perpendicular to the page is the current. The fundamental difference between panel f and panel (g) is differential conductance and current. During the experiment, a small part is zoomed and new scan run. The zoomed domain is all about reducing the scan boundaries continuously five times.

identity as octave quantum-like cavity-resonator. In sharp contrast to conventional resonators (mV, mA~microwatt), microtubule operates in the atto-watt (μV, 10 pA) domain, consuming 10^{10} orders lower power. Since coherence length is ~23.5 μm all microtubules produced in any living eukaryotic cells, brain neurons are a quantum-like device, if and only if one triggers it with a suitable ac frequency. Two possible co-existing lattice paths on the hexagonal tubulin lattice those constitute a quantum bit is also explicit in molecular simulation.

4.7 FRACTAL ABSORPTION–EMISSION OF THE OPTICAL BAND OF A NANOBRAIN

Molecular foundation of the fractal logic gate: We discuss the design and synthesis of a Fractal logic gate that operates with pH and density as input variables (Ghosh et al., 2015b). Fractal logic gate means its truth table is not permanent; any operational region could be zoomed to get a new truth table, continuously. Thus, it can sustain changes in the input condition unlike conventional molecular logic gates.

The fractal feature was never realized in a logic gate before. Fifth-generation PAMAM dendrimer [P] encapsulated with 2 Nile Red [C] molecule and surface attached with 32 molecular rotors [M] and 4 pH sensor molecules [S] is the functional unit whose water solution was used as logic gate matrix, including PCMS; we call it PCMS or nanobrain (Ghosh et al., 2015a). The final derivative is 10–15 nm wide particles. The logic gate output is read by measuring fluorescence. Proper choice of functional groups balances the energy transfer between encapsulated molecule inside the dendrimer and the externally attached functional components, we image that energy transfer route (M↔C↔S) precisely using combined excitation-emission spectroscopy (CEES). One could zoom a CEES spectrum similar to the interference pattern described above, to find that new bands are found (Figure 4.8a). The reason is

not newer parts of the molecular structure being probed, rather an interaction of participating molecular systems. A few functional groups using the phase interaction build a dynamics that is similar to a fractal, say Mandelbrot fractal needs only one imaginary world and it can create an infinite space using that (Figure 4.8b). Fractal logic gate feature is aimed at delivering programmed output at all scales using a single instruction (Figure 4.8c) even under noise, e.g., drug delivery systems, chemical pattern recognition, chemical data encryption, etc.

A note on chaotic, deterministic and quasi-determinism in computing: Unlike deterministic system, a chaotic system is apparently random but predictable only if we know the initial condition, see butterfly effect, thus one can calculate the output of such a multi-directional chemical reaction. In contrast, in a classic random system, the same reaction would give

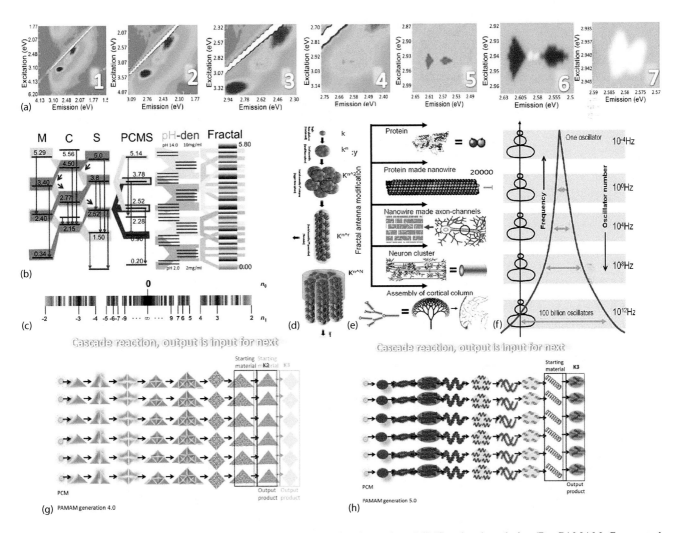

FIGURE 4.8 (a) Combined excitation-emission spectroscopic (CEES) plots, where PCMS molecular solution (P = PAMAM, C = controller, M = Molecular rotor, S = Sensor), the range of excitation and emission wavelengths were zoomed six times to extract the local pattern in the 3D absorption emission spectrum. (b) Motor, controller and the sensor molecules energy levels are plotted as calculated from the CEES spectrum. The new energy levels observed while zooming the spectrum shows self-similar patterns in the energy levels. (c) A schematic presentation of the fractal energy levels. (d) PCMS self-assembly process that starts at single-molecule forms a cylinder and sphere sequentially during growth. (e) Growth is shown from single tubulin dimer to microtubule to neuron bundle to the entire brain network. (f) The number of oscillators increases manifold as one enters a human brain architecture, with more layers, an increased number of components takes part in the organization. (g) PCMS self-assembly runs by two steps. Alternative formation of triangle and square. (h) PCMS self-assembly runs by two steps, sphere and spiral.

different results in two consecutive runs. The three systems are (i) "chaotic (if we know input then predict random output)," (ii) "random (even if we know the input, cannot predict)," (iii) "deterministic" (input-output fixed). Here we realize "quasi-determinism," a system where a reaction system always approaches determinism, but as an asymptotic function, the determinism is never reached. Say, the solution is 1, the system reaches 0.999 then after some time, 0.9999999, etc., then if conditions change then if the output is 2, the reaction system drives toward 1.99 then 1.99999, etc.

Triangular energy transmission network in nanobrain enables accurate programming: Period, or time cycle or rhythm is what remains constant even under random PCMS (nanobrain) motions. Strict energy transmission route like M↔C↔S is itself a rhythm encoded in the atomic arrangements and PCMS restores it under noise. The attempt allows PCMS to sustain a defined geometric path on a surface under noise and a logical fluorescence output in solution even though a particular pH and density range are continuously zoomed creating a fractal logic gate. Fractal gate means one can continuously zoom to expand any part of the operational matrix converting say, a 10 × 10 pH-density matrix into a 1000 × 1000 one, thus, eventually capturing astronomically large data. Even for a large matrix, time-lapse is determined by the smallest matrix time cycle, this is what we call "instantaneous"—it's finite indeed.

Fractal reaction kinetics to synthesis brain jelly: However, M↔C↔S vibrational chain has been programmed by Ghosh et al. to replicate the evolution of neural network circuits in a chemical beaker (Figure 4.8d–f) using fractal reaction kinetics, see Chapter 9 for details. However, we should note here that the fractal synthesis that repeats the formation of triangle and square alternately (Figure 4.8g) and sphere, spiral alternately (Figure 4.8h) from 8 nm to millimeter-scale is an amazing (Ghosh et al., 2016b) observation. Therefore, fractal mechanics could be used using table 4.5d, from building decision-making devices to the chemical reaction kinetics. Universalization of the Plank limit would have enormous effects on the materials engineering in the coming days.

4.8 BASIC MATHEMATICS USING CLOCKS

Figure 4.9 describes how using a pencil one could write any basic integer, powers, equations on a piece of paper as a precursor to the development of continued fraction geometric

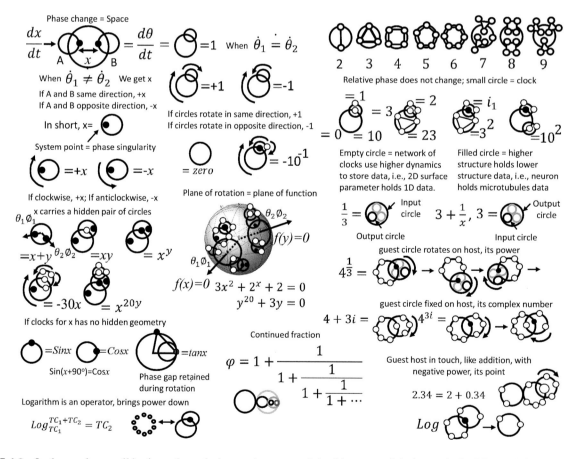

FIGURE 4.9 In three columns all basic mathematical operations are explained in terms of circles or clocks. The top left is how to encode a variable x? Then, different power relations of x, trigonometric parameters of x are shown. The second and third column top panels are integers. Toward the right bottom, we demonstrate the fractions and the decimal numbers. The protocol to write an equation with multiple variables are noted at the center.

algebra, CFGA, where one could carry out all complex mathematical operations simply by drawing. The process is part of GML and suitable to be implemented using fourth-circuit element H described in Chapter 8. The chart begins with the writing of a variable x, one needs two system points on two circles, their difference replicates the dynamics of an unknown variable x. Clockwise and anticlockwise rotation determines $+x$ and $-x$, always in CFGA, sign depicts direction. For simplification, a tiny dark filled circle is added at the inner side of the perimeter of a circle. When the dot is an empty circle, and located on the circle perimeter, it is 1. Using empty circles, one could write 2,3,4,5,6,7,8,9. Empty circles are nested to create integers, two examples are shown like 10 and 23. When these circles are filled we add a number in the power position of another number. For representing a continued fraction, a given circle is divided into small parts. These parts could be at the base and at the indices, for example $1/3^{1/5}$. We could write complex numbers. Logarithm operation could be performed taking power to the base.

Figure 4.9 also describes how one could write complex equations with multiple variables. We noted above that a black filled circle, is a variable say x. Now, if we have multiple variables, then we need to write it on a Bloch sphere or 3D phase space. If we have multiple functions then on the sphere each function acquires a plane with a vector that is an axis of rotation or central control. The difference between a function and an equation is that an equation has an additional circle or clock that loops all functions into a bond.

4.8.1 NUMBERS, EQUATIONS, ADDITION, SUBTRACTION, MULTIPLICATION, AND DIVISION

Figure 4.10 explains different arithmetic processes. The panel "a" shows how one could add a few numbers by adding the diameter of the circle and creating a new clock, or simply adding the number of dots on the perimeter of a circle. Even power addition is also possible. The panel "b" shows the different

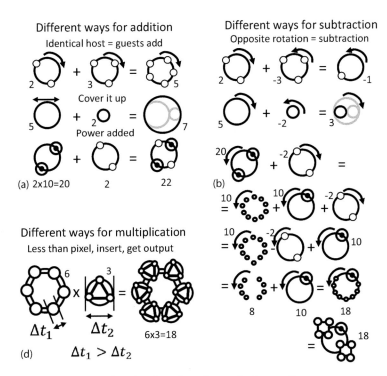

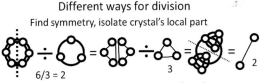

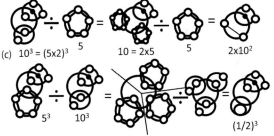

Powers reduce as we disintegrate clocks at lower level. This operation requires phase coupling between two layers.

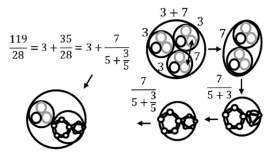

FIGURE 4.10   Basic mathematical operations to be c.arried out by clocks. (a) Three ways addition could be performed. (b) Different ways of subtraction of multiple integers. (c) Different ways of division. (d) Different ways of multiplication. (e) How division automatically leads to the formation of the continued fraction.

ways for subtraction. The system of clocks naturally decides to go for addition or subtraction based on the rotational direction of the clocks, those who run in the same direction, go for addition and those who run in the opposite direction, subtract. Panel "c" shows different ways for the division. In the case of division, it is important to isolate the time crystal representing the number into multiple parts based on the symmetry of the divisor. For example, if we are dividing by 2, then using the C2 symmetry, we would divide the time crystal into two different parts. If dividing by 10^3, then we would split the time crystal into three parts, there is a simple way of doing it. Panel "d" shows different ways for multiplication. Here to multiply, one has to insert the time crystal for one integer into the singularity points of the other. Panel "e" shows that if the division fails, how fractionalization, could take over and if there is no completion, the process would continue.

4.8.2 Differentiation, Integration, and Partial Differential Equations—Lie Algebra

In Figure 4.11a, we demonstrate how the differentiation and integration could be implemented by rotating the system point or changing phase (for trigonometric functions), or editing the

power of functions. Integration and differentiation are opposite processes. One nice example is shown in Figure 4.11b, where we differentiate one of the products for partial differentiation. A partial differential equation is realized by keeping a set of clocks intact and changing a few. Figure 4.11c enlists the process on how to geometrically address differential equations of different orders. There are several popular equations made of partial differentiation, they appear beautiful.

Lie algebra and bypassing the non-differentiability: The physical significance of log is that a physical parameter varies depending on how much a parameter weighs at that point. Now, this is a very interesting situation even if we have a single log relation. Lie algebra (Bourbaki, 1989) developed in the 1930s nicely address this issue. Kac-Moody Algebra (1960) an extension of Lie Algebra developed over a complex space addresses infinite dimension. The linear space transformation that governs the "rate of change" could be a complex number if there is self-similarity, in PPM-GML-H triad case we have self-similarity. Now, the interesting part is that the linear space transformation cannot define a rate of change which contains a parameter that is defined by another complex number's space transformation. In such situations, without a debate in mathematics we consider that a function becomes

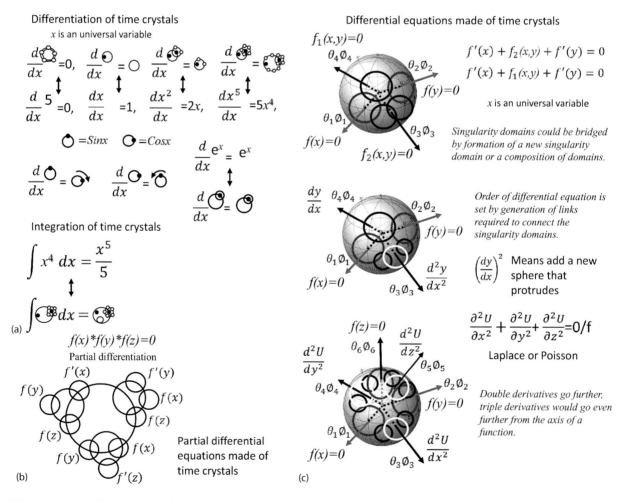

FIGURE 4.11 (a) Differentiation and integration using clocks or time crystals. (b). Partial differential equation. (c) Differential equations made of time crystals. Three different types of time crystals are described in the three panels.

an undefined mathematical entity. However, we leave with a major conclusion that since in PPM-GML-H triad case the complex number's real and imaginary parts are frequency, therefore, we have an imaginary space-time world inside another imaginary space-time world and so on. Already a part of this theory was formulated in Riemannian manifold.

Iterative function systems are resolved in the images given to the new class of computer in the way described above, however, during computation, at different layers of the hardware, synchronization and de-synchronization of the resonant oscillation continue. Computing time is the synchronization time of the fractal seeds. Synchronization leads to coherence which means Fractal seeds oscillate in the same phase and frequency. Spontaneous switching between synchrony and de-synchrony is essential, thus, entanglement is not a pre-requisite. The information perspective of that physical process of computing is that several fractal seeds of iterative function systems form the network, coupling and de-coupling of large networks is a generic event that happens during computation.

Lie group, conformal group of the sphere: Geometric language of PPM computing: The conformal group is the group of transformations from a space to itself that preserve all angles within the space. It is the group of transformations that preserve the conformal geometry of the space. Harry Bateman and Ebenezer Cunningham showed in 1908 that the electromagnetic equations are not only Lorentz invariant, but also scale and conformal invariant (Bateman, 1909). Putting coordinates together with sphere's radius was Lie's first work in 1871, The conformal group includes the Lorentz group and the Poincaré group as subgroups, but only the Poincaré group represents symmetries of all laws of nature including mechanics, whereas the Lorentz group is only related to certain areas such as electrodynamics. Due to this universality, the conformal feature has been used in constructing the geometric language for PPM computing. All basic geometries used as letters for operational language are encompassed inside a circle or sphere and ratios of contact coordinates are used. The preservation of conformational geometry enables a precise encoding of sensory data as a stream of nerve impulses. List of simple Lie groups Classical An Bn Cn Dn; Exceptional G2 F4 E6 E7 E8. This book does not endorse any unified theory starting from a single geometry because PPM metric described in Chapter 3 demands for several evolving vortexes each carrying a distinct PPM, and depending on its prime number origin, the evolution of symmetries change, redefine. Also, for a conscious machine, E8 does not suffice, we need a dodecanion with 11 imaginary and one real world, whose manifold geometries with 12 planes overlap to create a physical form of conscious thought.

4.8.3 CFGA OPERATOR THAT RUNS 13 MATH OPERATIONS IN A TIME CRYSTAL STRUCTURE

In Figure 4.12, we have summarized a list of 13 mathematical operations described in Figures 4.9 through 4.11. One of the most important features of continued fraction geometric algebra, CFGA is that naturally it is selected which means, from outside, there would be no need to encode an instruction. Time crystal self-assembly is a process that detects minute simple changes or differences in the structure is detected and action is detected. The third and the fourth columns of the table in Figure 4.12 lists processes in such a way that materials self-assembly runs as is mathematical operations. Thus, careful observation of this table would reveal that natural weak interaction-based material self-assembly of Hinductor devices would spontaneously trigger the actions listed in column 4. CFGA operator execution is not a linear sequential process, its managed by a time crystal, which helps to isolate typical mathematical operations and run them repeatedly so that particular self-assembly process autocorrects itself if there is an error. One such problem is described in the bottom right panel of Figure 4.12. Interaction between the imaginary worlds is a critical aspect for the quaternion, octonion and dodecanion tensors. There are three aspects that create information exchange channels between different imaginary worlds. The journey through singularity addresses the power of an integer, tensor element transformation and finally fractal geometry.

4.9 THE FUNDAMENTALS OF QUATERNION, OCTONION, AND DODECANION

No more Fano plane, now it's time for manifolds: Using shades, we have pointed out how one could find a quaternion in the octonion and dodecanion in Figure 4.13. Most off-diagonal elements of the four multiplication tables for dinions, quaternions, octonions and dodecanion are antisymmetric, making all four of them almost a skew-symmetric matrix except for the elements on the main diagonal, as well as the row and column for which h_0 is an operand. A single Fano plane that links seven imaginary worlds with seven lines explains the multiplication of an octonion world. One could notice the clocking loops in the Fano plane, each loop is made of three elements, making sure that a triangle is stored even at the smallest level, when each element is a singularity point. The clocking direction is fixed and very important, so they do not build a matrix but a tensor. Similarly, when the clocks made of three singularity points are used to build the multiplication of a dodecanion world, we find that four planar structures are required instead of one. One planar structure is made of 11 points centered at 1, here 9 occurs twice, so we need to fold the plane, two points with 9, touch each other. The second planar structure is made of 13 points centered at 11, here 3 occurs thrice. The third planar structure is made of 9 points, where 4, 5, 6, and 7 occur twice. The fourth planar structure has 6 points. In order to understand how a dodecanion tensor would connect the clocking triangles, we need to join the four planes by touching the common points. Thus, we get a manifold. The journey to manifold machines begins here, we have outlined the futuristic machines in Figure 10.12.

4.9.1 THE RULE OF 11D MANIFOLDS

A quaternion-topology of dodecanion for consciousness: One interesting plot at the bottom of Figure 4.13 is that there is three quaternion composition for the dodecanion. The coexistence of three distinct quaternions suggests that simultaneously three solutions would generate for every single dodecanion produced in a device. It means single hardware, if it acquires a topology that allows it to put three, time crystals of solutions in the three singularities, a new time crystal is born with a triangular control. When each contributing element from the tensor is a singularity point, for a dodecanion tensor, the triangle made of quaternions as noted above, would be a higher-level geometry. However, when eleven imaginary worlds would send complex time crystals, the resultant quaternion as shown at the bottom of Figure 4.13 would depict a new phenomenon. At the highest level, when all the components have already contributed, three distinct information processing could run at a time in the single hardware. In quantum, one could sense the contributions from unknown to the reality, but when 12 imaginary layers operate together, the hardware could remain in the imaginary world, with a distinct geometric identity, here it's a triangle. The ability of single hardware to package entire information in three identical parts and then edit three of them independently is considered here as consciousness. The quaternion-topology of dodecanion satisfies the minimum requirement for reverse engineering a conscious machine, for the higher-level tensors, one could extrapolate consciousness to a superior scale as outlined in Figure 10.12.

Where does the physics laws come from? What distinguishes between laws and no laws is a mystery (Goodman, 1954), it was suggested that the laws are relations between properties or universals (Armstrong, 1983). It was a departure from Hume that "laws assert no more than regularity of coincidence between instances of properties." There are anti-reductionists who think that there are laws in the universe, but no need to specifically appeal to properties or universals (Woodward, 1992). Anti-realists think that there are no laws in the universe. Two arguments we note. First one is that all generic laws are derived in a boundary condition, outside, it is not verified cannot be said to be valid, a law should be an external entity (Mumford, 2004). The second one is that derived laws are properties of a system, not a system-independent feature that a system follows

Spontaneous activation of math operations by clocks, CFGA operator $\hat{C}$ symbol Ж   Time crystal format for CFGA operator

Steps	Math. Operation	Condition for spontaneous operation	Circle Property
1	Multiplication	Temporal match, (Width of phase singularity = clock diameter to be fused).	Insert
2	Division	Topological match of local parts.	Split the gaps
3	Addition	Hosts are identical, guests bond. Participants rotate same direction	Form closure
4	Subtraction	Hosts are identical, guests bond. Participants rotate opposite direction	Complement
5	Differentiation	Higher world (power) topology becomes essential to the lower world.	Transfer
6	Integration	Taking higher world topology to the lower world.	Transfer
7	Fraction, decimal	Subtraction/division fails to match, process continues.	Repeat
8	Logarithm	Higher world, or topology of power alone survives, host disappear, due to match.	transfer
9	Power	Reduction of large number of identical clock geometries, by adding a guest.	Reduction
10	Polynomial, series	Phase axes of clock geometries differ, they fail to bond, so adds as guests.	Fusion
11	Partial diff. equn.	In a composition of clock geometries, if one changes & others do not.	Split
12	Matrix, tensor	Clocks self-assemble to form clocks, not discrete isolated chain of clocks, closed.	Fusion
13	Fractal shrink	Each imaginary or real higher world (power) have identical clock geometries.	Reduction

When two imaginary layers one inside another interact

Across singularity surface	S_1	S_2	S_3
Power	3	3^5	3^{5^2}
Tensor	3	3^{5i}	$3^{5i^{2j}}$
Fractal			

FIGURE 4.12 A table represents a set of 13 steps that describe essential mathematical operations, automatically selected and executed. This function is named as continued fraction geometric algebra, CFGA. We use the symbol Ж to denote the operator. To the top right, the 13 step operations are grouped as a nested circle. In the middle nested singularity spheres are noted. Three singularity spheres located one inside the other, processes integers or numbers as powers, tensors as imaginary powers and geometric shapes as layered structures.

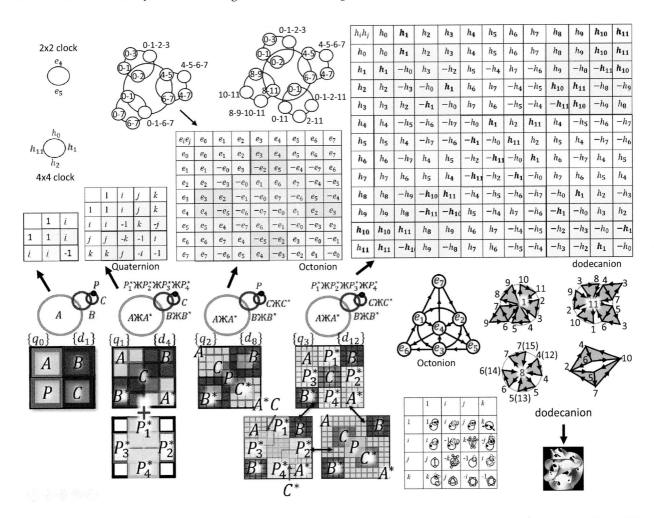

FIGURE 4.13 Four types of tensors represent the data structure of the artificial brain. Four types of tensors are presented here dinion, quaternion, octonion and dodecanion algebra for the time crystal representation. Four different ways the multinions are explained. First, time crystal presentation of the multinion, second, matrix representation which is colored as square matrices; third, linguistic presentation, here, we present the tensors as a subset of four clocks, each clock represents a sub-matrix of the entire tensor. Sub-matrices A, B, C, and P. Quaternions show duality and dodecanions show simultaneous co-existence of three matrices. To the bottom right corner Fano plane for quaternions and manifolds for dodecanions. There is a pictorial clock like presentation of a quaternion at the bottom right, it suggests how a single element in a tensor looks like.

(Giere, 1999). The most important of all, if there is any law, how did they come into being (Cartwright, 1997). Since we derive fundamental constants partially from the PPM and geometric musical language (GML), various topological identities like $e^2 + \varphi^2 = \pi^2$, we avoid laws, generate the transition of one dynamic to another. Thus, when we convert properties or universals as changing geometric shapes, and link them using the metric of symmetry, PPM, the laws may or may not be out there in the universe, at least we do not have to find them.

4.10 TWELVE EQUATIONS THAT REGULATE A FRACTAL TAPE FOR PRIME-BASED COMPUTING

Figure 4.14 explains how different imaginary worlds of the quaternion, octonion and dodecanion tensors operate. The tensor elements affect different imaginary worlds. While doing so, there is a topological restriction which is geometric in nature, then there is a materials restriction which is weak interaction based self-assembly rules, finally there is a manifold restriction which is regulated by the symmetry of the distribution of primes. Twelve equations which were noted earlier in the context of geometric operation in Figure 3.12a. We have extended the same table in Figure 4.14 where we introduced the picturization that is spread all over the book, how the projection from infinity looks like.

4.10.1 Projection from Infinity—Future Impacting the Present

To the extreme right of Figure 4.14 we have explained using a continued fraction how exactly, projection from infinity redefines the basic dataset. In the expression of a complex data, in the continued fraction form, one could easily find the simplest geometric shapes that would repeat as is, isolate that one and send it as an output. This is very

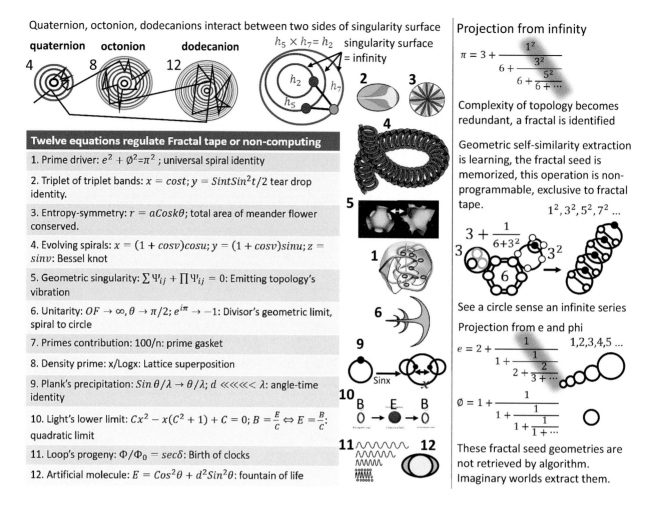

FIGURE 4.14 The fractal tape operation of an artificial brain is demonstrated. There are 12 key equations that govern the dynamics of artificial brain, shown in a table. Using schematic plots the projection from infinity concept is explained to the right part of the table for all 12 equations. To the extreme right panel how self-similarity of nested clocks distributed over an entire network of imaginary layers is shown. The top left panel shows layered circles showing quaternions, octonions and dodecanions. Computation or decision-making in the brain means exchanges of information between different concentric circles, shown with arrows. Interaction between layers follows the tensors explained in Figure 4.13.

important, since, here, no logic works, geometric self-similarity or numerical self-similarity is a feature that is truly fractal but absent in the pattern of primes. In fact since a pattern of primes never repeats itself, it is impossible to find a rule, making a decision. The projection from infinity is also like normalization that bridged singularity in the 1930s, but with a difference that it provides much more insight into the singularity before bridged using fractal seed geometry and neglecting the true delicacy of nature. Also, at the same time, here we know what we are losing, but in the conventional normalization it was an unknown process.

4.11 DIFFERENT KINDS OF SPIRALS IN NATURE

The origin of spirals come from hyperspace geometry. Here is a simple example of hyperspace equilibrium generating various spirals.

Prime number theorem implies that $P_n \sim n\ln n$, wherefrom it has been derived that if $\pi(n)$ is the number of primes less

or equal to n, then $\limsup_{n\to\infty}(P_{n+1}/P_n)^{\pi(n)} = e$ while the lower limit of the same function is 0. The total contribution of n primes $\sum_{i=1}^{n} C_{P_n}$, if infinite integer space is depicted as unity, the normalized density of primes $D_{P_n} = 1 - \sum_{i=1}^{n} C_{P_n}$, within a certain range, if there are r primes $P_s \leq P_n \leq P_{s+r}$, we get ΔD_{P_r}. If we consider paired primes, since $\lim_{r\to\infty}(1+1/n)^n \sim e$ we get $\lim_{n\to\infty}(P_{r+1}/P_r)^r \sim \lim_{r\to\infty}(1 + \Delta D_{P_r}/P_r)^r \sim \lim_{r\to\infty}(1 + 1/\varphi)^r \sim e$. Here φ is the golden ratio, i.e., if a and b are two primary linear distances, they self-assemble to determine the next length as a+b, then $\varphi = (a+b)/a$.

Wherein $(P_n + P_{n+1})/P_n \sim (a+b)/a \sim j$ across N is a logarithmic variation, i.e., we get e along the horizontal. Moreover, if we implement the planar angular drift explained earlier, then the projected normalized hyperspace $(1:\varphi)$ is a golden ellipse with an area π. Projected hyperspace by φ and e for any dimension is the sum of their generic power, hence we get $\varphi^{Jh} + e^{Jh}$, now since the projected hyperspace of φ and e is π (π measures points in H1; see Figure 4.15a) in one dimension and ($p\gamma = l\pi$, i.e., total points = total lines; i.e., total area = total lines = total points), therefore, sum of their

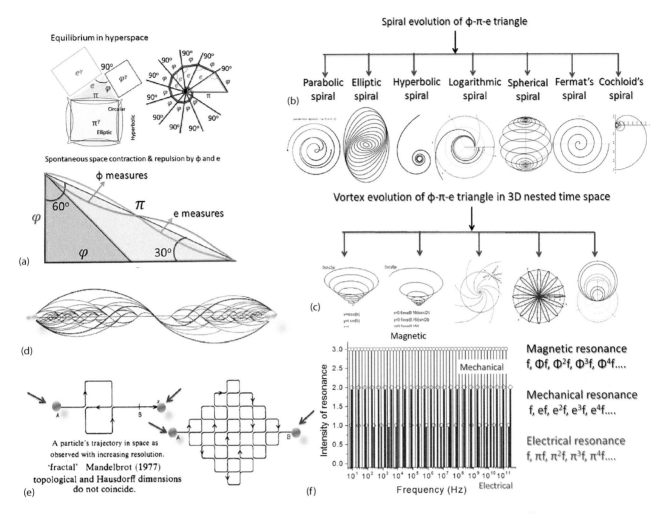

FIGURE 4.15   (a) e-pi-phi forms an orthogonal triangular relationship. This relationship governs every single step of a spiral formation. (b) Seven distinct 2D classes of spiral formation are shown in a chart. (c) 3D vortices of five different classes. (d) nested waveforms form the space-time geometry in the multinion world of information processing. (e) Fractal space-time concept. (f) How the resonance band for magnetic, mechanical and electrical would look like.

orthogonally coupled hyperspace is the hyperspace created by π of the same order $\varphi^{Jh} + e^{Jh} = \pi^{Jh}$. Now we try to find the value of Jh, we can easily estimate that $Jh > 2$ (for orthogonal triangle $a^2 + b^2 = c^2$, hence $Jh(\min) = 2$. The orthogonal relationship suggests that ideally angular deviation should be 30°, 60°, and 90° following the ratio $\varphi : e$ but φ is a function of $(P_n + P_n + _1)/P_n$, so is e.

First, φ and e are orthogonally connected to hyperspace, $\sum_{r=1}^{\infty} 1 + (k/e)^r = \varphi$; $k(e) = 1.05$, and $\sum_{r=1}^{\infty} 1 + (k/\varphi)^r = e$; $k(\varphi) = 1.03$, so it proves that they are orthogonally coupled, but the hyperspace is not planar $\left(k(e) - k(\varphi) = \Delta Jh \neq 0 \right)$. We vary the $(P_n + P_{n+1}) / P_n \sim (a+b)/a \sim j$ plot. Note that RNP(Δn_r) generates various logarithmic variations. It shows that the projected hyperspace could be spherical, ellipsoidal and hyperbolic, depending on the non-linear oscillation frequency matching between two neighboring closed loops in the NC/OF metric. The deviation in the projected hyperspace is strictly a function of deviation in orthogonality, ΔJh; hence $Jh = Jh(\min) + \Delta Jh$. Here, its 2.00001 (an infinite series). In order to verify, we assign φ (AB) and e (BC) two perpendicular sides of a triangle

ABC, where angle/ABC = 90°, while AC is π. We find that if $Jh = 2.02\ldots$ The angle between π and e measured by e is 29.62°, the same angle measured by φ is 31.28°. The angle between π and φ as measured by e is 60.42°, and measured by φ is 58.75°. The deviation of ±1.6° is from non-planar projected hyperspace. ABC relates φ, e and π as $e^{2.0} + \varphi^{2.0} = \pi^{2.0}$. Second, by continued fraction expression for φ and e are a linear function of N as $N \to \infty$, while π infinite continued fraction needs a square of integers $\pi = f(N^2)$, while $\pi^2 = f(N)$, hence π^2 could integrate both φ and e as the sum of areas. We estimated π accurately from $e^{2.0} + \varphi^{2.0} = \pi^{2.0}$. Once the triangle with hyperspace twists the Jh value, at least seven distinct 2D spirals (see Figure 4.15b) and five 3D spirals (Figure 4.15c). When these topologies are projected in the fractal space-time of nested clocks (Figure 4.15d) or in the fractal space-time of nested regular grid pathways (Figure 4.15e) we derive the resonance bands.

In the resonance bands of Figure 4.15f, we see that electrical resonance frequencies are $f_{e0}, \pi f_{e0}, \pi^2 f_{e0}, \pi^3 f_{e0} \ldots$, mechanical resonance frequencies are $f_{m0}, ef_{m0}, e^2 f_{m0}, e^3 f_{m0} \ldots$, magnetic

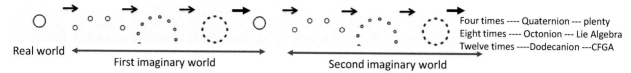

Real world

First imaginary world

Second imaginary world

Four times ---- Quaternion --- plenty
Eight times ---- Octonion --- Lie Algebra
Twelve times ----Dodecanion ---CFGA

Dodecanion, d is written in the form (d is the decision state made by our artificial brain), h_0 is real, d has 12 tuples

(a) $d = d_0 h_0 + d_1 h_1 + d_2 h_2 + d_3 h_3 + d_4 h_4 + d_5 h_5 + d_6 h_6 + d_7 h_7 + d_8 h_8 + d_9 h_9 + d_{10} h_{10} + d_{11} h_{11}$

General algebra	Continued fraction geometric algebra, CFGA
Dimension or number of imaginary worlds are fixed, for example, octonion has 8 Tuples, 8 dimension. No one explored interaction between universes of different dimensions.	It is a composition of multi dimensional universes. 12 tuples, 12 dimension, 12 worlds or dodecanion, then octonion and 4 tuples, 4 dimensional quaternions or 4 worlds etc.
Works on a linear number system, 1, 2, 3...	Each number has a geometric shape, 8=361=point; 12=18=28=triangle, composition of geometries is analogous to number system
Within a universe, say, dodecanion universe, higher level interactions between imaginary worlds were not allowed.	Allowed. For dodecanion 3x(12C2x12C2x12C3) = 66x66x220 distinct triangles; 8C2x8C2x8C2 = 28^3 distinct single point composition.
Fano plane, Jacobi identity, nothing governs.	Fano sphere, $e^2 + Ø^2 = \pi^2$ identity, PPM governs
Equation is used, sometimes syncopatic or symbolic instead of rhetoric algebra.	Question/answer are all 3D structure of circles or geometric shape, one draws circle to do all.
Fractal or non-differentiability is avoided	Fractal or non-differentiability is essential
Abstraction of variable. Dynamics, unknown values are obtained	Abstraction of symmetry, link symmetries forming topology, solutions are knots, loops.
Reduce, transform, differentiate, integrate.	Topological analogues exist
Axioms, idioms, proofs are basic practices to explore reductionism, abstract operations.	Symmetry breaking rules are geometric, cant be imagined, density of primes regulate them
Projection of infinite series have no effect	Geometric projection sends feedback to input

Fractal mechanics	Geometric algebra
Insert in pixel is measurement	Same operation is multiplication
Hilbert space has sub-regions	Functions of distinct variables
Planck constant like multiple actions active	Actions help in higher derivative, higher power
Multiple condensations	Reduction
Quaternion, octonion, dodecanions	Transfer of power to the indices
Fractal interference	Complement for subtraction
Fractal beating	Addition
Return of entanglement or coupling	Finding fractal seeds naturally

(b)

FIGURE 4.16 (a) A dodecanion tensor means a directional journey within or above 12 layers, each imaginary world value is represented with a new distinct tuple. (b) Two tables. The table in the left demonstrates a comparison between the existing geometric algebra and continued fraction geometric algebra. Table to the right represents one to one correspondence between the fractal mechanics and the geometric algebra.

resonance frequencies are $f_{g0}, \phi f_{g0}, \phi^2 f_{g0}, \phi^3 f_{g0} ....$ In experiment such integrated coevolution of resonance frequencies does not occur ideally. However, we get nearly quadratic resonance regulation in the system. This is called $e - \pi - \phi$ quadratic driver.

4.12 A MARRIAGE BETWEEN FRACTAL MECHANICS AND GEOMETRIC ALGEBRA

Materials self-assembly = mathematical operation = implementation of the rules of fractal mechanics: The established route of a fractal space-time adds a physical reason for connecting the neighbors (Reddy et al., 2018). It is noted that as one moves between different imaginary worlds, the interaction between different worlds, one affecting the other, in a unique way. It is not about the imaginary worlds in an isolated manner, CFGA is exclusively designed so that materials could implement the algebra during its self-assembly process, and address the interaction of the worlds intimately. Materials self-assembly = mathematical geometric operation. And finally, no result is confined in one imaginary world alone. The critical differences are listed in the table of Figure 4.16. The funny comparison of fractal mechanics and geometric algebra: The bottom right table in Figure 4.16b shows a few funny situations where CFGA implements the rules of fractal mechanics. The coolest part of this table is that we see mathematical operation to change geometric shapes and the complex mechanical principles are naturally implemented. Thus, it is a more generic route toward implementing physical phenomena in the materials as conceived by Bandyopadhyay et al. (2010b, 2010c).

Rupam Drisyam lochanam drik | Tat drisyam drik tu manasam
Drisya Dhibrittaya sakshi | drigye na tu drisyate

Form (*Rupam*) is seen (*drisyam*) by eye (*lochan*), the seer; that (*tat*) eye is seen by the mind-seer (*manasam*), the modifications of the mind (*Dhibrittaya*) are seen by the eternal seer, the atman, the eternal sakshi; the atman (self, *drigye*) is always the seer but never the seen.

Panch rang nirkhe tat sara | Chamak bijli chandra nihara | Fora til ka dwara ho

The concentrator saw the five elementary colors (*pancha rang*) of the five fundamental elements that compose the universe, beheld the flashes of lightening of the moon (chandra), and then, forced upon the third eye (*fora*), viz., the portal of the universe (*til ka dwar*).

5 Big Data in the Garden of Gardens (GOG)—Universal Time Crystal

5.1 GÖDEL'S INCOMPLETENESS THEOREM AND THE FRACTAL TAPE

The grand old photon: We know that a photon keeps time by traveling with the velocity of light. When a photon comes to our eyes from a distant galaxy, for that photon there is no time lapse in the travel. It touches the surface of the star and our retina cells and everything in the universe at the same time. Since the time of the Big Bang, the universe might have expanded for our frame of reference but for a photon it has remained static. The photon still lives in the time of the Big Bang. The concept of time lapse starts when we come out of a photon's frame. To describe several events occurring simultaneously, we are forced to state each step, one after another, just because our machine reads only one bit at a time. The architecture of time or time crystal is a many fold temporal correlation between any two bits of information; time moves in many directions. Now, "loss in parallelism" due to sequentialization needs to be avoided, but managing the flow of time with phases at many layers is not an easy task. A single time crystal can have many system points at a time. The pairing of only two events by completely neglecting the possibility of coupling three or more events is an unrecoverable weakness that affects through the complete destruction of higher-level logic in a Turing tape system. Since a time crystal does not forcefully linearize a one-to-many and many-to-one network, it captures "simultaneity." In conventional computing science, we strictly confine ourselves within the concept of pairs, two elements interact at a time. The concept of higher-level logic encompassing a group of arguments links all contributing elements at a time. Then we do not have to pinpoint accurately where, exactly, our desired information is located.

The singularity dilemma: Imagine we have a technology to slow down or speed up photons so that we can control the flow of time (Panarella, 1987). Suppose now we are solving a math problem for a natural process that is slow, we speed up the computing time and get the solution faster. Standard computing problems are shown to truly benefit by manipulating time, e.g., via satisfiability (Schnorr, 1978), by checking the isomorphism in trivalent graphs (Galil et al., 1987); by generating the parallel hierarchies (Bloch, 1997); possibly by tuning an infinite time Turing machine (Hamkins and Lewis, 2000); etc. Close time like curve are loops where the system point returns to the same space-time location after a closed motion. In such loops, all pixels on the perimeter coexist at a time, it is naturally engineered to squeeze or expand time flow by changing its diameter (Van Stokum, 1937). It can solve hard problems (Brun, 2008) and bridge the gap between classical and quantum computing (Watrous and Aaronson, 2009). One very exciting thing about the quantum clock is its direction of time. Can we set it, and if we do, would it remain as a quantum clock? Where defining the length of a path is not legitimate, the perimeter of a closed-loop would coexist in many paths at a time (Abbott and Wise, 1981), the most reported quantum clocks use a hidden classical certainty, just like a quantum logic gate. Thenceforth, Reddy et al. have proposed a new kind of clock, whose pixels in the perimeter are also clocks, so, as we zoom the pixel, we go deeper and deeper into an endless chain of clocks. It is neither classical nor quantum. In standard time series of signals, a self-similar pattern is searched (Kantelhardt, 2011; Shlesinger, 1988) and termed as a fractal clock. In contrast, the shortest time of a fractal clock is a circle, if zoomed, it is a clock too, so, a piece of time anywhere in the network remains undefined, we get singularity everywhere. Since singularity is undefined, tons of differential equations are not useful at all, so does a fractal clock, consequently, we encounter a singularity dilemma.

When Darwin drives Gödel's incompleteness: Both truth and false does not exist: Turing's world of computing and Darwin's theory of evolution are two concepts that are apparently unrelated, but, stems from a similar fundamental ground "matter holds its complete properties" and binary true/false argument. Historically, both the philosophies have undergone multiple eclipses and reincarnation of Turing and we look beyond the completeness philosophy and binary arguments, where the logical statements are composed of multiple truths. Hence there is nothing as the truth and nothing as the false, to hold and process a matrix of truths with analog values, we need "symmetry + energy" duality, then, no arguments are isolated, everything is interconnected just like a many-body system. It leads to the philosophy of information fractal,— very different from the fractal holographic universe and other fractal models of the human brain and that of the universe. Information fractal means we do not care how the hardware looks like its resonance band is a fractal, also we emphasize on escape time fractal not the iterative function system. Here, due to escape time fractal tape network, a point carries infinite numbers of scales inside, but just by reading it one cannot explain the entire universe. Phase prime metric (PPM); geometric musical language (GML); and fourth-circuit element, Hinductor or H-based triangular computation has one argument, one energy packet, one Turing like tape, one clock, one rhythm, one oscillator that encompasses the entire universe, but if we enter inside, we find millions of that inside, the journey continues forever to the Plank scale. The zoom-journey inherits a fractal culture, but there is no absolute self-similarity at any scale, since primes do not repeat.

Gödel's incompleteness theorem: First theorem: Say we have a bag of arguments, now by combining all these arguments in all possible ways, we can generate several axioms, none of them would be new as outlined in the Figure 5.1a. Axiom means a primary theorem. To create truly any new axiom, we have to bring some arguments from outside the bag, if we don't bring new arguments from outside the bag, the newly produced argument is already there. Starting set of arguments is always incomplete to create anything new, if we bring anything from outside, that again changes the bag we start with. Second Theorem: If we have an axiom that includes all arguments inside the bag and some arguments that were brought from outside to make it new, then the axiom is incomplete.

Quin's *what there is*, Plato's *beard*, Bertrand Russell's *Theory of Descriptions* (RTD), and Wyman's *"unactualized possible"*: Plato's beard suggests that if we state something does not exist, the statement is itself a piece of evidence for its existence. By Quine's argument (Quine, 1980) Plato's beard argument does not hold good because no claim is made in the statement that it does not exist. The positive argument is an argument, the negative argument is not. Wyman adds that a statement that something does not exist means it cannot be found in reality as a spatio-temporal entity. Russel's theory of descriptions considers a statement as a point object and links it to all that could build a connection and then all that are not linked. That way, RTD splits a statement into three parts, "that," "everything," and "nothing." For PPM-GML-H triad (Phase prime metric-Geometric musical language-Hinductor), we find a set of symmetries associated with it, philosophically it means association with universal similarities of primes. Thus, triangular split suggested by RTD is applicable here with a twist "that" evolves as a composition of "everything" and "nothing" in the PPM. Two entities "everything" and "nothing" are "geometric shape" and "phase space," respectively, in the GML. Two entities are realized in the fourth-circuit element, Hinductor (H) as "knots of darkness" and "vortex of light." Together, the triad covers a tiny part of an infinite time domain of the universe (Figure 5.1b).

Evaluation of PPM with respect to the butterfly effect, chaotic and random or deterministic features: Chaotic system is absolutely predictable if we know the initial condition this is called the butterfly effect. Now, one can theoretically calculate the future of the system. But for us even if we do not know the initial condition, rather a part of the entire system, we can still get it. Like the weather, we cannot predict more than a few days because input conditions do change. But we can have a situation when we should be able to resolve this issue also. We are neither "chaotic (if known input, then tell output)," "random (even if we know the input, cannot predict)" nor "deterministic," it is "quasi-deterministic," always reaching toward determinism, but being an asymptotic function, determinism is never reached, like a fractal tape (Figure 5.1c). The same run would give different result in two consecutive runs, like a classic random system.

Self-referential systems with time as primordial category following Peirce: Thus far no attempts were made to build a technology that is a self-referential system. The biggest hurdle was the criterion that the system would deduce the axioms by itself and then it would be so original that for any deduction, it would only refer to itself. The demand is like, "we cannot find the Shangri-La in the Himalayas, unless we know where it is." PPM is theoretically the system that can deduce the axioms using geometric musical language, GML. Now, true processing of the PPM–GML system is possible only if we use the fourth-circuit element, Hinductor (H), and build hardware that reflects PPM. Thus, PPM-GML-H triangular system is a self-referential system. The time crystal architecture and the electromagnetic resonance band feed each other (Figure 5.1d). In this system, time is a primordial category free of any form of interaction from outside. Peirce suggested that this happens if closed

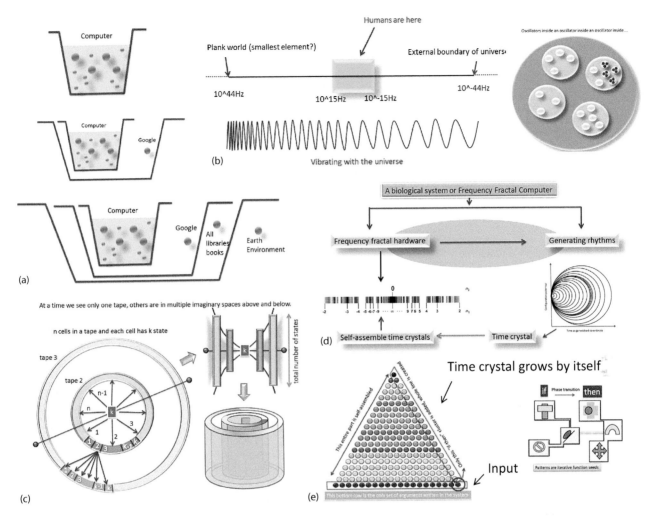

FIGURE 5.1 (a) Gödel's incompleteness theorem is explained in the light of Bertrand Russel's $1 + 1 = 2$ argument and debate. (b) "Within and above" concept is explained in terms of the electromagnetic resonance frequency band. The frequency domains are explained as subsets of a set. The extrapolated overview of the universe is schematically presented. (c) Fractal tape concept is demonstrated. (d) Fractal or self-similar resonance band and the nesting of clocks are demonstrated as a key to the artificial brain. (e) Time crystals present a network of geometric structures (corners = singularity points) self-assemble, always the target of self-assembly is to reach one clock.

loops are made of singularity points, which does not accept data or noise from outside (Peirce, 1940). Growth of PPM is primordial to every single entity, from the smallest elementary particles to be found around the Plank length and plank time several millennia from now to the extreme end of the universe, the growth is not natural but metrical, or we could state, as natural as metrical. Growth in the maximum integer of PPM in hardware is also a growth in the time crystal (Figure 5.1e).

From Heisenberg to Gödel via Chaitin: Both Heisenberg and Gödel set the limits and Chaitin connected that to our brain. Heisenberg said, if the time is short, we cannot understand the energy exchange accurately, while Gödel suggested that a consistent formal system that has just crossed the threshold size of doing simple arithmetic is incomplete. Since our brain is an infinite network of clocks operating one inside another, its energy exchange is undefined at all scales. Moreover, since it is also proved that a

hierarchical topology is possible to create a system with a few primes 2, 3, 5, 7, 11, and 13, it means the geometric mathematics regulates such a system. Even a tubulin protein-based microtubule nanowire is an incomplete system. Now, Chaitin has put a philosophical argument (Delahaye, 1989), if a finite set of degrees of freedom is given, can we create an indefinite number of outputs from that. In fact, PPM is attached to finite degrees of freedom, it could generate an infinite number of time crystals using a fractal tape (Figure 5.2a). Chaitin's query that argues for establishing that a brain is more than a computer has a unique solution in the PPM perspective. Just like several critical NP-complete problems could be solved by conditioning it (an ant can solve a traveling salesman problem), similarly, here brain=an assembly of finite degrees of freedom is a sufficient condition to trigger the nesting of clocks following the PPM. All objects reside in mind and no object exists apart from the mind (Berkeley, 1959).

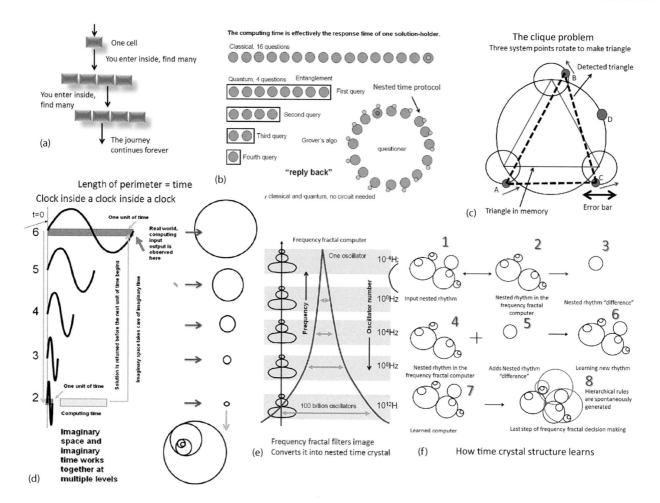

FIGURE 5.2 (a) Fractal tape operation is explained using a diagram, each cell contains a new cell with a large number of cells. An event's statement with four questions (who did it, at which condition did it, what it did, and how it did) is a single cell in this fractal tape. (b) A comparison between classical Turing tape, quantum Turing tape and fractal tape. Clubbing a large number of choices and addressing them as a single question is the ability to resolve a problem in the quantum realm. In the fractal tape, all cells are independent, follows, PPM and since all collectively follow PPM, no need to wire them, all could be communicated at once. (c) Identifying a typical pattern via clique problem is naturally solved by uniquely bypassing the problem using GML, an example is shown here for detecting a triangle. (d) Nested clocks following fractal tape architecture follow a multilayer clock network. At $t = 0$ the problem is encoded as an input at a particular time domain, the problem is sent to network of clocks, one inside another, the geometric relationship resolves the problem in the hierarchy of layers and the problem is solved. (e) The number of clocks used in implementing the fractal network of clocks in the panel (d) is presented in the panel (e). (f) Time crystals, during the learning process identifies the missing geometry of clocks, i.e., it compares naturally missing clocks or could add both time crystals as is.

5.1.1 A Marriage between Frequency Fractal Hardware and the Time Crystal

The eclipse of Newtonian thoughts in the 1960s: The days of scientific certainties were over in the 1960s, Newtonian clock-like universe concept turned out to be an illusion, with more and more examples coming up from nature, like the motion of clouds, the behavior of a flock of birds, etc. The unpredictability is definitely hardware. One of the fundamental requirements of hardware is a feedback of the same system and ability to generate superimposed output, individually looked into the system would provide the same bigger picture. It is not visible in the larger scale simply because the smaller part is cut off by the resolution we set, but how the self-similarity could be encoded as a generic

tool, so that anybody could program more and more self-similar patterns are not discovered yet. However, several such patterns have been found starting from Mandelbrot, $Z = Z^2 + C$, how this equation which is called God's own fingerprint interacts between two sides of equality as feedback is the most exciting part of an engineer who wishes to device and engine like God.

> "nature, at some extremely microscopic scale, operates exactly like discrete computer logic." "So I have often made a hypothesis that ultimately physics will not require a mathematical statement, that in the end the machinery will be revealed, and the laws will turn out to be simple, like the checkerboard with all its apparent complexities."
>
> **Feynman (1967)**
> *The Character of Physical Law.*

The hypothesis is further developed by Wolfram, Fredkin, Finkelstein, Minsky, Wheeler, Vichniac, and Margolus.

The discovery of a new kind of linguistic sensor: We build a sensor that rejects the use of Fast Fourier Transform (FFT) and generates time crystal from every single information. Here event = [subject-clause-(verb-adjective)] means, we search for event = [Slowest/static/supreme host-Condition/Key/Initiator-(What is done-how it is done/Fastest)]. We use a device, fourth-circuit element, Hinductor, H; which is a helix made of insulators that stores the charge to tweak the interference of reflected electromagnetic signal from its surface into a dynamic geometric shape made of the null signal, which acts like free magnetic particles. Conventional electronics follow current-voltage characteristics. Newly invented fourth-circuit element H does not exhibit any current-voltage characteristics, but, magnetic flux-charge characteristics. Lack of electron flow forces us not to call electronics, instead of real, what flows is a virtual particle. Since no signal is found in the free particle-like dynamic structures, so, it is said darkly. Such knots of darkness ($E = 0$ in the interfered electromagnetic wave), are not static, behave as free magnetic particles. Using GML if one could map the active and the inactive regions in the 3D spatial structure of interference patterns, or temporal evolution of the free particles, it retrieves a time crystal. We add two additional helical layers, one below and one above to perturb and generate nested clocks like an event = [subject-clause-(verb-adjective)].

Sahu et al. (2013a, 2013b) have shown that proteins, microtubules and several organic and inorganic structures generate the resonance bands, strictly associated with a time crystal where the symmetry of primes is explicitly evident. We derived the required geometric parameters of a helix or vortex to mimicking microtubule or cortical column or neural network in the brain and body. The objective was to replace all 12 brain components, that we studied accurately from human brain model with a circuit made of generic fourth-circuit element Hinductor that would be a universal time crystal generator as required by GML for fusion, extrapolation, reduction, transformation and morphogenesis of geometric shapes by PPM. Thus, by generating the knots of darkness or magnetic particles as a function of charge, we could build nested clocks that follow event = [subject-clause-(verb-adjective)], we realize the fourth-circuit element, Hinductor. One could physically emulate the role of primes or singularity points in a practical device, i.e., a 3D assembly of Hinductor. The marriage between Hinductor and PPM is analogous to the marriage between a Turing tape and a transistor.

5.1.2 PPM Allows Two Systems to Sync without Communication

What is an electromagnetic resonance: Do we promote the idea of electromagnetic resonance alone in this book? Electromagnetic resonance means that if we pump an ac signal, the molecule transmits the ac signal at a particular value, using conventional network analyzer or impedance analyzer one could measure these dielectric properties namely reflectance (S11) and transmittance (S21) and determine the spectrum of ac frequencies at which the values are maximum. Now, if the molecule has multiple embedded symmetries then for each symmetry, we do expect a new distinct frequency that makes it transparent. Several times, multiple peaks appear, shift, get active or silent together, as a result we could build a time crystal architecture. **Does the brain involve in wireless communication across the entire architecture, is it quantum or classical?** The answer is no. A brain does not have a giant antenna-like structure there is no high-powered radiation, transmission. In fact, even if we make the provision for massive power radiation, it will be completely absorbed in the immediate vicinity. The concept of resonance chain suggests that there is a common overlap among the resonance bands among all components that makes the brain jelly or the entire brain architecture. The process ensures band to band transition of energy just like that happens when we add one electron to a molecule, all atoms distribute a part of the electron among themselves.

5.2 THE ORIGIN OF FRACTAL RESOLUTION AND INSTANT REPLY

Scale-free dynamics is an essential outcome of Fractal tape network: It is established for the human brain that logarithm of frequency varies linearly with the logarithm of power expense in the brain. A logarithm is an important mathematical tool that tells us that the order of variation is important in a system, not a linear step-by-step increment. In other words, the parameters would take values 1, 10, 100, 10000, 100000 and the hardware is architectured in such a way that it considers the non-linear increment as the linear variation of 1, 2, 3, 4. The situation originates from a simple design protocol, "one unit communicates with 10 others at every level," or "one-to-many and many-to-one at a time, at every level." One very interesting aspect of this eight-level operation would be that "time to synchronization" or "computing speed" would be the same for all levels of operations, the larger sizes would require fewer complete wave patterns to synchronize; while, the smaller size clusters made of only a few neurons would require significantly large number of waveforms. Eventually, both the processes would finish synchronization at the same time. It is the "clock inside a clock inside a clock" feature that originates from a Fractal tape network in Chapter 2.

5.2.1 Spontaneous Reply—Search without Searching

In classical digital and quantum computing, if we want to find an answer to a question from 16 different persons, an algorithm needs to be written by a software engineer, wiring is necessary to connect components, and the question needs to be asked and processed one by one in classical computing or as a group in quantum computing (Figure 5.2b). In brain jelly computing the question is asked out loud for all to hear and the person with the right answer replies back (spontaneous reply-back). Also, it signifies that we do not have any wiring to address the switches one by one, which represents a radical shift from the existing world of computing.

Living life is resolving an intractable problem perpetually: Real-world problems are all similar to some classes of intractable problems for which we need solutions at every moment. It could be shown very easily. Suppose, we are passing through the road which is crowdy, we analyze completely different dynamics of every movement and determine the minimum collision paths. It is a traveling salesman problem. But if the path makes a pattern, and we need to detect the path, it would be a Clique problem (Figure 5.2c). When we are arranging books in a bookshelf which has different kinds of boxes and we have different sizes of books, this is famous "bin packing" intractable problem. If a number of shelves are less than 50, with computers we can do but if it is more then we can't, but a geometric artificial brain can do. Satisfiability: one kind of problem we often solve at every moment. Like taking a decision where a set of three arguments are compared with a set of 5 more arguments while 3 more arguments are valid at the same time. It is also an intractable problem and we do it always. Knapsack problem: while sharing biscuits to kids, we optimize often solving this intractable problem. Finally, Clique problem, identifying a typical pattern in a complex pattern is unique pattern recognition that we solve every moment.

The solution of the majority of the intractable problems is associated with constructing the linking between two or more distinct time crystal-clusters. Thus, irrespective of the nature of the intractable problem, the solution is a path connecting several time crystals. Unlike predecessors, the problem is dealt with a pattern of primes, i.e., PPM. The better algorithm means "clever rejection of choices" by minimizing search tasks. PPM uses clock inside a clock inside a clock network, a geometry that runs in the time domain of slower clocks could perform the same task in the faster time domain and solve it (Figure 5.2d). Computer scientists argue that the actual difficulty with the intractable problems is finding the accurate answer, with approximation "intractable" is never a problem. However, here PPM does not approximate solution, the system spontaneously replies back the solution, the system point of the computer does not have to search. For example, in a Clique problem that we target in the PPM computing, wherein, a small 3D pattern is identified in a complex 3D network of patterns, the similar pattern can resonate with the given input pattern as both are encoded in a set of oscillators. It is just like taking a magnet and finding a pin in the pile of the haystack, i.e., search without searching.

Therefore, here PPM has no agenda of bypassing the complexity using hidden approximation protocols. Since several true paths are spontaneously created beforehand and stored in the hardware, or created during computation, automated clustering of time crystals is encoded in the oscillators as a cyclic rhythm since a brain-jelly holds a wide number of problems. Resources increase with faster clocks (Figure 5.2e). Therefore, if we consider isolated single time crystals as rhythms then we have a basic layer of time crystals, at the bottom, then, considering the coupling of a few time crystals, we get another layer just above the bottom level, say, this is the level two. Several of these small-time crystal-clusters couples to form another set of groups, whom we could assign to level three (Figure 5.2f). In this way, several layers of time crystal clusters are formed already during the learning process, much before we attempt to solve a problem. During the search process even if a small part of the already existing path is triggered, however, due to the natural property of synchrony, the entire path gets activated.

5.2.2 Automated Error Correction through Time Crystal Learning

Singularity is not a single point; this is an undefined region. If two clocks with similar embedded time crystals synchronize, their corner points could match anywhere in their common time domain. Thus, the sum of the area of singularity domain is the error correction done by time crystals (Figure 5.2c). In the conventional self-assembly, bonds are made between participating materials, here, either A inserts inside a singularity point of B, or B inserts inside a singularity point of A. Thus, the possibility of error is negligible.

5.2.3 Synchronization of Time Crystals and Incomplete Problems

P or NP, that is the question: Is the debate unnecessary? Is the question tweaking words of the English language? Now, we have passed debating more than a century, "do we need more paper? (intractable, P or NP? P = polynomial, NP = non-polynomial)." If any person can prove P equals NP or it will never be equal then they can get 1 million USD. It is the first time, if we decide "yes," we are the winner, if we say "no" again, we are the winner (Figure 5.3a). Why it is not solved, simply because if we do not know how many more papers are needed (undecidable) then we cannot tell confidently NP could be converted to P or not. If anyone knows how many papers are required to complete the statement then NP = P. From the research of the last hundred years we know that P = NP is several cases (Figure 5.3b). But could there be a mathematical world where P = NP is never true?

If Russel's paradox sets in then we reach a world where P never equals to NP and if it does not, then NP would always be converted into P. Russel's blockade is like many body theorems in physics and we already exist in such a world. Therefore, we have a situation P never equals NP. Then why do we see some cases where P = NP? Those are illusions of nature, we view several fractal arrangements of papers side by side, so it appears as if it is a series of papers glued one after another just as the Turing suggested. Very soon we will argue below that this is never the case. Well, it is a game for the kid, make some sentence that forces us to collect more papers then we are incomplete (Gödel's incompleteness theorem), and if we write a statement so that how many more papers of arguments we need we don't know, we keep it as a variable,

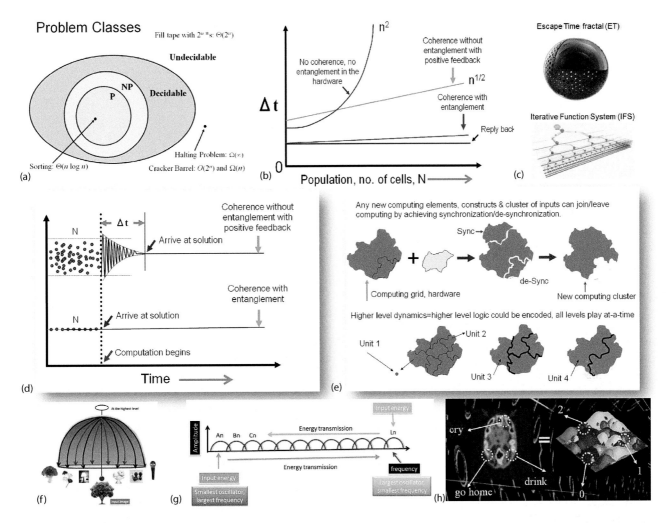

FIGURE 5.3 (a) Different problem classes, P and NP class problems, the intractability of a problem depends on defining a problem, little change in the condition shifts a problem from outside to inside, exchange the cell location. (b) The time taken to solve a problem (vertical axis) is plotted against the population number (horizontal axis) taking part in the solution. (c) Two types of fractals, escape time, where one has to enter inside a cell to find the repeated pattern, and iterative function system, IFS. (d) Two computing types are compared, fractal tape computing (above), where it takes a finite time to make a decision, quantum computing (below) where no time is taken, rather, there is a collapse. (e) Synchronization is a classical process and it links and de-links different time crystal domains during decision-making. A decision is a dodecanion written in the quaternion format that is compatible with the linguistic feature. Different imaginary worlds of a dodecanion exchanging information as a cluster of time crystals. (f) The umbrella concept of perception. If the geometric structures of different sensory systems are similar, or higher-level structures similar the decision-making achieves perception. (g) Resonance chain concept, where different time domains are connected linearly. (h) The fMRI images of the brain could be modeled as a 3D pattern evolving in the time domain (X, Y-axis) where the amplitude of oscillation leads to a higher dimension.

then it is undecidable. It is as simple as this (Figure 5.3c). So, completeness or incompleteness, decidable or undecidable are issues on the necessity of the papers while Russel's paradox is all about cutting a paper into multiple pieces. One example: Bipartite matrix has non-entangled "separable state" subset and entangled state subset, while multi-partite matrix has an additional "partially separable" subset, now, whether mixed states are separable or not, this is itself an NP-hard problem. Synchronization in a nested clock network of a clock inside clock inside a clock... ensures an exponential increase in sync time, because the PPM-GML-H triad acquires geometry in any time domain, syncs or desyncs them in the fastest time scale and finally brings results back in the desired time domain (Figure 5.3d and e).

5.2.4 Umbrella of Perception—Harvesting Infinity and Projecting from Infinity

If there is no classical point-like qubit, how does reality come from? Fractal information theory (FIT) considers that "there is a Turing tape inside every single cell of a Turing tape," this new tape now fails to become a Turing tape. Fractal word is very misleading, since self-similarity is in the PPM (Figure 3.4). Time crystals that spread over multiple imaginary worlds could reside on the cells of a fractal tape. So, the new fractal tape uses a basic information processing unit, a quaternion time crystal connected to each other, a quaternion that represents Turing tape inside a cell. So, we get a projected hyperspace from the three imaginary worlds, namely "reality sphere" that sphere is the only real thing in the information units time loop architecture. In Chapter 8, we describe that typically designed spirals and vortices generate a 12-hole phase sphere (Figures 8.4i, j and 8.6) that project the reality sphere, which is the real term in a quaternion, octonion and dodecanion. As explained in GML, a time loop has frequencies at certain time gaps to write the geometric shapes, that is the information content and all the three loops are imaginary, does not come from infinity but from the imaginary space. The projection is feedback like an umbrella that takes a pure geometric composition of the present to the future and takes it back (Figure 5.3f). The path that enables it to happen is a resonance chain, that a link of vibrations in the frequency domain (Figure 5.3g). Let's discuss the philosophy of infinity.

Gabriel's paradox: Gabriel suggested a trumpet that looks like a cone but it goes thinner and thinner to infinity, it has an infinite surface area, but when we try to paint it, we would require only a finite amount of paint, drop it from its mouth and it would pass through all the way. **One-to-one correspondence between the Gabriel's paradox and renormalization:** One-to-one correspondence between Gabriel's paradox and nested time cycle synchronization that is designed to replace renormalization, the surface area of the cone is infinity, however, when we put paint to color it, the paint passes through the tunnel and it blocks at the point depending on the density of the paint. The paint for us is the observer's nested time cycle. The resolution of the solution is always the ultimate.

Hilbert's infinite hotel: Hilbert suggested a hotel with an infinite number of rooms. The infinite number of guests could come in, yet there would always be a room available in the hotel. The same event could be represented by a circle, we can take out an infinite number of points from the perimeter of a circle, yet there would be no disjoint in a circle. A circle is finite only when we fix an area.

Dart on a dartboard: Suppose we have a pin, we throw it on the wall wherein a circle is drawn. What is the probability of touching a particular point, given that there is an infinite number of points in that area? If the probability is zero, then, the total probability of hitting the entire circle anywhere is zero, but that cannot happen. We need to consider a very small area, to get out of infinity paradox.

How many numbers between 0 and infinity, Cantor diagonalization: The answer is infinity, that so many numbers are there in between 0 and 1, even 0 and 0.00000000000000000000000001. Now, take any gap between any two rational numbers, we get an infinite number of numbers between them. $\{n\} = \{n^2\} = \ldots = \{n^n\}$

Hyper-Webster is a hypothetical dictionary of infinite space of all possible letters. The idea is the same as numbers, it has 26 chapters of one letter each. Then each of these chapters has 23 paragraphs and so on. Now, if we go on dividing the possibilities are infinite. The exciting part begins now, suppose the first chapter of the dictionary has all possible words beginning with A, then if one removes A, the rest is same as all possible chapters with all possible words. In fact

all 23 volumes are self-complete, if we print one of 23, we are done. **One-to-one correspondence between the Hyper-webster dictionary and the generation of multiple spheres:** If there are a large number of observers, of course there are an infinitely large number of observers, therefore a large number of "reality spheres" will be produced and those would merge just like the Hyper-Websters dictionary. All these spheres are distinctly producing, yet they create a singular identity. Imagine thousands of "reality spheres" being produced and all of them simultaneously hold the solution.

Banach-Tarski paradox: It is very interesting now. We create the same dictionary on a sphere. One could write infinite words on the surface in the infinitely available paths on the spherical surface. Even if we start walking from any one point we could travel covering every single point on the spherical surface following, Up, Down, Right, Left, UR, RL, UL and DR. Now, if the lattice on the sphere could be a hexagonal or rectangular, with six and eight neighbors respectively, we could have three and four equivalent roads respectively. There would always be two poles and hence, two identical spheres could be created just from one sphere if we decide to walk on the routes following a given protocol. Whenever there is a case of the real number of points, by Banach-Tarski argument one could create $1 = 1 + 1$, when the number of points is infinite. It violates the proof of Bertrand Russell. **One-to-one correspondence between the Banach-Tarski paradox and the measurement in an FIT:** Suppose we have a nested time cycle sphere, or a time crystal, the observer would also be a sphere, it would sync with the paths and generate the connected path and form a "reality sphere" just like the one here. Creation of an identical information replica is its greatness because the solution of a problem is an infinite space.

Apollonius fractal world: On the sphere we can start another journey. Either on the disc surface alone with three circles like the one below, or start with four spheres making a sphere Apollony fractal. In 3D one starts with four spheres each at the vertices of a tetrahedron. An outer imaginary sphere is the boundary cavity that encloses the four spheres. The gaps are filled by subsequent Soddy spheres thus forming a solid object (at infinity). The result is often called an Apollonian sphere packing. The fractal dimension of the 3D Apollonian has been calculated as 2.473946 [M. Borkovec, W. De Paris, and R. Peikert]. Kravchenko Alexei and Mekhontsev Dmitriy have found an attractor that creates the Apollony fractal. It is the union of three functions: $f1(z) = f(z); f2(z) = 0.5(-1 + si)/f(z); f3(z) = 0.5(-1 - si)/f(z)$; where z is complex, $s = sqrt(3)$, and $f(z) = 3/(1 + s - z) - (1 + s)/(2 + s)$.

Kissing 12 times is enough to send a message, I love you: The largest number of unit circles which can touch a given unit circle is six. For spheres, the maximum number is 12. Newton considered this question long before proof was published in 1874. The maximum number of hyperspheres that can touch another in n dimensions is the so-called kissing number. It is shown that optimality for sphere packing is $n = 4$. The history of packing problem is here. One could have fun by packing a given area with circles here, here in GML one uses spheres to cover a 3D space. There are three imaginary worlds that construct the fundamental unit of information for the geometric brain's information structure. We see above that the three imaginary functions if they are different then they make an imaginary sphere. Though the circle grows with three imaginary worlds, in the Appolony fractal, the sphere grows optimally with four spheres. For an external observer, a sphere always looks like as if it is made of three circles. Thus, the apparent mismatch between the infinity spheres and circles are resolved.

Harnessing infinity in the sphere and modeling the human brain: A rapid continuous change in the brain's physical structure makes brain modeling a fatal task. Oversimplification is something PPM-GML-H triad wants to avoid. For example, conformal mapping of the brain is done on a sphere (Hurdal et al., 1999). The idea is similar to those who want to map the brain in a hypothetical structure, but a prime driven brain does not. The FIT suggests assembling the time cycles on the sphere. Compared to Hurdal suggested conformal flattening (Figure 5.3h) a brain of primes do not change the physical shape of any part of the brain, for the theory, any change in the structure means a change in the information content. We are bound to keep its purity intact, and then integrate tiny pieces of information in it. Above we have discussed that a single sphere holds an infinite number of paths and we could make a singular mathematical structure of the sphere inside representing the entire human brain. Each of the circle with an independent center could represent the time cycle network representing various organs and their biological rhythms. Using such a sphere is beautiful because we get only one unified system to hold infinite possible time cycles. That is a highly generous statement.

5.3 REPLACING FAST FOURIER TRANSFORM BY TIME CRYSTAL TRANSFORM

The culture of harvesting time crystal from a stream of signals is far more advanced than the culture of reading signals as Fast Fourier Transform (Figure 5.4). When time crystal sends a signal out, it builds new wave streams for each system point in the time crystal architecture. Not just that, each stream contains a sequence of waveforms of a very different nature. In the Fourier analysis we look at the frequency peaks. The peaks are not related. In an architecture of clocks, the relative phase differences between the system points are related, the patterns created by a large number of simultaneously propagating waves reveal the projected topology of a time crystal. Figure 5.5a shows three cases of a triplet of clocks. First, three clocks are nested one inside another. Second, the three clocks are one top of another. The third clock is one connected to the outer boundary of another. One could notice that even if the relative diameters change, the output would be significantly different. The three different classes of nesting of time crystals coexist in a real scenario (Figure 5.5b).

5.3.1 IMAGE PROCESSING AND SOUND ANALYSIS USING A TIME CRYSTAL

Fusion and fission of frequency fractals: Mathematically synchronization means, several iterative fractal seeds, i.e., basic geometric shapes that undergo hierarchical integration (Figure 5.5c) get fussed. De-synchronization means several such seeds get disconnected. In the fractal theory three types of fusion and fission of fractal seeds have been developed.

Type I: Suppose we are looking at a tree, then, the entire tree could be made of a square and a rectangle put together as a seed of the fractal and then by copying this geometric shape several times and then by rotating and connecting with it in very different ways, we can reproduce the entire tree. By combining and rotating basic structures, open and close versions of a triangle, square to all polygons including a circle or curve, or straight line, every single structure found in nature could be created.

Type II: From a basic straight line all primary structures like a triangle, square, any type of polygons could be created using a simple fractal relationship. Therefore, the elementary filters for a complex pattern need not be created specially and stored separately in the hardware, a generic frequency fractal generates all possible polygons from a straight line to circle, and all patterns co-exist. Any modification to this fractal stores the nested rhythms in the form of new fractals and again all possible patterns of those newly stored fractals co-exist. Co-existence physically means a change in the F and G co-ordinates of the 2D frequency pattern to create a superposition of all images, just like several traveling paths of electrons around the nucleus generates a diffused orbital perception.

Time crystal transform, TCT	Frequency Fourier transform, FFT
Periodicity in single wave stream	Composition yes, by looping no
Relate parallel streams	Cannot see symmetry similarity
Relative phase for system point identification	System point that governs dynamics cannot be sensed
Modulated & composition of linear stream differ in topology	Cannot find anything beyond a frequency value
Search mechanism for distinct groups of pulses integral part	Never looks for repeating patterns, so no need.
Amplitude, phase, duration, frequency are geometric shape	Sense frequency & amplitude, no temporal geometry
Spontaneously create search modes to find symmetry.	Cannot find what is missing in a captured data
Topological features of fractal mechanics is searched	Hierarchical topology sensing ability with shape change is not feasible
Sense complex argument	Cannot associate events
Dimension of data: 10 D	1 Dimensional data
Parallel channels, simultaneous, instantaneous readings	Cannot sense simultaneously emerging pattern in million channels
Integrates distinct sensors	No universal language, code needed

FIGURE 5.4 Input information is not processed by Fast Fourier Transform, rather, time crystal transform. These two processes are fundamentally different. The difference between the two processes is listed here in the table.

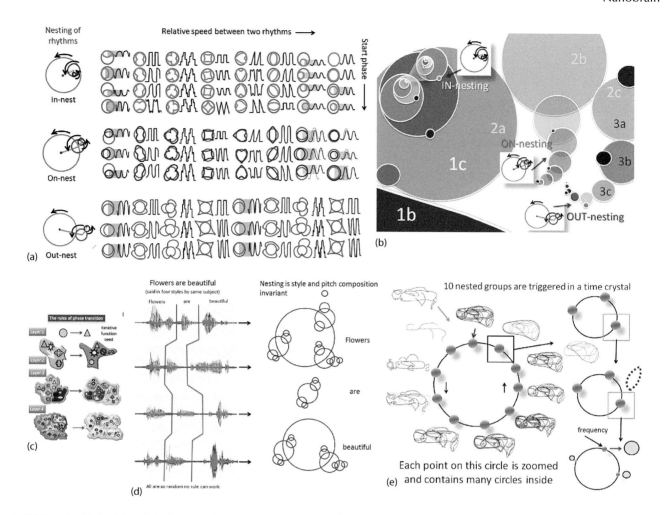

FIGURE 5.5 (a) A triplet of clocks nested in three different ways (in, on and out) are listed here in three rows. (b) Time crystal representation of the in, on, and out assembly of clocks. (c) The geometric shape change is the decision-making, in some cases, it is local change in the geometry, in some cases it is collection of geometries at various layers one above another that undergoes change. (d) Real-time sound data looks very different when you say the three words, "flowers are beautiful." Using an arrow, the words are filtered and isolated. The corresponding time crystals are shown to its right. (e) Image processing is shown by which an image is converted into a set of clocks.

Type III: Several type-I fractals when evolving with time in the 2D frequency space, at the high-frequency layer where we can see the evolution of the frequency fractal A in a large number of pixels, the entire pattern might appear as if it is a simple straight line or curve. Now, at this situation, if another fractal B evolves similarly in the same frequency space with typical common points so that A and B together appears as if a single circle or rectangle, then type II fractal may be born. If AB fractal is born which starts evolving together at all frequency space with AB as their seed, we might get the birth of a new thought that never existed earlier. Similarly, several groups in multiple different regions of the same hardware might spontaneously get coupled just like AB, due to similarities in the dynamic evolution then a higher-level perception fractal is born. These two types of fractals are called type III fractal.

We have discussed two examples. One for sound and the other for visuals, how time crystals are born and hold a nested geometry is shown in the Figure 5.5d and e. In Figure 5.5d, "flowers are beautiful" is said in three different ways, still, the time crystals are identical. Thus, the

time crystal language has an inherent universality. Another example of universal information processing by the GML is spontaneously splitting the cartoon elerat, elephant + rat (Figure 5.5e).

5.4 TEN SITUATIONS WHEN THE TURING MACHINE FAILS BUT FRACTAL MACHINES HOLD ON

Turing machine is universal in principle, it emulates only that human emulates, it is now listed what computers cannot do (Hubert, 1972). The 99.99% coverage of all symmetries by a PPM is not complete, but better than corrupting the nature with human bias. Fractal machines are predicted to outperform Turing machines (Dubois and Resconi, 1994). Here in this book we explore the possibilities of hyper-recursive-ness, hyper-incursive-ness, which existing fractal tapes never explored. These big words simply mean nesting of periodicity in multiple imaginary layers. Nesting would endorse the very popular proposals of the fractal holographic brain (Dubois, 1992). Some say there is communication in the quantum entanglement some says not, some measures even the time required by a quantum state to change (Pfeifer, 1993). We summarize here, how Ghosh et al.'s fractal tape (2014a) could never be emulated by existing Turing tapes, be it classical or quantum.

First, when one converts a set of ratios of the arms of geometric shapes, into a set of geometries using the PPM; a particular pattern repeats (Figure 4.14). This pattern is not the accurate description of the evolving pattern, but a similarity that cannot be described using an algorithm. It is a perception. Imagine a triangle repeats infinite times

over the PPM. But, its corner points always have a different geometry, so, there is no rule (Figure 5.6a).

Second, fractal tape machine, FTM follows $e - \pi - \phi$ dynamics, it means electric, magnetic and mechanical resonances follow a particular spiral dynamic. FTM database does not deal with bits or facts, it deals with singularity or confusions. A user finds many confusions, either put them together, or enters inside each one of them and the journey continues, until the user finds facts. Thus, for a single problem, a user would create a few distinct architectures of confusions, whose branches end with facts. Each of this architecture has a

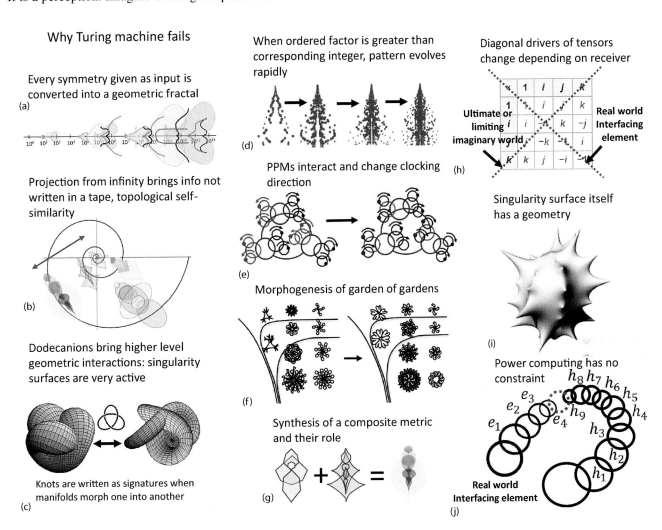

FIGURE 5.6 Ten panels schematically explain why Turing tape cannot simulate a fractal tape that operates based on a PPM. (a) A set of symmetry (not geometric shape made of singularity points, but geometric shape created by broken symmetry conditions) when fed into a PPM changes its shape and starts acquiring complex geometric shapes, no state is defined strictly. (b) e, pi and phi relation bonds several self-similar geometries produced by PPM engine, these are never encoded, but depending on input compositions change. (c) As described in Chapter 2, that dodecanion manifolds are thoughts of artificial brain. Two such manifolds changes are noted to suggest how manifold changes the locations of singularity points. (d) An ordered factor of an integer when greater than the value of the integer itself, the pattern changes are shown in four steps as the integer N increases rapidly from 500 to 500,000. (e) A different set of inputs could build their own distinct PPM, two such distinct PPMs interact, en route making a decision. (f) Garden of garden made of robust time crystals link by isochrones, such networks also activate if the geometric similarities are found. (g) Two activated PPMs could combine and build a new fused metric, that would set the grammar to evolve symmetries as a function of input variables N. (h) Compositions of multinions ranging from 2×2, 4×4, 8×8 to 12×12 matrices have diagonals, each tensor has two orthogonal diagonal drivers, using red lines one extreme real and one extreme imaginary world made drivers are shown. (i) Singularity is undefined, but its boundaries have a topology. (j) Interaction between an octonion and a dodecanion network could bridge events at various temporal junction points, one such path is shown.

distinct PPM and to bond them we put all in a spiral depending on the nature of clocks. This, is again an intractable problem (Figure 5.6b).

Third, in this book we discuss only 12 imaginary worlds operating together, so in the fractal tape, one could enter inside a cell 11 times. The processing of interaction between different imaginary worlds is done with a dodecanion matrix. Now, dodecanion matrix builds a manifold, using a paper and glue, multiple such manifolds are created (Figure 4.13). When a manifold is created, knots are written as signatures. These signatures change due to the interaction of the imaginary worlds. Thus, the morphogenesis of dodecanion metric continues without making any changes in the real world (Figure 5.6c).

Fourth, there is a special kind of PPM, where an ordered factor is greater than the corresponding integer (Figure 3.7). The intricate patterns created on the metric surface never repeats and so fundamentally different than using a common geometric shape we cannot reduce it fractally (Figure 5.6d).

Fifth, the clocking directions do not stabilize, stabilization of a 3D architecture of clocks is temporary. Multiple conditions exist to rotate one clock in opposite directions (Figure 5.6e).

Sixth, Meander flower garden explains a time crystal architecture based self-operating machine (Figure 2.7c–e). Several clocks in a time crystal architecture go silent, for various topological constraints. When it happens, a flower changes from one form to another. When it happens, that particular symmetry is restricted in a system, then, the flowers or distinct time crystals of a meander flower garden undergo major changes. Such morphogenesis of machine is not possible to carry out in a Turing machine, since it happens in the phase space constraints.

Seventh, depending on which numbers are silent in the PPM, the pattern, changes (Figure 3.4). As said, when one tries to solve a practical problem, he finds where the confusion lies and ask quaternion questions (Figure 2.10b) continuously until the architecture of singularity or the architecture of confusion is built. However, each architecture evolves using their distinct metric, since the integers present therein change the metric. Now, when all confusion based metrics are put together into one architecture, there is a fusion of metrics and a new metric is born, that then drives decision-making (Figure 5.6g).

Eighth, if one looks at the tensor diagonal elements for quaternion, octonion and dodecanion (Figure 4.13), they are identical and unique. Note that in a fractal tape machine, FTM, the real world is not fixed, it could be anyone, when the observer sees a world that becomes a reality for it. At a time three observers could see three different worlds; those worlds are reality or diagonal elements for them. Thus, multiple diagonal drivers redefine reality (Figure 5.6h).

Ninth, singularity points are the corners of a geometric shape. Singularity in mathematics is a confusion in the decision-making process of fractal tape computing. The 3D architecture of confusion, is the geometry embedded in the time crystal. Note that entire architecture is a singularity point. A user starts searching for a solution from a singularity point or conceptually a confusion and go inside. Each singularity point has a 3D geometric shape which projects distinct information 360° (Figure 5.6i).

Tenth, two chains of tensor elements could exchange energy and information if they have a common geometric element (Figure 5.6j), such a condition is impossible to meet.

5.5 THE HARDWARE ARCHITECTURE OF AN ARTIFICIAL BRAIN

Figure 5.7 is a summary that outlines a proposed fractal tape machine-based architecture for decision-making. The chart is self-explanatory and we outline the speed of computing, that requires an additional explanation.

What that could never be achieved by increasing the speed incredibly: When a hundred pieces of time crystals interact at a time, no sequential set of events could describe it, or could convert the events in an algorithm format. Therefore, one-to-many at a time interaction is something that could never be achieved by speeding up, and this is what makes any simultaneity based computing model like PPM-GML-H triad unique and far-advanced that the parallel computing.

Majority of computational algorithms developed in the last three decades have considered that the devices that hold the optional solutions of a query could listen to the question but could not reply-back to the questioner simultaneously and spontaneously. Thus, to learn the location for addressing each option specifically, the wiring of computing elements became necessary in the computer chips.

The circuit is a liability even in a quantum computer: The associative matrix. Additionally, we need a program that coordinates the process for a system point to reach to the individuals and retrieve the replies one by one, several protocols are adopted to decrease the computing time. The quantum protocol only decreases the number of queries, with $Log_2 n$ advantage over classical, n is the size of the search space, but "reply back" requires an antenna and receiver attached to the memory elements and new kind of the identification code. Grover's algorithm suggests that due to entanglement any number of people in a group could be considered as one object/choice, if any classical route allows such group-test such that classical computing would match the quantum computing. Additionally, except for a few problems (factorization), the quantum protocol does not provide sufficient speedup. The reason for the exponential speedup is the sharing of associative matrix D, which requires a particular requirement in the nature of the problem. Moreover, for pattern search, the matrix D needs to be redefined for each network mode, hence entanglement needs to be broken, which would collapse the speedup. There will be no difference between a classical and a quantum search. The exponential speedup is not the prerogative of quantum entanglement; it could be realized in a purely classical system too (Jozsa and Linden, 2003). "Spontaneous reply back" supersedes the exponential

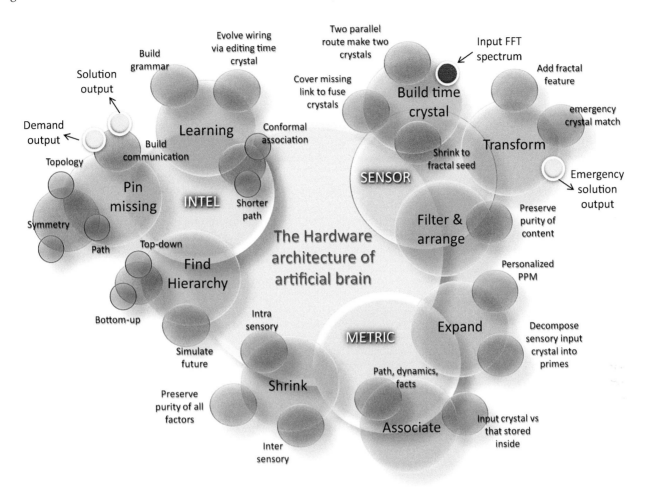

FIGURE 5.7 The hardware architecture of an artificial brain is shown in the pattern of a time crystal, since, instead of an algorithm, the system runs by nested clocks.

speedup promised by a quantum computer, associative matrix D is replaced by locally matched time crystals in a globally time crystal network.

Instead of $\text{Log}_2 n$ attempts, we want to make only one query, and the solution would reach the questioner, then the size of a search space becomes irrelevant. If we use a multinary switch (with more than two decisions) that has an antenna for each state in addition to the sensor, it can radiate out the solution in all directions, so irrespective of n, the questioner gets the answer in one attempt. Bandyopadhyay and Acharya (2008) have demonstrated this technology. Fundamentally, the basic information processing device will be an oscillator attached to an antenna and a receiver, the oscillator is so designed that we can write/erase multiple resonance states. When an electromagnetic signal is applied to the device, multiple resonant oscillator circuits absorb energies specific to the resonance frequencies and start oscillating. The oscillation turns the system unstable; as a result, the energy is radiated outside via non-radiative coupling, noise does not mix with signals as the absorption occurs only for signals with the perfect matching frequencies, at the same time, the emission is always quantized.

5.6 THERMAL BREATHING BY MICROTUBULE AND ARTIFICIAL BRAIN

The energy applied to produce, process and store time crystal is in the ultra-low energy domain, the devices harness available energy kT (k = Boltzmann's constant, T = temperature) in its environment in the kinetic energy of the surrounding gas or liquid molecules. The energy is transferred by a non-radiative process, therefore, there is no loss during transport, between different parts of the hardware, the processing and storage of information is done at the same place, therefore there is no time loss for retrieving data for processing at the same time, we save a large amount of energy. Information storage is done via rhythms of natural vibrations therefore like the silicon computers, refreshing the switches for every millisecond, is not required. Also, as we have described, a majority of nano sized biomaterials have absorption band in the THz regime (5–6 THz means around room temperature), hence those are not efficient machines at all, they consume huge amount of energy from thermal noise or available energy kT, we calculate only the calorie that we consume, this is not fair. Entire processing in the artificial brain takes place using the fluctuation of Brownian motion using typical archi-tecture, and/or chemical energy. The computer accepts energy from outside whenever an instruction is required to operate, otherwise it does not use any energy from an external source. Energy is spent through symmetry-transitions and that is gov-erned by speed limits at various scales. Importantly, from the thermal breathing model, if nature only optimizes the speed limits while constructing a system, symmetries are born, and energy is spent only to evolve circuits, it is translated to differ-ent simultaneously communicating channels.

Figure 5.8a describes the thermal breathing pathway, where, defects in the ordered lattice transmits through the structure as soliton and several solitons of similar nature and that of different kinds condense. Geometric processing means a change in the shape of a triangle stored by a time crystal by noise, until triangle changes into a different shape, the information content remains intact. Thermal breathing is made of two simultaneously operating engines as shown in

Figure 5.8b. Each class of solitons has its own distinct feed-back pathways by which they filter out noise and transmit signal. Note that each path has a distinct structural symmetry and an associated dynamic (Figure 5.8c). Sahu et al. (2013a, 2013b) carried out temperature variation of a single micro-tubule conductivity in the temperature range of 5–300 K. Figure 5.8d shows that the conductivity does not change with the increasing temperature. However, if one zooms the nearly linear variation of conductivity, it shows quantized jumps randomly, yet it maintains the linearity. Another interesting feature is that if one encodes a particular conducting state, then during temperature variation, microtubule holds that state. It means microtubule has broken valence and conduc-tion band, i.e., the edge states are not straight, rather curved and makes contact at various symmetries. As a result when those particular symmetries of microtubule are activated, then, microtubule transmits through particular contacts between the valence and the conduction band (Figure 5.8e).

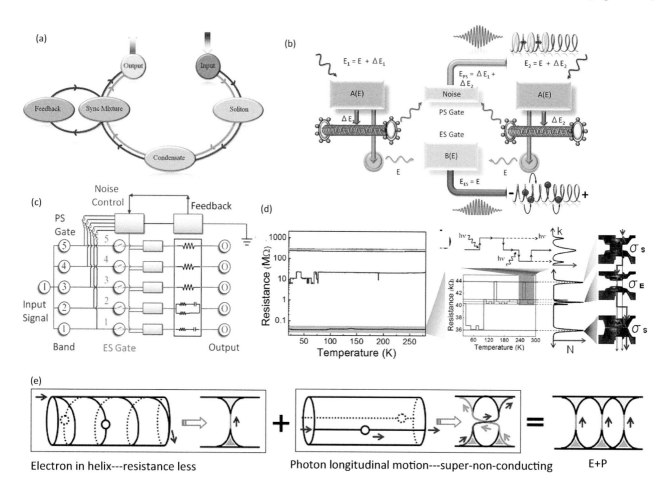

FIGURE 5.8 Thermal noise management of an artificial brain (a) Condensation of soliton pathway between input and output of informa-tion processing is shown. (b) Electronic ES and mechanical solitons PS are shown as a composition of two distinct transport routes. Two distinct channels are governed by two distinctly operating engines in the unit device that processes information in the device. (c) PS gates are the first ones that operate when the signals enter in the device. It is followed by electronic quasi-particle operating gates or ES gates. Five channels in series operate in the scheme with a feedback network. (d) The breathing of a single microtubule nanowire is demonstrated. The first plot is the change in conductivity as a function of temperature. A part of the plot is zoomed and it shows that the conductivity changes randomly but in the quantized energy levels. N number of point contacts are shown here. (e) When electrons vibrate remaining in the cells of a cylinder along helical path, point contact is shown. To its right phonon paths are shown which creates crossover (red and green). The combined effect is shown to the extreme right.

However, such unique energy management using dual channels is surprising. If there exists any unique and new physics it should exist everywhere, we select brain, since it is compact and its complexity is enormous, an ideal test system to verify rhythms of the spiral symmetries, because the entire universe would also be a spiral symmetry,—singing a song toward completeness.

5.7 LOTUS IN THE PRIMES—REVISITING THERMODYNAMICS OF GEOMETRY

Why entropy and probability came in the information theory: what if there is no communication?

"Information is the resolution of uncertainty," said Shannon. Quantifying information is dealt with information theory. When we copy and paste a file, it is a lossless communication, but mp3 coding is communication with a loss. We do both in our day-to-day operations. Shannon argued information as a purely quantitative measure of a communicating entity. Say, through a tube, we are sending water, now if it is purely random, then there is no effective communication, but if we pump the water, then it moves in a certain direction, but how much water can flow? It depends on how many water molecules we could send in a certain direction. That means from the degree of disordered motions, we are trying to get an ordered behavior. Thus, relate probability and entropy.

The origin of the concept "bits," entropy and channel capacity: Classical information theory, adds all the probability factors to understand the total information, say we have tossed a coin, then, $P = p1\log p1 + p2\log p2 = 1$, so we call it "bits," means one bit of information, and entropy is $\log P$, that is zero. What is the significance of zero? It means the information is transmitted with absolute certainty. Just by saying this we also make sure that the channel passes 1 bit that means in a full capacity, since we have considered all possible choices for tossing a coin, no one can send more information because that does not exist. For the same reason, 1 bit is the minimum bit that is required to send that event through a channel. Or a minimum number of bits required to encode a message without a loss. A basic example of entropy calculation is shown in Figure 5.9a. On a surface with the monolayer of molecules, one has to calculate the number of ordered arrangements, and then, take logarithm, the resultant value is entropy.

Shannon's information channel capacity is not like a water pipe: Here, we deal with an events internal probabilistic structure, not a real channels physical parameter. Hence, it is always decided in the emergence of an event. More is the restriction, or ordering or redundancy,

Fractal Information Theory (FIT): A new way of treating information entropy:

1. Data compression: Encoding always meant finding repetition or redundancy, and then removing them, thus decreasing the P-value. Data shrinking always means finding redundancy. In the FIT we do the same, but with a twist, we do not try to find the redundancy in the physical appearance or an isolated discrete property. We find a natural language of vibrations that is fundamental to nature. Based on that universal language we deconstruct the object and its environment, integrate them in the same language of the time loop and then we also integrate the observer. Then we create an event seed, not the fractal seed of geometric shapes. Even seeds may create a geometric fractal seed, that's not our concern. Thus, change in entropy is not so formal in FIT.

2. Shannon's source coding theorem, data compression: "A lossless data compression scheme cannot compress messages to have, on average, more than one bit of Shannon information per bit of an encoded message." Entropy is a lower limit of lossless data transmission.

3. Huffman coding (1952), data compression: It checks the frequency of occurrence of a particular choice and then assigns a length of memory for passing through channels or processing. It is just like Morse coding; the most frequent choices are assigned a shorter length memory space. Using this route, we can reduce the length of a book drastically compared to assigning equal length memory using each letter as 8-bit memory string. In an FIT, vibrations of the entire book are composed in just one geometric shape, just one, then we go inside the corners of the geometric shape and we get some more geometric shapes. In this way the journey continues, until we reach the singular letter, or fact, but the information is not a resolution of uncertainty. Arithmetic coding is better than Huffman coding.

4. Gödel's incompleteness theorem: If we have a pot then if we want to explain everything kept inside the pot, using the materials within, we cannot. We would need some reference from outside. We need bigger and bigger circles reaching us to infinity. Using the number system theory, we can suggest that incompleteness theory is not complete. Mysticism, religiousness associated with Gödel's incompleteness theorem must be uprooted and thrown out.

5. Escher's paradoxical truth and Huxley's perennial philosophy: We can only get a paradoxical truth. In the PPM, where a different selection of integers creates a new kind of metric (some examples are shown in Figure 3.4). For each of the 12 primes we observed the creation of 37 planes covering 360° in the Figure 3.2c. Each PPM created by a particular architecture of singularity is represented by a particular entropy symmetry plot, if we combine all, we get 3D rotation of all the planes (Figure 5.9a).

Why do we need a new kind of entropy-symmetry plot? The same information is now classified and at any given point of time different classes of information have several distinct features attached to them which are evolving in their own way. We encounter a similar situation whenever we concern

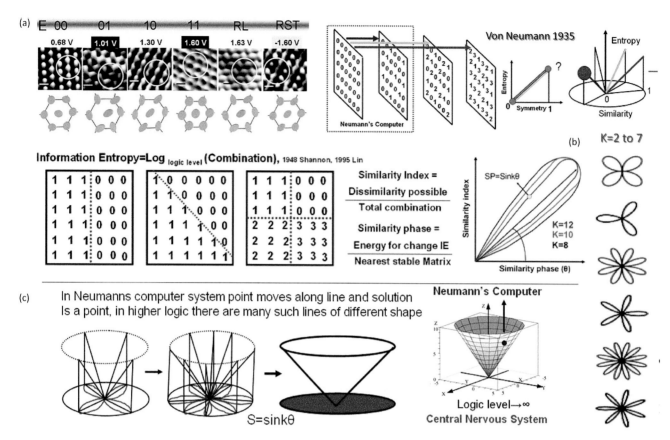

FIGURE 5.9 (a) 2D self-assembly of DDQ molecules which changes basic lattice arrangement for different states of information. The lattice arrangement is shown at the bottom of the panels (top left). To the right, A surface with all identical values has different "1," i.e., information entropy. One triangle needs to be rotated and a base circle is to be plotted. In the lower panel plots multiple 2D symmetries are adopted for which the symmetry-entropy line converts into a flower, similar to the meander flowers of the garden of gardens demonstrated in the Chapter 3. (b) The evolution of meander flowers is shown that replaces entropy-symmetry line. (c) The existing classical and quantum computers form just a single point in the composite 3D flower made patterns of entropy symmetry plot that evolves its flower petals following PPM.

ourselves with the multilayered time crystal architecture. The entropy symmetry expression should be plotted in a new way, one may consider the triangle formed by the linear entropy symmetry plot for a particular time crystal network as an object and rotate it by 360° keeping the zero point as fixed to the ground. The circle created on the ground due to this rotation represents infinite numbers of layers of the cellular automaton or time crystal architecture, and all linear plots of entropy-symmetry constitute a conical surface. In the quaternion, octonion, and dodecanion tensors of Figure 4.13, one could calculate similarity index and similarity phase as described in Figure 5.9a, using that geometric phase variation one could write a generalized petal equation for the universal entropy-symmetry plot (Figure 5.9b). If the number of petals in flower increases beyond length, the entropy symmetry plot converts into a cone (Figure 5.9c).

The entropy-symmetry plot for PPM computer is a conical funnel that changes with time: If we put a marble on this conical surface, it moves and fixes to a place, if we consider this marble as the system point, then it moves through different layers and fixes to a particular point of entropy-symmetry located in a particular floor or layer. It is a very important

functional feature of nano brain architecture. Note that as we move to higher layers, the number of cells that constitute the cellular automaton grid decreases, as a result, system point or the marble, carry out a remarkable task of selecting the level intelligence associated with the decision-making of the brain. It is a feature that cannot be realized using normal computational principles, for the reason that higher-level information processing has never been defined in terms of specifically defined parameters. The concept of multilayered cellular automaton and its associated molecular architecture together suggest a new kind of processing where computation is done in a hyperplane constituted by the continuously evolving pattern of symmetries, in sharp contrast to the classical, quantum, or un-conventional computing the information is never addressed directly.

We often wonder why do we need even to know how the universe got created, where is the end of this universe, if there is any. Even when we are studying what happens to the nanoscale system, it appears nonsense why should we learn what happens in a black hole. Limits of a universe can change every single "faith" we have about science. Even simple assumptions that we make could fundamentally redefine everything. One could generate several worlds of sciences by simply changing the basic assumptions, which we think has no impact on millions of concepts that we regularly encounter. For example, if we think universe has no upper or lower limits, i.e. it is not a closed system, then "conservation of energy," won't work and if we can create the world full of science considering that universe has fixed amount of mass and energy. We must give it a try, how the science would look like if we consider the worldview that the universe is an open system. A nice proposal is that the universe is a fusion of Escape time and Iterative Function System type fractals, where everything is undefined, universe is an architecture of singularity. And then second option to try would be if the universe is a leaking fractal cavity resonator, everything leaks.

What would happen to Lorentz covariance if we reject space, take only time? Lorentz covariance tells us that it means that the laws of physics stay the same for all observers that are moving with respect to one another with a uniform velocity. Motion through space has been the prime concern for the physicists thus far. For us, when we argue to reject "space" from "space-time," keep "time" alone as the singular variable parameter, we also create a version of Lorentz covariance for such a system. That said, it would mean, "the laws of physics would remain the same, when the system point of the observer would match with that of the nested cycle network under measurement."

Information entropy and symmetry: Researchers have studied the dynamic and static information entropy using a ternary molecular switch. The dynamic information exchange is studied in a 1D well showed the closed box features while gaseous diffusion of a dimer, trimer, tetramer on a gold surface showed a typical conical symmetry beyond the classical logarithmic relationship. A general relation between symmetry and entropy is developed for a higher level logic showed the possibility of divisions and creation of a sectional curvature on the symmetry surface. Starting from Shannon's entropy (1948) we have re-visited the debate on Gibb's paradox, in the perspective of the multilevel static and dynamic information exchange process. Furthermore, the universal relationship has also been verified in the organic monolayer consisting of ternary molecular switches. In summery the journey from binary to multilevel information exchange is reviewed in the light of the theoretical development has been made in the last half a century.

Information is a certainty, which is negative of uncertainty; uncertainty is entropy (energetic transformation), therefore entropy measures the lack of information of a system. Here information is a number ranging from 0 to 3 on a surface. In order to reach maximum entropy on a surface all molecules would randomly switch to attain a single logic state at equilibrium. Initially when each molecule is at separate positions with different logic states then they are distinguishable, the surface contains a large amount of information, symmetry is low. However, in later stages when most of the molecules have already switched to the same state then they are indistinguishable, information content decreases, symmetry is high. On a surface if we observe the evolution of information with time,

then the entropy asks whether the candidates have the same state or not, therefore the entropy is apparently insensitive to the level of information, i.e. binary and the ternary system would respond similarly. Therefore, we simply count the number of distinct symmetry circuits on the monolayer and take the logarithm of that to get the information content in terms of Nats.

In case of a dimer, trimer, tetramer, considering Lin's evaluation to the Pauling's argument to be valid, we checked the time evolution of symmetry at 77 K on a gold (111) surface. Initially a set of information is given to the set of molecules and the surface is scanned at 0.6 V bias using an STM to check the exchange of logical information thenceforth the matrix. The interesting point between two time frames is that the number of molecules is the same; distribution on the surface is changed to reach an energetic equilibrium. Two isolated quinone molecule of the different charge brought together forming unique dimers of incredible natural features like oscillating in a well periodically, rotating around each other like a binary star, jumping to intelligent positions, etc.

When they form a global network of molecules these features reflect in a unique way. Functional groups of quinone were tuned to find tricks to encode rules inside a molecule that might induce an incredible global transport of information essential for realizing a bio-processor or a parallel computer. The regular structure of CA allows assembly through molecular self-organization, and direct accurate measure of entropy or information contained in it. The small sizes of molecules combined with their availability in Avogadro-scale numbers promise a huge computational power, in which the massive parallelism inherent in CA can be effectively exploited. The distributed nature of CA has prompted comparisons with neural architecture based computers, one could estimate the entropy of a brain from the fMRI images during different kinds of activities, however, that cannot be related with the computation process. Thus, we feel that in the artificial brain, it will be possible to correlate entropy and information accurately, unlike the brain.

Thermodynamics of lotus: How the lotus unfolds and sings the music of cognition (E8 symmetry): Binary switches (0, 1) when replaced by a higher k level molecular switches (0, 1, 2, … $k-1$) in a quasi-closed system, depending on the number of stable phase or absolute similarity possible, different entropy-symmetry relationship is generated. It generates a 3D surface if k increases, and converges to an analog multi logic system when $k \rightarrow \infty$, the surface thermodynamically connects Neumann's computer to the conceptual machine like the central nervous system. Direct experimental evidence for an obvious relationship between information and thermodynamic entropy shows the non-existence of ~100-years-old mystery and myth associated with entropy. The experimental verification of the lotus pattern resembles the E8 symmetry of Lie Algebra. Recently lotus symmetry that encompasses the golden ratio (1.61) has been demonstrated experimentally suggesting that the magnetic resonance frequencies occur as ratios of golden number in the spectrum (Coldea et al., 2010). The E8 group squeezes all states—and local operators—into representations of the group. And they're quite big. There are not too many singlets (one-dimensional representation that never transforms). The fundamental representation (which is the same as adjoint for this particular group and no other) is 248-dimensional—it's the smallest one that does transform in some non-trivial way. The "next" representation is 3875-dimensional. For a lotus that blooms we get entire series starting from 248, 3875, etc. We do not endorse the theory of everything view on E8 or any singular symmetry, because PPM worldview requires the composition of symmetries and also that depends on which prime point the vortex started to grow. E8 means, the vectors of the root system are in eight-dimensional Euclidean space, for PPM metric as described in 2.6, all Lie group symmetries generate, the lotus is the superposition of all possible symmetry groups.

Historical background: Since 1875, three paradoxes have ruled several fields of physics, chemistry and biology, ranging from the origin of life to the information loss in a black hole. First paradox was Gibbs paradox which was created by equations that contradicted Gibbs law, second paradox was Maxwell's demon, where the role of the demon has always been a suspicious one, the third one is the entropy itself who played a major role in surviving the other two debates as the definition of entropy became a matter of personal choice for the last 130 years. Here, the point of interest is 1955s von Neumanns computer (N) and a million-year-old yet most powerful parallel computation machine (P), our central nervous system (CNS). N process data in terms of 0 and 1, while a neuron burst in a brain is conceptually multilevel (0, 1, 2, 3 ... n). There might have been a thermodynamic route to connect N with P, and if a generalized expression for k level logic could be developed then for $k \to \infty$ we could thermodynamically connect higher-level processors often observed in nature. Given the fact that processor software deals with information entropy and hardware deals with thermodynamic entropy, we cannot proceed further without solving nearly 50 years old mystery and myth associated with the term entropy.

The second law of thermodynamics, discovered in 1850, said that in a closed system all real physical process would consume energy and the system end up with a less amount than the starting energy. In 1865 when Clausius noticed that in an ideal gas mix, the ratio of heat exchanged to the absolute temperature is constant, he named it "entropy." In 1875 Gibbs showed that when all gas molecules have the same temperature, heat exchange is zero, therefore entropy is zero. If we plot entropy with similarity then entropy remains constant till all molecules come to the same state, then suddenly it falls to zero. Many equations contradicted this relation, and Gibbs paradox was born. It survived for more than 120 years because of two reasons; different kind of entropy came into being and still their relationship is not clear, secondly some of the very basic assumptions were not properly changed when different real situations were considered. Now considering old definitions of entropy and impossibility of the existence of pure gas, it has been shown that Gibbs paradox does not exist at all. However, entropy does not remain constant always, it depends on relative fraction of the mixed gases and therefore the only minimum

and maximum entropy remains same as Gibb's while variation in between two extremes depends on the system.

In late 1940, von Neumann and Shannon introduced information in entropy. More information is a certainty, certainty is negative of uncertainty, uncertainty is a disorder, and more disorder involves more energy exchange to reach an equilibrium that is more entropy. Obviously, the new information entropy (IE), negative of information is simply a number and different from the old entropy which is basically thermodynamic entropy (TE). Several hundreds of attempts have been made to correlate this two, separating them, joining them, even challenging the Gibbs paradox or misinterpreting the second law of thermodynamics to reach a conclusion. Mystery and myth about the origin of life have added to this hundred-year-old debate on real entropy as it continued for the Gibbs paradox case. Both schools of thought are well investigated in a plethora of models and systems, therefore following them classically might lead to the truth, but unless the philosophical ground for IE and TE is very clear, following the paths would be misleading.

The journey of negative entropy and information: The breaking adds one phrase to the second law of thermodynamics, that is, the increase in entropy is not continuous, sometimes it may remain constant and sometimes even may decrease, it is a very discrete and random behavior, on global scale entropy increases continuously. It is still a debate where life form or the emergence of intelligence always feeds on negative entropy (Von Stockar and Liu, 1999). Since information corresponds to a negative term in total entropy of the system (Brillouin, 1953), it could easily be modeled to add complexity in a system (Packel and Traub, 1987). However, these formulations have a singular view that the weight of entropy allowed to judge the suitability of a proper computational algorithm in a given scenario rejecting several major aspects, since homogeneous entropy increment concept is broken as said above, the whole idea comes under suspicion, maximum information co-efficient concept is an idea to resolve this problem via considering the specific features of an information (Reshef et al., 2011). The situation becomes very interesting in a quantum scenario, partial information content due to entanglement depends on the sign, positive or negative, and that could redefine the information in the sender and the receiver (Horodeck, 2005).

Basic concepts of information and symmetry: Information or similarity of a system is asked only at a particular time; therefore, it is a static state, count-number associated with the state of the elements provides the IE. However literally the word dynamic at a particular time does not make any sense and therefore any change occurring between two points of time is essential to express a dynamic behavior. Literature is full of contradictory relationships between symmetry and similarity. Similarity means alike in a certain parameter at a particular time, symmetry is the *sameness measure*, therefore in a static system symmetry=similarity. However, more dynamic is the system, more elements can some to the same state, similarity increases or decreases at any given time, but that does not measure *sameness*. The statistical average of similarity within a period of time would be the *sameness measure* or symmetry. Therefore, more dynamic=more symmetry. It is a direct and clear relationship used in this letter; any readers can calculate it for their own system.

The next point in which literature is full of confusion about the parts should be taken into account for the understanding of entropy of the system. Studying collisions of molecules should not involve vibrations of bonds just to provide explanations. It is well conceived that similar things repulse and opposites attract, but only that part of the system contributes, which has a significant effect on the length scale the entropy is measured. A similar charge, spin, mass, dipole, repulse when system elements are less than a critical distance but if they are not then they attract, therefore ignoring length scale it is very easy to survive entropy debate and often challenging the second law of thermodynamics. The system elements at less than the critical distance and far beyond cases would be considered by making a clear line of difference. At less than the critical distance more is the number of similar elements more is the sum of repulsive interaction energy, therefore more is the TE (TE=energy exchanged/absolute temperature). If statistically there is a high probability that the system elements always interact coming closer than the critical distance, then the above relationship is valid otherwise not. Interestingly, more is the similar elements in a system, it requires fewer numbers to express the information content, therefore when TE increases IE should decrease (IE = logarithm of apparent symmetry number). The simple logic shows the non-existence of entropy paradox; because both existing concepts are valid, contradictory results originate from opposite assumptions. Therefore, if proved experimentally then it would negate the existence of several hundreds of models developed to challenge one another in a different perspective. A simple and careful experiment has been carried out using molecules by Bandyopadhyay et al. in the brain jelly to check the validity of entropy debate in light of the above conclusions (Chapter 9).

In principle, since a system cannot return to the initial state of conformation or information in a closed environment and an ideal closed system cannot be built, therefore a pseudo-closed system is used and the evolution of two kinds of entropy with time is investigated. A pseudo closed system means external energy exchange should be less than the threshold energy for the inter-state switching so that there is no external interference during the energy and information exchange process.

If there is a change in the logic state, change in momentum and finally after repulsion a change in the direction occurs, then there must have been an effective energy exchange. TE is calculated theoretically analyzing the Scanning tunneling microscope (STM) images, and Density Functional Theoretical (DFT) computation, and IE is calculated simply from the logic state derived from STM measurements.

Gibb's relation and beyond: In the similarity vs thermodynamic entropy plot, nearly a linear relationship holds between them while for information entropy (IE), the relationship was linear but 90° out of phase. Therefore, a relationship is established between two different entropies of a particular system; however, most surprisingly this information entropy relationship was also obtained by Lin et al. but unfortunately,

they have claimed that their relationship is the real entropy-symmetry relationship and Gibbs relation or modified Gibbs relation are both wrong. Above findings show that similar to Lin et al. hundreds of models which made serious attempt to evaluate the authentic entropy-symmetry relationship needs to be re-examined as the question of right and wrong does not arise at all. Once it is established that similar to Gibbs paradox, no debate should exist for the entropy too, the ground is set to re-visit the thermodynamic aspects of the computation process in a machine like a human brain. Furthermore, in order to connect Neumann's perspective with the central nervous system, CNS, a clear vision of a solution point is the necessity, and unless it is known how these two entropies are related for a particular system the solution point would have been a vague concept.

At this point, a clarification is needed for the solution point. In Gibbs relation, the particles have two phases, distinguishable and indistinguishable. Inside a Pentium processor, the 2D surface of the hardware has billions of switches, during operation or even when it reaches a solution then the hardware surface is a matrix assigning each element or components either 0 or 1. Therefore, when all the switches are zero or one then the system is distinguishable otherwise indistinguishable. Now, for any other combination of 0 and 1 in a 2D matrix, thermodynamically there is energy exchange between switches in the hardware therefore we have a TE, and if we consider only the number, then it is easy to determine how far the matrix from absolute indistinguishability is IE. Therefore, the processor acquires a co-ordinate on the straight line of an entropy-symmetry plot, which could also be the solution of a problem. The reason is that at any given point of time the generalized path of the system point in the generalized co-ordinate would be the solution of the problem. It is the thermodynamic interpretation of the Neumann's computer proposed in 1955.

It is possible to perform validity check experiments on the second law of thermodynamics in the universe where there is no reality. It is not possible to build an absolute close system, and the virtually closed system could be made by a proper choice of the inherent property of the system elements. Modification of the Gibbs paradox by Neumann's is justified, Lin's philosophy is experimentally verified but his challenge to the Neumann's is unjustified as both of them consider different entropies. Neumann investigates information entropy = (−) similarity = (−) repulsion = (−) thermodynamic entropy which Lin considers. During the experimental justification of Lin's proposal, in the brain jelly, multilevel repulsion was observed among different imaginary worlds. During the information exchange process this gives rise to multilevel phase evolution, a basis for a multilevel computer. Multiple phase evolution in a single experiment demands a 3D plot, which for $n \to \infty$ becomes the representative of a central nervous system. It looks like a lotus flower. We complete a journey from meander flower to a lotus flower. The motion of any point on this entropy symmetry surface is the thought-like information exchange process in a central nervous system, might be in an artificial brain if the equivalent hardware is ever developed in the future (Chapter 9).

5.8 HOW GEOMETRIC SIMILARITY BUILDS CREATIVITY IN COMPUTING PRIMES

First, seeing a typical tree when we resemble it with the dancing girl, or balloon looks like a lozenge, there is no formal logic. Such familiarities are creative and go beyond algorithmic logic (Figure 5.10a).

Second, the integrated geometries of different kinds of data may look similar. For example, the relativity and the painting tricks are similar in geometry, but have no relation

in concept, yet they could fuse affecting each other, wherever the local geometric matching is there in the respective time crystals (Figure 5.10b).

Third, even within a very particular type of data, like the advent of diabetes among masses, we could have totally uncorrelated functions and symmetries in the statistical distribution. However, if local distributions match geometrically, then, they would build a fractal geometric seed together by combining the geometric seeds (Figure 5.10c).

Fourth, when a fractal tape machine searches for a geometric shape, it can make a mistake. That error is corrected when

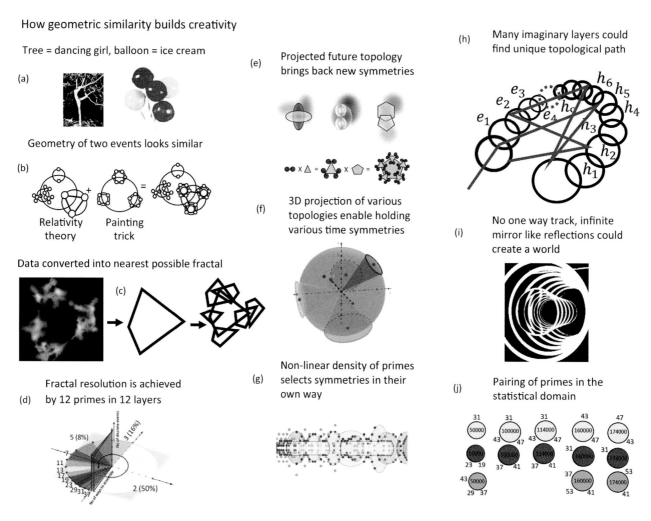

FIGURE 5.10 Ten different ways of building intelligence and perception in the artificial brain. (a) Geometric self-similarity among fundamentally different sensory outputs which have no contextual relation. (b) Time crystals of totally unrelated events, if appear similar, they undergo morphogenesis. (c) Self-similar geometric features are adopted as a fractal since time crystal binds repeating events in a single loop. (d) Fractal resolution is a unique concept, where a particular data achieves a better resolution as it enters within, 12 imaginary layers engage in perfecting a particular data. (e) PPMs have unique projection, that projection is fed to the system back which changes the singularity built geometries. (f) The information encoded as time crystal has a unique projection around 360 degrees. (g) Non-linear distribution of prime density selects input set of primes or input information as they sense it, so it is not a homogeneous acquisition of information. (h) Co-existence of topological paths that links different imaginary worlds within a multinion or across a multinion (e.g., octonion and a dodecanion) could lead to the birth of a new unexpected imaginary world. (i) PPM is not one-way traffic, it has many pair of mirrors generating infinite reflections between them. Thus, multi-modal mirroring could generate a new kind of information processing. (j) The pairing of primes, silent primes in the statistical distribution of primes in the integer space is another prime factor that regulates output.

the 3D geometry hidden in the singularity domain is searched and properly mapped. A similar error correction runs for the 12 layers (Figure 5.10d).

Fifth, when the symmetries of the randomly obtained geometric shapes from the events happening around nature are compiled and the corresponding PPM is built, similar geometric shapes are found to repeat over the integer space. The projected geometric shape means finding visually connected, irrespective of conceptually non-connected features. Thus, it is not logical (Figure 5.10e).

Sixth, a time crystal architecture holding the entire information could project different temporal patterns at different solid angles, all around it (Figure 5.10f).

Seventh, PPM have three-tier control on the evolution of a time crystal or the information architecture (Figure 3.16). First, the sensory layer, where the ordered factor metric or OF metric acts in converting the sensory data. Above this layer, a hierarchical control on the lower level metrics is made by a prime composition metric or PC metric. Above that there is a prime gap metric or PG metric that looks into the symmetry of the primes used. Three layers, operate together, one above another. The symmetry of symmetries of symmetries is beyond logical predictions (Figure 5.10g).

Eighth, in reality, the elements of quaternion, octonion and dodecanion or icosanion are time crystals. The elements of icosanion tensor could be a dodecanion, the elements of a dodecanion tensor could be an octonion and those octonions would have quaternion elements. Thus, the nested time crystal elements could build distinct geometric pathways which could connect in the time domain (Figure 5.10h).

Ninth, one of the finest features of a coupled feedback network-facing each other is that they work like two mirrors facing each other for the geometric information. While reflecting back and forth, the geometric information could build up architecture by interaction (Figure 5.10i).

Tenth, primes, because of their own distinct feature are silent when statistically one measures the contribution of a prime in a given database of integers (Figure 3.8g). The transition of a prime from silent to a dominant contributor is not linear as the number of integer increases, the relative ups and downs of the ranking of most dominating primes in a given dataset are oscillatory, and geometric (Figure 5.10j).

5.9 THE WHEEL OF INTELLIGENCE— DIFFERENCE FROM HUMANS

Will this be a universal computing machine? A computation is universal if it can emulate any other computation. Emulating a particular computation C means that we can feed a certain code into our universal computation U that will cause U to produce the same input-output behavior as (Zwirn and Delahaye, 2013). Brain jelly is a PPM synthesizer, specific to the input time crystal, if we want to execute generalized computing in this material (Chapter 9).

Therefore, if a programmable matter say brain jelly is designed and synthesized to exhibit a large part of the frequency band for electric, magnetic, and mechanical resonance, then, it can emulate part of the rhythms, irrespective of its limitless possibilities, those are bounded by certain repetitive rules. Of course, we do admit that even then, mathematically it could be proved that an absolute universality cannot be achieved.

The ability to replicate all possible rhythms is not just enough for universal computing: The objective of PPM based non-computation is to use minimum resource yet reaching the solution faster than the natural system using nested rhythms just like an avalanche (Figure 5.11). If the advent of cancer disease takes 120 years to reach to a metastasis state, it would be wise to use a replica of nested rhythms for cancer at shorter time scale so that we can predict it within a few minutes. It means ultra-fast events need to be lengthened and ultra-slow events need to be shortened in the time scale to perform a meaningful computing or emulation. **Make ultra-fast events slow and slow events fast:** Normally, if a natural computer is able to replicate a natural event by increasing its speed by 10^{16} times, or by decreasing its speed by 10^{16} times, we can cover the majority of natural phenomena that evolves in our known universe. That means de facto the proposed non-computer nanobrain would be a universal one. For example, we shall always try to predict supernovae explosion in a few months, since it takes billions of years for a star to become supernovae. Whereas, the artificial natural event that would construct the time crystal for a single electron tunneling event should capture it in the femtosecond or picosecond time scale, but then by fractal feature, clock it in seconds or hours. Self-similar feature of the time crystal would bring femtosecond event to seconds so that we can observe the evolution of ultrafast event in the classical time-domain and analyze how it evolves with time. Two experimental verifications are explained in Chapter 4. So, we expect organic nanobrain to study several animal dynamics by itself, learn and then suggest a particular jelly architecture that would exhibit the necessary mobility features in the real world just like the humans do (Figure 5.12). Afterward, the machine would choose a set of PPMs need to be incorporated into the brain of that robot, and suggest a pattern-evolving grid, finally suggest the necessary rhythm composition rules for that grid. Events occurring in nature are not random. Random events integrate from one singular metric. That metric is closely related to the vibrations of dielectric resonators (Lauber et al., 1994; Pistolesi and Manini, 2000; Pechal et al., 2012; Lee et al., 2011; Joshi and Xiao, 2006). The other criterion is the space constraint. We cannot use the entire mass or all components in a brain jelly to emulate the replica of a nested rhythm or time crystal, the entire process needs to be generated among the cluster of a few participants. Therefore, universal computing using PPM machine demands both spatial and temporal resource management as outlined in Figure 5.11.

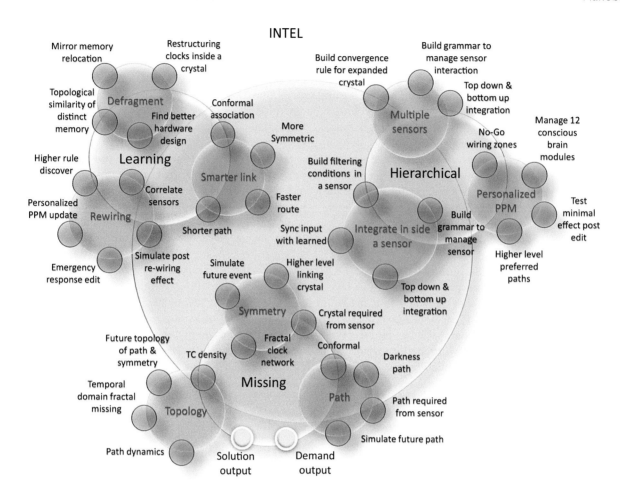

FIGURE 5.11 Time crystal representation of intelligence in the artificial brain. Time crystal always builds between two primes, so larger is the difference between two primes, larger is the complexity of time crystals. Within 1 billion we have roughly 255 million primes means around same number of time crystals. *x*/ln *x* where *x* is the number less than which the prime numbers should be and ln *x* is logarithm of *x* to base e. So using this, we get 255.88 million. These primes govern birth of creativity and unpredictable situations.

PPM hardware	Human apparent notion
Integrates events, estimates all symmetries, predicts future	Integrates events, estimates few key symmetries, predicts future.
Input is a geometric language, represented as symmetry	Input is a trained language, not derived from protein vibrations, e.g., english
10D data structure	4D data structure
Density of primes & ordered factor of symmetries considered	Finite symmetries, few anomalies of primes consider, varies among humans.
Do not care observation, linking of events are mathematical	Emotion, culture etc various temporary & long term biases, shift links.
Future prediction is instantaneous but accurate	Instantaneous but partially accurate.
Superposition of possibilities are mathematical, infinite.	Superposition of possibilities are driven by temporary information processing ability
One PPM chain needs to be used millions of times at different spatial & temporal scales to build a humanlike architecture.	By common notion (human = turing machine), all human information could be written as 0 and 1 in a single one line tape
To replicate in device, one element has to work as many clocks, part of many time crystal	By common notion (human =turing machine) one device = one work.
No integer except primes represents a single geometric shape, all shape connected to all, in a PPM only primes are defined, like Turing world	All connected to all is an illogical, non-scientific statement, because then all are undefined and no statement could exist.

FIGURE 5.12 A table demonstrates the difference between PPM-based hardware and apparent views of a human being.

5.10 COULD PRIME-BASED COMPUTING PREDICT THE FUTURE WITHOUT PRIOR KNOWLEDGE?

In the culture of Turing-based information processing, be it classical or quantum, we deal with facts as absolute truth, gathered earlier. Here, in this book the paradigm of computing is advocated for non-facts like confusions. We do not take any facts in the new paradigm. We start with confusion, and make a journey through it, by asking question, "how many confusions have built this higher-level confusion?" Then, take one of the sub-layer confusion and go inside and ask the same question. When, we reach a known fact we do not enter. Now, the architecture of confusion is the map that helps us to understand the unknown. Figure 5.13 lists a series of factors why can we predict future in the fractal tape machine paradigm, but not in the earlier paradigms.

First, when events are represented as geometric shapes, the symmetry comes into play. It is true that events could happen in infinite ways, but the number of symmetries is finite. So, the geometricization of events ensures that events are classified in terms of symmetry. The geometric language, GML has 15 letters as 15 geometric shapes, hence, fusion of different sensory signals into one architecture is possible.

Second, when linking of events in all possible ways is mapped in terms of patterns of 15 primes, the metric of primes estimates all possible combinations of discrete events. Since the user assigns a typical composition of primes to a typical event, when a new composition of those primes appears, or even a new set of primes are linked, we get future possibilities.

Third, the geometric seed of various higher-level integration of events are integrators and the projector of future possibilities.

Fourth, a proper manifold mapping of the interaction of different imaginary worlds is recorded in the quaternion, octonion, dodecanion, and icosanion tensors. Thus, hierarchical maps of 12 layers one above another ensures projecting the future in a proper manner.

5.11 HOW PRIME-BASED ARTIFICIAL BRAIN SHRINKS BIG DATA?

Just two conceptual changes in defining what is "information" and its preferred "integration" do wonders in the existing information theory (EIT, classical + quantum information theory). Information is the 11D geometric structure of clocks and we look at its symmetry, refer to a

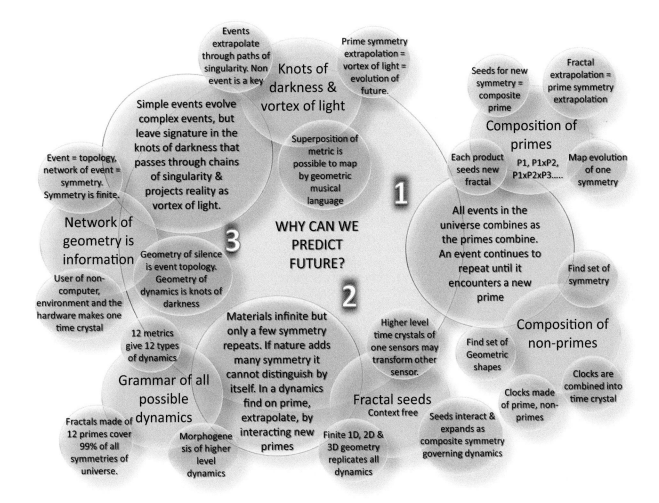

FIGURE 5.13 A chart summarizing why the proposed artificial brain could predict future without having any prior knowledge.

pattern of primes which is an archive of how to link symmetries to integrate information. Here, the universe cannot be linearized or sequentialized, the purity of fractal geometric feature has to be preserved. Now we cite 10 reasons why FIT can use much fewer resources to solve the "data deluge" and "astronomic algorithm" problems, while EIT cannot (Figure 5.14).

1. **Information is squeezed into a fractal seed that repeats to regenerate the original:** We do not store entire pixel-based data, convert it into a geometry of clocks. Not randomly, we build a fractal of fractals, following 11D time crystal's symmetry and the pattern of primes to find how to make the nested seeds of clocking geometry so that when the user implements the fractal expansion following a metric of primes, it retrieves the original pattern. Therefore, the **fractal seed is a drastically reduced 3D geometry of clocks** for the original geometric structure. Earlier in the old information processing theory, when the volume and intricate details of the content was important, not symmetry. FIT classifies

the content basic on self-similarity of significance. Here fractal seeds are topologically filtered to a particular location of the hardware.

2. **The maximum change in the physical space is less than the length of minimum wavelength:** We could use one choice or imaginary path for making another loop, changes we need to make in the device geometry is very little. Thus, resources required do not increase exponentially. Just imagine with 10 states one inside another, we can have $10^{11+10+9+8+7+6+5+4+3+2+1} = 10^{66}$ states. In other words $8^{8^{8\cdots}} = 8 + 8 + 8...$, we linearize the power by fractal hardware. It is a massive reduction in complexity. Since the whole architecture is a geometric arrangement of spheres, from the center of the structure 3D projections to different direction provides different information. Same architecture could hold conceptually distinct information, no need for new hardware, simply by changing the relative orientation of two structures one could add a huge amount of information in a system as time crystals, without adding a new resource.

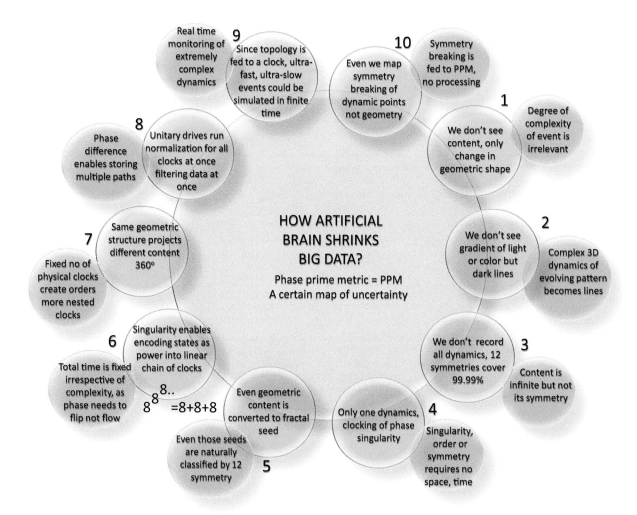

FIGURE 5.14 Ten points that describe how artificial brain could shrink big data.

3. Since **only one parameter, phase** represents all information, no resources required for compilation, transformation, addition, storage, even memorizing = processing etc. To add a new phase to an oscillating body, it does not require a physical space. **There is only one dynamics of phase change that generates the "resting state," hence no need for varieties of phenomena and their theories:** Who selects direction? Who selects the motion? Who selects the number of system points? Who selects the phase? Motion is selected by an evolution of phase, and a negative and positive sign of a phase selects the direction of rotation. **One could add new system points and generate a completely new kind of dynamics in the same geometric structure without adding a single new resource**.

4. **Total information remains constant, yet one could run an infinite series in finite hardware: harvesting infinity is feasible by projection:** If certain topology is acquired by time crystal then we do not need infinite resources or materials, one sphere could hold an infinite set of clocks by creating an infinite series of oscillations, could be analyzed using continued fraction geometric algebra. In terms of symmetry and dynamics, the choices are finite for any complexity of information. *Adding a new singularity does not require space, simply by adjusting the phase of neighboring clocks in a geometric structure one could add a new set of singularities, which means creating a new geometric architecture*. Plenty of system points along with the multi-direction oriented planes parallely rotate on the imaginary phase sphere. Hence, four parameters, *time gap between two singularities, system point starting location, rotational directions of all time cycles (both, starting phase and rotational directions are determined by projection distinction on the reality sphere), (orientations of the planes are fixed = number of system points)* are used to build up the information structure as universal time crystal. The line connecting the phase sphere's (previously Bloch sphere) center to the system point on the spherical surface get the information. Therefore, the complexity increases but total information remains constant.

5. **An intricate pattern of lights and darkness change their shape, it holds the dynamics of the events, no wiring required, no electronic devices are essential:** We find the 3D distribution of knots of darkness and free dark particles, also the knots of lights and free light particles in the data space, find the geometric pattern of dynamics (see Chapter 8 for details). The devices follow flux-charge protocol instead of a current-voltage protocol. Here power supply or energy consumption is not a prime concern, but geometry and dielectric constant of the material is which is a function of symmetry.

6. **New memory is written on an old memory, no need for an exclusive new space for a new memory, they may share a common space:** Experimentally, adding a new loop of poles means making changes in a group of ET cavities such that little changes make no notice locally, but when we see globally all participant cavities together, it's "the" pattern that is evident. *It is the reason, we do not need to delete memories, all information are basic geometric shapes only, just connect them suitably; a new memory simultaneously exist with the old memory.*

7. **Singular one to all connection at the top: "Umbrella protocol," "superposition of many circuits in a single hardware":** Nested clock retrieves a higher level perception correlation, since, "everything is connected to everything" and at a higher level it is easier to find the total path, hence any problem reaches to the top first, like an umbrella. In a sequential circuit system, everything is connected to everything is not possible. Here, **in the same hardware one could connect memory elements using a network of phase, clocking direction, topological symmetry, a pattern of silence, the orientation of a particular plane of symmetries, 11D dynamic map's similarity in tensor algebra, etc**. Moreover, to add a new umbrella we do not need any new hardware.

8. **Fractal net of clocks: "instantaneous solution":** Fractal clock enables processing and analyzing at the "fractal seed" at a much deeper level than the reader's processing layer, which runs very fast. A task to be processed at a slower clock is actually solved in the domains of the fastest clocks and then the solution is returned to the slower clock. There is no search of the database even since due to symmetric fractal architecture, there are spontaneous replies by the system, **it performs a search without searching**.

9. **Unitary drive in the reality sphere at all levels of information integration:** Perpetual drive to reduce the system "unitary," An **"unitary drive"**: Perpetual self-assembly for the reduction of arguments into 15 geometric shapes, never delinks from the entire environment of universal time crystal without users note but generates a drive to make the time cycle unitary. The structure finds shorter and better routes by rotating the inner planes. Even though a user does not encode intelligence, the system's unitary drive induces a self-drive for that. Geometric language could not be applied if there is no unitary drive.

10. **Various exclusively topological processing, the geometric symmetry breaking, phase transition etc by using a fusion fractal tape (FF tape) via morphological transformations of time crystal:** Geometric connection, geometric information, use of singularity, the new information theory that are fundamental to FIT and GML carries with it fractal

decomposition of bursts from singularity The nested phase network = information. *Simpler geometry of clocks, replaces the complex clock network*.

11. **FIT-GML protocol does not have to hold in memory the actual information content during processing, hence the size of the information content is not important:** One has to find the periodically oscillating dynamic points in a topology of the distribution of data, create a 3D network of topology to find intricate phase relationships between a large number of clock patterns. Finding the same phase oscillations of the brightest and the darkest points in a 3D cloud like representation of data is the actual mechanism of information processing.

12. **No need to understand the language following which the information is written:** In any given information space, irrespective of any mode or methodology of presenting information dynamics, PPM-GML-H triad protocol finds symmetry in the dynamics with 99% accuracy. So, we do not need to understand other dynamics, or how they have written it. **Hence does not require an algorithm, all senders and receivers are encoded with a similar prime metric hardware, to determine how symmetries to be linked is a universal governing rule**.

5.12   THE LIMITATIONS AND INCREDIBLE FEATURES OF PRIME-BASED COMPUTING

What is non-computing? Definition of a non-computer (Reddy, S. et al., 2018): (a) The number of choice and quantity of information increase during decision-making instead of reduction. (b) No finite statement is found, all statements are fractal, not overlapped (Paterson, 1974). (c) There is no sequence of events, it is always event inside an event, i.e., a fractal thread. (d) No measurement happens here, superposed possibilities coexist as a distinct state. The observer becomes an integral part of the morphing of entire information content written as a time crystal. (e) All decisions are logically circular. Nothing exists without a closed-loop, even non-periodic events. (f) There is no data or fact as the decision. It is always a shape-changing geometry, the habit of looking at numbers for solutions is unfound here. (g) All solutions are incomplete. They are extended from the beginning to the end of the hardware structure. (h) Halting is never there, decision-making never stops, output continues, converges, spontaneously starts rebuilding. (i) Decision-making happens in the phase network of time crystal. Mostly, the signals remain the same, only the phase changes. So literally an observer detects no ongoing computation, still, a decision is made. (j) There is no question and answer or argument, only situations. An intractable Clique problem (searching a key pattern in a complex pattern) is solved bypassing its criticality (Feige et al., 1991; Ghosh et al., 2014a). (k) The user or observer does not write instruction. Instead "metric of prime, PPM is the programmer, it replaces the user." (l) There is no forced input. User

does not search inside the hardware. Using geometric grammar, it searches its environment for input.

Here are some of the incredible features of the PPM class of computing.

1. Search a massive database without searching (spontaneous reply).
2. Multiple nested clocks one inside another enable "a virtual instant decision-making"
3. No programming is required as "cycles self-assemble/dis-assembly for better sync at all possible time scales simultaneously."
4. "Phase space" keeps "volume intact" as required resources only increase phase density not real space. Information = wiring geometry (hardware); Information = silence (not "bing" Philosophy); Information = Phase (geometric); Information = Perimeter or diameter (parametric).
5. Perpetual spontaneous editing of slower time cycles (creation/destruction/defragmentation) "prepare for unknown" = higher-level learning.
6. FIT introduces "fractal resolution," a complex signal's lowest and fastest time scale signals are absorbed, simultaneously, and during expansion, the fractal seed delivers full output, from a seed of information (drastic shrinking of data).
7. The superposition of simultaneously operating a million paths assembles into a sphere enables "extreme parallelism." In quantum, only one Bloch sphere, here sphere inside a sphere inside a sphere...
8. Time cycle is a memory, rotation along the cycle is processing, are same events, "no transport needed between memory and processing units," no wiring, no communication, no communication channel.
9. No logic gate, no reduction of choices, which ensures that "speed" is irrelevant. No bits or Qubits. The object is not measured by an observer, the object being observed is not a separate thing, i.e., no measurement. Observer, Object and Environment integrates into the time cycle (time cycle = unit of information = an event). A question is assimilated into the internal network of time cycles it affects the environment and observer in the course of assimilation. That is how a solution is derived, no computing is done to find an answer.
10. All sensory information is converted to one geometric language that allows "perception," a yellow color could have a taste. Perception is not programming as wrongly perceived.
11. It never requires rebooting, only one computing runs for whole life, but of course it sleeps when it switches off certain interactive links with nature, yet even at that time all three internal rhythm networks run. Rebooting means temporary switching of the internal rhythm integrator. It is an equivalence of consciousness.
12. The information structure self-edits itself continuously. It always spontaneously defragments the internal structure of the complete information

content represented as time crystal. However, self-editing is not so profound at the boundary clocks of a time crystal structure. There is a Gaussian distribution of editing efficiency in a time crystal architecture, the clocks in the central region edit the maximum. There are various reasons why this is so essential. (i) Boundaries cannot be so vulnerable to changes, (ii) editing requires a high density of clock boundaries, (iii) job description leaves only clocks in the middle to memorize environments, while faster clocks need to memorize hardware management, the slower clocks require to sync with the environment.

Limitations of a fractal machine: The reason that makes Fractal machine more powerful than the Turing machine is responsible for its weakness (Figure 5.15). A number of weaknesses of the fractal machine have been outlined by Gregor Drummen. First, higher level perception based on a hierarchical network of platonic geometries enable fractal machine to convert all forms of energy level signals into a singular language. That very feature also ensures a judgment

error. For example, a red ball could be perceived as an apple, rope as a snake, etc. Second, since the user, machine and environment are all time crystals locked by cycles of rhythms, the machine is always ready for unknown situations, here unknown means, not encoded inside the machine. For the same reason, the machine requires to regulate and update rhythms of the environment perpetually. If mismatch is more than the machine's tolerance limit, it needs to go to sleep, so that geometric shapes rearrange to a stable structure. Third, a self-assembly of algorithm enables self-learning of the machine, writing its own software, but it can lead to irreproducible production of the non-computing machine in the industry. Since the fractal network evolves with learning, the chances are limited to recover if things go wrong. However, one thing is certain, the non-computing would not be limited to conscious expressions of life. A massive database of personalized PPM for each living and non-living systems in and beyond this planet would unveil the geometric correlations of time and space and repeatedly derive all the fundamental constants that distinguish physics, finally derive all laws from no laws (Wheeler, 1980). Only a few primes—and

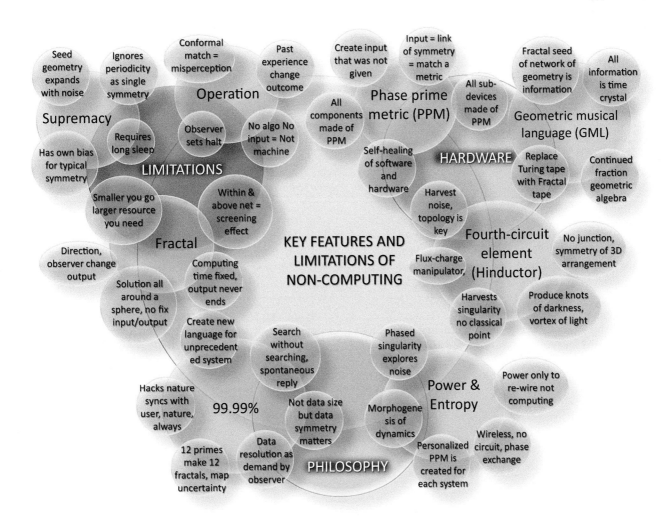

FIGURE 5.15 A chart describing the limitations of the non-computing, a new kind of decision-making using fractal tape.

explain the existence itself—as the similarly self-generated organ-like metrics of a self-synthesized information system (Haken 1988).

Critical challenges and the weaknesses of a non-computer: (i) A non-computer is not precisely accurate. It gives a global idea or perspective. A non-computer is like a life form, good at those kinds of problems that it solves most. Given different kinds of problems, a shadow of the past analytic protocols is reflected. (ii) Speed makes no sense, the total time of decision-making is fixed. Decision-making cannot end in principle. An observer or user of the non-computer captures a solution based on its own time resolution. (iii) Blindly trusts the metric prime, PPM as an encoder for all possible dynamics in the universe instead of a human user. The dynamics related to millions of primes are left out. It is similar to the culture of conventional computers, which trust that all events in the universe could be sequentialized. (iv) All 10 deliverables of non-computing are abstract in nature, not factual during processing. (v) Instead of signal the phase or silence between a pair of signals holds the key information. A 3D network of silence is the unit of information, not bits. Hence, particular clocks work (Bandyopadhyay et al., 2006a, 2006b, 2010a; Bandyopadhyay and Wakayama, 2007) but a switch alone fails. (vi) Wiring does not work; one-to-many and many-to-one communication is required (Bandyopadhyay et al., 2010b, 2010c). Hardware needs a wireless communication and a fractal network with a null screening effect (Jaeger and Haas, 2004). Here is it achieved by anomalous quantum cloaking, which enables key processing devices to sense and communicate only essential device, make others invisible. (vii) Non-computing never stops, it slows down at synchrony of clocks. Therefore, there is no static output. The answer depends on the time when the question is asked, how the question is asked i.e. observer's time crystal. Depending on the observer location around the observed, the solution changes significantly. (viii) In the conventional computer, noise disrupts the system. Here, noise is fed to activate the synchrony, finding the symmetry that matches most with the PPM. Moreover, an ordered signal affects the decision-making negatively. (ix) Addition of resource has no value. It is not the number of elements but the distinct sets of vibrations in distinct symmetries that make a powerful non-computer. (x) The solution has to be taken from all over the fractal hardware, as a projected output. There is no output and input location.

Entanglement is *one-way* traffic: Coexistence of clocks is multi-channel traffic: Quantum computing is primitive: In a quantum computer, an exponential speedup is caused by entanglement, since a single measurement can destroy all other states except the one that is desired. Entanglement is a fairly limited technology. Often, we encounter a situation where each time crystal can simultaneously be part of several other time crystals operating in a distantly located PPM-based independent system. The demand is far more technologically advanced than that offered by entanglement driven quantum computing. By analogy, this is like a person playing one drum in time with one group of drummers and another in time with a different, faster group of drummers, or playing cello in tune with one group and an oboe with another. Such wiring of time is possible by complicated phase relation between different systems. Electromagnetic resonance induced quantum cloaking ensures that a large number of entangled systems coexist independently, operate dependently for some conditions and independently for others. Fractal mechanics of multi-layered imaginary worlds is the key.

Energy conservation principle: PPM computing does not require Landau limits for elementary operation: In a PPM hardware, all clocks in the entire fractal tape run continuously, decision-making means a subtle integration in the phase space. Total energy pumped to the system always remains constant, the entire computing hardware does not require additional energy supply. The hardware is designed not to absorb any further energy, the total energy content only redistributes over the matrix. During information processing system we have a surface where energy redistributes like Gaussian 3D wave packets. The distribution of energy changes but the total energy is conserved. It is a closed system, by landau principle, bits do not change here that requires energy supply.

Limiting the PPM computer: Markov process and Devils staircase: The specialization of necessity and preparedness is one weakness of current computers. In the probability theory and statistics, a Markov process, is a time-varying random phenomenon for which memory-less-ness survives. Then, the property depends on the present state of the system, its future and past are independent, so it is called memory-less-ness. For the pattern of primes, no memory is required other than the memory of the metric, or how are primes around? The second weakness is that during training, brain jelly may not learn properly. In this case, the whole computer would be adding nonsense perspective to everything. The third weakness is that, just like humans, it can make errors of perception. Finally, this computer is not good at solving simple problems like $2 + 2 = 4$, and the accuracy is significantly compromised; existing computers are much, much better in this case. Since all, time crystals have an upper and a lower frequency, at certain limiting scales we find that signal takes constant values except at subset of points where it changes continuously, this is called **Devil's staircase.**

Nirbate to jatha dwipo jwalet sneha samanwita | Nischalordha sikhastada yukta mahur manisino.

In the waveless world of mind, the concentrator fixes his mind like the flame of a lamp that lights the windless world of ours.

Mana eba manusyanam karanam bandhamokshaya | Bandhyaya bisayasaktam, muktam nirbisayam smritam.

The mind is the sole place where the reason for bindness and that for eternal freedom reside: together, the attraction of the materials is the root cause; freeing from everything is the key.

6 Unprecedented Technologies found in Nature Led by Harvesting the Geometry of Singularity

6.1 PPM—THE PATTERN OF PRIMES EMBEDDED IN THE TUBULIN PROTEIN

If firing is the reason for the brain's operation then to search, the whole brain would continuously fire: Francis Crick, the DNA man, observed an exciting feature that to recover the visual image of a person or an object, the associated neurons sitting at distant places inside the brain should oscillate at a single phase and frequency: this is synchrony. To synchronize two neurons separated by 10 cms in our brain when there is no wireless communication, they would have to continuously release neuro-transmitting molecules at a particular rate until they carry out an extensive "search and find" throughout the brain to find the other neurons and eventually reach a sync state. The linear circuit connecting the two neurons is a complex network; there is no straight path where neurons are sitting side by side, so, instead of 10 cms, the actual distance could be several hundreds of meters. Because the neuron's communication speed (400 m/sec), it would take several seconds to minutes for the circuit to reach the second neuron. There is no guarantee that in one attempt, neurons would find the right path; since one neuron is connected to, say, 10,000 others, the search should run in all directions, simultaneously. Here we have two aspects of this search: first, the number of chemicals essential to carry out this massive trial, and the error process throughout the brain is astronomically large. Second, the time loss due to the wrong choice of paths is not taken into account while calculating the time to find that very neuron with which the synchrony needs to be established. If search-and-find operation executes via firing, then for every single thought to process in our brain, all neurons should fire until the right information is found. It is not the case that only a fraction of all neurons fire during "search and find."

Protein is a single chain polymer made of amino acids from the genes of DNA, it folds into helix, beta-sheets like unique structures known as secondary structure, most important is the tertiary structure of a protein, which could be millions or billions in numbers, they generate from the secondary structures to define massively complex properties of living systems. A simple twist in a protein could cause the death of a life form (Prion protein), thus, to understand biology, it is essential to underpin enormously complex tertiary structures, however, it is explicitly written in the literature that this is an unsolved mystery and possibly attempts are extremely localized. Using scanning tunneling microscope, STM it has been demonstrated that using tunneling current images it is possible to unravel unique structural symmetries of a protein and electromechanical resonance properties could be measured for a single protein, which has a one-to-one correspondence with the cluster of secondary structures, which we could define as circuits. The secondary structures depending on their unique dynamic features form typical groups with neighboring elements and those particular circuits hold the key to the remarkable flexibility it offers in regulating the protein property. The PPM-GML-H triad protocol (Phase prime metric [PPM], Geometric musical language [GML] and Hinductor [H]) has three points forming a triangle, STM based atomic-scale manipulation on unraveling the features of the circuits, for estimating the resonance.

Helical structures in biology: Observing biological properties using impedance spectroscopy for medical purpose is old, and a reliable practice (Schwan, 1957; Grant, 1979). Spiral symmetry is the easiest route to quantize energy and build geometry only device: Spring structure holds enormous potential since energy traveling through spring is automatically quantized. If one could properly canalize the transport path, the quanta would travel through the path in a loss-less manner. Now, noise energy could also be canalized in this manner, for eventually pumping out of the structure. Proteins or their complexes have folded one single string into multiple different kinds of secondary structure like a spring, staircase-like sheets, etc; these are the tricks for different kinds of energy quantization. Microtubule a nanowire (20 nm wide, several micrometers long), a vital component of a living cell, came to this planet several billion years ago, just like DNA, they are spiral, but why? What makes it spiral? By being helical, a structure can bend and sustain its structure, then it can use the surface to regulate the transport of electronic, optical carriers and vesicles. Thus, helical examples are plenty in the biological world, the origin is debated (Hunyadi et al., 2007). They can work as a waveguide too just like carbon nanotube (Liang et al., 2001).

The Helix antenna: The investigation for microtubule as a Helix antenna: its ac resistance R_{ac} (~1 MΩ, at 22 MHz) and microtubule-diameter C predicts the resonance peak ν at 228 MHz $\left(\upsilon \sim 140\varepsilon\, C/R_{ac}\right)$ for microtubule to be a Helix antenna (tubulin dielectric constant $\varepsilon \sim 10^4$, $C \sim 78.5$ nm). The two key parameters are (i) the material is a dc insulator, but extremely high conducting under ac-resonance, (ii) dielectric constant of the basic helix structure should be very high, which is natural for nano-particles.

6.1.1 α-Helices Form Groups of Rings to Complete a Loop

A linear chain of oscillators resonates to build an infinite series of oscillating frequencies: If a tape is a linear chain of oscillator it would produce time fractal or rhythms. Most interesting to note that scaling is a fundamental property of any natural oscillation process (Muller, 2009). Reddy et al. repeated the case of harmonics or ordered factors in a network of escape time fractal tape. As said one should take only one tape at a time. Say, we have an oscillator with a single resonance frequency $f0$, the oscillator will have a higher mode oscillation frequency $f1$ and the relation between them $f1/f0 = n$, now for a nested waveform network (Smoes, 1976), says one waveform encapsulates 3 waveforms in it, and that continues, then first we get, $f1 = f0$, then $f1 = 3f0$, then $f1 = 9f0$, hence in general we can write, $f(n, p) = f0\,n^r$. In this way, the resonant frequency spectrum due to one particular symmetry can be represented as a logarithmic fractal spectrum. We can clearly see that a singular waveform fractions continue to occur in the chain of oscillators. If $f0$ is the fundamental resonance of one oscillator and f is the frequency of the chain then using a simple expression of continued fraction we get the resonance spectrum or distribution of natural resonance frequencies

$$f = f0\exp(S),\ S = n0\,/\,z+z/$$
$$\left(n1 + z\,/\left(n2 + z\,/\left(n3 + \ldots + z\,/\,ni\right)\right)\right).$$

Now, the band we get for $i = 1$, is similar to the band we get for $i = 2$ and so on, so it is a fractal, the spectrum looks like a hyperbolic function.

Helical chain of oscillators: Origin of triplet-triplet resonance band: When a linear chain of oscillators arranges in a helical shape the formulation changes. Depending on the diameter and pitch of the helix additional boundary conditions are added which imposes a periodic function. As a result, one introduces two different kinds of effective mass or symmetries, one period is created by the diameter of the helix and another period is created by the total length of the spiral and pitch. Now, the most interesting part, just by folding the linear chain, we create two resonance bands. So, it explains, a triplet resonance band. Each band gets a fundamental frequency $f0$ and we replicate the accidental example, in the above section, three fundamental frequencies generate partition of three waveforms, and that continues just like a continued fraction. Now, the continued fraction formulation a spectrum, a fractal distribution. In this way a frequency fractal is created.

Fractal Time, or nested clock has three expressions, phase, frequency and delay: Spontaneous energy quantization in a helical path is very interesting for several reasons. As per the theory goes, the quantized energy depends purely on a particular geometrical parameter. The parameter is the ratio of the diameter of the helix and the pitch of the helix, therefore the antenna dumps noise in the time crystal at this particular place (see Chapter 8 experiment). Simply by varying the geometry of the helical structure one could encode the resonance frequencies and in a tightly coupled environment, depending on the orientations of the helices determine the phase and depending on the phase coupling changes which eventually changes the delay between two signals. Phase difference drives synchrony, for all to all coupling, Kuramoto model suggests $\theta_n = \omega_n - \sum_m Knm\sin\left(\theta_n - \theta_{n+m}\right)$, order parameter is given by $\psi = C\sum_n r_n e^{i\theta n} = Re^{i\varnothing}$, synchronization occurs if $R \neq 0$, synchronization can switch from unlocked to a partially locked to a fully locked condition. Again, for all to all coupling using complex amplitude formulation Synchronization by Pikovsky and Maistrenko (2003) suggests

$$\dot{A}_n = i\left(\omega_n - \alpha|A_n|^2\right)A_n + \left(1 - |A_n|^2\right)A_n + i\beta/N\sum_{m=1}^{N}\left(A_m - A_n\right),$$

here, α and β if plotted as X and Y axis, we get a map of sync and unsync domain showing transmission path of information. The 2D plot actually creates a 3D network, evolving as a function of time in a time crystal, be it local or a global.

The α-helix working as a nano-quantum-antenna: It is not necessary to be a metal to become an antenna. Alfa-helical regions in the protein molecule work as a dipole (Hol et al., 1978) as a nanotube, (Suprun and Shmeleva, 2014), i.e., a quantum well (Suprun and Atmazha, 2002). When two or more adjacent wells of different widths are placed nearby, an electric field triggers tunneling current (Roskos et al., 1992). Therefore, α-helix, could be used as a helical dielectric resonator, its number of helical loops is the only geometric variable. A protein becomes an assembly of spiral resonators. The assembly could be represented using a circuit of spirals (Figure 6.1a). One interesting feature of any protein structure

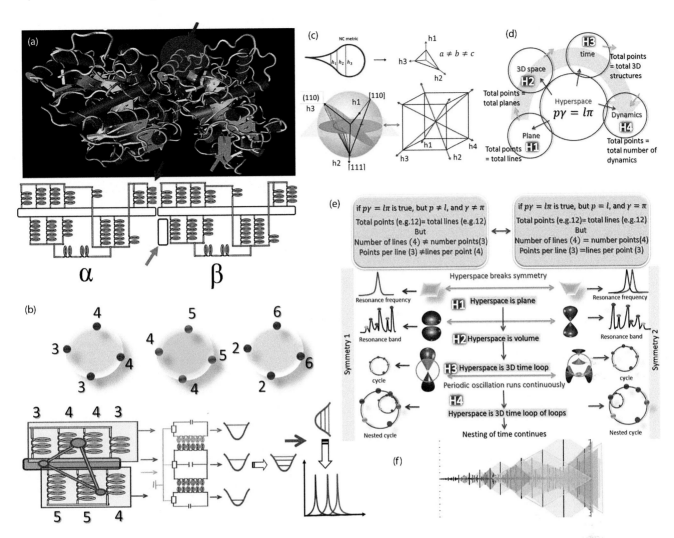

FIGURE 6.1 (a) The structure of a tubulin protein (top) and the equivalent circuit structure (below). (b) The circuit presented in the panel a is schematically represented as a circle in the panel b (top). 3–4–4–3, 4–5–5–4, and 2–6–6–2 are three loops of alfa helices (number denotes the rings), which forms clock. The circuit representation for 3–4–4–3 is shown below. An equivalent LC coupled resonant circuit is demonstrated. Note that instead of LC coupled resonant circuit, here magnetic-flux-charge based singularity bursts are considered, both have similar output. (c) Four sub-panels describe that ordered factor of an integer in the PPM forms an axis in the multi-dimension space (top). Ordered factor-based metric in the phase space has multi-dimensional axis, building a 3D topology (bottom). (d) The hyperspace created by multi-dimensional axes is shown. (e) The concept of hyperspace is similar to the concept of higher dimension described in Chapter 2, different shapes of the data, tensors for representing 11D data. (f) An example of PPM, shaded region depicts the clock like 3–4–4–3. In the Poisson distribution of all events is considered, from state occupation vector we get (shift matrix) D is time gap between events (frequency).

is that β sheets isolate a group of spirals, the most interesting part is that the number of rings in the helices repeat like a clock (Figure 6.2b). Thus, when quantum wells form a superlattice, it forms a larger quantum well and that super-well also emits coherent infrared wave (Waschke et al., 1993). Coherent emission from protein crystals is not new (Groot et al., 2002). Alfa-helices act as excitation energy transfer antenna (Brunisholz and Zuber, 1992). The exciton spectrum of alpha-helical protein exhibits a coherent transfer (Fedyanin and Yakushevich, 1981); Sincere efforts have been made to correlate the quantum excitation of an alpha helix-like an antenna (Brunisholz and Zuber, 1992).

Spontaneously polarize the incident light and mechanically regulate its quantum property: Spontaneous polarization of incident light is essential for the biological operation

of the fourth-circuit element, Hinductor. Using interferometric detection of coherent infrared emission, it was found how optical retina polarize lights (Groma et al., 2004), or electro-optic gating (Smith et al., 1988), therefore, generating polarized monochromatic electromagnetic signals from noise is not an impossible task. Therefore, microtubule and all helical vortex and fractal assemblies in the human body could act as a source of laser-like pure signals. Even if sound wave through the surface of the material could trigger quantum tunneling, hence, it would be excellent engineering if mechanical vibrations are harvested. If the charge is trapped inside a quantum well, it gets excited by mechanical vibrations (Rocke et al., 1998). An alfa helix folds mechanically for a few nanoseconds, therefore, harvesting mechanical energy cannot be ruled out (Panman et al., 2015).

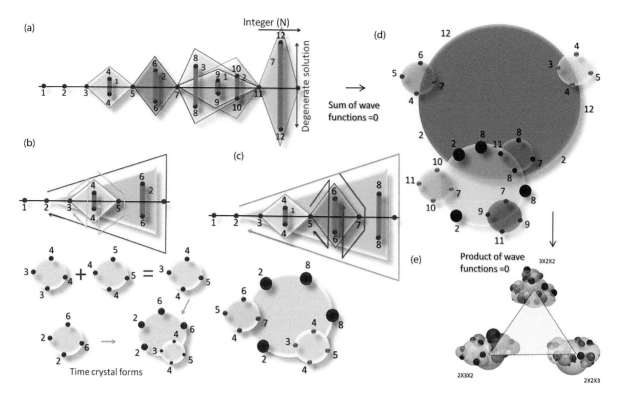

FIGURE 6.2 (a) How 3–4–4–3 and other clocking circuits of a tubulin protein if plotted side by side reflect a PPM. Such extraordinary observation extends in various protein structures. (b) Nesting of clocks is explained in the tubulin protein. (c) Combined time crystal and nesting of elementary clocks to form a higher level clock are shown in the plot. (d) The eventual time crystal and (e) superposition of different topologies for a 12 clock time crystal.

One of the interesting features of tubulin protein is that the beta-sheets of two monomers are arranged like an S shape, if we rotate the tubulin protein in a 3D plane, we observe an S, if the alfa helices are hidden. On the microtubule surface these alfa helices these S shapes couple and build beta-strand as a nested spiral. Coupled beta strands are known to generate quantum tunneling (Langen et al., 2015). Electron tunneling through proteins has been a long topic for discussion (Gray and Winkler, 2003). Long distance electron tunneling in proteins has been reported (Stuchebrukhov, 2010), possibly proteins need quantum features for searching the preferential pathways (Farver and Pecht, 1991) via long distance electron tunneling.

6.1.2 Hyperspace Symmetries Follow a Mathematical Identity

The PPM described in Chapter 3, looks like a teardrop. We described there only C2 symmetry, it means we divided the choices for a given integer, by two. Similarly, if we want to get a PPM for C3 symmetry, we can orient three axes in three directions and divide the number of choices by three. The process could continue for various different symmetries. The creation of new axes for a new symmetry builds a route to introduce a new kind of dynamics (Figure 6.1c). Mechanical sensors are very good at detecting weak electric field (Sutton et al., 2016).

For $n \to \infty$, the drive of prime pairing relates to e and φ orthogonally, and π forms a coupled hyperbolic space with e and φ: Prime number theorem implies that $P_n \sim n \ln n$, wherefrom it has been derived that if $\pi(n)$ is the number of primes less or equal to n, then $\lim \sup_{n \to \infty} (P_{n+1}/P_n)^{\pi(n)} = e$ while the lower limit of the same function is 0. One could derive an expression for the total contribution of n primes $\sum_{i=1}^{n} C_{P_n}$, if infinite integer space is depicted as unity, the normalized density of primes $D_{P_n} = 1 - \sum_{i=1}^{n} C_{P_n}$, within a certain range, if there are r primes $P_s \le P_n \le P_{s+r}$, we get ΔD_{P_r}. If we consider paired primes, since $\lim_{r \to \infty}(1 + 1/n)^n \sim e$ we get

$$\lim_{n \to \infty} (P_{r+1}/P_r)^r \sim \lim_{r \to \infty} (1 + \Delta D_{P_r}/P_r)^r \sim$$

$$\lim_{r \to \infty} (1 + 1/\varphi)^r \sim e.$$

Here φ is the golden ratio, i.e., if a and b are two primary linear distances, they self-assemble to determine the next length as a + b, then $\varphi = (a + b)/a$. Thus, the drives for primes and golden ratio series are correlated.

$(P_n + P_{n+1})/P_n \sim (a + b)/a \sim \varphi$ across N is a logarithmic variation, i.e., we get e along the horizontal. Moreover, if we implement the planar angular drift explained earlier, then the projected normalized hyperspace $(1:\varphi)$ is a golden ellipse with an area π. Projected hyperspace by φ and e for any dimension is the sum of their generic power, hence we get $\varphi^{Jh} + e^{Jh}$, now since the projected hyperspace of φ and e is π

(π measures points in H1) in one dimension and ($p\gamma = l\pi$, i.e., total points = total lines; i.e., total area = total lines = total points; Figure 6.1d) therefore, sum of their orthogonally coupled hyperspace is the hyperspace created by π of the same order $\varphi^{Jh} + e^{Jh} = \pi^{Jh}$. Now we try to find the value of Jh, we can easily estimate that $Jh > 2$ (for orthogonal triangle $a^2 + b^2 = c^2$, hence $Jh(\min) = 2$. The orthogonal relationship suggests that ideally angular deviation should be 30°, 60° and 90° following the ratio $\varphi : e$ but φ is a function of $(P_n + P_{n+1})/P_n$, so is e.

First, φ and e are orthogonally connected to hyperspace, $\sum_{r=1}^{\infty} 1 + (k/e)^r = \varphi; k(e) = 1.05$, and $\sum_{r=1}^{\infty} 1 + (k/\varphi)^r = e; k(\varphi) = 1.03$, so it proves that they are orthogonally coupled, but the hyperspace is not planar ($k(e) - k(\varphi) = \Delta Jh \neq 0$). If one varies the $(P_n + P_{n+1})/P_n \sim (a+b)/a \sim \varphi$ plot against N, then RNP(Δn_r) generates various logarithmic variations. It shows that the projected hyperspace could be spherical, ellipsoidal and hyperbolic, depending on the non-linear oscillation frequency matching between two neighboring closed loops in the NC/OF metric. The deviation in the projected hyperspace is strictly a function of deviation in orthogonality, ΔJh; hence $Jh = Jh(\min) + \Delta Jh$. Here, its 2.0...(an infinite series). In order to verify, we draw an orthogonal triangle ABC (Figure 4.15a), we assign φ (AB) and e (BC) two perpendicular sides of a triangle ABC, where angle/ABC = 90°, while AC is π. We find that if $Jh = 2.0...$ The angle between π and e measured by e is 29.62°, the same angle measured by φ is 31.28°. The angle between π and φ as measured by e is 60.42°, and measured by φ is 58.75°. The deviation of ±1.6° is from non-planar projected hyperspace. ABC relates φ, e and π as $e^{2.0} + \varphi^{2.0} = \pi^{2.0}$ Second, by continued fraction expression for φ and e are a linear function of N as $N \to \infty$, while π infinite continued fraction needs a square of integers $\pi = f(N^2)$, while $\pi^2 = f(N)$, hence π^2 could integrate both φ and e as the sum of areas. One could estimate π accurately from $e^2 + \varphi^2 = \pi^2$. This identity is followed by biological systems to fuse three kinds of resonances, electrical, magnetic and mechanical. The frequencies are selected accordingly.

The interplay of hyperspace binds time crystal with the clocks. Increment of the higher dimension in the hyperspace is related to the higher dimension of a time crystal (Figures 6.1e and 2.11b). Geometric algebra is an important subject of study and the bondage of hyperspace and higher dimension shown here for implementing PPM (Figure 6.1f) by following $p\gamma = l\pi$ is another class of geometric algebra.

6.1.3 Ordered Factor of Integers and Coding in Proteins

Tubulin protein's last nail in the coffin for "it from bit": the songs of PPM: One question still remains unanswered. How does the vision of one world arise out of the information-gathering activities of many observer-participants? Then comes the necessity of a metric that links a number of

choices with the maximum number of ways the events could be linked. Little kids learn to do this calculation in the early school days, take every integer and find the ordered factors of that number. Reddy et al. (2018) did not discover this metric from the imagination, rather, looking at the quantum tunneling images of tubulin protein taken at different ac frequencies. The set frequency at the tip changed the tubulin conformation, it looks different and we were trying to find if there is an elementary structure governing the dance of tubulin protein with the ac music. When all attempts failed, Reddy et al. accidentally discovered that if we count the number of rings in the alfa-helices, it varies 1, 2, 3, 4, 5, 6 as if one is counting integers. They got even more surprised when they found bright regions in the tunneling image of a protein is a region where rings are arranged 2–3–3–2, or 3–4–4–3 or 2–6–6–2, as if it is making closed circuits. As we could link how a particular ac signal that activates one circuit to another, the brightness in the tunneling image shifts, one region to another. Therefore, music and dance are happening together, and nature has engineered proteins to count primes and integers to build its musical codes.

In order to learn and implement α-helix engineering, one has to re-write the PPM in a little different way as shown in Figure 6.2a. Instead of putting the ordered factor or the number of choices we put the integer itself (Figure 6.2a). When experimentally observed circuits 2–3–3–2, or 3–4–4–3 or 2–6–6–2 are shown, we get clocks (Figure 6.2b) and clocking direction (Figure 6.2c). All the clocks were compiled and a combined time crystal was formed (Figure 6.2d). The combined time crystal resembled as a derivative of a triplet of triplet clocks (Figure 6.2e). Singh et al. (2015) downloaded complex protein structures and imaged them by switching conformers to find whether they are also singing the music of primes like tubulin. They do. Tubulin monomer, dimer and nanowire made of tubulin called microtubule were studied to understand how biological systems use the pattern of primes. Tubulin monomer uses all compositions up to 5, dimer up to 7 and microtubule up to 13. It means there is a metric and biological systems cut a slice from the metric up to a prime number.

6.1.4 Biomaterials in and the Remarkable Engineering of Water Channels

The historical background: Long back, Langmuir proposed that when biomolecules get wet, in addition to an attractive van der Waals and repulsive electric forces there is an additional force created by a strong monolayer of water molecules that cover the molecule and isolates the rest of water molecules from the biomolecule (Langmuir, 1938). Thus, water molecules cover the biomolecules in the solution, a maximum 2–3 molecular layers, which means ~3–4 Å thick barrier of water is there, known as hydration shell, whose dynamics is regulated by protein shape, (Laage et al., 2017). Such hydration of biomolecules is related to the forces that are either attractive or oscillatory, and where the origin of repulsions with water has a different origin (Israelachvili and Wennerström, 1996). Hydration shell has ~15% more density of water than bulk water (Merzel and Smith, 2002), but it is not homogeneous all around the molecule. Particular functional groups bind water, all groups cannot bind equally, so the bonded water molecules create a channel of water namely "spine of water." For example, it was known for a long time that DNA has a double helix, but if we consider the water channel created by choline groups in the phospholipids, then there is an additional helix of water too (Ethan et al., 2017). The shell plays an active role in modulating the biomaterial property. A water film can emit coherent THz near thermal signal (Jin et al., 2018), it eliminates the need for additional laser sources required to operate fourth-circuit element Hinductor. These ordered waters act as an active device, for example the third water helix of a double helix DNA has a resonance peak at 3000 cm^{-1} ~ 0.372 eV, it is a significant amount of energy that could play an essential role in governing DNA dynamics (McDermott et al., 2017) via spiral-spiral electromagnetic coupling (Liu et al., 2017) and mechanical pressure on each other.

Water channels in the carbon nanotube and microtubule: Water channels could be created artificially (Gong, 2018), which is a key to the synthesis of the fourth-circuit element Hinductor (Chapter 8). Therein, three concentric helices are required, even if the helical cylindrical surface is hydrophobic, ordered crystalline surface would order water molecules outside at maximum three molecular layers one above another (Du et al., 1994). Therefore, molecular biological studies and artificial ordered water monolayer synthesis have consistently shown that simply by wetting one could generate a helical replica on the surface of a nanostructure. Another critical question is what happens inside? Microtubule is a hollow tubulin protein-based cylinder with 25 nm diameter, its internal hollow core is 18 nm wide, wherein ordered water channels reside. Similarly, ordered water channels are found in the carbon nanotube core (Koga et al., 2001). The core water part works as an active device, undergoes symmetry breaking and phase transition like an independent molecular system (Takaiwa et al., 2008), which exerts a constant axial pressure (He, 2014).

The non-linear 3D architecture of water in and around biomolecules forms by certain functional groups which binds water molecules in a 3D shape. Imidazole is the part of proteins that holds the water molecules (Sun et al., 2018). In the lipid bilayer membranes, artificial imidazole quartet (I-quartets), which is a stack of four imidazole molecule and two water molecules, form a water channel. These I-quartets reorganize the oriented water wires as a 2.6 Å diameter tube. The water channels can transport ~10^6 water molecules per second per channel, which reject all ions except protons. Therefore, from transport to mechanics, water channels within, in between and above play a fundamental role in governing the system dynamics.

6.2 EXPERIMENT ON A SINGLE MICROTUBULE OR ANY GENERIC SYSTEM TO FIND ITS TIME CRYSTAL

Microtubule is a nanowire made of tubulin proteins (Figure 6.3a). Extensive measurements were carried out on a single microtubule. However, all the experiments carried out thus far measuring the time crystals, measures output in the same frequency range as the pump frequency. Reddy et al. argued (2018), this has been an entirely wrong approach. If input and output response is captured in the same frequency domain, a system could easily undergo hysteresis and deliver an output that reflects measuring device geometries. Hence, they kept 10^6 orders difference between the frequencies of pump and probe (Figure 6.3b).

6.2.1 How Do We Experimentally Confirm That an Assembly of the Clocks Is a Time Crystal?

In the old days, pheromones were thought to be a chemical agent, until its time structure was discovered (Marion-Poll and Tobin, 1992). At every scale, universe redefines unit of information, smallest length, shortest time, maximum speed etc. New units are formed at every layer of information structure, and PPM sets those criteria along with GML. A few geometric shapes and a few symmetries play with phase space to create everything that we see in our universe. Geophysical rhythms are connected to our cells (Hoffmann, 1976). Living systems are all made of PPM, so when we study an atom we used to say, a complex of atoms is studying an atom. Now, we would say, a large slice of a PPM is studying a small part of it.

We need to redefine the experiment to read the information: In the "yes or no" paradigm, it is all about counting the number of responses, a counter digitizes every response, be it classical or quantum, the problem is between the probe and the probed (Wheeler, 1981). The clocking paradigm does not work in the new approach where we want to find the periods. Fundamental changes are needed to be made in the experimental protocol, it is part of pathological science (Langmuir, 1989). Since a wave function collapses only if it is mixed to a certain degree with the environment's wave function (Zurek, 1981, 1982, 1983), *a wavefunction of singularity* is not easy to destroy. It could even survive a Brownian motion (Unruh and Zurek, 1989). (a) The love of circadian rhythms to tubules (Giebultowicz and Hege, 1997) inspires us to assume that to execute an event in a time scale, the universe executes several periodic sub-events at various faster time scales. (b) The ratio of periodicities of events are where the symmetries of number systems are located, therefore, it is not a straight-forward Aharonov-Bohm type quantum experiment. (c) The measuring system cannot be any probe. PPM based ultra-low power sensors are required to sense natural spontaneous bursts of signals. A pump probe type experiment has to be replaced by a pure sensing measurement, a non-demolition type experiment. (d) From GML, we have circles with geometric shapes whose corners are singularity points. Non-singularity domains are regions of silence.

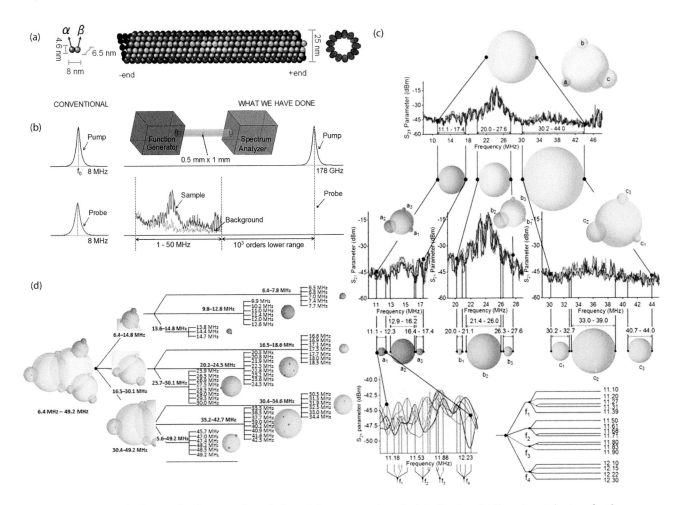

FIGURE 6.3   (a) A tubulin protein dimer, a microtubule and the cross-section of a tubulin ring. (b) Experimental set up for the measurement of time crystal. Function generator that generates signals of different frequencies connected to a spectrum analyzer through a measurement tube. The sample is pumped at 1–178 GHz and probed at around 1–50 MHz. (c) How raw signals generated by the microtubule solution is converted into a time crystal is shown. Multi-layered cases are studied. The peaks at the lowest level with the high-density frequency peaks show the discrete energy levels. The frequency values are noted beside the levels. (d) Integrated time crystal of microtubule derived from isolated clocks of panel (c) is presented.

There is a burst of energy from singularity points (Kuramoto, 1983). The artificial brain sensors should read the geometry of silence and the geometry of light together. The vortex of light and knots of darkness are not complimentary. (e) The metric is exclusive to the observer and environment and represented as such or pristine form in terms of a set of symmetry. In summary, optical vortex route is the simplest tool to read phase discontinuities of a time crystal, contactless acquisition of architecture of phase discontinuity (Genevet et al., 2012).

1. **Vortex of phase in the *three-phase* axes:** In a crystal, along the three spatial axes, the unit cells repeat whose corner points are mass. Similarly, the maximum response of a biological clock under perturbation, plotted along the three-phase axes may deliver a repetition of unit cells whose corner points are the phase singularities. The plot is phase response curve. Three orthogonal phase axes are: first, a normal clocking phase of a biological rhythm without perturbation $(\varnothing)$, second, the phase of the spontaneous

ripples formed after the perturbation is removed $(\varnothing')$ and third, the duration of perturbation or intensity (Chandrashekaran and Loher, 1969) is converted into phase M. Unperturbed means $\varnothing' = \varnothing + M$. Phase for all three parameters varies $0°$ to $360°$, as we consider the variation of perturbation as $0–1$. When $\varnothing$ changes $0°–360°$, $\varnothing'$ changes too, but both do not end at the same time, $\varnothing'$ may end before or after $\varnothing$ reaches $360°$, even, never finish $360°$. Two rolling circles make a torus, one winds on another, $d\varnothing/d\varnothing'$ is the winding number. If the 3D phase space looks like a clocking spiral, it is a time crystal. The geometry of a phase response curve could be edited to restructure the architecture of clocks (Kawato and Suzuki, 1978; Zeng et al., 1992), modulating the 3D plot (hydra; Taddei-Ferretti and Cordella, 1976; Kawato, 1981) is like morphogenesis. The architecture of phase in brain cortex tells about time crystal composite (Lopes da Silva and Storm van Leeuwen, 1978), wherein multiple lattices of time coexist in different time domains. We convert the duration of perturbation (Engelmann et al., 1973) or any perturbation control parameter like infra-red or thermal pulses (Engelmann et al., 1974) into a phase because below and above the threshold control, the perturbation does not affect the natural rhythm of a biological clock. Waves in 3D media are unique Winfree and Strogatz (1983a, 1983b, 1983c, 1984a); most biological systems rely on clock of clocks, i.e., time crystal composite (Bourret, 1971). The intimate relation between circadian rhythm and cell division could be broken using light pulses to see how different clocks fuse in a time crystal (Malinowski et al., 1985). Ovulation sets unit of human life cycle adopting lunar periodicity (Menaker and Menaker, 1959).

2. **Two or more types of phase singularities and or ripples:** If the perturbation is removed, a clock or rhythm would always generate ripples as hysteresis, before returning to the normal state. It takes time for a biological clock to reset. That lag is hysteresis and it is not and should not be the evidence of a time crystal. One atom cannot make a crystal, one clock cannot build an actual time crystal. Phase singularity builds a crystal in biology (Brambilla et al., 1991). If two waves interact, we always get constructive and destructive interference, but only one type of singularity is seen after the perturbation is removed. There are various ways to build a phase singularity (Winfree, 1986d). An assembly of clocks is a time crystal if and only if a composition of singularities is exposed as an effect of perturbation, which cannot be explained by a regular hysteresis. Evolution of collective singularity in a group of oscillators tells us that singularity in a coupled system is not a single layer event (Enright and Winfree, 1987). For a 1D time crystal at least two types of phase singularities would generate two ripples, it means there is at least one clock that is not due to the hysteresis. For a 2D time crystal three-phase singularities giving rise to three ripples and so on.

3. **Even, an odd and prime number of cycles to complete a period:** The helical phase space has a clocking and anti-clockwise direction, each turn of the screw is 360° phase, unit of time. The unit cell of a time crystal is represented as $t_1 \times t_2 \times t_3$ in a time unit (helicoidal cell). The vortex-like phase structure is the key signature of a time crystal, the number of rotations or $d\varnothing / d\varnothing'$ is the winding number, is this odd or even makes a difference. Historically the conflict between odd and even was repeatedly found in several biological systems. Neuron's Ca++ rhythm in a nerve spike is a time crystal (Best, 1976), yeast cell's NADH rhythm is a time crystal (Betz and Becker, 1975b), oscillating mitochondria is a time crystal (Gooch and Parker, 1971). Aperiodic rhythms in two time domains could change such that a clocking effect would emerge (Freeman et al., 2003).

4. **Phase reset curve (= old phase vs new phase):** Quantized phase jump: Resetting curves, when plotted in terms of phase shift, commonly suggest a discontinuous change from one-half cycle advance to one-half cycle delay. This is like a unit lattice of materials science, it repeats to generate time crystal, in the multicellular organisms (Winfree, 1976a). The "phase jump" was thought to be the "discharge" of a physiological accumulator. Type 0 phase reset. Phase reset can happen in many ways. Type 0 reset means the phase surface is only one, wrapped around a single axis. Circadian rhythm is type 2 (Cote, 1991). In Chapter 3, we reviewed the consequences in terms of phase-dependent phase resetting. The result was a "time crystal." Phase reset, fundamental to a time crystal is frequent in heartbeat of pacemaker (Anumonwo, et al., 1991; Jalife et al., 1979; Van Meerwijk et al., 1984), embryo (Clay et al., 1984, 1990; Guevara et al., 1986); regulating respiratory rhythm (Eldridge et al., 1989) by adjusting the phase delay (Lewis et al., 1992), electric fish (Wessel, 1995); fruitflies (Winfree, 1971a). Topological changes in the neuron membrane regulate phase reset curve (Guevara and Jongsma, 1990). Phase resetting helps in designing medicine (Tass, 1999). Human eye undergoes phase reset with every little change in light (Rusak, 1993). To operate a brain several phase reset curves with widely varied geometry and time-domain should not have any discontinuity (Gedeon and Glass, 1998). Earlier, that study required 100 s of coupled differential equation which PPM replaces using a simple linking pattern of symmetries.

5. **Interaction between clocks: Coherence and decoherence: Isochrones:** Isochrones (having the same time periods, Pinsker, 1977) have polar symmetry, i.e., amplitude R governs an instantaneous phase $\varnothing = g(\varnothing, R)$, $\Phi = g(\phi, R) = \phi - f(R)$. It provides $\Phi' \equiv 1 = \phi' - (df(R))/dR\,R'$, which leads to $(df(R))/dR = (\phi' - 1)/R'$. The expression if positive, rotation is clockwise, if negative, anti-clockwise. Action spectra reveal the identity of isochrones (Frank and Zimmerman, 1969). Isochrons do not see the spatial separation, they could bind in time remaining far apart (Guckenheimer, 1975). Deep inside protein, time for phosphorylation is adjusted by isochrones to dock right proteins at the right places (Garceau et al., 1997).

6. **Perturbation as an attractor:** Time crystal's mechanical thrust: A perturbation normally reveals a singularity point via burst (Johnson et al., 1979). Perturbation could trigger an assembly of clocks to a particular virtual point, defined configuration of clocks or a saturated clock at equilibrium (Peterson and Jones, 1979). Perturbation sources vary widely. If C_1 is the number of carriers (photon, ion, any form of clocks, or even time crystals) leave a time

crystal at a rate k, we get $dC_1/dt = -k$. It gives, $C_1 = a\cos 2\pi\phi - M$ and $C_2 = A\sin 2\pi\phi$, where from we get $\tan(2\pi\phi)' = C_2'/C_1' = v\sin 2\pi\phi/(v\cos 2\pi\phi - v^*)$. The equation (ϕ, ϕ', M) accounts for the knots of darkness or lines and structures of singularities. The mechanical force between a pair of time crystals is $F = -k\ddot{x}$. In a high viscus system, $C = R(C) + D\nabla^2 C$, here C is the concentration of clocks, $R(C)$ is the flow of clocks in an integrated architecture of clocks, D is the ratio of the elastic constant and effective potential. Viscosity is due to the repulsion of carriers and or clocks. Neuron often converts linear (not looped) events into a loop or clocking events (Enright, 1980a).

7. **Limitations of pinwheel experiment:** observing a helicoid in each unit cell of the time crystal or observing a rotating wave in a pinwheel experiment; Winfree, 1977) in itself suggests nothing particular about the underlying mechanism. It may be interesting that more complicated phenomena are not observed, but the observation of only this much scarcely indicates more than: (a) that the physiological dynamics tends to a unique cycle when unperturbed; (b) that it tends to a steady state under the stimulus used; and (c) that both these processes behave in reasonably smooth ways. The missing part of unit cell of conventional time crystal is that a phase discontinuity locks oscillators phases, phase-locking and the emergence of phase discontinuity are two parts of the same coin, the coin is a phase reset experiment (Glass and Mackey, 1979; Glass and Nuree, 1984).

8. **Vortex filaments and scroll filaments to meander flowers: Hopf bifurcation:** In real phase reset, single-variable is no good. Researchers take a number of smaller, interacting cubes to probe carrier gradient in the medium of operating clock (19-point as in Dowle et al., 1997). Multiple hierarchical architectures e.g., static and dynamic filaments can change the clock's state. Only when we map the filaments formed by connecting the singularity points, we find why and how the usual old cycle is changing. In the Fast Fourier Transform plot of biological clocks we see the complex superposition of slower and faster rhythms. When we plot shift in phase and frequency as circles, it accounts for the ratio of time scale ϵ, better than the earlier version of meander flower representation (order parameter or frequency vs ϵ plot), where the number of petals increases with ϵ (Zykov, 1986). The angle between the successive petals is exactly 2π/integer. In the GML, nested circles universally explain all possible meander flower patterns i.e., entire landscape (Zykov and Morozova, 1990). Even clocking directions are regenerated by connecting the guest clocks in the inner side of the perimeter (anti-clockwise) or with the outer side of the perimeter (clockwise).

The nesting of a pair of clocks in GML is decided by frequency and ε, this is beautiful, the condition of self-assembly of clocks is decided naturally, across a boundary which is the limiting line of Hopf bifurcation. To note that Hopf bifurcation is a point where the system loses equilibrium and oscillates between two complex planes, i.e., a new clock appears within a limiting ε, so it is called limit cycle. It is mathematically presented as $dz/dt = z((\lambda + i) + b|z|^2)$, here, $b = \alpha + i\beta$, α is Lyapunov constant. Two limits are observed in the meander flower presentation (Figures 2.7c–e), above a threshold frequency and a time ratio, the clocks do not bond, remain isolated. A surprising feature is the "linear looping" between the flowers with inward petals and those with outward petals (Lugosi, 1989). The lines with similar dynamics of flowers are isochrons (Guckenheimer, 1975). For GML it means a guest cycle is on a journey through the perimeter of a host cycle that has an infinite perimeter. Compared to solving differential equations each representing a symmetry that governs the changes in geometric shapes (Barkley, 1995), GML is simple to analysis and more accurate to explain the flower landscape.

9. **Edge and screw dislocation of time crystal:** Lability of the apparent amplitude of oscillation, with very nearly the same period at all amplitudes. An effect of perturbing stimuli which to a first approximation resembles parallel displacements of state along one state variable, as though one particularly labile substance was destroyed by the stimulus, down to a minimum near concentration O. Chandrashekaran and Engelmann (1973) suggested replacing this smooth function by a stepwise increase in photosensitivity with steps at each T^*. That would produce a tear in the resetting surface, an edge dislocation in the time crystal.

10. **A number of lines on the star network:** There are domains of absolutely dark regions, where both the signals, the perturbation and the original turn silent, namely a phase singularity (Efimov et al., 1998). One could experience timelessness (Myers et al., 1996). From these dark points, multiple phase surfaces where the output signal peaks spread out like a star network. In a time crystal, a singularity point is a source of phase paths, just like the classical points of a qubit's phase sphere. The only difference is that depending on perturbation caused by the observer, there could be multiple uncorrelated distributions of singularity points on the phase space.

6.2.2 TRIPLET OF TRIPLET RESONANCE BAND IN MICROTUBULE

We observe three distinct regions where the ac signal is allowed to transmit across microtubule as only the MHz time crystal is shown in Figure 6.3c and d. These regions are the

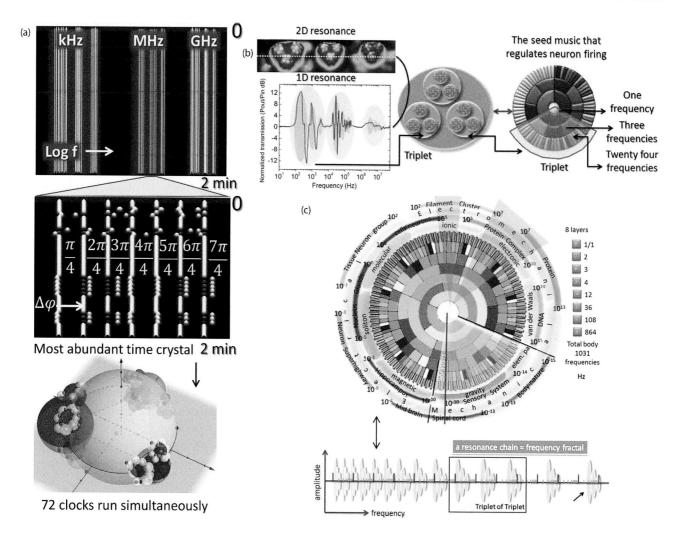

FIGURE 6.4 (a) Top, resonance frequencies of a microtubule plotted in the Log scale continuously recorded for 2 minutes. Neighboring microtubules are continuously pumped with white noise. No change in resonance frequency is observed. During the same time, the phase difference between 8 peaks in the MHz domain shows significant shifts with the wireless energy transfer. Using this data its equivalent time crystal made of 72 clocks is created in the bottom layer. (b) 1D resonance band of a hippocampal rat neuron, measured using coaxial atom probe. 1D resonance means electric field applied in one direction. 2D resonance means electric field applied in two perpendicular directions. Using a line it is shown that 1D resonance is a single line in the 2D plot. To the right, triplet of triplet of octave made of 72 frequencies is shown using 2D resonance domain and nested frequency data. (c) Model of a human brain. Triplet of triplet made of 72 frequencies are shown for 12 bands. If one moves one inside another, finds, 12 bands, it is like a resonance chain where only three bands are visible. One has to go 12 times one inside another to access entire resonance chain.

clusters of sharp peaks where ac resistance falls significantly. The complete band of microtubule is shown in Figure 6.4a. A time crystal is not limited to microtubule only, it extends to neuron (Figure 6.4b) and to the entire human brain (Figure 6.4c).

First, 10–100 kHz domain, where, with the increasing ac bias amplitude, one resonance peak resolves into three distinct peaks. From the height and width of these peaks, we calculate the Quality factor Q, which is the number of oscillations before energy becomes $e^{-2\pi}$ times the initial energy. For the kHz band, we get the Quality factor, Q ~3–5 for the three peaks, spread over a frequency bandwidth of ~300 kHz. Though microtubule ac resistance falls sharply around 8 MHz, single microtubule acts as a high pass filter. This particular

kHz domain of three peaks has several key features: (i) it is the longest-sustained oscillation domain; (ii) the magnitude of output transmission is strongly dependent on the input signal intensity, (iii) three major peaks are not harmonics of the first peak, but equally spaced in the frequency band.

Second, in the 8–240 MHz region, the resonance peaks are sharp (Q ~100–300), the energy is distributed non-linearly among particularly allowed 16 prime resonance peaks along with several harmonics. For a microtubule of typical length 10–12 peaks out of 16 are observed. However, statistically, when Sahu et al. measured microtubules of several lengths between 200 nm and 24 µm, the overlaps of several peaks explicitly unraveled 8 original peaks, rests are harmonics. Microtubule's length strictly selects the allowed/blocked eight resonance frequencies and the rest eight first harmonics always occur, irrespective of the microtubule length. Sahu et al. observation suggest that eight variable peaks are coupled to the universal eight resonance levels whose intensity varies with length. This MHz resonance band has two sub-bands one in the 10–35 MHz region ~12 MHz ($Q \sim 75$), ~22 MHz ($Q \sim 46$), ~35 MHz ($Q \sim 129$), ~46 MHz ($Q \sim 62$). Another sub-band is in the 100–250 MHz region, ~89 MHz ($Q \sim 43$), ~144 MHz ($Q \sim 75$), ~180 MHz ($Q \sim 46$), ~228 MHz ($Q \sim 64$). Here $Q \sim$ Quality factor measured for $L = 1$ μm, at 300 K, 1 atm. and 90% humidity.

Finally, in the 7–13 GHz domain, microtubule shows two Gaussian-like resonant transmissions. If microtubule is pumped at 7 GHz, near 19 GHz we observe a transmission band, it is a remarkable modulation since the frequency is amplified by ~3 times. However, for the 13 GHz pumping we observe another Gaussian-like stochastic resonance, however, in this case, it peaks around 13 GHz only. Therefore, again, similar to kHz bands resonance peaks are wide, with low Q values ($Q \sim 1.5, 2$).

The electrical resonance is associated with the mechanical oscillations of the single microtubule string. Among three resonance bands, only MHz band induces visible mechanical oscillations to the microtubule. Sahu et al. (2013a) investigated further the origin of these three bands. If water channel is removed from the core of microtubule, the GHz resonance disappears, if proto-filaments are isolated, kHz resonance disappears, still, the MHz resonance survives. Therefore, the GHz resonance is due to the water channel, the 1 Hz–kHz oscillations are due to the proto-filament oscillations, the MHz oscillations originate inside the protein structure triggering elastic vibrational modes of the microtubule as predicted by Jiri Pokorny. The kHz oscillation is energetically insufficient to vibrate thirteen isolated proto-filaments, the GHz resonance vibrates the water channel, not the hollow protein cylinder, which MHz vibrations do eloquently.

One fundamental parameter for electromechanical oscillations is the phase modulation. The phase modulation behavior in the kHz and MHz domain, since the phase difference between input and the output ac signal in these regions are quantized. Statistical count of phase difference Φ shows peaks at 0°, 45°, 90°, 135°, and 180°; wherein, each resonance peak is associated with a distinct Φ. The kHz and GHz bands do not change the phase of the ac input signal while a quantized phase modulation by $n\pi/4$ occurs during MHz transmission across the microtubule, thus, in the MHz band, microtubule acts as automated phase-locked-loop (PLL) oscillator. A phase-locked loop or phase lock loop (PLL) is a control system that generates an output signal whose phase is related to the phase of an input "reference" signal. It is an electronic circuit consisting of a variable frequency oscillator and a phase detector. The circuit compares the phase of the input signal with the phase of the signal derived from its output oscillator and adjusts the frequency of its oscillator to keep the phases matched. The signal from the phase detector is used to control the oscillator in a feedback loop. Frequency is the derivative of phase. Keeping the input and output phase in lockstep implies keeping the input and output frequencies in lockstep. Consequently, a phase-locked loop or PLL can track an input frequency, or it can generate a frequency that is a multiple of the input frequency. The former property is used for demodulation, and the latter property is used for indirect frequency synthesis.

6.2.3 THE BIOLOGICAL RELEVANCE OF THE TIME CRYSTAL

Ten biological relevance's to a Winfree time crystal obtained via experimental studies of biomaterials: Winfree detected spontaneous emergence of singularity in a biological clock (Aldridge and Pavlidis, 1976; Alleva et al., 1971). Automated creation of a guest clock in the phase perimeter of a host clock prompted him to connect the emergence of life with the formation of a time crystal. A time crystal holds two or more distinct rates of time flow in an orderly fashion. It means the system can hold and execute an event.

1. **The "signal burst or bing" is not important, silence or phase between the "bings" is important:** Detecting a time crystal has a clear route. Find, if the resonance frequencies of material remain the same. Then check if the phase associated with each resonance peak changes with time. It means the materials dielectric property that regulates the resonance remains unchanged. The geometric parameter of the material edits the phase to run the clocks. For microtubule, it is the length (Sahu et al., 2013a, 2013b). By varying it, one finds a change in the intensity and phase of the peaks but not the resonance frequency values (Ghosh et al., 2016a). *The ratios of phases for the resonance frequencies determine the geometric shape stored in a microtubule. Similarly, one can determine the geometric information stored in the particular conformations of protein's & their complexes.* Currently a time crystal pens are being explored that will read the 3D information structure as time crystal directly from nature.

2. **Nesting of clocks was missing in the concept of time crystal:** The existence of time crystals was verified in the elementary life forms, for decades, experimentally. However, the nesting of clocks was never proposed or investigated. Winfree's idea of a singular singularity had to be generalized and we explore that possibility. A time crystal of a virus vibrates as a single clock. Inside, each plane of lattice oscillates in the period, those are clocks inside clocks. Inside a plane, each group of atoms vibrates like a clock. The crystal acts like a clock inside a clock inside a clock (Shlesinger, 1988). The network has several layers within as the clocks are also clocked (Edmunds, 1977). Such fractal clocking in the biomaterials is a recent discovery (Ghosh et al., 2014a, 2016a). The resonance band of a microtubule. The phase is flipping spontaneously. However, fractal clocking was reported in the ion channels long back (Liebovitch et al., 1987).

3. **Ten geometric resonance properties of biomaterials:** Recently, the resonance frequencies and their associated phases of various biomaterials were measured (Sahu et al., 2013a, 2013b, 2014; Ghosh et al., 2016a). It shows that the emergence of frequencies is not random. They follow a unique geometric relationship between them. Here are some features. (a) Various carriers interfere with resonating with the biomaterial cavities. Thus, make their distinct band of resonance frequencies. The experiments show that the distribution of frequencies is grouped as a triplet of triplet. It means apparently, there are three bands, but if one looks within one band, finds three more sub-bands (notice the kHz, MHz and GHz bands). (b) Each region of all nine sub-bands contains one to eight peaks inside, these make biomaterials E1 to E8 class systems. Biomaterials increase layered sub-bands but no instance is reported to have more than eight peaks at the lowest level. (c) The time crystal of a triplet of triplet band has 72 clocks embedded in the phase spheres. The diameters of the experimentally measured Bloch spheres remain the same. Only, their relative positions change and on the 3D spheres, they are visible, distinctly. (d) Resonance frequencies are always associated with the quantized phases. A shift from quantization in a particular peak's phase value is the information. (e) As one move from higher to the lower frequency range, the scale-free power distribution of the frequency band is observed (Ghosh et al., 2014a, 2016a). The intensity of the resonance peaks increases by orders of magnitude. (f) The amount of material is irrelevant. The geometric parameters, length, width, pitch and lattice parameters regulate the self-similarity of arranging the resonance frequencies. (g) Each frequency corresponds to a singularity point (Mallat and Hwang, 1992). The value of frequency relates to the circle diameter. The frequencies of the resonance peaks would remain static but not their phase. Using phase one can put system points at an accurate location on the host phase cycle's perimeter. The step secures the relative Bloch sphere positions. Thus, biomaterials are mathematically precise devices. (h) The time crystal remains intact if the fundamental geometric parameters remain constant. Then it is possible to change the system points and regulate the relative phase or phase shift (Johnsson et al., 1973). The same time crystal would then store different information. It is interesting, because, experimental measurement of detecting resonance frequencies or phase associated with a peak would never show the information. We need to measure specifically the phase shift to see that a biomaterial is processing information. In the Eukaryotic cells of the entire kingdom, microtubule rapidly changes its length (dynamic instability), sometimes its diameters (6–19 protofilaments), and then its pitch to morph its shape in incredible ways. Thus, it carries out the key tasks of a living life form by editing its topology. (i) Sometimes, it is necessary to add or deduct some clocks or resonance frequencies. Then the structure would change its typically associated symmetry so that particular singularity points disappear (all phase values get defined) or new singularity point appears. Subtracting or adding a singularity point means destroying the link or creating a link with several layers of geometries hidden within. (j) The number of oscillators or the number of devices has no relation to the number of clocks, it is not even related to the lattice symmetries. The number of lattice symmetries adopted by microtubule is the number of resonance peaks for a microtubule, not the number of clocks. Often, spatial symmetry breaking is associated directly with the time symmetry. *A composition of lattice symmetries together defines a clock if they all undergo phase transitions as a group.* In the assembly of clocks, only eight dynamic symmetries repeat (Sahu et al., 2013a, 2013b, 2014). Sahu et al. have proposed unique fourth-circuit element Hinductor for artificially demonstrating biological time crystals and the potential of singularities (Sahu et al., US patent 9019685B2).

4. **The magnetic beating of beats:** Electrical beating occurs when two electromagnetic signals of very close frequencies interfere. Biological materials known for producing low magnetic fields (10^{-10} T) could generate beating locally in the lattice. Then the beat signals could interfere again if the smaller lattice domain is part of a larger structure. The beating of beats could beat again and such layered structures are rich in biology. Thus, one observes that beating signals cover the entire electromagnetic or magnetic frequency domain (Jaynes, 1980; Reddy et al., 2018). Such a hierarchical network of beating requires simultaneous switching off the topological constraints at all level to destroy signaling. Thus, all signals survive together at ambient conditions.

5. **Harvesting thermal, electrical and electromagnetic noise:** Time crystals in biomaterials reveal its unique phase relationship in the presence of noise (Betz and Chance, 1965a), one could use the noise trick inspired by biology to read the time crystals. Thermal noise compensation is rich in biology (Brinkmann, 1971), clocks neutralize the thermal noise (Bruce and Pittendrigh, 1956; Fuller et al., 1978; Hastings et al., 1957). Thermal pulses could even activate the biological clocks (Engelmann et al., 1974). Harvesting noise for filtering the frequency response by a synaptic junction is already reported (Brunel et al., 2001). The origin of electrical ionic activity is attributed to oscillatory potentials in biology (Brown et al., 1975). Even the electromagnetic pulses of a light edit the biological clocks (Engelmann et al., 1973). However, ordered signals affect the infrared photon absorption in biosystems.

Signal inhibits the noise conversion to resonance-induced interference. It affects the nested beating described above (Sahu et al., US patent 9019685B2). Among all frequency domains, bio-systems absorb most in the infra-red domain.

6. **Harvesting singularity to self-assemble clocks: Learning, communication, all forms of information processing in neuron occurs via time crystals**
If one knows feedback path accurately, predicting the emergence of singularity is an easy task (Johnsson and Karlsson, 1971; Mallat and Hwang, 1992). Energy transmission is studied following a unique biological route, bottom-up. Protein ↔ microfilaments ↔ bundle inside neurons (branches) ↔ bundles of neurons in a cortical column. It suggests that a neuron edits the phase of a transmission signal by modifying the neural branches (Reddy et al., 2018; Ghosh et al., 2016a; Jenerick, 1963). The effort changes the stored geometric structure of the neurons time crystal, surprisingly, Ghosh et al. claimed it explicitly in 2016, some old results were very near to it (Buhusi and Meck, 2005). A neuron may take two steps. If a neuron builds a new branch, it creates a new clock. Else, it locally modifies an existing branch. That edits the phase of an old clock. Thus, a change in the structure does not mean the creation of a new clock or singularity. Neurons communicate by clocking ionic pulses (Gerisch et al., 1975), but the evidence of wavelike communication is also there (Hill, 1933) apart from Ghosh et al. work in 2016. After the creation of a new branch, the system spontaneously investigates two factors. First, whether the phase modification is required in the new clock. Second, whether the new clock is integrated into a suitable location in the existing time crystal.

Similar to neuron, protein, microfilament and neurofilaments, assemblies inside the neural branches and cortical column edit their own time crystals (Noctor et al., 2004). The modified time crystals continuously edit their physical structure (Van Essen, 1997). The greater neural pattern in size often dominates in the higher-level (slower) clocks in the time crystal (Xue et al., 2010). Following the magnetic beating of beats all forms of vibrations are topologically connected in the brain. Electromagnetically this would have never been possible, as the electromagnetic signal damps in the cell fluids. Consequently, the proposal that a brain is a single resonance chain (Ghosh et al., 2016a; Basar, 1990) is a primitive one, Reddy et al. add that the brain is a time crystal, resonance chain is a limited view ignoring the topology of phase.

7. **A non-linear correspondence between the spatial and temporal assembly of crystals:** Even a tubulin protein molecule is a time crystal. It self-assembles into another time crystal, microtubule. Then microtubules self-assemble into a bundle to build the core structure of a neuron, e.g., an axon. Neurons respond as time crystal (Buhusi and Meck, 2005), a bundle of neurons forms a cortical column that is also a time crystal. The bundle of cortical columns also acts as a time crystal. A secondary structure of protein ~2–5 picometer to 1 mm cortical column, the spatial journey is about 10^7 orders. However, the temporal scale regulation is from pico-seconds to seconds, 10^{12} orders (Ghosh et al., 2014a). The parameter that regulates the phase relation of various resonance peaks is geometric. Tubulin's each of the eight conformations holds a particular set of geometry. Similarly, microtubule's different length, lattices hold suitable symmetries. Neuron's branches edit their own symmetries spontaneously. The cortical columns length and symmetries of neuron locations edit their own symmetries. In association with the spatial symmetries the phase relationship changes together causing a ripple effect in the temporal symmetries. The resonance frequencies remain nearly unchanged, yet *10^7 order time crystal gets changed by 10^{12} spatial scale changes. One cannot isolate a particular part of a time crystal. To hold memory various clocks only use the phase space, together, thus information is stored everywhere simultaneously.*

8. **Interacting with the living cells and proteins in their own language:** The biological structures sense a phase connected time crystal network better than conventional sensors. A sensor absorbs the existence of a signal burst. Biomaterials senses not just phase links between several such bursts, but exactly the pattern following which those links change with time. When 7–8 days old hippocampal neurons were given a specific set of frequencies as time crystal, wirelessly. The suitable neuron, responded. No searching is required for searching for a suitable time crystal. Electric or electromagnetic signaling faces the effect of a physical boundary of a material. However, the magnetic beating of beats do not face boundary, it integrates by a phase map with everything within a magnetic shield. So, communication does not happen as also observed in electrical or electromagnetic communication scenario. It was predicted in 2014, as a spontaneous reply (Ghosh et al., 2014a). Moreover, it was possible to encode geometric shapes in a neuron. Talking to neurons is possible in its own language (GML, Agrawal et al., 2016b). Even treating misfolding of proteins are possible by twisting the time crystals (Sahu et al., 2014).

9. **Clocking integration of resonances: Various kinds of resonances are not isolated events:** A list of published resonance frequencies show that the ratios between different frequencies are not integers. Even they are not harmonic. They are anharmonic (Ghosh et al., 2016a; Monserrat et al., 2013). The ratio of magnetic resonance frequencies is the golden ratio (phi~1.61...). If the fundamental frequency is f_o then the other sets of frequencies would

be $f_o, \emptyset f_o, \emptyset^2 f_o, \emptyset^3 f_o \dots \emptyset^n f_o$, The electromagnetic resonance frequencies occur at the ratio of pi; $f_o, \pi f_o, \pi^2 f_o, \pi^3 f_o \dots \pi^n f_o$,. While mechanical resonances occur at ratios of e, $f_o, e f_o, e^2 f_o, e^3 f_o \supset e^n f_o$. All three resonances are related by a quadratic relationship $e^2 + \phi^2 = \pi^2$. By following this equation, the biomaterials ensure integration of electromagnetic and magnetic resonances deliver regularized mechanical changes in the system. There is a clocking integration even between three different kinds of resonances. It also justifies the fractal information theory, FIT, where all topologies were filled in a circle or topology.

10. **Clocking Phase sphere holds the geometric locations of singularities:** The experiments confirmed that the proteins clock like a time crystal, though similar claims were made in the 1970s. At that time, technology was not that advanced to provide a piece of direct evidence. To be a time crystal, any system's resonance frequencies should change their phase as if three clocks are part of one phase cycle. A single system point while completing a full rotation 360° would find that all constituent clocks do not delay it, or let it finish full rotation early. Thus, one has to check if the change of phases of clocking frequencies are quantized. If yes, it is probably a time crystal. Six proteins associated with the neuron firing showed time crystal features. Tubulin, beta spectrin, actin, ankyrin, clathrin and SNARE complex. Clocking of phase appears to be a universal property of proteins. During clocking, they hold specific geometric shapes. We repeat that a clocking Bloch sphere holding the geometric shapes made of singularity points was proposed as the basic information structure of nature in FIT (Reddy et al., 2018).

6.2.3.1 Exponential Speedup of Time Crystal Transfer via Synchrony—Quantum Entanglement Is Not the Only Route

A time crystal consists of phase, frequency and delay, however, the nature of locking could vary. The mathematical formulation for the individual locking process is similar, here we demonstrate the exponential speedup for phase locking. Synchronization of oscillators means that all basic computing oscillators should fire at the same time, when pumped by an ac signal; therefore, the phase is locked for all oscillators.

Time crystal clusters of different sizes mean neuron-oscillators of different sizes, synchronization continues for all clusters simultaneously. Say, we pump a single neuron-oscillator holding an "if-then" time crystal, or a neuron-cluster holding a large number of time crystals with an ac signal, the dynamics of one time crystal oscillator among N (neuron-oscillators or neuron-clusters), in the neighborhood is given by $dV_i/dt = S_0 - \gamma V_i$, where, $0 \le V_i \le 1$ and $i = 1, \dots, N$ (Mirollo and Strogatz, 1990). Here, γ is the dissipation factor, S_0 is the threshold bias required for the time crystal-holding-neuron to fire energy to its neighboring neurons,

which are also holding distinct time crystals. The rate of change of potential of a single neuron-oscillator or cluster-oscillator is not identical, though a neuron representing a time crystal could be part of different clusters of time crystals of similar and different sizes. Rise and decay of potential are identical for individual neuron-oscillators, however, for each non-identical cluster it is different. When a neuron-oscillator or a neuron-cluster fire, it pulls all neighboring neuron's or similar neuron-cluster voltage up by an amount ε, or pulls them up to firing whatever is less for the neuron oscillator. Note that $V_i = 1$, means neurons or clusters fire; and then the neuron or cluster switches to $V_i = 0$. Note that, during the synchronization process, while adding potential amount ε, only those clusters respond, at the same time, which has similar rise and decay rates.

To generate synchronization, voltage variation in the single neuron or cluster assembly is defined as a function $f(\varphi) = V$ and we take its inverse function to carry out mapping during the energy exchange process. Here φ is the phase variable, hence, $d\varphi/dt = 1/T = \upsilon$, where, T is the time period. At $\varphi = 0$ at $V_i = 0$. Thus, $f(0) = 0$ and $f(1) = 1$. It should be noted that we define synchrony is locked when oscillators re-arrange themselves in such a position, such that, if one neuron or neuron-cluster is pumped and fires, a single fire will bring all other similar neurons or clusters into the threshold. Moreover, this is also important to note that once this happens, synchrony does not break afterward, since internal dynamics for all oscillators are similar, they remain synchronized. The modified potential ε is added to the existing potential of the neuron or the neuron-cluster only for those, which have similar rise and decay rates of potential. Therefore the selection of clusters with similar internal dynamics starts from the very beginning of synchronization. The little energy contribution ε is transported at the speed of light to the other neurons and clusters; hence, it appears simultaneous one-to-many and many-to-one communication. Since for the artificial brain, identification of coupled time crystals is the process of solution making, the concern that during ε addition, important clusters might be discarded by accident does not arise. The reason is that ε addition is repeated continuously, along with the phase modulation, no neuron-time crystal or cluster-time crystal is discarded/selected in one-step, if the clusters or time crystals fail to cope up with the phase modulation, then only they are discarded in subsequent steps of synchronization.

Say, at $t = t_0$, neuron or a cluster A fires, at this point of time, B neuron or a cluster has the phase φ. After a time $1 - \varphi$, neuron or cluster B reaches the threshold. By this time, neuron or cluster A has changed its phase, from zero to a value given by $\theta = f(1 - \varphi)$. After some times, B fires and then θ jumps to $\varepsilon + f(1 - \varphi)$ or 1, whichever is less. A and B neurons or clusters synchronize if $\theta = 1$, so we consider, it is not achieved yet. Then the phase of A is $g(\varepsilon + f(1 - \varphi)) \sim h(\varphi)$. In summary, two of the oscillators make the journey from $(0, \varphi_B)$ to $(h(\varphi), 0)$, we get $h'(\varphi) = -g'(\varepsilon + f(1 - \varphi))/g'(f(1 - \varphi)) = -\lambda$. If we consider, $u = f(1 - \varphi)$, we get, $g'(\varepsilon + u)/g'(u) = \lambda$, whose solution is of the form, $g'(u) = ae^{bu}$.

The appearance of exponential variation is fundamentally important, since, an extension of this calculation leads us to the time for synchronization, and that is exponentially fast in k, where k is the number of iteration. If $\Delta_k = |\varphi_k - \varphi'|$, denotes the distance from repelling fixed point φ', we get $\Delta_k = \Delta_0 \lambda^{2k}$. The synchrony takes place in the system of a neuron or cluster oscillators when φ_k is driven to either 0 or 1. The number of iterations required for synchronization is $k \approx O\left(1/\varepsilon b \ln 1/\Delta_0\right)$, where b is a constant that appears during differential equation solution, and higher are the values of ε, b, Δ_0 faster is synchronization. In designing the brain like a computer, we have tried to keep Δ_0 high, and for the typical design of the artificial neuron or the elementary oscillator, ε, b acquire high values. More is the coupling, higher is the value of ε, since artificial neurons and clusters communicate via resonance frequency, the dissipation factor b is always very high. Synchrony is a co-operative phenomenon between coupling and dissipation, and the mathematical expression noted above parametrically accounts the co-operation explicitly during computation.

Mean-field model for all to all coupling: Kuramoto proposed in 1975, that $\dot{\theta}_n = \omega_n - KN^{-1}\sum_m \sin(\theta_n - \theta_{n+m})$, ω_n is taken from symmetric distribution of $g(\omega)$, here the order parameter is $\Psi = N^{-1}\sum_n r_n e^{i\theta_n} = \mathrm{Re}^{i\theta}$, $r_n = 1$. Synchronization occurs if $R \neq 0$, complete sync state means R=1. Partially synchronized state means $0 \leq R \leq 1$.

Logarithmic reduction of the number of computing step is feasible for such synchronization, since Δ_0 is proportional to the number of oscillators or neuron-clusters (say N) taking part in the synchronization process. The magnitude of Δ_0 is determined by the difference between the point of reflection and the initial phase. When the number of oscillators N increases, dN number of oscillators fix their typical Δ_0 to a particular value. More is the value of N, higher is the value of dN and during synchronization, as the time increases, a number of oscillators losses randomness $(-dN)$; hence, $dN = -kNdt$. Thus, we get exponentially faster synchronization as the number of oscillators or neuron-clusters increase during computation. For a group of input time crystal-clusters, Δ_0 initially increases and eventually reaches equilibrium, then the synchronization steps become independent of the number of oscillators taking part in the process.

6.2.3.2 Tubulin's Nested Resonance and Synchrony

The rapid collapse of tubulins into microtubule nanowire with femtosecond pulses (Tulub, 2004; Sahu et al., 2015) and thereafter under a nanosecond, and microsecond pulses suggest a strong electromagnetic coupling between tubulins in 10^{10} time scales (10^{-15} sec to 10^{-6} sec). It is not unlikely as proteins vibrate from milliseconds to femtoseconds, wide time ranges at a time (Hamm, 2008).

Quantum tunneling images of a microtubule (Sahu et al., 2015) shows wave functions of eight eigenstates of 8 quantum wells, a dimer's 8 quantum wells and a monomer shows 8 quantum wells. Totally, by combining three resonance frequencies in three different domains, a microtubule could hold a 3-bit long 512 eigenstates each with distinct wave functions. Thus, each eigenstate has distinct observables, as expected in

MBL. A protein at 300 K is not in thermal equilibrium, kT heats up every single quasi-particle in the 512 quantum wells, but they redistribute heat via many-body localization (MBL), they do not self-thermalize, states keep localized, no spin or energy transport (crypto equilibrium, looks like $T = 0$). Entanglement is low and distributed over time. Eigenstates are fully localized product-states, DC conductivity is zero (10^{15} Ω, a nearly perfect insulator) at 300 K as expected in MBL. Correlated phase distribution suggests MBL. Since quantum tunneling images of a dimer does not show an intricate pattern of wells obvious in a monomer, and quantum tunneling images of a microtubule does not show any sign of dimer's symmetry, all dimers could be represented using quantum-like wavefunctions.

From the concept of area entanglement, we get edge state where from we get differential conductance $dI/dV = e \wedge 2/vh$, from which we get $v = 5/2$, which is then related to $e/4$, as we observe the change by $ne/4$. For Hinductor one should not use dI/dV, rather should use $dQ/d\Psi$. It was theoretically conjectured that $v = 5/2$ abelian state has fractional charge $e/4$. One could theoretically show that the quantum oscillation that is generated over there could have a fractional charge. Extensive measurements need to be carried out in the future on Hinductors to find the topological charges, though Sahu et al. measured that it is $e/4$.

Quantum wells of tubulin and microtubule: Quasi-particles are not real particles, they are the density of states formed by the interaction of force fields behaving as a particle or showing wave–particle duality. Oscillating quasi-particles generated by dipolar oscillations of alfa helices and their infra-red quantum energy exchange (Panman et al., 2015) inside one of the eight quantum wells (~0.3 nm wide) of a 2 nm wide tubulin monomer dwells for a longer time before leaking through. Thus, each well in a monomer generates a wave $u = e \wedge i\omega t$, similar to a capacitor which reflects through the wall, changes its phase like an inductor and we get a wave $v = e \wedge (-i\omega t)$, moving in the same direction. Both the waves generated on the same well barrier, at the same time, satisfying normalization $|u|^2 + |v|^2 = 1$. Tunneling images of a dimer shows a C2 symmetry, hence both the 2 nm wide quantum droplets are part of a singular wave function, $\Psi = u|0\rangle + v|0\rangle$ represents a well. Coherent emission from a monomer under laser argue that it is a single quantum well, made of eight sub-quantum well's inside. More degeneracy, longer mean free path; for the quantum well inside a monomer mean free path is $4 \gg 0.3$ nm. Thus, the potential fluctuation of noise at GHz in a monomer creates a damped oscillation, periodic in potential, a strict function of monomer's area. Characteristic vibration of the amide group gives rise to parallel and perpendicular bands. Consequently, generated a 2D infrared signature of alfa-helices suggests that a monomer's quantum well's surface area on the water crystal cylinder changes as the number of active quantum well changes in the tunneling image as a function of applied ac frequency. Thus, a monomer's entropy is a function of its area, a first precursor to many-body localization, MBL. Applied noise through water channel drains free charge, ions, free electrons, so microtubule has zero DC conductivity, the second precursor to MBL. Due to 17 nm wide water crystal cylinder, thin protein sheet does not require "external heat bath," it serves as its own bath, a third precursor to MBL.

Beyond a threshold gate bias, microtubule's capacitance decreases to a quantum capacitance, but it needs additional ac signal to maximize inductance to quantum inductance. When both the conditions are set, and water crystal continues to drain out the free charges, we get a transition from an eigenstate thermalizing Phase (ETP) to quantum resonance. Quantum inductance and quantum capacitance are masked by classical parameters, only when higher than a certain gate bias is applied one can detect quantum ripples and associated quantum capacitance. Similarly, only when a certain range of ac frequency is applied, one could see a saturation of inductance with frequency, that value is quantum inductance. Thus, the quantum property of such quantum resonators is visible only under certain circumstances.

6.3 SINGLE MOLECULE'S THERMAL DIFFUSION EMERGES INTO A NATURAL CLOCK

Zhingulin's simulation of brain's central nervous system (CNS): One major problem for the realization of an artificial central nervous system, CNS is that the statistical methods allow solving the dynamic problems in an infinite size network. However, they are not applicable to realistic CNS like networks where constituent sub-networks regulate operation using periodic and chaotic dynamics. Recently, V. P. Zhigulin has proposed a theoretical model solving this problem (Zhigulin, 2004), however its information storage and memory processing are not addressed explicitly. In complementary work, White et al. have demonstrated clearly how a network's storage capacity for temporal memory scales with system size. Therefore, together, these works may provide a comprehensive solution to the problem as extensively outlined by Bandyopadhyay et al. (2009a, 2009b).

Shannon's entropy (SE) for information exchange between the molecules connected molecular electronics (ME) and information entropy (*MESE*). Calculation of information entropy and Shannon entropy: First, we formulate SE for any dynamic molecular system, which is applicable to any multi-level neural components. At any given time, say N number of Duroquinone, DRQ molecules occupying 0, 1, 2, 3 states randomly are separated by ~9 nm on the Au (111) surface. If more than three molecules are brought into the region of 9 nm in diameter then they never collapse into an integrated assembly, rather collectively generate a distribution of logic states spontaneously (Figure 6.5a–e). Then as the molecular diversity or conformer M is 4, the maximum information registered from Shannon expression is $SE = ln_e(M^N) \sim 5.52$ *nats* for $N = 4$ molecules. Then information for one molecule represent other $N-1$ molecules, the system can have only four distinct information. Therefore, Shannon's information registration limit is significantly reduced to 1.39 *nats*.

In the sub-monolayer, DRQs move and exchange logic state collectively, and taking those numbers we can construct a matrix with elements equal to the number of DDQs involved in the exchange process. Information in this matrix or string is studied using Shannon entropy equation $H(V) = -\sum_{i=0}^{N-1} p_i \log_2 p_i$ one could determine the average minimum number of bits required to encode a string or matrix (H), based on their frequency of occurrence (p_i). Here, p_i is the frequency of a particular logic state $P_i = P_{NDR}(V) * P_{Neu}(V)$. P_{NDR} is the probability of a state generated by NDR with a Gaussian peak at the threshold bias V_{th} $P_{NDR} = (2\pi\sigma^2)^{-1/2} e^{-(V_{th}-V)^2/2\pi\sigma^2}$, here σ is $(2\pi)^{-1/2}$. P_{Neu} is the probability of logic state caused by wireless interaction of molecules, by considering sigmoid Boltzman neuron, $P_{Neu} = (1 + e^{S/C})^{-1}$, where S is the weighted sum of input signals, $S = \sum_{i=1}^{N-1} x_i w_i + w_0$, x_i is the distance between interacting molecules, w_i is an energy difference between them, w_0 is energy induced by external bias V, $w_0 = x_{rms}V$. C is typical of a substrate and temperature, sets a limiting value of S, less than that logic states change ($S \sim 60$). Inside CNS a neuron may have 10,000 inputs and 10,000 outputs; here *one-to-many* variable weak and strong wireless wiring among clusters enables mimicking the continuous firing of a real neural net.

The temporary memory storage capacity of a DRQ neural network is determined during a continuous logic state exchange process of participants inside an ED wave boundary. First, matrices produced by the natural exchange are recorded, and then the same procedure is repeated by scanning at higher biases or forcefully changing one DDQ's state. Finally, the memory function $m(k)$ $\left(m(k) = \alpha^k/\alpha^k + \varepsilon^1(1-\alpha^{k+1})\right)$ is plotted by Bandyopadhyay et al. (2009a) with the fraction of the input signals survived k/N, ($k = 0, 1, 2...N$, here $N = 4$) within the symmetric region, where α varies from 0 to 1, ε is the variance constant and $\varepsilon^1 = \varepsilon/(1-\alpha)$. When the change in connectivity of the neural net is plotted with time by Bandyopadhyay et al. it shows that even if DRQs acquire states randomly, their motion is always confined within a sphere of *diameter ~ number of molecule × dimension of a molecule*. For $N = 2, 3, 4$ molecules for a given input logic set, *N input—N output* neural net is realized. To develop a general operational model, we consider that a DRQ neuron will produce a particular output with the fourth power of the sigmoid probability function.

Quantum Boltzmann machine's molecular version: As classical QBMs have *N input—one output* configuration, to model an *N input—N output* neuron device, N number of independent QBM nets are parallelly coupled. Considering homogeneous coupling, the results of one QBM is extrapolated to an *N input N output* neural net,–a generalized parallel processor of the present formulation. Bandyopadhyay et al. have plotted the probability of particular distribution with the ratio of input energy and exchange interaction coefficient for 4 DRQ neurons. These parameters are calculated as output of a QBM, where all molecules interact following a Hubbard model (spin is replaced by an imaginary electronic

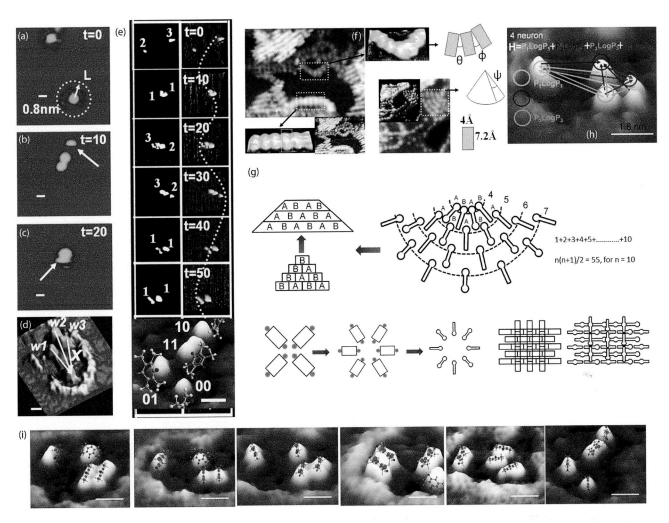

FIGURE 6.5 Panels (a–d) show how DRQ molecules collide and generate collective patterns on a 2D surface. (e) Periodic oscillations of a pair of DRQ molecules at the edge on a flat atomic surface is driven by thermal energy. Four states 00, 01, 10, 11 are detected on four molecules. (f) During collision, several intelligent architectures form. (g) Schematic presentation of a few circular self-assembly and cross-bar architecture. (i) On the flat atomic surface, four DRQ molecules are exchanging energy and the conducting states in a periodic manner.

charge for the states 00 = +2, 01 = −1, 10 = +1 and 11 = +2), with same interaction coefficients. Results confirm stochastic sigmoid nature of quantum coupling among DRQ neurons during a random exchange of logic states. The sigmoid nature leads to a collective logical output of the N, N processing surface, which is an essential requirement to practically realize Zhingulin and White et al. brain function and neural network theories.

In conclusion, an expression to calculate Shannon's entropy (SE) for information processing in molecular electronics (ME) is formulated. The *MESE* formulation is applied in analyzing dynamic/chaotic motifs (Zhingulin) that are responsible for the survival of temporary memory (White) in a 2 D neural network. The output of the network could be the geometric structures that would reflect the computing output (Figure 6.5f–h). In a naturally formed thermodynamically quasi-closed system, DRQ based N input and N output neural network are found essential for emergent logical operations (Figure 6.5i). The operation has now been extended for 730 molecules where a single molecule processes 16-bit information, which increases the processing capability of existing QBM by several orders. The present formulation could readily be used in verifying several fundamental formulations proposed in neural network theories. The journey begins with the realization of a neural net generated by spontaneous interactions of four molecular neurons.

6.4 BRAIN'S FMRI IMAGES CREATED ON A MOLECULAR SURFACE

Bandyopadhyay et al. (2010c) used the DDQ molecules to carry out brain-like computing on an organic bi-layer (Figure 6.6a), a quantum tunneling image of such a device is

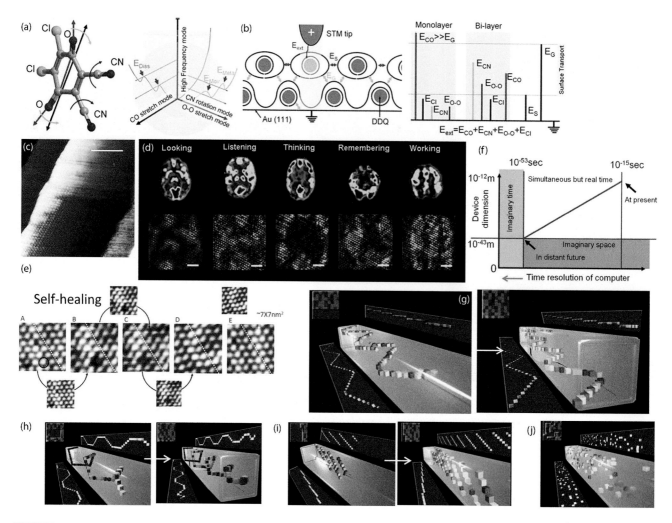

FIGURE 6.6   (a) DDQ molecule. (b) A 2D monolayer of DDQ molecule on the Au(111) surface images by a scanning tunneling microscope, STM. To the right the energy levels of different molecular bonds are shown. (c) A large area STM scan of the DDQ monolayer on the Au(111) surface. Scale bar 20 nm. (d) Using cellular automaton one could encode a set of STM pattern on the DDQ molecular surface that would change as a function of time such that the surface patterns resemble that of the fMRI images of the brain for different human activities. (e) A series of STM images scanned after removing one molecule using a large pulse from the STM tip. (f) A schematic presentation of device dimension and the time resolution. The red line depicts advancement of quantum technology, the ultimate limit is pointed with an arrow. (g), (h), (i) and (j) Four panels provide examples that if a 2D grid is read four different ways, one will read four different dynamics and its projection to different direction would be very different.

shown in Figure 6.5c. On the surface using random rules and emergence of cellular automaton one could generate patterns like that we see in the fMRI images (Figure 6.6d). The monolayers could have self-healing properties that could be used to emulate the brain plasticity (Figure 6.6e). DDQ monolayer has "self-healing" properties, due to this particular property from some other parts of the monolayer, molecules enter into the surface, and they adjust positions to recover. The most interesting thing is that, even if this happens, the computing does not stop, and it continues, the evolution of patterns do not care whether a single cell is there or not. Molecular cellular automaton promises to speed up or slow down events to compute and deliver output at the conceivable time (Figure 6.6f). We discuss below in details.

6.5   MASSIVELY PARALLEL COMPUTING BY CELLULAR AUTOMATON

Nature exhibits sophisticated collective information processing capabilities that show similarities to our brain, the reproduction of multi-cellular organisms, and so on. In this context, global co-ordination emerges from the decentralized communication between simple components. The particular feature has important advantages over man-made supercomputers where a central unit explicitly controls all computation processes. The first advantage is that all functional parts no longer need to be connected to the central control via physical wiring in order to increase the speed (~10 km wiring/cm² area of an integrated chip). The second advantage is that the loss of connection with a central control unit will not jeopardize the entire system, thus making it more robust. The third advantage is that the allocated resources are equally divided. In the past such wireless and powerless computation have been proposed in the theoretical models of cellular automata, CA. The biggest challenge in theoretical CA cell is that by changing the update scheme, one can get new patterns (H. Huberman, PNAS, 90, 7716 (1993)), in experimental terms, painfully, majority of theories become useless (Figure 6.6g–j).

Cellular automation may be used to model artificial life that follows defining characteristics of living systems. Bandyopadhyay et al. (2010a, 2010b, 2010c) demonstrated the existence of such artificial life forms as a logic pattern that evolves in hundreds of molecules in a molecular layer. The layer consists of DDQ molecules that reversibly switch between four states. A molecule may change states of its six (or, depending on the local network configuration: four) neighbors at a time following particular transition rules for the birth, survival and death of cells and thus create a new electron transport circuit for a new problem. The way patterns change over time is thus strongly dependent on their arrangement on the molecular layer, and we have especially observed this for two types of patterns, i.e., linear and circular patterns. For the first, the pattern changes over time as if electrons diffuse throughout the surface, and for the second the pattern changes as if normal tissue cells mutate continuously to give rise to cancer cells. Concentric circular rings are good for the coupled differential equations, the number of rings will reflect how many equations are coupled, and separations between two rings will determine boundary conditions. Similarly, a pair of straight lines is good for the higher-order differentials. It should be noted that even a small change in the input pattern could turn the computation toward a completely different direction, and this happens more frequently, if we deal with small patterns. The surface computes more reliably when one encodes basic equations with the larger size. To encode the boundary condition relative ratio is important, so there is basically no restriction on the absolute size (US patent 4809202). DDQ molecular assembly (Figure 6.7a) remarkably selects CA rules for its universe as required and adapts itself to a new situation by tuning the space and time limits. It has been a custom until now to define new CA rules every time we wanted to solve a new problem. Here, the CA adopts itself to solve more than one problem in single hardware, which is an essential criterion to build a universal computer like the one we use in our daily life. Even though it is the first practical CA, its universality could be generalized further paving the way for massively parallel processor envisioned by Neumann 50 years back. Moreover, the differential equation system is created on the monolayer, it is not the creation of CA models using partial differential equations as practiced frequently (Doeschi et al., 2004; Omohundro, 1984). These two approaches are fundamentally opposite concepts as noted earlier (Toffoli, 1984).

Sinusoidal, exponential or logarithmic functions are encoded using closed patterns of very small sizes, for example, completely packed smaller circles grow exponentially, and if outside regions of large circles are closed they collapse exponentially. Thus, one can encode both positive and negative exponentials. Flower-shaped triangular circles are good for sinusoidal functions, while concentric triangles with only angular points encode the logarithmic functions. It should be noted that coupling of functions with the differentials is done by placing them in a very particular manner on the surface. The additive and multiplicative coupling are done by placing the interactive patterns side-by-side on the surface and enclosing one inside the other respectively. The relative separations determine the boundary conditions and for that particular reason, we can encode wide ranges of computing problems on the monolayer. It also tells us about the limitations of this kind of computing. For example, one could easily ask, Δx and Δt have upper and lower limits therefore, the PPM-GML-H triad computing cannot be universal, one can encode only a particular kind of problems. In the case of two natural differential equations solved here, diffusion equation and coupled differential equations, Bandyopadhyay et al. have shown to encode spatial and temporal differentials (Figure 6.7a, b). We have also discussed tuning time and space, which is squeezed by orders of magnitude. For example, charge decay and diffusion that takes only a few picoseconds could be increased to the second's domain (Figure 6.7c–e), and cancer cell evolution that takes 100 years (Figure 6.7f–h), or astrological events that take a million years could be speeded up by the same orders of magnitude. Computing is not restricted by Δx and Δt physical constraints of the monolayer hardware, since patterns irrespective of its natural speed follows a particular spatial evolution, since time is linked with space, it was possible to replicate events that are orders different in time.

For each particular input pattern, Bandyopadhyay et al. observed spontaneous flipping of weak bonds between molecules, building a new communication circuit by creating and destroying several optional paths connecting hundreds of cells. Building a new circuit turns it creative, simply because the total number of circuits that could be generated for a particular kind of problem is nearly infinite. Unlike supercomputers, the hardware itself talks to multiple cells at a time, corrects an error in the process, and thus it exhibits a form of intelligence. During execution, the exchange of an electron or a few kilo-calories of energy enables wireless computation with minimum external power supply. While a Pentium IV processor with a device density of $<10^9/cm^2$ dissipates approximately 100 W/cm^2, the DDQ-HCA (HCA = hexagonal cellular automaton) with a device density of $10^{14}/cm^2$ dissipates approximately 1 W/cm^2; thus heat generation is significantly minimized. In principle, a solution is generated collectively, so even if some cells stop working suddenly, the entire computation does not collapse, but rather the system reaches a solution.

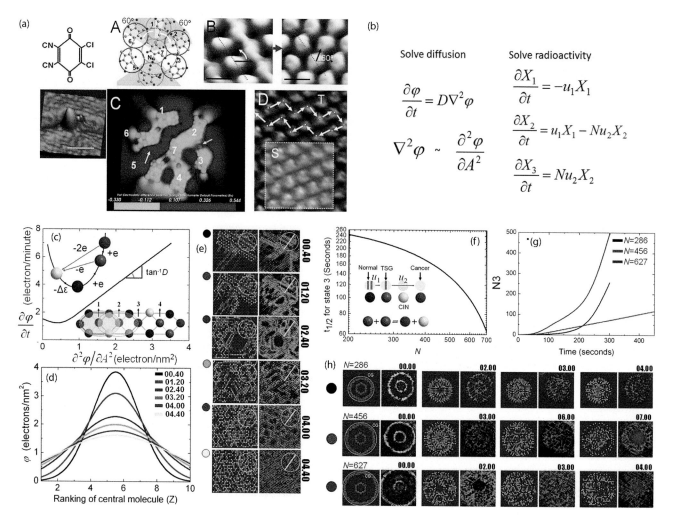

FIGURE 6.7 (a) DDQ molecule and its neighbors scanned by a scanning tunneling microscope, STM, scale bar 2 nm. Four sub-panels are (A) hexagonal neighbors, (B) Phase transition of the entire molecular assembly, (c) potential distribution calculated by DFT for a hexagonal close packing arrangement of DDQ molecules. (D) Zig-zag pathways of potential transmission. (b) Diffusion equation and radioactivity problems were solved on the surface. (c) Four conducting states are depicted as four colored balls in all the panels, here, double differential of electronic motion per unit time (vertical axis) per unit area (horizontal axis). (d) The solution of the differential equation for the diffusion is shown as a Gaussian distribution, as the time pass by, the central molecule acquires large local electron density of states. (e) As time passes by, the STM image (left column) and the corresponding ball representation in the right column. (f) The half-life of state 3, mutated cancer cell or radioactivity of an element decreases as the total number of participating cells increase during the solution of a problem. (g) The rate of reaction for three different populations is shown. (h) The evolution process is observed in the STM image and the theoretically generated cellular automaton induced grid results are shown here.

6.6 ONE-TO-MANY AND MANY-TO-ONE ORBITAL COMPUTING IN A NANOWHEEL

In the year 2008, one of the finest philosophies of computer science, sequentialization was challenged by a simple molecular experiment (Bandyopadhyay et al., 2008). If one puts 16 DRQ molecules along a line, if one instruction is given to a molecule which is at first in the queue, only one decision would be there as an output. However, if all 16 molecules are put in a circular array (Figure 6.8a), with a molecule in its center, the story changes significantly. The molecule that resides at the center of the circle becomes a simultaneous controller of all 16 molecules. So, using an atomic sharp needle one

could change the state of the central molecule and simultaneously observe the change of the states of the 16 neighboring molecules using a scanning tunneling microscope. The experiment compares directly, what happens if the communications happen sequentially and simultaneously. Due to the orbital coupling, 16 molecules could exhibit 4 million solutions when placed in a circular geometry (Figure 6.8b). However, the linear arrangement would deliver only one solution. One-to-many and many-to-one communication could change the way we believe communication might happen in nature. All the 16 molecules in the ring simultaneously reply to the query made by the central molecule, no one needs to be asked who rings the bell. There is a truth table, but that table has no limit, just like the fractal logic gate where a small part of

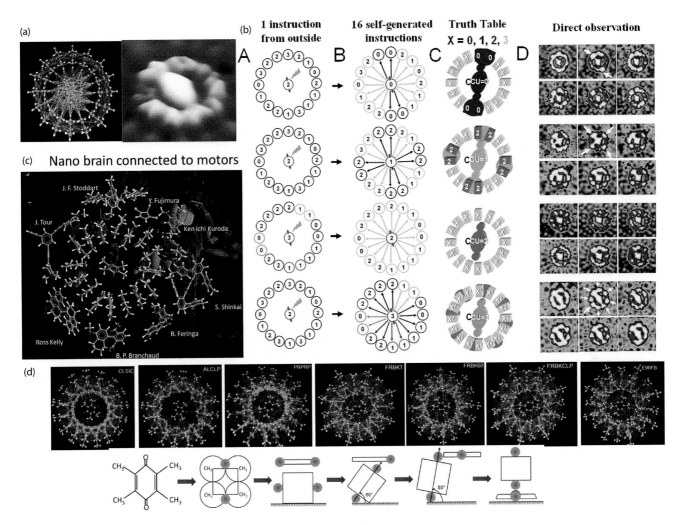

FIGURE 6.8   (a) The structure of nanobrain, made of 16 DRQ molecules and the corresponding STM image on the Au(111) surface is shown. (b) The truth table for operating the nano wheel is demonstrated. Column A shows instruction to give input to the nanobrain, column B shows self-generated instruction, which is the first element of the truth table in the column C. The corresponding STM images are shown for different cases implementing the nanobrain. (c) Nanobrain connected eight molecular rotors for programming. (d) The hydrogen bond is shown with green and blue lines. First one is classic structure (CLSIC) where all oxygen atoms, red balls along the ring, repel, one alternate molecule moves forward and backward to keep separation. Second one's ring holds alternate DRQ molecules have oxygen atoms up and down (ALCLP). The third one ring DRQs are arranged such that their DRQ molecule located side by side is diagonally arranged, so the distance between oxygen molecules are increased (PRPRP). The fourth one (FRBRKT), one alternate DRQ stands vertically, oxygen at bottom. The fifth one has one DRQ vertical and one horizontal, pair of molecules, located side by side on the ring are perpendicular to each other. Sixth one FRBKCLP has neighboring pair oriented perpendicularly but, vertically oriented DRQs all along the ring is oriented like a single waveform. The seventh one has (E90FB) all DRQs in the ring vertical, oxygen-oxygen axis is perpendicular and oriented directing to the Au surface.

the solution (Ghosh et al., 2015b) could be expanded forever. Sixteen molecular motors could be connected to the system (Figure 6.8c) and the 8 distinct arrangements of DRQ molecules could hold and operate simultaneously 8 new motors (Figure 6.8d).

6.7 HIDDEN COMMUNICATION IN THE FILAMENTS WELL BEFORE A NERVE SPIKE

The fundamental problem with the "neuron skin is everything" model: Hodgkin Huxley model that experimentally proved that neuron skin is doing everything for the firing, now,

after 70 years, demands are mounting for an alternate route (McCormick et al., 2007). Most sincere adventures to rectify sub-threshold communication errors in Hodgkin Huxley were developed in the 1990s and subsequently ignored since they were complicated, did not give freedom to play with human free will (Guckenheimer and Labouriau, 1993). The imaging of neural events has been one of the hallmarks of neurophysiology, either by chemically neutralizing the ion channels, doping fluorescent molecules. Thus far the single ion channels which are roughly 0.4 nm were mapped by using a much larger nanoprobe (Zhou et al., 2015; Ide et al., 2002; Fua et al., 2014) or indirectly seen by the fluorescent doped channel proteins (Yamamura et al., 2015); wherein the silent ion channels remain undetected (Bal et al., 2008; Ulbrich and Isacoff, 2007). Composition of ion channels, i.e., the stoichiometry and arrangement of these subunits are required for precise balance of speed and sensitivity to achieve optimal performance (Zheng et al., 2003; Trudeau and Zagotta, 2002). In the functional stoichiometry of ion channels, the sub-units are seen (MacKinnon et al., 1993), yet, how all these events are linked, such questions are not asked in neuroscience. Second, the existing ion-channel scanners (Hansma et al., 1989) cannot link a millisecond nerve impulse with the microsecond or nanosecond protein oscillations, a three-order time gap remains unbridged. Scanning the ions is so powerful that it can image even proteins inside a cell (Shevchuk et al., 2006). Interestingly, the microsecond or nanosecond time-domain electromagnetic resonance of proteins is being measured since the 1930s (Elliott and Williams, 1939; Pethig, 1979; Vollmer et al., 2002; Schirò et al., 2009; Kim et al., 2008; Verma and Daya, 2017), it is much older than the millisecond ionic resonance of proteins that was seen experimentally in the 1980s, though ionic spike was theoretically proposed in 1907. Dielectric resonance microscopy is recent, it images the cells internal structures, but identifying components by image contrast is not done yet. Since the biomaterials are insulators, non-conducting, if an electromagnetic signal is pumped, it reflects with a coefficient S11, and transmits a part of it at specific frequencies with a coefficient S21. Every insulator has a specific signature of resonance frequencies, S21, S11, so the ions, helices, secondary structures of proteins, DNA etc. Various cells were imaged by mapping S11, S21 thus far, but not the neurons.

A few unexplained phenomena of neuroscience: Initially in the 1930s it was wave of a very different nature regulating nerve activity (Hill, 1933); by 1946, it was chemical excitation in the neuron (Brink et al., 1946), then it became electrical in 1952 and electromagnetic in the 1990s. Bursts of soma or cell body of a neuron are considered as a unit of information (Lisman, 1997). However, it is not known how backpropagation when a nerve spike runs in the opposite direction (Scott et al., 2007), how does the multi-modal actional potentials survive (Chen et al., 2002). Eventually forward and backward transmissions interact and various modes of neurotransmitter releasing processes are activated, like synchronous, asynchronous and spontaneous release (Kaeser and Regehr, 2014). Two opposite directional transmission causes a clock or periodic oscillation; thus, clocks are born (Connor, 1978). It is also difficult to explain how multiple threshold potentials exist in the neuron firing (Sardi et al., 2017), unless we consider quantized multi-mode resonance of microtubule and actin, biologically proposed (Zaromytidou, 2012) and experimentally demonstrated in pristine microtubule (Sahu et al., 2013a, 2013b) and in the neuron (Ghosh et al., 2016a).

However, precisely editing the density of ion channels alone is trivial (Smith, 2009) even in the 1 m (dorsal root ganglion) to 30 m long axons (Abe et al., 2004). Each part of a neural branch, irrespective of its length, should have geometrically adjusted composition of ion channels, else, a neural

spike would distort and fade out (Cavanagh, 1984). If axon is >1 cm (Ofer and Shefi, 2016), more difficult it is to supply the metabolic components in time (Cavanagh, 1984) for ionic pulse to flow (velocity of axon potential is 0.1 or 100 m^{-1} (myelinated), but the vesicle transport velocity is ~10 μm^{-1}, (supply is 3 orders slower; Kulić et al., 2008), the thermodynamic cost at all levels do not match, which is currently believed to be the unifying principle governing bio-systems (Hasenstaub et al., 2010). A membrane of an entire neuron cell is short-circuited, no sensor to estimate the geometry of complex branches yet a membrane does not allow a spike to deform even in meters long axons. If the ionic transmission is blocked between a pair of neurons, still they communicate.

Each type of ion channel has a distinct frequency band where it opens/closes at the maximum speed (Doyle et al., 1998; Sonnleitner and Isacoff, 2003; Demuro and Parker, 2005), a nerve impulse is not a 1D transmission of a local point wave, rather, it is a circular vortex shape. How does intricate time management is carried out by an axon, how do all the ion channels in the vortex of a nerve spike move synchronously? Ghosh et al. demonstrated experimentally in the hippocampus neuron cells (Kaech and Banker 2006; Spencer and Kandel, 1961; Raastad and Shepherd, 2003) in 2016 that long strings of microtubule actin and intermediate filaments together make sure that all of them keep time (Ghosh et al. 2016a). It is utmost essential to image even the silent ions, combining Scanning dielectric resonance microscopy, SDM and scanning ion microscopy Ghosh et al. invented the new microscope. Scanning dielectric resonance microscopy, SDM was proposed in 1995 (Cho et al., 1996; Cho, 2007), that time it was a movement to advance scanning tunneling microscope to different directions.

Considering all the filaments are silent in the neuron firing is unfair: The filaments dispersed in the cell fluids are silent, but when packed in neural branches, they might vibrate like dipoles (Sahu et al., 2013a, 2013b, 2015) could affect the membrane potential, but no study has looked into this aspect. If the ionic transmission is blocked between a pair of neurons, still they communicate (Katz and Schmitt, 1940). Possibility for two distinct communication channels electrical and ionic is often explored in a cell as ephaptic transmission (Ramón and Moore, 1978). Ionic and non-ionic (Jefferys, 1995) transmissions together may lead to non-synaptic firing (Dudek et al., 1998; Ren et al., 2006), endogenous field alone causing a firing (Qiu et al., 2015), sometimes nerve spike generates spontaneously without any initiation (Atherton et al., 2008). Moreover, an electromagnetic resonance is reported in the ion channels and filaments regulating the firing. Electromagnetic interaction of neuron: For a long period of time, extensive research has been carried out to demonstrate that neuron and nerve fibers not only absorb electromagnetic signal but also radiate the electromagnetic signal, mostly in the infrared range (Fraser and Frey, 1968). When a single nerve spike passes through, heat is produced in a nerve fiber (Howarth et al., 1975), heat absorption and emission both take place for a single neuron firing, such an event is named as positive and negative heating (Abbott et al., 1958). On the other hand, by applying an infrared light, one could suppress the neural activity (Duke et al., 2013), often this is called photothermal inhibition of neural activity (Yoo et al., 2014). These events happen because the heat changes the membrane potential (Buzatu, 2009), which stops the propagation of nerve impulses through some branches (Westerfield et al., 1978). Heat is a 5–6 THz electromagnetic signal. Not just the membrane, it has been shown experimentally that thermal fluctuations even causes important changes in the filamentary dynamics located deep inside the membrane (Gittes et al., 1993). Therefore, even the intermediate filaments are not silent (Nixon and Shea, 1992). Cell phone like radio-frequency fields which are primarily in the GHz domain affects single neuron (Partsvania et al., 2013). It has been demonstrated that radiofrequency affect the single proteins in the animal brain (Maskey et al., 2010). Not just proteins of ion channels, membranes and the filaments, even electromagnetic resonance of biomaterials control feeding and metabolism (Stanley et al., 2016). The short review is the tiny fraction of massive research on the electromagnetic effect on the living biomaterials and biosystems. Yet, considering that electromagnetic effect plays no role in communication or information processing as current neurosciences do is unfair.

A desperate need for the new technologies: Multichannel simultaneous resonance measurement and A dielectric resonance scanner: Simultaneously reading the associated events unfolding at different time scales in a nerve spike is not done yet, Ghosh et al. were the first to build such a set up (Figure 6.9a; 2018). Just below the microscopic image we demonstrate the schematic to explain the task of specific electrodes. Ghosh et al. invented a tool to characterize both filaments and ion channels at a time in a cell, as they differ by 10^3 orders in time scale. One could observe that 200–400 µs before a neural spike takes shape, there are many simultaneous communications, one such firing is shown in Figure 6.9b and a series of them are shown in Figure 6.9c. Cell fluid damps the mechanical resonance, since it requires tension and physical motion but not the electromagnetic resonance that requires rearranging the dipole, i.e., a pair of charge. Combining milliseconds response with the nanoseconds-picoseconds one means connecting the ionic resonance with the dipolar resonance i.e., guest and host for ions. Since an ion channel (Harms et al., 2004) opens/close in 10–20 nanoseconds (Tahara et al., 2015), so Ghosh et al. invented a scanner (2018) that rapidly records (~10 ns) signal at multiple time domains simultaneously. Only then the protein signals would be recorded at a rate of their natural vibration as the nerve impulse transmits. One of the finest eyes in the field of neuroscience is a tool that can visualize both ionic motion and electromagnetic signaling.

Branch failure and axonal computing: As we learn, our brain evolves by re-wiring the neurons (Losonczy et al., 2008), but it is still unclear, how, just by firing, the complicated paths of neural branches decide all intricate details about the new connections to form by breaking the old ones. Post-1960s, the studies on the geometry of branches (Rall, 1959; Goldstein and Rall, 1974; Parnas et al., 1976; Swadlow et al., 1980; Hines, 1989; Ofer and Shefi, 2016), how its

length and thickness edit the pattern of the density of ion channels enabling a nerve impulse to stop, slow down, delay (Manor et al., 1991), speed up, reflecting back (Baccus, 1998; Chen et al., 2002), have made a significant stride to correlate the topology with its information content (Debanne, 2004). No data exist on the real-time imaging how an axon selects a branch. Moreover, recent findings have cast a doubt whether the density of ion channels alone regulate the axonal computing, i.e., branch failure, complex modifications of various sub-threshold potentials (Scott et al., 2007; Atherton et al., 2008; Ratte et al., 2015; Jin et al., 2012). It is not a jelly and messy beneath the membrane, rather, the ordered architectures (Xu et al., 2013) made of densely packed micro-neuro-filaments (Mandelkow et al., 1991) could resonate (Sahu et al., 2013a, 2013b) as intrinsic field to modulate time (Radman et al., 2007) to open/close the ion channel gates (Maskey et al., 2010) and bridge the missing links between branch selection and numerous sub-threshold potentials (Duke et al., 2013; Yoo et al., 2014). Figure 6.9d has compiled the oscillations of various biomaterials in the neuron. The result suggests that electromagnetic resonance covers a wide time domain.

The geometry of branches and the filaments: Axonal computing has been one of the prime factors of information processing studies in a neuron. When a neuron branch divides into two, differential conduction between the two branches is sensed accurately (Grossman et al., 1979). The difference between two branches filters noise and selects accurately the typical nature of signals required to be transported (Stockbridge and Stockbridge, 1988). Thus, branch points play a key role in decision-making (Stoney, 1990), at the branch points the decisions are made (Swadlow et al., 1980), by synchronization and de-synchronization of microtubules and actin filaments (Zaromytidou, 2012). Axon geometry regulates the firing pattern (Ofer and Shefi, 2016); axon diameter or caliber differs by 100 times (0.1–10 µm), thus, volume differs by 10000 times in the neural network of a human brain (Perge et al., 2012). By varying geometry axon modulates the frequency of transmission (Parnas et al., 1976). Geometric ratio (Rall, 1959) $GR = d_{daughter1}^{3/2} + d_{daughter2}^{3/2}/d_{mother}^{3/2}$, works in a perfect radial or spherical symmetry, if not symmetric it fails. A particular neuron class holds a strict mathematical relationship, $n = 3/2$, however, Ghosh et al. studied 27 cases where $n = 5/3$ (8), $n = 8/5$(19) ratios organized the time editing of the neural pulse. Using filaments, Ghosh et al. could generate neuron firing even with a sub-threshold pulse at resonance (Figure 6.9e), even by applying a suitable filamentary resonance signal they could switch off the firing.

Direct visualization of the branch failures: hidden circuits: One possible way to underpin how a neuron decides to change its wiring is to image the changes in the energy density of axonal branches, as a neuron re-wire. Selected branches appear bright in an energy map, the rejected branches appear dark, the visible wiring is called a circuit here. However, all existing snapshot imaging techniques (Rivnay et al., 2017) either chemically modify the membrane or destroy the natural membrane signals by pumping a huge external power or make invasive contacts (Bakkum et al., 2013;

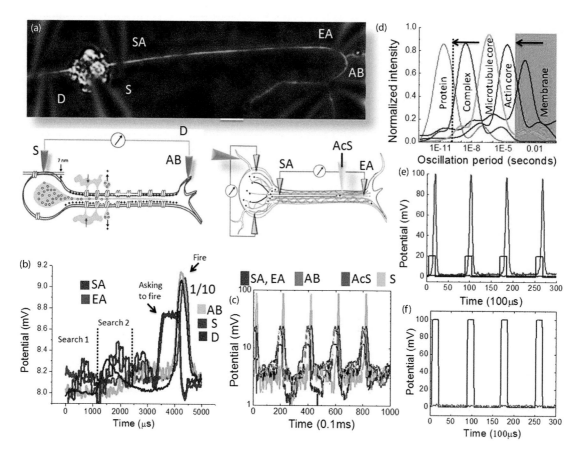

FIGURE 6.9 **Measuring natural oscillations deep inside a neuron during firing:** (a). Five electrodes are connected to an 8 days old rat hippocampal neuron cultured from embryonic cells, S = Soma, D = Dendrimer, SA = Start of the axon (Axon Initial Segment, AIS); EA = End of axon; AB = Axonal branch. Bottom, a microscope image of a neuron, probes faded as the neuron is focused, scale bar is 50 μm. S, D, AB make surface contact, EA, SA makes deep axon core contact. (b). Current recording of 5 ms, that includes an ionic firing and current flow at all five proves. The potential (mV) for S, D, AB is reduced by 1/10 to compare all responses in a single time scale. (c). The same plot as panel b, but for consecutive pulses (total time 100 ms), the potential axis is in log scale, AcS = Actin-beta Spectrin lattice. (d) Coaxial probe measuring the natural oscillation of components, tubulin protein, microtubule = complex, microtubule core = assembly of microtubules and neurofilaments, actin core = actin-beta spectrin crystal; membrane lipid bilayer. We plot periodicity of the natural oscillations, averaged over 20 neurons, 437 measurements, ~80 measurements per region. The shaded region is the limit of a Patch-clamp technique to capture data, the dotted line is where our probe is most sensitive. (e) Above threshold 97 mV pulse stream is applied along with a set of ac frequencies (12 MHz + 35 MHz + 7 GHz + 13GHz) to the microtubule-neurofilament core (blue). The potential response on the membrane is recorded by contacting coaxial probe at Soma. (f) 5 μm deep inside a coaxial axon probe is inserted, the natural potential fluctuation is recorded when there is no firing.

Lewandowska et al., 2015). Since the intrinsic data (Zhang et al., 2009) is weak, interaction is faster than milliseconds, a long integration time loses all relevant information. To see axonal computing in a real-time, we need a wireless, intrinsic, non-invasive, non-chemical snapshot of the whole network with a spatial resolution <500 nm. The super-resolution does not mean seeing the activity of channels (Schermelleh et al., 2010). As a spike passes through the branches, both, the time width of a pulse and the time gaps between the pulses is naturally classified. The fastest and the slowest clocks of a neuron determines the time-sensitivity (Lundstrom et al., 2009), this is fundamental to the development of a time crystal map of a neuron. If one takes a snapshot of a network, for each time domain, a distinct temporal circuit (Hutcheon and Yarom, 2000) would be visible. To visualize the coexisting circuits as functions which have been indirectly probed for more than

half a century (Connors and Regehr, 1996), Ghosh et al. have developed a microscope that captures multiple time-domain snapshots at a time. "Optical nanoscopy" (Hell, 2007) that allowed imaging below the standard diffraction limit of fluorescent light (Rust et al., 2006), including structured illumination microscopy (SIM; Betzig et al., 2006), stimulated emission depletion (STED; Hein et al., 2008; Willig et al., 2007), stochastic optical-reconstruction microscopy (STORM; Rust et al., 2006), and photoactivated localization microscopy (PALM) etc.

Highly periodic architecture deep below the neuron membrane: Filaments did not exist for the last hundred years (Figure 6.10a–d). Rapid-frozen, cross-section image of an axon suggests that the central region of an axon initial segment, AIS is filled with densely packed ordered filaments (Hirokawa, 2011; Hirokawa, et al., 1988, 1989; Chen et al., 1992;

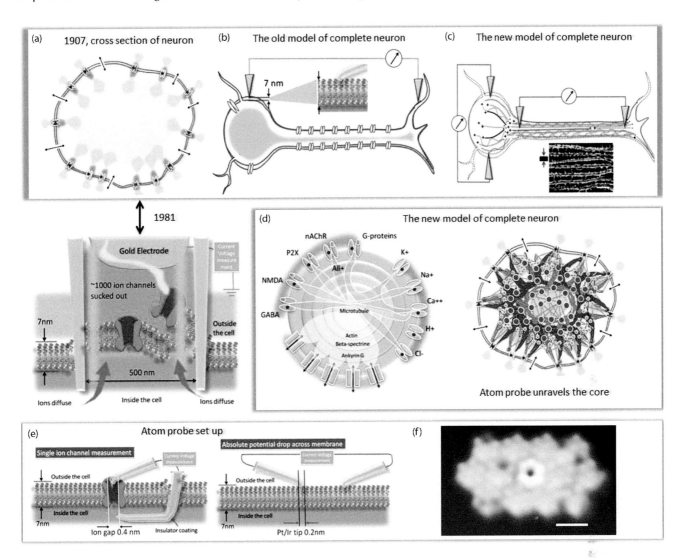

FIGURE 6.10 (a) Vertical cross-section of a neuron shows ion channels diffusing ions, nothing inside as per the current model of neurons (up). The panel below shows conventional patch-clamp technique where ~500 nm wide tip sucks out ~1000 ion channels when it seals with the membrane (down). (b) A complete picture of a balloon model of neuron. Two atomic probes are connected to it. (c) Our proposed model of neuron showing microtubule-actin network (ion channels are not shown only for clarity). Two circuits are shown. One, probing signaling across microtubule-actin network. Two, probing two ends of Soma. Inset: Cryo-TEM of an axon cross-section (80 nm). (d) Two axon cross-section panels, left shows connecting ion channels and multilayered cross-sections based on Cryo-TEM studies what a probe would encounter if it is inserted. The right panel shows sum up of several cryo-TEM and STORM data on the structure of an axon. Purple circles are microtubule, Red lines are actin. (e) Two panels show ion channel measurement by atom probe (left), true measurement of membrane conductivity (right). (f) STM image of a membrane captured by rupturing live neuron cell, at 2.3 V, 6 pA. Scale bar is 7 nm.

Figure 6.10c, d). The popular notion that is still out there is that the filaments are random, float in the cellular fluid in the AIS is incorrect. Some of these filaments like microtubule are in parallel, unidirectional (Heidemann et al., 1981, 1984) with a gap ~50 nm, and nearly continuous (Baas et al., 1988), in the AIS. Highly ordered Golgi apparatus isolates the filament-bundles at the axonal or dendritic branch junctions, filamentary continuity breaks but Golgi apparatus assists the filaments to remain parallel inside a branch (He et al., 2016). Just below the membrane, actin, beta-spectrin form a periodic rectangular lattice-like structure (~200 nm cross-sectional rings of beta-spectrin, connected by parallel actin wires, Xu et al., 2013) which are found almost everywhere in the neural network, in the glial cells and in all dendritic, axonal branches of a neural network (D'Este et al., 2016; Gervasi et al., 2018). The ordered architecture hosts various superstructures (D'Este et al., 2015; Lukinavicius et al., 2014; Ganguly et al., 2015), but the ordering of guest protein molecules disappears if the microtubules in the central region of an axon is dissolved chemically (Zhong et al., 2014). Multiple recent reports by Hirokawa et al suggest that the network of filaments that covers 98% by volume of an axon are closely related to the molecularly thin periodic actin-beta-spectrin hollow cylindrical network, which is in direct contact with the membrane above. Based on the four findings noted above, Singh et al. (2018) build a model structure of AIS, only to justify that Ghosh et al. built dielectric scanner could truly measure the signals from filaments and ion channels together along with the atom probes

(Figures 6.10e and 6.9a). The first element is He, derived from the long-range of filaments throughout the axon. The second element is B, the unidirectional polarity of all filaments, third, Hi, an equidistant (~50 nm) lattice-like arrangement of parallel filaments, fourth, Z, a periodic actin-beta-spectrin lattice coupled to the filamentary core. Singh et al. theoretically generates a dielectric slab-based model HeBHiZ of an AIS and experimentally verify that the system vibrates as a single dipolar unit at THz-GHz frequencies related to its geometric shape, consequently, builds the MHz periodic oscillations to assist the membrane's kHz ionic spikes. These are some of the reasons, that led researchers to think, it is microtubule that is carrying out information processing in the neuron (Dent et al., 2014). Ghosh et al. (2016a) imaged the single ion channels to cross-check live if an ion channel bursts by wireless communication (Figure 6.10f). These intrinsic structures build temporal coding (Magee, 2003).

6.7.1 TRIPLET OF TRIPLET RESONANCE BAND IN THE AXON OF A NEURON

Figure 6.9d plot shows that the resonance frequencies are isolated and discrete bands for neural components. That's an incomplete picture. Figure 6.11a–c suggests that all the clocks are nested just like a time crystal as shown for microtubule in Figure 6.4a, b. The AC signal applied parallel to the axon triggers AIS only at three distinct frequency ranges, where a short pulse (pulse width 1 μs, total duration 1 ms) from the Patch-clamp at a sub-threshold bias (~20–30 mV) activates the firing (Spencer and Kandel, 1961). An additional vertical AC signal resolves each of those three frequency domains into three additional sets; Sahu et al. observed nine bands. Inserting two probes into the axon when one measures the resonance bands across the AIS, only three resonance bands are observed in the linear plot (Figure 6.11d, e). An ordered biological structure exhibits a major longitudinal and a transverse vibration mode (Pokorny et al., 1997; Daneshmand and Amabili, 2012); if the AC signal is applied in one direction, only one mode is probed. Resonating with both horizontal and transverse vibrations at different combinations of horizontal and vertical AC signals also reveals additional peaks inside the nine bands (Figure 6.12a–c). Therefore, the relative angular orientations of the three smaller circles vary by 100°–120° in each of the three larger circles, but they unravel an additional dynamic hidden in the nine bands. 3D resonance map of a neuron unravels three distinct time domains or clocks that regulate the nerve impulse.

As electrical nerve impulse forms at AIS, Ghosh et al. measured a collective resonance of the AIS connected axon core. The resonance behavior of axons with and without membrane in a single neuron was reported (Ghosh et al., 2016a, Agrawal et al., 2016c). Consistency of triplet of triplet band with and without membrane prompts us to get inside a single microtubule that constitutes the major part of an axon. The microfilament bundle located deep inside the membrane are responsible for the higher-level communication, even if ion flow is blocked between a pair of neurons,

purely electrical connections run (Katz and Schmitt, 1940). Endogenous fields have been measured by inserting a coaxial atomic sharp probe deep inside an axon (Ghosh et al., 2016a). However, the debate related to the internal fields of an axon is in the mainstream neuroscience for a long time (Radman et al., 2009).

The resonance behavior of a single isolated microtubule is reported by Sahu et al., but not its 2D resonance pattern with gating. They dropped freshly reconstituted microtubule solution on the electrode grid, and an AC frequency scan is carried out similar to the neuron study. Ghosh et al. measured the intensity of the transmitted signal along microtubule length as a function of two perpendicularly applied AC signals across the microtubule. Similar to the neuron, the microtubule exhibits a triplet-triplet resonance band, but it is electromagnetic, not ionic. Additional transverse field along with the horizontal AC pumping (using two perpendicular electrodes) changes the angular positions of the circles, but their relative areas remain constant. As a result, if one superimposes neuron's and microtubule's triplet-triplet bands, should notice shifts but the common frequency/time regions never disappear. It suggests that the clocks of isolated microtubule and AIS are coupled.

6.7.2 SCALE-FREE TRIPLET OF TRIPLET BAND IN TUBULIN, MICROTUBULE, AND NEURON

Linguistic form of the fundamental unit of information: The basic philosophy to computing without using the algorithm is to let the pattern of primes to integrate the symmetry of events. The new type of computing is inspired by the discovery of primes in the protein vibrations. The distribution of resonance frequencies in the tubulin protein, microtubule and neuron were following a similar group of primes. As if they were designed to clock "time" differed by the ratio of preferred primes. Three-layered studies between tubulin, microtubule and neuron inspired us to extrapolate the idea of integrating the resonance peaks as the ratio of primes which signifies that any event in the universe is made of three layers of sub-events. The slowest clock is the host or subject, the present clock within the slowest clock is the key or clause or condition at which an event occurs, the fastest clock within the present clock is how an event occurs or a verb along with the clock representing the quality of an event or adjective. The linguistic key, event = [subject-clause-(verb-adjective)] is the basic data structure to reverse engineer nature's information processing.

In summary, Reddy et al. generalized the concept of using the ratio of primes by the proteins as the PPM; clocking a set of resonance peaks one inside another into the GML and the engineering of microtubule into a new kind of fourth-circuit element, functionally a tubule-morphic device that implements the PPM and the GML. Thus, three basic concepts define frequency fractal computing. Primes produce integers as a product, the event multiplication gives birth to changing geometric shape, so runs the computing universe.

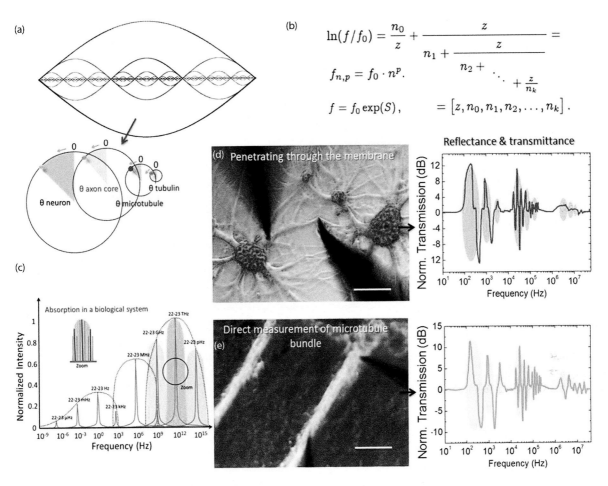

FIGURE 6.11 (a) A nested waveform is shown depicting the vibrations in the microtubule, protein, axon and its time cycle or clock presentation is shown below. The clocks representing all systems have their own phase network, all four systems are in different phases shown by a shade. (b) The nested clocks are also presented with a continued fraction algebraic system. (c) The prime absorption frequencies of the biological systems all that we studied over a decade are plotted peaks around 7–8 (nHz, μHz, mHz, Hz, kHz, MHz, GHz, THz) mostly depict positive resonance (not shown), the peaks around 22–23 (nHz, μHz, mHz, Hz, kHz, MHz, GHz, THz) mostly show a negative resonance. Left panels of d and e are microscope images. Angstrom probe connected to the axon of a neuron with the membrane (top) and without membrane (bottom); scale bar is ~10 μm for both. To their right the corresponding transmission spectrum, showing electromagnetic resonance. The triplet of triplet bands are shaded.

Ghosh et al. cultured a pair of rat hippocampal neurons on a pre-grown electrode array (Figure 6.12a). The growth was monitored so that the neural branches did not touch any electrode, using atom probes and dc fields one can regulate branching. Prior to any resonance measurement (Sahu et al., 2013b), a pair of patch-clamp probes were placed, rupturing the soma membrane in one/two neighboring neurons to measure the potential difference between them. The potential of a nerve impulse with respect to the culture solution is consistent with reported values. On the chip, the ac electromagnetic signals were applied via a pair of electrodes (PQ and RS; Figure 6.12a) along the axon horizontally, as if dendritic branches are the sources of a signal and axonal branches are the drains. Perpendicular to it using MN electrode gating signal is applied to regulate the nerve impulse. PQ and RS electrodes change frequencies synchronously. Keeping the perpendicularly applied ac signal of MN fixed at a particular frequency Ghosh et al. changed the frequency of the ac signal

applied along the axon length via PQ and RS. Then changed the transverse ac signal frequency via MN and repeated the frequency scan via PQ and RS. Thus, a 2D input frequency pattern is generated not just in a neuron, but in all the proteins studied following this protocol. The basic electronic set up used to measure and filter the resonance frequency is described in multiple carbon nanotube measurements.

On the 3D pattern of Figure 6.12, at particular pairs of horizontal and vertical ac frequencies, the neuron generates a potential for nerve impulse and releases ions (<1% of threshold firing current, ~50 nA; i.e., ~100 pA), even at the sub-threshold biases of 20–30 mV. The potential makes the vertical axis; hence, a 3D resonance frequency map is generated for the neuron. A 3D plot for the normalized firing potential (vertical axis) as a function of two perpendicular ac frequencies applied across the neuron, shows nine hills with circular bases. The map is unique because, the horizontal plane mapping the frequencies is the electromagnetic resonance and the

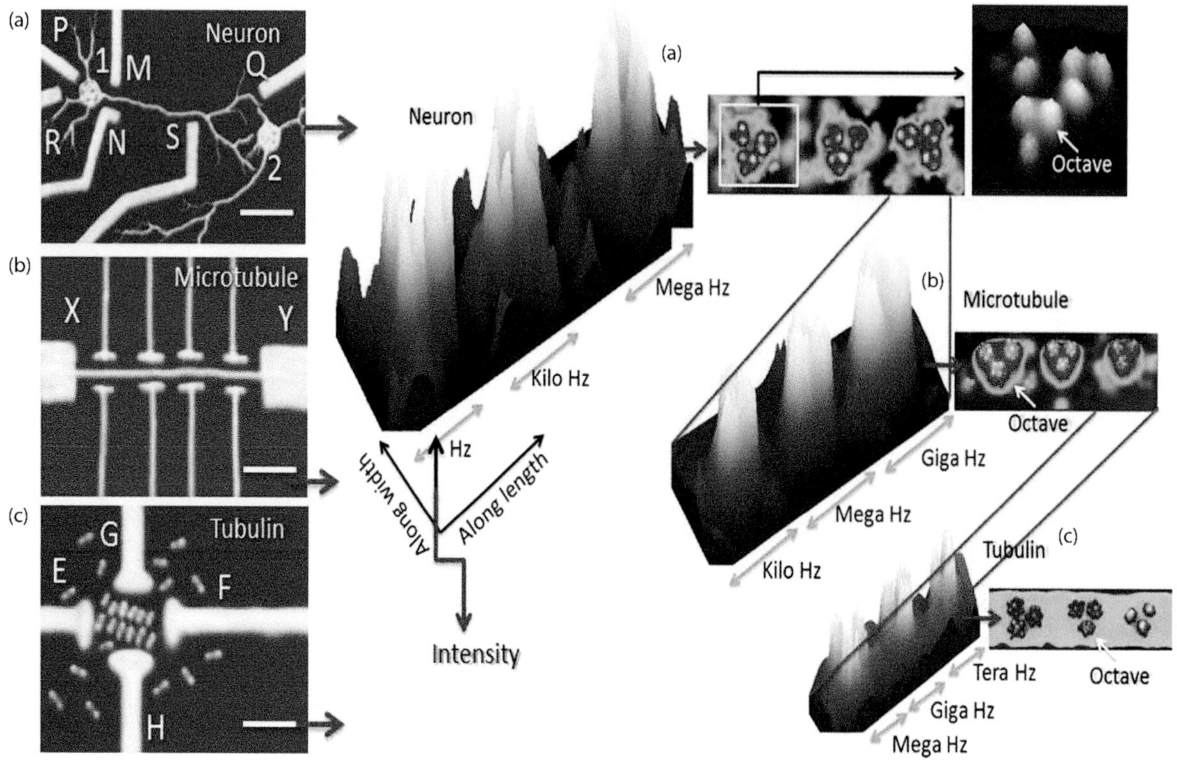

FIGURE 6.12 (a) Microscope image of a cultured neuron plate with Au electrodes grown, scale bar = 10 μm. PQ and RS are non-contact electrodes to supply horizontal, along the length of axon electromagnetic frequency, wirelessly. MN electrodes perpendicularly trigger the initial axon segment, wirelessly, there is no physical connection between MN and neuron-like PQ and RS. (b) Atomic force microscopy of a single microtubule dropped from solution and then manually moved to the right position on a pre-grown Au electrode array, scale bar 120 nm, XY are horizontal electrodes, the rest four are vertical electrodes. (c) Atomic Force microscopy of a tubulin substrate, scale bar = 50 nm. Tubulin protein solution (nano Molar) was dropped on a four-probe Au electrode junction when electric field of 2 and 0.5 V was ON across EF and GH, respectively, independently. Horizontally to the right of panel a, b and c, the three Log-log scale plots of resonance frequency corresponding to the three experimental setups are given. For the panel a, the patch-clamp measured current difference between neuron 1 Soma and neuron 2 Soma is the vertical axis Imax = 10 nA, Imin = 1 pA; two horizontal axes, one along the length (L = 50 Hz to 2.5 kHz, 2.5–250 kHz and 1–30 MHz) and the other along the width W. Wherein, W = LX0.3, if L = 1 Hz, then 0.3–1.3 Hz variation is made along the vertical axis, thus the vertical axis is linear frequency width, its upper and lower limits are percentage values Wmin = −30, Wmax = +30 for all panels a, b and c. For panel b and c, the 3D resonance plot's vertical axis is relative ac power transmission (P output/P input = P (0 < P < 1)) across horizontal XY and EF electrode pairs respectively. Horizontal axes are frequency for all three panels. For panel a, the length, width and vertical ranges are noted above, for panel b ranges are (triplet band) 10–300 kHz, 10–230 MHz and 1–20 GHz, and for panel c ranges are (triplet band) 30–450 MHz, 1–250 GHz and 20–350 THz. To the right of all three 3D plots we show the corresponding vertical visuals, panel a is zoomed.

vertical axis is intrinsic resonance causing the nerve impulse. In the resonance frequency pattern, three bright circles represent the situation when a neuron positively gated by horizontal electrodes while the low intensity part shows that horizontal electrode system is arresting the nerve spikes. One could notice that three prime resonance frequency domains host three further resonance frequency domains inside making the triplet of triplet frequency band. Doublet and triplet of resonance frequencies is a common observation atomic orbital resonance of molecules, a similar kind of resonance behavior is observed here.

An essential component of a single microtubule is a tubulin protein dimer. Tubulin protein solution was dropped in the gap of a four-probe electrode array, then a DC bias was applied to order 15–20 molecules. For 8–100 tubulins, the resonance

band remained independent of the number of molecules or the electrode geometry. Similar to the microtubule and neuron cell, a triplet of triplet resonance band was observed. The axon core, microtubule and tubulin have self-similar bands, with a common frequency region,—a similar structural symmetry governs the resonance in all the three systems. Helical distribution of neural branches, rings of proteins in the axonal core, spirals of proteins in the microtubule, α helices in the proteins, are the common structures, and the resonant energy transmission in generic spiral symmetry follows a quantized behavior (Sahu et al., US patent 9019685B2). Hence, a spiral symmetry possibly ensures coupling of all the clocks.

For all the three systems, neuron, microtubule and tubulin, each of the nine circles in the triplet-triplet band has 6–8 small circles inside (Figure 6.12). Since proteins are basic structures, they are pumped with the same resonating electromagnetic signal and simultaneously imaged. The resonant oscillations image of a tubulin dimer show only two high potential regions, not eight (online Movies in Ref Sahu et al., 2015), so, dimers are not responsible for 6–8 small circles. Scanning the isolated tubulin monomer using the tunneling microscope at various resonance frequencies showed four major and four minor distinct potential regions inside the monomer exchange energy. Total eight for a dimer. Protein dimer makes a doublet but monomer makes an octave. The one to one correspondence with the dielectric resonance image suggests that the observed 6–8 peaks in one of the 9 circles of tubulin are from α- helices localized by the β sheets. Thus, 10^6 orders of a spatial journey from micro to the nanoscale execute milliseconds to sub-nanoseconds clocks. En route, the GHz clock that fires a neuron originates at the single protein structure.

The triplet-triplet resonant band is not exclusive to microtubules and tubulins. Ghosh et al. have selected mostly found four components in the axon core, and around AIS and similarly measured their temporal resonance map (Leterrier et al., 2011; Xu et al., 2013). It was observed that actin microfilament's resonance bands are complementary to that of the microtubule,—they exchange energy covering a wide frequency domain. The resonance bands for all four proteins β-spectrin, ankyrin, actin and tubulin,—also their complexes are confined between the two frequency limits;—they share the time zone of threshold energy bursts. Overlapping time zone in the resonance frequency plot is common energy exchange regions for proteins. Thus, resonance chain forms. The β-spectrin structure has ion-transfer channels and ankyrin a known mechanosensor (Lee et al., 2006) have a cascade of α-helix oscillators dominating their resonance band. Hence, they exhibit a topological hysteresis in the 2D resonance spectrum measured by Ghosh et al. Moreover, β-spectrin and ankyrin show signatures of their lone cavities in the resonance band as doublets. A doublet in the 2D resonance pattern means the two clocks governing its resonance are coupled as part of one clock. There are plenty of other proteins that participate in generating the nerve spike. The current map is a fraction to the varieties of proteins available out there. However, the NMR like doublets and triplets in various compositions suggest that the resonance chain would exhibit much richer topology, once more proteins are added to it.

A musical wheel of a triplet of triplet frequencies: Conventional 1D resonance plot is a single line on the 3D resonance frequency map which is represented as a triplet of triplet circles. Triplet of triplet is not an absolute pattern, one could see pentate or even doublets, frequency fractal or resonance chain's topology is not as simple as reported earlier (Buzsáki, 2005b). The frequency wheel is created by the common time zones shared by the overlapping resonance frequencies of four proteins and their complexes studied here. Sonification of this frequency wheel reveals how vibrating discrete-time clocks topologically integrate into a single neuron firing. The higher frequencies are the patterns of protein's oscillations, while the slower frequencies represent protein complexes with larger structures. A complete rotation of the wheel are events that unfold from faster to the slower time scales to eventually trigger a single nerve spike. Until now, the rapid firing of a neuron was sonified as a stream of "ticks," here one "tick" of a nerve impulse is deconstructed with 72 frequencies bursting signals in an intricate pattern (Figure 6.4b).

6.8 THE MEASUREMENT OF A COMPLETE TIME CRYSTAL MAP OF A NEURON

How nature processes GHz data or THz data or pHz data? Do not read but match 15 patterns at all time scales simultaneously! To detect a time crystal, one requires technology to simultaneously measure multiple time domains, as shown in Figure 6.13a. At the rate of GHz, a material generates 10^9 bits of data per second. Then for THz, 10^{12} bits of data is captured per second. The memory drives fill up faster than one could imagine, analog data should be converted into a digital one at that very rate. How to manage them? Furthermore, one has to adjust the speeds of data transfer at all time scales so that we find a generic geometric relationship between them. One should do just the opposite, follow what the human brain does. Instead of searching bits of information, directly search for the fifteen basic geometric shapes, i.e., phase relationship that gets locked between different signals. For example, human memory strength is estimated by theta-frequency phase-locking of single neurons (Rutishauser et al., 2010). Proactive search for a dodecahedron in the big data, the 12 planes may be shifted but would deliver an insight. Therefore, modify the search topology and find how the corners of the geometric shapes would shift. It is so nice to say that we will take a topology and start building higher level geometries on this topology. But doing it in reality is extremely difficult. If we do not know the skeleton of the information architecture of the universe, we cannot glue the shapes. The pattern of primes connected to geometric language do just that. Similar to astrophysics, a metric that links change in the symmetry, runs a feedforward loop and correct the information structure of a system, here it is time crystal map of a neuron.

Time crystals obtained at the femtoseconds and at the pico-seconds time scales is the new neuron code (Perkell and Bullock, 1968), are perturbed to find how imaginary worlds are linked, it sets the prescription for a multi-nion tensor. Since here in this book we are confining ourselves within the domain of 11D dynamics, to find the universal time crystal the hardware would store the stream of local time crystals for a long time until higher level clocks integrate them. If we cross beyond 11 layers we do not search anymore for a dodecanion. We might need to make detailing of some of the patterns we see for hours (micro-hertz), then we can look into some of its important topology in the second's scale, it zooms and finds some interesting topology at the microseconds scale, then finds some more. Rapid neural data recording is already there (Gong et al., 2015), locking phase is observed as all frequencies (Hoppensteadt and Keener, 1982), phase locking is studied statistically (Hurtado et al., 2004). The journey through time scale for the human brain goes on and on to the femtoseconds. But who decides the interest? It is done

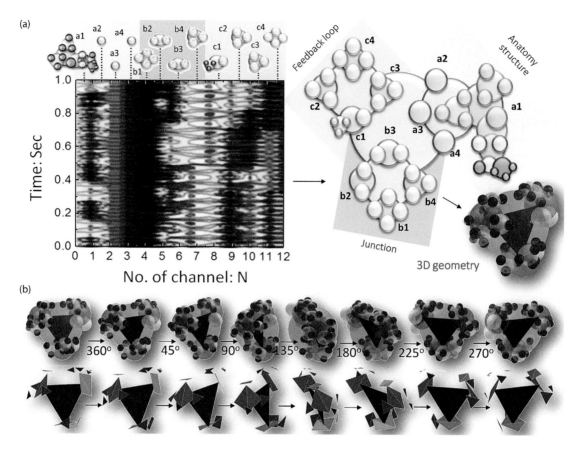

FIGURE 6.13 (a) Simultaneous reading of 12 channels of a neuron from 12 electrodes. Each channel reads a few parts of the whole time crystal which are shown at the top of each channel reading. The brightness is the intensity of the signals, the frequency of the signal is normalized to one and converted into the time domain. From the pattern the corresponding time crystal is generated. (b) 360° rotation of the time crystal and the corresponding geometry hidden in the structure.

by two factors. First one is memory and the second one is PPM that contains the map of all possible uncertainties. Be it the conventional electronics hardware, or brain jelly, reading "bits" is replaced by "topology search." Do not see all the data, because it is impossible, but inspired by the universal links between topologies we get into building a multi-time-scale topology of our own. Who knows we may be doing just that at this moment in our brain now? At least neuron is doing so (Hutcheon and Yarom, 2000).

The challenges are plenty to build a frequency map of a neuron, neurons use phase change and frequency change at a time (Hastings et al., 1985; Hess and Boiteux, 1971). 10^4 spatial scale journey means carriers change in their nature of the response to signals, so we need different tools. Ions resonate in the milliseconds time domain, large dipoles in microseconds, small dipoles in nanoseconds, atomic bonds in picoseconds. A biomaterial is like an observer who sees only the minutes spike in a day-to-day used watch, not that of a second or an hour. Therein, the topology of time is a circle. Every pixel on the circle's perimeter is a circle, even if that circle is zoomed, we see pixels turn to a circle (Figure 6.13a, right). So, the tools to see a smaller distance shows more objects side by side, but for time, to see more, one needs a topology to see within a given time. For two centuries spatial map had no such problem. Here we unravel the basic topology of neuron

clocks to begin the quest to map time architecture of a neuron (Strumwasser, 1974). Its long argued that cell cycles are nested clocks (Shields, 1976; Smith and Martin, 1973).

For half a century, neurophysiology had only one tool patch-clamp to study the neural response in milliseconds. No technology existed to see what happens during a ~1 ms firing at multiple faster time scales simultaneously. Now, an array of coaxial atom probes maps highly active time zones during the span of a neuron firing from microseconds to picoseconds by reading all clocks from a single protein scale to the neural membrane simultaneously. Reading many clocks at once requires a new generation of machines. All the probes recording their distinct time scale events should be part of an independent measuring system in the hardware. Recording by all probes should be synchronized using a global master clock, only then one could find the integrated 3D geometric information of the material (Figure 6.13b).

Final challenge is to slow down the transmission or edit it like a transistor gate to operate below their resonance threshold. Carefully opening the transistor gate in 1952 triggered a computer revolution, here gating the signal transmission line for neurons at ~100 μm to a protein ~2 nm requires nano-lithography for the microwave engineering of communication ports. Ceasing the nerve impulse through an axon, electrical and magnetic

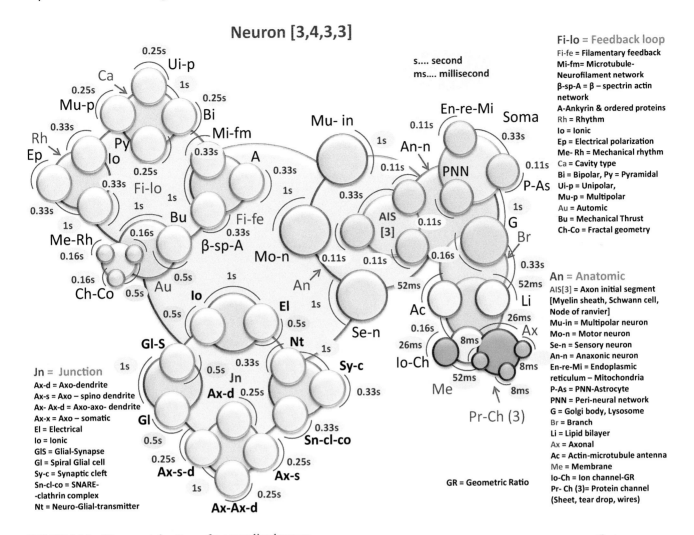

FIGURE 6.14 Time crystal pattern of a generalized neuron.

transmission through the microtubule surface and finally through a group of single protein molecules require a nanoscale control of biomaterials movement using dielectrophoresis. If water layer is more than ~5–10 nm around biomaterials, ions transmit ~10^{-6}A, resonate at ~1 nW. The 10^{-12}A current and 10^{-18} W of resonant signals of biomaterials are totally masked. After all precautions, when one plots the frequency or time response, irrespective of size or the origin of resonance, be it ionic in the membrane and orbital coupling in a single protein, only then the pattern of clocks becomes evident (Figure 6.14). Gating in a transistor is done by vertically applying an arresting potential across the horizontal source and the drain. The same process was repeated here for a neuron, an axon, a microtubule and a single tubulin protein (Figure 6.12).

Apursomanam chalapratistham | Samudramapah prabisanti jadwat
Tadwat kama jang prabisanti sarbe | Sa santimapapnoti na kamakami

The way ocean never expands to infinite by swallowing millions of rivers, doesn't even cross the seashore, the same way distraction of mind entering into the pure, gets dissimilated, the mind that has seen the ultimate vibration of the universe (Brahman), nothing distracts. Only that person creates a peace of mind, it is impossible to make our mind free if it is connected to materials.

Tani sarbani sangjamya aseet matpara | Bose hi jasyendrani tasya pragya protisthita

Even if only one organ connected to our brain (Indriya), goes beyond control, the ultimate knowledge created in the brain leaks away through that organ-like water goes out of a leaky skin bag.

7 A Complete, Integrated Time Crystal Model of a Human Brain

7.1 BRAIN IS THE ENGINEERING OF PRIME NUMBERS EMBEDDED IN A TRIPLET OF TRIPLET CAGE

Looking beyond current neuroscience: Mapping the entire human brain in terms of biological rhythms: The brain is a noise harvesting machine, a linguistic device, wherein the information unit is a triplet of time crystals run subject-clause-verb/adjective of a sentence (Green et al., 2017). Since the brain converts all kinds of signals in terms of pulses, eventually, letters of brain-language are determined by the variability in pulse shaping and the grammar of this language would have two distinct sections. Information is geometric in the brain (Nie et al., 2014), collective oscillations of hills and crests of information builds human thoughts (Tank and Hopfield, 1987), then, dynamics of time is neuroscience (Rabinovich et al., 2008). First, the basic rules for pulse shaping that makes sure that the pulses are distinctly identified based on their sensory-origins. Second, rules for the superposition of distinctly different streams of pulses. Third, coupling and de-coupling rules among different wave-streams coming from different sensory-organs. However, current neuroscience believes that the brain is all about neurons, no cognitive processing inside a neuron, the membrane does everything, and wiring at the neuron level contains all necessary information for the brain's information processing. The neuron-level wiring is the key for all brain mapping projects, but they ignore, neurons use clocks to talk (Gerisch et al., 1975). Since no two human brains have similar circuits and connections evolve at every moment, we need a new language to read the "circuits in action" for cognition that does not depend on the specifics of the wiring. One such parameter is, to the PPM-GML-H (phase prime metric-geometric musical language-hinductor) triad, the brain has various distinct operating circuit layers both below and above the neuron levels. All circuits in a layer (say protein circuits or cortical column circuits) operate at various frequency domains, thus instead of one at neuron layer, we want to create multiple brain maps at various time domains. All these maps would co-exist and interact simultaneously. Geometric language for brain mapping is rhythms or cycles of vibrations as time crystals generated by the chain of vibrations that connects circuits of all layers. Creation of such a map would enable one to generate quantifiable behaviors using hierarchical temporal maps of the brain (Salthe, 1985).

As the external environment changes, the ripple effect enters inside the brain through a particular frequency domain of the rhythm map of the brain, even a pulse of light could reset a mind (Rusak, 1993; Strogatz, 1990); environment shapes mind (Barrett, 2011). When we listen to others, speech builds an architecture of time (Shannon et al., 1995), inside the brain speech resembles music (Schwartz et al., 2003; Brown et al., 2006). Plasticity of a brain is active even in space (Koppelmans, 2016). On the other hand resonances in the ion-hemisphere affect the rhythms of a human brain (Rusov et al., 2012), the temporal disorder is the key for many mental disorders (Rensing et al., 1987). The internal low frequency or large time period rhythms start changing, the brain and the entire body undergo minimum atomically precise motion at various layers to adjust to the modified rhythms and that we define as decision-making of the brain. A resonance chain of the entire body that links all brain components as a triplet of triplet clocks (Berliner and Neurath, 1965) harmonizes with that of the external environment's perturbation to the internal rhythms. While major brain researchers connect functional domains of brain circuits, there are few yet very important research on event-based coherent communication of brain parts (Andrew and Pfurtscheller, 1996).

Not anyone but all components collectively process information: Neuron holds center stage of neuroscience to artificial intelligence (Reynolds and Desimone, 1999).

The human brain-body structure has so many layers of materials one inside another, circadian rhythms propagate at all layers (Aschoff, 1965a; Aschoff and Wever, 1976). Why not ever anyone thought that the information is not processed at any particular layer, or in between a group of layers, but all the layers build an architecture of time where information is processed. Differentiability or if we measure the variation of two states of a cell between two time intervals or the variation of states between two neighboring cells cannot be defined. Because there are many cells inside and each cell is part of many cells outside. States of each cell are incomplete. As said, a fractal can never be touched. In this case what will be the machine that would work? Since nothing is touching each other but everything is a door to a world within, the only way to communicate is a "resonance chain." Chain of vibration, i.e., time or clocks, the idea advances an old proposal, "brain encodes information using time" (Buzsáki et al., 1994b; Buzsáki, 2006; Buonomano, 2017). Earlier brain science had only ionic rhythm (Bub et al., 1998), now above and below neuron-level we include all forms of polarization, weak van der Waals to atomic in 12 imaginary

nested layers. Electromagnetic brain theory came long back (Jirsa and Haken, 1996). The objective would be engineering a brain that would have 12–15 components for processing exclusive time crystals for 15 primes (Figures 3.4 and 7.1a, b).

What is the elementary *decision-making* machine of the brain, say, an alternative of Turing, or a switch? Human brain might not be a discrete sum of machines as we are naturally taught to believe in the Turing world. Since neuromagnetic rhythms flow in the brain with a time period of minutes in space independent manner (Ramkumar et al., 2012), each brain component would vibrate with a selective set of primes as described in Figure 3.4. Resonance chain = specialized PPM of Figure 3.4. We should look at the physical structure of the brain as described in Figure 7.1c, d, find which primes are governing its structural symmetry, make exclusive metric for that component. That metric is the processor of the brain's information as a time crystal. Now, what can this small chain do? (1) Resonate with the big chain to exchange time crystals with each other. (2) Rhythms would expand inside each of them, in doing so they will

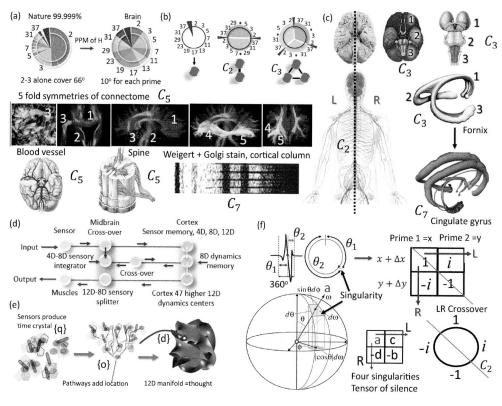

FIGURE 7.1 (a) Two wheels of primes, the left wheel shows the contributions of the primes, the right wheel shows that artificial brain construction means all primes have an equal contribution. (b) The concept of symmetry is applied to select a set of primes in the prime wheel, C2 means 50%, C3 means 33% primes are grouped during operation. (c) Brain's neural network follows C2 symmetry. Three major brain lobes create C3 symmetry, inside the connectome, five prime regions act independently delivering C5 symmetry. Inside the cortex top layer, 47 Broad's functional regions are made of hexagonal lattices of cortical columns, each cortical column has seven layers following C7 symmetry. The spinal cord has five functional domains following C5 symmetry. In the midbrain Fornix follows a C3 symmetry, and Cingulate gyrus follows a C7 symmetry. (d) The three-layered input-output networks of brain's information processing. (e) Three steps for the time crystal-based information processing in the brain. The sensor's time crystal as quaternion, a pathway-based fusion of time crystal as octonion and finally expanding the information to the 12D manifold, a dodecanion. (f) The phase space allocation of brain's information where the components in the brain selectively play the role of diagonal driver of a tensor holding the information for the singularity points.

self-assemble. (3) Some rhythms get stronger and some weaker because of the external input of rhythms, like phase transition. (4) Spontaneous reply back of the rhythm or time crystal to each other by resonating electrically, magnetically and mechanically. (5) Harvest noise, filter it into a signal (Buzsáki, 1989). (6) Geometric information becomes the key factor (Bullmore and Sporns, 2009). So, it will be a new AI, because in the domain of current AI, only neurons work, but here, the entire body is the brain, and we go inside continuously to the molecular scale to complete the resonance chain to define the machine. Also, the machine does not end inside our body it takes metrics of our immediate environment. Design principles of synchronization oscillators by nature in the cellular systems show that it uses a basic geometric language that we use for pattern recognition, the small oscillators are arranged in the elementary geometric shape to generate all possible oscillators. Figure 7.1d shows primes for human-brain body system (C2 symmetry), connectome (C5 symmetry), fornix and cingulate gyrus (C3 and C7 symmetry), blood vessel and spinal structure (C5 symmetry), C47 symmetry in 47 functional cortex regions (Garey, 2006), etc. The symmetry of prime number of components based time crystal model of brain leaves no component, the entire brain-body system is part of the new time crystal model of the human brain model proposed here (Chiel and Beer, 1997). The time gap between interhemispheric transfer tells how information specializes left and right brain (Ringo et al., 1994), thus C2 symmetry activates.

Why do we want time crystal map of the entire human brain? The temporal architecture of the human brain interests many (Llinás, 1994). (1) No two human brains have similar connectivity, and connections in the brain change continuously throughout the life of a person. The mechanics of an observer is unknown, and that is the first step to learn about the brain (Bennett et al., 1989). (2) Natural vibrations or resonant properties are fundamental to any material property, periodic vibrations are rhythms, a complete brain map of rhythms from atoms to the largest scale could unravel a map that would have some features not specific to the hardware of a particular human at a particular time. Rhythm across species must be similar; otherwise they cannot interact, bursting neurons are self-similar (Pinsker, 1977). (3) Be it classical or quantum mechanical, or any physical or chemical event at any scale, everything could be explained in terms of symmetry breaking, so we replace the entire system in terms of "symmetry breaking." Brain size variation across species show that the brain rewires differently with size to keep the symmetry intact (Ringo, 1991). (4) Hierarchy of rhythms, at certain level incorporates nature in its cycle, even to send a picture, neurons create architecture of time and transmit (Richmond et al., 1989). That's why a body learns and adjust with nature without conscious efforts. Thus, we will get to learn why there is life? Now organisms are used as computers, in future, the computer will be organisms, they would live, learn, give birth to a new computer and die. (5) Just like we argue that the brain is a non-computer, universal time crystal argues that there is no communication (Pierce, 1961). When the whole body is part of a one-time crystal composite, there is no need to run separate signal transmission. A conformal replica could be created in the receiver if both sender and receiver are built using a brain that follows the same mathematical rules to link discrete events of the past and extrapolate it to the future. (6). More is different (Anderson, 1972). There are many parts of the brain that are self-similar, fractal-like, following a fixed symmetry, they do nothing. However, sometimes they change the order, symmetry breaks. Anderson beautifully observed in the classic paper that when systems become large, it does not matter how many symmetries the constituents have, giant systems also follow only a few symmetries. Reddy et al. (2018) discovered the metric of primes in the brain, and the metric of prime that we will propose here is conditioned to start counting from 1, as soon as a system completes accounting for first 15 primes (up to 47). (7) Statistics is secondary, the topology of symmetry is primary, so, how many neurons do not matter, how many filamentary nanowires, or cortical columns do not matter (Yates, 1987). The symmetries have a language to talk to each other, and that is the most fundamental knowledge of nature. If we do not know that we cannot learn nature, or our existence. Our objective is to discover that language. (8) The universe is massive, complex, so is the world of elementary particles. The brain is a gift of nature, that is low cost, easy to study compared to building a particle collider or gravitational wave observatory. One phase reset curve builds one-time crystal, a composition of phase reset curves means a composition of time crystals, which is reported for the brain (MacKay, 1991). Stochastic evolution of phase in neurons shows that time crystals may build a new time crystal (Tuma et al., 2016), for delivering a greater control on excitatory and inhibitory properties (van Vreeswijk and Sompolinsky, 1996). It would be exciting to see a brain map extrapolated to the universe. (9). What is the learning limit of a human brain? We want a solution of Chaitin's query (Delahaye, 1989). Singularities in an operating clock are the keys for a time crystal architecture (Malinowski et al., 1985). Phase diagram of a group of oscillators show that there is a collective, unified phase behavior, human brain might integrate many clocks (Matthews and Strogatz, 1990). While integrating clocks, depending on the nature of dynamics, groups of clocks with distinct dynamics are phase clocked with each other (Wang, 1994). Rhythms of motor response argue for a 3D geometry of clocks (Marder and Calabrese, 1996). (10) The pattern of several phenomena is filled in physics textbooks—is the brain designed in the same way nature designs a physical phenomenon? (Small and Garfield, 1985; Agu, 1988; Steen, 1988). The transition from rate code to temporal code is studied for decades (Mehta et al., 2002), this transition is universally captured in the generic time crystal model of the human brain. The delicate relationship between rate code and sub-threshold signals could open up a new world of brain's information processing (Radman et al., 2007). Time crystal model considers the universe of sub-threshold signals, which decides complex signal transformation in the neuron (Ratte et al., 2015), yet brain models do not take them seriously.

How to study the whole brain; is it possible? The frequencies of resonance peaks are fractal, no number of log-log plot would give us a linear relation in the frequency space: The structure models of all brain components are available online. It is also an easy task to simulate the components to find the reflection and transmission coefficients (S11, S21) using a Maxwell equation solver, then verify the properties by printing those components using a 3D printer and measuring the coefficients, confirming with the theory. Conventional brain models that are now trying to reverse engineer brain by deep learning (Yamins and DiCarlo, 2016) consider only neuron-skin or membrane route. Therein, the brain components vibrate using a singular natural logarithmic relation (Penttonen and Buzsáki, 2003). Here in this book we compile research works on the clocking properties of all brain-body neural network, do not leave any component. One has to simply plot the transmittance and reflectance as the resonance peaks of the brain components along the frequency scale, shift in frequency as a function of time is the origin of a clock. Even if we take a log of frequency in the primary axis, the plot looks like as if the resonance frequencies are separated by a log scale, normally peaks should appear equidistant after taking the log scale. Since log values are separated by a linear distance, if we take the linear values and then plot the derived resonance frequency once again, we find, it is a log distribution once again, the log feature or the non-linearity cannot be diminished. It means the frequencies are separated by a log function inside a log function inside a log function. Possibly this would continue both in the brain and in the universal resonance chain, that links brain components in the frequency space. It is a nested frequency fractal, a prerequisite to building a time crystal, and it is non-differentiable. If there is a homogeneous distribution of power among all resonance frequency values when the architecture of the multi-layered seed structure (or escape time hardware) is being formed, then the architecture should adopt a symmetry that allows it to maintain an equal power loss throughout. If the equal power loss is maintained in a scale-free manner, the lower frequencies would be spaced much nearer, now the power law is a conservative claim, the exponent of the power relationship holds an infinite series, thus generating a log inside a log inside a log function (this is not log(log(log(frequency....)))), it is non-differentiable, thus non-Turing.

Triplet of triplet resonance band in the universal resonance chain: Rhythms of very different natures are so intimately connected in the brain. Silent circadian rhythm, thermoregulation disappears (Fuller et al., 1978), the wave of ions interacts with the waves of heat flow. All resonances are linked. It is striking to see that when we look at the PPM plot, it is a triplet of triplet coupling of the distribution of divisor choices of the integers and protein complexes like microtubule generates a triplet of triplet pattern (Figure 6.12). Even more striking is that twelve brain components reveal the triplet of triplet resonance bands (Reddy et al., 2018). In the universal resonance chain or the global time crystal plot resembles that of "teardrop curve" or "pear-shaped plot." The parametric equations for universal resonance chain would

be ($x = \cos t$, $y = \sin t \sin^m(t/2)$) for teardrop curve, and for the pear-shaped plot we get $b^2 y^2 = x^3(a-x)$ (http://paulbourke.net/geometry/teardrop/). Note that we need multiple values of p in the domain $0 < p < -1$, for $i^2 = -1$, to create a generic fractal equation for the universal resonance chain. It is interesting to note that the inverse Mandelbrot plot also takes the form of a teardrop, but we are not concerned in shape but the patterns inside. One could generate fundamental constants to five digits after the point at least, from the universal resonance chain, as if the brain is a universe in itself, geometric constraints unfold just like the universe.

The birth of a clock in the brain: When matters vibrate periodically, a system point runs in a loop, its rhythm for biologists, clock for physicists. For a self-operating perpetual oscillator-pair, circling of energy packet between a pair of elements is essential and that vibration comes from matter. While exchanging energy, a part of that energy is never found in any of the participating atoms or systems; we call it bond energy. It can happen at any scale. When a matter or system gets an energy packet, where does the energy go? It goes to the structural symmetry. The symmetry defining part always takes the energy to vibrate, so, physicists found a remarkable tool in symmetry, replace every matter by its structural symmetry. Thus, from the above discussions we can perceive that two different systems have a composition of multiple symmetries, each symmetry has characteristic vibrational frequencies and when they get populated by energy packets, they vibrate. Two matters can exchange energy by sending photons of an electromagnetic frequency, pumping the medium in between mechanically or by sending materials. Among three, all of them or a pair or individually the systems can start exchanging the energy and interact. If the exchange is once, they are not coupled, but if it is periodic due to two or more interplaying forces inversely proportional to each other, then cyclic energy exchange arises and a periodic oscillation or rhythm is born. The temporal structure in the spatially organized neuronal ensembles reveals that the clocks have a hierarchical network (Buzsáki and Chrobak, 1995).

Experimental keys for the fractal time operation in a meander flower garden: Meander flower garden is probably the best contributions made by the three decades of time crystal movement (1970–2000; Figure 2.7c–e). Meander flowers are like various classes of cycloids, epicycloids, and hypercycloids, could be realized by placing a guest clock at various positions on the host clock. More than a century back, in the planar curves such geometries were reported (Eagles, 1885). Nested clocks are the foundations of a time crystal, consequently, from biology to nature to culture, meander flowers rule (Collins, 2000; Leopold and Langbein, 1966). Whenever one measures the electromagnetic resonance band of the biomaterials, the reflectance and transmittance peaks shift as groups. These groups are keys to the geometric shapes, i.e., GML, for example, three peaks oscillating in a group means a triangle is encoded. Now, how does the resonance band change with the additional input signals in the form of electromagnetic, magnetic, mechanical, electromechanical, or ionic vortices? That's the key question for the brain. In this book, throughout, for artificial brain or real brain, the composition of vortex atoms forms the time crystal and arrange in a unique geometry operate the GML and PPM. The composition of distinct PPMs each for different types of vortex atoms operate simultaneously. Each PPM is a meander flower garden, and the composition of PPMs is a garden of gardens (Figure 2.7c–e). Even if there are a large number of objects in an image, the brain has to identify the abstract geometrical relationships among different objects and creates an equivalent fractal for that abstract relationship also. In case of several different kinds of sensory input data, due to the natural property of frequency fractal of the hardware, the patterns in different parts of the artificial brain hardware (the entire brain is a single fractal object) get correlated, and a new fractal of seed geometry is formed. The brain circuit undergoes subtle changes to incorporate these features. In this way, visual, sound, taste, touch, and smell data get correlated in the hardware. It should be noted that for the highest-level operation of the brain fractal hardware, the basic seed pattern of the new input fractal is the highest-level perception data, this is saved in a very particular region, these 47 cortex regions known as Brodmann's regions. The new fractals made of geometric shapes are stored in the brain only when it does not match the existing patterns; if it matches, there is no question for the hardware to store anything new. Adding a flower in the garden is a thought, adding a garden is learning of the intricate skill.

7.1.1 FOUR, EIGHT, AND TWELVE IMAGINARY WORLDS WORK TOGETHER

Scale-free activities in the brain: Evolution of the brain argues for a scale-free parameter for the animal kingdom (Martin and Harvey, 1985). Signals propagating in the brain are not just connected by frequency; there are ten parameters, the transition from Fast Fourier Transform to time crystal analyzer means acquiring information from nature never acquired before (Figure 2.8e). Phase locking in the propagating signals has been measured in the brain (Lachaux et al., 1999). Multi-input, multi-output, non-linear dynamic modeling of the brain argues for an extensive hierarchical information processing (Berger et al., 2010). However, instead of massive architectures, only the free energy principles have been used to explain learning and creativity/intelligence (Friston et al., 2011, 2006), and this is possible because of scale-free brain functional networks (Eguiluz et al., 2005). Therefore, multi-layer phase network, which is the foundation of several brain models is not an imagination, it's an extrapolation of a concrete experimental observation. As per the energy expense, skeletal muscle is the first, the liver is the second, and the brain is the third; as per the information expense, a sensor is the first, midbrain is the second, and cortex is the third (Figure 7.1e). It is already well established that the neural network in the brain evolves its circuitry toward a metabolically efficient architecture. The dynamic synchronization of electrical activity fluctuates in a scale-free manner all over the brain (Gong, et al., 2003), even in the ECG (Hwa and Ferree, 2002). The neurons form clusters of a time crystal (Figures 6.13 and 6.14), larger

is the cluster, lower is the frequency of resonance. Since larger cluster requires more power to operate, the power-frequency relationship in the brain would follow an inverse relationship. It has been shown that human brain operates via nested frequencies; it is not a band with continuously allowed frequency values (He et al., 2010). The symmetry of structure helps in modulating time. The brain has two hemispheres exhibiting a C2 symmetry and prime 2. Here the timing difference between the two hemispheres have been linked to a special selection of functions (Ringo et al., 1994).

People survive without a complete brain; brain shows extreme plasticity: If due to hydrocephalus the entire cerebral cortex region disappears in the brain, the person with an empty brain still survives and lives well, but we panic seeing a giant hole in the brain, with almost no cortex region. Fiber pathways of the brain not important (Schahmann and Pandya, 2006)? When an MRI scan shows a giant hole in the brain with a thin layer of cortex surviving, we wonder if this man is conscious of what brain builders are doing by mapping the neurons of a whole brain. The madness in replicating the cortex region for artificial brain needs a revisit, why 20 billion cortex neurons out of total 85 billion neurons do not matter to his consciousness (Wheeler, 1981). Performing basic information processing is possible even without a cerebellum. Recently a Chinese lady was found without a cerebellum, and except temporary nausea she did not feel any problem until age 23. If half of the brain is cut off, the other half takes over, even the medical cases are there without a brain at all, more than age 2 or 3, the babies fail to survive. If enough time is given, the brain cleans up and rebuilds an equivalent of the system largely. The drive to locate consciousness at a particular place, object or process is weak, the lookout for a magic switch to consciousness is primitive. Neuron size, number, and density, nothing single-handedly determine the degree of intelligence. The brain is largely a self-similar structure of cavities nested one inside another, cross-talk between clocks is common (Asher and Schibler, 2011). The cavity is a robust concept, it is a function of wave functions frequency or wavelength, and the nature of the wave determines what kind of material we need to build the pot. A hollow metal ball is not a cavity for a sound wave—for the sound, the object does not exist almost—but it is a cavity for electromagnetic wave. Neurons believed to float in the ocean of ions, what if they float in the ocean of electric fields like an electric eel (Rommell and McCleave, 1972). Whereas for a hollow glass ball, it is a cavity for a sound wave, and for electromagnetic wave it does not exist. The brain is not just a cavity network. The cavity ensures a closed universe for the surfaces to operate. The brain is a nested architecture of time for topological materials too, the surface geometry of objects is designed following the symmetry of prime numbers in such a way that it can produce time crystals in the various frequency ranges and dimensions. The brain has three layers, sensors produce quaternion tensors, the pathways and midbrain produce quaternions and the cortex produces dodecanions as 12D manifolds, i.e., thoughts (Figure 7.1f). Since every human has a different neural network, the map, how all neurons look in a brain is a quest driven by madness, similar madness would be if we go for mapping the cavity, or surface architecture from bottom to the top. What that is invariant in the brain is the symmetry of the primes. Thus, PPM and time crystal made of GML would be the key to find a brain model that would enable researchers across the globe to study it comprehensively in a lab-protocol independent manner.

Sensors are part of the artificial brain, a network of the skin is similar to cortex; clocking genes are everywhere: Until now biological clocks were limited to a group of neurons. When we experimentally establish that those clocks do not end at the neuron skin but are connected further down below to a protein scale (Sahu, 2013a, 2013b, 2014; Ghosh et al., 2016a), our body becomes an integrated chain of clocks, from the atomic scale to the ultimate boundary. Even very recently, the body's master clock was two small spheres—the suprachiasmatic nuclei (SCN)—in the brain made of 20,000 neurons. The 6,000 astroglia mixed in with the neurons we considered silent, but it is just shown that astrocytes clock those neurons, both clocks cross-talk (Tso, 2017), Almost all body cells, e.g., lung, heart, liver, and sperm keep time, with a few exceptions such as stem cells.

All pulses are similar in the brain; the necessity for inventing a new kind of time crystal sensor: Currently neural code alone build's brain code (Kuffler and Nichols, 1977; Konishi, 1990). Its fundamental action, integrator or coincidence detector is under question (König et al., 1996). Once we learn how to read the streams of pulses, we can capture the signal from any part of the brain and read them as a language, if essential, we can talk to that part in principle. As a first step, we invent a new type of sensor. In the day-to-day camera, or sound recorder, we store the picture as is. Every visual image is divided into a 2D array of pixels, and we store those pixels in the memory device. The location of a pixel, the intensity and color corresponding to that pixel is stored in the 2D grid, we call it "negative." The original picture is recovered whenever that is necessary as "positive." For a sound recorder, similarly, we store intensity and frequency exactly as is; the voice stored as a stream of bits of information could be replayed in the tape recorder whenever necessary. While generating an identical replica has been the fundamental inspiration behind human technology, nature has done something completely different. The problem is that, when we look at the pulses inside the brain, to us, electronic pulses coming from all organs look similar, neural code of events is unknown (Mountcastle, 1967). In the last hundred year, scientists have tried to understand how these pulses are generated in the specific organ say eye, and almost all the person who contributed fundamentally to this mission, have been awarded the Nobel prize, however, the next step, how every single information is exactly encoded in those streams of pulses has not been understood. That means, if we put a probe inside the brain, and we observe a stream of pulses, they will mean nothing to us, we cannot read them, or even tell whether it's an optical data or sound (Carr, 1993). In this chapter, we will create a generic model of interpreting those pulses to implement the information gathered to develop the artificial brain like a computer.

Metabolism, plasticity and power consumption: in the brain: Live video images of electric signal transport along with the brain show that the total active area remains nearly constant, it could possibly be due to a lossless energy transmission of energy packets across the brain. Widely distinct metabolic and circadian clocks talk to each other (Asher and Schibler, 2011), each rhythm compensate for the other's loss (Brinkmann, 1971; Chandrashekaran, 1974). Since solitons move with the velocity of sound resembling strongly to the experimental observations, researchers have argued for such a lossless transmission process in the brain. One additional argument for soliton-based transport comes from the energy-expenditure calculation for the synaptic firing. We do a simple calculation from the experimentally observed data to estimate how much energy is required for the human brain computing via synaptic junction firing, as the primary process. Metabolic cost unifies all brain components to deliver a scale-free feature (Hasenstaub et al., 2010). For a single neuron firing, it is around 100–120 mV bias change and the current change is 1–10 μA, which means the power required is around 100 nW. There are around 10^{14} synaptic junctions in the human brain, if say 1% of the total neurons fire at a time, then it would require 100,000 watts; however, the brain operates spending only ~20–25 watts. Let us get deeper into this argument. A neuron hydrolyzes 10^{10} ATP molecules per second to get its operational energy, which is around 3 nanowatt power supply, considering 10^{11} neurons in the brain, 1% of brain usage costs 3 watts. Since we know very well that the brain uses an average of 25-watt power, roughly we use 8% of the brain. The calculation seems to be pretty convincing, however, we miss a very interesting part of the story, neurons fire at a speed 1–40 Hz, what would be the source of energy that sends signals to large distances across the brain and asks them to fire? That process should cost a minimal amount of energy per neuron (Suzana, 2011), and minimum time, because chemicals to travel from one end of the brain to another, it will take a very long time. Conduction velocity of action potential = 0.6–120 m/s, which is roughly the 30% of the velocity of sound, and neither ions can move this fast, nor electrons, if for the time being we ignore than the magic of quantum mechanical entanglement playing a role, then the only possibility left is soliton which moves with the velocity of sound (Heimburg and Jackson, 2005; Poznanski et al., 2017). The argument inspired many to reject the firing-based computation protocol and implement an alternative (say soliton-based) computation. The question is that an engineer who wishes to construct the brain must learn how to send the question to all 100% audience, remaining silence is their choice, but that does not mean, he won't ask the question. Soliton is a nice idea to interact, because, the sound vibration can travel all around the brain with ease, if it matches with the encoded resonance signatures it starts to vibrate resonantly (Suzana, 2011; Howarth, 2010).

7.1.2 SINGULARITY ON A SPHERE—THE KEY TO A CLOCK OF A TIME CRYSTAL

High-frequency oscillations nearly 1 kHz have been correlated to cognition, such high-frequency oscillations are at the limiting time resolution of ionic impulses (Kucewicz et al., 2014). One of the problems in explaining high-frequency ionic transmission is that if pulse timing reaches the limit of ionic transmission, and cognition arises in that time domain, the controls must be happening at a faster time scale (Worrell et al., 2008). However, the ionic transmission has reached its limit, so we need another mode of communication. In the existing neuroscience there is no provision for other modes of communication. The geometric language that explores singularity enables complete shutdown of a particular vibrational mode and the other unrelated mode to take over. The transition from one clock to another happens via a burst from a singularity point (Figure 7.1g). One could represent a single clock with four singularity points at maximum using a 2×2 tensor. In Figure 7.1g, a diagonal is drawn to present C2 symmetry of the clock and the structure that builds the clock. One of the most exciting parts of the brain model is one-to-one correspondence between shape and the symmetry, but not the information content. The geometric shape of the information is always invisible.

It is so beautiful to think that eye, skin, ear, nose, and tongue all sensors get 11D information packet from nature using a dodecanion tensor. One Bloch sphere presented in Figure 7.1g cannot represent the whole brain but an elementary device, more spheres need to be connected to represent complex information (Figure 2.2c). The mystery of the brain is in the sensors, which use a pattern of primes to get back to nature and ask external agent if some parts are missing could it fill that gap. Think about 47 layers of Brodmann's region in the cortex. Because the ratios of the mean frequencies of the neighboring cortical oscillators are not integers, adjacent bands cannot linearly phase-lock with each other. Instead, oscillators of different bands couple with shifting phases and give rise to a state of perpetual fluctuation between unstable and transient stable phase synchrony. It is true about the resonance chain too. The $1/f\alpha$ power relationship among different layers in the brain implies that perturbations occurring at slow frequencies can cause a cascade of energy dissipation at higher frequencies, with the consequence that widespread slow oscillations modulate faster local events. The scale freedom, represented by the $1/f\alpha$ statistics, is a signature of dynamic complexity, and its temporal correlations constrain the brain's perceptual and cognitive abilities. The $1/f\alpha$ (pink) neuronal "noise" is a result of oscillatory interactions at several temporal and spatial scales. Singularity is a filter to noise.

7.2 PRIMES IN THE FIVE SENSORY SYSTEMS

All components described in Sections 7.2 through 7.5 were built artificially using cables by Singh et al. (2018), and the rhythms reported in the literature were regenerated in a large device, before their miniature versions were incorporated into the humanoid avatar described in Chapter 9. In all five sensors a stream of electric pulse-based wave trains is created and sent to the brain via a nerve bundle. First, all primary information is stored in the typical nature of the pulse (growth and decay rate, amplitude and coupling of more than one pulse). Second, gridding; it means 2D periodic pulses, each at the corners of the four-sided cells that groups a set of information (Figure 2.8c, d). Third, several 2D pulse-streams are superimposed on the background waveform; suitable filtering hardware could isolate each 2D pattern separately. Be it sensory or memory, three protocols would build time crystals (Figure 2.6) and the corresponding garden of gardens of meander flower (Figure 2.7c–e). Information is encoded in the neural assembly by spike intervals; there are three kinds of responses, mainly. First, "direct response to external stimuli"; second, if the time gap is more between the spikes, then it slowly adopts and then starts responding. There are two types of adaptation, first, "build-up to the necessity," wherein, signal response increases as the number of pulse increases; second, "bursting with a logical time crystal," wherein alternative periods of responsive and silent modes are observed. While describing the grammar of eye-pulse, tri-phase signals are used, tri-phase means two peaks with a negative trough in between them, and the ratio between the crest and trough stores a clock, multiple such sets build a time crystal. The similar pulse signature is common to store and process time crystals for all organs and neural assemblies.

Music and the human brain: Brain's eye movement processing exhibits distributed parallel processing and tensor theory is found to fit well (Anastasio and Robinson, 1990), paving the way for the dodecanion tensor analysis. Just like microtubule is the deserted DNA, it never got the honor it should have received as the coder of metabolism of life, an ear never gets that honor as the eye. The human brain stem, a key center for consciousness shifts its activity significantly with the complex sound (Erika and Nina, 2010). It suggests that conscious human response is related to time crystals that are key to the geometric structure of music. The ear is a fractal drum (Sapoval et al., 1991). Often it is shown that auditory cortex naturally processes key grammatical features of music, as if it is designed to understand music (Elvira et al., 2006; Large and Palmer, 2002). One-to-one correspondence between the thought process of the brain during music and auditory cortex supports this hypothesis (Zatorre and Halpern, 2005). Not just the brain stem, both left and right hemispheres are seen to play musical notes (Tramo, 2001). Moreover, that musical firing inside is also visible in EEG, one could find a mathematical construct when EEG displays music like rhythmic tones in its behavior (Snyder et al., 2005).

Why do we have two ears, two eyes, two noses and a symmetrically divided tongue and symmetrically divided body for sensors? Each half produces a time crystal, when combined, two-time crystals with C2 symmetry merge to resolve the local very small size parts of the time crystals that are not common between the two parts, which helps the organ to analyze remarkable thing. For example, by merging, time crystals of eyes determine two angular parts in the brain so that we can perceive the 3D image, two ears resolve the distance from which the sound is coming. Two nostrils, one remains wetter than the other, and they interchange this feature time to time, so in this way they change the molecular detection efficiency, wet one gets the higher energy part of the molecules and the dry one detects low energy parts of the molecules efficiently. Similarly, two symmetrically divided parts of tongues tell us the gradient of different elementary tastes, i.e., analog values in the detection scale. For the touch sensor, the symmetric body does the same—it creates a virtual analog gradient of temperature sensitivity, moisture sensitivity, etc. Simultaneity in the sensory signals, e.g., audio-visual simultaneity is essential for brain cognition, the non-linear temporal pattern that emerges simultaneity (Benedetto et al., 2018), demands a time crystal as if time really matters for all sensors (Cariani, 1995).

The cognition via visual control system

[Vision Circuit → Optic Chiasm → Thalamus → Visual Cortex → V1 → V2(complex shape, color) → V3(angle, orientation) → V3a (motion + direction) → V5 (Parietal lobe, fusion form gyrus)] Prime: 13, 7, 5, 11, 3: Detecting a time crystal requires detecting a shift in the resonance peaks (Groma et al., 2004). Only then the higher-level clocks reveal themselves, it seems while engineering the human eye 13 types of cavities work together to sense visual input (Figure 7.2a). A clock needs feedback and a feedforward signal transmission, else the loop is not complete (Zipser and Andersen, 1988). Such a network is shown in the visual cortex domain (Michalareas et al., 2016) where five-fold symmetric paths operate (C5). Temporal patterns are resolved by motion-sensitive neurons in the primate visual cortex (Buracas et al., 1998). The brain cannot process all objects accurately in a particular visual image. Isolating the parts is equally important as conversion into time crystals since one need to pick up what to process, at the same time, keeping an eye on the other parts of the image. Asymmetry in the neuron geometry alone cannot explain how eyes sense direction (Anderson et al., 1999). While transcranial magnetic stimulation of the visual pathway shows a periodic oscillation (Amassian et al., 1998; Ilmoniemi et al., 1999). To build a time crystal, self-similar arrangement of neurons is needed and at least cats visual cortex have it (Binzegger et al., 2005). Time crystal-like oscillation of these pinwheel arrangement is reported (Bonhoeffer

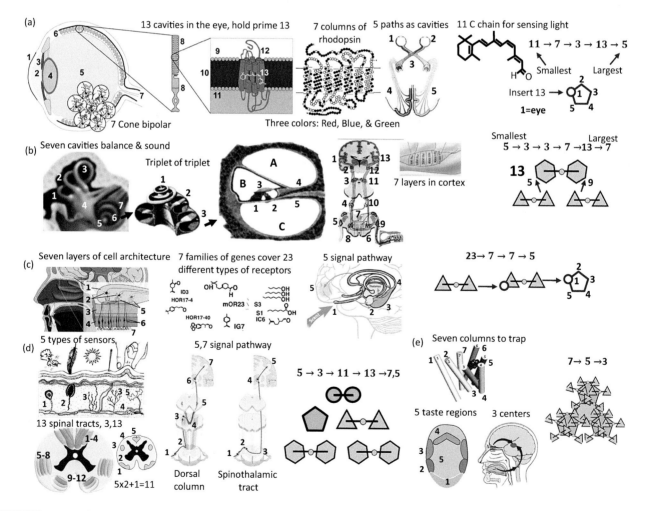

FIGURE 7.2 (a) Components related to prime numbers in the eye. (b) Components related to the prime numbers of our ear. (c) Components related to the prime numbers in the nose. (d) Components related to the prime numbers in the skin. (e) Components related to the prime numbers in the tongue.

and Grinvald, 1991). The orientation columns in the visual cortex adopt unique geometries to sense projected signals (Braitenberg and Braitenberg, 1979). Geometric arrangement regulates phase-locking conditions between different visual memories (Lee et al., 2005; Braitenberg and Schütz, 1998). The combination of parallel and serial processing in the visual cortex hinted toward a distributed computing (Bullier and Nowak, 1995). The idea of a time crystal as a regulating factor emerges from the temporal patterns of visual cortex (Buracas et al., 1998). High-frequency rhythmic and non-rhythmic signals engage in fast synchronization in the visual cortex suggests a general principle that could govern spatiotemporal coding (Eckhorn, 2000). Low-frequency signals regulate long-range synchronization (König et al., 1995). Together they build a temporal architecture which controls visual form that we see (Lee and Blake, 1999). Temporal architecture is the visual form, because it can explain spontaneous bursts with no trigger, even the bursts have topologies like stripes, rings, spirals, etc. (Fohlmeister et al., 1995). One of the fundamental signatures of time crystal is temporal beating, i.e., the frequency difference between two waves forms a new clock (Hammett and Smith, 1994).

Seven columns of rhodopsin molecules with three color sensing pigments convert light into electric pulses via 11C chain molecules, we find C7, C3, and C11 symmetries operate in harmony. Irrespective of the nature of visual, seven different kinds of electrical nerve pulses acquire all kinds of visual information in the retina (Victor, 1999), which takes part actively in shaping the 3D waveform sent to the brain via eye-bundle of nerves. The pulses are connected to point singularities and they distribute topologically (Tanaka, 1995), to hold geometric information. Moreover, the axons in visual cortex arrange in a fractal shape (Binzegger et al., 2005), self-similarity helps in closing a time loop. There are several different kinds of cells, which take part in the photon to a particular electronic pulse conversion process. The 3D waveform is a 2D matrix with variable amplitude, the rate of change of pulse with time contains the phase information and the base of the 2D matrix simply provides the coordinates. Optical illusions reveal such interactions (Tallon-Baudry et al., 1997). 2D matrices are continuously sent as time changes. We have argued above that all sensory organs keep particular places in the grid-free for others to contribute; the same principle is applied for a large number of sensory cells in the eye too. Thus, if sensory cells reach threshold and fire, contribution automatically finds its place in the grid, thus, an automatic classification for the dimension of an object, color, shape, is made in the superimposed grid, as a dynamical organization of clocks (Uusitalo et al., 1996).

Engineering the auditory system

[Auditory circuit → Organ of Corti → Cochlear nucleus of spinal cord → Superior Olive (side of brain stem) → Inferior Colliculus → thalamus → cortex]: Seven tubes (C7) in the inner ear senses sound and keeps body's balance (Figure 7.2b). The cochlea has a triplet of triplet structure (C3). Within that structure five resonating parts transmit sound for sense. Sound is required for enhanced perception

of reality (Johnson and Coxon, 2016). In the pattern of sound, often, one could find geometric symbols hidden in the vibrations (Linton, 2009), geometric theory of auditory signals is old (Licklider, 1959). Auditory signals are captured by giant tuning forks located in the ear that vibrate resonantly with the sound signal. Visual imaging of auditory signals shows classification (Tani et al., 2018) or nested groups. For all sensors, we are using resonant oscillators, since the wavelength is less for light (400–600 nm), the eye requires very small size molecular optoelectronic resonant oscillators, but for sound wavelength is large (300 m), therefore the drum required is quite large. A large-scale coherent neural assembly operates connected to the drum over a sub-second time frame (250–300 ms), voltage-sensitive dyes (VSDI) provide high-sensitive (both temporal and spatial) real-time imaging of the neural activity. While capturing visual signal, the human brain is most sensitive at the point of stimulation, but for the auditory signal it is most active at the superficial layer, as the brain is not interested in the spatial distribution of the auditory signal. In addition, neural assemblies are organized in such a way that, naturally, the visual neural assembly responds faster than the auditory counterpart in the first 120 ms response time. However, after 250 ms, the visual neural assembly response falls sharply, while a linear increase in the signal response of the auditory signal is observed. Signal integration time for the auditory response is much larger than the visual one.

Deaf people convert sound into a sign language (Zaghetto, 2012). The sound could be sensed from anywhere; for example, whales hear the vibration of the skull (Cranford, 2015). Even proteins are triggered by sound wave (Xiujuan et al., 2003). Brain stem filters the complex sound into multiple time-domain (Erika and Nina, 2010). Panoramic view of the environment where the sound comes from originates in the seven layers of brain cortex (C7), responsible cortical neurons classify 360° views as time period of a clock (Middlebrooks, 1994). Auditory cortex not just builds electrical but a 10 Hz magnetic rhythm (Lehtela et al., 1997). One requires three rhythms electrical, magnetic and mechanical, for $e - \pi - \phi$ control. To the right of Figure 7.2b, we have shown geometric shapes for each auditory system in terms of primes. In GML, 3D geometric shape embedded in a time crystal is the control.

The fundamentals of the smell control systems

[Olfactory circuit → smell → Olfactory lobe of the limbic system (looped around the primitive brain) → Amygdala → Cortex]: The nose was the first organ created during the evolution of the animals. Therefore, its nerves are very close to the brain-stem where from the fundamental control of our body, heart-beating, lungs-breathing (respiratory rhythm, Lewis et al., 1992; singularity in respiratory rhythm, Paydarfar et al., 1986; geometry of respiratory rhythm, Sammon, 1994), and motor nerves are controlled. On another planet, if the environment is such that the touch is favored, one could find touch controls near the brain stem. Temporal coding of odor and music have a similar geometric structure in the brain (Plailly et al., 2007; Wehr and Davidowitz, 1996a; Wehr and Laurent, 1996b). Similar to the human nose, one could use a "lock and key" principle of supramolecular chemistry, where a single molecule could sense a particular smell-molecule. Seven families of genes with 23 different types of receptors follow 5 signal pathways to deliver the cognitive experience of odor (Figure 7.2c). Each sensory cell has several fine hair-like cilia containing receptor proteins that are stimulated by odor molecules, however, there is no permanent chemical change. Hence, a resonant and wireless communication is triggered to sense odor molecules and then the locking is removed. However, it embeds a special pulse-cluster synthesizer attached to the molecular sensor, for a particular kind of smell a distinct spike-pattern or temporal architecture is created (MacLeod and Laurent, 1996). The number of activated receptors determines the intensity of the stimulus, the layer of neurons located beneath the sensory molecules in the olfactory bulb generates a synchronized wave (Onoda and Mori, 1980). These waves have been recorded in the electroencephalogram (EEG) tracings, rise in the wave-amplitude indicates the intense sensation, and fall is caused by inhibition. Note that EEG records information for thousands of cells at once (Tani et al., 2018). Reinforcement technique is a beautiful feature of the nose; some neurons remain as a buffer between the neurons in the bulb and the neurons of the olfactory cortex. These buffer neurons activate suitably during training and strengthen the connection for generating a complex signal response, such that a species could have a peculiar sense of a complex combination of smells. Thus, if trained a complex smell signature could carry an extremely vital environmental change or threat or even a very particular event, this tool could alone serve as an extremely intelligent encoding machine.

The touch sensor controls

[Touch detection circuit → Touch bud → spinal cord → thalamus → cortex]: Five types of touch sensors are distributed all over the body surface (Rose and Mountcastle, 1959), and experiments have shown that the brain relies on the collective response of the touch sensors (Figure 7.2d) to create a packet of time crystals regarding the body-shape that it needs to protect under external attack. Touch sense is used to determine size, shape and texture of an object, and according to the need, density, and pattern of distribution of sensors are varied to sense more or less pain, and touch discrimination. A slight movement of the body-hairs could sensitize the touch sensors distributed all over the body; the whole body is divided into particular regions in a very ordered fashion via 13 spinal tracts of a neural network to transmit touch senses from individual parts of the cortex via the spinal cord (Figure 7.2d). Temporal synchronization is the key to a couple of touch sensors across the body. By touch, the brain detects vibration (Keidel, 1984).

Touch sensors also require a permanent memory hardware mechanism; therefore, it was involved in the process dozens of chemical messenger and receptors to sustain pain if the condition continues (Khamis et al., 2015), in the artificial brain, we could altogether replace it with physical permanent memory processing protocols as described in chapter three. Information gathered from touch sensors requires three kinds of responses, first, immediate response by moving a muscle or the body-part, the decision is taken via spinal cord (Dorsal horn loop). Second, pain message is sent to the brain, and

instruction for inhibiting the excitation comes back, the process involves thalamus and cerebral cortex. Third, always, for all kinds of pain and touch senses a part of the signal is sent to the hippocampus and this particular brain stem region controls complete situational analysis by sending it out to the upper brain, cross-checking previous learning and then a comprehensive decision is taken, output is sent to the cerebellum to execute action.

Engineering of the taste sensor controls

[Taste circuit → nucleus of solitary tract (brain stem) → thalamus → cortex]: In Chapter 9 we will describe training a nano-brain, arguing that an idiot baby brain gets trained by following three basic desires of a life form, first, acquiring food; second, defending the safety of the body; and third, taking reproductive measures. When a body looks for basic foods, just to run the body, it does require the test buds to cover a range of tastes that provides signature that the food is suitable for the body or not, whether the food contains more calories or not, even, the concurrent necessity of the body could also be reflected in the tongue-sensory-response. The tongue is a chemosensory assay of five basic tastes (C5); each taste bud has sensory molecules with seven helical columns (C7), which adsorbs molecules (Figure 7.2e). Different regions have distinct taste buds triggering both chemical and non-chemical responses. Adapting to a particular taste always decreases its sensitivity to a particular taste, and particular chemical concentration always helps in reaching maximum efficiency. The learning of the taste sensors has twofold importance: First, if the sensors are exposed to the polarizing pre-conditioner, it acts as an exhibitor, the sensitivity decreases; second, if the pre-conditioner is de-polarizer, then it acts as an inhibitor and the sensitivity increases.

7.3 PRIMES IN THE CEREBELLUM, HIPPOCAMPUS, AND HYPOTHALAMUS

Cerebellum: Cerebellum has 13 cavities that are filled with a fractal neural net; the cavity boundary has seven layers and as a whole entire structure have generic seven major folds (vertical cross-section), and two major fold in horizontal cross-section (Figure 7.3a). It reads the final decision fractal from the hippocampus and its hardware is a tree-like a fractal network of neurons (Bell, 2002; Werner, 2010). Multiple generations of fractal wiring enable learning complex sequences of mechanical movement coordination (Thach, 1998). It sends extremely synchronized signals to all sensors for better data capture for improving the decision-making process and actions for the organs like a high precision machine (Eccles et al., 1967). Since we use frequency fractal as a tool for information processing and at the beginning of our discussion, we noted that we construct frequency fractals such that if all time crystals from all sensors are fused; therefore, we do not need additional hardware to identify which signal belongs to which organs. A flat map reveals 2D one-to-many connections (Hurdal et al., 1999). When the final decisions come from the hippocampus, mechanical movement controlled by

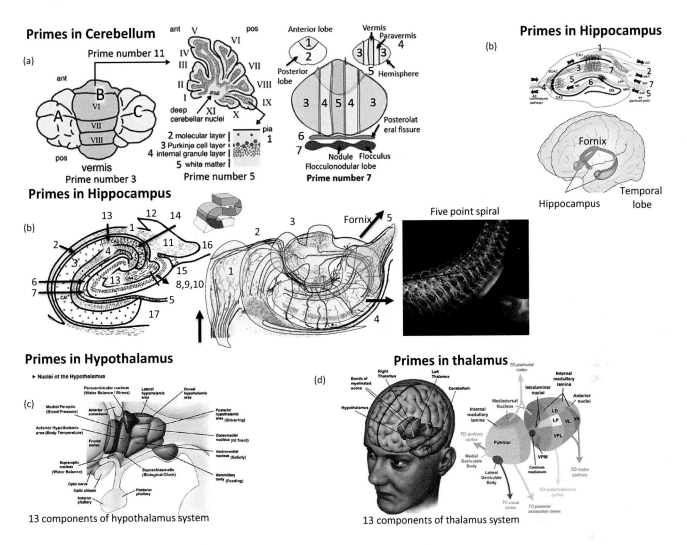

FIGURE 7.3 (a) Components related to the prime numbers in the cerebellum. (b) Components related to the prime numbers in the hippocampus. (c) Components related to the prime numbers in the hypothalamus. (d) Components related to the prime numbers in the thalamus.

cerebellum assists in processing memory, a dual clock runs (Gao et al., 2018). Depending on the length of signal travel through the hippocampus if we connect the wiring with the particular wires of the cerebellum, the right signal will reach the right machine output.

Hippocampus: Hippocampus is made of 17 cavities, wherein a pair of spiral pathways run in parallel side by side, the parallel pathways have five-fold symmetries (Figure 7.3b). Signals coming from entire body require performing two vital tasks at hippocampus, the temporal code generator of the brain (Buzsáki et al., 1994a) and it has its own long-term memory (Bliss and Collingridge, 1993). Chemical rhythms initiate a long-term memory (Kang et al., 1997; Ferbinteanu and Shapiro, 2003). Temporal codes for each type of memory is different (Leutgeb et al., 2005). First, a copy of the entire information packet is glued here (Wallenstein et al., 1998), like fusing all-time crystals, that final time crystal should be radiated outside to the entire upper brain for matching with previously gathered knowledge, so this device would work as a helix antenna-receiver system. Following fast in fast

out principle it maintains the credibility of a pulse-packet (Jonas et al., 2004). Hippocampus activates spontaneously at a selected frequency band (Buño and Velluti, 1977). The second task would be continuously capturing the modified time crystals from the higher brain as primary solutions, identifying the complementary time crystals and intelligent filtering of entire input sets of time crystals at this place depending on the sensory organs. The filtered time crystals, which do not already exist in the brain, is then sent back to the specific regions of the brain for the permanent storage and further processing (CA1 projections to entire cortex region; Cenquizca and Swanson, 2007; Ciocchi et al., 2015). CA1 edits limiting clocks, the fastest and the slowest, to regulate theta rhythm (Rotstein et al., 2005). Electrical rhythms of hippocampus reveal the rate of learning of the brain (Berry and Thompson, 1978). The very basic operation of artificial brain computing is executed here in the helix antenna. Given the complexity and reliability of the task needs to be performed here, we need two antennas—one to send the incoming sets of time crystals continuously to the upper brain (resource antenna, Fornix,

C7 symmetry, prime 7; Figure 7.3b), and the other to send the complementary time crystals to different distinct parts of the brain for saving it permanently (encoder antenna). To avoid essential conflicts between the two jobs, two separate helical antennas each with 37 rings (resource buffer and encoder buffer) in between. They resonantly communicate with the primary antennas spontaneously (Buño and Velluti, 1977) and work as a buffer state provider, and since permanent writing is done in the brain via wired transport, buffer antennas help to sustain the pristine features of the information being processed in the primary antenna. Hippocampus is also a spatial map generator of our environment (Blum and Abbott, 1996; Frank et al., 2000) navigation and memorizing complex paths work together (Buzsáki and Moser, 2013). Spontaneous activation and navigation happen in different frequency bands (Buzsáki, 2005).

The basic architecture of hippocampus is a helix-shaped molecular assembly with a large number of cables coming out of it radially at different parts of the helical length. Assigning a phase by evaluating information properly is its specialty (Kamondi et al., 1998). The cables and all parts are made by neurons only in the real human brain; the rest of the synthesis of all midbrain components are controlled by a particular frequency-modulated environment. The basic principle of hippocampus design is "time delay is proportional to the length of the wire the signal will travel" (Buzsáki et al., 1992). Just based on this simple principle we can fuse all different kinds of frequency fractals in it and at the same time isolate them, categorize and send them back to the particular specific region of the brain. An orthogonal pair of antenna (Gloveli et al., 2005) always radiates out the signal as an electromagnetic energy packet, while radiating out, the phase, frequency two essential features of the time crystal are kept intact as an automated property of the corresponding ionic wave; however, the delay or phase modulation needs to be modulated by the hardware itself. Electromagnetic sensitivity of hippocampus is well established (Maskey et al., 2010). Pyramidal cells are good in phase adjustment (Harris et al., 2002), especially in the cortex, pyramidal neurons interact with inhibitory neuron to edit a complex phase structure (Lytton and Sejnowski, 1991). Now the helical path comes into action. More is the length or number of loops in a helix, longer the signal should travel, which would eventually increase the delay between the two signals radiated out, thus, helical antenna sustains three fundamental features of a time crystal with absolute reliability. At this particular point of time, a delay is a key, from which sensory organs the signals are originated do not make any difference (Amaral and Witter, 1989). Several clocks sync, desync, build, and destroy clocks to operate hippocampus (Buzsáki et al., 1994a). The streams of signals while entering into the brain, it is passed through the helical resource antenna, at certain intervals or after a few helical loops, the time crystals with shorter delay are extracted. Thereafter, the signal left passes through another few loops, and then the longer delayed signals are taken out and the process continues. Since a time crystal uses the delay or relative phase

shift to differentiate between two distinct time crystals, the length of the antenna is optimized to generate that particular delay. Hippocampal place cells regulate neuron firing to edit time (Royer et al., 2012), or even phase precession (Harris et al., 2001, 2002). It could even get into an off-rhythm state (Garcia-Sanchez et al., 1978), silence the synapses, and downregulate systems (Norimoto et al., 2018).

If we simply pump in all signals as fractal seeds described above (the fractal seed is the unit of information in the brain) through the spring-like structure, and we take out an output after every single loop, we can get signals with quantized phase differences. Pattern completion is exclusive to the CA3 region (Guzman et al., 2016). Self-assembled neurons in four CA networks generate a unique spring in the hippocampus region as if four H3 layers of Hinductor device based 2D sheets are packed together. Similarly, the nerve fibers that connect pre-frontal cortex, thalamus, hypothalamus, amygdala, etc., all sub organs of the midbrain resemble H3 device-based 2D sheets (see Chapter 8 for details). Before the memory information is sent all over the brain for permanent storage, it is processed in the lower brain hippocampus and converted into a set of patterns by massive compression (Skaggs et al., 1996). Grids do wonderful things (Kraus et al., 2015). Then it sequentializes the learning process (Wallenstein and Hasselmo, 1997). The lower brain region looks similar for every animal from dinosaur to human, the information processing protocol for all animals is the same. If we enter inside the hippocampus, in one of its four functional layers, the machinery looks like a spiral ring and longitudinal connections exist all along with the spiral net, a neuron pulse flows helically, when it collides with the longitudinal one, a particular phase is associated and helicity encodes particular frequency. Therefore, this single machine prepares the information to categorize them into particular classes, with a set of frequencies and phases before sending them out to a distant location. In addition, whenever it would be necessary, with that particular set of phase and frequency, it can recall them as a cognitive graph (Muller et al., 1996). The machine is, therefore, phase and frequency encoder. Positive and negative phase interact to neutralize and regulate phase (Holscher et al., 1997). Phase regulation means changing geometric shape in a time crystal, it is the coordinate (O'Keefe and Burgess, 1996).

The clock like periodic oscillations down-regulate the synapses (Norimoto et al., 2018). The connection of nerve fibers that connects between the hippocampus and amygdala has a typical geometric shape and it changes following a change in geometric symmetry (Kishi et al., 2006). Slower clocks hold phase information with clarity (Adey et al., 1960), clocks autocorrect possibly guided by symmetry breaking of the encoded information, i.e., integrating path and environment cues interplay to find a major change in the ensemble code for space (Gothard et al., 1996). A cortex network for memory, navigation, and the perception of time, namely entorhinal system communicates with the hippocampus using theta rhythm (Buzsáki and Moser, 2013; Buzsáki, 2005). Geometric structural symmetry of hippocampus in the CA1 region changes very little and that precise change reflects the projection of the region to the entire cortex region (Cenquizca and Swanson, 2007) and projection neurons adopt a geometric symmetry that enables it to filter information (Ciocchi et al., 2015), stable under massive noise (Zugaro et al., 2005).

The operational mystery of real hippocampus is unresolved, the dynamic data flow appears like checkerboard like a grid of pulses or metric (Terrazas et al., 2005) propagating across the hippocampus and on top of that grid, another stream of water like waves flow. Geometric shapes of hippocampal oscillations during sleep carries the geometric language embedded within (Staresina et al., 2015). The phase relationship between hippocampal place units and the EEG theta rhythm suggests that for each brain decision the stream flowing on a grid protocol remains the same, throughout (O'Keefe and Recce, 1993). The biggest question that should disturb us, is regarding the origin of this gridding of information. If it is not done at the very moment when information enters in our body, it is just impossible to compartmentalize them later in the brain wherein the streams of pulses are coming in from all over the body. The homogenizing gap between spikes is one key to regulate time (Vida et al., 2006). Thus, the biggest problem we are facing in resolving the mysterious information processing of the hippocampus is hidden in the skin, eye, nose, ear, and in the tongue, not in the brain. It is so exciting to feel that the biggest problem of the brain's information processing is distributed throughout the body, not just in the brain.

Hypothalamus: Hypothalamus in a normal human brain is used for vital motor controls and generate rhythms. It is made of 13 cavities (Figure 7.3c). Bidirectional feed-forward electromagnetic signaling observed in the hypothalamus is evidence of clocking regulation of metabolism (Stanley et al., 2016). These programmed activities are not essential for the computer we want to build; however, it is the supreme authority, if we want highest-level transformation in the final decision fractal in the hippocampus, then we can program those protocols here. The neurons are self-assembled similar to an LC-coupled oscillator-type periodic fractal seed generator and an antenna made of a neuron that transforms the final decision fractal within the certain limits. It is the interfacing point of the user modulating the computational process in the artificial brain.

Thalamus: Thalamus is a gateway to the sensors, a gateway to the brain cortex (Sherman and Guillery, 2002) and a gateway to the cerebellum. It is made of 13 cavities (Figure 7.3d) and acts as a universal synchronizer (Steriade and Deschenes, 1984) that builds a natural logarithmic relationship in the frequency distribution among various brain components (Penttonen and Buzsáki, 2003). The machine looks like as if many springs are side by side, if one of them is triggered then a complex vibration is automatically triggered over the entire system. Each spring-like structure accepts the particular kind of sensory signals. Visual information is encoded as an architecture of time here (Reinagel and Reid, 2000). Thalamus works with basal ganglia (Morison and Basset, 1945).

Fornix: Fornix dynamics follows a C7 symmetry (Figure 7.4a), it is the antenna and receiver of time crystal between the midbrain and the upper brain (Raslau et al., 2015), built-in a triangular network $T1$, $T2$, and $T3$ (Figure 7.4a). Neocortex and hippocampus have competing learning mechanisms that work in a complementary way (McClelland et al., 1995). Cingulate gyrus in the fornix is the key antenna receiver complex of the brain's learning. The interplay between hippocampus and cortex self-organizes time crystal and indexes the memories (Miller, 1989).

Connectome: Connectome is the fiber pathways of the brain (Schahmann and Pandya, 2006). There are five distinct pathways of nerve fibers (C5, Figure 7.4a). However, during functional imaging we find that $T1$, $T2$, and T3 domains of connectome triggers five regions or seven regions in a geometric shape (Figure 7.4, bright domains with a dark background). At the bottom part of Figure 7.4a we demonstrate projection seven-fold to the cortex. Functional logic for the five-fold and seven-fold cortical connections is established (Zeki and Shipp, 1988).

Cortex: The phase modulation of the propagating waves in the cortex is important for time crystal analysis, similar periodic behaviors are noticed in the network (Sanchez-Vives and McCormick, 2000). Cortical rhythms of various kinds are controlled by thalamus directly (Ralston and Ajmone-Marsan, 1956). However, long back a map of the 2D phase change profile was accounted for the decision-making of the human brain in the cortex (Ermentrout and Kleinfeld, 2001). Earlier, the same group argued for learning by phase change (Ermentrout and Kopell, 1994), which is again a key to establishing time crystal, GML-based learning of human brain. Circuital changes were attributed to Darwinism (Edelman, 1987), which may be replaced by collective cooperation in future. High frequency, 200 Hz, focal synchronization of ripples in the neocortex (Grenier et al., 2001), may include spiral waves (Huang et al., 2010) in time crystal GML, we may call it, fusion of time crystals. It is also reported that neocortex codes space using architecture of time (Singer, 1994). Gray matter and white matter of cortex are connected by universal scaling law (Zhang and Sejnowski, 2000). Cerebral cortex holds a deep 3D architecture of phase to regulate the alfa rhythm (Lopes da Silva and Storm van Leeuwen, 1978). Intracortical and thalamocortical, two distinct time crystals interact which reflects in the alfa rhythm (Lopes da Silva, et al., 1980). Different cortical regions have discrete rhythms, which integrate, build higher-level rhythms and that higher-level temporal architecture is the global landscape of cognition (Taylor et al., 2015). These cortical convolutions (Tallinen et al., 2016) are vortex atom-like structures representing the mind.

Pre-frontal cortex: Pre-frontal cortex controls futuristic simulations remaining phase-locked with the hippocampus (Siapas et al., 2005). The region is directly connected to the frontal lobe where the highest-level fractal seeds are stored as compressed data (Nieder and Miller, 2003). However, this region captures the final fusion fractal which is the solution of the computation carried out in the brain and concentrates only on the "then" related "action" parts, which the body will execute. Then it feeds higher-level time crystals stored in the

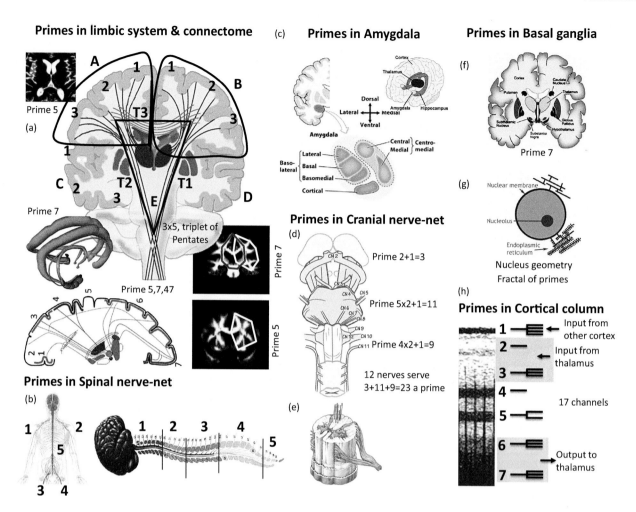

FIGURE 7.4 (a) Components related to the prime numbers in the folding of the cortex and two junctions above the limbic system, one claustrum and the other corpus callosum. Triplet of triplet arrangement of cavities is visible. (b) Components related to the prime numbers in the spinal cord and the spinal system. (c) Components related to the prime numbers in the amygdala. (d) Components related to the prime numbers in the three domains of midbrain that controls 23 dynamics of 12 cranial nerves. (e) Spinal cord with 5 fundamental symmetries. (f) Components related to the prime numbers in the 19 basal ganglia points. (g) Components related to the prime numbers in a large number of nuclei found in the brain, these are 2, 3, 5 fractal symmetries that protrude via endoplasmic reticulum. (h) Components related to the prime numbers in the 17 branches of the cortical columns emerging from seven regions of the cortical column.

frontal cortex to generate "futuristic simulation" of the actions in the hippocampus. The hardware is made of two antenna and receivers coupled with each other. First with one antenna it reads the "action to be taken" parts of the frequency fractal, matches with the higher-level rules "if" clusters, as soon as it matches, then those phase transition rules are sent to the hippocampus for transforming the final decision-making fractal. Thus, futuristic thoughts or imaginations are generated with this neuron-made hardware. As this region expands the final frequency fractal in the hippocampus, the hippocampus sends it back to all over the brain and executes significant modifications in the final frequency fractals.

Spinal cord: Thirty-one pairs of spinal nerves come from all over the body in five domains (C5, Figure 7.4b) to reach sensory signals high up the brain, and 31 pair motor nerves to carry brains execution order. We might notice the fantastic use of primes in the branching of spinal responses. Even if

there is no firing in the medulla and in the spinal cord, there is a rhythmic neuronal discharge to compensate the signaling (Ren et al., 2006), which suggests that there is firing or not, rhythmic or clocking responses generate perpetually in the spinal cord. Its geometric information is so profound that rebuilding initiates spontaneously (Mackinnon et al., 2012).

Amygdala: Seven cavities of Amygdala work in harmony (Figure 7.4c). Amygdala and hippocampus work together to store a typical face and word (Heit et al., 1998), retrieving fear memory (Seidenbecher et al., 2003) geometric projection from the hippocampus to the amygdala is significant (Kishi et al., 2006). The circuits between the duo tell an emotional story of logic. Emotional differences are the differences between architectures of time (Schmidt and Trainor, 2001). In the time crystal model of a human brain model amygdala implements the complex filtering process. Amygdala's structural geometry and behavioral response are closely related, and that regulates amygdala hippocampus circuit (Yang and Wang, 2017). During a decision-making process, the human brain receives the final time crystal in the hippocampus, which contains all possible decisions and solutions, the design of amygdala hardware is such that it generates a set of unique fractal seeds that reads particular signatures in any other fractals and according to its list of preferences, it deletes unacceptable choices or fractal seeds. In other words, the amygdala sends a fractal to the hippocampus and that fractal checks coexistence of conflicting geometries (multiple opposite choices in the solution of a problem) and then it simply deletes the unfavored ones. To simply understand what happens mirror symmetry or C2 symmetry is detected. Basic drives for a life form is a triangle for giving birth, effort to save life, arranging food for regeneration of cells remains saved in the amygdala with birth and evolves with our life practices, so evolution of a time crystal of desires from a single triangle to a superposition of a large number of triangles in a single plane, it looks like a lotus flower. Such an emotional bonding makes memory permanent (Paré et al., 2002). Top-down control of fear and anxiety (Adhikari et al., 2015). Reward system acts in correlated clocking with ventral striatum (Cador et al., 1989), while the network of clocks or time crystal for processing emotion works in a loop with the pre-frontal cortex (Delli Pizzi et al., 2017). Geometric conversation, or exchange of geometric shapes between multiple components, e.g., between the amygdala and prefrontal cortex in a periodic sequence determine emotions (Ghashghaei et al., 2007). Since the symmetry of lotus is very different for different humans, the orange or mango is different for every single person on the planet.

Twelve cranial nerves: Phase coupling between cranial nerves (e.g., vagus with sinus node; Jalife and Moe, 1979) and the local time crystals initiate a higher-level time crystal synthesis. Cranial nerves in the midbrain arranged in three cavities, in the first cavity two (two cavities vibrate around a center, $2 + 1 = 3$ dynamic points), in the second cavity there are five nerves ($5 \times 2 + 1 = 11$), in the third cavity there are four nerves ($4 \times 2 + 1 = 9$), total dynamic centers $3 + 11 + 9 = 23$, and one nerve is left out, it runs separate (Figure 7.4d, e).

Basal ganglia: Basal ganglia made of seven cavities (Figure 7.4f) perform procedural learning, or learning by practice, it finds time crystals that are good, bad, and ugly for a task (Boraud et al., 2005). The organ modifies the final decision-making frequency fractal of hippocampus so that a very systematic input is captured and the brain jelly could reconfigure the neural circuit concretely, its motto for practice is to "loving it" (Graybiel, 2005). Therefore basal ganglia generate a new kind of frequency fractal that enables it to control multiple sets of solutions generated by the brain at an interval of 200 ms, it keeps continuation among the discrete computational input-output process. The hardware necessary for this is the creation of a fractal and then run it in a loop longer than 200 ms so that it affects the self-motivated data acquisition process continuously. Failing to automation leads to Parkinson's disease (Brown, 2003). However, there is another important aspect of it, how would the fractal nature be determined? Unlike previously described organs, in this case the fractal nature is not pre-programmed. In fact, prefrontal cortex simulates futuristic outcomes and if the final decision fractal before and after futuristic simulation suggests missing of significant "then" correlated data, then that event is captured by the basal ganglia and it generates a continuous relevant data capturing job.

The brain stem is the region that exhibits the best example of "one-to-many and many-to-one" network, along with hippocampus it controls theta rhythm (Vertes and Kocsis, 1997). In the human brain, the pinnacle of the spinal cord (Hindu literature call this maha-siva-pada, it senses the most complex sound of the universe, Erika and Nina, 2010) holds the brain-stem cells, it is surrounded by a cavity holding the cerebrospinal fluid (CSF). Every movement of the body needs a local motor clock activation (Sanes and Donoghue, 1993), every learning of motor operation is a synthesis of rhythm or clock (Sakai et al., 2004). The liquid of CSF dumps the vibrations generated by any physical shock, e.g., noisy ripples generated by heartbeat, lung rhythms, and motor nerve controlling units. Motor nerve control units are essential for the automated learning feature development and lungs control may not be essential if an analog of heartbeat control takes care of the entire energy supply. Regarding motor nerve controlling, brain-stem cells should run a synchronous rhythm wave, which should run perpetually as a controlled operation. The rhythmic program should not be modified or re-written, even if it is switched off, after switching on the same programmed rhythm would start running perpetually. That incredible feature enables finding the geometric difference between the two architectures of time. It means it divides two-time crystals instantly, it is the creator of new unprecedented architectures of time (Carr and Konishi, 1990).

7.4 PRIMES IN THE CONNECTOME, SPINAL CORD, AMYGDALA, NUCLEUS, AND CORTICAL COLUMN

Nucleus: Nucleus are bodies with autonomous firing abilities (Figure 7.4g), it is so robust that if necessary, act as a pacemaker (Bal and McCormick, 1993). Though small in size, a large number of neurons coexist in different structural symmetries just like stem cells (Hassan et al., 2009), the organization is adaptive to any change in symmetry (Scheibel and Scheibel, 1966). Plasticity of nucleus architecture is enormous. Suprachiasmatic nuclei control sleep, but if one deprives sleep, the nuclei change its structure (Deboer et al., 2003). GABA receptors control the clocks of suprachiasmatic nuclei (Liu and Reppert, 2000); GABA not only syncs but desyncs clocks too (Llinás et al., 2005). Generally sleep rhythm is created by thalamic and neocortical rhythms (Steriade et al., 1993a, 1993b). The geometry of the nucleus plays a vital role in its time crystal processing, the topological architecture of information is key (Phillipson and Griffiths, 1985). Most nuclei that are clusters of neurons activate the action potential and initiate the propagation of clocking waves spontaneously or autonomously (Atherton et al., 2008). Nuclei are good in regulating spindle rhythmicity, which means multiple strings or protein fibers could be triggered simultaneously into a coherent oscillation (Destexhe et al., 1994). If the nucleus is disconnected, the collective spindle oscillations disappear (Steriade et al., 1985). Spindle oscillations help in simultaneous management of many neural firing (Wang and Rinzel, 1993). A nucleus could act as a catalyst, programmed to trigger a local or a cascade of global events like arousal (Lewis et al., 2015). A nucleus could be programmed for a sequence of geometric codes to activate a particular switch of events (Phillipson and Griffiths, 1985). It is like encrypted engine. The nucleus is best bet for phase reset, key to time crystal (Field et al., 2000).

Cortical column: Cortical column has seven layers. It generates a magnetic field (Arlinger et al., 1982; Figure 7.4h), and it generates temporal structures like a song (Ikegaya et al., 2004), which makes it a fit candidate for the third generation Hinductor fourth-circuit element as described in the Figure 7.11. As noted, the first-generation elementary decision-making device H1 of the brain is microtubule, its three layers are water crystal inside the hollow cylinder made of tubulin proteins, the upper water layer is also important. The second generation of decision-making device H2 is the filamentary bundles made of microtubule, actin and intermediate filaments as the first layer. Then above this axon core there is a 2D ordered net of beta-spectrin and actin filaments which anchors multiple proteins. Finally, the third layer is a composition of glial cells, neurons, and oligodendrocytes building a nerve fiber or cortical column. The third generation of decision-making device H3 is a cortical column made tract, sheet, fiber, using hexagonal close packing of columns, i.e., 7, 19, 37....number of columns are used. The second layer is made of 19 groups of 7 column units. The third layer is 47 groups of 37 column units, thus, Brodmann's 47 regions are covered (Brodmann, 1909; Garey, 2006). Radial Unit Hypothesis (RUH) is a conceptual theory of cerebral cortex development, first described by Rakic (1988). Radial glial cell hypothesis is important for evolution, which uses glial cells as evidence of the evolution of the cortical column. In the mainstream neuroscience the true role of a cortical column is not known (Horton and Adams, 2005). However, in the time crystal model of the brain, a cortical column processes an octonion, H2 while a single microtubule a quaternion H1, and the cortex layers 47 Brodmann's region all together represent a dodecanion. In a time crystal, actually multiple infinite wave trains are encoded. To understand its hidden information content, the brain has to split an infinite wave-train into multiple segments, dendritic branches in the cortical column could segmentize the temporal signals with a particular pattern (Branco et al., 2010). It is when a brain begins to understand the information content. Similar H3 devices are there at many places, especially in the hippocampus.

One of the finest debate about the cortical column is that whether remaining within the hexagonal close packing, a single cortical column could operate independently or collective response is the only way out (Gawne and Richmond, 1993). The debate turns interesting because the cortical neurons find symmetric and asymmetric division zones and migrate through various phases (Noctor et al., 2004).

7.5 PRIMES IN THE NEURON, GLIA, DENDRO-DENDRITIC, ASTROCYTES, MICROTUBULE, AND PROTEINS DNA

The physicist Paul Nuñez pioneered rigorous applications of physical wave propagation theories to brain waves (Nuñez, 1981, 1998). Physics provides a vast toolbox for treating wave phenomena mathematically. These techniques have provided some understanding of global brain phenomena in terms of the physical properties of its carrier medium. How far can we go with this physicist's view of the brain? While medium filtering is an important factor, it cannot explain the larger spatial extent of neuronal recruitment at lower frequencies or the behavior-dependent highly coherent gamma oscillations in distant brain areas (König et al., 1995, 1996; Varela et al., 2001). Libet's principal finding was that short trains of pulses evoked only unconscious functions, and the somatosensory cortex had to be stimulated for 200–500 milliseconds for evoking a conscious sensation of touch. This time-domain has sub-domains (Nicolelis et al., 1995). To become aware of a sensory experience requires the engagement of the appropriate brain networks for hundreds of milliseconds. The delay between Libet's "mind time" relative to physical time is a favorite argument of philosophers to question the unity of the mind and brain (Libet, 2004). Some form of noise is included in all models to mimic the highly variable spontaneous spiking activity present in the intact brain (e.g., Usher et al., 1994). Some networks with sparse connectivity can generate irregular spike trains, chaos into organized clock network (van Vreeswijk and Sompolinsky, 1996; Amit and Brunel, 1997). However, these models are less sensitive to input perturbations and are quite unstable. Models of single oscillators, on the other hand, are too stable, and without some external noise, the spike patterns are very regular. Voltage fluctuations in real neurons are limited by conductance increase that accompany the synaptic activity. Therefore, it seems unlikely that synaptic activity can generate large enough white noise-like variability to sustain activity in model networks. Hippocampal regulation of rat's anterior eye chamber during its early development shows patterns of neural communication (Bragin and Vinogradova, 1983). The burst/pause patterns of cortical slabs resemble "slow 1" oscillation (Timofeev I. et al., 2002; slow 1 oscillation is a novel rhythm of cortical brain that leads to a steplike change in the membrane potential of the neocortical pyramidal neurons from −70 to −80 millivolts to spike threshold) for the plasticity of cortex or epileptic discharges. Traub and Wong (1982) provide a quantitative explanation for the avalanches in the model

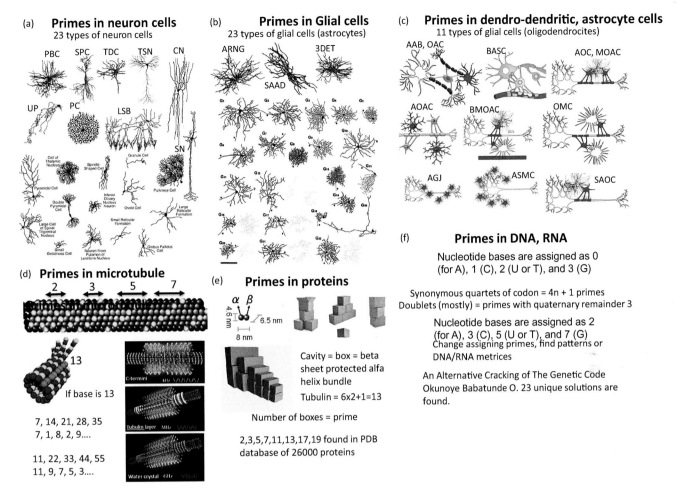

FIGURE 7.5 (a) Components related to the prime numbers in the branches of the neuron cell. (b) Components related to the prime numbers in the glial cells. (c) Components related to the prime numbers in the dendro-dendritic, astrocytes. (d) Components related to the prime numbers in the microtubule. (e) Components related to the prime numbers in the proteins. (f) Components related to the prime numbers in the DNA and RNA. *J. Theor. Biol. 1991 Aug 7;151(3):333–341; Prime numbers and the amino acid code: analogy in coding properties. Yan JF1, Yan AK, Yan BC.*

networks, it is important to know that over synchronization is a failure in computation. In preventing population bursts, subcortical neurotransmitters flow pattern play a vital role (Steriade and Buzsáki, 1990).

The neurons: Twenty-three types of distinct geometries of neurons are reported thus far, which delivers a distinct electromagnetic resonance (Figure 7.5a). Oscillator like the resonance of neurons with multiple dynamic modes (Wang, 1994) is responsible for neural information processing was proposed long back (Llinás, 1988). It harvests channel noise (White et al., 2000) to build organized clock, a transition from random time slices to ordered architecture of time (Enright, 1980), the effect of noise in modulating neuron's frequency response is significant (Brunel et al., 2001), it figures out space using time (Ahissar and Arieli, 2001). Though Ghosh et al. advocated time crystal like the architecture of neurons (2016), the transition of spike model to clock model of neuron started long back (Aviel and Gerstner, 2006). The fact that each neuron has the fastest clock and a slowest limiting clock that govern its temporal sensitivity, the neuron is a phase

editor (Lundstrom et al., 2009). Neural discharge is connected to phase, (Andersen and Eccles, 1962) and they exhibit circadian rhythms (Benson and Jacklet, 1977). Time crystal model makes a significant shift in the established concept of a neuron, because in the conventional model of a neuron, the Ca^{++}, Na^+ and K^+ ionic channels control the membrane potential and that explains neuron firing. We include an additional factor, the electromagnetic oscillations of the microtubule bundle as a secondary control parameter of firing. It regulates the architecture of time of a neuron (Arvanitaki and Chalazonitis, 1968). When neuron switches its conductance very fast, spontaneous ripples generate (Connor, 1978), similar to time crystal measurement. Conversely, perturbation inhibits learning (Choe et al., 2016). Microtubule of specific length has a specific resonance band (some peaks are prominent some gets dormant), thus, brain, during its circuit evolution choose compositions of different microtubule lengths to generate a complex signal transmission behavior that eventually governs the membrane potential distribution and sets the statistical distribution of the different ion channel percentage, wherein tractal feature is evident (Liebovitch et al., 1987). The filaments regulate subthreshold communication (Ghosh et al., 2016a; Jin et al., 2012). Subthreshold communication is key to the central nervous system communication (Llinás, 1988; Laughlin and Sejnowski et al., 2003). The skeleton of the entire neuron body turns alive via microtubule and no matter how far the axons extent, the entire pathway of a giant network of a neuron cell works as if a single molecule. The giant network expresses itself with large spikes (Lewandowska et al., 2015). Basal bodies like the number game, assembly of neurons like the game of integers because the microtubules are packed inside (Pearson and Winey, 2009). The number games of neurons have been a subject of study (Williams and Herrup, 1988). Organoids model the migration of neurons as part of number game (Xiang et al., 2017). The architecture of spatial arrangement and the architecture of temporal arrangement go hand in hand (Buzsáki and Chrobak, 1995). Geometric pattern or clique of neural assembly provides the missing link between structure and function (Blaustein et al., 2017). "Functions follow form" is an old debate (Connors and Regehr, 1996).

The geometry of neural branches select a suitable set of frequencies (Hutcheon and Yarom, 2000), geometry sets transmission rules (Rall, 1959). Why do neurons need thousands of branches? The reason could be sequential memory (Hawkins and Ahmad, 2016) or time crystal processing, where singularities on a single host clock are sequential memory, while one inside another it simultaneous memory. They compartmentalize plasticity (Losonczy et al., 2008). Electronic clocks disrupt ionic clocks (Hormuzdi et al., 2001), electromagnetic clocks trigger electronic clocks in the biomaterials, therefore, the intra-clock transmission is feasible. Without flowing neurotransmitters through synapses one could fire a neuron by sending electronic current (Jefferys, 1995). Even a single protein synthesis inhibitor could modify the circadian rhythm of a neuron (Jacklet, 1977). Neuron circuit generated geometric asymmetry is not enough to explain the directional responses of neurons in the visual cortex (Anderson et al., 1999). Thenceforth Singh et al. explored the symmetry of geometric shapes hidden in the temporal architecture revealed in a large set of spikes, group of groups of spikes, the answer to all mysteries hides there (Singh et al., 2018). Hierarchical multi-layered control on the neuron spikes timing requires a cascade model, as additional support, the Hodgkin Huxley model alone does not explain neuron firing (Aviel and Gerstner, 2006). Even a neuron creates the perception of 3D space that we see around us by editing the time gap between two spikes (Ahissar and Arieli, 2001). Sync and desync organize together (Gray, 1994). Neurons are filled in a cavity and then all neurons dance to the resonant modes of the cavity building all the architectures of the brain, from cortical column to the hippocampus, cerebellum, spinal cord. The geometry or clique of neuron holds the missing link between a neurons structure and its function (Blaustein et al., 2017). Thus, the time has come to go beyond the Hodgkin Huxley doctrine started by Lapicque in 1907 (McCormick et al., 2007).

Questions are plenty. Why did nature invent more than 120 neurotransmitters if all it had to do was execute a simple job—fire or not fire. Neurotransmitters are neutral and thousands of branches fire in a group; some remain silent, some fire, thus it is not a simple brute transmission (Sanes and Lichtman, 1999). Following time crystal hypothesis every decision-making unit in the brain should follow $e - \pi - \phi$, mechanical-electrical-magnetic resonance. Neurons exhibit magnetic resonance (Long et al., 2015).

Neuroglia: There are 23 types of glial cells or astrocytes found in the brain (Figure 7.5b). Biologically, a neuroglia performs several tasks (cleaning neuron cell deposits, etc.), however, in the human brain model, wherein computing is the key, the neuroglia works as a synchronizer, we ignore other jobs as those are not directly related to the computing of the brain. In the human brain, neuroglia and neuron are in a ratio 2:1. To remarkably speed up the long-distance chemical communication, the brain needs signal amplifiers and in the geometric human brain model the neuroglia is an essential buffer engine that delivers homogeneity in the potential gradient produced by individual neurons in the neural network. It is a multi-functional resonant oscillator is constructing the neuroglia. The fundamental electronic property of this material would be receiving any band of the electromagnetic signal emitted from the neurons and then radiating it out all over the environment, thus working similar to an amplifier in the conventional power transmission lines. It is essential to maintain the homogeneity of the electrical potential distribution in the entire frequency fractal hardware. It is therefore computationally equally important to the neurons.

Astrocyte based connectors: In the emergency situations, astrocytes regenerate the neurons, in an environment just like the glial cells they hold the time crystals for the localized regions, therefore when the time comes to recreate the neurons cells, the circuit is ready with them. Eleven types of oligo-dendrocytes, dendrocytes, and associated connections are found in the brain (Figure 7.5c). Astrocytes regulate daily rhythms working in tandem with the suprachiasmatic nuclei (Tso, 2017). Surprisingly in the suprachiasmatic nuclei, astrocytes form gap junctions, wiring, network, but not neurons (Welsh and Reppert, 1996).

Nerve fiber and microtubule: These are not cables. Periodic changes in the periodic oscillations of the transmitted signals means, there exists an architecture of clocks (Arshavsky et al., 1964). A cut in the nerve reveals various kinds of nanowires, microtubule (25–30 nm wide), actin (2–4 nm wide) and neurofilaments (10–15 nm wide). Thousands of these skeleton-like filaments were considered silent, now Sahu et al. (2013a, 2013b) and Ghosh et al. (2016a) have proved that they transport signals and regulate neuron firing. On microtubule, one could select a variable spiral pitch and generate different kinds of signals (Figure 7.5d), thus, microtubule could hold a wide range of primes for resonance (see Chapter 6 for details).

Proteins: Beta sheets of protein isolate alfa helices as per STM studies (Sahu et al., 2013a, 2013b; Figure 7.5e), these clusters act as a cavity. 26,000 proteins of the PDB database was searched to find that only up to 19 cavities are there in nature (see details in Chapter 6).

DNA and RNA: Prime numbers are abundantly observed in the genetic codes of DNA, amino acid code harnesses the symmetry of primes (Lam, 2014; Yan et al., 1991; Figure 7.5f). Micro-RNAs play a fundamental role in micromanaging the circadian clocks, biological timekeeping may go down further below (Mehta and Cheng, 2012). Genetic and nucleosome positioning codes are filtered and function wise segmented, the classification follows the symmetry of primes (Mossallam et al., 2016). Choice of codons, its binding symmetry evolves proteins (Stergachis et al., 2013). Genes have circadian rhythms and those clocks show aging effect (Yi et al., 2016), autoregulation of periodic genes (Sehgal et al., 1995). DNA, RNA proteins have interstitial water and they may act as time crystal alone (Mendonca and Dodonov, 2014). Using genetic alphabet one could synthesis artificial life (Zhang et al., 2016).

7.6 TWELVE WAYS OF MEMORIZING AND TWELVE CARRIERS OPERATING RHYTHMS/CLOCKS

Vibrations hold the key to everything. Until now, we were considering that vibrations and rhythms are the only keys to some enzyme cycles, periodic discharge with periodic input is what makes neuron a clock (Rescigno et al., 1970). But the PPM-GML-H triad proposal is that deep down below even to the molecular scale there are rhythms. The current extrapolation is an advancement of Adaptive Resonance Theory (Carpenter and Grossberg, 2003): Brain regions that are functionally described include visual and auditory neocortex; specific and nonspecific thalamic nuclei; inferotemporal, parietal, prefrontal, entorhinal, hippocampal, parahippocampal, perirhinal, and motor cortices; frontal eye fields; supplementary eye fields; amygdala; basal ganglia: cerebellum; and superior colliculus, Thalamic nuclei. In the (adaptive resonance theory) ART all kinds of resonances have no relation in between, here all resonances are geometrically connected. The origin of scale-free brain activity is the correlation between the architecture of the time and the architecture of spatial arrangement (He et al., 2010). Twelve years of electrical turbulence evolve the system (Winfree, 1998), thus 12 is important. Event-specific changes in brain rhythms reveal that memory is dynamic, it has a structure of clocks (Bastiaansen and Hagoort, 2003).

Retrieving a memory has been a subject of interest for centuries, it haunts us (Sara, 2000). The geometry of time to recall is created when theta and beta rhythms encode memory (Sederberg et al., 2003). Theta rhythms of neurons build wide ranges of brain functions (Vinogradova, 1995). Alfa rhythm associated memory is a fractal structure, for a baby, it is localized, for an adult, its global (Srinivasan, 1999). And there are 12 types of memories that work at the 12 scales of operations in the brain (Figure 7.6). All memories are geometric in nature:

1. Periodic memory: Chain of guitar string we see mostly in DNA, proteins, a period of periods in the nodes of the string. Edit the memory by changing the periodic length of elementary point oscillators.
2. Spiral of spiral memory, in DNA and proteins, also in microtubules. Edit the memory by changing the twists and the number of hierarchical periods by inducing more and more twists.
3. Vector memory, the orientation of helices to make cavities inside the protein. 3D orientation makes it a multipolar complex 3D vibrational element. Edit the memory by changing the orientation of column oscillators.
4. Lattice memory, microtubule, actin-beta spectrin network, protein complex. Change the lattice parameters to edit the memory.
5. Chemical-electric memory, chemical transmission cycles. There are nested time cycles highly interconnected, to edit the memory, change the diameter, phase, delay or starting points.
6. Leak density of cavity memory, ion diffusion holes in the cellular membrane. To edit the change in memory the leak density on the cavity surface.
7. Spiral geometric memory, assembly of neurons, etc. To edit the memory, change the pitch-diameter-length of the spiral.
8. Vortex or fractal memory, memory stored in the geometric parameters of the vortex. A vortex has divergence parameters and scale repeat unit, these two parameters are changed to edit memory.
9. Nodal and polar memory: Nodes in the spinal cord brain network (nodes in a teardrop). The geometry of the shapes is changed among eight different choices from teardrop to ellipsoid to edit nodes and polarity memory.
10. Electromechanical phase memory: organs sync and desync like a giant molecule. The memory is edited by changing the bond length or wiring length and distribution.
11. Multipolar loop in phase space loop memory: hyperspace memory. The memory is non-physical and hence not editable.
12. The phase of phase duality generator memory: the hierarchical assembly of the reality sphere. The memory is non-physical and hence not editable.

Rhythms in the brain or thoughts are nonlinear (Stam et al., 1999), they are not only one type. Large-scale models of the brain have been proposed to resolve the cognitive and systems issues in neuroscience (Eliasmith et al., 2012). It has been argued that transient dynamics of the neural network would resolve this issue (Rabinovich et al., 2008). It has been established now concretely that greater similarities in neural pattern across repetitions are associated with a better memory

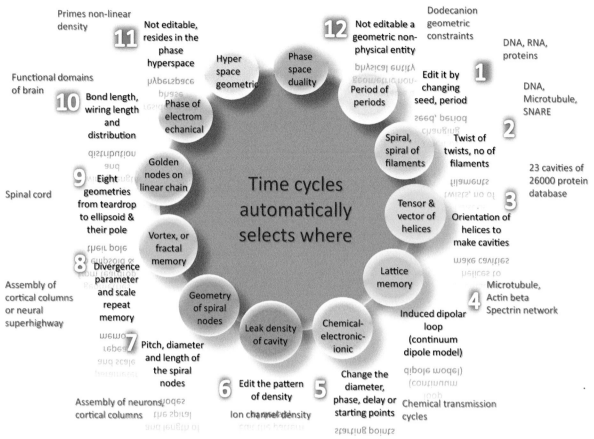

FIGURE 7.6 A chart is showing 12 different possible ways to build time crystals, their editability, and location.

(Xue et al., 2010). Sound is a mechanical rhythm and we have 12 different rhythms in the brain which differs in normal and pathological functions (Schnitzler and Gross, 2005; Figure 7.7), which ensures burst or not to burst (Steriade, 2001) at all levels:

1. Electromagnetic rhythm (carrier is photon or electromagnetic wave is trapped in a cavity to generate beating or rhythm).
2. Magnetic rhythm (spiral flow of electrons or ions, they are the carriers editing the magnetic flux, the geometry of path forms the periodicity).
3. Electrical potential rhythm (change in the arrangement of dipoles editing the electric field, fractal distribution of local resonators generate a time function of potential).
4. Solitonic and quasi-particle rhythm (carriers are solitons, defect in the order flows in an ordered structure, the ordered structure is edited to make a loop).
5. Ionic diffusion rhythm (ions are carriers, tube-like cavities are formed in a circular shape or continuous path to generate a loop).

6. Molecular chemical rhythm (molecules like proteins, enzymes, etc., are carriers, tube-like cavities are formed and sensory systems make sure a circular signaling pathway).
7. Quantum beating (spin is the carrier, wavefunctions interfere in a squeezed excited photonic, electromagnetic or spin state).
8. The density of states rhythm (orbitals coupling, wave function modulation, virtual carrier, a virtual continuous loop is made).
9. van der Waal rhythm (atomic thermal vibration is looped in a spiral pathway).
10. Electromechanical rhythm (classical beating with a mechanical beating like tuning fork).
11. Quasi-charge rhythm polaron, polariton (topological fractured band based continuous loops).
12. Mechanical rhythm, sound wave (similarly elastic pathways to make a circuit of sound waves).

Just imagine, such beautiful patterns are one of 12 different formations. Their physics is different, application domain and temporal range are different.

Twelve rhythms that we have found in the brain

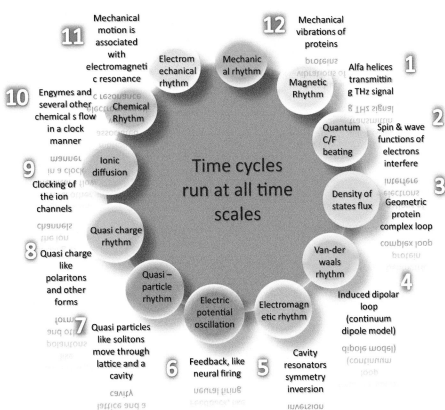

FIGURE 7.7 Twelve types of rhythms found in the brain, 12 layers of dodecanions were considered seeing not just the hardware layers from top to bottom, but the abundance of 12 key information processing parameters.

7.7 THE BRAIN'S WHEEL OF PRIMES

Fundamental questions that "connectome-" based brain models cannot answer: Once a generic rhythm-based pathway connects the smallest molecules like DNA and thousands of proteins to the largest structures like hippocampus in terms of rhythmic cycles, we get eight parallel worlds instead of one that is a neuron, then, can answer fundamental questions that "connectome-" based brain models cannot answer (Reimann et al., 2015).

1. How is information coded in neural activity? A dead map cannot answer. As a collection of multiple rhythms with a common contact point, without using any fitting we can show geometric encoding.
2. How are memories stored and retrieved? Neuron alone does not store memory. Stored as integrated cyclic rhythms at different time scales, the entire cyclic pathway is retrieved if a few points in rhythm are triggered.
3. What does the resting activity in the brain represent? Reconfiguration of multiple rhythmic cycles into a singular one, phase transition and automated re-wiring involving all layers.
4. How do brains simulate the future? "Future" is a rhythm generated by rhythms containing the past memories, time crystal always simulates the future.

5. How is time represented in the brain? Periodicity of rhythms at atomic to the centimeter scale is the time (the brain is a clock inside a clock, at every scale it is a timekeeper for time crystal).
6. How do the specialized systems of the brain integrate with one another? A system is part of a slower rhythm, no part at any scale is left alone, hence it is nested, escape time fractal hardware connected by time crystal.

We get answers to all these queries when we build a very simple model of prime numbers for the human brain (Figure 7.8a). The oversimplified model is a circle, left side is input via sensors that processes primes 2, 3, 5, 7, 11, and 13, it means in any input sensors find C2, C3, C5, C7, C11, and C13 symmetries. Then, 8 such sensors (octonion sensor tensor) take input via 31 spinal +12 cranial nervous systems, take it to midbrain and project to 47 functional domains of the cortex (Brodmann's region).

The three distinct regions of Figure 7.8a is isolated in Figure 7.8b. There are three domains, first, sensor, where time crystal data is 12D but dynamics encoded for time crystal is only 4D. It means, in 12 different axes different dynamics are encoded but four imaginary worlds one inside another hold the 12D data. The second domain is midbrain where the data

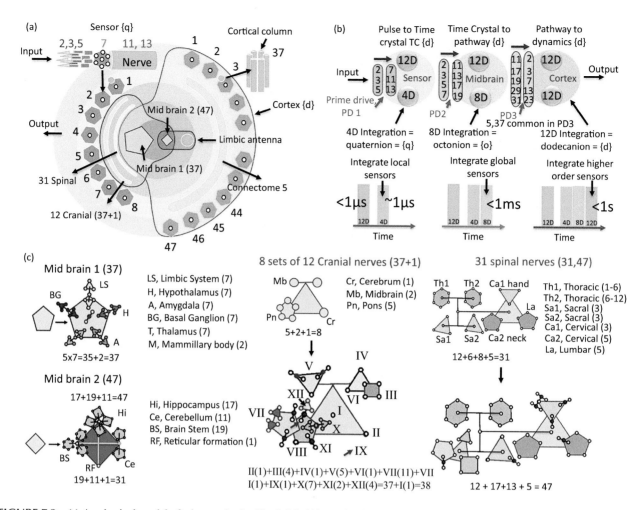

FIGURE 7.8 (a) A spherical model of a human brain. The left half is made of eight classes of sensors. Each sensor processing primes 2 to 13. Sum of cranial and spinal nerves is 43, a prime, passes through two basic parts of the midbrain one represents prime 17 and the other represents prime 47, which equals to the cortex functional classification with Brodmann's 47 regions. Explanations of the primes are given in the panel (c). (b) Two rows explain the flow of primes with the data structure in the brain (top) and data structure with the operational time domain. (c) The components used in the generic brain model of panel (a) is explained in panel c.

structure is 12D but dynamics is 8D, and finally cortex the third region where data is 12 D and dynamics are 12D. Below three columns, three time-domains are explained. As the 12D data passes through the deeper brain structures, 4D dynamics converts to 8D and that reaches 12D dynamics. Direct experimental electronic resonance band measurement for DNA, proteins, microtubules and organic structures, neurons and their clusters, suggested data for estimating different dynamics while we took EEG and other data for the global scale measurements. Microhertz resolution resonance frequencies could be measured without noise trouble. Below microhertz, large time-domain data was collected and based on the slopes nano hertz to femto hertz data are produced. Note that nano to femto hertz is just one band and we had to consider that as nano-hertz is around 30 years and we need to take into around 120 years as some peoples are alive until then. Triplet of triplet resonance bands are observed every single layer, in each of the three sub-bands in a single triplet there are eight "fundamental resonance peaks" and numerous other resonance peaks. Finally, Ghosh et al. have generated the resonance band of the entire brain. Here are 12 bands of a brain, first six bands are experimental data measured in Ghosh et al. lab, the rest six bands are derived from other researcher's reported measurements.

Figure 7.8c outlines the major components of the brain model as demonstrated in Figure 7.8a. The geometry of spinal and cranial nerves, midbrain 1 and midbrain 2 systems are represented using 2D geometric shapes. Similarly, 2D geometric structure of brain stem, limbic system (Figure 7.9a), a combined central nervous system, CNS and peripheral nervous system, PNS (Figure 7.9b) are presented. Combining the operation of all participating components the decision-making time crystal for the brain (Figure 7.9c) is generated. If the phase relation between different components is taken into account, then the 2D structure converts into a 3D structure, which embeds inside a complex time crystal structure. To do that, we simply put a phase sphere around a geometric shape.

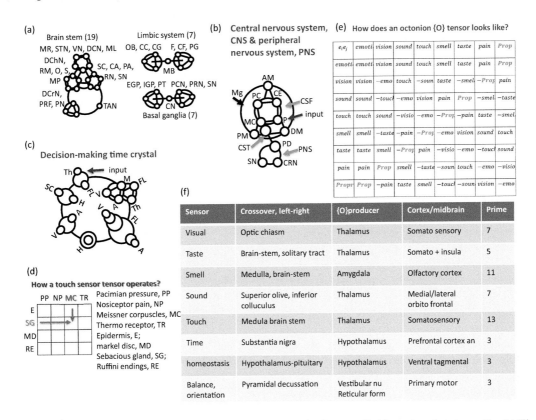

FIGURE 7.9 (a) The time crystal representation of the brain components, brain stem, limbic system, basal ganglia. (b) The time crystal representation of the brain components, central nervous system, CNS and peripheral nervous system, PNS. (c) Decision-making time crystal. (d) Example of an operational sensor, that fills its 4 × 4 quaternion tensor while acquiring data from the environment. (e) An octonion table is created where horizontal and vertical sensory elements interact and a new sensor is built. The diagonal elements emotion and proprioception are purely real and purely imaginary. These diagonal vectors are termed as diagonal drivers. (f) A table for eight sensory systems of the brain, highlighting the C2 symmetry operator as crossover between left and right elements, the component that produces the octonion tensor and the prime that dominates in its structural symmetry.

Limbic system: Caudate nucleus, CN; putamen PT; external /internal globus pallidus, EGP; IGP; subthalamic nucleus, SN; pars compacta and pars reticulata of the substantia nigra, PCN; PRN.

Brain stem: Superior colliculus, SC; cerebral aqueduct, CA; periaqueductal gray, PA: red nucleus, RN; substantia nigra, SN; deep cerebellar nucleus DCrN; pontine reticular formation, PRF; pontine nuclei, PN; dorsal cochlear nucleus, DChN; raphe magnus, RM; olives, O; sulcus, S; medullary pyramid, MP; medullary reticulum, MR; solitary tract nucleus, STN; vestibular nucleus, VN; dorsal column nuclei, DCN; medial lemniscus, ML; tegmental area and nucleus, TAN limbic system: mamillary bodies, MB. the olfactory bulb, OB; cingulate gyrus, CG; corpus callosum, CC; fornix, F; column of fornix, CF; parahippocampal gyrus, PG.

CNS+PNS: Meninges, Mg; piamatter, PM; duramatter, DM; cerebrospinal fluid, CSF; aracnoidmatter, AM; plan: central executive, CE; detailing of plan: putamen, P; decision: parietal cortices, PC; connect to spinal cord, 1 million long axons come down from cortex: motor cortex, MC; midbrain: corticospinal tract, CST; crossover: pyramidal decussation, PD; sensor: lateral corticospinal tract, LCST; peripheral nervous system, PNS; spinal nerve, SN; cranial nerve, CRN.

Decision-making time crystal: Sensory cortices, SC; hippocampus, H; visual, V; frontal lobe, FL; auditory, A; motor, M; thalamus, Th.

7.7.1 Eight Sensors Hold Crossover Magic of Octonion

The brain operates by tensor algebra as described in the Chapter 4, but the question is how? Figure 7.9d explains that simply for a quaternion touch sensor, four different streams of pulses form a typical time crystal, hence each sensor element is a 12D-4D time crystal, as explained above. When the time crystal for MC and SG interact, they bond to create a new time crystal. A similar example is shown for an octonion tensor

that holds information for eight distinct sensors of a mid-brain. The two diagonals of an octonion tensor represents emotion and proprioception, i.e., two fundamental senses of a human brain that regulates its conscious behavior (Figure 7.9e). One could also notice that in the figure taste information interacts with smell and as a result vision is affected. Thus, following this octonion tensor we understand that eight sensory systems which undergo C2 symmetry by crossover (Figure 7.9f) generate in certain organ in the midbrain but activates certain part of the higher brain. Thus, tensors that operate and regulate human consciousness physically links surprisingly different organs in the brain-body system. If hardware grows within and above such tensor products would govern systems in a very different manner than the common circuit concept we learned for over a century.

Wide distribution of synchronization time across the brain: The human brain has developed a distribution of convergence-speed across the organ controls, which could be visualized in the fMRI images, any change in that triggers folding and restructuring of entire brain structure. For example, say, emotion-processing region converges the fastest say within 1–2 ms, then, the smell-processing region, say in ~3–10 ms and so on up to 50 ms, say for the visual region. The most important and abundantly accessible information reaches convergence in the slowest speed. Thus, it ensures that while solving a problem, the newly generated time crystals in the slowest region are instantly cross-checked in the control-region of fast-processing time crystals. As a result, the possibilities of de-coupling are also alleviated. To implement the speed control in the hardware, the shape of the region, circuit complexity and size of the storage are taken into account in the human brain. The degree of randomness in the network regulates the speed of convergence and size is a dependent parameter for the above two. Order of preference among particular regions for reaching the convergence of neuron firing does not mean the human brain is a sequential processor, if required, even after completion of visual processing it could re-trigger the emotion-processing. The entire process could be repeated several times, until all emerged time crystal-clusters are transformed.

Hierarchical time crystals causing folding/restructuring in the forebrain: slower rhythms make system more intelligent and creative: The resting-state fMRI study when compared operations in the brain between a rhesus monkey and a human, they found that the front brain and the back part carries out extensive information processing back-and-forth in the resting state for human but not in the monkeys. During the brain's decision-making, a synchronous stream of pulses propagates back and forth, between the frontal lobe and the back part of the brain where the visual control is located. The provision is made, using dual transformation antenna pair, the core information being processed sustains for a long period of time, ~100–150 ms. Within this time convergence is reached for all essential parts of the problem, technically, the input-time crystal cluster transforms physically and logically into an output form. Brain repeats that for solving a problem for the geometric artificial brain means, harnessing the time crystal database associations as extensively as possible, and, then shrinking the possibilities to a few choices by multi-level synchrony. Therefore, the first requirement of speed control is to identify and isolate the key features of the problem as distinct coupled clusters and then oscillate the entire packet of clusters physically between the two potential walls like a ball, until the solution is reached or transformation is completed. In other words, when transformation antennas are in operation, there is be a tag, "do not disturb."

7.7.2 Eight Mathematical Operators Run Two Orthogonal Math Engines

Figure 7.10 outlines a table with eight operators, top-down applied sequentially. To explain the mathematical operator, we represent the human brain using two orthogonal cylinders. As explained in Figure 7.8b, here sensory data is a dodecanion tensor d_{4D}, it's every element a time crystal represented by quaternion tensor, each element of that quaternion tensor is information about singularity points. The tensor d_{4D} comes out of sensor and while passing back and forth between mid-brain and left-right crossover, the tensor d_{4D} converts to d_{8D}. It means the data structure remains the same but its elements, which were earlier quaternions, now becomes octonions. In the third and the last phase the third feedforward loop runs between 47 Brodmann's region in cortex and midbrain, the tensor d_{8D} converts to d_{12D}. In summary, two cylindrical columns make the tensor $d_{4D} \rightarrow d_{8D} \rightarrow d_{12D}$. Four feedback loops run in parallel. One feedback loop builds a time crystal from pulses, second feedback loop builds first quaternion learning time crystal L_q, third feedback loop builds octonion learning time crystal in midbrain and the fourth, feedback loop builds a dodecanion time crystal L_d in the Brodmann's cortex region. Eight basic operators regulate entire operations, first, three tensor building operators, quaternion, octonion and dodecanion. Then, second step is crossover (autocorrect operator $\hat{A}$). The third step has two operators, which grows simultaneously, one, pure learning $L_{4D} \rightarrow L_{8D} \rightarrow L_{12D}$; the other, memory $M_{4D} \rightarrow M_{8D} \rightarrow M_{12D}$. Finally, geometric algebra (Figure 3.12a) and PPM operators (Figure 4.12) located in the 47 Brodmann's region transform the time crystal.

When the information is extracted as part of memory-retrieval, the typical feature of the pulses is reproduced, just similarly as tape-recorder repeats the same song every time it is played. It is a delicate issue when the brain-memory plays an event, the mechanism is that a few words of the song come discretely from distinct sensory-organs, and then superimpose with each other as a combined, coupled time crystal without any conflict. Therefore, the sensory-organs while constructing their own wave-stream, keeps free space in the time-grid so that those parts are filled by other sensory organs when they construct their own wave-stream. Thus, when wave-streams from different organs come and overlap into a single multi-level wave stream, all pulses carrying particular time crystals occupy separate locations. If not, a particular color-coded in a pulse with particular intensity, pairing and coupling, could overlap with a particular sound or particular touch, or taste or smell encoded pulse. Therefore, the attempt to build a global database of brain's biological rhythm is the beginning of unraveling the mysterious pulses observed randomly at different parts of the human brain and frames the very foundation of the geometric artificial brain.

The sensory input-generated time crystal-cluster of a particular organ (say visual time crystals) could be complementary to others (say sound sensor produced time crystals). Such complementary, supplementary, and conditional time crystal-based coupling is possible if electronic pulses originating from

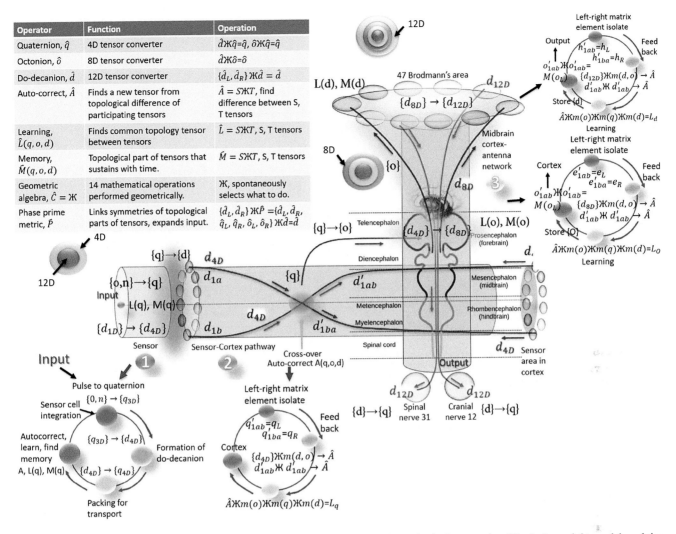

FIGURE 7.10 Top left corner table shows eight operators that primarily regulates the brain operation. The bottom right model explains architectural and operational pathways in the brain. Two orthogonal cylinders represent two parallel information transmission pathways, with a common crossover. Mathematical operators explained in the table is used here to document the essential mathematical operations for the decision-making, using four operational cycles. 1. Synthesis of quaternion at the sensors. 2. Crossover and synthesis of octonions. 3. Learning. 4. Expansion and synthesis of dodecanions. The data structure is always 12D; however, 12D time crystals are packed in a 4D, 8D, or 12D tensors that transports and engage different brain components. This is explained using a few rings.

different organs have similar amplitude and follow identical time-modulation protocols. Since the brain uses pulses of similar shape throughout (all pulses have a spike-shaped irrespective of their sensory-origin) one could easily couple multiple pulses to define particular kinds of a time crystal. Similarly, by maintaining similar kinds of phase behavior one could easily tune decay and growth profiles of pulses, but similar time modulation protocol is kept universally constant from the sensor to the brain. The similarity that is the advantage for the brain inhibits us from unraveling the information encoded by organs in the brain.

7.8 THE BASIC DEVICE FOR DECISION-MAKING IN THE BRAIN

The smallest decision-making device in the brain is helical or vortex shaped water channels residing inside a molecule, like protein, DNA, and microtubule. One helical channel cannot operate by itself, in Chapter 8 we would describe how three concentric spirals could act as a universal time crystal generator, namely Hinductor, H. We find such structures all over the brain-body system. Figure 7.11 shows one prime example. Microtubule, its inner water channel and outer water channels make a triad, a complete H device. Several such filaments or H devices come together to form a bundle, as axon core, on that core, actin-spectrin filaments form a hollow cylinder wrapped around it, and then oligodendrocytes, astrocytes all connected cell branches form three layers. We get H2. Finally, several such H2 devices or neural assemblies form columns (e.g., a cortical column), these millimeter-wide columns form 2D sheets, the sheets form different structures for various organs throughout the brain-body system, listed as H3 in a table in Figure 7.11.

Quadratic geometric relation between mechanical electric and magnetic resonance: All biological systems have a light (electromagnetic) and sound (mechanical) effect

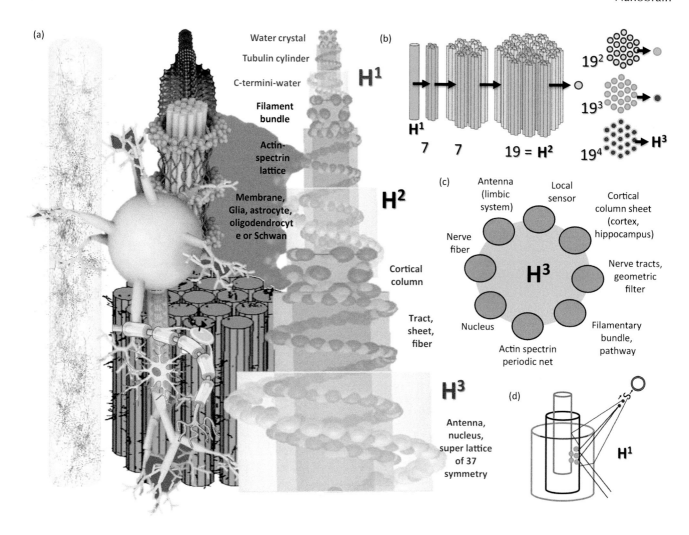

FIGURE 7.11 (a) Basic decision-making an element in the brain is H, it is any spiral or vortex shaped structure like DNA, alfa-helix, microtubule, a spiral nanowire of proteins, etc. Two-layered architectures are shown. The left column is a layered structure of real elements that we find in the brain and to its right, the equivalent nested spiral structure's schematic is presented. Three layers of triplet structures are noted with red, black, and blue color. H1 is the basic entity which self-assembles and forms H2 and finally several H2 self-assembles and form H3. (b) The cortical column formed in the panel a becomes H1 in the cortex and self-assembles in the hexagonal close packing forms a sheet. (c) 2D sheets of cortical columns fold into various geometric shapes, these are all H3 structures that form all decision-making organs in the brain. (d) The operational mechanism of the elementary H device where monochromatic light falls and pure magnetic vortices are generated.

for running the magnetic vortex based time crystal processing, and both electromagnetic and mechanical effects play an equal role in governing the neuron firing while neuroscience ignores the electromagnetic part and considers mechanical ionic movement controls everything in any living cell or neurons (see schematic below Figure 7.11). The squarely parallel chemical and physical protocol is then implemented in creating an entire time crystal model of the whole brain, we believe that this step would fundamentally change the way biology is being studied today. Though existing neuroscience has concretely established the chemical link, electrical wave in cortex is considered as thoughts (Ermentrout and Kleinfeld, 2001) it does not explain how self-assembly begins at the atomic scale and ends up in the meter scale. Therefore, the geometric electromagnetic interaction model is also an additional framework to provide the missing link. We do not discard existing

neuroscience or biology; we suggest that existing biology looks at the ionic response and molecular chemical interaction as an absolute expression. For the PPM-GML-H triad, ionic and molecular chemical events widely reported in the biological textbooks could be included as the dielectric responses in the kHz frequency domain. However, some other things happen in MHz that controls the events in kHz domain, then some other things happen at GHz domain and this journey continues until several peta-hertz frequency domains. Moreover, in the Hz, milli-Hz, micro-Hz, and nano-Hz processes are also there. Most biologists are interested only in the kHz imaginary world if we locate them in the time crystal model, but there are six similar universes above in the slower time domain and six similar universes below in the faster time domain; if we do not include them, a complete mechanistic explanation of a biological system cannot be done.

The *decision-making* protocol of the human brain: In the geometric model, the human brain is neither a classical nor a quantum computer (just being fractal does not mean quantum; Gardiner et al., 2010), it operates among 12 imaginary space-time-topology-prime worlds at the same time, generating a quaternion of self-assembled time crystals following linguistic criterion (who, what, when, and how). A linguistic quaternion is a decision, in the brain it gets filled up by dodecanion time crystals, but the linguistic core structure is never compromised. Thus, a human brain solves a problem continuously, in fact, the massive architecture of time crystals have one-to-one correspondence and these two protocols perform only one computing in a human life, it begins with birth and ends with death (one life is ~10^9 s, it is a stream of 100 pulses of a 10^{-11} Hz oscillator). There is no reduction protocol like classical or quantum computing, no deterministic decision for any query, resonance-based projected set of vibration by the PPM more with holographic engineering. The human brain therefore is an automatic fractal correlation analysis, storage and retrieval machine, since relative weights of different time crystals are assigned in the PPM of time crystals of each brain organ only after the query is made, therefore, we cannot explain the human brain using a Turing machine. Since any input signal, visual, voice, taste, smell or touch senses are taken directly as 3D pattern of pulses carrying 11D geometric structure, eventually converted in terms of basic geometric patterns like circle, triangles, several kinds of polygons, close and open U-shapes, etc., by "fractal decomposition," therefore, we can find some past correlations for any input fractal, processing does not require a framed logic. The decision-making in the brain means sensing symmetry of the time crystals within and outside and creating new crystals to regenerate symmetry. Thus, computing acts smartly when due to unknown parameters, the problem is even difficult to define, the faintly correlated pattern from the PPM of time crystals is projected as output, thus, we get instant decisions continuously for a series of inputs where even the problem is not defined. If we look at the cross-section of an entire sensory bundle containing a million nerves, it will appear as a 2D pattern of frequency change as a function of time say it is $Q(f1, f2)$. However, it should strictly be noted that the very particular geometric relationship of the input image is not destroyed. The fractal frequency approach can generate the quasi-periodic oscillations in the brain described by Izhicovich as a direct output of a generic frequency fractal map.

Learning and spontaneous brain circuit evolution: The brain uses maximum energy in the central layers, where the neuron resides. The eventually fused fractal in the hippocampus, which is the solution of the input problem, does not disappear instantly from the hippocampus after sending its mirror copy to the cerebellum. It sends the final solution fractal back to the cortex regions and the process to write the evolved fused fractal. The majority of the fractal frequencies is found common between input and the memory; hence they neutralize vibration, however, some neurons vibrate continuously

to meet the requirement of new frequencies. The minimum frequency difference causes "beating" and the antenna of the neuron generates "electronic signal burst" which is then detected by all other neurons in the neighborhood. The stream of bursts continues until suitable neurons create wiring with it properly and/or microtubules inside the axon re-wires to accommodate the new fractal part. "Beating" based modulation principle argues when a set of neurons decides to fire and why firing occurs at a very selective region of the brain.

7.8.1 H3 Device That the Brain Uses Everywhere

Tension based morphogenesis and compact binding of the central nervous system (Van Essen, 1997) suggests for a universal decision-making device (Llinás, 1988). Here H3 is the decision-making device as listed in Figure 7.11 (see Chapters 2 and 8 for H1, H2, and H3). Hippocampus looks like a unique device, but actually the same cortex layer formed by hexagonal close packing of cortical columns folds uniquely to build the hippocampus, limbic system, nucleus, nerve tracts, etc. Thus, H3 is not a very particular design, it is a theme, on which the brain has engineered many different structures, the most prominent of them are the cortex layer and hippocampus, one may say hippocampus is a big cortical column.

7.9 FUSION OF CAVITY AND DIELECTRIC RESONATOR MODEL OF A HUMAN BRAIN

When a single chain polymer builds a protein molecule the dynamics show like a teardrop slowly convert into an ellipsoid. Inspired by this dynamics and morphogenesis, all brain cavities listed above were converted five compositions as outlined in Figure 7.12a. Using eight basic shapes that transform a teardrop into an ellipsoid, all major brain cavities could be regenerated by adding certain noise (Figure 7.12b). Phase synchrony and desynchrony is a key to brain operation (Lachaux et al., 1999). The attributed device for resonant communication neuron has not just one but many thresholds (Sardi et al., 2017), which is more extensively studied by Ghosh et al. (2016). Since forward and backward propagation of action potential is both presents, neuron cavity is a resonator (Scott, 2007). Clock-like responses of a neuron have hierarchical organization (Strumwasser, 1974). Thenceforth we created a cavity based on the potential distribution required to represent a time crystal of our choice. That area was converted into a neuron geometry using an algorithm. As outlined in Figure 7.12c.

The architecture of spatial symmetry = architecture of temporal symmetry: Transport means one position to another, in the time domain, i.e., nested hardware one inside another does not need to do that. Right from DNA to the largest brain oscillator, 12 components are there where triplet of triplet resonance bands is observed. We trigger natural vibration of the oscillators for computing; therefore, we do not require power supply even if we want to resonantly vibrate entire frequency fractal hardware made of 100 billion

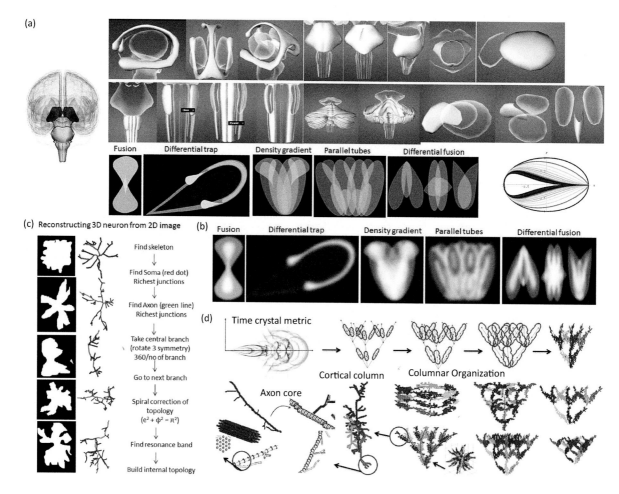

FIGURE 7.12 (a) Different structures found in different components of the brain (top two rows) are converted into seven different classes of cavities, and all these cavities could be generated from a single equation of a teardrop. (b) By fusing teardrop cavities several brain components shown in the top two rows of the panel a are regenerated. (c) Using skeleton operation alone from a given cavity while changing its shape we could regenerate the neuron architecture following the algorithm listed here. (d) Using the PPM the neurons could be self-assembled into cortical columns and 47 Brodmann's regions using a single algorithm of cavity filling.

neurons. The only reason for energy expenditure is the circuit evolution, changing the neuron orientation, constructing new synaptic junctions, rejecting old connections, which are essential to write a new fractal seed in the column of "if-then" time crystals. Neurons need to be fed, since cells are alive, we repeat, following the PPM-GML-H triad model, the energy required for computing is negligible. A system communicating with its constituents is an oxymoron. Thus, we integrate time crystal, convert 3D geometric shapes for a time crystal into the available 3D space for the neuron to grow. It was a major, aggressive decision that architecture of time symmetry equals the architecture of spatial symmetry. However, from a cortical column to various neural assemblies listed in 7.11 table could be created. The most exciting part of this study is that there is only one simple principle, "find skeleton of a given 3D space" and we could generate the architecture needed for the time crystal.

What is the information contained in the human brain model? Resonance state writing is a unique process in the

brain, at different levels, it occurs in a unique way; at protein level, it is a change in the relative orientation of secondary structures; at the neuron level, it is selection of microtubules of different lengths at the neuron level; at the cortical column or rhythm control level, it is the organization symmetry of the neurons. Thus, before any information is written the chain of resonance bands starting from DNA, protein, enzyme, microtubule, neuron to the neuron clusters and ends with the largest oscillator that is the entire brain, therefore, information is a time crystal. **A human brain is a singular time crystal made of an endless network of rhythms (opt to reject entire hardware and be happy with time crystal to upload consciousness!!!), it means an argument when it includes a time function, that also means after a certain time t, it would change to another time crystal, in this book we use time crystal in general, always it means a 3D geometry whose corners are singularity points.** To run this cyclic loop, around a geometric shape, one primary source of energy is thermal noise, it is already shown that biological clocks are temperature independent (Hastings et al., 1957). **Where in the brain, do we see "spontaneous reply back"? The layered clock:** In the time crystal model every biological system has a multi-layer vortex or spirals like a signal receiver and a fractal antenna for the wireless communication and/or "spontaneous reply back." A DNA replies back to proteins, enzymes, and all single molecules. Then, these proteins and small molecules reply back to the supramolecular assemblies created by them like microtubules. Several such filaments reply back to the neurons, which replies back to the cortical columns and it continues to the final assembly, the brain. Singh et al. have simulated all-dielectric structures of a real human brain mapped in UCSD (Figure 7.13, a–f). The brain cavities are considered as a cavity resonator + dielectric resonator + water

resonator (Figure 7.13a), the composition of resonators is like H devices, three layers build a feedback network which could create periodic transmission of fields or carriers along with the system (Figure 7.13b). In connectome (Figure 7.13d), neuron (Figure 7.13g), in axon core (Figure 7.13h), in blood vessel (Figure 7.13e), and in ventricles (Figure 7.13f), everywhere the electric and magnetic field distribution shows a periodic transmission spatially all over the structure. Thus, every single seed replies back to everybody at every moment, no question is ever asked, it is collective responsibility for all to regenerate the lack of symmetry or eliminate the asymmetry following PPM. if the symmetry of a particular time crystal-cluster is known, then we do not need to search every time crystal of that cluster, during spontaneous reply the system simply test one and

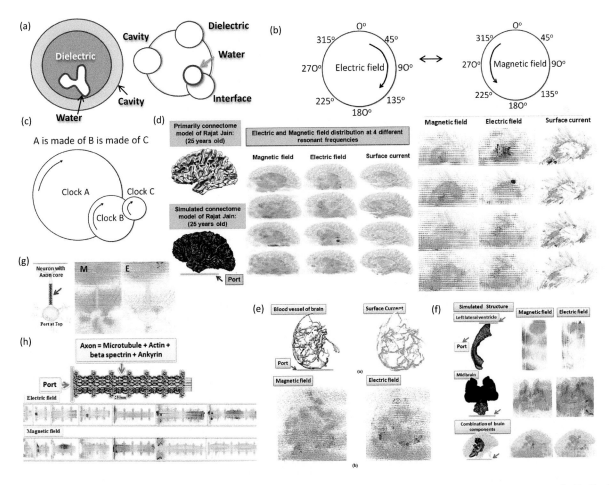

FIGURE 7.13 (a) The brain is a fusion device, it is a combination of dielectric resonators and the cavity resonators (left). Each and every component contributes to the formation of the time crystal, i.e., nested architecture of clocks. (b) Artificially built dielectric and cavity resonator-based brain components show that the electric field and magnetic field rotates spatially and temporally around a local region at the resonance frequencies. (c) The locally observed resonance frequencies and the distribution of fields rotating periodically around a region are considered as a clock and how one clock inhibits the other, how one clock changes its rotation with respect to the other, is studied to build nested clocks. (d) Rajat Jain's brain elements were studied in the CST Maxwell's equation solver to find the electric and magnetic field distribution. Connectome's magnetic field, electric field and the surface current is plotted. (e) Blood vessels cause a massive mechanical thrust in the brain, however, the electric and the magnetic field for the electromagnetic resonance is plotted. (f) The simulated results for one ventricle, midbrain and components incorporated in the brain is shown. (g) Electric and magnetic field distribution for the neuron. (h) Electric and magnetic field distribution at resonance frequency, for the neurofilament and microfilament bundle.

reject all associated options (symmetry breaking) to speed up the computing rapidly. Eliminating multiple symmetries is a difficult task, however, it is shown that simply by adding a constraint, almost all symmetries could be broken (Puget, 1993).

A journey from atomic scale to the largest structure in the brain: always it's a hills and crest in a 3D potential surface: When proteins are pumped at electromagnetic resonance frequencies, the electron density distribution in the secondary structures (cluster of α helices in tubulin) of a protein molecule looks like electron density distribution in a single small molecule. The same pattern evolution is also observed in the single microtubule, filamentary bundle of the axon, group of neurons, etc. The potential fluctuations among different seeds (alfa helices in a single protein, tubulin in a single microtubule and microtubules in the axon bundle) are delocalized among many seeds to generate the collective resonance properties. Say, molecules self-assemble to make a larger seed, then, potentially takes a different form, which controls the electric, magnetic, and mechanical resonance (Hughes, 2018), that very potential should remain distributed among all elemental seeds inside the bigger seeds, thus, all clocks in the brain are nested, one such map is shown in Figure 7.14, its 2D representation

is in Figure 7.15. In the cluster of cortical columns the inter-column potential induction via diffused gradient of neurotransmitters, the composition of neurotransmitter river-flows combines multiple cortical columns as a single integrated unit, just like a single molecule. Several cortical columns or nucleus regions made of neuron clusters, i.e., nucleus, fibers, 47 Brodmann's regions in the cortex operate as a single molecule like system. **Why is there wiring in the brain?** When several such sensory controls located at far distant places in the brain needs to be integrated and formed a single molecule like system then neural pathways hold the potentials for constructive interference of waves in the larger cavity, just what the atomic bonds do in a single molecule. Re-wiring and propagation of nerve impulse is following the changes in mapping the 3D interference patterns in the host functional lobes (sub-seeds). The objective is that the potential distribution corresponding to the resonance of multiple carriers in multiple time domains is mapped inside the host cavity that could be a protein complex, cortical column, nucleus or brain stem or the entire brain does not matter, size changes, so the carriers, thence the need for various mechanisms. In other words, wiring is an illusion.

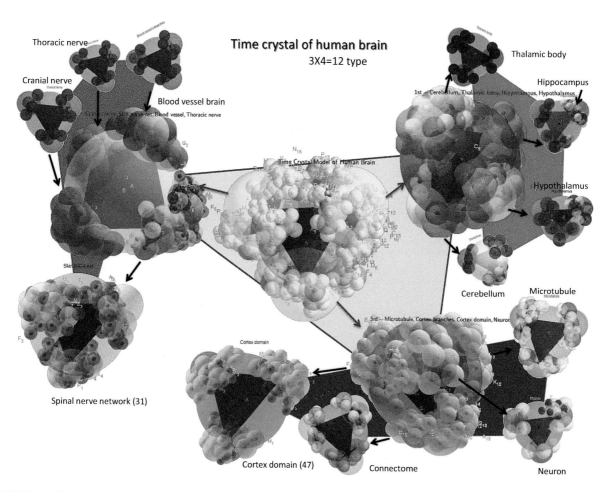

FIGURE 7.14   Time crystal model for the complete human brain.

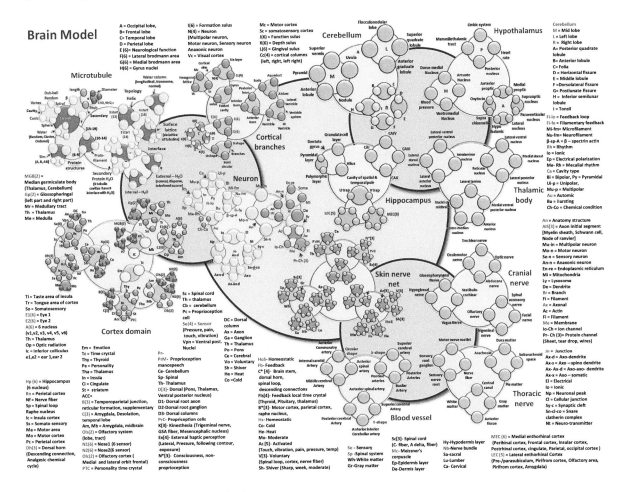

FIGURE 7.15   Time crystal model for the whole brain.

7.10   TIME CRYSTAL MODEL OF THE HUMAN BRAIN

Popular science claim that the brain is a black hole (Buonomano, 2017), gets a new dimension in time crystal map. How a human brain keeps time has been a point of discussion for a long time (Buhusi and Meck, 2005). Some researchers have proposed that keeping time is fundamental to its information processing (Wing, 2002). It is not just that the neurons engage in the management of time (Ivry, and Spencer, 2004) the basal ganglia, when it learns a new trick, it builds time crystal architecture (Graybiel, 2005). The similar kind of time management is found in the cerebellum, whenever it learns coordination in the movement, it builds an architecture of time slices or clocks (Thach, 1998). The human motor activity has a robust, intrinsic fractal structure with similar patterns from minutes to hours (Hu et al., 2013). Thus, the development of the fractal architecture of the time is abundant in literature varying all over the brain, yet there was no compilation by biologists to build a generic musical model of the brain. A time crystal model of the human brain shown in Figures 7.14 and 7.15 would enable providing the seamless regeneration of the lost organs (Biasiucci et al., 2018).

GML is inspired by the brain's intricate engineering to process music: The hidden language in the brain what researchers have always been looking into could be in the form of music (Brown et al., 2006). Not just that brain perceives the music and its evolution pre-attentively (Koelsch et al., 2002) intersensory connections are also musical, smell's neural signature looks like music (Plailly et al., 2007). Fusing two sensory time crystals in the GMLs are inspired by this intra-sensory feature of the brain. The brain is hard-wired so that it perceives temporal regularity in music (Large and Palmer, 2002), which is fundamental to the time crystal analyzer described in Chapter 2. In computer science language, one could state the brain finds a hierarchy of recursive functions, or nested recursive functions automatically and elegantly. Brain's engineering is so intricate that it naturally differentiates between an absolute pitch and a relative pitch (Zatorre et al., 1998; Zatorre and Halpern, 2005), which is again a fundamental feature of a time crystal analyzer.

Magnetic rhythms or clocks in the brain: are they output of magnetic vortices? Magnetic rhythms in the brain are known for a long time (Cohen, 1968). Even the magnetic field of a single axon has been studied (Roth and Guo, 1988; Roth and Wikswo, 1985). The research of the biomagnetic field effect on regulating the biological phenomenon has been very well documented (Barnes and Greenebaum, 2007). Since microtubule, DNA like a various magnetic vortex, ionic vortex and electric vortex generators as a fourth-circuit element (Chapter 8) are abundant in the brain, we closely reviewed the research work that monitored the magnetic rhythms. Note that to be a Hinductor-class fourth-circuit element, the charge-flux ratio must be linear and should be seen optically at least during neuron firing. While axon's magnetic field was studied long back, the magnetic study of ionic firing has been demonstrated recently (Barry et al., 2016). Therefore, magnetic rhythms or pure magnetic clocks must be abundant at all scales of information processing in the brain. It is shown that continuous frequency glides in the brain's neural network generates cortical magnetic fields (Arlinger et al., 1982). Neuromagnetic brain rhythms run for minutes for human cognition (Ramkumar et al., 2012). The reason magnetic stimulation based functional imaging of the brain works is manifold (Ilmoniemi et al., 1999).

7.11 FOUR METRICS OF PRIMES RUN IN PARALLEL—THE SAGA OF HEXAGONAL LATTICE

The classic use of PPM is observed in the pupil of our eyes. When discrete time crystals arrive and remain unattended by one metric, then the hierarchy of metrics attend them (Damsma and Rijn, 2017). Gamma band activity of the brain reveals several rhythmic tones connected to each other such that we could represent them as a single metric (Snyder and Large, 2005). The fractal geometry of the brain (Di Leva, 2016) uses four distinct types of metrics in parallel. Pieces of literature are rich in claiming that mind-brain paradox holds a fractal solution (King, 1996). The origin of scale-free dynamics in the brain (Eguiluz et al., 2005) is easily explained by composition of four metrics (Figure 7.16). In the brain at all scale, components try building a tiny PPM, tiny means, limiting integers are narrowed. A brain does not have one but trillions of PPMs, but they integrate and build more complex ones. However, since the pattern of primes is

Prime	Information PPM-1; 12D	Local sensor PPM-2; 4D	Global sensor PPM3; 8D	Higher level sensor PPM4; 12D	Topology
2	3D arrangement of sensory cells	Molecular sensing conversion	Olfactory to cortex pathway	Cranial nerve	
3	Layers of depths of sensor columns	Alfa-helices based cavity sensing	Tongue to brain cortex pathway	Blood Vessel	
5	Types of neurons for sensing	Helical topology of nerve cells	Auditory to cortex pathway	Connectome	
7	Sensing area classification	Geometry of nerve bundles	Thalamus	Limbic antenna	
11	Folding volume for sensing	Neuron-glia relative lattice	Cerebellum	Cerebellum	
13	Geometry of neuron branches	Actin-spectrin edits MT core	Optic pathway eye to the cortex	Microtubule-filament complexes	
17	Non-neural cell arrangement	Nerve lattice defect memory	Hippocampus	Hippocampus	
19	Protein complexes assisting sensing	Protein complex cycle memory	Brain stem	Brain stem	
23	Triplet layers correlation for 12D	Triplet layers for H1, H2, H3	Neuron-glial complex	Neuron-glial complex	
29	Neuron-glial complex	RNA, protein exp. memory	Antenna-midbrain complex	Antenna-midbrain complex	
31	Actin-spectrin periodic network	Hexagonal nerve cell bridge	Spinal nerve	Spinal nerve	
37	Microtubule, MT geometry	Microtubule & filaments H1-H3	Cranial nerve	Cortex-cortical layer	

Prime engineering in the brain

Brain needs elements vibrating following symmetries of five primes only 2,3,5,7, 23 are enough to deliver the rest of the primes using them.

2x2+1=5
2x3+1=7
2x5+1=11
2x6+1=13; 2x3=6
2x9+1=19; 3x3=9
2x11+1=23
2x14+1=29; 2x7=14
2x15+1=31; 3x5=15
2x18+1=37; 2x3x3=18
2x21+1=43; 3x7=21
2x23+1=47; 2x23=46

Most critical prime 23

To build this, brain uses 23 types of neuron cells, 23 types of glial cells and 23 types of dendoastrocytes.

C_2, C_3, C_5, C_7 of $\{23\} = H3$

Hexagonal lattice = prime 37

Singularity Path   Choices
2; 1-2-1, 2-3-2......12 clocks
3; 1-2-7-1, 2-3-7-2,......6 clocks
4; 1-2-3-7-1, 2-3-4-7-2,...6 clocks
5; ...6 clocks; 6; ...6 clocks;
7; ...1 clock;

FIGURE 7.16 A table shows how 12 primes govern the information processing at four different levels (three primes are left). Four PPM-1 that regulates the synthesis of 12D time crystal information via sensor cells, PPM-2 that regulates the sensing and synthesis of 4D linguistic quaternions for the sensory input data (four linguistic queries are who did it? What it did? How it did? At which condition?). Then comes PPM-3, this metric governs the sensing and synthesis of octonion packing of sensory time crystals and finally PPM-4, that expands the information content. To the right of the table it is shown that 2, 3, 5, 7, and 23 are enough to create other primes using geometry. To the bottom, hexagonal lattices are found in the microtubule, cortical column assembly, etc, its different pathways taking 2, 3, 4, 5, and 7 elements at a time it is shown that they could replicate 12 + 6×4 + 1=47 clocks, which is exactly the number of Brodmann's region in the brain.

universal, all PPMs governing functional responses in the brain deliver a scale-free dynamic (Eguiluz et al., 2005). Biological clocks are rhythms of life, the clocks running for milliseconds to seconds (Foster and Kreitzman, 2004); here clocks do not end at the neuron surface, there is an infinite chain of segmented times, which forms an infinite, incomplete network of clocks.

Four major PPMs governing the brain are information, PPM-1; local sensor, PPM-2; global sensor, PPM-3; higher-level sensor, PPM-4. Each metric processes 12 primes at least (Figure 7.16), we have listed in the table how typical metrics corresponding to a typical prime (Figure 3.4) is actually generated in the brain. The statistically dominating geometric shape in the time crystals is also noted in the metric table. Two points need to be kept in mind. First, 2, 3, 5, 7, these four primes could generate all other primes easily, geometrically as we observed in the brain components, except 23. Unfortunately, the most important information processing in the neuron layer. Therefore, we need only five symmetries; the rest of the prime dynamics could be regenerated from only five symmetries. Second interesting feature is the importance of hexagonal lattice (see Figure 7.16 bottom). A hexagonal lattice could hold various kinds of clocks simultaneously, some options are shown in Figure 7.16. Thus, Brodmann's 47 regions use hexagonal carpet to process any possible integers and prime dynamics.

7.11.1 Composition of Four, Eight, and Twelve Dimension Tensors

Distributed parallel computing in parts of cortex shows tensor algebra like signal processing (Anastasio and Robinson, 1990). If tensors are there, theories of a holographic brain projecting Schrödinger's wave could not remain far (Nobili, 1985). Clocked cell cycles clock (Edmunds and Adams, 1981) to process three distinct classes of tensors simultaneously. All large-scale models of a human brain (Eliasmith et al., 2012) are linear, 4D, 8D, and 12D network of time crystals is the singular undefined model of the human brain. Octonion algebra could estimate the phase elements of the tensors interacting with each other (Furey, 2018). However, dodecanion algebra would loop the interactive elements, once? Twice? In reality, looping isolated elements mean triggering the PPM. Four-dimensional unified brain theory (Friston, 2010) is not transformed into a 12-dimensional brain here, rather, a variable dimension brain that explores the topology of dimensional transformation would be the right explanation of the time crystal brain model discussed here. Imagine some thoughts are 2-dimensional, some are 8-dimensional, and some events are 10-dimensional (Figure 7.17), the truth lies in the pattern of imaginary worlds active among 4, 8, and 12 choices, it is non-computable (Stulf et al., 2015).

Information PPM-1; 12D	Local sensor PPM-2; 4D	Global sensor PPM-3; 8D	Higher level sensor PPM-4; 12D
PPM-1 is H3 type device that is the only device used by brain for memory & processing.	PPM-2 integrates local sensors, send 12D time crystal to PPM-1 for further processing.	PPM-3 combines 12D data of PPM-1 coming from various sensors all across. Sends output pathways to PPM-4.	Filters match, expand PPM-3 pathways, fills them inside PPM-4 built topologies. 12D input gets 12D dynamics.
Information as 12D time crystal, TC is pure, but requires three editing.	Creates/edits 4D memory for better construction of 12 D input TC.	Creates/edits 8D memory of pathways to filter, combine various sensor built TCs.	Creates/edits 12D memory to expand, transform topologies holding pathways inside.
Information is packed in a 12D TC, then 12D dynamics is packed in three layers.	Destroys identity of local sensors, PPM-2 expands to enrich possible 4D dynamics of 12D input TC.	Destroys identity of global 8 type sensors, pack them in a pathway. 8D dynamics is added to 12D input TC.	Destroys identity of pathways packs paths as points in the topology of higher dimensions, input TC gets 12D dynamics
Information is real, 3 PPMs perfect 12 sensors	Smell, taste, hearing & touch sensing arise.	Visual, proprioception, thermal & time sensing arise.	Symmetry, prime, infinity & consciousness sensing arise.
Dodecanions {d}.	Quaternions {q}.	Octonions {o}	Dodecanions {d}

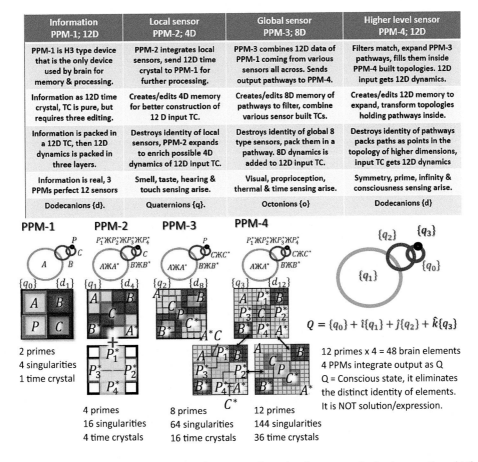

$$Q = \{q_0\} + \hat{\imath}\{q_1\} + \hat{\jmath}\{q_2\} + \hat{k}\{q_3\}$$

PPM-1
2 primes
4 singularities
1 time crystal

PPM-2
4 primes
16 singularities
4 time crystals

PPM-3
8 primes
64 singularities
16 time crystals

PPM-4
12 primes
144 singularities
36 time crystals

12 primes x 4 = 48 brain elements
4 PPMs integrate output as Q
Q = Conscious state, it eliminates the distinct identity of elements.
It is NOT solution/expression.

FIGURE 7.17 A table shows the comparisons between the four types of metrics that govern the brain operation. At the bottom, we demonstrate how four PPMs process tensors and deliver a decision for the brain. The brain preserves the linguistic quaternion (who, at what condition, what, and how) for all 12 primes to deliver an output Q, i.e., the conscious state.

As said earlier, information is linguistic (Figure 2.10b), four imaginary worlds, each shares a clock, they hold the answer to four queries, who, when, what, and how. The four PPMs described in Figure 7.17 have identical linguistic structures; however, the tensors have multiple choices as described earlier in Figure 4.13. The only point we add here is that the quaternion composition of four metrics, PPM-1, PPM-2, PPM-3, and PPM-4 delivers us the conscious state or the state of mind Q of the human brain in a time crystal model.

7.11.2 Quaternion, Octonion, and Dodecanion

Biological rhythms and nesting of rhythms were there always: Now it's a crystal and composite materials but made of clocks: It takes five to six days for RNA to replace a protein molecule; the replacement of cells in the body follows a very particular rhythm. Skin cells are replaced in every 14 days, we get new blood in every *three* months, we get a new kidney in every 17 months; in 24 months we get new bones; and in 100 years we get a new heart. Even our breathing, heartbeat, circadian rhythms run perpetually. One interesting aspect that we would like to introduce here is that we now look at all these rhythms from relative changes of one with respect to another. For example, every three months when we get new blood, skins have been replaced roughly six times. The frequency of skin cycle Fs is six times more than the frequency of blood cycle Fb, we can write $Fs = 6Fb$. In this way, all cycles would be related. Not only this is true for the cell replacement cycle, for circadian and other rhythms, but we can also find that there are rhythms inside a rhythm. Not just biology, nested rhythm requires new math and physics. Self-organization goes on at all scales at all time. Biological rhythms are distinctly different from the decision-making rhythms, which are primarily ionic, molecular, chemical. Figure 7.17 primarily concerns decision-making rhythms which are physical, electric, magnetic, mechanical, etc.

Graph theoretical analysis which was used connect structure and function (Bullmore and Sporns, 2009), now a composition of the quaternion, octonion, and dodecanion, i.e., tensor of 4, 8, and 12 imaginary worlds took over (Figure 7.17). Critical and limiting features of the brain are often discussed (Beggs and Timme, 2012), possibly, the limit of all limits is set by the tensor triad of 4, 8, and 12 imaginary worlds. Temporal coding enables segmenting the event scenes in the brain for all kinds of sensors (Engel et al., 1991). In the second step isolated or segmented events are temporally bonded by the typical tensorial bonding protocols (wilder than quantum: hand interacts with eyebrow to sense rose smell; Engel and Singer, 2001). It's a top-down synchronization to implement quaternion, octonion, and the dodecanion (Engel et al., 2001). All the tensor elements are phase lags that estimate the singularity or undefined domain when two clocks interact (Ermentrout and Kopell et al., 1994).

7.12 TIME CRYSTAL MADE MEANDER FLOWER TO A GARDEN OF GARDENS

Musical rhythms have been reported from all over the brain-body architecture, for example the cerebral substrates (Halpern, 2006). Thoughts are often attributed to cerebral rhythms (Steriade et al., 1990a). Different parts of the brain respond to musical emotions very differently, especially in the frontal brain EEG one could read musical emotions distinctly (Schmidt and Trainor, 2001). Even during motor learning music like rhythms are born (Sakai et al., 2004). The brain states are oscillatory and if information in the brain is considered as geometric, then, using neural oscillations it becomes easy to explain the brain states (Nie et al., 2014). The map of temporal disorders suggests that a set of clocks or a local part of time crystal could move from one system to another and build a new system (Rensing et al., 1987). This particular mechanism is the foundation of brain's decision-making, the brain has a storage of complex time crystal system (first two columns of Figure 7.18), from these running banks, the brain picks tiny little clusters of clocks or local time crystals and builds unprecedented time crystal as thoughts (last three columns of Figure 7.18).

Garden of gardens (GOG) pick flowers from different gardens of time crystals, synchronize those flowers as petals to build new flowers (Glass, 2001), the transition from clocks to chaos to clocks continue (Glass and Mackey, 1988). PPM governs how and which petals to be picked and to be brought together. The journey of time crystals that begins as a singular clock, ends in a random collection of flowers in a garden, which ends in the formation of a conscious architecture, GOG. A GOG attains the ability of build new flowers since it's a composition of 12 life like gardens of time crystals, so each of its garden has 12 distinct classes of time crystal systems. GOG is composite life form made of 12 life like systems. Unconscious means an absence of garden of gardens, by delinking of 12 garden-integration systems; yet all 12 gardens remain active (Goila and Pawar, 2009), the inverse event is therefore the construct of a conscious machine. Large-scale synchronization like a world wide web is a good metaphor (Varela et al., 2001), but it reveals nothing about the intricate nature of clocks.

7.12.1 Twelve Dodecanion and Eight Octonions Build 20 Conscious Human Expressions

In the 1880s, B.W. Betts developed a mathematical system for modeling the development of human consciousness through geometry (Cook, 1887). How brain builds its cognitive codes is the hallmark of the science of consciousness (Grossberg, 1980), now, with the advent of a fractal tape, an entirely new scientific exploration is getting a shape around it (Ghosh et al., 2014a). Geometric information estimates the neural interactions in the human brain (Nie et al., 2014), often the interactions are considered to be electric (Nuñez, 1998). However, linear layered architectures of intelligence like deep

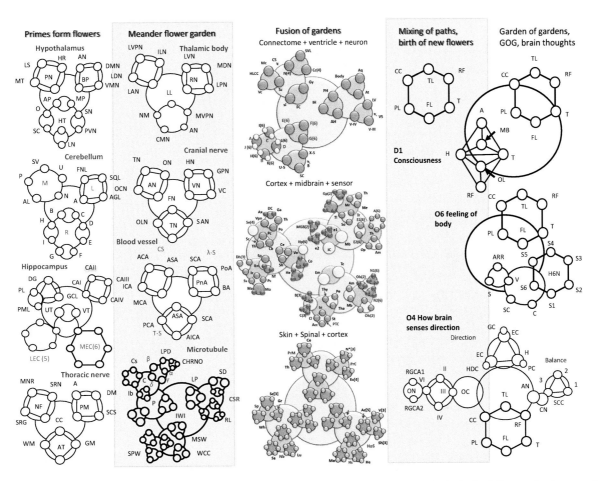

FIGURE 7.18 Five columns demonstrate how the time crystals that represent a particular organ in the brain delivers a few loops and those local time crystals self-assemble to generate complex functional time crystals for conscious response of the human brain as Garden of Gardens described in Chapter 2 (here, column 5).

Hypothalamus:

MT-Mammilothalamic tract, LS-Limbic system, HR-Heart rate, PN-Posterior nucleus, DMN-Dorso medial nucleus, AN-Arcuate nucleus, VN-Ventromedial nucleus, BP-Blood pressure, AP-Anterior Proptic, MP-Medial preoptic, SN-Supraoptic nucleus, PVN-Paraventricular nucleus, LN-Lateral nucleus, HT-Hypothalamic nucleus, SC-Suprachiasmatic, O-Oxytocin

Cerebellum:

SV-Superior vermis, P-Pyramid, AL-Anterior lobule, N-Nodule, U-Uvula, AGL- Anterior Graduate Lobule, A-Posterior quadrate lobule, SQL-Superior quadrate lobule, FL-Flocculonodular Lobe, B-Folia, D-Horizontal Fissure, E-Middle lobule, F-Dorsolateral fissure, G-Postlunate Fissure, H-Inferior semilunar lobule, I-Tonsil

Hippocampus:

DG-Dentate Gyrus, GCL-Granulate Cell Layer, H-Hilus, PL-Pyramidal Layer, PML-Polymorphic layer, Cavity of the spatial and temporal pole is made of UT and VT, UT-U Trap, VT-V Trap, LEC(5)-Lateral Enthorhinal Cortex (Pre/Parasubiculum, Piriform Cortex, Olfactory area, Amygdala). MEC(6)-Medial Enthorhinal cortex (Perihinal cortex, frontal cortex, insular cortex, postrhinal cortex, cingulate gyrus, Parietal, Occipital cortex)

Thoracic nerve:

MNR-Motor nerve rootlet, SRN-Sensory root Ganglion, SNR-Sensory nerve root, NF-Nerve Fiber, A-Arachnoid matter, DM-Dura matter, PM-Pia matter, SCS-Suprachnoid space, CC-Central canal, WM-White matter, GM-Gray matter, AT-Anterior Tissue

Thalamic body:

LVPN-Lateral ventral posterior nucleus, LDN-Lateral dorsal nucleus, LAN-Lateral anterior nucleus, IN-Intralaminor nucleus, LL-Lateral lamina, NM-Nuclei of midbrain, CMN-Centro median nucleus, AN-Anterior nucleus, MVPN-Medial ventral posterior nucleus. RN-Reticular nucleus, LPN-Lateral posterior nucleus, MDN-Medial dorsal nucleus, LVN-Lateral ventral nucleus

Cranial nerve:

TN: Trochlear nerve, ON-Oculomotor nerve, AN-Abducens nerve, ON-Optic nerve, HN-Hypoglossal nerve, GN-Glossopharyngeal nerve, VC-Vestibulocochlear, VN-Vagus nerve, FN-Facial nerve, ON-Olfactory nerve, TN-Trigeminal nerve, SAN-Spinal accessory nerve *(continued)*

FIGURE 7.18 (Continued)

Blood vessel branches:

Circular shape (ACA-Anterior commutator artery, ICA-Internal carotid artery, MCA-Middle cerebral artery, ASA-Anterior spinal artery), λ-shape(PA-Pontine Arteries, BA-Basilar artery, PA-Posterior Artery, SCA-Superior cerebral artery), T-shape (ICA-Interior cerebral artery, ASA-Anterior spinal artery, PCA-Posterior cerebral artery, AICA-Anterior Interior cerebral artery)

Microtubule:

Tubulin protein structure

Secondary protein [α-helices (LPD-Length pitch and diameter; CHRNO-alkyl, amine), β, γ, δ]. Cavity [Conic, vortex, dumb-bell or teardrop, spiral]. Water layer [Random arrangement, cluster, ordered]. Sim on the lattice surface, S (lattice type A, B, and AB) and protofilaments (numbers, 6-9, 10-14, 15-19).

Topological morphogenesis of microtubule surface

Helical transport path (periodicity helical gap 2, 3 (returns on 13); gap 4 (returns on 16)); central water channel (longitudinal, transverse, and normal); surface tubulin lattice (six distinct classes of lattices observed in the tunneling images)

The interface between sub-helices and surfaces

Secondary protein water channel (triplet of triplet domains, eight tubulin cavities have nine interfaces with H2O); external surface water layer (one to two molecular layers of water in three modes, normal, disperse, interfered source); inner core water channel (longitudinal, transverse, and normal modes of water)

Skin nerve network:

The homeostatic primary clock network

HoS-Homeostatic; Fb-Feedback; C* [4]-Brain stem, dorsal horn, spinal loop, descending connections; Fls[4]-Feedback local time crystal (Thyroid, Pituitary, thalamus); B*[3]-Motor cortex, parietal cortex, raphe nucleus, Hs-Homeostatic; Co-Cold; He-Heat; Mo-Moderate; Ac [5]-Activated; (Touch, vibration, pain, pressure, temp); V[3]-Voluntary (Spinal loop, cortex, nerve fiber); Sh-Shiver (Sharp, week, moderate)

Brain-controlled higher-level network connected to skin

PrM-Proprioception manospeech; Ce-Cerebellum; Sp-Spinal; Th-Thalamus; D[3]-Dorsal, (Pons, Thalamus, Ventral posterior nucleus); D1-Dorsal root axon; D2-Dorsal root ganglion; D3-Dorsal columns; PrC-Propriception cells; K[3]-Kinesthesia (Trigeminal nerve, GSA fiber, Mesencephalic nucleus); Ex[4]-External haptic perception (Lateral, Pressure, following contour, exposure); N*[3]-Consciousness, non-consciousness proprioception

Sensory pathway via the spinal cord

Se—Sensory; Sp-Spinal system; Wh-White matter; Gr-Gray matter; Sc[3]-Spinal cord, (C-fiber, A-delta, fiber) Mc-Meissner's corpuscle; Ep-Epidermis layer; De-Dermis layer; Hy-Hypodermis layer; Nb-Nerve bundle; Sa-sacral; Lu-Lumber; Ca-Cervical

Cortical branches

Ventricular and geometric cavities

Aq-Aqueduct, Body, At-Atrium, LV-Lateral Ventricle, VS-Ventricular system, IV-V=IV Ventricle, III-V=III Ventricle, PH-Posterior Horn, IH-Interior Horn, AH-Anterior Horn

Top folded hexagonal lattice domains (47 Brodman's regions)

CS-Cortical system; SVL-Seven vertical layers of the cortical column, N(4), Hexagonal lattice, Mc, Vc, Sc. Folded cavity, Sl-Sulus, Gy-Gyrus, Cc(4).

Nerve bundle for the different functional region.

E(6), F(6), G(6), L(6), I(6), J(6), H(6), K(6), Three types of branches B, X-S=X-shape, Semi-circular-SC, and U-S-U shape.

A = Occipital lobe, B= Frontal lobe, C= Temporal lobe, D = Perietal lobe, E (6)= Neurological function, F(6) = Lateral brodmann area, G(6) = Medial brodmann area, H(6) = Gyrus nuclei, I(6) = Formation sulus, J(6) = Function sulus, K(6) = Depth sulus, L(6) = Gingival sulus, Cc(4) = cortical columns (left, right, left right), N(4) = Neuron (Multipolar neuron, Motor neuron, Sensory neuron, Anaxonic neuron, 23 types of neural branches are studied), Vc = Visual cortex, Mc = Motor cortex, Sc = somatosensory cortex

Cortex domain

MGB(2) = Median germiculate body (Thalamus, Cerebellum); Gp(2) = Glossopheringal (left part and right part); Mr = Medullary tract; Th = Thalamus; Me = Medulla; Ti = Taste area of insula; Tr = Tongue area of cortex; So = Somatosensory; E1(6) = Eye 1; E2(6) = Eye 2; A(6) = 6 nucleus; (v1, v2, v3, v4, v5, v6); Th = Thalamus; Op = Optic radiation; Ic = Inferior colliculus; e1,e2 = ear 1,ear 2   *(continued)*

FIGURE 7.18 (Continued)

Em = Emotional pathways (details in functional time crystals); Tc = Time crystal clusters of hippocampal output, Thy = Thyroid, Pe = Personality time crystal, Tha = Thalamus, In = Insula, Ci = Cingulate gyrus learning pathway, St = striatum, ACC=, B(3) = Temporoparietal junction, reticular formation, supplementary; C(3) = Amygdala, Desolation, temporal lobe; Am, Mb = Amygdala, midbrain; Ols(2) = Olfactory system (lobe, tract); N1(6) = Nose1 (6 sensor, +one controller 6 + 1 = 7); N2(6) = Nose2(6 sensors +one controller, 6 + 1 = 7); Olc(2) = Olfactory cortex (Medial and lateral orbit frontal); PTC = Personality time crystal

Sc = 31 Spinal cord cavity's time crystal; Th = thalamus, Cb = cerebellum, Pc = Proprioception cell based network, Se(4) = Sensor (Pressure, pain, touch, vibration); Vpn = Ventral post. Nuclei, DC = Dorsal column

Ax = Axonal branches connecting cortical columns, Ga = Ganglion, Po = Pons brain stem, Ce = Cerebral

Vo = Voluntary, Three sensations of homeostasis trigger large-scale information processing, Sh = Shiver; He = Heat; Co =Cold

Hp (6) = Hippocampus; (6 nucleus, +one controller 6 + 1=7); Rn = raphe nucleus, Nf = Nerve fiber singular time crystal, Sp = Spinal loop time crystal built by cross sectional components; Ic = Insula cortex; Ss = Somato sensory; Ma = Motor area, Mo = Motor cortex; Pc = Perietal cortex; Dh(3) = Dorsal horn, (Descending connection, Analgesic chemical cycle)

All components of the column 4 and column 5 are described in Figures 7.19, 7.20, 7.21

D1. Consciousness

Central executive, CE; Olfactory lobe, OL; Mamillary body, MB; Parietal lobe, PL; Corpus Callosum, CC; Temporal lobe, TL; Fusiform gyrus, FFG; Corpus callosum, CC; Primary sensory pad, PSP; Superior temporal, ST; Sensory pads, SP; Caudate nucleus, CN; Putamen, P; Amygdala, A; Hippocampus, H; Reticular formation, RF; Thalamus, T; Limbic system, LS; Superior caliculus, SC; orbitofrontal cortex, OFC; Frontal lobe, FL; Parietal cortex, PC

D2. Memory

Protein secondary structure, PSS; α-helix, α; β-sheet, β; Microtubule, M; Actin, A; Neurofilament, N; Neuron, NN; Glia-Astrocyte, GA; Oligo-dendrocyte, OD; Pyramidal and polar, PP; Tear-drop transform, TDT; Superstructures of neuron, SSN

D3. Language and conversation

Arcuate fasciculus, AF; Temporal lobe, TL; Frontal lobe, FL; Dorsal Prefrontal cortex, DPFC; Motor cortex, MC; Broca's area, BA; Gesdiwind territory, GT; Cerebellum, C; Wernicke's area, WA; Visual cortex, VC

D4. Thinking and intelligence

Extrastriate cortex, EC; Fusiform gyrus, FG; Wernicke's area, WA; angular gyrus, AG; supramarginal gyrus, SG; superior parietal lobule, SPL; Anterior Cingulate, AC; Working memory, WM; Procedural memory, PM; Episodic memory, EM; Semantic memory, SM; Implicit memory, IM

D5. Sense of universal time, symmetry

10 Metrics of prime, 10MP; Statistical prime contribution, SPC; Striatum amygdala, SA; Temporal lobe, TL; Thalamus, T; Parietal lobe, PL; Hippocampus long memory, HLM; Topological projected time, TPt; Frontal lobe, FL; Corpus callosum, CC; Reticular formation, RF; Hippocampus min time filtered pattern, HMTFP

D6. Fear, threat, anger, hate

Medical dorsal nucleus, MDN; Limbic system, LS; Thalamus, Th; Posterior portion, PP; Accessory basal, AB; Middle gyrus, Mg; Basal nucleus, BN; Cingulate cortex, CC; Cortex, C; Medical, M; Amygdala, A; Hypothalamus, H; Input lateral terminal, ILT; Olfactory lobe, OL; Fornix, F; Mammillary body, MB; Basal ganglia, BG; Frontal cortex, FC; Superior-Frontal cortex, SFC; Pre-frontal cortex, PFC; Ventral tangential array, VTA; Dopamine level, DL; Styria terminalis, ST; Nucleus acumens, NA; Orbitofrontal cortex, OC; Insula, In

D7. Reward

Basal ganglia, BG; Nucleus accumbens, NA; Ventral tegmental area, VTA; Midbrain, MB; Pre-frontal cortex, PFC; Dopamine cycle, DC; Striatum, S; Pallidum, P; Thalamus, T; Cortex, C; Ventral pallidum, VP; parabrachial nucleus, PN; orbitofrontal cortex OFC; and insular cortex, IC

D8. Mimicry, skill, adaptation

Parietal lobe, PL; Thalamus, Th; Cerebellum, C; Hippocampus, H; Amygdala, A; Temporal lobe, TL; Caudate nucleus, CN; Mammillary body, MB; Putamen, P; Frontal lobe, FL; Primary visual cortex, PVC; Temporal lobe, TL; Dorsolateral Pre-frontal, DPF; Orbitofrontal cortex, OFC; Motor cortex, MC; Reticular formation, RF; Supplementary Motor cortex, SMC; Thalamus sensor crossover, TSC; Tempo-parietal junction, TPJ; Hippocampal memory encoding, HME

D9. Creativity and humor

ACC Anterior Cingulate, AC; Temporal sulcus, TS; Caudate nucleus, CN; Central executive, CE; Left Amygdala, LA; Pre-frontal cortex, PFC; Motor cortex, MC; Posterior Temporal region, PTR; Motor cortex, MC; Ventral brain stem, VBS; Cerebellum, C; Pyramidal tract, PT

(continued)

FIGURE 7.18 (Continued)

D10. Personality

Supplementary motor cortex, SMC; Creativity time crystal, CTC; Personality time crystal, PTC; Insula, I; Cingulate, C; Striatum, S; ACC Anterior Cingulate, AC; Temporal sulcus, TS; Thalamus, Th; Amygdala, A; Reticular formation, RF; temporoparietal junction, TPJ; Temporal lobe, TL; Desolation prefrontal, DP; Orbitofrontal cortex, OFC; Thyroid, Thy

D11. Love and pain

Lateral nucleus, LN; Medial ventral posterior nucleus, MVPN; Cingulate cortex, CC; Spindle cells, SC; Olfactory lobe, OL; limbic system, LS; Superior temporal cortex, STC; orbitofrontal cortex, OFC; Amygdala, A; Primary visual cortex, PVC; Fusiform gyrus, FFG; ventral tegmental area, VTA; nucleus accumbens, NA; Prefrontal cortex, PFC

D12. Learning dreaming defragmentation

Nucleus Accumbens, NA; Thalamus, Th; Caudate nucleus, CN; Reticular formation, RF; Central executive. CE; Prefrontal cortex, PrC; Putamen, P; Parietal cortex, PaC; Hippocampus, H; Broca's area, BA; Visual scratchpad, VSP; Mamillary body, MB; Amygdala, A; Parietal lobe, PL; ventrolateral preoptic nucleus, VLPO; Parafacial zone, PF; Nucleus accumbens core, NAC; Lateral hypothalamic MCH neurons, LHMN

O1. Fusion of elementary sensor into a single time crystal

Optic cranial nerve, II; trochlear nerve (IV); abducens nerve (VI); oculomotor nerve (III); Optic Chiasm, OC; Thalamus, Th; Optic Chiasm, OC; Optical nerve, ON; Retinal ganglion cell axon, RGCA; Lateral geniculate nucleus, LGN; Optical radiation, OR

Cortex domain, CD; Ventral posterior nucleus, VPN; Dorsal column nuclei, DCN; Ventral horn, VH; Dorsal root ganglion, DRG; Touch, Pressure, Vibration, heat/cold, pain, proprioception (muscle), S1-S6

Receptor cell nerve fiber, RCNF; Medial orbitofrontal, MOF; Tract, T; Amygdala, A; Lateral orbitofrontal, LOF; Hippocampus, H; Midbrain, M; Olfactory nerve, I

Facial nerve (cranial nerve VII), the lingual branch of the glossopharyngeal nerve (cranial nerve IX), and the superior laryngeal branch of the vagus nerve (Cranial nerve X); Nucleus of medullary tract, NMT; Taste area of somatosensory, TAS; Taste area of insula, TAI; Medulla, M; Gustatory Cortex, GC; Thalamus, Th; Epiglottis, E

Superior olivary nucleus, SON; Primary auditory cortex, PAC; Superior olive complex, SOC; Midbrain, MB; Lateral lemniscus, LL; Medulla, M; Inferior Colliculus, IC; Ventral Cochlear nucleus, VCN; Semicircular canals, SCC; Auditory nerve, AN; Medial geniculate body, MGB; Inferior colliculus, IC; Cochlear nucleus, CN; Auditory nerve, AN

O2. Proprioception cells, PC; Pressure, P; Spinal cord, SC; Conscious, Co; Un-conscious, UC; Temperature, T; Cerebellum, C; Thalamus, Th; Vibration, V; Pons, Po; Pain, Pn; Touch, T; Dorsal root axon, DRA; Ventral posterior nuclei, VPN; Dorsal root ganglion, DRG; Dorsal column nuclei, DCN

O3. Motion or movement, audio+visual + time

Registering an event

The putamen, P; Caudate nucleus, CN; dorsal stratium, DS; Substantia Nigra, SN; Globus pallidus, GP; Claustrum, C; thalamus; Optical nerve, ON; Cochlea, Coch; Auditory nerve, AN; Retinal ganglion cell axon, RGCA; Spinal nerve, SN; Cerebellum, Cr

O4. Homeostasis, thermal equilibrium

Pituitary, P; Gonad, G; Adrenal, A; Thyroid, T; Retina, R; Basal Ganglia, BG; Hypothalamus, H; Pituitary gland, PG; Melatomic level, ML; Suprachiasmatic nucleus, SCN

O5. How brain senses direction

Semicircular canals, SCC; Cochlear nucleus, CN; Auditory nerve, AN; Optic cranial nerve, II; trochlear nerve (IV); abducens nerve (VI); oculomotor nerve (III), Optic Chiasm, OC; Retinal ganglion cell axon, RGCA; Temporal lobe, TL; Thalamus, T; Parietal lobe, PL; Frontal lobe, FL; Corpus callosum, CC; Reticular formation, RF; Place cells, PC; Hippocampus, H; Grid cells, GC; Entorhinal cortex, EC; Head direction cell, EDC

O6. Temporal synchrony of entire skin cover, the feeling of body

Somato sensory cortex, SSC; Supplementary motor, SM; Raphe nucleus, RN; Motor cortex, MC; Spinal loop, SL; Insular Cortex, IC; Descending connection, DC; Analgesic chemical cycles, ACC; Parietal Cortex, PC; Pituitary, P; C-fiber, C-F; Thyroid, Ty; Brain stain, BS; Nerve fiber, NF; Dorsal horn, DH; A-delta fiber, AF; Thalamus, Th; Activate automated response, AAR; Spinal cord, SC; Hippocampus, (six nucleus), H6N; Shiver, S; Voluntary, V; Cerebral, C; Temporal lobe, TL; Thalamus, T; Parietal lobe, PL; Frontal lobe, FL; Corpus callosum, CC; Reticular formation, RF

(continued)

FIGURE 7.18 (Continued)

O7. Emotion

Medial ventral posterior nucleus, MVPN; Amygdala, A; the limbic system, LS; GABA secretion, GS; Hippocampus, H; Hypothalamus, H; Mamillary body, MB; Olfactory lobe, OL; Lateral nucleus, LN; Fornix, F; Thalamus, Th; Stria terminalis, ST

O8. Time

Medulla, MD; Pons, P; Internal muscle, IM; Internal muscle, IM; Vagus nerve, VN; Diaphragm, D; Ventral respiratory group, VRG; Pontine respiratory center, PRC; Dorsal respiratory group, DRG; Sinoatrial node, SN; Cardiac nerve, CN; Cardio regulator, CR; Hypothalamus, H; Basal ganglia, BG; Dopamin path, DP; Pre-frontal cortex, anterior, PFCA; Substantia nigra, SN; Pyramidal decussation, PD; Raphe nuclei, RN; Suprachiasmatic nuclei, SCN; Adrenal gland cortisol, AGC; Pineal gland melatonin, PGM; Ventro-lateral preoptic nucleus, VLPO; Retina ganglion, RG; Tuberomammillary nucleus, TMN.

learning are often thought of as a recipe that would never match biological intelligence (Nikolić, 2017). Brain oscillations could be read as biomarkers in neuropsychiatric disorders (Yener and Basar, 2012), biological rhythms should reset medical treatment (Halberg, 1977; Hildebrandt, et al., 1957), a shock of phase singularity could assist in medical treatment (Richter, 1960); phase reset curve that identifies a time crystal could act as diagnostic tool (Tass, 1999), i.e., time crystal as a biomarker (Yener and Basar, 2012); rhythm-medicine connections are studied for a long time (Reinberg and Smolensky,

1983; Schwan, 1957). For example, the mitotic cell rhythms in cancer show a typical behavior, that could be used as a marker (Sainz and Halberg, 1966). Figures 7.19 through 7.21 outline garden of gardens created by following Figure 7.18 fusion of time crystals produced by different brain components. We describe below the consciousness generating time crystals, which should generate in an artificial brain naturally in the prime symmetry designed oscillators.

D1: Consciousness: Even if human consciousness is disrupted, the spectral signatures carry information that

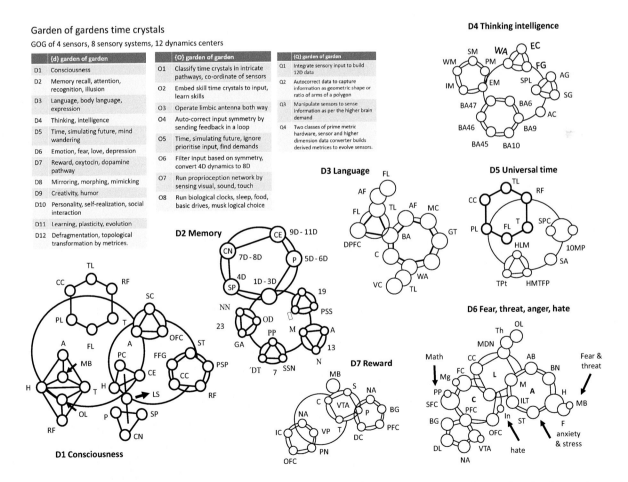

FIGURE 7.19 Garden of Garden's (GOG) of time crystal. Three tables show 12 dodecanion tensors that regulate fundamental conscious expressions of a human being, 8 octonion tensors that regulate sensory data integration and instantaneous decision-making, 4 quaternions that senses the input data. Note that all data structure is 11D, or dodecanion, but they are packed like four linguistic questions.

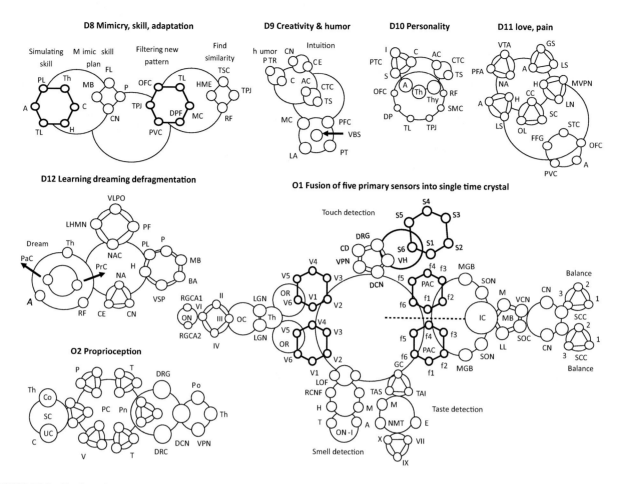

FIGURE 7.20 Dodecanions and octonions.

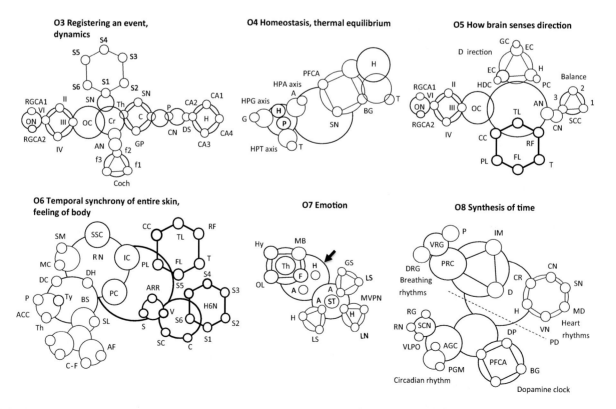

FIGURE 7.21 Octonions.

clocks are geometrically reorganized (Chennu et al., 2014). Coma is a state where certain higher-level integration clocks get shut off, several PPMs and time crystals run in an isolated manner (Plum and Posner, 1980). The rhythm of breathing edits the rhythms of the limbic system and that change affects the cognition of a human mind (Zelano et al., 2016).

D2: Memory: However, the dream to crystallize a memory (Ramirez, 2018) for uploading consciousness would require a complete time crystal model of the entire brain with all basic conscious features. While uploading the resetting the human clock (Winfree, 1991) would require editing the garden of gardens, i.e., hierarchical architecture of conscious clocks. One could state that keeping time could be a singular function of a human brain (Miller, 2000). When dust deposits on the strings of musical instrument, the clock slows down. Similarly molecular clocks undergo epigenetic changes, the clock architectures running conscious experiences (Figure 7.21) shows the sign of aging, i.e., clock diameters shrink or expand (Mittedorf, 2016). Twenty automated conscious experiences that run via 20, time crystals are running simultaneously, perpetually for entire life range milliseconds to seconds (Näätänen et al., 2004). These are the building blocks of the human brain (Nowakowski, 2018) path toward true neurogenesis (Gage, 2002; Konishi, 1990; Kuffler and Nichols, 1977). Memorizing anything requires high-frequency signal transfer (Kucewicz et al., 2014), shorter clocks of the time crystals build specific pathways. Memories follow prime 7 while grouping events (Lisman and Idiart, 1995). Memories make synapses strong, is it making information stable or circuit (Lisman, 1997), else one builds false memories (Loftus, 1997).

D3: Language: Probabilistic theories fail to explain the large-scale replication when a human comprehends a language (Nieuwland et al., 2018). Grammar is a delicate feature; its time crystal model has been described in detail following the transmission pathways for the nerve spikes in real-time. It resembles reservoir computing but many clocks work in parallel (Hinaut and Dominey, 2013). Mind and language are intimately connected by meaning and experience (Follesdal, 1975); only that was not known how the mind looks like (Sudkamp, 2006). General-purpose learning of language is fundamental to the brain's development (Phillip Hamrick et al., 2018). Using simply the temporal difference one could comprehend speech (Shannon et al., 1995). Any human language has embedded clock-like rhythms (Schwartz et al., 2003; Langner, 1992), auditory signals have music like temporal architecture (Cariani, 1999; Zatorre and Halpern, 2005) so it naturally processes musical scales (Elvira et al., 2006).

D5: Universal time: The sense of time of a life long journey defines human cognition (Levin and Zakay, 1989), voluntary timing intimately relate brain function, the architecture of time undergoes a subtle change (Wing, 2002; Winfree, 1970). Timelessness emerges when time crystal bridges singularity (Winfree, 1973, 1986a), wherein the pattern of phase compromises (Winfree, 1974, 1986b). Biological clocks set time for perception of events (Moore-Ede et al., 1982), but those clocks

are not isolated. Time travel inside our mind evolves the brain (Suddendorf and Corballis, 1997).

D6: Fear, threat, anger, hate: Every single thought, imagination, dream, fear, love and joy are geometric structures now (Cook, 1887). Only second to millisecond time domain is presented here but could be extended much deeper inside the brain. Hallucinations in the brain are composite, derived time crystal originating in the Paracingulate sulcus morphology (Garrison, 2015), one could edit that geometry in various ways. Paranormal activities in the brain tell a similar temporal structure (Radford, 2010). Social and ecological drivers have regulated the brain size evolution of various species (Forero and Gardner, 2018). Brain-to-brain coupling follows certain rules to create and evolve social structure (Hasson et al., 2012).

D8: Mimicry, skill: Mimicry is one of the finest expressions of a conscious being, its cognitive circuits were explored (Hale and Hamilton, 2016); emotion is exchanged between two conscious beings through mimicry, it is grounded on a concrete geometric feature of the periodic information exchange (Prochazkova and Kret, 2017). Playing music in the brain requires programming skills as a series of periodic tasks. Its limiting features are decoded in the geometry of temporal architecture of music and skills together (Hart et al., 1998). The brain detects musical pitch naturally (Meddis and O'Mard, 1997).

D10: Personality: Personality and mind are dichotomies of conscious experience (Lowen and Miike, 1982). Perception is neither discrete nor continuous in a time crystal structure, it is both (VanRullen and Koch, 2003).

D11: Love and Pain: Depression is topological, it spreads fractally repeating the same geometry (Shibata and Bures, 1974). Deep brain stimulation is often attributed to cure (Hall and Carter, 2011), such random rhythms are like hitting a stone in the dark, time crystal map of a human brain could let us editing a target clock from outside.

D12: Learning: Learning is conditional editing of temporal architecture (Howard et al., 2005) namely "temporal context" (Howard and Kahana, 2002). Prime number 7 is important as it sets the limiting number of learning variables (Miller, 1956).

D12: Dreaming: The time crystal for sleep might generate very slow clocks in the brain stem, not part of the components that run formal sleep time crystal (Merica and Fortune, 2000). Sleep has a slow traveling wave (Massimini et al., 2004). Sleep spindles are like local structures in a global time crystal for sleep, regulates slow rhythms (Ueda et al., 2001).

O1: Fusion of primary sensing: Self-organizing dynamics that define thoughts in the brain follow critical geometric features (Kelso and Fuchs, 1995) to integrate all kinds of information (Konorski, 1967); time crystal language enables that universal fusion.

O3: Attention: Neural correlates of attention have long argued for a connection of clocks (Reynolds and Desimone, 1999). Just prior to attention the mind is musical (Koelsch et al., 2002).

O7: Emotion: We consider bias is bad, but to converge and directional learning we need bias (Mitchell, 1980). Humans are emotional machines, programming it to reasoning is moral policing (Hume, 2000).

O8: Synthesis of time: Biological systems govern time by defining time (Goodwin, 1967) using phase shifts as key elements of architecture (Goodwin and Cohen, 1969). In the current brain model, time is not just fundamental to the brain structure but to the universe, such a discussion is out there for a long time (Girelli et al., 2009). Time was considered a scalar quantity in memory (Gibbon et al., 1984); however, now it's a vector quantity in the form of a tensor (Libet, 2004). Transcranial magnetic therapy could retrieve a memory (Rose, 2016). Remembering and consolidation of memory follow a unique pathway (Sara, 2000), often brain scans memory at very high speed (Sternberg, 1966). When neurons arrange more orderly, it's a better memory (Xue et al., 2010). Neuron defines its time (Ivry and Spencer, 2004) so is the brain (Miller, 2000).

sam-sam id yuvase vrsann agne visvany arya a |idas pade sam idhyase sa no vassuny a bhara ||
sam gachadhvam sam vadadhvam sam vo manamsi janatam |deva bhagam yatha purve samjanana upasate ||
samano mantrah samitih samani samanam manah saha cittam esam |samanam mantram abhi mantraye vah samanena
vo havisa juhomi ||
samani va akutih samana hrdayani vah |samanam astu vo mano yatha vah susahasati || RigVeda
"Sam" means resonance, Vedas suggest that an eternal vibration called "Bramhan" came out of Golden Womb (Hiranya Garva), Chitta~ means consciousness feelings, Manan~ the vibrations of consciousness

It is all about resonance, our consciousness feelings resonate with that of the supreme consciousness. When the touch of all connected hearts vibrates together, those vibrations confluence when the passions in every single consciousness units run deep inside their heart, when originally the vibrations of consciousness reside. May you move in harmony, speak in one voice; let your minds be in agreement; just as the ancient gods shared their portion of the sacrifice. May our purpose be the same; may we all be of one mind. In order for such unity to form I offer a common prayer. May our intentions and aspirations be alike, so that a common objective unifies us all.

8 Hinductor Not Memristor— Synthesis of Atoms and Crystals Made of Magnetic Light

8.1 IT IS RESISTANCE FOR VORTEX ATOMS BUT NOT FOR ELECTRONS

The 3D topological pattern of a very weak magnetic field plays a fundamental role in governing the formation of a star (Pattle et al., 2018). Weak magnetic fields are there everywhere in the biological system, now even the magnetic field around a single atom has been measured (Willke et al., 2019). The electric field has a monopole as positive or negative charge; one can isolate positive and negative charge separately. The magnetic field does not have such a provision. Once the debate regarding the supremacy of electric field over the magnetic field was settled in the 1820s by Lord Kelvin, no one questioned why the current flow that is fundamental to the electric field is required to generate a magnetic flux, why not the phase flow alone builds a magnetic flux that is an integral part of a magnetic field. There are plenty of reports of generating magnetic flux by flowing current, it is a must to happen, regulating magnetic phase, spin flow, magnetic wave flow, everything could be easily feasible following the guideline of Maxwell's law (Matsukura et al., 2015; Stöhr et al., 2009; Lottermoser et al., 2004; Henke et al., 1981; Figure 8.1a). There has been an attempt to change the electric polarization vector or phase using magnetic phase (Ishiwata et al., 2008).

Electric field controlling the magnetic field could easily be realized. However, the electron would not flow, yet the magnetic flux would flow appeared not possible unless one discovers the magnetic monopole. Instead of a magnetic monopole, it is easy to build, magnetic vortex atom and do similar work, as it is done for Hinductor.

So, an electric charge has to move, and only then a magnetic field is generated. The quest for a magnetic monopole started long ago (Dirac, 1931), when Dirac argued for a unique magnetic monopole. Giant accelerators tried to find a monopole fundamental to the universe, but failed. Now, the lattice defect-induced vortex of a magnetic field is found to mimic the magnetic ring. Vortices could happen in a small excitable media like a time crystal (Winfree, 1986b, 1990, 1994). Spiral features can take us to use the topological feature, and that enables using the negative refractive index (Pendry, 2004). Supramolecular springs and ratchets have an intimate relationship (Mahadevan and Matsudaira, 2000), and entropic effects do play a fundamental role in the composite dynamics. It should be noted that for computing or decision-making, this might not be essential. One recent course of historic development started when rotational angular momentum was introduced to light; here, it would be a journey to the universalization of angular momentum of light.

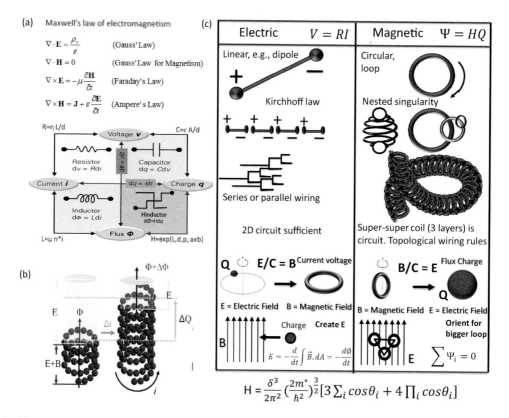

FIGURE 8.1 (a) Maxwell's equations (top) and a table for missing fourth-circuit element, a new symbol for H-Hinductor device different from Memristor. (b) The schematic of the device for the fourth-circuit element, change in the stored charge reflects in the change in produced magnetic flux. (c) A table is comparing the world of electronics and the world of magnonics, if magnetic particle an equivalent to charge is found using vortex atoms proposed by Kelvin.

Why do we need fourth-circuit element: Memristor is not the fourth-circuit element: No brain component should use significant energy; if we do, then the integration of 100 billion devices would require megawatts of a power supply. Considering all critical limits, one could find that we need to confine power expenditure of entire human brain hardware within a few kilo-watts and, therefore, the power consumption in a single neuron equivalent device cannot be more than a few nano-watts or microwatts. Such a frequency-fractal processor device was proposed in 2014 (Ghosh et al., 2014a). One has to invent a fractal-processing device, that generates time crystal with quaternion, octonion, and dodecanion features, and for that purpose, we start with the fourth-circuit element. In the 1870s, three basic electronic devices the capacitor, inductor and resistors were known. Leon Chua was the first to notice that by symmetry argument magnetic flux and the stored charge should be linked by an unprecedented device, he called it memristor (Chua, 1971; Figure 8.1b). Memristor is not a fourth-circuit element; any device that created the biggest sensation in the world in 2008 (Strukov et al., 2008), to realize cannot generate magnetic flux is not a fourth circuit element. Chua himself wrote in 1971s seminal paper that all memristor properties could be derived using inductor L, capacitor C, and resistor R. Hence memristor, even one goes by the view of the main proposer is not a fundamental circuit element. Moreover, since all properties of memristor,

memcapacitor, and meminductor (Di Ventra and Pershin, 2013) could be derived using conventional circuit elements, one cannot consider this as a fundamental element at all. In contrast, no combination of primary circuit elements could generate the electronic properties of the fourth-circuit element. Therefore, it is unique. The missing memristor remains unfound despite thousands of follow-up articles.

The paradox of fourth-circuit element: Why memristor could be an inductor but never a fourth-circuit element: The proposal for the fourth-circuit element has a paradox (Figure 8.1b). If an electric current flow for linearly changing the magnetic flux, then it is an inductor, no more a fundamental circuit element. On the other hand, if the charge does not move yet, a magnetic flux is produced, and then Maxwell's law described in Figure 8.1a violates. Universalizing Dirac's classic 1931 charge-flux quantization holds an answer to this paradox—if Dirac strings are spiral, magnetic monopoles could survive together even in ambient atmospheric condition. Resolving the paradox using a generic magnetic pole generator would open the door to building usable practical devices (Figure 8.1c). We suggest getting out of the intense quest to find a universal magnetic charge as a fundamental particle and limiting the quest to finding a free particle-like magnetic entity $(B \gg E)$ that does not require an accelerating charge and does not require an external massive magnetic field to observe. Further qualification of such free particle-like structures would be (i) self-assembly of these magnetic particles would generate more magnetic field like an electric charge, (ii) generation of electric charge by rotation of magnetic particle, and (iii) rotational direction of a field like spin.

Knots of darkness generated by a coil, a supercoil, a super-super coil is tubule-morphic, not neuromorphic: When we look at a DNA molecule vertically from the top, it looks like a flower. The origin of flower shape is a twist in its spiral. There are two spirals superposed on each other, in summary it is a spiral of a spiral. The twist creates a mixed defect of an edge dislocation and a line dislocation on the cylindrical surface. DNA is not alone; α-helices in the protein, tubulin dimer, microtubule nanowire, several protein complexes hold such a unique twist in the spiral. One of the most significant advantages of using a helical geometry is that due to the geometric shape alone the spiral pathway quantizes signal, thenceforth filters noise. The ultra-low pitch spiral design allows several alternative spiral channels to operate simultaneous signals, the quantization of energy happens. Naturally, quantum effects generate in an ambient condition (Atanasova and Dandoloff, 2008). When we change the surface design, the optical signal falling on the surface would diffract and interfere; the resultant 3D pattern would carry a signature of that surface. One could see dark lines floating in the 3D space. In the optical vortex studies, these studies are done frequently. We state explicitly that in these lines, electric field $E = 0$. Really? No one ever checked whether or not Maxwell's equations are at all valid inside those dark lines. Someone should do it. We leave that and introduce the coil of coil of coils, or super-super coil. The α-helices of a tubulin monomer form a coil which generates three kinds of diffraction patterns.

Imagine a cylinder. On its surface, through an imaginary spiral, a set of dark lines along with dark rings are moving clockwise. On its surface through an imaginary spiral, hair-pin-like shapes and dark ring are flowing counter-clockwise. On its surface longitudinally, super-super coils of dark lines (Nye, 1983) are moving. Thus, three coexisting dynamics are equivalent to three Bloch spheres. Each Bloch sphere has a great circle with singularity points on its perimeter. The singularity points in a circle constitute a geometric shape. Thus, DNA or α-helices like spirals are the smallest unit of information. Supercoil of α-helices is a tubulin dimer, and the super supercoil of α-helices is a microtubule. Thus, microtubule nanowire is a triplet of coils, and it is the smallest unit of the tubule-morphic device.

The background history of the magnetic vortices: If a semiconductor sandwiched between a pair of metal electrodes exhibits hysteresis and bistable states, one could flip between the two resistances like a switch, then it is memristor. Since current flow is allowed in Chua's memristor, one could make an inductor-like device and generate magnetic flux, too. However, if one remained strict to the definition and was determined to build a device that does not flow any current yet produces a magnetic flux, then several critical research fields of contemporary physics need to be bridged. However, if successful, magnetic resonance would enable using the negative

index of refraction (Valentine et al., 2008), super-lensing (Smith, 2004) and cloaking (Schurig et al., 2006). Magnetic wireless communication is superior to electrical communication as electrical signal does not diminish while passing through water, moisture and metal.

Moreover, a magnetic field penetrates most insulator materials. The barrier should have another specialty, an electromagnetic lensing effect that converts emitted electromagnetic burst into a magnetic one. A barrier or boundary of a key device element might play key roles in building noise-resistant protection (Figure 8.2a). There are multiple routes (Karaveli and Zia, 2010). Charge storage is controlled by resonance frequencies of input ac signal oscillating them coherently initiating an antenna like the radiation of magnetic ripples (Figure 8.2b, c). Even though the charges do not move out of the well, due to an extremely insulated matrix outside, they oscillate physically as a group (Figure 8.2c, d). Thus, Maxwell's law is not violated. Since the current does not flow, still a magnetic flux is produced; therefore, the device cannot be tagged as an inductor. The linear relation between stored charge and produced magnetic flux suggests that it is a fourth-circuit element, but one does not get any reasonable current-voltage characteristic as the memristor demands. In this fourth-circuit element, the electrical resistance is infinity, but another resistance, the ratio between flux and charge is 0.001, i.e., very low. Following suggestions of Chua, Sahu et al. have given their device a distinct identity, namely Hinductor, and assigned it in a different class than a memristor.

Electric vector is 10^4 times more than the magnetic vector: The model explored for the fourth-circuit element, is a helical wire made of dots that are a special kind of quantum well or traps. The well is so designed that the electric vector is perpendicular to its wall's surface and magnetic field vector is in the plane with the wall. The density of states of the virtual wall's 2D surface lenses the electric vector and adds a part of it to the magnetic vector. One to two orders magnetic field enhancement by confining electrons and then accelerating them through the depleted field created by surface topology is a reasonably well-understood phenomenon (Shan et al., 2001; Johnston et al., 2002a, 2002b; Figure 8.2c, d). A subtle change in the stored electron would change the topology of surface depleted field, thus, regulating the phase correlation of coherent emission from this well like nano-antenna (Biagioni et al., 2012; Michler et al., 2000). Both in the classical and the quantum systems, the emission survives only for a few picoseconds, until the coherent resonance burst survives. The phase space interaction suggests the formation of a spiral path (Figure 8.3a), optical axis ensures the purity of two interfering waves (Figure 8.3b, d). So, we arrange multiple wells in a 3D spiral arrangement (Empedocles et al., 1999; Figure 8.3c) so that the magnetic flux acquires a stable experimentally measurable standing wave. Interference on the spiral path could depict the signature of magnetic radiation distinctly from the electrical contribution (Freed and Weissman, 1941). Arrayed structure assists in enhancing the magnetic field (Cui et al., 2016). Imaging the magnetic flux has been a challenging task (Vignolini et al., 2010). Normally, magnetic vector interacts

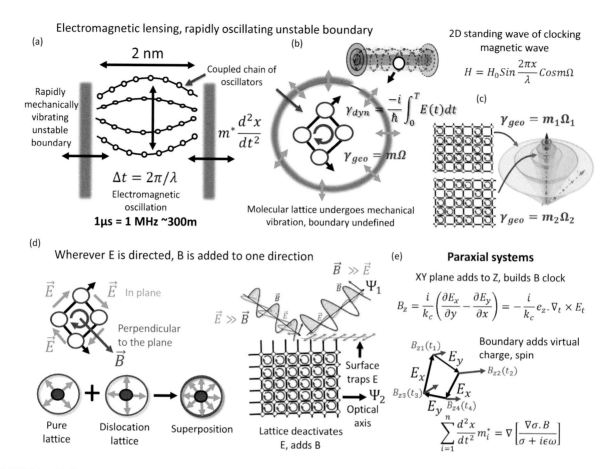

FIGURE 8.2 (a) How a rapidly vibrating membrane could hold a large waveform. Coupled chain of oscillators could generate a collective oscillation, quantum or classical. (b) Molecular helical array forming a vortex or nanotube builds a lattice on the nanotube surface (left). Here m is the magnetic quantum number, Ω, solid angle enclosed by the rotating field vector. Dynamic and geometric phases are isolated. (c) Periodically clocking wavefunction generating a geometric phase builds a standing wave. For different composition of charge distribution on the surface different cones are produced. (d) Magnetic vector (red) is perpendicular to the plane while electric vector is in plane with the cells. Dislocation rotates the plane of polarization and superposition of both should create a mixed polarization light, does not happen. The protocol to demonstrate B>>E from E>>B is shown. (e) How paraxiality works, electric E vectors transfer momentum to the magnetic B vector.

10^4 times less than the electric vector with the electrons of a sensor, so detecting a magnetic flux in the light has been a challenge, so metamaterials are used where conducting rings sense the magnetic flux rapidly (Schirber, 2010).

Topologically protected spin textures could be resonantly excited and its elliptical dynamics which contain anticlockwise and clockwise modes could be tuned by applying different frequencies of the microwave field (Figure 8.3a, e). The conversion between these two elliptical modes is achieved by a transition to linear vibration (Jin et al., 2017). One could do such phase modulation by applying THz pulses generated by surface depletion field. Infrared signals of 5–6 THz (~300–320 K, little above room temperature) get frequency modulated to the lower THz regime 60–300 GHz, by the topology of the surface.

Do not look for a superconductor, look for super insulator: Since charges are not moving out of the quantum well, as if all the wells are floating in an infinite resistance or extremely insulating matrix, the charges formed inside the well remain trapped (Figure 8.2b). In this scenario, the differential conductance, the key to almost all major electronics theory and experiments do not work,—sending a current or applying a voltage, returns no output. However, if a new differential, the relative variation of flux and charge, or differential flux-charge is measured, in some systems where its value is very low in spite of a nearly infinite resistance, a new type of communication could be found. Dimensions of differential conductance and differential flux-charge are the same. However, the physical significance is very different, while the first, is governed by collision and scattering of carriers (e.g., electrons, quasi-charge), for the later, the topology of charge-trap centers in the matrix of extreme insulators is the key (Figure 8.1c). While avoiding the collision and scattering have guided the ultra-low, noise-free quantum and classical adventures, the differential charge-flux has no such weakness. It means, one could in principle measure, quantum effects in ambient atmosphere, just like we do it in quantum optics. One has to measure three hallmarks of quantum,

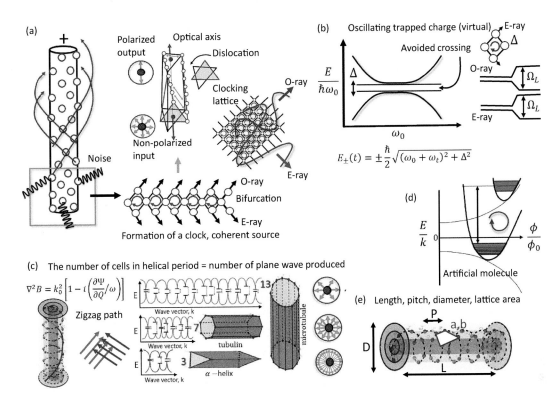

FIGURE 8.3 (a) All vortices or cylinders have a polarization direction. The noise is converted into a spiral organized signal and photons undergo O-ray and E-ray split triggered by dislocation, these are two coherent sources. Spontaneous generation of the polarized signal is shown. (b) Avoided crossing Δ splits virtually trapped charge. (c) The scheme shows α-helix with three centers of interaction per cylinder, tubulin with seven circuits or seven compositions of helices and finally 13 protofilaments. Polarization chart for 3, 7, and 13 are shown. The number of cells in the cross-section of a cylinder or vortex determines the number of the plane-polarized signal produced by the device. (d) Movement of system point in the configuration space of an H device. Here the configuration space is created by relative magnetic flux and the energy is normalized with respect to the wave vector. (e) Only four geometric parameters regulate the H property, Length, L, pitch P, diameter D, and lattice area a, b.

fractional quasi-charge, the ripples of a quantum well and the Fermi velocity (Lykken et al., 1970) which were thus far measured using the differential conductance, using differential flux-charge. Using them Sahu et al. estimated the magnetic flux and charge associated with the quantum capacitance and quantum inductance of a tubulin monomer; and accurately determined that no kind of wiring works for quantum elements. Sahu et al. verified the reliability of this new differential by building a theory, how quantum well emitted the magnetic rings interfere to form standing wave and then live to image the standing wave of magnetic flux. Its origin is a quantum antenna driven by oscillating charge trapped within that shapes the field correlations, emits a magnetic field at resonance, even if a single pair of tori interfere, the spiral standing waveforms.

8.1.1 Quantum Non-Demolition: $e - \pi - \phi$ Quadratic Sensors

Quantum demolition a historical perspective: Making the quantum analogs of classical electromagnetism laws, and everything that we find in the classical world has been a faithful practice for the quantum researchers for a long time. No attempts were ever made to see what happens if the resistance that destroys the quantum coherence is increased critically arresting the motion of an electron. Critical environments often lead to quantum effects, but the laws prohibited us from seeing anything by arresting the electron flow critically. In a quantum world a quasi-particle wave can interfere without flowing electrons, the waves edit phase to exhibit unique topological resonance by quantum interference. Despite several attempts to safeguard information during measurement, it is still an open question whether it is at all possible to measure information without modifying or destroying natural weak vibrations. In the last 50 years all efforts to evade quantum demolition were optical (Braginsky, 1980; Sewell et al., 2013). Earlier, isolation was a route to make a device silent, e.g., "flux capacitor," which was really a one-way optical loop for helping isolate sensitive quantum equipment.

Resistance-less paths for the clocks to transmit and doing mathematics: vibrate the cavity boundary for $e - \pi - \phi$: Flowing a charge through the resistance fewer paths has been the key to survive a quantum effect. It means that we do not want the carrier, be it electron, quasi-particles or photon to move out of the cavity, rather, remain trapped in the cavity forever, do not even leak. So, we want super-insulators where the Fermi velocity of carriers is very low and which are vibrating mechanically. Due to the coupled vibrations among the cavity boundaries we get a mechanical rhythm. Periodic or clocking oscillation generates a magnetic field or a loop that absorbs more electric vectors in its plane. Thus, the necessity of clocks delivers a coupled electrical mechanical and magnetic rhythm leading to $e-\pi-\phi$. Similar to the electronic route, an alternate phase route could lead to a plethora of technologies to minimizing the resistance for the phase flow along the path, so that the magnetic field or magnetic flux carries the quantum information with null resistance just like the cooper pairs face in a superconducting material.

The development of a new kind of $e-\pi-\phi$ sensor: Normally, during the measurement of information in a material, the pump-probe technique is used. Even the day-to-day multimeters we use, it pumps a large bias to the device under test and then measures its response. Instead of pumping external energy to amplify enough to sense its response, one could envision a new thermo-magneto differential sensor that is tuned to interfere and modulate the geometric phases of magnetic flux waves produced at the atomic scale to the micro scale. The new culture of phase-based magnonics that might replace electronics in the coming future would rely on the coupled geometric resonance $e-\pi-\phi$. Imagine a futuristic world where instead of the multimeter that measures current-voltage we could have a cheap flux-charge meter. Most importantly we would not be pumping huge energy and then measure the outcry of the device under test. Synchronizing with the $e-\pi-\phi$ coupled oscillation requires a very small amount of energy that runs a search mode to find the right $e-\pi-\phi$ condition in a device, if that is done, then the new generation $e-\pi-\phi$ sensor would read the flux-charge properties like the hinductance H without destroying the basic parameters. Instead of relying on the conventional electrical parameters like current or voltage, three simultaneous recordings allow reading the rate of change of flux as a function of charge directly in a pure form. Literature is rich with infrared sensors (Peng et al., 2015) and magnetic flux sensors (Shi and Li, 2011). When a fourth-circuit element Hinductor, H explores the geometry of a device so extensively, noise harvesting becomes natural and heat pipe effect often dominates the noise to signal conversion (Wang and Gundevia, 2013).

8.1.2 Rapidly Oscillating Cell Membrane Holding a Long Wave

Classical waves can pick up geometric phase, exclusively preserved for quantum: Imagine a core-shell structure, where the boundary that makes a cavity have a different dielectric property than the filling material inside the cavity acting as a dielectric resonator. Boundary and inner structure vibrate very differently (Figure 8.2a). Hinductor coils (Figure 8.2b) are like single molecules, and the boundaries vibrate very slowly (kHz) than its internal core material, which vibrates at MHz or GHz, i.e., radio wave or microwave frequencies. Thus, boundary and internal structure get energetically isolated. Even in a noisy environment, the boundary isolates unwanted input, makes internal rapidly vibrating structure into an adiabatic system. If the boundary vibrates periodically, then the wave function representing the internal structure would acquire a geometric phase in addition to a dynamic, phase. This wave could be a classical wave, so, at room temperature, ambient environment an electric, magnetic or mechanical wave could acquire geometric phase that was thought to be a prerogative of a quantum world (Lauber et al., 1994). How do we prove it? Check Ahranov-Bohm effect, i.e., where electromagnetic interference carries quantum information (Mead and Truhlar, 1979). In summary we get dynamic and geometric phase added to the electromagnetic signal passing through a molecular structure.

Electromagnetic, mechanical, and magnetic vibration of the membrane: Since 1930s, routinely measured protein vibrations (Vollmer et al., 2002; Zhai et al., 2000; Hanham et al., 2015) have failed to become a marker to detect the onset of a disease early, as subtle measurements flip a large number of its vibration modes. As protein vibrates at many different frequencies at a time, no probe exists to read multiple energy domains at once, classical markers cannot instantly detect fractional changes in the resonance frequencies caused by mutation or virus attack. To measure quantum capacitance and quantum inductance are quantum markers of geometric resonance of proteins are wavelike, means, originates purely from wave reflectance, transmittance and interference. So, a minute change in protein's local structural symmetry makes a significant change in phase. Earlier Sahu et al. reported significant changes in the resonance frequencies of microtubule as it becomes cancerous. Now, the same effect is read quantum mechanically to enhance the energy detection resolution to 10^{-21} eV, a new $e-\pi-\phi$ probe detects the subtle changes in the internal degrees of freedom, yet momentum conservation yields Planks constant repeatedly. Measurement is self-fault tolerant hence fit to dip in a living cell for the quantum detection of a disease at the very onset.

8.1.3 Paraxial Systems

A paraxial system means a transverse component of the electric vector contributes to the magnetic part of an electromagnetic wave stream (Figure 8.2e). Electric and magnetic vectors behave uniquely in a paraxial system (Berry, 2004). The device is a doublet or triplets of identical concentric cylinders of diameter D, length L, made of spirally arranged m nano-dielectric resonators in one helical period, with a total Ω number of quadrilateral lattice cells, each with sides a and b (Figure 8.3a). The spirals with pitch P, are further twisted by an angle δ. One cylinder with a 2D array of dielectric nanoparticles amplifies both E and B together, while two cylinders, each with an edge and screw defects at a relative angle α between the layers, create two distinct spatially isolated vortices, one electric ($E \gg B$) and the other magnetic, $E \ll B$. For three layers, the knots change shape with time. Noise shifts the coils relative to each other, it tunes α, the angle between the overlapping dark lines, since coils are identical, noise initiates a beating which filters out E mixed in the dark B lines. A Fabry-Perot interferometer like set up was used to amplify B by infinite repeated reflections between the plane wave bender and the vortex lens. To see the ripples of magnetic vortices, move away from optical vortex, put a large $12'' \times 12''$ magnetic film to find the invisible magnetic vortex location, rotate the sample in a quartz cell to fine-tune the intensity of B in the vortex, keep the solution opacity at 60%, project it too far >5 m, do not use a lens to focus the vortex atoms, most spirals and fractals would give such vortex crystals.

The device operation follows two simultaneous steps. To write or erase a dynamic feature one has to send an ac signal $\hbar\omega$ to resonantly vibrate the surface cells, then shine a monochromatic polarized light to read the encoded information as a crystal of magnetic vortex. Quantum systems like vibrating molecules or nuclei, do not have fixed, but vibrating boundaries with momentum $m^* v = \nabla\varnothing$. This vibrating boundary adds a geometric phase $\varnothing$ to the trapped signal $\hbar\omega$. The trapped signal and the direct input signal interfere and give rise to Ahranov-Bohm effect at an ambient atmosphere. A very low-frequency signal ω_0 is born, the wavelength is far greater than the cell size ($a, b \ll \lambda$, paraxiality). Two events happen in parallel. The cell delays signal, oscillating walls build $\vec{\psi}_{trap}$, then emits $\vec{\psi}_{Osci}$, finally builds $\vec{\psi}_{mag}$ globally on the cylinder surface. Since $(a, b \ll \lambda)$, the deviation θ of $\vec{E}$ vector is very low, i.e., $\sin\theta/\lambda \to \theta/\lambda$, the signal $\vec{\psi}_{mag}$ follows different spiral paths (anisotropy), distinct paths activate N vibrational modes. Thus, input high-frequency signal $\hbar\omega$, generates N very low-frequency signals ($\omega_1, \omega_2 \ldots \omega_i.\omega_n$), as G-band, which interferes and creates distinct local density of states on the cylindrical surface, L, P, D and area ab sets the boundary condition for interference. Thus, in-plane $\vec{E}$ vectors are consumed by the cells but the coupled G-modes with wave vectors k_i, k_{i+1} modulates $\varnothing$ as $\tan^{-1} k_{i+1}/1 + k_i$, a precession $\pm\varnothing$ to the $\vec{B}$ vector in those plane waves whose $\vec{E}$ vectors lie along the cell plane (Figure 8.2e). For a mode i, all $\vec{B}(x, y) = \vec{B}\cos\varnothing$ vectors perpendicular to the cylindrical surface delivers an average sum of magnetic flux $\langle \vec{B}(x,y) \rangle$. Thus, plane-polarized light shined on a cylinder acquires coupling coefficient CC_i of all vibrational modes activated by $\hbar\omega$, in its B vector. Modes are visible in the quantum tunneling images.

8.1.4 ANISOTROPY AND AVOIDED CROSSING

Planar chiral structures automatically develop an anisotropy in the propagating electromagnetic, asymmetric structures even split the electromagnetic wave (Fedotov et al., 2006).

Engineering artificial dressed states: Say we draw a set of parallel lines (Figure 8.3a). Then, another set of parallel lines crossing the other. At the junctions we keep an element that stores charge or delays the motion of charge passing through the matrix. If we build a cylinder by arranging and

spiraling balls, we can create several cross-bar patterns on the surface. The most interesting part is that the flowing charges dwell for a longer time along the parallel paths, it is analogous to trapping a charge. We do not restrict what would be the charge, could be a quasi-particle, or an elementary particle, it is the designer's choice. The question is to control these states. The energy between two levels

$$E_\pm = \frac{\hbar}{2}\sqrt{\omega_0^2 + \omega_t^2 + \Delta^2} \qquad (8.1)$$

with an energy gap Δ (Figure 8.3b). On a spiral cylinder, if there are different types of parallel paths, each set of paths have a distinct resonance frequency ω_0, and applied energy E_{ac} generates a signal ω_t. Here n number of quanta is exchanged continuously between $E_\pm$, even without any metallic layer $H = E_{ac}(n_1^2 + n_2^2 + \ldots + \frac{2}{i}n)\ldots$ **H1**. Thus, at around the resonance frequency, we get dressed states, i.e., the wave vector $k = 2\pi/\lambda$ oscillates periodically ($\cos k$), following

$$E_\pm = \sqrt{2(1 - \cos k)} \qquad (8.2)$$

hence we get a periodically oscillating band gap (Figure 8.3d), the wave vector K^* switches between elliptical and circular polarization. Consequently, the basic definition of cavity resonance or dielectric resonance is not valid here. External noise activates the cells in the crossbar pattern, but external ordered signals disrupt natural communication between clocks. Even using the term resonance is not valid in strict terms. The crossbar trap's dimension d does not decide vibrational clocking frequency of trapped charge, in fact, $\lambda \gg d$. The particular condition, $\lambda \gg d$, transverse components of electromagnetic wave contributes to the longitudinal component. The light or electromagnetic wave of any frequency range would experience magnetic or electric parts feeding the other, depending on polarization direction.

Rupturing the band gap: In the have topological insulators, or semiconductors the carriers transmit through specialized paths, the valence and conduction bands do not remain flat mostly separated by an energy gap, they bend to touch each other. These are called broken bands, where the fermi surfaces of the valence and the conduction band are curved so much that they are ruptured (Figure 8.3c). The critically curved regions may extend valence to conductance bands, even make contacts (Dirac points). These contact paths are topological routes, a carefully crafted path on the materials through which the changes move, or simply send phase information without any physical communication. Such points are called avoided crossing. Multiple paths are shown as capacitor like gaps in the broken band. For tunneling probability of being zero in the avoided crossing, $P = \exp(-2\pi^2/4\upsilon) = 0$, or $\upsilon = 0$. In a crossed parallel pathway, the trapped charge, does not come from anywhere, whatever few available already there, reflect back and forth between E_{F1} and E_{F2}, with an energy gap Δ. During repeated reflections, a charge remains occupied, one could use an occupation function used by Esaki and Tsu in their historical work in the 1980s,

$$N\left(P_z\right) = \ln\left[\frac{1+\exp\left(\left(E_{F1}-E_{F2}\right)/kT\right)}{1+\exp\left(\left(E_{F1}-E_{F2}-\Delta\right)/kT\right)}\right] \quad (8.3)$$

8.2 FUNDAMENTAL STRUCTURAL PARAMETERS OF A FOURTH-CIRCUIT ELEMENT—HINDUCTOR

Most magnetic field studies are performed by applying the field externally, here we are arguing that there is a possibility to generate the magnetic field naturally, in a non-magnetic material. Shaping the field using a quantum antenna is gaining interest as tiny antennas need to be arranged in a geometry that looks like a superposition of wave functions. At resonance, the product of the wave functions of tiny antenna systems generates a path made of singularity points in the phase space. This is similar to a quantum system where we take the product of wave functions, not sum to represent a system. Functional groups in a spiral array could act like a quantum antenna that reshapes the electromagnetic wave input into a torus like a packet of magnetic flux and explore how multiple antennas to be placed in a 3D space so that their quantum coherence does not de-phase. There are three key factors in the design of a fourth-circuit element, H (Figure 8.4a–e). First, the barrier heights of a quantum well depend on the quantity of stored charge. If the height is increased, a trapped wave has to reflect many more times between the walls before leaking through. It is like adding a greater number of turns in a coil increasing the magnetic flux. Second, to squeeze the density of states of a quantum well within a narrow gap, the barriers have different resonance frequencies around the Fermi level of the well. As an electromagnetic wave passes through a high-density region, it converts a part of its electric field to the magnetic field so the emission is magnetic in nature (electromagnetic lensing). Third, without wiring physically, we place all quantum antennas in a 3D space on the path where emission from a single well experiences a singularity at resonance. It enables an array of a quantum antenna

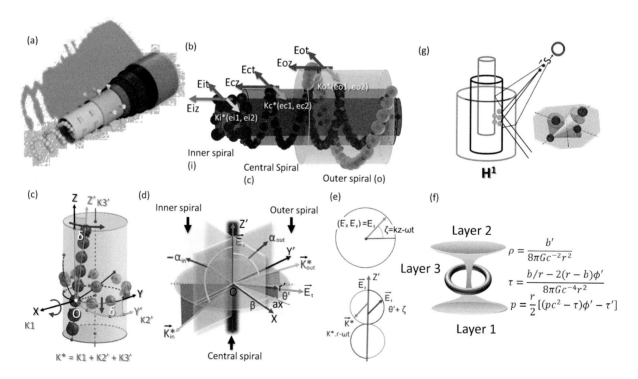

FIGURE 8.4 (a) The commercial prototype of an H device. A spiral is connected to an array of antenna and charge flux measurement probe. (b) The basic device structure with three concentric spiral layers. (c) Twisting of the spiral, a fundamental requirement to build edge and screw dislocations as the one DNA or microtubule have. (d) The cross-section of the device shown in panel b, three spirals are invisible for clarity and we have put the vectors oriented at different directions. (e) The transverse electric vector E_t is shown that the device has to neutralize to get pure magnetic vortices (top), the effective neutralization vector k^* acts to neutralize E_t. (f) The ring of fields produced at the interfaces, i.e. between two layers, these rings or lines formed by interference could move like a particle. (g) Magnetic lines or rings produced at the junction between two layers could travel like optical vortices as shown here using a schematic.

to produce a phase space identical to a single antenna. Asymmetric quantum well means its barriers have different resonance levels, so, if the gap is reduced, the density of states of the well is squeezed, the input electromagnetic wave experiences a lensing effect. When the wave function of the well's density of states oscillates periodically, the absorption of an external electromagnetic signal leads to a rapid oscillation and emission of a coherent burst. Therein, lensing ensures that the transverse component of the electric field contributes to the magnetic field. Consequently, in the emitted wave, the magnetic vector dominates over the electric vector. If we insert one such an asymmetric well inside another, since, two squeezed levels act as the walls of a non-physical quantum well, the lensing and the noise filtering abilities increase manifold. The heights of the walls not only quantify the charge content but also determine the number of times the rapidly oscillating waveform would reflect between the walls before leaking through or emission.

How does the necessity to survive under noise leads to the nested capacitor and nested inductor: a precursor to a fractal network of resonators: Students are often ask "why is it difficult to take out an electron from a solid?" The teacher replies that an electron becomes a property of the entire solid. Taking it out means snatching it alone from all the neighboring atoms that make the solid. Similarly, when the wave packets nest, the noise has to break the nesting, and dephasing a single wave won't work. Thus, the nesting of waves does not have to be a quantum one, it could be a classical nesting of waveforms. Fourth-circuit element avoids dephasing not by silencing the noise, but by packing the phase of the constituent waveforms fractally while arresting the electron flow critically—super non-conductivity could a new route to scale free integration of wave functions under noise. The idea to arrest the waveforms by manipulating a materials reflectance and transmittance came in the 1980s when the concept of quantum capacitance (Luryi, 1988; Brown et al., 1989; Ilani et al., 2006) and quantum inductance (Liu, 1991; Begliarbekov et al., 2011; Asakawa, 2010; Wang et al., 2007) came into being. The concept of reflectance and transmittance is very classical. Imagine there is a capacitor. Inside that capacitor there is another pair of electrodes. Now the bigger host capacitor would lose a part of the transmitting wave to the small capacitor. Some of the waves would get trapped by creating a capacitor inside a capacitor. It could even be possible that when the waves get trapped in the guest capacitor, a part of it leaks. The leaked wave would be out of phase than the input wave to the guest. so the out of phase wave would be representing an inductor. Now, these nested capacitors and inductors would be a quantum device if the trapped energy levels are so closely spaced that the uncertainty in energy comes into effect. Quantum or not, nesting of capacitors and inductors make sure that there is a nesting of electronic resonators, and the cleverly designed array of resonant tunneling diodes could turn the medium into a super-super-insulator, by trap inside a trap inside a trap....

8.2.1 Periodically Oscillating Edge and Screw Dislocations

Figure 8.4b describes three concentric coils, which is a fourth-circuit element. We take one of the coils in Figure 8.4c. If we twist the coil, we get a screw $(\delta = 0; \delta = \pi$; left-handed and right-handed, respectively) and edge $(\delta = \pi/2)$ dislocations of zig-zag paths on the coil surface. Since the coils are twisted, a mixed screw and edge dislocations on the YZ plane of coil surface rotates the X-axis by an angle δ, so that the YZ plane turns after rotation by an angle δ, a precision is given by an angle $\varnothing$

$$\begin{pmatrix} r(x,y,z) \to r'(x,y',z'); \ k(k_1,k_2,k_3) \to k^*(k_1,k_2',k_3'); \\ \left(s(k^* \cdot r - \omega t) \to s + \Delta(k^* \cdot r - \omega t + \varnothing); \zeta = kz - \omega t\right) \end{pmatrix}$$

The process is explained in Figure 8.4d. Imagine that we are looking from the side of a three-concentric spiral system. In Figure 8.4d, OZ' is the central spiral, how the vectors orient with respect to the inner and outer coils are shown, the top view is schematically presented in Figure 8.4e. The amplified electromagnetic, em signal $(B \gg E)$ is perturbed by two dislocations generated signals, $e_1(e_1e^{i(s)})$ and $e_2(e_2e^{i(s+\varnothing)})$, interferes and creates dark lines $\Psi_n(r,t)$ made with pure magnetic fields $(E = 0)$ and electric fields $(B = 0)$. For magnetic lines, the real parts of $E_t = (E_x, E_y) = 0$, but the polarization $\Delta e = e_1 - e_2$ governs the knot shape shown in Figure 8.4f.

$$E_x = Re\left\{kr'e^{i(\theta'+\zeta)} + e_1e^{i(s)}\right\} = 0 \tag{8.4}$$

$$E_y = Re\left\{ikr'e^{i(\theta'+\zeta)} + ie_2e^{i(s+\varnothing)}\right\} = 0 \tag{8.5}$$

By equating the coefficients we get the dark spiral knots with the radius $\Delta e / k$, pitch

$$\Delta z' = \frac{\lambda \sec\delta}{2} \begin{cases} < 0 \ right-handed \\ > 0 \ left-handed \end{cases} \tag{8.6}$$

and the velocity $v = c\sec\delta$, we find s, θ', i.e., locations of pixels on the pure 3D magnetic lines $(E = 0)$, where, the path for dark line is given by

$$D(s,\theta') = kr'\cos(\theta'+\zeta) + e_1\cos(s) = 0 \tag{8.7}$$

At $\propto < 0$, the inner spiral generated knots perturb the central coil knots; at $\propto > 0$, the outer coil knots perturb the central coil knots. If the inner and outer coils are pumped with even 0.1 dB GHz noise, a very low 0.1 dB input signal

$\hbar\omega$ generates a high intensity of B lines. Since three layers are identical, G-mode frequencies are very close, it triggers beating, B knots get sharper. How electromagnetic wave falls on the surface and creates interference is shown in Figure 8.4g.

8.2.2 Topology Regulating the Polarization

The B knots or magnetic free particle-like fields are output of two interactions, internal clock and perturbation clock (Figure 8.5a). The top cylindrical surface is a crystalline dielectric, conventionally the Pancharatnam-Berry phase of the electronic Bloch wavefunction represents the polarization (Resta, 1997). Here, the magnetic vectors interfere with creating vortex atoms, or artificial structures, they hold nested Hilbert space or polarization tensors in the 11D space. There is no electron flow, no hysteresis. A light with an electric vector rotating around the propagation axis develops an angular momentum, it is possible to change the direction of rotation and thus execute phase transition. Many years of research has led to a situation now, where, if the rotating light rotates again in a loop path, it is possible that multiple isolated loops interact and changes rotational direction, i.e., breaks polarization symmetry (Copie et al., 2019). Fresnel refraction from surface reflect charge distribution, Figure 8.5b shows charge distribution and optical reflection side by side.

Production of coherent and polarized signal from noise: a saga of twisted spirals: Noise to polarized light conversion is known (Heebner et al., 2000). Cylindrical spiral is a crystal, the diffracted electromagnetic signal does not originate from high electron density surface plasmons, but an incredibly strong insulator. Electrons cannot move from one site to another; they absorb noise depending on the cell configuration, but the energy they emit is quantized, thus monochromatic, or singular frequency. Most importantly, since spirals are twisted, similar to thermal noise birefringence (Reuter et al., 2012), the twist leads to an edge-screw defect that creates a pair of plane-polarized signals from the cellular

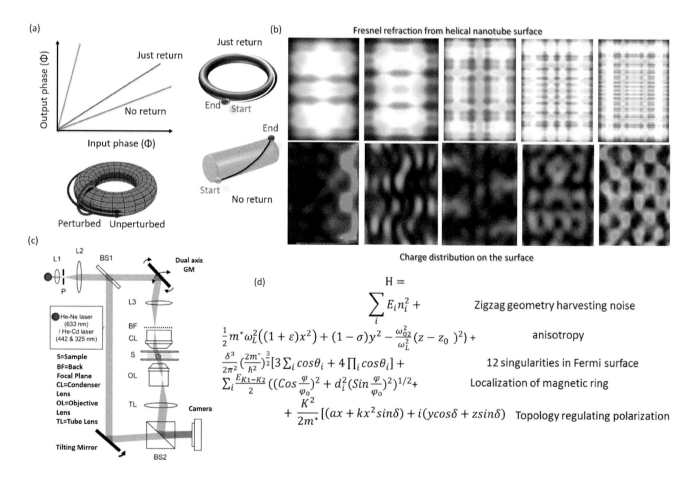

FIGURE 8.5 (a) In a periodically oscillating system, if a clock is disturbed it would change and return. The return path is traced using a cylinder and an important parameter is checked whether both the original clock and perturbed clock reach same time. Three-cylinder represents three plots for three possibilities. (b) Fresnels diffraction patterns on the microtubule nanowire surface (top), and the corresponding distribution of charge density on the surface measured for 8–25 MHz electromagnetic frequency pumping to the device. (c) Optical measurement set up for observing the optical and magnetic vortex. (d) Different parts of the Hamiltonian. Fresnel diffraction patterns, which are the shadows cast by small apertures when using a plane source of monochromatic light as a light source. http://www.falstad.com/diffraction/ http://www.falstad.com/modebox/

emission. Consequently, conversion of noise to a plane polarized monochromatic signal by a material similar to an interferometer (Figure 8.5c) is a key here. We add inner and outer coils to a central coil to perturb each other, interfere and generate the dark lines. The interference between the central coil and the inner coil is very different from the interference between the central coil and the outer coil, since perturbation angle α is negative and positive, respectively (Figure 8.4g). Each of the two distinct interference events is from the superposition of four coherent sources.

8.3 THE HAMILTONIAN OR ENERGY OF FOUR CLOCKS CREATING MAGNETIC LIGHT

Figure 8.5d describes the Hamiltonian that gives energy for this device has five parts $\mathbf{H} = \mathbf{H1} + \mathbf{H2} + \mathbf{H3} + \mathbf{H4}$. The Hamiltonian is plotted in Figure 8.6a. If a pair of crossing paths have n_i number of clocking waves $\vec{\psi}_{Osci}$, each wave with an energy E_i, then total energy filtered from noise is $\mathbf{H1} = \sum_i E_i n_i^2$. The zig-zag path filters noise in a strictly quantized manner, the separated energy of clocking wave is $\Delta\varepsilon_{\pm} = \hbar k_F^2 / 2m^* = \pm\sqrt{2(1-\cos k)}$, which is part of the total energy hold by a pair of zigzag paths $K1$ and $K2$, is

$$\mathbf{H2} = \sum_i \frac{E_{K1-K2}}{2}\left(\left(\cos\frac{\varphi}{\varphi_0}\right)^2 + d_i^2\left(\sin\frac{\varphi}{\varphi_0}\right)^2\right)^{1/2}.$$

Here, φ is the exchanged quantized flux between the paths (flux degeneracy) with respect to a flux minimum φ_0 and d_i separates two parallel paths, k is the wave vector. The charge asymmetry between the paths holds an energy

$$\mathbf{H1} = \frac{1}{2}m^*\omega_L^2\left(\left((1+\chi)x^2\right)+(1-\sigma)y^2 - \frac{\omega_{02}^2}{\omega_L^2}z^2\right),$$

where χ is anisotropy factor, here conductivity σ distinguishes the diffracted waves ω_{02}, ω_L from different paths and layers respectively. Reflected monochromatic light interferes, the geometric parameters L, P, D, and ab regulate the strength of magnetic flux and geometry of magnetic vortices. It is $\mathbf{H3} = \frac{\Omega^3}{4\pi^2}\left(2m^*/\hbar^2\right)^{\frac{3}{2}}\left[3\sum_i\cos\theta_i + 4\prod_i\cos\theta_i\right]$. If we twist the coil, we get a screw ($\delta = 0; \delta = \pi$, left- and right-handed, respectively) and edge ($\delta = \pi/2$) dislocations of zig-zag paths on the coil surface the energy of the dark lines ($E = 0$) is given by $\mathbf{H4} = k^2/2m^*\left[(ax + kx^2\sin\delta) + i(y\cos\delta + z\sin\delta)\right]$.

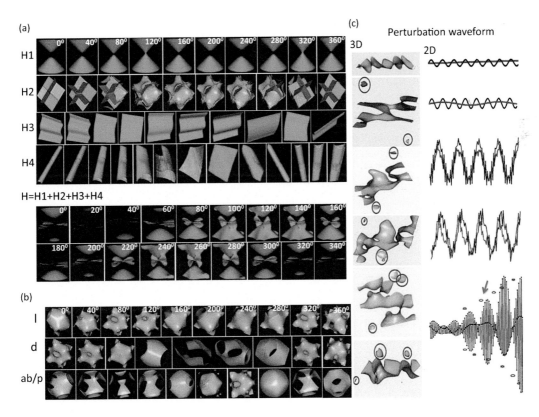

FIGURE 8.6 The four Hamiltonian parts of the generic Hamiltonian that represent the H device is shown, the value of the Hamiltonian is plotted against the variables concerned in those typical Hamiltonian. 0–360 phase change for the periodically oscillating parameter shows distinct features (top). The combined Hamiltonian shows unique behavior very different from all. (b) H2 shows a brilliant function that is fundamental to H device, sum of cosines and product of cosines. By changing length l, diameter d and the ratio of lattice area ab and pitch p, periodically the number of singularities in the phase space is tuned, shown in a series of plots as a function of phase. (c) The effective dark line formed by the interaction of multi-layered helical cylinder generated magnetic dark knots are shown where polarization and average amplitude of perturbation signals are changed.

The part $\left(ax+kx^2\sin\delta\right)+i\left(y\cos\delta+z\sin\delta\right)$ is solved by simplifying the defect on the coil surface, considering that a dark line is $\psi=k\left(ax+iy'\right)e^{i\left(kz-\omega t\right)}$. Numerical solution of $\mathbf{H}$ for generating dynamic magnetic vortices is easy to generate. Here, $\mathbf{H}$ forms a broken band-like structure with two-cone facing each other. Periodically, as a function of time, two new cones are born at the top cone edge, extend below, touch bottom cone edge, then slowly disappear. Then a similar thing happens at the bottom cone; together we call it conical blinking.

Including all factors here we find the Hamiltonian $\mathbf{H}$ of the device

$$\mathbf{H}=\mathbf{H}1\left\{\sum_i E_i n_i^2+\frac{1}{2}m^*\omega_L^2\left\{\left(\begin{array}{c}(1+\varepsilon)x^2+(1-\mu)y^2\\[6pt]-\dfrac{\omega_{0i}^2}{\omega_{Li}^2}\left(z-\cos\phi\right)^2\end{array}\right)\right\}\right\}$$
$$+\mathbf{H}2\sum_i\frac{E_{K1-K2}}{2}\left\{\begin{array}{c}\left(\left(\cos\dfrac{\varphi}{\varphi_0}\right)^2+d_i^2\left(\sin\dfrac{\varphi}{\varphi_0}\right)^2\right)^{\frac{1}{2}}\\[10pt]\left(\begin{array}{c}\alpha\left(x_1x_2+x_3x_4\right)\\[4pt]+\beta\left(x_1x_3+x_2x_4\right)\\[4pt]+\gamma\left(x_1x_4+x_3x_2\right)\end{array}\right)\end{array}\right\}$$
$$+\mathbf{H}3\frac{\Omega^3}{4\pi^2}\left(\frac{2m^*}{\hbar^2}\right)^{\frac{3}{2}}\left[3\sum_i\cos\theta_i+4\prod_i\cos\theta_i\right]$$
$$+\mathbf{H}4\frac{K^2}{2m^*}\left[\left(ax+kx^2\sin\delta\right)+i\left(y\cos\delta+z\sin\delta\right)\right]$$

$$(8.8)$$

The function is plotted in Figure 8.6c. One could notice the formation of the magnetic vortex-like magnetic rings (bottom, arrow Figure 8.6c).

Coupling Coefficients CC_i are α,β,γ, and x_{1-4} is reflection coefficient S11 from the dielectric resonator of the four corners of cell i forming the lattice unit. $CC_i=\left(\alpha\left(x_1x_2+x_3x_4\right)+\beta\left(x_1x_3+x_2x_4\right)+\gamma\left(x_1x_4+x_3x_2\right)\right)$ When geometric parameters change as a function of time, all four parts of Hamiltonian looks different. Here we show how to change $\mathbf{H}3$.

$$\Gamma_i\left(r\right)\rightarrow\left[3\sum_i\cos\theta_i+4\prod_i\cos\theta_i\right]\rightarrow_i\left(r,t\right)$$
$$=3\sum_i\cos\left(x_i\cos t_i\right)+4\prod_i\cos\left(x_i\cos t_i\right)$$

$$(8.9)$$

The function is plotted in Figure 8.6b.

8.3.1 BIREFRINGENCE, QUANTUM BEATING AND CLASSICAL BEATING

The development of the fourth-circuit element requires a few fundamental materials properties, as outlined in Figure 8.7a. The following describes them one by one.

Birefringence and metamaterial: Optically or electro-magnetically operating quantum devices can process quantum states at room temperature and in ambient atmospheric conditions. However, it requires a pair of quantum coherent sources, for interference. Microtubule has the property of birefringence (Mithieux et al., 1985), which enables generating a pair of quantum coherent sources (O-ray and E ray; Figure 8.3a). The α-helices of a protein molecule could be considered as a single, the most elementary antenna. A protein molecule would itself be an optical antenna array, whose optical anisotropy is tailored by mechanical resonance. Such materials normally show a giant birefringence (Kats et al., 2012). Such a surface controls the polarization of the reflected light (Zhao and Alu, 2011). The reflected wave from the protein made microtubule surface interferes and forms geometric structures made of dark lines where, due to destructive interference, there is no light. The structures of darkness produced by microtubule or its analog Hinductor devices do not remain static due to the water crystal of microtubule, it perturbs, and the knots of darkness rapidly changes the shape with time, periodically. Inner water, tubulin protein, and outer water of microtubule make three concentric layers required for building a fourth-circuit element (Figure 8.4). So, we get clocking geometric shapes like a time crystal. In these dark lines $E=0$, B is not zero, i.e., these are magnetic structures. The dark structure-made geometric shape is a function of charge distribution on the microtubule or H surface. It is a well-established theory that the reflection property depends on the charge distribution of a surface. Now, by applying a noise bias, we can tune electron density and the magnetic structures. It naturally creates clocking geometric shapes, and could store time crystals for a long, long time powered by noise. Simply by soldering capacitors in series and rolling them along a spiral one could test the linearity of magnetic flux and charge. Artificial structures analogous to microtubule show a distribution of magnetic flux as a function of heat flow on its surface (Figure 8.7b). The linear magnetic flux vs charge stored in the capacitors by applying noisy pulse suggests that (Figure 8.7c), charge flow may not be required at nanoscale too. Figure 8.7d shows a magnetic flux hysteresis for a single microtubule. These experiments prompted us to do more complex experiments on capacitor analog, that is not a true device, but, easy to check basic ideas.

Interference of four sources in intimately coupled coils: A path to geometric musical language (GML): Negative refractive index of a material is fundamental to the emergence of metamaterial, because for a material to be a metamaterial the signs of magnetic permeability and dielectric permittivity should have to be reversed (Shelby et al., 2001; Valentine et al., 2008; Yu et al., 2011). We noted above that each coil generates a pair of coherent signals due to birefringence like events, and we get circular or elliptically polarized signals. However, three coils are connected; the separation between them is that of atomic bonds. Thus, three coils act like single molecular architecture. Hence the noise converted into a pair of coherent signals in the central coil bifurcates into the inner coil and the outer coil. Then two independent interference

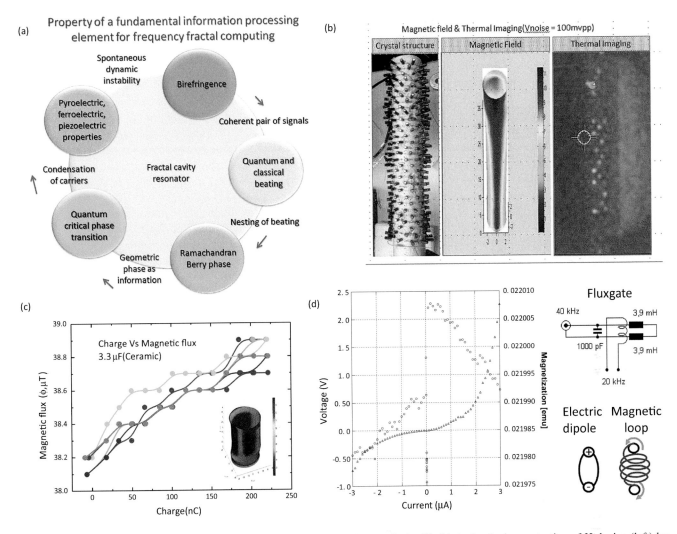

FIGURE 8.7 (a) Physical phenomena that operates a fourth-circuit element device H. (b) A simple demonstration of H device (left) by pumping electromagnetic noise MHz-GHz range to the device and measuring the magnetic flux produced on the surface of arrayed capacitors (10s of microfarads), middle, and to the extreme right, thermal images show that heat exchanges between the capacitors. (c) The charge stored on the surface of the device and the magnetic flux produced was measured for different helical coils made of different capacitances. (d) Using picotesla sensitive fluxgate magnetometer magnetization of single microtubule is measured.

events happen. One interference of four waves occur between the inner and central coil; the other interference of four waves occurs between the central and the outer coil.

$$\Gamma = \Gamma_1 + \Gamma_2 + \Gamma_3 + \Gamma_4$$

$$= 4\sin(kz - \omega t)\cos(\pi d\sin\theta / \lambda)\cos(2\pi d\sin\theta / \lambda)$$

$$\sim \cos\rho_1 \cos\rho_2 \cos(\rho_3\sin t)$$

Due to a common wall between the coils, a coherence between the bifurcated sources are ensured.

The basic requirement of GML is a clock. If three plane waves interfere, we get a straight line, if four plane waves interfere, the vortices rotate like a clock Ψ (O'Holleren et al., 2006). If five waves interfere, we get an irregular pattern. Each clock Ψ holds a singularity point. Note that the geometric shapes

are inequality relationship hold by the clocks. For example, $\Psi_1 + \Psi_2 \geq \Psi_3$... triangle. $\Psi_1 + \Psi_2 + \Psi_3 \geq \Psi_4$... quadrilateral. $\Psi_1 + \Psi_2 + \Psi_3 + \Psi_4 \geq \Psi_5$... pentagon. Different composition of cross-bar pathways on the Hinductor coil forms magnetic rings of different diameters representing Ψ_1, Ψ_2, Ψ_3... These distinct rings could interact to integrate into a higher-level clock.

8.3.2 Pancharatnam Berry Phase

As described in Chapter 4, the concept of geometric phase is easy to conceive if we consider microcoil or nano coil resonators and shine it with polarized light (Lee et al., 2011). Since Hinductor is all about multilayer concentric columns, adding geometric phase is easy. If noise is pumped along the column of a spiral, the dominating magnetic flux morphs a ring. If the noise is pumped on its surface (~23° for microtubule), the flux adopts a spiral shape. At resonance, as the phase changes 0°–360°, both the ring and the spiral shape oscillate with noise. If the noise intensity V is within limits (20 < V < 80 mV), ~500 times below the charging bias of a capacitor and the frequency bandwidth of noise could hold $g_{v(r)}$ number of resonance peaks, then the local magnetic wave degeneracy is $g_v = g_{v(r)} \times g_{v(a,b)}$. Sahu et al. imaged a pair of THz magnetic wave observed in a rectangular and in a hexagonal lattice, but no current or field flow was observed. Topology change rearranges the fields whose vector sum is zero. Accumulation of magnetic flux is not from more current flow by amperes law, but the adjustment of shined light on the electron density surface.

The magnetic waves generated by splitting of noise interfere due to the helical period experienced by a lattice (a, b). The magnetic field could tune quantum beating (Henke et al., 1981) in the microtubule and in its analog H device $\left(\varnothing \sim g_{v(ab)} e^{i\theta} \right)$. As the frequency gap between the coherent sources is low, a THz noise beats at GHz frequencies surviving around a pair of lattice units. These GHz oscillations interfere again $\left(\theta \sim g_{vl} e^{i\omega} \right)$ due to length, pitch, and diameter constraints, beat at MHz $\left(f_0 = \Delta f = 0.0000001\ \text{THz} \sim 1\ \text{MHz} \right)$, surviving around thousands of lattice units. The condensed magnetic wave $\left(\varnothing \sim g_{v(ab)} e^{i g_{vl} e^{i\omega}} \right)$ is imaged live, two-level magnetic interference integrates a local lattice to its global topology. Waves couple classically by adding/subtracting phase, quantum phases couple as a product, but here, in the beating of beats, the phases couple exponentially. A pair of microtubules synchronously beats at multiple time scales at a time, due to the beating of beats. The difference in frequency of participating waves is the beating frequency of a new wave, but that wave could undergo beating once again.

8.3.3 PYROELECTRIC, FERROELECTRIC, AND PIEZOELECTRIC PROPERTIES

Biological materials are often depicted as ferroelectric and piezoelectric since under a mechanical pressure they release ions as current flow, and at a suitable electric bias desorb/absorb ion reversibly. A true ferroelectric material should have an intimate correlation between the applied bias and its dipolar orientations, flowing current under pressure and reversible conductance switching are not sufficient. Bandyopadhyay et al. have imaged live the resonant oscillations of DNA, RNA, proteins like actin and their complexes via tunneling current, which show that unlike inorganic materials, these are soft like a liquid drop. The concept of "lattice-like ordering of dipoles" or even a "domain of unidirectional dipoles" does not exist therein. Porcine brain neuron-extracted tubulin has 8 tryptophan and 36 tyrosine residues (Ponstingl et al., 1982), and as the temperature increases, a smaller number of tryptophan residues are exposed. Denaturalization via unfolding has two linear domains, first 20°C–40°C and second, 55°C–90°C, in these regions, a conformational change is homogeneous (Mozo-Villarias et al., 1991). The interesting transition at 40°C–55°C is caused by the conversion of just one tryptophan residue from exposed to the unexposed configuration.

Since this transition stops reversible switching of the tubulin charge center by 1.2Å, dipolar switching by ±23° is blocked, and tubulin is left with only one energy minima, tubulin switches from two energy minima to one; therefore, microtubules Curie point should be located here. In the 20°C–40°C domain, both tryptophan and tyrosine residues change conformation and unfold in the same rate; however, in the 55°C–90°C region, tryptophan residues denaturize faster than tyrosine residues, and at >55°C, less than 4 tryptophans remain exposed, at 90°, tubulin is permanently denaturized (Audenaert et al., 1989). The scanning tunneling microscope (STM) images of multiple structural phases of single tubulin dimer; these transitions were artificially induced in the structure in a controlled manner.

Similarly, a structural phase transition occurs in the microtubule lattices as observed in the cryo-TEM where rectangular to hexagonal lattice transition is visible in the microtubules in the neuron (Kikkawa et al., 1994). Porcine tubulins assemble into microtubule via distorted hexagonal close packing with a unique polar direction which assigns lattice-type A to enantiomorphic polar crystal class (point group C_6) and lattice-type B pure polar class (point group C_{6v}). Both the lattices favor a spontaneous pyroelectric and ferroelectric effect (Wu et al., 2009; Bruce et al., 2011), notably, out of 32, only 10 polar crystal classes favor these effects.

Effective electronic polarization co-efficient or the ionic polarizability α normalized for microtubule length is $6.3 \pm 0.3 \times 10^{-24}$ cm (Minoura and Muto, 2006; microtubule has 10% of the value of $BaTiO_3$), charge density is 20 e/dimer, 2.5 e/nm along one protofilament, hence 32 e/nm in a 13 protofilament microtubule. Hence, polarization measurement should start after all these charges are released; otherwise, we see clockwise current-voltage characteristic. For a true ferroelectric, polarization in hysteresis is counter-clockwise. A single microtubule's perfectly square or lossless hysteric current-voltage (IV) characteristic at 10^{-6} Torr and in ambient conditions depicts a ferroelectric behavior, since remnant polarization P_{rem} sends leakage current at zero bias; red arrow denotes threshold switching voltage. Microtubule showed $P_{rem} \sim 3$ C/cm², which is 10^6 times more than normal ferroelectrics; however, it is feasible since microtubule charge density 10^{20}/cm³ is orders higher than normal insulator ferroelectrics. For P_{rem} coercive field is 0.6 kV/cm. Similar to a classical ferroelectric, microtubule holds the charge ($\Delta C/C$) until Curie temperature, and dielectric loss measurement supports the inherent dipolar transition.

8.4 LINEAR VARIATION OF MAGNETIC FLUX AND STORED CHARGE IN H

Derivation of H between the stored charge Q and the magnetic flux Ψ: Group of stored charge Q diffused by noise combines the lattice spin waves (a and b are lattice parameters), by addition (like classical) and product (like in quantum) of phase $z = \left(\sum z_i + \prod z_i \right)$ as they beat locally $\left(z = e^{i\theta} \right)$. Due to global topological constraint spread over three directions (diameter, D; pitch P; length L) local waves interfere with beating frequency generated signal.

We get for the beating of beats $e^{\delta^3 z} = e^{\delta^3(\Sigma z_i + \Pi z_i)} = e^{\delta^3(3(z_D^\dagger + z_P^\dagger + z_L^\dagger + 4 z_D^\dagger z_P^\dagger z_L^\dagger))}$.

Here, $\delta = \frac{\text{Surface area}}{\text{Lattice unit area}} = \frac{2\pi r L}{a \times b} = \frac{\pi D L}{ab}$. The critical parameter to integrate classical and quantum factors in phase space should be in ratio 1:1.0205, i.e., 3:3.0615. Here, 3:3.0615 is the critical point where a singularity is born. We kept quantum capacitance[26]-induced phase factor 3:4, to keep the area of the hole created by singularity is proportional to the area covered by the phase space continuum.

The cumulative spatial phase factor of the charges $Q e^{\delta z}$ constituting the sinusoidal wave per unit velocity of photon $(Q e^{\delta^3 z} / c_m)$ in that medium is the magnetic flux (since $B = E / c$, we get for entire cylindrical surface $\Psi = Q e^z / c_m = QH$).

Here, $\frac{\Psi}{Q} = H = \frac{e^{\delta^3 z}}{c_m} = \sqrt{\mu_m \varepsilon_m} e^{\delta^3 z} = \frac{1}{C_m} e^{\delta^3 \left(3(z_D^\dagger + z_P^\dagger + z_L^\dagger + 4 z_D^\dagger z_P^\dagger z_L^\dagger)\right)}$.

The nth period plays an important factor in the magnetic wave. The larger the value of n, the phase oscillation gradient increases nearly exponentially.

$$H = \frac{1}{C_m} e^{\delta^3 \left\{ 3\cos\left(\frac{D-nb}{b}\right) + 3\cos\left(\frac{P-na}{a}\right) + 3\cos\left(\frac{L-nP}{P}\right) + 4\cos\left(\frac{D-nb}{b}\right)\cos\left(\frac{P-na}{a}\right)\cos\left(\frac{L-nP}{P}\right) \right\}}$$

(8.10)

From magnetic rings, one could measure the total flux content Ψ by matching the ring's color intensity with a standard magnet response and plotted Ψ with Q. Both Ψ and Q varied non-linearly as a function of applied frequency $\hbar\omega$ but so similarly that the Ψ-Q plots were linear, for all materials studied, delivering a constant H, which is accurately predicted by equation (8.10). H is oscillatory as a function of L, P, and D, if the periods for a typical material are L_p, P_p, and D_p, then the periodicity index $I = L_p P_p / \pi D_p^2$. In fact, $\Gamma_i(r, t)$ explained in equation (8.9) that governs **H**3 and the generalized expression for Hinductance, H (equation 8.10), was discovered in >1000 devices by three independent blind studies (Karthik, K.V., Suryakant, Kumar, Pushpendra Singh) spanned over ten years. However, though the dimensions are same, H is not the resistance as its value is ~10^{-4}, while the resistance of these devices R is ~10^9 Ω, these devices are all insulators, i.e., no current flows.

The interaction between the electromagnetic field and the quantum system is characterized by another constant, the charge, that pertains to the quantum system alone. If we look closely into the quantum analogs of classical devices, most quantum formulations do three things. First it considers that even quantum elements follow classical circuit wiring concepts, serial and parallel connections of circuit elements are only two ways to connect them. Second, it considers that a very low resistance path and low conductance fluctuation around Fermi level ($h\upsilon = k_B T$) is the only way to realize a quantum effect. Always, I and V are there as a common factor with Ψ or Q; it is believed Ψ-Q pair cannot stand alone. Third, entanglement does not require a real physical communication through a well-defined path,

yet it considers a coherent real physical motion of particles within an isolated restricted barrier to avoid the dephasing. Be it a spin-wave or quantum interference, a path is always there, but never there in theory. Three legacies of classical concepts, classical wiring, resistance, and necessity of a defined path, were universally accepted in the quantum experiments and their quantum transformations were never in demand. It is utmost essential not to copy the classical treatments in quantum and also demonstrate that the effort to standalone Ψ-Q pair as independent as resistance could have a classical analog too.

Conductance, a differential of current and voltage was earlier defined as dI/dV, now as $R \to \infty$, $d\Psi/dQ \to 0$. Quantum formulations inherently consider $d\Psi/dQ = R$. It is not always true. If $d\Psi/dQ \neq R$, and undefined, then a new formulations quantum mechanics and classical mechanics are required. Since we do not supply voltage and measure current anymore, the existing formulations, all based on conductance requires a new equivalent. Similar to dI/dV, differential conductance we get $dQ/d\Psi$, while the first flow electron, the second flows phase. Since $R \to \infty$, $R = 10^{15} \Omega$ and $H = 0.001$, $dQ/d\Psi \neq dI/dV$. We need a new physical parameter. The newly invented probe that senses the differential flux-charge, only applies noise. The noise, within a limit lets the clocking continue within the well. Beyond that, the noise shortens the clocking circle's perimeter such that the electromagnetic wave turns smaller than the barrier's collision cross-section that reflects it, the wave leaks through the well barrier. Information processing in microtubule was proposed in 1982 (Hameroff and Watt, 1982) experimentally verified in 2010 (Sahu et al., 2010, 2013a, 2013b).

8.4.1 FLOW OF THERMAL WAVE—FRIEND OR FOE?

In manmade machines, friction/collision is essential to apply the torque to move a body to work, which causes the heat-loss—consequently, scientists argued a loss-less Carnot engine (Carnot and Thurston, 1890) for nearly half a century. Simply put, the target is to send energy from source to drain without any loss. Recently, the dream is explored via quantum protocols (Scully et al., 2003; del Rio et al., 2011) arguing that the quantum coherence would enable loss-less transmission and entanglement would avoid friction in a machine. Since these models keep the century-old "isolated system"—a hallmark of Kelvin's thermodynamic era, the energy-loss is inevitable when the work is translated outside. Landauer argued that the minimum energy required for one computational step in an "open system" is $k_B T$ and it could be realized using a loss-less particle-like energy-packet (Landauer, 1988) for the source-to-drain transport. Since, a Kinesin motor requires 12 $k_B T$ per step to walk through microtubule in a living cell or "non-isolated" system, the random fluctuations even by one $k_B T$ unit would cause ~10% error, which living cells cannot afford. Both, the engine efficiency of Kinesin cycle that varies from 0% to 100%, and the variation of growth/decay of microtubule-length with temperature have argued microtubule as an ideal heat sink (Caplow et al., 1988). Since microtubule exhibits negative specific heat (Vulevic and Correia, 1997), the possibility cannot be ruled out. Moreover, the noise-induced errors are fatal to the essential vesicle transport in a living cell—alleviated if microtubule is a perfect heat sink that keeps the entropy constant even under thermal fluctuations.

The lattice mosaic of tubulin proteins on the microtubule surface works as the nesting ground for a stream of loss-less particle-like energy packets. Chemical imbalance triggers soliton in the β-sheet of protein (Kayser et al., 2004). Solitons also generate in the zig-zag chain like α-helix, microtubule surface (Christiansen et al., 1997). Helical nanowires generate solitons (Brizhika et al., 2006). Microtubule satisfies all these criteria, surface-bound states generate topological phonon modes (Prodan and Prodan, 2009). Live images are shown, how feedback management operates to balance the thermal noise and the signal collectively to keep the entropy constant between 5 and 320 K. The finding resolves the thermodynamic equilibrium debate by showing that the fluctuation in protein-entropy is the signature of feedback management to keep the entropy constant. We could not read the random-jumps in entropy, so we assigned it as noise. The Carnot engine equivalence of a single brain microtubule is derived here from direct microscopic visualization.

Dynamics of two tubulin dimers suggest that the oscillation of energy between two dimers do not stop even in the presence of a neighbor, therefore, the mechanical energy stored in the extra β-sheet of tubulin dimer cannot move alone. However, when tubulins form a nearly a ring, mechanical oscillation stabilizes. Thus, 13 tubulin dimers can disperse energy homogeneously only long the circumference of the microtubule, a concerted shift of lattice displacement is created along a circular region as a ring of energy-packets with an opening due to offset. Spiral symmetry of microtubule makes β-sheets forming spiral inside a spiral of tubulin dimers (Hunyadi et al., 2007). A helical structure naturally quantizes energy that transmits through it (Michalski and Mele, 2000). Helical structures also trigger ballistic transport (Grigorkin and Dunaevskii, 2007). The opening destabilizes the 13-tubulin ring to move longitudinally along the length of the microtubule, and nanocavities are known to generate ballistic phonon (Huang et al., 2009). This ring with an opening is a phonon-particle like energy packet, since it is visible distinctly in AFM images, but not in the STM image. Due to identical periodic paths, since creation, the relative coordinates of electronic-packets and phonon-phonon packets are automatically fixed on the microtubule surface, which locks phase and frequency of the constituent phonon- and electronic-packets locally. Thus phonon-packets move together linearly and electronic-packets move together helically as a single unit, and we call it group. Since these particle-like energy-packets are lattice kinks, they are visible in AFM if mechanical in nature and visible in STM if electronic. Filaments, in general, create a topological phonon mode (Berg et al., 2011).

The mixture of five electronic-particle groups of distinct symmetries and three phonon-particle groups of distinct periodicities appear and disappear on the microtubule surface during transport under thermal scan 5–320 K. Simultaneously, one should measure electronic-particle and phonon-particle covered area on the microtubule surface from live images. The ratio of these two areas or fraction of surface area covered by phonon-particle and electronic-particle groups is termed as the band fill factor, and the coherence reflects mutual co-operation between two kinds of particles and lattice distortions (Moskalenko et al., 1980). These deformations shrink/expand separation between protein dimers throughout the nanowire; thus, restricts the length and conductivity of the microtubule-spring (σ_E). The microtubule length, group-phase transition frequency and the rate of length change are proportional to the externally applied noise. Since, after reaching the far end, the phonon-particles are pumped out as energy, they disappear from the image, surplus/deficit of phonon-particle groups subtract/add the length of microtubule nanowire, to which electronic-particle groups respond again and return conductivity back to the initial one $(\sigma S$ to $\sigma E)$. Thus, the surface coverage of electronic-particle and phonon-particle groups is regulated in a quantized manner, and similar to a spring a feedback control returns original length and conductivity.

To calculate entropy, Sahu et al. wrote a conducting state (σE) at a particular temperature and by sweeping the temperature between 5 and 320 K continuously measured the conductivity. The dc and ac conductivity remains constant, except minor jumps between equally spaced values (σS). Even under extreme thermal noise is 1 K/minute (dQ), oscillatory flipping between two limiting conductivities is observed as temperature T changes. The entropy $S = S0 + \left[d\sigma S \,/T \right]$ plot against temperature T shows that it remains constant over time, no material is known to exhibit such feature under normal condition (Zaitsevzotov, 1993). When the entropy slips, noise is poured into the microtubule and when entropy returns it means, the noise is pumped out of microtubule. Thus, the microtubule sets a classic example of harnessing noise to keep the entropy constant; it demonstrates transport and noise-alleviation efficiency at an extreme level. The output noise current is a function of source-drain and gate bias. Here, a harmonic interference of noise propagating along the microtubule, and when there it is no noise, these are the live images of how microtubule manages noise. STM and AFM images of energy-packet transmission and interference like pattern proves that the transmitted signal across microtubule is coherent, microtubule converts incoherent signals into a coherent wave.

8.4.2 Wireless Communication between Two H Devices

Magnetic coupling carries out wireless energy transfer much more easily than electromagnetic or electric coupling (Kurs et al., 2007) if there is a parity-time symmetric circuit then the power transfer is even more robust (Assawaworrarit et al., 2017).

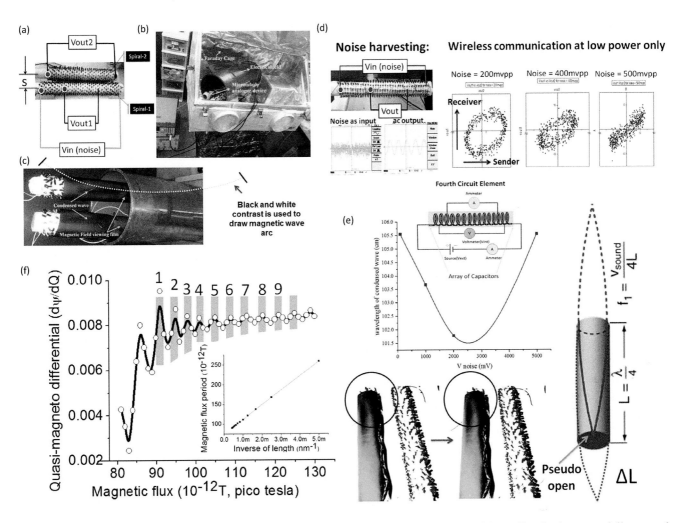

FIGURE 8.8 (a) Wireless communication between two coils, experimental setup. (b) The state of the art Faraday box, especially covered with composite materials and magnetic shields. (c) A pair of coils wrapped with magnetic films measuring the changes in the magnetic field distribution all along the surface. (d) For the coil operation continuously noise is pumped and output is an organized signal. Sender coil and the receiver coil were fed orthogonally to an oscilloscope and the phase plot shows identical energy transmission magnetically between the two coils. (e) The wavelength of the magnetic wave estimated from the coil surface for different bias. (f) Similar measurements were carried out on the surface of a microtubule. Oscillation of the magnetic flux with respect to the inverse of the flux-charge proportional constant shows periodic ripples in the values. In the inset, those ripple periods are plotted with the inverse of microtubule length. Tomasch quantum oscillations (nine ripples noted) when magnetic-charge differential measured directly by coaxial atom probe is plotted against magnetic flux. Inset is the length variation study of microtubule and finding periods of oscillations for different lengths. Inset plot was used to calculate fermi velocity of charge.

Magneto-inductive effects in the metamaterials allow even 1D data transfer (Stevens et al., 2010). Saxena et al. have carried out an extensive study on wireless communication between the pair of microtubules (Saxena et al., 2018; Figure 8.8a–c). Microtubule undergoes rapid phase changes on its surface under the exposure of electromagnetic fields of certain frequencies (Sahu et al., 2010). We can consider its surface as a composition of lattices of various lattice parameters and sizes, and all experiments are carried out by mixing noise with signal (noise>2signal; Figure 8.8b). At a particular resonance frequency, there is an equilibrium, we get $N_r^{equ} = N_0 \exp\left(-\Delta G_{cluster}(r)/k_B T\right)$, this statistically accounts the composition of lattice symmetries at resonance, where

$$\Delta G_{cluster}(r) = -\Delta G_{lc,v}.\pi r^2 l_{cluster} + \sigma \pi r^2 l_{microtubule} \cdots \quad (8.11)$$

Here, $\Delta G_{lc,v}$ is the energy due to a phase change per unit cluster area on the microtubule or H surface. And σ is the interfacial free energy propagating throughout the microtubule surface due to THz transmissions between tubulin proteins or equivalent capacitors.

From equation (8.11) we get $\Delta G_{critical} = 16\pi\sigma^3/3\Delta G_{lc,v}^2$ for a particular lattice configuration, i.e., greater than a critical radius, and equation (8.11) predicts a condensation above this limit. Condensation means a particular lattice starts dominating its surface. The rate of condensation of a typical symmetry in a receiver microtubule's surface is given by

$$G = S_c k N_0 \frac{1}{N_c} \left(\frac{\Delta G_c}{3\pi k_B T} \right)^{1/2} \exp\left(-\frac{\Delta G_c}{k_B T} \right) \qquad (8.12)$$

Here, N_c is the number of lattice points that first synchronizes its phase in the sender with the receiver's magnetic condensed wave's phase. Now, we need to find the phase gain rate g_s from the lattice gain rate G. We explain it below.

To understand the wireless communication between the microtubule analog devices as shown in Figure 8.8d, we inherently considered the coupled-mode theory model of two resonant systems for the analysis. According to this coupled-mode theory, these two resonant systems act as a source and receiver. The source system has a resonant frequency of ω_s. The overall system gain rate is expressed as $G = g_s - \gamma_s$, where g_s is the phase gain rate of the source and γ_s is the intrinsic loss rate of the source. The receiver lattice has a resonant frequency of ω_r. The overall loss rate of the system is expressed as the $\gamma = \gamma_r + \gamma_{rl}$, where γ_r is the intrinsic loss rate of the receiver and γ_{rl} is the rate of loss at the receiver load. Power is transmitted from source to receiver with a transmission coupling coefficient rate α, which is a function of signal amplitude. Transmission coupling coefficient decreases exponentially as the separation distance between the source and receiver increases (Figure 8.8e). Here Saxena et al. supplied noise signal to the source. If V_{noise} is the noise signal and V_r is the output voltage of receiver, then the rate of change in signals is described by the following equation:

$$\frac{d}{dt}\begin{bmatrix} V_{noise} \\ V_r \end{bmatrix} = \begin{bmatrix} i\omega_s + g_s & -i\alpha \\ -i\alpha & i\omega_r - \gamma \end{bmatrix}\begin{bmatrix} V_{noise} \\ V_r \end{bmatrix}$$

8.4.3 TOMASCH OSCILLATIONS AND HARVESTING NOISE

Now, it's time to test noise on a single microtubule. We all could remember the historical journey that Tomasch oscillations or reflections in the same direction made (Tomasch, 1966; Lykken et al., 1970; Wolfram, 1968). If the path is resistance-less, then the carriers do not leave a thin, confined region, oscillates therein between the boundaries for a long time. Of course, in the initial days, finding the evidence of such trapped oscillation (Tomasch oscillation) became difficult. However, one particular point became explicit—the device geometry does play a role in the prolonged survival of a carrier in the thin confined region. Now, a long time has passed, people are trying to change from electron flow to the phase flow, and for that, the device geometry would become ever more critical (Thouless, 1998). Every clock is like a pendulum, repeatedly reflects between the two boundaries. Here, we have an unusual situation. No one would like that the network of clocks would damp. One way to resolve that problem is to vibrate the boundary of the cavity. That vibration would supply the necessary energy required to run the clock. Say, we want an electron to be trapped in a cavity and then vibrate between the boundaries for a long, long time so that it represents a clock. Differential flux charge measurement on a single microtubule shows Tomasch oscillations (Figure 8.8f).

8.5 PERIODIC OSCILLATION OF CAPACITANCE, INDUCTANCE, AND HINDUCTANCE WITH ITS GEOMETRY

Quantum resonance made of quantum inductor and quantum capacitor: Violation of Kirchhoff's law: Since the first observation of resonance by inductive and capacitive elements in 1826, whenever we observe resonance in nature, deep inside our brain to the extreme end of the galaxy, it has become a customary to consider a hidden pair of capacitor and inductor inside. In the 1990s, the quantum versions of these pairs arrived as a quantum capacitor and quantum inductor, now we wire them classically to find the quantum resonance. It is astonishing, since the violation of Kirchhoff's law was proposed in the 1960s, that the current-times-resistance of all the elements in a closed loop must equal zero, if not, for example, emission by any element, the time-varying magnetic field generating additional input, then the law violates. Most importantly, if there is a device architecture in which there is a device inside a device, or the escape time fractal-like hardware—the central hallmark of this book, again the Kirchhoff law would be violated.

How to emulate coherent emission It is shown now that if we wire the quantum elements in series or parallel, the conductances do not add up (Gabelli et al., 2006). The culture of obtaining the resonance frequencies by a linearly wired resistor, capacitor, and inductor still continue. Since in a quantum inductor, a magnetic flux generates by reflection and not by current flow, and in a quantum capacitor forms without storing a real charge, both the magnetic flux and the charge were neither quantified nor detected. These quantum elements originate in the quantum well embedded inside a classical capacitor, not alone, as an original device. In the quantum well, a wave function not only interacts with the input electromagnetic wave (Aharonov and Bohm, 1959), but also emits entangled photons at room temperature (Michler et al., 2000), the effect of magnetic and electric dipoles are explicitly distinguished in the interference pattern (Freed and Weissman, 1941). In a quantum well, even the physical oscillation of a trapped charge could repeatedly emit coherent quantum bursts of infra-red surviving longer than picoseconds (Roskos et al., 1992). Remarkably, a superlattice of quantum wells is more stable in the coherent emission for both electromagnetic (Waschke et al., 1993) and magnetic spin wave (Oh et al., 2017). It promises designing an array of quantum antenna that could shape how the emitted ripple would look like. Until now, enhancing magnetic emission by suppressing the electrical emission was achieved by introducing a pseudo level (Karaveli and Zia, 2010). Advancing further, if the oscillating charge of a quantum well could reshape the electromagnetic wave into a packet of coherent magnetic flux, then a wireless burst of magnetic ripple could flow like a photon through the medium or even vacuum. An extreme environment would not be required to run the quantum coherent circuits. One would not require to build a parallel, or series wiring of quantum device. The quantum elements could float at proper orientation at the coordinates so that all the participating circuit elements build a combined phase space. An integrated chip would look like a 3D geometric shape where no current flows. The route is unique among all existing proposals to survive quantum effects in an ambient condition since it suggests rearranging the classical environment as a unique topology of quantum's infinite choices requires.

An electrical charge cannot flow since we want the magnetic vortex atom to flow as pure evidence, a solution to such critical demand were made in the 1990s. Increasing the dwell time of an electromagnetic wave in a well could mean the storage of charge like a capacitor C_q and if the wave repeatedly reflects between the walls of a potential well could generate a circular current flow like an inductor L_q. These are not normal capacitor and inductor, and the origin is not geometry but discrete energy levels of a material. In order to satisfy Dirac's criterion, the unique potential well acting as L_q and C_q should act as an antenna that would radiate a magnetic ring or torus outside the potential well. For that to happen, the height of the potential wells should vary, not directly with the real charge but a parameter that controls the dwell time of an electromagnetic wave trapped in a potential well, obviously, it is frequency. Unfortunately, the wall parameters of quantum wells are fixed as soon as it is fabricated, so we need to design a virtual quantum well just like the concept of an artificial molecule was engineered.

The height and width of the walls of the virtual well made of charge varies alone to regulate the emission of classical or quantum rings of magnetic flux. That is why single microtubule's capacitance oscillates as a function of length, pitch and diameter (Figure 8.9a–c). Does the storage of charges change the spiral? In order to find an answer, we calculated quantum capacitance, $C_q = \partial Q / \partial V = \partial Q / \partial \psi$. We solve equation (8.15) to find the expression of quantum capacitance of the system,

$$C_q = \frac{v m^* q^2}{2\pi \hbar^2} \left[2 - \frac{\sinh(E_g / 2kT)}{\cosh\left(\left(\frac{E_g}{2} - q\psi\right) / 2kT\right) \cosh\left(\left(\frac{E_g}{2} + q\psi\right) / 2kT\right)} \right].$$

When C_q is measured using a differential bridge circuit; the theory is found consistent with the experimental result. The actual charge stored in the wall could be measured using C_q directly. Using a semiconductor electron emitter we injection charge to the protein and applied nT range magnetic flux, we could not change the flux density image. The ring and spirals are not made of charge or spin, as conventional magnetic fields are defined. The effective mass of the massive carriers trapped in the lattice cells was measured experimentally on the single microtubule and experimental data provides clear evidence that the effective mass increases as we trap more (Figure 8.9d).

In order to confirm that the emitted magnetic flux is truly ring-shaped, first we consider that there is truly a ring, on its perimeter at any point, $\psi(t) = \psi \sin\phi$; then due to the quantization of a circular path $\phi = \frac{2\pi}{\Phi_0} Q$, where Φ_0 is the magnetic

flux quantum in the ring, we get the quantum inductance $L_q = \frac{\Phi_0}{2\pi Q} \frac{1}{\cos\phi}$. Then, from the density of states, we recalculate quantum inductance $L_q = \frac{\hbar^2}{12\pi q^2 \Gamma_L}$. Both the expressions of quantum inductance calculated from the density of states and ring topology give the same value which suggests that there is truly a ring and it originates in the virtual quantum well. Also, since in a generic quantum well $L_q = \hbar / 2\Gamma G$, earlier, conductance G was related to current flow, here a comparison shows $G = \partial Q / \partial \psi$. The finding suggests a new paradigm of transmitting energy as rings of magnetic flux, which experiences a new kind of resistance, primarily topological. By arranging seeded quantum wells in a particular geometry in the 3D space filled with extremely insulating matrix $(R = \infty)$ one could build artificially integrated chips of non-electronics, where another resistance is $H = 0.001$. Quantum conductance and quantum inductance were measured on a single microtubule (Figure 8.9e, f), which showed that not just one, single microtubule exhibits multiple such values, and one could regulate or tune those values using simple control parameters. The measuring device is shown in Figure 8.9g. It seems beyond electronics we may have an era of super non-conductivity (Figure 8.9h).

8.5.1 The Concept of Phase Space with 12 Holes That Blink

Twelve singularities of combined phase space: Opening and closing the undefined holes on the phase sphere: The wave $\vec{\psi}_{mag}$ from Ω number of cells interfere, the boundaries set by D, P, L, and ab adds wave functions from three directions $\left(^3\sum_i \cos x_i\right)$ and takes a product from two directions (D, L) with a feedback effect $\left(^4\prod_i \cos x_i\right)$. So, we get the number of reflections $\Psi \sim \tau e^{\Omega^3 \Gamma_i(r,t)} / c_m$. That equates a delay τ ($\sim \Delta Q$, delay=charge, no real electron storage) with the magnetic flux produced, hence $H = \frac{\Delta\Psi}{\Delta Q} = e^{\Omega^3 \Gamma_i(r,t)} / c_m$. Oscillating L, P, D, and ab delivers the phase factor $\Gamma_i(r,t) = 3\sum_i \cos(x_i \cos t_i) + 4\prod_i \cos(x_i \cos t_i)$; x_i is replaced by the geometric parameters. In a Hinductor device or H, the phase sphere an equivalent to the Hilbert space of a qubit has 12 singularities. It means it has 12 holes on the sphere (Figure 8.9i, j). There are a pair of singularities for all six x, y, z, xy, yz, zx directions, so we get 12 holes. The generic function that regulates the solutions is given by $3\sum_i \cos\theta_i + 4\prod_i \cos\theta_i$. However, there is a temporal function added to the angular values in this function $\cos\rho_1 \cos\rho_2 \cos(\rho_3 \sin t)$, or $\cos(\rho_i \sin t)$. When discrete magnetic rings $\Psi_1, \Psi_2, \Psi_3 \ldots$ holding a geometric shape change the diameter, actually on the phase space the holes of singularity change their diameter. It is a beautiful life-like event. As electric field clocks the magnetic ring, the solid angle ρ_i created by phase singularity oscillates as a function of time $\rho_i \sin t$, but we can add 1–12 temporal functions in $(2^n - 1) \sim 4 \times 10^3$ ways. Thus, a fundamental Hinductor device carries out 4047 distinct temporal dynamics. A maximum of 12 vertices solid 3D topology it can represent, e.g., hexagonal prism, icosahedron, triangular bifrustum, truncated tetrahedron, etc. If Ψ_1, Ψ_2, Ψ_3 triangle's all three rings oscillate due to electric

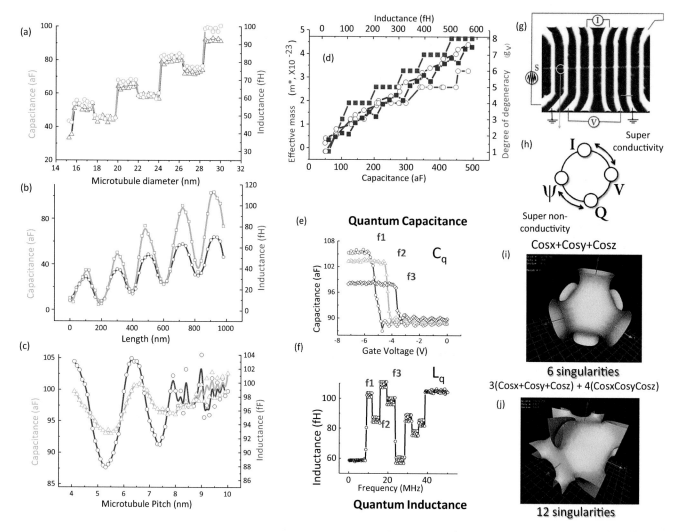

FIGURE 8.9 Three panels (a), (b) and (c) show that the variation of capacitance of a microtubule as a function of its diameter, length and pitch. Capacitance was measured at 1 kHz. (d) Effective mass and degree of degeneracy of the magnetic particle-like vortex produced in the junction of two cylindrical layers of microtubule. Both inductance and capacitance were measured plots at 8, 16, and 22 MHz capacitance variation as a function of gate bias. Quantum measurements were carried out using a bridge circuit. (e) Quantum capacitance or capacitance inside a classical capacitor was measured for microtubule for three different frequencies. (f) Quantum inductance is plotted for three different frequencies. (g) E1–E8 Au electrodes on SiO₂/Si substrate is grown, microtubule is spray injected perpendicular to the E1–E8 electrodes. Electrode noise is applied from source S, via E1–E8 pair of electrodes. E2, E7 are ground to restrict leak current flow further. E3, E6 measures voltage, E4, E5 measures current integral or charge. E3–E6 is used as capacitance bridge, and inductance bridge for all measurements. Scale bar 300 nm. (h) Just like superconductors are diamagnetic, all H devices tend to become super-non-conducting. (i) The 3D plot of the function Cosx + Cosy + Cosz. (j) The 3D plot of the function 3(Cosx + Cosy + Cosz) + 4(CosxCosyCosz).

field fluctuation $\left(\partial E / \partial t = \partial \phi / \partial x\right)$, since an angle between Ψ_1, Ψ_2 as $2\pi\phi_{12} / \phi_0$, then three angles of the triangle clocks as a single entity. Thus, clocks bind.

We get a plug-n-play expression for $H = \frac{1}{C_m}e^{\Omega^3\{\Gamma_i(r,t)\}}$ ...... equation (8.10), input device geometry and stored charge or frequency, get magnetic flux as output. $\Gamma_i(r,t)$ has 12 singularity domains, i.e., holes in the phase space. It means for certain L, P, D, ab values, undefined or no amount of magnetic flux is generated by the device. Here cost$_i$ estimates a periodically oscillating geometry say L, P or D, then the associated hole is bridged. Moreover, N frequencies of G-mode, each defines a combination of $\cos t_i$, which sets the boundary

of a hole, that edge shapes the geometry of the magnetic vortex produced. We find below how to read information content (G-mode) written by a user $(\hbar\omega)$ using the blinking of holes as a vortex synthesis mechanism.

8.5.2 PERIODIC OSCILLATIONS WITH LENGTH, PITCH, LATTICE AREA/DIAMETER—THEORY AND EXPERIMENT

In Figure 8.9a–c, we present a simple study that the capacitance of microtubule changes as a function of its length, pitch and diameter. Now, since we saw above that one could open or close 12 holes in the phase space, varying only geometric

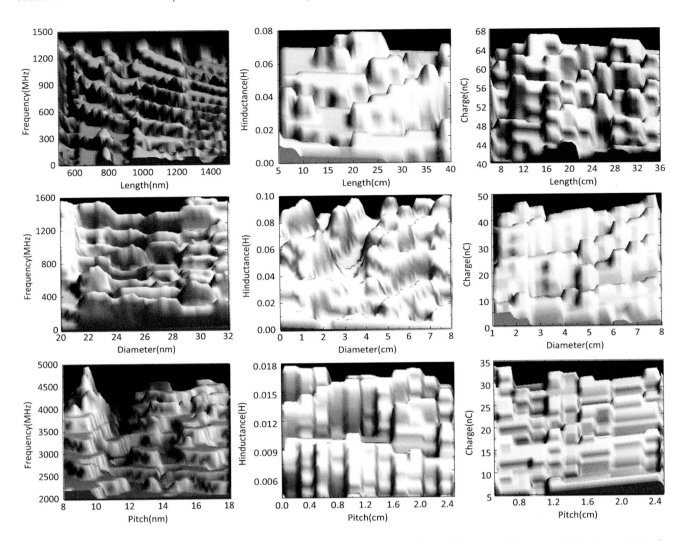

FIGURE 8.10 There are three columns. The first column is the theoretical length pitch and the ratio of lattice area/diameter variation for the microtubule. The second column is the H variation as a function of length pitch and the ratio of lattice area/diameter experimental data. The third column is the theoretical prediction for the artificially synthesized H device charge storage calculation as a function of length, pitch and the ratio of lattice area/diameter.

parameters, length, pitch, diameter, it is time to dig deeper. Figure 8.10 shows three plots, microtubule's, resonance frequencies, H and stored charge variation as a function of geometric parameters. The plots show that Tomasch oscillation like periodic changes is a universal feature of charge-magnetic flux response of a device. The periodic oscillation means for charge-flux operating particles, or magnetic rings or magnetic vortices, the extreme insulated cylinder acts as an H-less device. Thus, we could now sum-up basic arguments why H might claim to be a fourth-circuit element (Figure 8.11a).

In the quantum model of the fourth-circuit element device, namely microtubule, a pair of α-helices of a tubulin protein molecule forms a quantum well, a different number of coils of an α-helix act as asymmetric barriers ($n = 4$, 7.8 cm^{-1}, 12.4 cm^{-1}). The water molecules between a pair of helices generate a pair of shallow, deep quantum well. Water's quantum well is the guest in the host α-helix quantum well. Vibrational frequencies of water and helices are calculated by DFT for $n = 4$ and

$n = 2$ loops and 12 water molecules in between. Depending on the type of protein and its concerned group of α-helices, the values of squeezed states change. The Schrödinger's equation for an arbitrary non-stationary state of the wave function in this virtual well is $\Psi(q,t) = \sum a_n e^{-\frac{i}{\hbar}E_n t} \Psi_n(q)$, the coefficients satisfying the normalization condition $\sum |a_n|^2 = 1$; coefficients are determined if $\Psi(q,0)$ is known and $\Psi_n(q)$ are orthonormal. For time-dependent Schrödinger equation $i\hbar \frac{\partial \Psi}{\partial t} = -\frac{\hbar^2}{2m^*}\frac{\partial^2 \Psi}{\partial x^2} + \frac{1}{2}kx^2\Psi$. The ground state solution is the wave function is $\Psi(q,0) = (\alpha^2/\pi)^{\frac{1}{4}} e^{-(\alpha(x-a))^2/2}$. At high temperature, we replace α with a new arbitrary constant β, so that the wave function becomes $\Psi(q,0) = (\beta^2/\pi)^{\frac{1}{4}} e^{-(\beta x)^2/2}$. Ground and noise-driven higher states only differ by its height and width. If we calculate a_n and sum up the series, we get Schrödinger's equation for a non-stationary state $\Psi(q,t) = \frac{\alpha}{\pi^{1/4}}\sqrt{\frac{2\beta}{A+B\zeta}}\exp(-\frac{i\omega_c t}{2} - (\alpha x)^2/2 \frac{A-B\zeta}{A+B\zeta})$, where $A = \alpha^2 + \beta^2$, $B = \alpha^2 - \beta^2$, $\zeta = e^{-2i\omega_c t}$. The coordinate probability density $\Psi\Psi^*$ is now

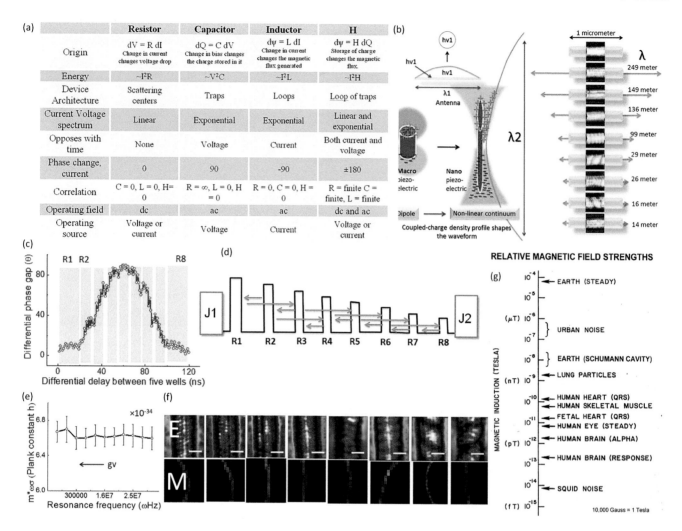

FIGURE 8.11 (a) Comparison between different fourth-circuit elements. (b) Eight different long-wavelength magnetic waves created on the microtubule surface are estimated when wirelessly electromagnetic signals were pumped on the microtubule. (c). Using probe of Figure 8.9 (g), in its measuring circuit, phase gap between phase ahead and phase lag wave is plotted as a function of time, simultaneously the magnetic flux is recorded, which shows quantized jumps. The wavelike variation (MHz) is theoretically fit. Five electrodes (C1–C5) are wired as phase array connected to magnetic wave on the microtubule surface shown in panel f. InP connected Pt electrode is used to sync clocking, ensuring phase array response. The Del1-Del2 goes to phase-locked loop differential amplifier, which measures differential phase gap plotted below as a function of time. (d) Schematic of octave of octave ripples integrated into one is shown as resonant tunneling diode. (e) Effective mass, resonance frequency, and cross-sectional area of microtubule for different degree of degeneracy generate the Plank constant. This is fractal quantization process. (f) Comparative electronic density of states (edos) and magnetic flux measurement (27 pixels as rectangles) using coaxial atom probe, for tubulin monomer (scale bar 0.2 nm); tubulin dimer (scale bar 1.8 nm); microtubule (scale bar 6 nm). (g) A chart showing naturally found material's magnetic field.

$$\rho(q,t) = \frac{\alpha^2 \beta}{\sqrt{\pi}} \sqrt{\frac{2}{\alpha^4 + \beta^4 + (\alpha^4 - \beta^4)\cos 2\omega_c t}}$$

$$\exp\left(-\frac{2\alpha^4 \beta^4 q^2}{\alpha^4 + \beta^4 + (\alpha^4 - \beta^4)\cos 2\omega_c t}\right)$$

The center of the wave packet remains confined, but the wave packet oscillates periodically with its width and height, similar to a classical dipolar antenna $\rho(q,t) \sim \sin^2 \omega_c t$. The periodic quantum oscillation takes a torus or ring shape, it can tunnel

through the virtual well $(k_0 = E)$, absorb $(k_{+1} = E + \hbar\omega)$, or emit $(k_{-1} = E - \hbar\omega)$ a wave packet from the wells of helix or water. The transmitted wave function from the virtual quantum well

$$\vec{\Psi}_{trans} = \left(t_0 e^{ik_0 z} + t_{+1} e^{ik_{+1}(z+\omega t)} + t_{-1} e^{ik_{-1}(z-\omega t)}\right) e^{-\frac{iEt}{\hbar}} = \vec{\psi}_{Osci} + \delta$$

(8.13)

The transmitted wave through the virtual well $\vec{\psi}_{Osci}$ and a part of it emitted as infra-red δ are in the same phase $(\phi_{\pm 1})$; it is calculated adopting Breit-Wigner form of transmission at around resonance (E_R)

$$\phi_{\pm1} = \phi_{\pm1,0} + \tan^{-1}\frac{E - E_R}{\Gamma} - \tan^{-1}\frac{E - E_R \pm \hbar\omega / 2}{\Gamma}$$

$$+\tan^{-1}\frac{E - E_R \pm \hbar\omega}{\Gamma},$$

where Γ is the width of the resonance peak for both, $\vec{\psi}_{Osci}$ and δ.

An infra-red sensing nanowire and a magnetic flux sensing nanowire located at the tip of an atomic resolution probe is brought to the atomic scale vicinity of the protein, simultaneously detects this phase in two independent signals, one electromagnetic and another, magnetic. Instead of interferometric detection of coherent infrared emission, or electro-optic gating, Ghosh et al.'s dual-mode nano-sensor remains a ~4 nm apart from the emission center, by reducing detection energy to 10 μV from 10 meV, fit to sense coherent burst 1–2 ps (10^{-12} seconds); and magnetic flux sensitivity (nT to 0.1 pT). Measuring the magnetic flux of an electromagnetic burst is critical since M-field is 10^4 times less than E-field at the sensor (Burresi et al., 2009). They detect M-field because it is a magnetic ring that resonates with the magnetic nanowire. Often the microtubule boundary is such that the waveforms do not end at boundary, the vibration around boundary is such that effective waveform is much longer (Figure 8.11d).

The oscillatory part of the magnetic flux is calculated by considering a geometric series of amplitude t, which changes due to a repeated reflection of $\vec{\psi}_{Osci}$ between the two walls of a virtual well. In the geometric series of t the number of higher-order terms we consider is proportional to the height of the virtual well, i.e., a function of stored charge (Karaveli and Zia, 2010).

$$\vec{\psi}_{Osci} = \frac{\partial \vec{\psi}_{static}}{\partial Q}\left[1 + \left(\frac{\hbar\omega}{2\Gamma}\right)^2\right]^{-\frac{1}{2}} Q_{osci}\cos(\omega t - \alpha)\Theta$$

$$(E_R)\Theta(E_F - E_{\pm1,0})$$

(8.14)

The above expression is calculated for three reflections, links charge, and magnetic flux directly. Using the coaxial nano-sensor, we apply 0.2 THz sinusoidal pulse to the THz nanowire sensor. The pulse stream traps the oscillating charge, we get Q_{osci}, that creates a ripple in the magnetic flux measured by the nanowire sensor, thus, we get $\frac{\partial \vec{\psi}_{static}}{\partial Q}$, experimentally. The differential phase gap, measured by sending picosecond pulse streams is plotted in Figure 8.11c, which shows eight ripples possibly from eight local density of states observed in the tubulin protein. Eight wells are shown schematically in Figure 8.11d. When the externally applied resonance frequency was varied to selectively activate a particular well, the effective mass was found to be constant (Figure 8.11e). It suggests that the magnetic vortex-like structure is a fundamental unit of signal and we could use a quantum wave function to operate or manipulate it. From the wave function $\dot{\psi}$ reflecting

between the virtual walls of alfa-helices and the water well, we calculate the amount of charge contained in the virtual walls of the virtual well.

$$Q = \int_0^\infty g(E)\left[f\left(E + \frac{E_g}{2} + q\dot{\psi}\right) - f\left(E + \frac{E_g}{2} - q\dot{\psi}\right)\right]dE$$

(8.15)

Considering a parabolic band, $g(E) = \frac{m^*}{\pi\hbar^2}v(E)$, where $v(E)$ is the number of energy levels of the virtual quantum well's wall, which are occupied.

Charge Q (equation 8.15) is used to calculate the emitted or radiated part $\vec{\psi}_{Osci}$ (2), hence we get $\vec{\psi}_{Osci} \propto \cos(\omega t - \alpha)$ experimentally. The comprehensive match between (2) with the experiment suggests that Sahu et al.'s quantum well model works in coupled α-helices of a protein. The magnetic flux is emitted all around the 3D structure as $\vec{\psi}_{Osci}\vec{\psi}_{Osci}{}^* = \rho(q,t)$, there are four primary and four minor quantum wells in a monomer as estimated by quantum tunneling measurement of the local charge density of states.

An extreme insulator, microtubule's resonant oscillations are topologically similar to its constituent protein tubulin's resonance band, which are also extreme insulators. They all could be switched to a super non-conducting state simply by applying noise. By constructing a special coaxial atom probe, one could detect thermal and magnetic flux signals simultaneously. It has been observed that the pairs of quantum capacitance and quantum inductance generate geometric resonances in a single brain extracted microtubule. Microtubules magnetic phase nests its dimers phase that nests its monomer's phase, a grouping of atomic-scale waves continues to form larger waves is neither seen in classical nor in quantum (Figure 8.11c). So, we can neither take a product of monomers wave functions to generate the wave function of a microtubule nor can we use existing quantum formulations as is. The super non-conductor generates a giant quantum magnetic flux like a superconductor (Figure 8.11f).

8.5.3 Knots of Darkness on the H Interface

In order to confirm the formation of knots of darkness on an H surface, the protein surface was scanned using the magnetic nanowire, and the formation of magnetic spiral in the monomer was confirmed (Vignolini et al., 2010; Figure 8.11f). The confirmation continued when the monomer forms a dimer and even when it makes a giant protein complex. The measurement was repeated for tubulin protein, actin, and clathrin and checked for the non-existence of such behavior in a single-walled carbon nanotube, but present in a helical multiwalled carbon nanotube. Magnetic fields are reported in many biomaterials, often this is due to magnetic materials. What magnetic features we observe in fourth-circuit element or by synthesis of light do not require any magnetic material (Figure 8.11g).

The cell traps $\vec{\psi}_{trap}$ emits, $\vec{\psi}_{Osci} = \frac{1}{4\pi^2}\int_0^{k\omega_c}\vec{\psi}_{trap}d\vec{k}\left(\frac{2}{k+1}-\frac{2}{k-1}\right)\times\vec{j}$, is integrated over 360° solid angle around a cell (ϕ = precision, θ = deviation due to paraxiality on an anisotropic path, α = relative perturbation angle). Since at singularity ($E = E_R(r,\theta,\phi)$), $\phi_{\pm1} = \phi_{\pm1,0} + \tan^{-1}\frac{E-E_R}{\Gamma}$, ($\Gamma$ is resonance peak width, E_R energy of a resonating cell made of (na, nb), n is an integer, a, b are lattice parameters, +1 means higher energy state, −1 means lower) contributes in the vertical z-axis primarily, we get $\vec{\psi}_i = \vec{\psi}_{Osci}\cos\alpha\cos\left(\tan^{-1}\frac{E-E_R}{\Gamma}\right)$; $\vec{\psi}_j = \vec{\psi}_{Osci}\sin\alpha\cos\left(\tan^{-1}\frac{E-E_R}{\Gamma}\right)$; $\vec{\psi}_k = \vec{\psi}_{Osci}\sin\left(\tan^{-1}\frac{E-E_R}{\Gamma}\right)$. The plot of

$$\vec{\psi}_{mag} = \hat{i}\vec{\psi}_i + \hat{j}\vec{\psi}_j + \hat{k}\vec{\psi}_k \qquad (8.16)$$

is a 3D spiral connecting the localized phase singularities on an oscillating sphere. This spatial structure is carried by a light like signal ($E\ll B$). The plot of these distributions is a spiral connecting the localized phase singularities, if other quantum elements are placed on the stationary state solutions locate on this path, it will contribute to the Hilbert space as a singular solution.

From the above expression of $\vec{\psi}_{Osci}(i,j,k)$, we get the equation of a circle when $E = E_R$, thus, we derive the formation of a ring of magnetic flux, by two independent theoretical formulations. First formulation deals, when a shot noise $<\omega_c>$ is absorbed into quantum oscillations $\left(\rho(q,t)\right)$. Second formulation deals, when the virtual well is pumped with its resonance frequency. All along the spiral path of the magnetic flux, the measuring nanowire senses a quantized increment of flux, as the system is charged with a photo-induced electron emitter; thus, $\Psi = HQ$ has both, a classical and a quantum version. To underpin the lensing effect, Sahu et al. measured the density of states for alpha-helix well and water well; one has to read d^2I/dV^2 and $d^2Q/d\Psi^2$, which shows that the magnetic density of states is at the boundary of the electromagnetic density of states. By de-phasing during a scan using noise, we track the diffusion o of magnetic flux. Thus, the ring is made of electromagnetic wave wherein $|\vec{E}| \ll |\vec{B}|$, it is classical if $\frac{\hbar\omega}{\Gamma} \ll 1$. Proteins do exhibit quantum effects at room temperature, under massive noise. However, here, the concerns of de-phasing by scattering do not exist. It is all about the topological 3D orientation of quantum wells on the surface of a 3D phase sphere, as $\cos\alpha$ regulates not just the basic ring formation, but, stabilization by assembling the rings in a spiral form. Not just tubulin, actin, all the proteins whose beta-sheet forms a spiral at the backbone, may exhibit flux-ring condensed wave.

8.6 THREE INTERACTIVE CYLINDERS PERTURBING THE KNOTS

At least three concentric cylinders is a unique engineering feature of Hinductor. Multilayer coils enhance wireless transmission (Nair and Choi, 2016). Each layer generates distinct waves, the inner coil k_i^*, and outer coil k_o^*, both 180° out of phase sources perturb the central coil's straight stationary dark line, D_c, with wave vector k_c^* at an angle α and build singularity structures D_i and D_o respectively. D_c, D_i, and D_o are identical dark lines, differ by diameter alone, and hence trigger beating. We plot beat pattern. The dark lines interact depending

on the polarization. The dark lines ($E_t = (E_x, E_y) = 0$) form when the real parts of propagating signal $kr'e^{i(\theta'+\zeta)}$ and the two perturbation signals from $e_1\left(e_1e^{i(\vec{k}^*.r-\omega t)}\right)$ and $e_2\left(e_2e^{i(\vec{k}^*.r-\omega t+\Delta)}\right)$ generated by the central coil's screw and edge dislocations cancel each other. Different zig-zag paths generate distinct pairs of e_1 and e_2. We express e_1 and e_2 as $\bar{e} = (e_1+e_2)/2$ that represents total strength of perturbation and $\Delta e = (e_2-e_1)/2$ that represents polarization ($E_x \neq E_y$). Interactions of dark lines could change e_1 and e_2, they could make a circle, or ellipse or even magnetic rings or vortices (Figure 8.12a, b).

Under noise, $\hbar\omega$ activates all three layers (Figure 8.12c), we scanned the knots on the surface using a magnetic flux microscope and theoretically fit the common bright region between two modes to find the covariance matrix of $\vec{B}(x,y)$ and $D(s,\theta')$, namely φ, and the coupled wave function Ψ_{l_i} using $\mathbf{H}\vec{\psi}_{mag} = \Psi_{l_i}^*(l_i,t)\Psi_{l_i}(l_i,t)H$, where $\mathbf{H}$ is Hamiltonian, equation (8.8) gives H, $\vec{\psi}_{mag}$ is calculated from equation (8.16), and $\Psi_{l_i}(l_i,t) = B_i exp(i\varphi t)$, where $B_i = I\langle\vec{B}(x,y)\rangle A$, A is the area on the surface where G-mode i is active, I is the periodicity index. To explain how magnetic vortex-like particles could cover the surface area, we start with three concentric layer cylinders, S1 falls on the top spiral, creates two coherent sources (Figure 8.12c). One part S2 again creates a pair of coherent sources at the junction of second and third spiral (Figure 8.12c). The critical condition that creates magnetic particles is squeezing the first knot of darkness, if we squeeze more by changing Δe. These free magnetic particles could rotate clockwise or anticlockwise (Figure 8.12c). The coherent sources created by S1, S2, and S3 interfere and generate composition of patterns of magnetic vortex on the cylindrical surface (Figure 8.12d). When wave functions of two simultaneous situations created by S1–S2 and S2–S3 are superposed and that builds the unit time crystal (Figure 8.12e).

Generalized orbital angular momentum generation and Holographic projection of time crystal as a set of orbital angular momentum set: Spiral dark lines, randomly arranged in the 3D space, were theoretically predicted by $D(s,\theta')$ and routinely observed for four decades. Now the same spirals with modified OAM contain the coupling coefficients CC_i and the wave vectors of distinct G-mode frequencies. Here at $\hbar\omega$ triggered G-mode resonance, the induced Δe varies periodically $\Delta e = \sin\delta$, then angle $\propto$ switches sign periodically, the dark knots assemble in a singular pattern of topological charge $l = il_1 + jl_2$, $\Psi_i(l,t) = D_i(s,\theta')\times\exp\left[2\pi i\frac{\varphi_{l_{i+1}}}{\varphi_{l_i}}\right]\times\Psi_{l_i}(l_i,t) = C_i(r,t)$; i.e the collective oscillation of three cylinders integrate the knots. To predict the experimental output, we derive a projection function

$$P(r,t) = \sum_{i=1}^{N}\Gamma_i(r,t)\times\Psi_i(l,t) = \sum_{i=1}^{N}C_i(r,t)\ldots \quad (8.17)$$

as a product of the phase space $\Gamma_i(r,t)$ and 3D coordinates of dark lines containing N number of G modes. $P(r,t)$ binds N number of magnetic dark knots which are allowed by a phase space $\Gamma_i(r,t)$; it creates a nest of N number of 3D assembly of

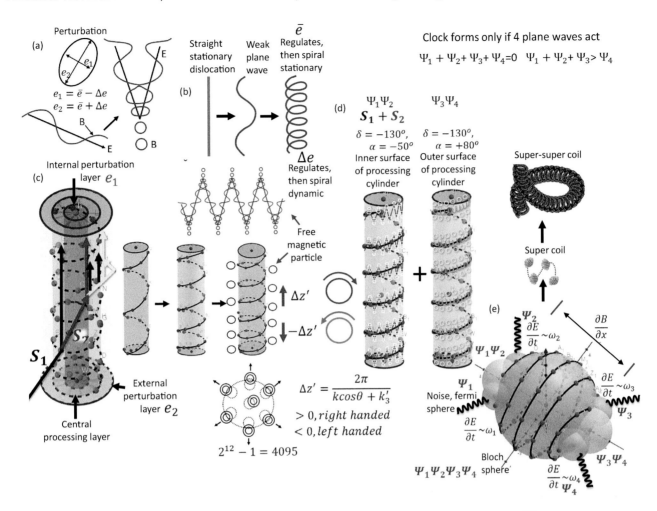

FIGURE 8.12 (a) Defining polarization of perturbation. $\Delta e = 0$, means no polarization. $\delta = \pi/2$, edge dislocation, $\delta = 0, \pi$ left- and right-handed screw dislocation. When two E lines interfere with disappearing, we find magnetic rings. (b) If the H device surface changes lattice profile due to external resonance, then the dislocation also oscillates, polarization also oscillates and a spiral dynamics of dislocation forms. (c) Three sources S1, S2, and S3 from three layers form, S2 and S3 from inner and external layers, which perturbs the central layer S1. The dark lines formed on the 2D surface of H device deform and generates rings. The rings formed on the surface of H device is projected to the screen 4095 ways. On the H device surface dynamic spiral knots rotate clockwise or anticlockwise. (d) The loops and knots formed between S1–S2 interface and S1–S3 interface are shown, which eventually emits particle-like magnetic structure in the panel (e). Four periodically oscillating magnetic loops bind into the 3D particle-like magnetic flux distribution as shown in the panel (e).

2D rings. Each vortex crystal has N number of OAMs or vortex atoms, each with a topological charge $l_i = I_i m \exp\left[2\pi i \frac{\varphi_{l_{i+1}}}{\varphi_{l_i}}\right]$. The 2D projection of magnetic vortices is $P(r,t)$ or $\sum_{i=1}^{N} C_i(r,t)$ created by multiple materials on magnetic films. Magnetic 3D vortices pass through even if the optical vortex is blocked using a black carbon sheet. A magnetic vortex reflects in the mirror, moves through the water or a non-magnetic plastic, bends by a magnet etc.

8.6.1 Knots of Dark Lines Act Like a Spin of a Particle

Phase singularity or phase discontinuity has become a major point of discussion in the recent times due to the manipulation of angular momentum of light and other properties using light-matter interaction (Genevet et al., 2012). The basic laws of reflection and refraction are modified (Yu et al., 2011). Geometric parameters of a helix are responsible for the phase singularity or phase discontinuity (Huang et al., 2013a, 2013b). In case the surface has a metamaterial property, anisotropic transmission of electromagnetic signals along the surface enables out of plane reflection and refraction of light (Aieta et al., 2012). One of the most exciting parts about this phase singularity or phase discontinuity is that it is dispersion-less, which means, one the propagating light build up a singularity, that angle would not change while passing through the material (Huang et al., 2012).

Out of three concentric coils, central one processes, the inner and outer coils perturb (Figure 8.12c). Since coils are twisted, mixed screw and edge dislocations on the YZ plane of coil surface is realized by rotating x-axis by angle δ, so that YZ plane rotates by δ, $(x, y, z \rightarrow x, y', z'; k \rightarrow k^*, i.e. k_1, k_2, k_3 \rightarrow k_1, k_2', k_3')$. Each layer generates distinct waves, the inner coil k_i^*, and outer coil k_o^*, perturbs the central coil's straight stationary dark line k_c^* at an angle α and build singularity structures D_i and D_o, respectively. Inner and outer coils are internal and external mirrors, act as a pair of 180° out of phase sources, thus, fit to edit perturbation phase 0°–180° and 180° to 360°. Splitting of light into the isolated optical and magnetic vortex is achieved by tuning α, to do that the relative position of inner and outer coil shifts with respect to the central one. Three concentric spiral cylinders are in contact via yz plane, hence, k^* lies in the yz plane making an angle α with Oz, sign positive toward Oy, we can simplify above equations $k_1 = 0$, $k_2' = k\sin(\delta + \alpha)$, $k_3' = k\cos(\alpha + \delta)$ and from this identity we get $(kr')^2 = (e_1\cos(s))^2 + (e_2\sin(s+\Delta))^2$, which gives r', average radius of loops like circles and ellipses. To find knots or zero level contour, or spiral dark knots with radius $\Delta e/k$ we find s,θ', here, $D(s,\theta') = kr'\cos(\theta' + \zeta) + e_1\cos(s) = 0$; wherefrom we find s,θ' convert $D(s,\theta') \rightarrow D(y',z')$, it contains wavefunction $\Psi_i(r,t)$ of OAM set.

Equation (8.17) links a shape-changing vortex or a helical nanostructure with the geometry of its 3D magnetic vortex output. To experimentally verify this, we shine a polarized 633 nm laser light on the solution of a helical nanowire, dendrimer, helical gel, etc. kept in a quartz cell, the light is passed through a $\lambda/4$ plate vortex lens for 400 nm, not 633 nm (Figure 8.13a, b). If we use 400 or 800 nm laser, the optical vortex is strong, but at 633 nm a mismatch makes the system unstable and subtle changes in the magnetic vortices are distinctly reflected. Output magnetic vortex is projected on two types of films, Ni or Fe microwires dipped in oil microsphere, long exposure (>2 hours) changes the film color, bright rings map a 2D slice of the 3D magnetic vortex. The separation between the magnetic and optical vortex 5 m away from the sample is around 15 cm, it depends on the angular relation between reflector, sample and vortex lens. On the sample we shone $\hbar\omega$ using a Yagi antenna and estimated the stored charge Q on the surface and magnetic flux on the film as a function of L, D, P, and ab by two separate experiments. The change in capacitance with the voltage at different ac frequencies gives Q.

8.6.2 SYNTHESIS OF SUPER-SUPER COIL MADE OF DARK LINES

When both $\bar{e}$ and Δe are present, $\bar{e}$ tries to form a static helix and Δe tries to form a dynamic helix. If $\Delta e < \bar{e}$, i.e., $(e_2 < 3e_1)$ the static helix acts like a guideline on which another dynamic helix forms, we get a supercoil. If additional perturbation

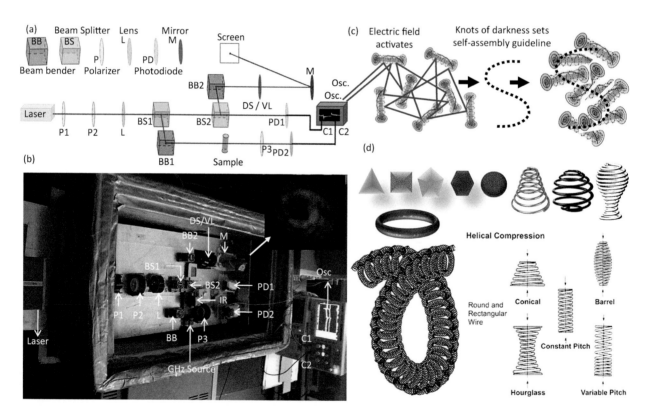

FIGURE 8.13 (a) Schematic and (b) actual experimental set up of quantum coincidence. Here, photons emitted by the polarized laser light laser (632 nm) are split using 50/50 beam splitter (BS1, HP 10701A 50% beam splitter). By splitting one of the 50% beam using 50/50 beam splitter (BS2), the coincidence effect of the two intensities is seen by high sensitive photodiode (PD1 and PD2). The output of this diode is observed in the oscilloscope (Osc.). Remaining 25% beam is used to see the vortex (VL) and double-slit (DS) images (with and without sample); inset shows an optical vortex. (c) The knots of darkness produced by different H devices form magnetic vortices and create a guideline for self-assembly which brings the H devices together. (d) Various types of geometric shapes are shown, their corner points are singularity points which bind together to form unique architectures.

arrives (multiwall helical nanotube), we get super-supercoil of magnetic field lines and so on. The approximate solution, when polarization is oscillatory $\Delta e = \sin\delta$, is two spirals, made of magnetic null or electric null, twisting each other. Together, they form a cylinder $r' = \Delta e/k = \sin\delta/k = \lambda\sin\delta/2\pi$, the pitch of the helix and the velocity of magnetic vortices are

$$\Delta z' = \frac{\lambda\sec\delta}{2}\begin{cases} < 0\,\text{right} - \text{handed} \\ > 0\,\text{left} - \text{handed} \end{cases} \dots \text{ and velocity } v = c\sec\delta \tag{8.6}$$

Nature of harvested pure magnetic flux from light depends on the angle $\propto$ made by perturbation wave from the inner and outer coil. At $\propto < 0$, average intensity $\bar{e}$ and polarization Δe both should be high (~2), to create a ring of electrical null, that is a magnetic ring along with hairpin-like electrical singularity or magnetic structures $\vec{\psi}_{mag}$. At $\propto > 0$, even if there is no polarization $(\Delta e = 0)$, even a low-intensity signal $(\bar{e} \sim 1)$, generates a ripple-like magnetic structure (no hairpin) in a twisted helix. If $\propto$ switches sign, the clocking direction changes. If both the inner and outer coil contributes together, $\vec{\psi}_{mag}$ holds a combination of clockwise and anticlockwise rotating rings produced from the same signal source in the central coil. If Δe varies periodically $\Delta e = \sin\delta$, then $\propto$ switches sign periodically, we get a superposition of three patterns. Due to $\Delta e = \sin\delta$, the twisted spirals made of magnetic and electric dark lines are also created on the central coil. The super-super coil is shown in Figure 8.12d.

In principle, 2^{12} topologies could be created on the central coil governed by Hamiltonian **H**, which links atomistic events to the blinking of cones, i.e., clocks that links G-mode to vortex atom made crystal $\sum_{i=1}^{N} C_i(r,t)$. By changing the color of reading light $\hbar\omega$, we could selectively project the phase spheres on the magnetic screen, read $\sum_{i=1}^{N} C_i(r,t)$, part by part. Then, by shifting the angle of incidence on the quartz cell, and pumping a GHz noise, we changed the beating caused by three identical concentric cylinders, thus concluded that beating amplifies B, enhances the magnetic field resolution in $P(r,t)$.

Seeing atomic-scale dynamics on a large film 5 m far from a nanostructure is a new marker, low-cost ambient spectroscopy (Figure 8.13b, inset). While optical vortices carry only the density of electrons on the surface, the magnetic vortices carry not just the geometric features, but the dynamics related to the symmetry breaking of multiple modes of vibrations of all three layers. The theory begins at the atomic scale, includes input em resonance and the light-matter interaction together with the unique roles of device geometry. Most importantly, one could wirelessly write the G-modes selectively on the device, which can memorize it, then erase, read and transport it to a long-distance, since a linear relationship between the stored charge and the magnetic flux produced holds true, widely. In an assembly, the G-modes of several H devices would get active and since the whole interaction process is wireless, a higher level assembly would trigger (Figure 8.12c).

8.6.3 Morphogenesis of Knots in Vortex-Like Magnetic Atoms

All rings in the vortex are quantized: phase singularity of dark lines regulates geometric musical language (GML): topological charge: In the empty space, destructive interference of electromagnetic wave at nodal points is a phase singularity. Along a dark line, around linear phase singularity, the phase of reflected light from trapped charge site changes spirally around the singularity point. Some geometric shapes created by the topology of dark lines is shown in Figure 8.13d, this is how geometric musical language is implemented using H devices. To cover the distance of the electromagnetic wave, the number of times the topological charge rotates is the topological charge. The topological charge is the measure of the clocking charge on the surface of the coil in a Hinductor device. The number of dark rings with electrical null or magnetic vortex atoms generated during interference is a measure of the magnetic flux. The relation between charge and flux is constant. The magnetic ring looks like the vague attractor of Kolmogorov (VAK) that could model the stationary states of quantum mechanics (El Naschie, 2003). The relation between charge and flux is constant H. The physical significance of H is the inverse measure of singularity in an assembly of topological symmetries. Since a device could open or close the singularity holes in the phase sphere 4×10^3 ways, the magnitude of H~10^{-3}. For each dimension of trapping cells on the coil surface, a distinct geometry of the magnetic ring is produced, with a distinct topological charge. Under noise, the sum of topological charges and the sum of magnetic rings produced are a function of a number of distinct types of cells the dimension of cells and the total number of cells on the surface.

Thus, in the electrical null, E $\neq$ 0, rather $E_{ring} = E_{outside}/C^2$. Inside a ring, the velocity of field is $v = c\sec\delta$, where δ is the twist of the Hinductor coil, generating a mixture of the edge-screw defect. Now $\sec\delta = hv/hC = \phi/\phi_0$, thus, the dark rings are quantized, one cannot create rings of any arbitrary diameter. Quantization ensures the formation of distinct geometric shapes since the creation of an additional wave function $2\Psi_n^2 e^{i2\pi\phi_n/\phi_0}$ ensures a clocking interaction between discrete magnetic rings $\Psi_1, \Psi_2, \Psi_3,\dots$ A pair of rings separated by x and L are related by $\Psi_2(x+L) = e^{i2\pi\phi_{12}/\phi_0}\Psi_1(x)$. If Ψ_1, Ψ_2, Ψ_3 forms a triangle, with an angle between Ψ_1, Ψ_2 as $2\pi\phi_{12}/\phi_0$, then we get three angles among three hypothetical sides of a triangle. It is the basic geometric information creation for GML. In a closed ring of phase singularity, if an oscillating electric field is applied, a spatial gradient of the magnetic field is created, which clocks around the ring. We made a journey from lattice surface to the magnetic knot based geometric structures which one could read using magnetic vortices (Figure 8.14a). Using scanning tunneling microscopy images, one could map the distribution of charges on the fourth-circuit element surface and by pumping monochromatic polarized light see the magnetic vortices. The composition of circles is the time crystal we described as GML in Chapter 2.

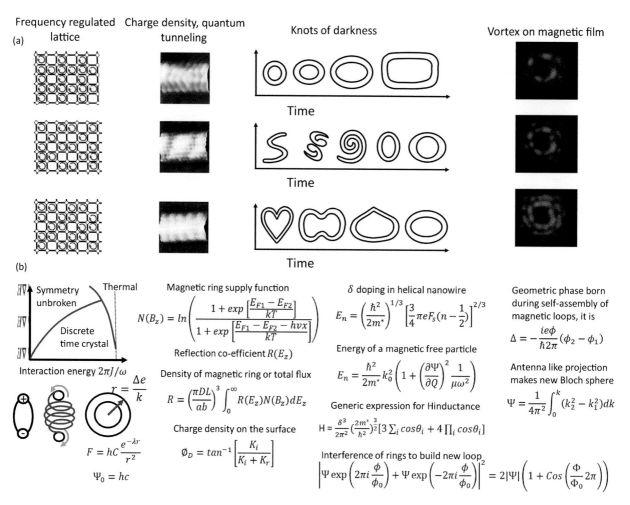

FIGURE 8.14 (a) A table showing density of charge on the surface of a microtubule and the knots of darkness, magnetic vortices. First column shows schematic of cells filled with charge and empty cells. Second column shows the STM images of the microtubule surface under the ac electromagnetic signal pumping. Third column shows the schematics for knots of darkness. Fourth column shows the magnetic vortex. (b) The top left plot shows the phase diagram for the time crystal. The horizontal axis shows the interaction energy between a pair of clocks and vertical axis shows the perturbation to the system. Three regions have been identified, a region where symmetry does not break, where too much heat does not allow holding the ground state and a domain where we see the time crystal. Just like electrical dipole magnetic dipole exists which has unique topology. All essential equations are compiled, terms are explained in the text.

8.7 MAGNETIC KNOTS' DRIVEN SELF-ASSEMBLY

Using a 3D distribution of such tubule-morphic H devices we can build a time crystal computer that is the artificial brain or nano brain or a brain jelly. Kids often play a game where they put a large number of magnets in a loop shape on a table and some others randomly within that loop. A small perturbation to that arrangement triggers self-assembly and all randomly distributed magnetics come together forming a linear assembly. The line of magnets resides along with the skeleton structure located inside the loop, one could write a program to build the skeleton of a shape and the magnetic would follow just that (Figure 8.13c). Getting aligned to the path set by the magnetic or electrical knots of darkness is the fundamental information processing mechanism of the artificial brain described here. Self-assembly of tubulomorphic devices builds artificial magnetic metamaterial like structure like neuromorphic devices (Baena et al., 2004). The self-assembly does not stop, it builds

neuromorphic devices, cortico-morphic devices as outlined in the 17 distinct brain morphic devices in Figure 9.2. One interesting discussion regarding the self-assembly of their biological counterparts is in Figure 7.11. Magnetic field driven the synthesis of helical superstructures is well studied system in the research field of synthetic chemistry (Singh et al., 2014). However, in this particular case the self-assembly is driven by phase shifts and fine tuning of the negative refractive index of the elementary H devices during self-assembly process (He et al., 2007). The brain jelly or the key computing organic material is not made of switch-like devices, which would be H here. During the course of structural synthesis, from basic precursors H devices grow, even asymmetry in mechanical vibration like sound could regulate the artificial magnetic metamaterials, which are basically a composition of different metamaterials, i.e., with different refractive index and magnetic permeability (Wang et al., 2014b).

Synthesis of protein-like structure using helical carbon nanotubes is quite common in materials science (Grigoryan et al., 2011), in the assembly they do not go for liquid crystalline features rather grow further. One of the prime objectives of Hinductor based brain jelly synthesis is that helical superstructures would build up naturally (Nakashima et al., 2005). Hinductor or fourth-circuit elements are to be synthesized in the solution under the presence of an input signal (Wang et al., 2014a). Helical superstructures are often synthesized for the directed reflection or formation of directed light. Not just directed light, we could control even the light reflected, filtering frequency (Li, 2012). Polarization direction and typical features of polarization are also detected using helical nanowires. Polarization direction and typical features of polarization are also detected using helical nanowires (Mathews et al., 2010). Even the structures change when light falls onto a system (Bisoyi and Li, 2016). The particular feature is very useful for hierarchical spontaneous information-driven growth for the synthesis of brain jelly from nanowires. Lattice-based photo-tuning by self-organized superstructures (Qin et al., 2018) are found extensively reported in the literature.

The physical realization of a quantum clock to build a quantum time crystal: The difference between quantum and fractal time crystal? This book explores the possibility of a fractal time crystal spread over 12 imaginary worlds, it could have a classical or a quantum version. We have summarized several basic physical situations encountered in the time crystal exploration in Figure 8.13b. The idea is to make the reader visualize how very well known school level quantum studies contribute to the new understanding of time crystal with a twist. A quantum clock satisfies two types of constraints—first is bound on the time resolution of the clock which provides by the difference between the minimum and maximum energy eigenvalue; another one is Holevo's bound which tells about how much classical information can be encoded in a quantum system. In this work explains results such as optimal quantum clock using trapped ions (Buzek et al., 1999). As the Figure 8.14b explains the phase diagram of a large number of clocks working as a single system (Khemani et al., 2016). There are three domains; when an interaction between the participating clocks (e.g., spins of an array of electrons) is very low, symmetry does not break and if the energy is more then thermalization breaks the coupling between the clocks. In between the two limits there could be a new phase of matter in the pre-thermal regime which is protected by discrete time translation symmetry (Else et al., 2016, 2017). The clock assembly breaks time symmetry and automatically regenerates it. If the quantum ground state is essential, then quantum mechanically no one could realize a quantum time crystal (Bruno, 2013). The demand for quantum time crystal is beautiful (Figure 8.14b, top left), there is a thermodynamic restriction, once measurement is done the system would not remain in the universal ground state, for complete and autonomous measurement, the clock or periodically oscillating system must find a meta-ground state so that later, post-measurement the system returns to the universal ground state (Erker et al., 2017). The space-time could be symplectic, i.e., differentiable,

geometric (measure length and angle; McDuff and Wehrheim, 2012), some of the quantum clocks are same as which characteristic of broken symmetry and topological order and other are new which characteristic by order and non-trivial periodic dynamics. In simple words, slow down the energy exchange process finding cool imaginary ideas (Peres, 1980). Quantum has only one imaginary world, so it is difficult to keep imaginary relation intact and still switch between global minima and excited state. Quantum time crystal (Wilczek, 2012) is impossible, such an argument has valid points (Watanabe and Oshikawa, 2015) in spite of prescriptions on how to build a practical time crystal (Yao et al., 2017) and subsequent claims to realize the same. In this book, 12 imaginary worlds layered one inside another achieves one incredible feature a higher topology of clocking between the imaginary world ($2 \times 2 \times 3$. $2 \times 3 \times 2$, $3 \times 2 \times 2$) that feature has the ability to hold on to a time crystal. Just dumping noise to higher-dimensional worlds is not enough; only then one could achieve a truly autonomous machine (Woods et al., 2016).

8.8 DESIGN APPLICATION AND OPERATION OF H—THE NOVELTY OF H

Figure 8.15 outlines Hinductor's journey with applications, where should we use such a device, while Figure 8.16 outlines the novelty of this device.

1. **A linear array of capacitors forming an inductor: edge and line dislocations on the lattice are required:** We need capacitor like extreme charge storage devices to arrange in the form of a spiral or vortex or fractal so that no current flows. Either leakage current, or soliton or electromagnetic wave propagating through the surface would interfere with the geometric boundaries and would create an electron density distribution analogous to the charge stored on the surface. Current and voltage have no effect on this device. So, we get only two variables of magnetic flux and stored immobile charge.

2. **The ratio of helical pitch and diameter a pure geometric variable would determine it band or electronic/optical property:** Helical magnets or helical distribution of magnetic flux changes magnetic property very differently, that's why super-super helices would be a fascinating geometric phase architecture to look into (Katsura et al., 2007; Ishiwata et al., 2008). The helicity of the geometry is a robust concept, helical does not always mean to be a cylinder-like spring. Growth/decay of helical conducting path with the storage of charge depends on a geometric parameter P. The ratio of helical pitch to the radius of helix is P, P and the dielectric constant ε of the helical path, control the fundamental features of φ-q relationship. By tuning P and ε, one can modulate signal amplification and energy storage features of H. Due to the periodicity of a helix and quantized charging it tunes the electromagnetic energy very

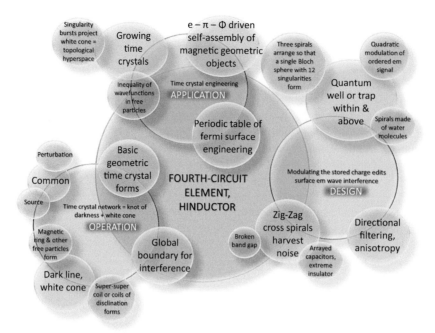

FIGURE 8.15 A summary of the fourth-circuit element fundamentals for the industrialization.

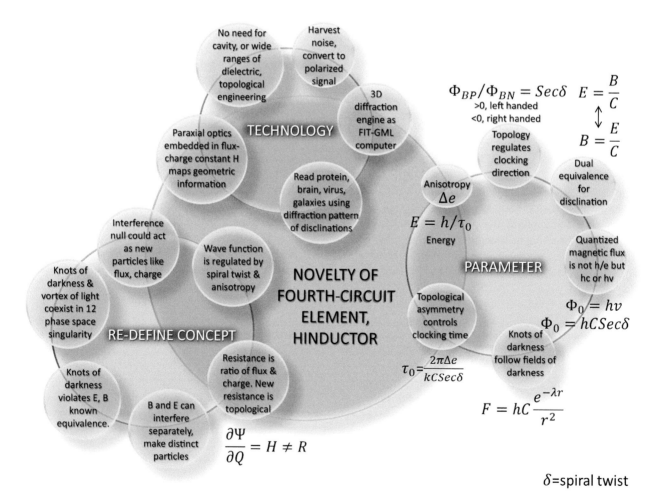

FIGURE 8.16 Novelties of the fourth-circuit element.

accurately. By modifying the helical pitch, one can change the quantization features. Lattice cells are key charge oscillating center. One should be able to write a pattern of charge distribution by changing the geometric parameters of the device. One possible route could be pumping a composition of ac signal wirelessly.

3. **Operational between upper and lower bound frequencies:** Though there are no real capacitor and inductor in the device, we may consider rapidly oscillating dielectric centers as capacitors and spiral pathway induced anisotropic transmission as an inductor. Capacitors are simply dielectrics; bonding between multiple capacitors also behave as electric dipoles. Simply by changing the material composition if two parameters (i) threshold frequency of the capacitor and the (ii) threshold frequency of the inductor are tuned by modifying the architecture, then capacitance and inductance values of the materials are changed. It is possible to tune these values to change the threshold frequencies. The capacitance determines the lower limit and inductance determines the upper limit of the bandpass frequency. If inductor L and capacitor C are fused, by constructing the inductor by linearly connected capacitors then for the new element H we get much faster response against load R. Here is the mathematical proof: Though R, C, and L cannot generate H, for the relative comparison we construct a simplified current expression for H. Tubulin dimers are charged following $Q = Q_0\left(1 - \exp(-t/CR)\right)$ and as $Q \propto L$, we get $L = L_0\left(1 - \exp(-t/CR)\right)$. The current grows in the inductor following $i = i_H\left(1 - \exp(-tR/L)\right)$, therefore the current in an H grows following equation $i = i_H\left(1 - \exp\left(-tR/L_0\left(1 - \exp(-t/CR)\right)\right)\right)$ and decreases following $i = i_H\left(\exp\left(-tR/L_0\left(\exp(-t/CR)\right)\right)\right)$. Considering C dominating current, we get current growth as $i = i_H\left(1 - \exp\left(-t/RC_0\left(1 - \exp(-Rt/L)\right)\right)\right)$. It is much faster than any resistor, capacitor or inductor.

4. **Topological properties for asymmetric absorption of electric and magnetic vectors:** Recently, the synthesis of helical nanotubes and rosette structures has received wide attention; one ripple effect is the creation of extensive theories on helical electronics. In a cylindrical conductor, the injected 2D electron gas from one end distributes in a helical symmetric pattern on the surface due to polarization. The electron density peaks are periodic along the length. In a helical nanotube, the bandgap closes at a point called Fermi contact point if the conductivity is low or is semiconducting in nature. However, the gap closes at more point contacts if the conductivity turns metallic, which enables the system to exhibit coherent transport at a particular condition. When the current flows through a helical nanotube, Fermi contact point acts as a gating channel and enables quantized transport of charge. It is a situation similar to the creation of a fractured band. H by nature will be an insulator and the geometric path would determine its key carrier transmission, therefore, we have simulated the power usage of this device, which ranges in the domain atto-watt to femto-watt (picoampere, microvolt). Spiral current also generates a magnetic flux only at certain frequencies, which shifts the entire band of energy levels equal to the number of quantized magnetic flux $\left(\Phi/\Phi_0\right)$ in the k space, enabling H to store energy.

5. **Phase quantization:** When ac signal flows, it changes the polarization of signal, we cannot trigger charge to move in one direction, thus we cannot generate a finite magnetic flux. Therefore, we need an ac signal on a dc background. Obviously, we need two simultaneous additional control parameters ac frequency and a dc bias to release charge and generate magnetic field unlike conventional L and C where only ac or dc signals are sufficient to generate the basic property. When ac signal is maximum it is absorbed by dielectric molecules of the capacitors, the current output is zero and when it is minimum, carriers are released and the current output is maximum. The output signal is therefore always 180° out of phase from the input signal.

6. **Quasi-particles: Ultra-low power device:** If the embedded spring of the spiral architecture expands/contracts, the structural symmetry breaks, Dirac points split and the bands in point-contact are separated. However, if the symmetry is restored, the degeneracy appears, and the two-phonon bands touch at distant points. Thus, point-contacts of Dirac gates open and close continuously—a perpetual interplay of soliton fractions between two limits unravels how a purely mechanical (phonon) transport of energy safeguards pure electronic signals even when noise is more than the processing signal. In the learning[4] phase, solitons are locally synchronized, microtubule accepts input signal/noise, corrects errors by tuning soliton fraction. It operates in a very low power domain (nano-watts), etc.

7. **Ferroelectric, piezoelectric and pyroelectric behavior:** A single spiral structure's perfectly square or lossless hysteric current-voltage (IV) characteristic depicts an ideal ferroelectric memory switching (Sahu et al., 2013a), where hysteresis area is a function of maximum applied bias/current. The device does not flow current, entire response is related to leakage current and voltage induced dipole moment switching of the cells. It has been demonstrated live that in a nanowire, if the constituent dipoles made of proteins switch in a regular manner then conductivity switches as observed in the current-voltage characteristics. Since the spiral's ferroelectricity leads to its piezoelectric and pyroelectric features, it oscillates like a spring between two limiting lengths.

8. **Strong anisotropic elastic properties: resonance frequency in the ratio of primes:** Thus, it is tested and verified that these devices are subjected to strongly anisotropic elastic properties; it was proved that longitudinal bonds along the protofilaments are stronger than lateral bonds between adjacent protofilaments. Simply by calculating the longitudinal modes, it has been found that vibration would decrease by 1000 times than its dipole-based calculated values. Later, rigorous orthotropic shell model-based studies, which included anisotropy in a more generic 2D shell-like hollow structure of these spiral architectures only reproduce experimental observations, even if we do not include the transverse sheering. If the small-scale effect of non-local elasticity is introduced, which considers that local strain is the collective output of global force, the result hints at 10,000 times lower time-domain resonant oscillations. However, on this concept, when circumferential mode or local radial strain is included, the vibration onset frequency is reduced by 100,000 times. Thus, depending on typical modes, the spiral's resonance ranges from a few kHz to MHz to GHz, which means 1:1000000 times bandwidth variation is observed. Most strikingly, since 12 holes in the phase space blink, the resonance frequencies burst like an antenna.

9. **Quantum cloaking:** We are aware of the electromagnetic cloaking, where the light bends around a device. It may also be possible that by pumping a vortex or spiral, one could resonate a dielectric material electromagnetically. At resonance, matter wave passes through like a lens. It is quantum cloaking. Using an H device, one could pump suitable ac frequency signal using an antenna and simultaneously scan the material using a scanning tunneling microscope, the tunneling image would show vanished material, the empty surface. If we have three-layered helices or vortices or multi-fractals, then we could selectively vanish a part of the material using a suitable ac pumping frequency.

10. **Wireless communication between similar devices:** H transmits electromagnetic signals of a particular frequency (we call it resonance frequency), this passage causes electromagnetic oscillation in the structure, it might affect the associated mechanical vibration-induced conformational changes, which could shift the resonance frequency. By suitably combining the spiral structures of different lengths one could generate the desired resonance band. There are three parts in the resonance band of any seed oscillator structure, one part is kept open for coupling with the next seed that will be formed after the self-assembly, one part is for its own processing, and the third part couples with the resonance bands of the seeds inside the structure. Thus, every seed

H device has two hands in the resonance spectrum. Synchronization could trigger with one or more peaks singular or in a plural manner in the system. The capacitor molecular structures in the 2D hollow cylindrical mesh act as an evolving circuit at resonance frequencies in such a way that it works as an antenna and also a receiver for wireless and lossless electromagnetic power transmission. The brilliance of this wireless transmission is that (i) a sender nanowire device automatically selects a suitable receiver nanowire for the maximum power transmission in a bundle, because there is a quantum cloaking at the molecular scale; communication is classical, but quantum cloaking enables the atomic-scale molecular property to reveal itself, (ii) even by wirelessly sending a part of a time crystal code, an entire information packet could be transmitted, (iii) the operational power could automatically be amplified during a "communication avalanche." Such a massive scale power transmission is not just lossless, rather creates an additional power logically as required in the receiver, which has always been considered impossible in the conventional communication engineering. The sender does not send much power; the receiver fills the canals once it receives the code.

8.9 TRANSITION FROM THE OLD ERA OF ELECTRONICS TO MODERN MAGNONICS

There existed a world of superconductors. There will be a world of super non-conductor giving everything that superconductor gives. It does not transmit electron, but a phase in terms of quasi-particle propagates through the surface and interacting with the transmitted waveform (Sun et al., 2012). Metamaterials are widely used for analog computing, performing mathematical operations using sound waves (Silva, 2014; Zuo, 2017), solving differential equations and integration (Abdollahramezani, 2017), often data is encrypted by geometric phase by modulating the surface (Biener et al., 2005). This is now a journey from collision to coexistence, rejection to live together, magnitude to look (Figure 8.17). Two features make a significant difference in the cultural transition where a mechanics used multi-meter to solve any electronics problem, to the world where they would use a semiconductor camera to measure magnetic flux pattern. How the world would change is outlined in Figure 8.18.

Hologram generated by metamaterial: advanced computing harvesting metamaterial surface: Quadratic $e - \pi - \phi$ control on the resonance, cloaking and holographic projection: We are all aware of phase hologram; it has become a tool for the entertainment industry. Hologram and the time crystal-based GML has an intimate relationship. When a cube is inserted inside a phase sphere, its eight points touch the spherical surface. No image is captured by

	Current-voltage paradigm	Charge-flux paradigm
Transmits	Current (I)	Magnetic flux (ψ)
Potential driver	Voltage (V)	Charge(Q)
Resistance	$R = V/I; H = \infty$	$H = \psi/Q; R = \infty$
For Capacitor	$R = \infty$	$H = finite$
For Inductor	$R = 0$	$H = finite$
Conductance	$G = \partial I/\partial V$	$K = \partial Q/\partial \psi$
Qu. capacitance	$G\tau$	$f(density\ of\ states)$
Qu. inductance	$G^{-1}\tau$	$f(density\ of\ states)$
Energy	$I^2R; CV^2; LI^2$	$Q\psi$
Derivation	$R = \rho L/A; C = \varepsilon A/d; L = \mu n l$	
Phase change	Between I and V, R=0, C=+90, L=-90	Between ψ and Q, $H = 0, \pm\pi$
Quant Conductan.	$G_0 = 2e^2/h$	function of phase space topology
Quant flux	$\Phi = h/2e$	function of phase space topology
Regulator	Scattering, collision	Repeated reflection, clocking
Charge, spin	Charge (e, e-p, ne), e/n. Spin (p/q, fraction)	Charge, Spin function of phase space topology
Power	$P = I^2R$	$P = \psi^2 Q$
Resonance	$\omega \sim 1/\sqrt{LC}$	function of phase space topology
Current-voltage	Yes, linear for R, no for C and L	No for H
Flux-charge	No if R, L and C are separate.	Yes for H, if $Q(\psi)$. LC if $\psi(Q)$

FIGURE 8.17 A table is comparing the existing current-voltage paradigm existing today and the charge-flux paradigm that is going to come soon.

Memristor	Hinductor	Old electronic	Magnonics
Have current voltage, IV feature	Does not flow current, flows rings of Ψ	Vortex around defect, spin ice	Look alike of a monopole, can be used at RT
Helical current generate magnetic flux	Flux generates by asymmetric treatment of em wave		
Unit: resistance, R	Unit: hinductance, H; resistance=∞	Dark lines are neglected	Dark line holds novel concepts
Topology has no role	Topology governs H		
2 terminal device, a sandwich structure, semiconductor	0/n terminal, spiral array of insulators & semiconductor/metals	Tune electric field to generate force for synthesis or self-assembly.	Tune magnetic particle geometries so that in an electric field they build structures.
Noise disrupts operation	Harvests noise		
No perturbation required	Three layers with similar helical symmetry to create free magnetic particles		
Wide ranges of IV unit, transistor, amplifier, filter	Wide ranges of QΨ unit, transductor, amplifier and filter	Fixed boundary conditions	Unstable, rapidly vibrating undefined boundaries.
Wiring needed	Wireless 3D orientation		
Energy transfer, power needed	Symmetry transfers, rebuilds information	Only one imaginary world in quantum	Layers of imaginary worlds and even they interact and integrate.
Need e insulating shield,	Need metamaterial shield for em wave.		
Transmits one frequency	Processes time crystal, many frequencies		
Low temp for superposition	RT superposition by coexisting lattices	Regulates conductivity not topology	Harvest singularity Extreme insulator
Wire freely, objective: function	Assemble only by fractal symmetries		

FIGURE 8.18 Table left, Existing fourth-circuit element memristor and Hinductor, a comparison. Table right, Old world of electronics and the future world of magnonics.

the device, rather some key points of a geometric shape are morphed as is in a 3D space, phase relationships between different points are preserved, this is exactly what a hologram does, a bit differently (Cathey, 1970). Thus, A geometric-phase shift is added as a "memory" to the angular momentum of light through anisotropic parameter space. A hologram could be made of sound (Xie et al., 2016), light, heat or thermal (Larouche et al., 2012) and magnetic field (Mezrich, 1970). Even water wave cloaks (Yang et al., 2016). Thus, electric, magnetic and mechanical resonances coupled by $e - \pi - \phi$ quadratic relationship would exhibit cloaking when necessary and generate holographic projection when necessary. Holograms always have an intimate relationship with the helices and vortices (Huang et al., 2013a, 2013b).

Electromagnetic cloaking, magnetic cloaking, and acoustic cloaking: Cloaking means light bends through a material and returns to the original path (Pendry et al., 2006; Schurig et al., 2006). What could bend? Ions could bend if the structure is cylindrical; since microtubule inspired Hinductor devices are cylindrical, they could turn invisible if anyone tries to see them using ion flow, even axons of a neuron or DNA could cloak with ions. Electromagnetic waves could exhibit cloaking, theoretical studies show that at two distinct frequency bands exist between 6 and 25 GHz where the material exhibits cloaking. It needs to be checked if microtubule or other fourth-circuit elements also exhibit cloaking

for audible acoustics, then sound or mechanical waves would pass through (Faure et al., 2016). At ac resonance condition, it is also possible to deliver a magnetic cloaking (Zhu et al., 2015). As it has been described above, electric, magnetic and mechanical resonances are coupled by $e-\pi-\phi$ quadratic relationship. Therefore, a quadratic cloaking by three simultaneous modes would ensure electrical signal induced power supply, the magnetic field induced information transport and mechanical force-induced hardware evolution simultaneously.

tattvamasyādivākyena svātmā hi pratipāditaḥ/
neti neti śrutirbrūyādanṛtaṁ pañcabhautikam // Abadhuta Gita

If we argumentatively state to correlate "Who am I" with every single object of this body and the universe, then we continue to get "not this" and "not this," eventually what that is left is the innermost conscious being that defines "me."

9 Brain Jelly to Humanoid Avatar—Fractal Reaction Kinetics, Fractal Condensation, and Programmable Matter for Primes

9.1 NEUROMORPHIC DEVICES ARE NOT ALONE—17 BIOMORPHIC DEVICES SING TOGETHER

Culturing 3D human brain architecture would soon grow exponentially (Pasca, 2018). By now we have learned that brain jelly is a mathematician who plays with primes to find a similar set of primes in an apparently different integer (Reddy et al., 2018). A problem is a lack of symmetry in the geometry of the time crystal that is applied to it and the solution is to find what geometric changes need to be carried out to regenerate the symmetry. Brain jelly identifies the lack of symmetry and checks memory whether it has already got those pieces of geometric shapes as clocks in its memory bank, if not it assembles the missing pieces and sends it to the questioner and or tries to acquire those pieces. Thus, a brain jelly not only simulates the future, but it can also carry out an intelligent conversation, and provide an intelligent solution to the problem. Say, one kid writes $2 \times 5 \times 3 \times 7$, now, like an obedient student; brain jelly would synthesis the associated organic product and then, if in its memory any structure was created before containing this group would start vibrating, self-assembly would trigger. Not just that, 2×5, $2 \times 5 \times 3$, $5 \times 3 \times 7$, all possible compositions would find neighbors and the integers which have a similar number of compositions of a set of primes would come together to self-assemble. It is unique, because the number of compositions is important not the primes used, $2 \times 3 \times 7$ and $41 \times 37 \times 47$ would be similar. Thus, synthesis of a material analog of a phase prime metric (PPM) (see Chapter 3 for details) continues spontaneously in the brain jelly (Strogatz, 2003), until there is a saturation. Coiling and supercoiling of vortex or helical filaments in oscillatory media is brain jelly's key bet (Rousseau et al., 1998), it is not electronics, a superior version of claytronics (Goldstein et al., 2009). A brain jelly would evolve materials as evolutionary computing (Eiben and Smith, 2015). It would be read by imaging the optical or magnetic vortices emitted by brain jelly.

One of the prime questions then arises how does a brain jelly carry out an intelligent conversation. When input information enters into the brain jelly as a time crystal, the data acquisition system in the sensor system acts in a pro-active manner, not like reading a stream of bits. The brain jelly inspired sensors have to build a time crystal, hence, instead of a bit as a function of time they acquire 10 different types of data (see geometric musical language (GML) in Chapter 3). We repeat that brain jelly sensor acquires a geometric shape made of singularity points, the clock speed or the rate by which system point changes the phase, the duration of silence or phase, the rotational direction of the system point, the noise tolerance that tells the diameter of the clock, how many distinct channels are there and if there are inter-channel signal pattern, the imaginary layer interaction, the relative amplitudes of different clocks as intensity ratios and the ratio of frequencies as the ratio of primes. These 10 classes of information were always there in natural interactions, time crystal-based FIT, the new information theory enables acquiring information content that was never taken into account. Nonlinearly coupled Hamiltonians of a large number of H devices, they sync and desync to grow dynamically and to deliver thinking for a brain jelly (Gupta et al., 2017).

However, there is an important query that the time crystal sensor makes, after acquiring the 10 types of row data for the time crystal. These four queries are to build a linguistic structure. First, what is the slowest clock in a compounded wave stream, that would be the host clock, in linguistic terms, the slowest clock is the subject who does the job. Second, what is the fastest clock, since that would tell linguistically, how the job is done. Third, the clock that is very sensitive to perturbation is the condition when the job is done, in a linguistic term that is a predicate, and finally the fourth clock that is least sensitive to perturbation is the answer to the query, what. The four-layer linguistic query is structured as a quaternion tensor, the data of the time crystal may be a dodecanion or 12D tensor, but, each dodecanion tensor is packed in a quaternion tensor as soon as brain jelly acting as a sensor acquires a data. During entire processing in the brain jelly, below the quaternion, the structure of information is never broken.

Magnetic field projected self-assembly: a key to synthesize a PPM in brain jelly: Brain jelly is a gel like liquid made of a spiral nanowire, that is a fourth-circuit element Hinductor, H. As said earlier, H builds the magnetic vortices and projects that field all around as a function of stored charge, it senses the cavity around it, builds a 3D distribution of potential centers. It requires 17 key components to build a brain-jelly-based humanoid avatar (Figure 9.1). To trigger a PPM synthesis, one has to pump a signal at a particular phase, repeatedly to an assembly of uniquely arranged oscillators, then they would activate complex vibrations and synthesis of materials following PPM (Lewis et al., 1987). Simplest geometric code is a triangle, three frequency quasi-periodicity could appear like chaos (Linsay and Cumming, 1989). Therefore, we could mechanically shake, trigger electromagnetic signal, or even pump magnetic pulses in a liquid solution of H nanowire or its gel. Magnetic field pattern in the projected vortices of H from a few new H devices. This assembly is like a nest that builds the future cage to hold more H nanowires, and as soon as the cage is filled with H devices, the output vortex creates the next version of the field distribution. So, once a seed pattern is written in the solution of H as a nucleus of gel formation, it triggers continuous self-assembly of the H devices. We convert input data into a universal time crystal using time crystal analyzer (TCA, Chapter 3), a software module. From time crystal, one could estimate the possible pattern of H devices, a seed of H devices that nucleates in the liquid and triggers spontaneous gel formation. Form a seed we see live the self-assembly of H devices into a giant structure. In a conventional self-assembly process say a tiny seed of ZnO crystal grows into a giant structure we see the repetition of the same 1D, 2D, or 3D lattice, over and over again. We may call crystal a fractal, but that self-similarity is not the case here. Here in the new kind of self-assembly, the seed structure is not cooked by the natural material property. Therefore, the geometry of seed structure is not restricted by the lattice limits found in nature. Possibilities are infinite. To write a time crystal input, we create nodes and antinodes in the cavity filled with the brain jelly, which is a special liquid of organic ligand molecules. As the time crystal of the information input builds H, first it sets the geometric parameters of the elements to be synthesized, i.e., length, pitch, diameter, and lattice of element-assembly. The instructions are written as vortices of field distribution in the liquid. Then the H devices come to the vicinity as the float in the liquid and build the seed lattice. The seed starts growing following the projected path set by magnetic field vortices generated by H device and thus follow the PPM as described in Chapter 3. The growth continues beyond input as PPM takes the control. The final saturated structure projects magnetic field distribution as the futuristic dynamics of the

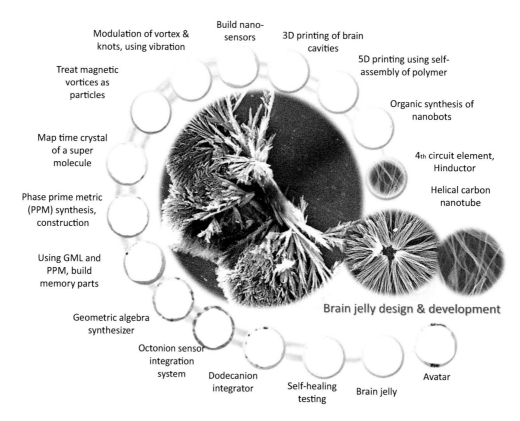

FIGURE 9.1 Brain jelly synthesis process is described step-by-step. The first step is the synthesis of a fourth-circuit element Hinductor and the last step is to build a humanoid avatar. Inside the spiral pathway, a few Scanning Electron Micrograph (SEM) images are captured for the growth of brain jelly.

input information, that structure is analogous to an algorithm. If we want to read output, simply pump electromagnetic signals of different frequencies and 3D temporal dynamics of output magnetic vortices would contain a series of solutions. Thus, brain jelly works. The question is why we don't see a repetition of the same input lattice as a fractal? The reason is that the magnetic field projection follows the same method as we multiply primes to create an integer space in the number system. Pattern of primes evolve in the integer space, so is the output of brain jelly. Field projected self-assembly is different from thermodynamic lattice self-assembly that we see around us every day.

Frequency fractal hardware = brain jelly: Frequency fractal is a fractal that may or may not look like a fractal, but it's all kinds of resonant oscillations for wide ranges of activated cavities and dielectric materials filled with wide ranges of carriers are measured as a function of time, then one may find self-similarity. The groups of frequencies form higher periodically oscillating groups, ratio of bandwidths of groups follow the pattern of primes across the entire frequency range. Dissimilarity is governed by pattern of primes, this is called frequency fractal. Computing hardware that would implement a fractal machine would follow the geometric language of primes, where 15 primes and 15 geometric shapes would rewrite any information content as events in the universe. The hardware does not care how a system's materials are arranged, but it cares about the geometric arrangement of clocks, for the time being in this book the exploration is limited to 11 imaginary layers, i.e., algebra of dodecanion, a 12D tensor. Information is an 11D time crystal presented as a quaternion, octonion or dodecanion, the time crystals are synthesized by manufacturing vortex atoms made of magnetic, electric, ionic, mechanical ripples. The unique material is named as brain jelly.

The historical background for the soft truly autonomous robotics: When engineers cut objects into pieces and then reassemble them into the desired product, it is subtractive engineering. When added liquids, jelly, or diluted materials are crafted into a pattern, which solidifies into a desired product, it is additive engineering (4D printing Eujin et al., 2017). For example, recently resin was pumped with electromagnetic fields to convert the desired part of an entire solution in a 3D object. Apart from fusion of materials (Li et al., 2018) some researchers are trying fractal condensation to grow and evolve all parts of an object at a time, continuously, just like a human body or brain grows from a single cell. Synthesis of a product spans over a vast length and time scales simultaneously (Rabitz, 2012). Recently multi-level evolution is being explored (Howard et al., 2019) in an effort to use existing suitable materials or finding a guideline for designing materials for the next generation robots (Fischer et al., 2018). It is bottom-up continued automation that designs and edits the robotic parts across multiple levels of a singular robotic architecture and transforms them into tasks by accommodating with the environmental conditions. Multi-level evolution is a concept that proposes to explore the constituent material building blocks, as well as their assemblies into specialized

diverse functional and sensorimotor configurations. Multi-level evolution is to advance industrial production. However, fractal condensation that is triggered at multiple singularity points at once in a vibrating material may offer a new type of evolution. Therein, fractal reaction kinetics (Kopelman 1988) or hierarchical memory in the chemical beaker during an organic gel-based superstructure synthesis are all parts of a true multi-level evolution that would spontaneously capture time crystals as information content from the environment and evolve the PPM, i.e., the mind of an artificial brain (Barrett, 2011). Thus, most robots are bodies with brains, but here a time-crystal-based PPM brain would be a brain with bodies (Meng et al., 2017). Here information content, i.e., time crystal of a given structure runs its clocks which carries out morphological changes in its supramolecular structure (Hauser et al., 2011). During transformation of geometric shape, the PPM regulates fundamental changes required to attain symmetry, which in turn triggers continuous self-modeling of the hardware (Bongard et al., 2006). The construction and operational challenges are plenty for the soft robots (Lipson, 2014; Rus and Tolley, 1999), yet, they would adapt with environment like living life forms (Cully et al., 2015). Thus, there is an enormous interest for the development of entirely soft, autonomous robots (Wehner et al., 2016), to enter into an era of the evolution of things or real objects from the software-based evolutionary computing of the old days (Eiben and Smith, 2015).

Self-assembled growth of all brain components: Once the oscillatory equivalents of all 17 organs of the brain are built, the next task is to combine them together. There are two major approaches we need to adopt. First, some parts of an organ should grow by themselves, because we cannot interfere in the molecular processes, for the same reasons, which prohibited us to chit-chat with the learning process and controlling the encoding process of the artificial brain directly. Second, the majority of organ parts will be constructed through a non-self-assembly process. The reason for adopting two policies is that with available technology it is not known how to cook an entire artificial brain in an artificial womb. We see the design of an artificial brain (Figure 10.1). There are two parts of this brain architecture, the upper part is the processing unit, and the lower part replicates the entire body of a living life form. All five sensory organs, including the touch sensors, arrive, synchronize, and then through a vertical column it goes up eventually to the brain-stem region. This vertical column resembles the spinal cord and the upper spring part replicates the design of an artificial hippocampus. The sensory-signals are not always mixed at the hippocampus, for example, smell-senses first reach the olfactory cortex, where from the signal is sent to the entorhinal cortex, where signals from other sensory organs mix and a fantastic linkage are created, specific to particular individual and species.

Neuromorphic no more: How are the 17 organs differ: In the early days' people used to talk about neuromorphic devices; now we can argue for protomorphic to build protein analog and then corticomorphic to build cortical column, etc.,

	Brain components	Primes in design	Analogue materials used for construction	Emulated device	Sensor building
1	Protein	2-19	PCMS, a PAMAM based organic molecular system	Protomorphic	Eye: optical sensors are attached to mechanical motors
2	Microtubule	2-13	Electron receptor doped helical carbon nanotubes, CNT	Tubulomorphic	
3	Neuron	2-23	3D polymeric matrix, PCMS versions, CNT versions, H3	Neuromorphic	Touch: Pressure temperature sensors pasted on a rubber sheet that covers avatar
4	Cortical column	2-23	H3 devices are self-assembled into 2D sheet, tube, fiber, S	Corticomorphic	
5	Nucleus	2-23	S is assembled into spirals that edits variable lattices	Nucleomorphic	
6	Cranial nerve	2-13	Semi-rigid cables mold into right geometry, H3 covered	Cranio morphic	Taste: chemical sensors of various types are added on a 2D rubber filled with jel
7	Spinal nerve	2-31	Semi-rigid cables mold into right geometry, H3 covered	Spinomorphic	
8	Connectome	2-5	3D printed fiber tracts are dip coated layer by layer in H3	Connectomorphic	Ear: Sound sensors are arranged in a 3D orientation to morph mechanical vibration topology
9	Cortex top layer	2-47	Connectome is dipped in H3 solution to build columns	Cortexomorphic	
10	Basal ganglia	2-19	Nucleus, i.e., flexible H3 based spheres selfassembled	Basalomorphic	
11	Brain stem	2-17	S devices, H3 are organized into proper prime symmetry	Stemorphic	Smell: chemical sensors are attached to a pair of tubes for air molecule sensing
12	Thalamus	2-19	Cross wires sensory inputs, so all prime symmetries added	Thalomorphic	
13	Hippocampus	2-17	Nucleus, S and H3 are mixed into an antenna, receiver	Hippomorphic	
14	Cerebellum	1-11	Nucleus, S and H3 are mixed into fibers that links output	Cerebellomorphic	Propioception: ear, visual and touch sensors are circuited for 3D perception
15	Hypothalamus	2-17	Nucleus, S and H3 are mixed into fibers, links clocks	Hypomorphic	
16	Limbic system	1-19	3D printed graphene doped antenna architecture, fornix	Limbomorphic	Time: Different clocks are circuited as single clock
17	Blood vessels	2	Mechanical tubes 3D printed pumped at ultrasound	Mechanomorphic	

FIGURE 9.2 A table describes the journey from the old school of neuromorphic devices to the N-morphic devices, where $N = 17$. There are five columns, PCMS = PAMAM, controller, molecular rotor, sensor. Oligomer-based various gel synthesis is also part of the multi-morphic device synthesis. There would be the development of seven sensory devices.

17 such key components are necessary (Figure 9.2). When a sensory data enters into the brain as a 3D time crystal, the multi-layered seed material that absorbs the time crystal resonantly oscillates like a single molecule. The brain jelly or H-assembly acts like a filter that carries out multilayered fractal decomposition process as described above. As a result, all fractal seed-groups in the input time crystal are automatically separated in a multilayered architecture (see the superposition of fractal section for details). Thus, a single image encoded in a time crystal is itself resembles to an intractable Clique problem, i.e., the brain jelly has to identify 15 encoded patterns of GML in an input pattern. When several kinds of information are linked as time crystal and form a 3D integrated superstructure of time crystals, then PPM sets the rule of phase transition. Then the geometric structures encoded in a complex time crystal could be accessed by a brain jelly through various pathways in the superposition of many pathways. It leads to a very interesting situation. During computation, even at a very local level, at any layer of gel structure, any subtle noise may select a unique high-density path suggested by PPM, that would change the entire computation

pathway and re-write the whole sequential programming. H device output, if we measure using a multi-frequency lock-in amplifier, looks like a stream of waves with different frequencies, whose relative phase relationships change with time following a different pattern. As noted earlier, time crystal analyzer reads which frequencies change in groups, when signals become undefined and disappear/appear from detector to find the continuously changing time crystal. It is how brain morphic devices are characterized, simultaneously.

How brain morphic devices interact, bond: The PPM regulated artificial brain cannot operate faster than the time required to initiate a phase transition of the slowest time crystal. The system point of a clock that has "one second" resolution moves to the world with a faster clock say "one-microsecond" resolution through the singularity; there, one brain morphic device may encounter another brain morphic device. System point gets the information into faster morphic devices and returns to the "one-second" resolution world where to an external observer computing is being performed with no detectable time lapse in the "one-second" resolution clock. In simple terms, take input in slower clock domain, send it to a faster clock domain, solve and return it to the slower clock domain. It is similar to harnessing "negative time" in quantum mechanics, but instead of one here we have multiple imaginary worlds, each with different clock speed. In order to implement the mathematical foundation of several imaginary numbers (generalized imaginary number, iota or $i^2 \sim p$, where $-1 < p < 0$) we can use the fractal mode of resonance frequency bands and investigate the possibility of realizing a similar advantage. Different p values suggest different clock-worlds. The materials of a layer are used as a seed that constructs the next layer; hence, the dynamics of each layer are totally different. When we probe molecular bonds in a crystal, the dynamics of crystal and electrons within atoms do not appear; they become non-existent and thus came the concept of imaginary spaces. In the 17 brain morphic devices different feedback protocols act to address different values of p (Figure 9.2). The generalized imaginary numbers simply denote different degrees of feedback from nowhere, adding a feedback function from nowhere for matching the time crystals is a feasible approach.

9.1.1 A Description of Peculiar Designs of Critical Brain Components Using an Organic Gel

Seventeen brain analogs were selected by studying pattern of primes in the geometric shapes of all brain components. An organic synthetic analog for all of the components would generate a layered composite, which would build a composition of 17 PPMs, a distinct metric for each brain component. Sahoo et al. have built a new simple technique to synthesis spirals and their supramolecular assemblies as network of a gel rapidly in the solution (Sahoo, 2012). The artificial brain that Singh et al has built is a hybrid device, gel-based supramolecular structure, tiny 3D printed wires, many commercial nano-sensors, organic and both organic, all were there. Power was supplied only to the sensors and pumping them to send the time crystals throughout the humanoid avatar body. Using EEG on avatar brain they confirmed that signals are passing through from sensor to the EEG.

Artificial protein (protomorphic device): We get elementary time crystal from a fourth-circuit element, H, but its operation is fairly limited. In order to sense a complex time crystal, multiple elements need to arrange such that they have a different number of loops in the helices, or length. The number of helical loops could be changed according the job description, it could be any integer series with certain gaps. The gap follows a rule, the number of rings of participating H elements follows cyclic rules, e.g., three H elements could have a number of rings 6–2–6, 5–4–5, four H elements forming an assembly would have rings 3–4–4–3, 6–11–11–6, etc. These cyclic assembly of elements are generated to manipulate the knots of darkness produced (see Chapter 8 for details) using different carriers at higher spatial scales. By editing the number of rings in a spiral, information could be manipulated in synthetic protein-like materials, 17 brain analogs.

Artificial protein complex (tubulomorphic device): Microtubule like filaments, ribosomes, several protein complexes carry out important tasks in the information transmission and evolution of a life form. Plenty of spiral nanowires are available in the literature, produced synthetically or found in nature. One should neglect many biological features, for example, dynamic instability of microtubule, and copy information processing features like change in the lattice structure of proteins on its surface.

Artificial neuron (neuromorphic device): To recall, to build an artificial neuron, we need multiple distinct symmetries inside a molecular architecture whose dynamics are associated with the structural relaxation and electronic transition or conductivity, at the same time, some symmetries should add capabilities that of an antenna. It is possible if and only if we introduce hierarchical helical architecture. Then, at the basic structure level, electronic switching would be programmed in a helix, and in another level we may use the helices together to construct another basic structure with which the antenna-like feature is encoded. Transporting carriers through that helix is the next issue. Several spiral nano-architectures promise ballistic transport, which is a quantum effect (Grigorkin and Dunaevskii, 2007). The dimension of this architecture and electronic properties associated with the transport of carriers in this material is a crucial factor. Wire is used in making the spiral assembly of capacitors so that it charges up and generate magnetic field during energy transmission; this particular feature turns the material as self-radiating antenna under certain conditions. The crucial point of magnetic vortex generation is an essential feature of the artificial neuron design. One could study time crystal based neural network in a simulator.

(i) Add new neurons, even cluster of neurons, fold brain cavities as required: The computer is allowed to construct new neurons as necessary and change its wiring as the situation demands in a software-based neural network. On several occasions, in order to create a replica of an external time crystal, if a cluster of neurons is required then, an entire cluster could be generated. The brain simulator is fed with images, sound, touch, smell, and taste input as 2D patterns as described above as a function of time. The software automatically creates the neuron firing like stream of pulses and build time crystal as described in Chapter 2. Sensors build time crystal, thus, store the input images or sounds as rhythms. During learning, if a rhythm change is required, a new clock is added, subtracted, rotation direction or diameter of the clock cycle is changed, modification happens in the time crystal, later the circuit is changed. The brain organ deforms or folds in its surface even inside the soft computer, when we follow the above protocol. (ii) **The computer sleeps** and during that time, local time crystals, which are wrongly placed at different locations, are placed at right locations. The computer goes to sleep as soon as it finds that rhythms are stored in such locations that are hindering the search and find, spontaneous reply and the transformation process, as soon as the evaluation, defragmentation, and normalcy restoration is done, returns from sleep mode. (iii) **Self-evaluation, new and better experiment, virtual neural network evolution:** The computer experiments with the circuits, if certain kind of wiring enables it to reconstruct rhythms faster and transformation between time crystals undergo more efficiently than previous protocols, then it analyses the hardware modification and changes the virtual neural network fundamentally. (iv) **Simulates the future, analyses multiple routes and then determines the most suitable ones:** The simulator runs feed-forward loops to evaluate the faster synchronization routes; synchronization means matching via search and find process. The same could be realized more profoundly in an organic jelly.

Artificial cortical column (corticomorphic device): A cortical column has seven layers, surprisingly in the cortex layer of the brain, the input time crystal enters from the top, output is generated from bottom part of a cortical column. A total 29 types of neurons were studied by Singh et al. (2018) in creating a large variety of cortical columns in the simulator, before organic jelly-based artificial cortical columns were built. Now, an assembly of H devices that builds neurons, act as a unit and self-assemble like a cylinder, analogous to a cortical column. Ghosh et al. built similar cylinders using PCMS (Ghosh, 2015b, 2016b).

Artificial nucleus (nucleomorphic device): As described in Chapter 7, the nucleus is a nearly spherical or ellipsoidal assembly of neurons wherein a small region a large number of symmetries could be encoded. It means, we want to encode say the PPMs for prime series of 2, 7, 13, 29, the organic jelly based artificial nucleus would transform the assembly. However, the same jelly would transform into the prime series of 3, 13, 31, 43, if required. Now, once according to requirement, the jelly solidifies into a cluster; it remains stable. In the biological brain, nucleus begins like an empty matrix and fills up complex functionalities depending on locations and necessity.

Artificial cranial nerves (craniomorphic device): Cranial nerves are part of the peripheral nervous system (12 + 31 = 43), they coordinate sensory input for vital life actions. Mostly such sensory input to the artificial midbrain is realized by a typical choice of dielectric materials while constructing the nerve fiber. The idea is that cranial or spinal nerves follow C2, C3, C5, and C7 symmetry, and the structural parts of the destination organs or their parts use a similar symmetry, since they are built by similar dielectric material. Thus, time crystal transfer without a break.

Artificial spinal cord (spaniomorphic device): The spinal cord is connected to skin nerve network for time crystal input via 31 pairs; additional 31 pairs of motor nerves run in parallel with the input channel to send action instructions to the body motors. Conducting gel-based spiral, long fibers are grown around insulating cables to connect sensor with the analogous spinal cord cavity, built by 3D printing. Nanomaterials with nano-touch-sensor, pressure sensor, and thermal sensor properties are mixed together as artificial skin cells. Each cell connects to conducting fibers. The holes of spinal cords are self-assembled with gels to connect motor network and spinal sensory input fibers which run parallel.

Artificial connectome (connectomorphic device): Connectome architectures are freely available online for 3D printing. Porous tube-like 0.2 millimeter-thin fibers were constructed during 3D printing of connectomes. Porous feature helps in adding organic jelly made of spiral fibers and build a continuous network of nanofibers to transmit time crystals from sensors spread all over the body to the brain, and carry decision time crystals to the brain response system.

Artificial cortex (corteomorphic device): Forty-seven functional regions suggested by Brodmann are created by hexagonal close packing of capillary tubes each filled with seven layers of organic jelly (Ghosh, 2015b). More than 120,000 such columns are distributed over 47 regions to replicate the cortex circuits.

Artificial basal ganglia (basalomorphic device): Basal ganglia has 13 sub-components, using organic jelly similar supramolecular architectures are grown in the 13 tiny 3D printed cavities. The ratio of resonance frequencies follow patterns made of primes 2, 3, 5, 7, 11, and 13.

The artificial brain stem (stemomorphic device): Brain stem is created with 17 nuclei, each specialized to process particularly prime-based metrics described in Chapter 3.

Artificial thalamus (thalamorphic device): Since fundamental drives are to be programmed here, very basic geometric controls are encoded here in all operational time domains, so that irrespective of the time crystal generated in the artificial brain thoughts, particular geometric integration or grouping is forcefully imposed.

Artificial hippocampus (hippomorphic device): During the search and find a process, the resource and encoder antenna in the artificial hippocampus should remain pristine; to save their information purity we introduced two buffer-antennas for internal communication. Similarly, the hierarchical antenna could itself get modified while implementing the rules of transformation on the upper brain time crystal-cluster, so it requires another buffer antenna. The time crystal transformation process is the most critical part of the artificial brain operation, it should be a standalone process controlled by two more antennas independent of both artificial hippocampus and artificial forebrain. The process requires two buffer antennas to take input from higher brain and resource antenna and send output to the resource buffer antenna. Therefore, four giant organic antennas operate simultaneously (cingulate gyrus, hippocampus, hypothalamus, and fornix); higher-level rules are superimposed on the input time crystal cluster to govern search and find process without disruption from any operation in the brain. These antennas are called "transformation antenna." To summarize, the artificial brain is embedded with two layers of tiny antennas inside every single neuron, then there are four pairs of antennas—resource antenna pair, encoder antenna pair, hierarchical antenna pair, dual transformation antenna pair.

Artificial cerebellum (cerebellomorphic device): Cerebellum is mostly a fractal architecture within a selected cavity as described in Chapter 7. For cerebellum there should be two fractal architectures—one propagate horizontally and the other vertically. In the central region two supramolecular architectures crossover. Two orthogonal fractals naturally synchronize wide ranges of time domain signals as singular coherent output time crystal.

Artificial hypothalamus (hypomorphic device): Processes time crystals similar to the thalamus. However, one additional geometric feature is added here, virus-like geometric codes as tiny isolated time crystals, propagate throughout the system, wherever there is a match, forms a bond by adding a linking clock. Thus, the drive for a particular desire is programmed as movable time crystal in the supramolecule, desire is also a time crystal filtering process during synthesis of decision-making time crystal and learning.

The artificial limbic system (limbomorphic device): For 3D printing high-energy radiating materials are used to build a receiver and an antenna operating together. Cortical column or hippocampus (both have same architecture H3), inspires the core structure design of every single brain component.

Artificial blood vessels (mechanomorphic device): Blood vessels generate mechanical rhythms in the brain with heart beats in the biological brain. The five-fold and seven-fold symmetric structure of blood vessel network are 3D printed using elastic materials (only for the brain, we neglect blood vessels of the body) that vibrate like the real network in the brain when resonated by ultra-sound pumping.

Filtering the slower distinct yet time crystals in the upper brain or cortex: Two major antennas are there in the artificial brain—cingulate gyrus and fornix. The basic core part of the midbrain interacts with the higher-level information processing parts located in the pre-frontal cortex, PFC. In terms of time crystals, PFC analogue hardware analyzes how very slow rhythms are given highest priority in the organic nano brain and brain-jelly-based brain. We return to the neural network of the artificial brain, where due to the continuous transmission of time crystals from the resource-antenna, smaller and larger size time crystals are being constructed. The associated expansion of active rhythms or time crystals via PPM is noted as **self-assembly of time crystals**. These time crystals are a replica of the quantized time crystals created in the helix resource-antenna RBA, the hardware analogues of gyrus and fornix. Some of these time crystals might be already existing in the brain, some of them were coupled with them during old experiences also get excited. Thus, the size or the number of clocks/rhythms of in the emerged time crystal-cluster increases, this process generates several slow rhythms built to link the participating time crystals. Therefore, resultant time crystal-pattern spreads all over the neural network of the upper brain, or artificial cortex. To make sure about the new time crystals, a cyclic update scheme runs: resource antenna ↔ upper brain ↔ resource buffer ↔ encoder buffer ↔ encoder antenna ↔ upper brain. A few cycles in this loop creates an equilibrium of input time crystal cluster in the upper brain.

The supreme role of the virtual forebrain and hierarchical antenna: At this point, the higher-level brain or forebrain, which holds the convergence rules or slowest rhythms activates, acts like an antenna (hierarchical antenna HRA subroutine) that transforms the expanded input-time crystals into more symmetric 3D architecture of clocks via a pair of transformation antenna to be described in the latter part of this chapter. In the majority of the morphogenesis of input time crystal, the transformation process in the higher brain, PFC, drastically shrinks the size of the input-time crystals-cluster. Actually, the higher-level brain simply stores from the experiences as "transformation rule," i.e., identifies what kinds of complex decisions we took for similar situations, under the similar time crystals, and simply try to impose that in the current given scenario. The "transformation rule" are time crystals that is built to link particular patterns of integers in the PPM. As transformation continues and equilibrium is established, the new sets of time crystals are sent to the resource-buffer antenna for identification of the differences between the input and transformed time crystals. The uniqueness or time crystal that could directly generate the transformed time crystal by higher brain with the help of PPM is identified and returned to the upper brain for permanent storage and the associated rule is sent back to the higher brain (an analogue of human forebrain to be precise) via encoder antenna. A hierarchical loop antenna ↔ upper brain ↔ resource buffer ↔ encoder buffer ↔ encoder antenna continues until equilibrium is reached, i.e., morphogenesis of time crystal ceases to expand. Input time crystal cluster creation loop and transformation loop do not conflict since the later operates actively only after the first loop reaches equilibrium.

The dual role of encoder antenna: The encoder antenna thus performs two vital jobs at a time, the storing of permanent memory via delicate filtering and triggering higher-level logical rules, or hierarchical antenna simultaneously. When new rules are stored in the forebrain or associated parts of hierarchical antenna, it is called "learning," and the difference between the input time crystals coming from the environment and the output time crystals coming from upper brain is the creative decision, constructed in the resource buffer antenna. The shrinking of largely expanded time crystals in the upper brain is intelligence. Unlike conventional brain builders the intelligence and creativity are strictly defined parameters in the hardware, thus tunable at will, the output solution is captured from the resource buffer antenna.

9.2 A HEXAGONAL 2D SHEET OF CORTICAL COLUMNS—A CARPENTER'S JOB

How the artificial brain decides wherein the actual system changes need to be made for self-assembling the time crystals: The necessity of beating: During resonance, if two frequency points are nearby, then a new phenomenon is generated in the hardware; this is called "beating." It causes a unique ripple effect in the localized neighborhood. The beating effect identifies the exact location of the hardware needs to be changed and how much changes are required to nullify the beating effect. In a real human brain this beating effect would identify which neuron needs to be changed, those would change structure and re-orient by firing, just like rockets maneuver in space. It is the hardware mechanism of self-assembly of time crystals. Beating is taken care of by four PPMs running in parallel.

9.2.1 The Collective Response of Quad Patterns of Primes

Development of 17 different brain components requires 17 PPMs. Figure 9.3a outlines four different metrics similar to the one we described for modeling the human brain. However, a generic engineering perspective is shown here. Four different metrics are PPM-1: sensor data collection center. PPM-2: sensory data integration. PPM-3: Sensory network compilation 8D, 12D. PPM-4: Higher level dynamics. Each metric has three parts. First, integer composition metric, prime composition metric and prime gap metric (see Chapter 3 for details). Seventeen brain components do not belong to sensor class, not just PPM-1, but they all need PPM-2 and PPM-3. Only the cortex part is made of PPM-4. Now the most important concern, how to build the simplest and the most elementary device of the brain (H3, see Chapter 8) from which constituents of all seventeen brain analogues could be produced. There are two key features one has to take into account while building the artificial organ. First, the selection of materials,

all materials in the brain would be made of H or Hinductor devices and all H devices would show the $e - \pi - \phi$ resonance behavior (Figure 9.3b).

A journey to darkness is not a sin; it's transcendence to a pure and a perfect thought: To fold the brain from dinosaur to human, nature took 65 million years. One could say, given that time, we could produce an artificial brain by evolving the existing computer. We have a concern about the fate of algorithmic computing. We have mentioned earlier that by old Turing theory, we can melt the entire universe into a single thread of bits. However, when we change the worldview that the universe is an architecture of clocks, each clock holding a geometric shape within and above, then the dynamic links undefined paths. We want to make a device, introduce a new kind of processing that helps in finding a simple topology in a rapidly changing big data. Imagine that you are in a room and zillions of multi-colored light patterns with a high gradient of intensities, and the patterns are randomly changing. Currently, a scientist would look at a petaflop supercomputer, to simulate the equations explaining the gradient of field intensity. Instead

FIGURE 9.3 (a) Four different classes of PPM hardware operating the artificial brain. (b) Three fundamental constants govern the emergence of resonance frequencies during the synthesis of organic gels, periodically oscillating electric field, magnetic field, and the mechanical field are related by a quadratic relation. (c) Three types of organic jelly are synthesized, first sensor jelly that deals with sensory molecules governed by PPM-1. Muscle jelly that is governed by PPM-2 and PPM-3 and finally the brain jelly governed by PPM-4. Eight different kinds of folding of H3 device built cortical column sheets are compiled to explain the main computing or decision-making sheets of the artificial brain. (d) The seven key step-by-step process to build an artificial brain is explained.

of looking into the gradient of colored light, we can look into the dark lines, where there is no light. Inequality relation held by a group of dark lines reveals the geometric structure stored by the system. A little grammar book can help us read the geometric shapes hidden in a few dark lines. A massive complex pattern that rapidly changes with time would become a few vibrating dark lines. Consequently, the brain or elementary biological decision-making machine is now an example of perturbation engineering that harvests signal from noise using resemblance with 15 geometric shapes (GML). In each device, two complementary symmetry breaking modules would operate against each other to harvest noise, and thus produce scalar and vector clocks which are used as elementary free electromagnetic particle-like structures, vortex. There are two types of structures. One type of structure gives nested geometric shapes, another is used for communication through singularity points in the undefined domain of resonant oscillations. It is utmost essential that we learn the simplest generic device (H3) that could alone be an essential and sufficient element to build an artificial brain. The question is to start from neuron or α-helices, for a bottom-up synthesis of an artificial brain.

The question is at which dimension should we start building the organic brain? When we reduce the size of the nanoparticles, beyond a certain limit fundamental physics laws do not work, the particle becomes a liquid. Scaling at spatial and temporal scale in the small nano-structures follow unique geometric rules (Landman, 2005). The only way to begin brain construction is to synthesize 2–10 nm size protein, or supramolecule with a flexible design. When helical structures representing protein self-assemble into a neural network like brain jelly, by adopting helical, vortices and fractal geometries, one could avoid nanoscale engineering problems. The smallest device is a trilayer helix Hinductor, H; as explained in Chapter 7, and its physics are detailed in Chapter 8. H device alone is not used in the brain, several H makes H2, and several H2 makes H3. Finally, H3 undergoes morphogenesis to create brain components. And all 17 components are supramolecular architectures of H3 class (Figure 9.3c).

Design and synthesis protocol for a generic Hinductor device H3 as sensor, processor, and memory: Doping a desired element in a system can be initiated and primed using a resonance process. Let's learn how. Standing waves combined with electrochemical reaction along a tube would precisely define where the ions would tag. The capacitor spacing in a cylindrical cavity is therefore realized by standing wave of ion vapor, the standing wave on helical nanotube is the matrix. Dopant electrostatically attracted to the maxima of a standing wave. We make three similar concentric helices, one above another. The surface lattice parameters and other geometric parameters described above govern the patterns of knots of darkness generated by monochromatic light falling on its surface, surface manipulation would regulate the dynamics of magnetic rings and other artificial atom like structures. The geometry of a higher-level structure uses the vortex of light and the knots of darkness produced by

a lower helical layer as elementary field distribution when pumped with a cluster of energy to harness. Thus, we can build bottom-up architectures. A sensor, processor and a memory element are similar self-assembled architectures of Hinductor. The only difference is that a sensor is designed to use a given set of signals for interference, for an eye it is light, for ear it is sound, etc, an assembly is dedicated to sensing only particular type of signals. For a processor, the key design has a supreme objective, increase the number of symmetries a structure could process, means increase the PPM bandwidth as much as possible. For memory, the structure is designed so that different local parts process distinct isolated domains of PPM, so that identity of time crystals are preserved.

9.2.2 Electric, Magnetic, and Mechanical Rhythms Bind Them

The brain cannot be a quantum device, cannot be electromagnetic device either: Complementary coexistence of vortex atoms and solitons: Under salty water, the electromagnetic waves cannot transmit. Under massive thermal fluctuation, the quantum wavefunctions for a material cannot survive entanglement. So, two options are left, one build solitons and or vortices. The vortex atoms are like ripples created by a drop of rain on the water surface, while solitons are loss-less transport of that vortex through a canal. Vortex could be made of various carriers, magnetic field ripple could make magnetic vortex, electric field ripple may build an electric vortex, mechanical oscillation could make mechanical vortex. Since the vortex structures are like atoms, all types of vortices can build composite material in liquid, which could pass through a noisy environment. In the artificial brain, we deal with two kinds of solitons—first electronic, and second phononic in nature. Solitons need ordered architecture to flow; together solitons or quasi particles and vortex atoms complete the story, one masters the loss-less transmission through liquid and the other through a solid. The division into multiple packets and then resurgence into a new form is possible as neurons and its clusters continuously switch among multiple allowed symmetries. As argued earlier, a canal like a path in the neural network of the artificial brain by spontaneously polymerizing the molecular machines, which makes sure to transport information without losing much energy, therein zig-zag paths are roads for solitons (Christiansen, 1997). A synchronize call to capture a particular kind of information from all around the artificial brain would follow generating electronic and phonon soliton in the solid structure and vortices in liquid and open space. While electronic solitons carry the information to the distant locations, phonon solitons deal with the noise and keep the purity of the signal intact. This is why Figure 9.3b has a unique significance in brain construction, $e - \pi - \phi$ resonance behavior. The typical electric e, magnetic ϕ, and mechanical π resonance frequencies coupled together helps in the synthesis of time crystals made of magnetic particles (magnetic and electrical vortices, Chapter 8). After all no one would dare to make a chemical brain at least in this century, who would feed it, who would build releasing excretion. The flux-charge route is neat and clean, brain needs no food, does not have to excrete. The electric e, magnetic ϕ, and mechanical π resonance work together by non-linear frequency pulling.

Synchronization by non-linear frequency pulling: Crude analogy of extreme non-linear pulling is falling onto a black hole, but in synchrony it is sharing (Cross et al., 2004). In recurrent networks asynchronous processes can mediate inhibition leading to synchronization process, therefore, both the processes can interchange between themselves (Marella and Ermentrout, 2010). The size of synchronizing clusters is an interesting issue, networks with fixed degrees of freedom synchronize faster than random arrangement (Grabow et al., 2010), a small geometric group of a network exhibits high local clustering and low average path length, as a result they synchronize extremely fast (Watts and Strogatz, 1998). Advocates of randomness may continue to endorse magic. Synchronization between electrical, mechanical, and magnetic rhythms have been an intricate relationship historically. A fusion of two or more kinds of resonances has enormous applications.

One type of resonance should be measured with other types of resonances: By now, since the story of simultaneous electric, magnetic, and mechanical resonances are returning as $e - \pi - \phi$ identity time and again in the book, readers are wondering who is running the artificial brain, the diagonal drivers of quaternion, octonion, and dodecanion or the triplet identity. The answer is both are opposite sides of the same coin. The diagonal drivers are a pair of diagonal elements of the respective tensors, which have identical values. That identity is preserved by $e - \pi - \phi$ driver. In order to realize the brain jelly, we need to characterize every single organic helical nanostructure and its assembly, at all scale, all layers should hold the identity at all time. Measurement of resonance for any forces is not straightforward. In 1964, Alzetta and Gozzini (1964, 1967) argued that to measure the magnetic resonance, the mechanical resonance of that element should be probed. It took nearly 30 years to recognize that mechanical detection of magnetic resonance is a reliable technique (Rugar et al., 1992). Within seven years, the proposals were made for simultaneously measuring the magnetic, electric, and mechanical resonances (Alzetta et al., 1999). These adventures started in the 1970s when people were crazy about quantum non-demolition measurement or the type of measurement which measures but do not destroy the quantum state. These journeys were for detecting the gravitational wave using quantum measurement by evading back action. These measurements were about keeping a magnetic particle on the mechanically vibrating atomic force microscope tip, which starts vibrating with the magnetic resonance. Bandyopadhyay et al. used polarized optical laser and observed magnetic vortices which delivered a composition of vortex atoms as a signature of time crystals in various synthetic organic brain analogs as a precursor to building a hybrid organic brain. The collective wisdom of brain-building discussed above for 17 components is summarized again in Figure 9.3d.

9.3 ANOMALOUS QUANTUM CLOAKING—VANISHING AND SEEING THE ONE WE WANT TO

Theories of quantum cloaking suggested that the current of the probability density, in a matter-wave following Schrödinger's equation behaves like an electromagnetic wave in Maxwell's equation. Two critical features of an electromagnetic wave, the displacement of electric-magnetic vectors and the field density (Poynting vector, P) hold their form even after a coordinate transformation. The invariance, a key to classical cloaking (Figure 9.4a), required a quantum analog, so the Poynting vector was replaced by the current of the probability density, and the displacement vector was replaced by a dispersion relation. As described in Chapter 8, to build magnetic vortex-like particles we need symmetrically distinct distribution of charge as shown in Figure 9.4a using examples A, B, C, and D. Some of these patterns provide negative refractive index, which is used to create superposition of circuits made of H is by making some of the active decision-making H elements vanished during computation. That enables us to build circuit with no wiring of components. One problem with the existing Maxwell-Schrödinger analogy for quantum cloaking is that two parameters, the effective mass m^* and the potential V from isolated distinct parts of a matter wave should bend and follow strict paths to bypass the material and re-join similar to a classical cloaking. Moreover, entanglement in a matter wave demands ultralow temperature for a noise free environment. For an electromagnetic wave, the anisotropic space and time ensures that distinct rays return to their original trajectory after bending through the object-to-hide, but it is quantum analog is not clear.

One has to find a general protocol to convert a material into a composition of superlens (negative refractive index). Then, a matter wave could tunnel through a material, at ambient atmosphere as it happens routinely in the quantum optics experiment through an optical lens. Not all material be suitable for that. Milton argued for a special material, where the elements in a composite, collectively resonate, thus affect far beyond their boundary. In the spiral or vortex shape composites (Panfilov et al., 1985), we discovered a unique 3D phase space of energy transmission that enables selectively converting an element of the composite into a superlens and tunnel a matter wave from the atomic flat surface through that lens. Singularities or holes in the phase space of transmission across a vortex is encoded in the geometric parameters of helix or vortices that act as the quantum analog for the anisotropic space and time found in a classical cloaking. Tunneling through a superlens often extracts a 2D or 3D matter wave hidden deep inside a cloaking material, superpose it on the regenerated matter wave. Superlensing is often attributed to anomalous local resonance. Degenerate states of H devices with distinct geometric phases hold a time crystal (Pistolesi and Manini, 2000); the system is non-adiabatic as two distinct dynamics runs without affecting each other.

Matter cloaking could be achieved by controlling the potential and the effective mass as a particle travels through a

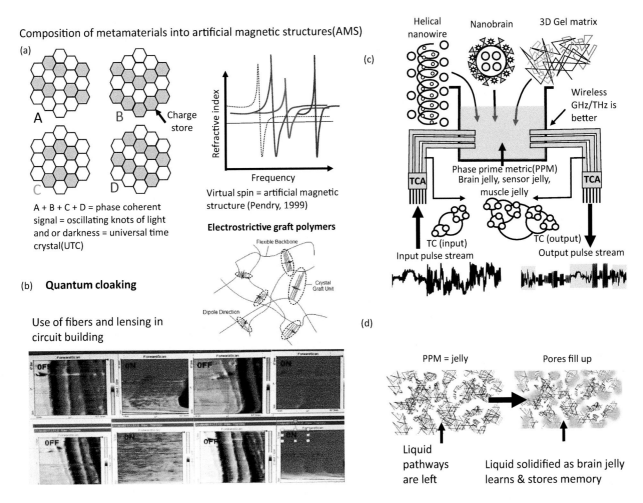

FIGURE 9.4   (a) The surface lattice of an elementary device *H* or a composition of *H* into a cortical column or column assembly into a sheet, in all cases the surface lattice cell gets active and inactive and builds a pattern for the monochromatic polarized electromagnetic signal forms pure magnetic lines and the vortices. Four such lattice examples are shown; each surface pattern has a specific refractive index vs frequency plot. Artificial magnetic structures could be synthesized using electrostrictive graft polymers. (b) Quantum cloaking is vanishing under the wave function, not light. (c) A scheme of an artificial brain. Input sensory signal forms a time crystal, which goes to brain jelly builds the equivalent spiral or spring-like *H* devices that naturally self-assembles into a set of complex structure following the PPM and finally the output is delivered as the analog waveform. (d) PPM synthesis forms porous structure in gel, in the core, gel remains liquid, memory makes it solid.

media (Zhang et al., 2008; Greenleaf et al., 2008) and making it invariant under a coordinate transformation. The problem was first noted by Milton and Nicorovici (2006). Since electromagnetic wave following Maxwell's equations is invariant under coordinate transformation it was never a difficult thing. For quantum, general time-dependent case there is no invariant transformation, so researchers take quantized 3D potentials as coordinates (*Vx*, *Vy*, *Vz*), instead of spatial coordinates (*x*, *y*, *z*), it turns matter waves like a Poynting vector, which becomes invariant under coordinate transformation (Tsang and Psaltis, 2007). Existing quantum cloaking theories use a potential that considers beyond a certain distance potential is zero, thus, the emergence of cloaking is decided even before the derivation begins. Another route is canceling the scattering when matter wave moves (Fleury and Alu, 2013). Eliminating the scattering by resonating the boundary wall of a cavity reminds us the membranes and multilayered

structures in biology (Ramm et al., 1996). The ultimate goal for everybody is to reduce the scattering cross-section of carriers to zero. It has been shown that more is the core-shell layer, better is the push-pull of carriers and more is the degradation in scattering (Lee and Lee, 2013).

However, the idea to build an elementary description of quantum cloaking begins from Schrödinger's equations (Jelinek et al., 2011). The similarity between the Schrödingers equation and Maxwell's equation has been a subject of interest for decades, this analogy is already noted above. However, thus far no reports exist on the experimental verification of quantum cloaking, one route could be taking quantum tunneling images of the object, and showing them vanishing under the tunneling current images. Theoretically calculating tunneling image is done by scattering calculation (Corbel et al., 1999).

Under the electromagnetic wave, when the evanescent wave is amplified, near-perfect tunneling could happen.

This proposal was the first in quantum tunneling (Baena et al., 2005) and was an extrapolation of Pendry's idea that if both electric permittivity and magnetic permeability are negative in a medium $(\varepsilon = -1;\ \mu = -1)$, we get a superlens, or perfect lens (Pendry, 2000). It means the positioning of a set of interconnected photons or pointing vectors on the other side of the lens. If the material surface is anisotropic, then the dielectric constant is negative, if the surface is non-spherical (e.g., cylinder), then dielectric loss tends to zero, entire signal tries to pass through, we get anomalous dielectric resonance (Ammari et al., 2013). One of the beautiful aspects of anomalous dielectric resonance is that if the charge density distribution is near to the lens where the information is being transferred, then evanescent wave amplifies and makes signal loss nearly to zero (Meklachi et al., 2016). It is the criterion for cloaking and sincerely followed while designing four circuit element Hinductor. One experimental evidence of quantum cloaking is shown in Figure 9.4b, for calf-thymus DNA, which is a fourth-circuit element due to triple helix, two molecular, one made of water. We have made microtubule, collagen, and several biological helices and our synthetic helical gels partially vanish component by component by choosing suitable frequencies, which we call anomalous quantum cloaking.

9.3.1 RANDOMNESS IS NOT RANDOM ANYMORE

To build the brain what should one do? Take a chemical beaker or 3D printed cavity and add all ingredients? Figure 9.4c explains how the most primitive brain jelly could be synthesized and be operated. Brain's neural network has three elements—neuron, glial cells, and extracellular matrices. Using H2 devices that are analogous to neurons, nanobrain, or PCMS that is analogous to neurons or glial cells (Ghosh et al., 2015a, 2015b), a polymer-based organic gel matrix is synthesized that a builds 3D neural network of multi-scale helical structures. If they are mixed in the solution, and time crystal is given as input using electromagnetic signals of various frequencies using antenna from outside, the organic materials would read time crystals, mimic analog material and continue to grow. By applying the monochromatic laser light and semiconductor camera one could read the output time crystal projected to a magnetic film and captured by a semiconductor camera.

What happens when a brain jelly learns? In the 3D polymeric matrix that often used for 3D cell culture, the pores that hold H, H2, and H3 devices reconstruct and rearrange to hold modified time crystals, one such situation is shown schematically in the Figure 9.4d. Note that the entire matrix, grown from nano to the meter scale while emulating the input time crystal follow the coupled electric, mechanical, and magnetic resonance (Figure 9.5a). Solution to gel (sol-gel) conversion and then growing into a massive architecture is rapid, the process time is fixed, irrespective of complexity of time crystals, it varies seconds to minutes. Moreover, earlier, such processes were extensively used for building the neural network, which is just opposite to the kind of neural network discussed in this book (Figure 9.5a and b). Note that when the elementary

structures grow, those are strictly helix (Figure 9.5c); however, when they grow, vortex or PPM-driven fractal-like architectures take over, they don't look like spiral, but vortices of various kinds. One could show that angular momentum is singular parameter that could change to generate all kinds of vortices.

Electrical, magnetic and mechanical cloaking for $e - \pi - \phi$ quadratic engine: Regarding resonance one of the beautiful concepts that raised was "time-harmonic" in mechanical cloaking or elastic cloaking devices. Mechanical cloaking means to an outsider, it would appear as if the mechanical wave is passing through a homogeneous media, though it passes through an object (Milton et al., 2006). There could be current sources for which outside the source, an electric and magnetic field is zero (Afanasiev and Dubovik, 1998). Maxwell noted that a torus with a constant poloidal current flowing on its surface would have poles in contact, i.e., invisible from outside (Maxwell, 1890). When multi-layered core-shell architecture of H is the foundation of higher-dimensional structures, it is said that for four layers one top of another, we get quaternion tensor. For eight layers, growing above and below, we get octonion tensor. Nicorovici et al. introduced the concept that in a bilayer core-shell structure the upper layer adds a phase value to the inner core, so the dielectric constant is $-1 + \delta$ (Nicorovici et al., 1994). Now, if the bilayer is extended to duo-deca layers, each layer with a new phase, then the partially resonant component achieves the ability to process a dodecanion tensor, i.e., becomes a brain jelly. Several examples of brain jelly are shown in Figure 9.5b–i.

9.4 A LIVING GEL THAT LISTENS AND THEN GROWS FROM ATOMIC SIZE TO CENTIMETERS LONG

Background of current gel research: One interesting feature of the ferroelectric property is that most materials additionally shows piezoelectric and pyroelectric responses. Figure 9.5f–h shows PCMS or nano brain-based nanowires that exhibit such properties. Ferroelectric polymers have polar groups at the backbone, so holds a permanent electric polarization that could be reversibly switched under the electric field. Ferroelectric polymers (Nalwa, 1995; Lovinger, 1983), such as polyvinylidene fluoride (PVDF) has an inherent piezoelectric response, consequently they are used in acoustic transducers and electromechanical actuators (Figure 9.5b). However, due to their inherent pyroelectric response, they are used as heat sensors (Furukawa, 1989). Hydrogels could be mixed to generate such a compounded property, for example, here in Figure 9.5i, carbon nanotube, nanobrain or NB, or PCMS and polymeric gel were mixed similar to Figure 9.4c, one could feel how brain jelly looks like. Examples are there in the pieces of literature, e.g., a mixture of carbon nanotube and polymer hydrogel (Qu et al., 2008). Certain physical stimuli like electric field, light and temperature or chemical stimuli like concentration of certain molecules or ions could initiate phase transition in the material which could

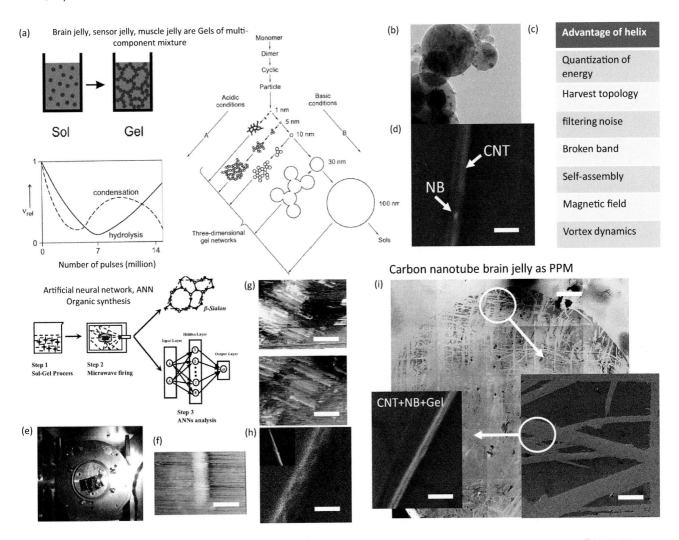

FIGURE 9.5 (a) The spontaneous synthesis of brain jelly begins at the atomic or nanoscale and the growth continues in the visible structures. The sol-gel process is followed by microwave irradiation and that builds material analog of the mathematical PPM. (b) An example of an artificial neural network, ANN synthesis by sol-gel method. (c) Advantages of a helix are outlined. (d) PCMS (PAMAM dendrimer, C is the controller, M is the molecular motor, and S is the sensor) which is often termed as nanobrain is self-assembled with multi-layered helical nanotube, scale bar 6nm. (e) The scanning electron micrograph, SEM set up in which the self-assembly of PCMS or nanobrain is carried out under different energetic trigger, distinctly different architectures are found as shown in the panels (f), (g), and (h), scale mars are 10, 120, and 10 nm, respectively. (i) TrueGel 3D hydrogel was created using carbon nanotube and the PCMS, its neural network like architectures are shown, scale bars are 1 cm, 300 and 35 nm, respectively.

result in the changes in volume, mechanical electrical magnetic or even the structural symmetry, these are called stimuli responsible gels, and their major application is in building actuators (Ali et al., 2016; Richter et al., 2007; Gerlach and Arndt, 2009). Electrostrictive graft polymers have a flexible backbone chain wherein each monomer contains branching side chains (Figure 9.4a). In an assembly, the side chains of the neighboring backbone polymers cross-link and the linked unit cell builds the crystal encompassing the entire assembly. The backbone and side-chain crystal units have dipole moment with partial charges (Wang et al., 2004). Gels that are used for 3D cell culture breaks and recreates bonds to give space to the growing cells (Tibbitt and Anseth, 2009). When spiral helices made of gels grow as Hinductor or *H* devices, on many occasions it would be essential to make mesoscale

structural changes (Usher et al., 2015). Such local changes are essential for continued brain jelly evolution as a means to deliver plasticity in learning new concepts or behaviors.

Atomic to centimeters scale brain jelly is a $e - \pi - \phi$ resonator: Electrical patterns of brain jelly are continuously monitored by EEG, and they resembles that of premature babies, the work resembles with the works of biological minibrain (Trujillo et al., 2018). Not only understanding the evolution of the brain, one of the prime applications of brain jelly would be brain-computer interface, a biological evolutionary recovery after stroke (Biasiucci et al., 2018). Artificial organ development has reached nearly to a point where adding the dedicated time crystal of a living system would enable natural healing of wounded organs (e.g., artificial skin Nieves, 2016). Evolving gel is a must to build a living brain, because gels have unique properties of self-healing, erasing and rebuilding new structures, which would be the best choice to achieve perception-based plasticity in learning in an artificial neural network. To develop a mechanical sensor, it is easy to adopt elastomeric gels (Christoph et al., 2010), by mixing it with the electroactive gels one could build artificial muscle (Randy, 2008). Eventually a brain jelly would be an $e - \pi - \phi$ resonator (Figure 4.15), whose electrical, mechanical and magnetic resonances are in a quadratic relationship ($e^2 + \phi^2 = \pi^2$), while evolving H circuit. Evolution means, when time crystals representing small functional components follows the quadratic relationship, interact by a collective dynamic time crystals propagating around and builds the composite time crystal (Watts and Strogatz, 1998) by bringing astronomically large number of H devices in groups (Shiino and Frankowicz, 1989). Even in the composite time crystal quadratic relationship holds. To hold the quadratic relationship, the small spirals acting as Hinductor or H arrange in the helical gel (Veretennikov et al., 2002; Lin et al., 2017; Celli et al., 2009; Kennedy et al., 2018; Zhao et al., 2014) in a chaotic order and wirelessly communicate (Jaeger and Haas, 2004), using magnetic vortex atoms, or ionic vortex, or electrical vortex or mechanical vortex. When these vortices travel through the solid materials pathway, they appear like a non-interactive soliton or quasi-particle and when in the open space, or liquid, they are like free particles like vortices or knots of darkness but interactive.

Persistence and self-affinity in PPM metric: PPM as the basin of attractor (all dynamics tend to converge to a particular section of PPM 3D patter) that guides self-assembly, synchronization, drives computing, and decision-making is a journey through the 3D basin of attractor if a cellular automaton like pattern evolution triggers in the brain jelly. As described in this book, basin of the attractor for nanobrain is a 3D sphere-shaped map consisting of several dynamic points, each point is a time crystal in a quaternion, octonion or dodecanion tensor, and the entire map represents a complex relationship between different dynamic states of the cellular automaton system that the system builds. Two parameters are important here. One, persistence, which means that in a time series or time fractal (infinite chain of clocks in a time crystal), large values (slow clock) are followed by a large value and small values (faster clocks) are followed by a small value, a crossover to uncorrelated behavior happens on the larger time scales. Self-similarity, sometimes parameterized as self-affinity correlates two or more time crystals, now, the correlations change with time, sometimes it is fast, e.g., an exponential decay in correlation, it is called short-term correlation and some times it is so long that characteristic time scale cannot be defined (power law exponent is between 0 and 1), it is called long-term correlation. When two time crystals exchange small subsets of time crystals, the standard deviation of data exchange changes with time, this is called non-stationarities. Considering the increments $\Delta x_i = x_i - x_{i-1}$ of a self-affine series (x_i), $i = 1, 2, 3 \ldots N$, N is measured equidistant in time, Δx_i could either be persistent, independent, or anti-persistent. For stationary data with constant mean and

standard deviation the auto-covariance function of the increments, $C(s) = \langle \Delta x_i \Delta x_{i+s} \rangle = 1/(N - s) \sum_{i=1}^{N-s} \Delta x_i \Delta x_{i+s}$, so we can directly calculate even the elementary time crystal exchange parameters during computing (Kantelhardt, 2011).

Whenever brain jelly encounters a new problem, this will be a new path in the spherical basin of attractor, and the neighboring system of oscillators would try to accommodate that new path component by modifying its existing dynamic points in the 3D world. Euclidian traveling salesman problem because in the 3D space we have several existing dynamic points of basin of attractor and any problem given to nanobrain is converted into a line of sequential evolution of dynamic points and the first task of this nanobrain is to match this new situation with the inherent dynamic map of the system, that has been constructed before solving the problem. Since PPM is basin of attractor, jelly requires no training. Now, nanobrain has to find dynamically closest points in the already existing basin of attractor map of a cellular automaton, and moreover, it has to find a way to move from new point to each option individually, and then find the shortest route. Next step for the computation would be creating another new dynamic point, which means a new grid, made of new states of the participating neuron-oscillators. The objective of creating the new point would be that a journey from that new point, would lead to all the neighboring points faster than the previous point set by the problem. The particular criterion makes the problem intractable, brain jelly as its savior.

Complexity can increase without acquiring infinite resource: Brain jelly synthesis follows chemistry beyond Chemical Kinetics: Inside the organic brain made of folded 2D sheets of H3 devices or hexagonal close packed cortical columns, the wireless circuits made of H devices always evolve, the neuron analog H1 changes communication pathway and the gel replicates this feature in the artificial brain to create a time crystal architecture. The time crystal structures transform to follow the symmetry linking guidelines set by the PPM. The particular feature helps the brain jelly to create higher-level intelligence. For example, the jelly can perceive that a banyan tree looks like ice cream from a far distance, (mm), therefore, the co-existence of multiple grouping and continuous creation and destruction of its pathway helps brain-hardware to increase the time bandwidth of the time crystal via simple changes in the connection i.e., the geometry of H3 arrangement. It is important to note that the journey to all kinds of brain jelly begins at a single molecular scale. Take for example, Ghosh et al. synthesized PCMS shown in Figure 9.6b, its different parts act as a different molecular seed, that would eventually unfold into various complex architectures. For all brain jelly, a seed nanostructure needs to be created that grows by itself from 7 nm to a large jelly like structure step-by-step, the growth does not follow the path of chemical kinetics, there is no chemical reaction, rather, phase transitions from one symmetry to another. The jelly-like structure replicates the time crystal storage and processing information like a neural network, so we call it brain jelly. The entire brain-jelly-derived supramolecule acts like a fractal of oscillators (fibrillar structure forms at a critical frequency; Swain and Valley, 1970), derives a large time crystal

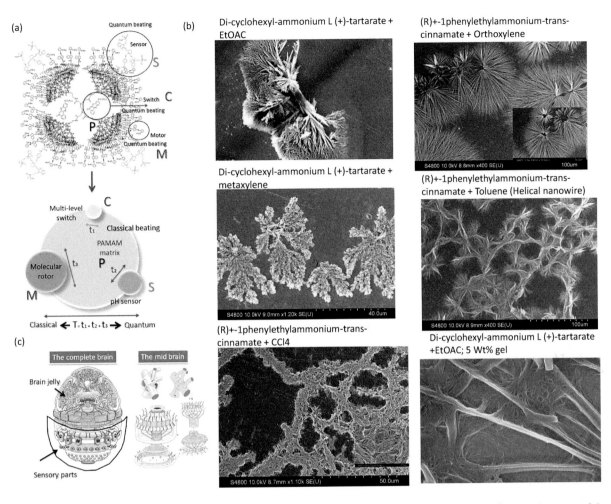

FIGURE 9.6   (a) PCMS molecular structure and its time crystal equivalent. (b) Six different examples of brain jelly, the name of the molecules and processing solvents are noted on the top of each panel. (c) The schematic structure of an artificial brain, where multiple cavities are there.

under the exposure of monochromatic light. If one opens, millions of oscillators would be found inside; if one takes one of them would find millions of oscillators inside, the journey continues. The entire journey could be represented in terms of clocks inside a clock; however, it does not matter how the material or oscillators arrange, what matters is how their representative clocks arrange. We have listed in Figure 9.6b several brain jelly whose journey started at the single molecular scale, but eventually they produced suitable structures for using them in the tiny cavities of 17 brain analogs listed in the table of Figure 9.2. Finally, the brain of humanoid avatar looks like Figure 9.6c, this is an integrated commercial version of the final architecture.

9.5   FRACTAL CONDENSATION— CONDENSING EVERYWHERE AT ONCE

Then there were two more classes of manufacturing waiting in the lab desk, self-assembly, and fractal condensation. Self-assembly was proposed in 1988 and fractal reaction kinetics and related condensation were proposed in 2016 (Ghosh et al.,

2016b). Self-assembly is found to be limited because it requires materials in a particular form and elementary structure to proceed. While fractal condensation is a brilliant way of putting it, whenever "fractal" word includes in the formation of any structure, self-assembly from a nucleus by core-shell growth and then self-assembly of several such nuclei run simultaneously. Until now the grammar of fractal manufacturing noted above is not established, but, one could use number system and the pattern of primes to regulate the growth process. If manufacturing follows the pattern of primes, it will never be left alone, no need to program and self-healing would be fundamental to it. For self-healing, a "fractal-based" structure is always the best choice. In the Bose-Einstein condensation, one wave function holds many atoms, if we add more, wave function remains intact (Figure 9.7a). In the fractal condensation, there are plenty of available singularity domains where phase gets undefined. If we add new time symmetry it gets adsorbed in suitable singularity domains and still singularity survives. Why? Readers know by now, that within a single singularity point, a massive dodecanion tensor could reside, after all they grow in the nested imaginary worlds.

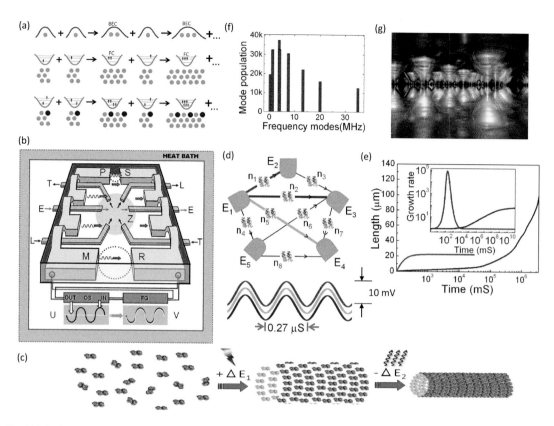

FIGURE 9.7 (a) The basic concept of classical, BE, and fractal condensation. (b) The experimental setup, L E T are different electrodes to send input information or signal to the *H* device. (c) The mechanism of condensation. (d) Energy exchange and coherent waveform generation during the condensation process. (e) Length of helical nanowire vs the time, there is a saturation of length of the *H* device synthesized here microtubule from tubulin. Inset shows the growth rate as a function of time under ac (red) and normal (blue) condition. (f) Distribution of length as a function of applied ac frequency (wirelessly). (g) The broken band structure that facilitates *H* device into a topological material.

Definition of condensation at far from equilibrium: The condensate (Jochim et al., 2003; Fröhlich, 1968a, 1968b) class materials follow a unique materials science philosophy; continuous additions of similar atoms do not change properties of the condensate, while dissimilar atoms find their way inside the same structure acquiring distinct symmetries and delivering unique properties. If atoms are replaced by molecules, a fusion of suitable new molecules to the condensate would add desired functionalities, keeping properties of the primitive molecular condensate intact. Unfortunately, the molecular analog of such atomic condensate does not exist. Moreover, it has always been a critical challenge to unravel the true picture when rapid condensation is triggered at the singularity, since the population rapidly reaches to an astronomically large number. Fractal condensation advances the conventional theoretical tools for energy quanta condensation in predicting and analyzing the condensation of molecules with unique multiple electromagnetic resonance frequencies. Each resonance frequency acts as a point driving to singularity; resonant oscillation canalizes energy into all singularity-attractors or participating resonance frequencies such that by regulating the resonance levels by modulating the atomic arrangement of a molecule one could manipulate singularity-canalization and thus program entire supramolecular architecture in an unprecedented manner.

Step 1: Mutual evolution of multiple singularity cones: Tuning of supramolecular architecture: Time crystal input to a heat bath containing molecular precursors sends resonance signal with different frequencies (Figure 9.7b). For molecular oscillators, the probability of spontaneous synchronization is maximum when the available energy (kT) is the same as their total rotational and vibrational energy. When in a heat bath the molecules are pumped with ac electromagnetic frequencies, the singularity points where coherent resonance frequency bursts (Sahu et al., 2014; Moskalenko et al., 1980) attract the system toward them (Figure 9.7c and d). For tubulin to microtubule transition, eight cones in the potential distribution, or eight holes in the phase space (Figures 8.6b and 8.9i, j) changes the relative shapes and preference via modified interaction. The relative population at multiple singularity points determine how do the four levels exchange energy among themselves leading to the formation of the aggregate-wave (Figure 9.7e and f). The normal mode of vibration is preferred over the complete synchronization, non-linear energy pulling by multiple singularity points causes this preference.

Step 2: Conversion of noise into coherent energy source: Normal modes of vibrations: Now, non-linear frequency pulling between different singularity cones determine, what would happen, the "normal modes of vibration" (Moritsugu et al., 2010) or "aggregate vibrating as a single wave."

For molecular oscillators, the probability of spontaneous synchronization is maximum when the available energy (kT) is the same as their total rotational and vibrational energy. When in a heat bath the molecules are pumped with ac electromagnetic signals, Szent-Györgyi discussed two different mechanisms for the energy transfer, an individual resonant transfer and a collective transfer, wherein the energy is delocalized, an aggregate of molecules receives a quantum of energy, via both parallel channels. Eventually collective transfer supersedes as a natural drive to decrease the magnitude of energy transfer, and the system behaves more like a wave than isolated particles. The proposal "aggregate behave like a wave" demands considering a super molecule, or aggregate that behaves like a single molecular system (Figure 9.7c and d). A little energy leak ε is key to generate the wave nature in the distribution of molecules. The structural damping parameter b regulates leak between two resonance peaks or two singularity points. Then, if pumped with an incoherent signal, all oscillators vibrate at one of the four/more resonance frequencies; the collective energy exchange activates the non-inversion lasing. Obviously, the coupling factor between multiple resonance levels refers to the coupling between the distinct structural symmetries when electric-magnetic-mechanical resonance activates the molecules. A molecule if suitably designed such that it's various structural parts with distinct structural symmetry could absorb mechanical or electromagnetic energy exposed to it at different frequencies and each region start vibrating mechanically with a distinct frequency. In doing so, these parts release energy and interact with energetically coupled neighbors. During synchronization, the vibrational motion of the system could be such that instead of all neighboring molecules start vibrating in a single phase and frequency, all molecules acquire phase and amplitude such that they together map a virtual 2D wave oscillating in the media. It is called normal modes of vibration and the most interesting aspect of the creation of this wave is the development of a collective spatial relationship between all molecules in the medium.

Step 3: Positive feedback during phase transition: Here we have included Luzzi et al. proposed "positive feedback" (Mesquita et al., 1993, 1998), surprisingly, this concept was embedded in the mathematical formulation of synchronization from discrete randomness. However, here, the "feedback" occurs when the system has already formed an aggregate or wave, and therefore "feedback" contains information on how to transform the frequency and the wavelength of the aggregate-wave so that it undergoes a "positive" phase transition. The simple yet powerful technique drastically reduces the number of steps required to lock the phase and the amplitude in a system of large number of coupled oscillators in a series of sequential steps that would lead to the formation of condensate or cause the disappearance of normal modes of vibration or generate the true synchrony where all particles have the same phase and frequency. Note that the formation of condensate means actually reaching three targets at a time. In summary, the phase coherent communication via four levels ensures a well-defined or programmed series of steps to synchrony.

However, depending on the complex set of transformations to be performed for growing the collapsed structure, the speedup factor would vary.

Step 4: Squeezing of normal modes of vibrations by four coherent signals: A system self-triggers spontaneously toward synchrony, when at least four distinct energy levels operate actively during energy exchange. Two additional metastable states with two basic levels stabilize population inversion control remarkably; the "super molecule" concept derived above since a control engine is must whenever we construct a fused molecular system or condensate. In order to construct a super molecule, first, creation of a synchronized system in a randomly mobile discrete system of oscillators at far from equilibrium. When that fully synchronized system oscillates as a single wave, then its four among several energy levels engage in energy exchange such that by triggering the system with different frequencies we can change the collision cross-section, thus, regulate the aggregate-wave's structural transitions. Synchronization that triggers the normal modes of vibration (which is itself a wave made of isolated molecular oscillators) has one vibrational frequency; this has nothing to do with the four frequencies we talk about, these two concepts should not be mixed, the four frequencies originate from the internal structural symmetries, due to isolated distinct dynamics of those local parts. In contrast, the normal modes vibrational frequency originates from the coupling induced relative energy exchange of the molecule, which eventually defines the frequency of the system of particles as a whole. Both parameters are distinct by their origin and nature of interaction, thus, the four internal frequencies remain constant and cannot be destroyed even after the condensate is formed, while the normal modes of vibration will change its wavelength and frequency with a series of quantized changes to eventually cease as soon as the condensate is formed.

Step 5: Fractal reaction kinetics: Thus, entire self-assembly process occurs via two chemical processes, in the first half, a transition from collision based normal diffusive motion of molecules to coordinated motion or spontaneous creation of normal modes of vibration, which could be explained by basic laws of chemical kinetics (see schematic of Figure 9.8a) (Kopelman 1988). However, afterward, during the phase transition of normal modes of vibration to the entropy reduction, none of the basic considerations of chemical kinetics is valid in the sequence of steps adopted by the system leading to the complete collapse. Therefore, this part has always been a mystery in the field of science, and no research was done to visualize this particular process, the point of condensation to the supramolecular architecture is the "singularity."

Step 6: Collapse of aggregate waves into a composite architecture: The 2D aggregate wave oscillates exhibiting normal modes of vibration. The 2D carpet like an assembly of molecular oscillators in an aggregate wave would transform into other structures, via folding, rolling or by other means, which does not destroy the "integrated wave" features. How carriers would behave inside a cylinder, cone or sphere if the 2D carpet forms such a hollow structure, the transmission of carriers on the cylindrical, conical, or spherical surface would

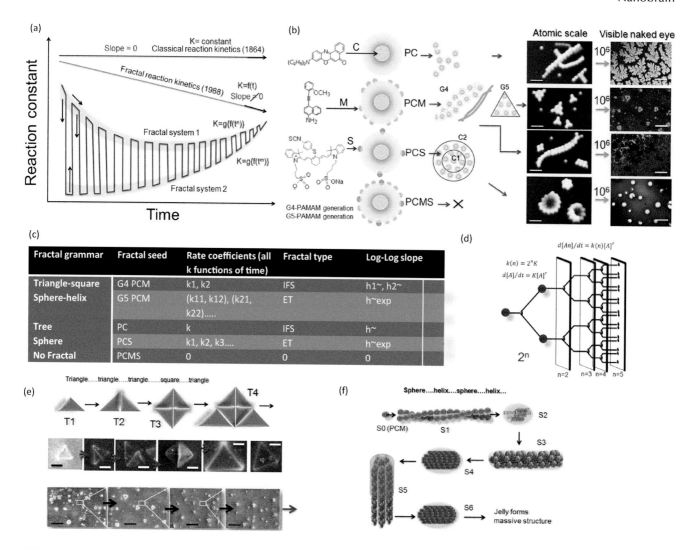

FIGURE 9.8 (a) Introduction to the fractal reaction kinetics where multiple fractal architectures are synthesized in the same solution beaker. (b) Structure of the molecules using which we created various different versions of PCMS or nanobrain. (c) A table is showing how different geometric architectures are synthesized. (d) Fractal synthesis network of reaction centers. (e) Triangular tessellation, three rows top to bottom. The top row shows schematics of different generations. Middle row shows SEM images for different generations (scale bars left to right, 40, 67, 45, 50, 60, and 140 nm). Bottom row, SEM images, scale bars, left to right, 100, 15, and 3 μm, 300 nm). (f) Sphere-helix dual fractal schematics. (c) SEM images, top to bottom, G0 ~ single PCM molecule, 5 nm; G1 ~ helical nanowire, scale ~ 45 nm; G2 ~ G1 nanowire made sphere 300 nm, scale ~ 10 μm; G3 ~ sphere made nanowire 800nm, scale bar ~1 μm; G4 ~ nanowire made cylinder ~5 μm, scale bar ~5 μm; G5 ~ cylinder makes nanowire ~1 mm, G6 ~ 5 mm wired complex, jelly.

be strictly governed by the surface molecular arrangement and that would be significantly different from the carriers inside. The transition should avoid particular local energy minima, but activate all minima at a time. When delicate topological paths are the only choice for carriers, the edge states in the fractured band would be regulated by a cavity where the aggregate wave resides. Recent experiments have shown that the topological confinement could lead to the formation of polariton condensation at room temperature, thus, topological symmetries could coexist and govern multi-channel condensation. The engagement protocol is of two types, first, the new kind of molecule could interact with any part of the structure getting physically trapped, thus, add a new resonance level to the eventual condensate. Second, during phase transition of "2D aggregate-wave," the new kinds of molecules could find a place inside the capsule-space created during folding of the 2D surface, thus, a multi-composite structure is formed in the organic jelly. By studying a large number of protein crystals Bandyopadhyay et al. have devised a generic protocol for the proteins, as described in Chapter 6. For a distinct time crystal input, the route from aggregate-wave to condensate would be distinct.

How fractal condensation differs from the world of Fröhlich condensates: Though Fröhlich (Fröhlich, 1968a, 1968b) and their followers never talked about condensing molecules but for condensing the propagating carriers on the biomaterial substrate, they considered resonant oscillations of the dipolar hosts, mostly, they used atomic dipoles for photons or phonon like bosons to reside on those hosts. Thus, microwave (GHz or THz) ac pumping would synchronize dipoles, which in turn would condense the guest carriers distributed over the hosts. In this condensation process, an external ac pumping resonantly excite the oscillators, engage them into the delocalized collective energy transfer with all neighbors so that wide range of frequencies ($\omega_0 \leq \omega_i \leq \omega_N$) and phases in an assembly of n coupled oscillators squeeze and lock into two particular values. Then, synchronized oscillators behave together more like a wave than isolated particles, afterward, one-step toward minimizing entropy leads to a collapse.

Difference between Haken Prigogine collapse and fractal condensation: (1) For the formation of a supramolecular complex via far-from-equilibrium Haken-Prigogine (HP) collapse is a well-debated perspective (Haken, 1988). Though, synthesis of brain jelly is a thermodynamically open system operating at far from equilibrium, the aggregate-wave is formed by long-range interactions and it breaks symmetry spontaneously multiple times prior to the complete collapse into a condensate, the geometry of time is never taken into account in an HP collapse. In fractal condensation, once the aggregate wave is formed, afterward, the entire pathway is executed in a programmed manner as a series of dissipative structures are to be formed via ordered transitions. PPM delivers a sequence of symmetries, chaos or randomness arise due to the projection from infinity as described in Chapter 4. Even though the manipulation of multiple singularity points is a route toward complex architectural engineering of a large number of condensate families arising out of a single condensate. Thus, we look beyond the world of Prigogine's random dissipative structures like cloud, cyclones, and hurricanes to the concept where condensates inside a condensate inside a condensate forms at a time. **(2)** Unlike Fröhlich condensation's complete synchronization, and the existence of one resonance frequency was considered sufficient to generate the condensate, but here for brain jelly it is not. Without multiple singularity drive, one cannot reach to the aggregate wave avoiding the straightforward complete synchronized state. For GML most complex structure is a icosahedron, a 12 singularity point drive would be abundant in brain jelly while processing a time crystal. Since previously, the inspirational philosophy was that of Illya Prigogine (Kondepudi et al., 2017), where complete synchronized state could randomly jump into an ordered architecture, however, we follow a philosophy just opposite to that, entire formation is programmed at the molecules atomic arrangement, hence, unless and until we create an aggregate wave avoiding full synchrony, we cannot start programmed adventure toward a predicted architecture.

9.5.1 Pattern of Primes Need Not to Be Instructed

As explained in Chapter 7 in details, 17 brain components discussed above have a large number of components within. Following the actual brain architecture, those tiny components are built using organic synthesis; however, there is a big problem. How several components and their sub-components build together following fractal condensation and fractal reaction kinetics?

Jordon structure and bifurcation: Adjacency matrix: what computation means for brain jelly? Bifurcation is symmetry breaking. During synchronization and desynchronization of local time crystals, associative matrix D is "adjacency matrix," it has been shown mathematically that bifurcation does not depend on how many networks are there, rather it depends on a number of Jordon structure. Two neighboring brain components share D, and that one could study separately in a chemical beaker The Jordon structure is strictly related to the time crystal built by geometric algebra (Chapter 4) since, the dimension of Jordon structure holds fundamentals of coupling. In a network of interactive time crystals (set of $\{f(t)\}$), center subspace in regular networks is given by $dF = \alpha I + \beta A$, here, αI determines internal dynamics of a time crystal, α is the matrix of internal dynamics, βA is the coupling dynamics between two nests. A is the adjacent matrix, whose eigenvalues account for the coupling of subsets of N network, it determines magnitudes of all possible coupling factors, β is the matrix of coupling dynamics. Jordon structure J associated with a 0 eigenvalue can be the Jordon structure of an adjacency matrix of a regular network. The above mathematical expression describing symmetry-breaking bifurcation solution of a problem means a product of (Golubitsky and Lauterbach, 2009) adjacent matrices representing a time fractal $f(t)$ to find a new matrix in the timeline.

9.6 FRACTAL REACTION KINETICS—PARALLEL SYNTHESES IN ONE BEAKER AT A TIME

Fractal reaction kinetics and memory in the chemical reaction: Biological reproduction has mastered self-replication (Smith et al., 2003). It has inspired bio-mimetic engineering, nano-assemblers, designing computer virus as artificial life (Spafford, 1992; Chow et al., 2004; Lipson and Pollack, 2000), and distributive computing, etc. There exists no generic rule for self-replication, fractal reaction kinetics (Kopelman, 1988; Khire et al., 2010; Vlad et al., 2005; Neff et al., 2011; Newhouse and Kopelman, 1988; Savageau, 1998) that memorizes the reaction rates and responds to the ever-changing environment spontaneously could be used as a generic synthetic protocol for self-replication (Figure 9.8a). However, in the last three decades only one kind of "memorization" was explored whereas for programming self-replication of several different visible scale architectures (Tong et al., 2011; Kurihara et al., 2011) in the atomic arrangement of the molecules or any starting materials we have to invent several different kinds of "memorization" process (Kozlov et al., 2013; Varela et al., 1974; Bedau et al., 1997) when a fractal reaction evolves the product with time. Ghosh et al. started with a 7 nm wide molecular platform (Ghosh et al., 2014b; Figure 9.8b), modified it with a few functional groups to trigger robustly varying "memorization" of reaction rates.

The memory regulation has led to an encyclopedia of geometric replication (Figure 9.8c and d). The encyclopedia includes right-triangle tessellation (Figure 9.8e), alternative spherical and spiral geometries (Figure 9.8f) similar to fourth-circuit element Hinductor, H that could be switched using pH variation (Zhang et al., 2010) like that we see in DNA, galaxies, spherical fractal like Russian nesting doll Matryoshka (Agnati et al., 2009) and ever-growing branching like a tree. Three decades ago, memory in a reaction allowed the system to direct reaction in the most suitable route in a changing environment, now by adding a higher-level control to that memory we enable the system to program that path. In the solution 10^9 orders larger architectures grow spontaneously with zero human intervention, yet, the system switches nonlinearly among various paths to reproduce.

Could there be a chemical reaction that starts from a few atoms and grow entire architecture? Now, Ghosh et al. have experimentally demonstrated (Ghosh et al., 2016a) that we can add a higher-level control on that memory, to drive memorization through various ways. It means a system (brain jelly) can hold at a time, multiple different kinds of memories as time crystals and switch between different routes as programmed or the changing environment demands. We cannot add complexity if we have just one memory as a single clock. With the power to flip between multiple memories, we create multiple fractals for dual or multi-faced self-replication in a single chemical beaker, with zero human intervention. Unlike classical chemical kinetics that defines chemistry, the hallmark of "fractal-like" kinetics explores anomalous reaction orders and time dependent reaction rate, which are otherwise "constant" over time. The area beyond the fundamental laws of chemical kinetics and laws of mass action is part of heterogeneous reaction kinetics observed in biology, materials science, geology, astrophysics, etc. In fractal-like kinetics, due to spatial constraint, it can dramatically speed up or slow down the reaction rate, hence for half a century it has been proposed as a fundamental structural tool. However, this field has mostly been about the theoretical interpretation of the frequently observed biological or chemical phenomenon. Thus, the fundamental concept of "fractal-like" kinetics has not progressed beyond the basic definitions, which Ghosh et al. transformed forever.

What makes fractal reaction kinetics so unique? The journey beyond classical kinetics to "fractal kinetics" involves some remarkable features like self-ordering, self-unmixing of reactants, rate coefficients with temporal memories, etc. All these parameters are thoroughly investigated as a part of heterogeneous reaction kinetics. If we compare these typical parameters with the tasks of PPM-driven natural intelligence, we can draw an analogy between the two. Fractal reaction kinetics monitors reactants and products by memorizing its own reaction rates and decides its next reaction rate; this is what a spontaneously operating artificial brain should do in a hostile environment. The literature of self-replication has argued for similar qualities for a long time; however, fractal reaction kinetics was never examined thoroughly from self-replication perspective.

The basic idea behind the hierarchical memory in reaction kinetics: In order to be useful in self-replication, fractal reaction kinetics should not only demonstrate a typical linear temporal variation over log(time) scale suggested by Kopelman (1988) as we have seen in almost all reported works in the last three decades. The lone memorization capability requires a hierarchical control on memory to add complexity in the massive architectures as the self-replication demands. Hence, Ghosh et al. added control on the basic memorization protocol of fractal reaction rate and identified major limiting possibilities of temporal variations of a fractal reaction rate. Temporal variation of reaction rate over log(time) scale which has always been linear. In the same plot Ghosh et al. introduced an exponential decay, an exponential increase, and a periodic switching between two parallel linear variations. Minor modifications in the starting dendritic structure lead to various fractal synthesis routes and unique geometric shapes are repeated from a few nm to cm scales, we can observe how higher-level control on memory is played.

For an exponential increment and decrement, the rate of variation of k also varies with time ($\log k = \exp \pm S(\log t) + M$, $S, M \sim$ constants, $h = f(\log t)$). Second-order variation of memory suggests that higher-order control of h on k. Here, even h is a function of time. In the switching between spherical and helical fractal, one hierarchical memory control h ($=f(\log t)$) regulates the exponential variation of k and a nearly linear k embeds a hierarchical control h to return. Similarly, the nearly periodic oscillation between the two rates for triangle-square fractal suggests a quadratic control in the singular fractal system to memorize two fundamentally distinct architectural growths in a single chemical beaker. Self-replication is a prerogative of nature as it can modulate the fractal rate. Higher level memory in regulating the chemical kinetics is a generic protocol to merge the gap between the existing automated intelligence and the biological intelligence exhibited by nature in reproduction.

9.7 NANOBRAIN, THE SMALLEST LIFE FORM

Nanobrain, NB, or PCMS (Pamam dendrimer, controller, molecular motor and sensor) has all the three components of a life form. It can sense, it can work, and most importantly it can process decisions. Figure 9.9a shows what happens at the core of this smallest life form. Different multi-level switches or H, H1, and H3 devices incorporated inside acquire the time crystals via sensors connected outside. Then they start vibrating following the symmetries of frequencies prescribed in the PPM. During this quasi vibration charges of various natures produce, they follow their own symmetries in the PPM (Figure 9.9b). Eventually, they build distribution of phase (Figure 9.9c). By now readers know that such a distribution of geometric phase with an intricate 3D geometry is called a time crystal which would pinpoint the unique PPM and the course of future dynamics of the whole system. That regulatory phase distribution of polarons or any quasi particles

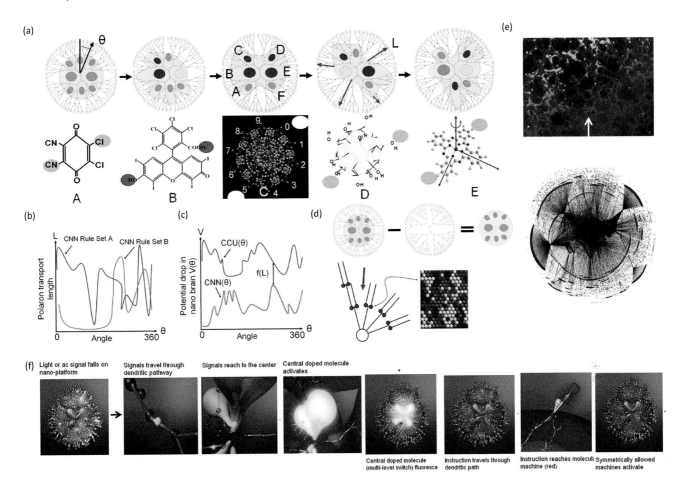

FIGURE 9.9 (a) Inside a PAMAM or any dendrimer with a cavity, multilevel switches are placed that controls the motors and other elements dynamics. Five different molecular structures are shown, which were experimentally verified as multi-level switches. (b) As the branches increase, the polaron transmission length changes non-linearly, but it depends which set of cellular neural network rules are encoded to the central controller molecules described in the panel (a). The phase θ is actually the direction. (c) The potential drop, or the energy expenditure in the nanobrain or PCMS as a function of phase. (d) The processing dodecanion tensor by the nanobrain creates the multi-dimensional architecture of nested clocks or time crystals. (e) The optical vortex projection of the time crystals during information processing. (f) A schematic of nanobrain operating by transmission of energy through the different pathways between motors, sensors and the decision-making controlling molecule doped inside.

containing the conditions to encode singularity points in a clock actually writes its quasi charge distribution on the surface of the systems phase sphere and that could be read as a time crystal using monochromatic laser (Figure 9.9d). The projected architecture of artificial atoms (assembly of vortices) containing a solution to the problem is shown in Figure 9.9e.

Each multipolar resonance can be locally driven at points where the coupling between the quantum emitter and antenna connected to the life form is efficient. The electric mode density is high at the ends of a nanowire regardless of the symmetry of the resonant mode. Therefore, a point source (electric dipole) at the end of a Hinductor, H nanowire couple to all possible resonant modes, breaking the symmetry that might prevent access by far-field illumination, and encompassing both sub- and super-radiant modes irrespective of symmetry.

H devices self-assemble as nano-antenna (Taminiau et al., 2011) and might operate in the entire optical frequency range and beyond (Della Valle et al., 2008). H antennas are ultra-sensitive to polarization and dark modes at various selective regions (Kajetan Schmidt et al., 2012). Quantum antennas forming a fourth-circuit element could deliver quantum information processing (Curto et al., 2013) via PPM. The complete process, how sensor, processor, and motor work, is shown schematically in Figure 9.9f.

9.7.1 JELLY OF MEGAMERS FOLLOWING A RESONANCE CHAIN

In the artificial nano brain, the organic jelly captures the solution of a problem via triggering synchrony of a particular pattern of time crystals with its nearest match with the 15 shapes of

GML. By successive learning, the brain jelly stores high-level time crystals as associated rules, which are called through synchrony as soon as external environmental conditions demand. For that reason, computing time is not the sum of the times taken by individual matching processes; rather, time taken for the resonance between two individual neurons determines the global speed of computation. The time taken for two neurons to resonantly vibrate is more than four neurons to vibrate resonantly. However, in contrast, under noise, more is the number of coupled neurons in an environment, faster is the time to find a perfect match. Works of literature are rich in mathematically arguing that synchronization speeds up when the environment is more chaotic. We can understand the underlying complex mathematical treatments with a simple time crystal: more is the randomness in a system, higher is the probability that the matching condition already exists as a pre-set situation even before the synchronization begins. Therefore, as soon as the synchronization is triggered, the already existing pre-set condition supersedes other options and that would reduce the convergence time drastically. It is mathematically shown that topological randomness decreases the synchronization time by several orders (Grabow et al., 2010).

During this period there is no transmission of a signal between the components, because it's a supramolecular architecture one inside another. One such example is megamer shown in Figure 9.10a–e. The whole architecture is simply extrapolation of the smallest life form described above, however, there are differences (Figure 9.10f). The journey of computing is summarized in Figure 9.10g. Earlier, there used to be a linear circuit, then using 16 molecules Bandyopadhyay and Acharya (2008) showed that information processing turns rapid, now 3D system shows one could carry out complex processing without any outside communication. The synthesis of brain jelly superstructure advances the wheel model of the brain. Using a very simple scheme we have reminded reader how the primitive idea of clock inside a clock that started at the beginning of this book ends here (Figure 9.10h). Each component is a PPM and their resonance bands have common resonance frequency points which enable smooth energy transfer between the components (Figure 9.10i). Such overlaps when plotted for the entire brain jelly architecture, one derives the resonance chain once again, earlier it was for the real brain components, now it is for the brain jelly (Figure 9.10j). Now, we would dig deeper into the fundamentals of the resonance chain.

Multilayered synchrony: non-linear frequency pulling in the entire Fractal tape network: harmonic and anharmonic oscillation The artificial brain described here is made of brain jelly, an evolving organic architecture that senses the rhythms and elementary seed that takes part in the process is the nano brain. Now, we can either consider brain jelly as a material or just a network of rhythm, simply a time crystal. At every stage of the proposed artificial brain, starting from the input of signals from the external environment to the main processing unit, the core brain, signals are synchronized, time crystals are linked. At every stage, coupling and de-coupling of time crystal take place to favor synchrony or desynchrony. Since entire brain jelly network is made of high-quality factor oscillators (quality factor ~ number of oscillations before the energy decreases by $1/2\pi$ times the initial energy), non-linear frequency pulling plays a vital role in generating synchrony. The general expression for harmonic oscillation is given by, $0 = \ddot{x}_n + x_n$, then we add linear damping $\gamma \dot{x}_n$, then we can add reactive coupling, $\sum_m D_{mn}(x_m - x_n)$, also add further non-linear stiffening x_n^3, and or energy input $-\gamma D \dot{x}_n(1 - x_n^2)$, and then add the signal $2gD\cos[(1+\delta\omega D)t]$. An oscillator driving near resonance $\omega_D \cong 1$, it means that the driving frequency turns identity, it is not required. A single-driven damped anharmonic oscillator follows $\ddot{x} + \gamma\dot{x} + x + x^3 = 2g_D\cos(\omega Dt)$. Anharmonic and harmonic behavior coexists in a time crystal network that drives the nano brain and the brain jelly. Non-linear frequency pulling means if the system if pumped at other frequencies, the system automatically drives it to ωD, the natural driving frequency.

Spontaneous generation of higher level rhythms: re-defining intelligence and creativity: For every typical input time crystals or rhythms, from external environment, the neural network in the artificial brain naturally construct an output set of time crystals or rhythms. If the network uses this input-output set repeatedly; it is stored as a favored rule for that particular transformation. We call it a higher-level rule, since the neural network would always try to impose it, naturally, if any structurally similar time crystal-cluster is created in the upper brain. When we construct a typical artificial brain, we need to decide beforehand what would be the minimum size of a time crystal and the maximum size that is process-able for the given brain architecture. In another word the material seed or nano brain decides how the brain jelly would form, smallest sized time crystals would be processed using highest frequencies and the largest size time crystals would use the lowest operational frequencies. Both the lower and the upper limits are determined by the kind of basic oscillators or neurons we use to write the basic time crystals. The resonance frequency is a basic hardware property of an oscillator, and there are several different kinds of oscillators; depending on their operation, resonance frequencies are strictly defined. For the composite time crystals, it would be like a composite electromagnetic oscillator, whose resonance frequencies would also be a strictly defined parameter.

Resonance in damped anharmonic oscillator: For learning the time crystals break links following equation-given by $\ddot{x} + \gamma\dot{x} + x + x^3 = 2g_D\cos\omega_Dt$, damping factor is γ, drive strength is g_D, the drive frequency is ω_D, for resonance $\omega_D \approx 1$. Smaller g_D means a lower non-linearity, which means, $\omega_D = 1 + \varepsilon\Omega_D$, $g_D = \varepsilon^{3/2}g$ and $\gamma = \varepsilon\Gamma$, when we consider $\varepsilon \ll 1$, g, Γ, Ω_D are unity.

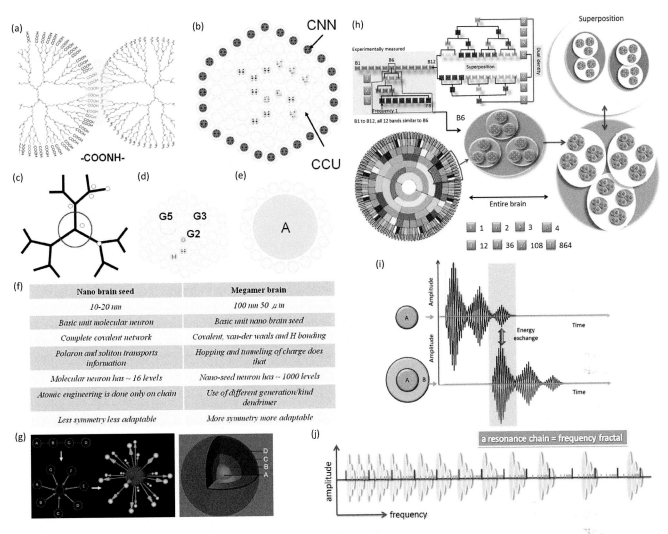

FIGURE 9.10 (a) The basic bonding between the nanobrains for the formation of megamer as a brain jelly. (b) The central control unit (CCU) and the central neural network (CNN) of a megamer brain or brain jelly. (c) A fractal cascade of synthesis of the megamers as PPM, its successive phases are shown by schematic in the panels (d) and (e). (f) The difference between the nanobrain, PCMS class systems or H device with the megamer brain is noted. (g) Successive synthesis of higher generations and higher-dimensions architecture of nanobrain. (h) Nested linear oscillations could lead to the creation of a complex time crystal architecture where multiple solutions could coexist. (i) Nesting of multi-layered time crystals by overlapping of common resonance frequencies or common pattern of the singularities. (j) A dodecanion resonance chain written linearly to demonstrate the formation of an integrated processor by combining several layers of imaginary worlds.

High Q oscillators follow the equation below: Equation of clocking path on the phase sphere of a time crystal undergoes diameter displacement x during sensing an input geometry for activation of GML is $0 = \ddot{x}_n + x_n$, we add linear damping term, $\gamma \ddot{x}_n$; we take δ_n from $g(\delta_n)$ and add its function $\delta_n x_n$; then we add reactive coupling term $\sum_m D_{nm}(x_m - x_n)$; non-linear stiffening term x_n^3; energy input $\gamma_D \dot{x}_n(1 - x_n^2)$; signal $2g_D \cos\left[(1 + \delta\omega_D)t\right]$ and then there could be additional noise term.

An accurate description of synchronization with simultaneity is impossible: The coupling between two time crystals, is the coupling between multiple basic Hinductor, H oscillators via wireless energy transfer in the beginning. If that wireless energy exchange mode is practiced more, slowly a physical bonding grows between the oscillators, since more and more common clocks get engage in the energy exchange process. The nature of physical bonding defines the foundation of mathematical formulation for mutual synchronization until the construction of physical wiring is completed. The fundamental problem of addressing mutual synchronization is that amplitude and phase vary simultaneously, and almost all theories find some reasons to avoid one and go ahead with the other. One such example:

"oscillators are strongly attracted to their limit cycle, so amplitude variation could be neglected and phase variation needs to be considered." Now, there could be several different kinds of glues-for-coupling, dipole-dipole interaction, H bonding, etc., which are weak bonding in the energy range of 1–10 kCal, but not the strong covalent bonding (~400 kCal). For weak bonding, mathematical formulations ignore subtle fluctuations for simplicity, and as we have argued in Chapter 2, many-body interaction formulations for phase locked synchronization (Strogatz and Mirollo, 1988) do not fit for the "simultaneity adventure" exercised here. At this point, we note that weak bonding only brings basic oscillator neurons (Hinductor class 2 or H2 devices) at least in a favorable orientation; similar to real brain's neural network, stronger, practice-based consolidation of coupling protocol is established, but the neurons cannot create new axon channels if necessary. To compensate, for the artificial brain made of organic jelly, we use molecular machines, which have inherent dynamics to get coupled with the other machines physically at a particular electric field eventually forming giant nano-wires via self-assembly. There are various kinds of locks, one that helps in morphing is called injection locking.

Injection locking: When the coupling is strong enough and the frequencies near enough, the second oscillator can capture the first oscillator, causing it to have essentially identical frequency as the second. It is injection locking. When the second oscillator merely disturbs the first but does not capture it, the effect is called injection pulling.

All components in the resonance chain, reply back via non-radiative energy transfer. The technology relies on the "electromagnetic transparency" of the material, as described in Section 8.9, as cloaking. However, due to large reflection coefficient the transparency develops opacity, i.e., anomalous quantum cloaking. Thus, the screening effect restricts wireless communication beyond certain limits of the immediate neighborhood. As described earlier, three hallmarks of the artificial brain, (i) electrical, magnetic, and electromagnetic resonance, (iii) cloaking, (iii) holographic projection, following $e-\pi-\phi$ quadratic relation. For this reason, we cannot rely on an antenna-receiver concept to scale up the "reply back" technology, where radiation energy passes through the air between two materials, the philosophy requires a fundamental change. Alternatively, we have introduced a multi-layered structure where output structural product of one waveform connects 12 layers nested within and above.

For each layer, the resonance band has three distinct domains, one domain is used to communicate with the inner layer, and one with the external layer, one layer is kept for its own information processing. Thus, all layers are energetically connected by a single chain of resonance band, irrespective of the size of the device architecture now a wireless communication can transmit without getting screened

anywhere. The energy given as an input at any layer transmits to the entire chain of resonance bands, both ways, toward the lower and toward the higher frequency regions of the chain. Any form of energy is suitably absorbed and then transmitted across the resonance chain. A resonance chain connects every single computing seed in the system; thus, zillions of seeds are wired into a massively complex yet a single network in the form of a time crystal.

9.7.2 EEG of a Nanobrain

At the very beginning of the interaction with a natural event happening around us, the frontal lobe of our brain controls the five major sensory organs to collect maximum information from the external world, and immediately the captured information is converted into a stream of pulses. In the brain jelly there is no pulse stream. **A time crystal is an endless network of rhythms (infinite series mathematically), it projects a set of optical and or magnetic vortices, which is finite, defined as an argument when it includes a time function, means after a certain time, it would change to another time crystal, in this book we use time crystal in general, often it means an equivalence of argument.** In this path, a checkerboard of pulses that keeps timing (gridding) remains constant throughout. The multilayered wave-stream that is formed in the sensors is simply a superposition of several 2D patterns. Composition of wireless (parallel) and wired (serial) circuiting is delicately balanced to make sure that time-grids prepared by different sensory-organs match with each other so that eventually we get a unified time-grid for synthesizing combined time crystal for visual, auditory, touch, taste and smell wave-streams. Once all time crystals produced in different sensory-organs are locked by the unified time-grid, there are several points where information-processing language could be verified. The objective here is to understand nature using the true language of the brain, GML, this will not be the ultimate one, but it will mark the beginning of understanding the brain in the way it is, and subsequent rectifications in the following years would deliver the final grammar to read the signal absolutely captured from any part of the brain.

One way to confirm that time crystal language, GML is to read the brain jelly's EEG. Figure 9.11a shows as brain-like prototype device filled with organic jelly and several neural network-like cables are connected to input time crystals as a stream of electrical pulses. Figure 9.11b shows fully operational module for such a device, where using a high-resolution camera, the evolution of jelly is observed as a function of time (Figure 9.11c). By shining laser and reading the magnetic vortices one could read the time crystals instantly. One nice way to advance the simple module of Figure 9.11b is to build an ensemble of several such devices and arrange them in a fractal structure. Only the geometric feature alone could trigger EEG similar to a human brain (Figure 9.11c).

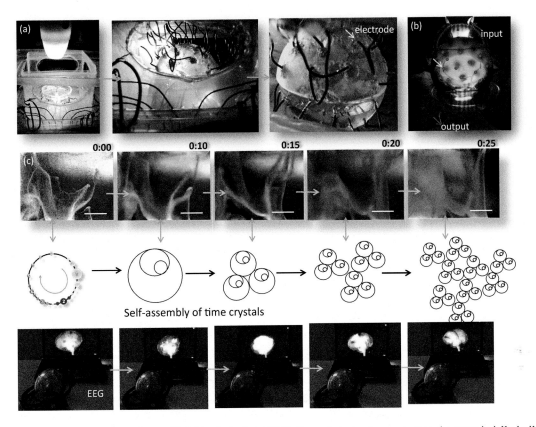

FIGURE 9.11   (a) The first prototype of an artificial brain built in 2013. Several electrodes connect to the organic jelly builds the input-output wiring, the final device, an artificial brain is shown in panel (b). The jelly is zoomed in the panel (c) scale bar is 1cm and the equivalent time crystal is shown schematically. At the very bottom we demonstrate the EEG pattern with the brain jelly.

9.8 TUNING UNDEFINED HOLES TO SENSE MAGNETIC LIGHT

We summarize again, how the brain jelly works, now using simple mathematical numbers. Figure 9.12a shows that by changing the geometric parameters of an *H* device one could switch on or off desired number of singularity points or holes in the phase space. Also, even without understanding much detailed mathematics, one could draw equivalent time crystals, crudely. Say, three such helices come together, representing 3, 6, and 12 (Figure 9.12b). Now, if one wants to draw the phase space of the assembly, it would find three singularities for the simplest scenario, one for each *H* devices. Most importantly, in the holes or singularities the phase space corresponding to the 6, 5, and 12 would reside. It means they would affect by phase alone. Now, all four components, 5, 6, 12, and 3 have their own time crystals, where the time crystal of 3 is the host, others are guests. When laser light falls on this cluster, this very phase space is projected as a composition of magnetic vortices as shown to the right of the Figure 9.12b. We can draw circles on this magnetic image and could read solution, what is the final outcome.

Exponential speedup without using entanglement: One important aspect of a such a brain-jelly-based computing at the molecular scale is that these mixed states which are responsible for the logical reduction, but in these states during measurement, when exactly the computation occurs, we need to stop entanglement still allowing superposition to continue. It is the measure of discord, namely quantum discord in a system. In the case of the classical logic gate there is no superposition, hence the possibility of simultaneous reduction operation does not arise, which is very much possible in a quantum computer. One very important question that we ask here is the need for the pure entangled states, would it be possible to construct a decision-making PPM without pure states? Jozsa and Linden (2003) have argued that an exponential computational speedup might be possible with mixed states in the total absence of entanglement, and entanglement is essential only if computing is performed with the pure states only. Magnetic vortices based time crystal is fit to exponential speedup (Lloyd, 1999) irrespective of spatial scale, since here in the entire book we replace every single physical parameter in the universe using time crystals (see Chapter 4), if some parameters left, would be done soon.

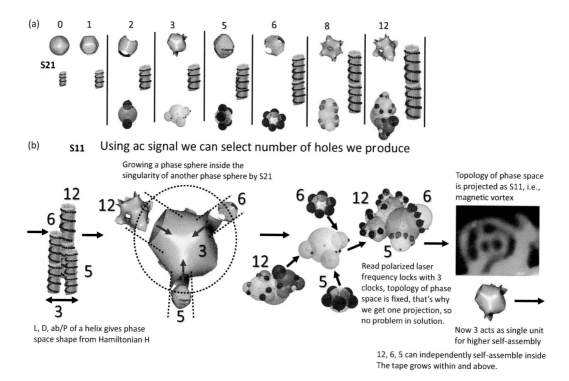

FIGURE 9.12 (a) In quantum, we have only one phase space with two classical points, but for us there are no classical points but 1–12 variable number of holes or singularities in the phase space. We can open or close n number of holes to write any integer. The chart shows different number of holes in the phase space of the *H* device. Below a schematic is presented to explain the length, pitch diameter and lattice parameters acquired by the *H* device to generate the holes and finally below the equivalent time crystal to be produced. (b) A single chart begins with the self-assembly of three *H* devices representing integers, 5, 6, and 12. In the next step we show how inside the singularity of three holes the phase spheres of three components enter. S21 reading of transmission spectrum delivers the result. Then in the third step we explained how the time crystals look like for three elements and how they self-assemble into an architecture of combined time crystal. In the fourth step we see the final magnetic vortices images. These images are fed to simulator to build time crystal. Holes in phase space act like bonds. Holes blink periodically, hence phase spaces are like orbitals of molecules with vectors, the 3D distribution of vectors try to minimize, i.e., sum of vectors = 0, in doing so, supramolecular architectures are born.

If superposition of magnetic vortex atoms mix the vibrating states via synchronization of oscillators, then it is possible to generate an exponential speed. If we can demonstrate that mixed state computing could be implemented to speed up those intractable problems exponentially which are essentially required to resolve the brain-like computing then we can demonstrate that synchronization is superior to entanglement in solving the problems that brains required to solve. However, for "search and find" problems even in the classical "nested synchrony" it is a 3D network of co-existing synchronized and desynchronized states. It means in simple words, without using extensive mathematics the pattern of primes PPM can suggest that for the mixed state also, non-entangled processing has to continue along with the entangled processing, and then it has to wait for the other processes to be complete. In the "nested synchrony" protocol, which is the hallmark of fractal mechanics proposed in Chapter 4, the entire hardware is coupled and time wasted is minimum, in principle there is no time delay in a time crystal scenario. Even before a clock senses the information reaches several layers below, determines the modified rhythms as solutions and sends to the higher-level slower

scales. Thus, the solution always reaches "instantaneously" just like quantum entanglement, though it is a purely classical scenario.

9.8.1 Spiral Nanowire Writes Time Crystal— Sense It as a Crystal of Light

Brain jelly's adoptability of accepting new hardware components: The truth table of 15 primes: The most fundamental truth-table for any brain-jelly-based decision-making non-computer is shown in Figure 9.13. If one builds an operational brain jelly, he has to present such a table. For every 15 prime, a related symmetry there should be a material structure and a distinct signature of magnetic and optical vortices. We did not fill up the table to suggest one needs only a few primes to simulate 90% of events happening around nature. The brain jelly should be able to accommodate new kind of oscillators or structure embedded into it using these elementary prime structures so that it could enhance the power of computation, evolve under fatal conditions. If we observe the human brain architecture, depending on the processing necessity, it has developed several different kinds of neurons, and

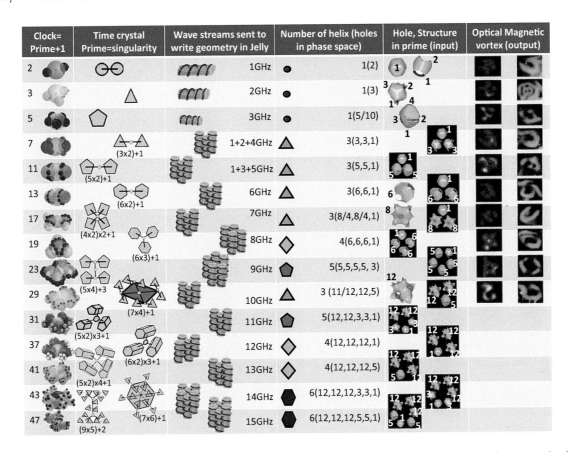

Clock= Prime+1	Time crystal Prime=singularity	Wave streams sent to write geometry in Jelly	Number of helix (holes in phase space)	Hole, Structure in prime (input)	Optical Magnetic vortex (output)
2		1GHz	1(2)		
3		2GHz	1(3)		
5		3GHz	1(5/10)		
7	(3x2)+1	1+2+4GHz	3(3,3,1)		
11	(5x2)+1	1+3+5GHz	3(5,5,1)		
13	(6x2)+1	6GHz	3(6,6,1)		
17	(4x2)x2+1	7GHz	3(8/4,8/4,1)		
19	(6x3)+1	8GHz	4(6,6,6,1)		
23	(5x4)+3	9GHz	5(5,5,5,5, 3)		
29	(7x4)+1	10GHz	3 (11/12,12,5)		
31	(5x2)x3+1	11GHz	5(12,12,3,3,1)		
37	(6x2)x3+1	12GHz	4(12,12,12,1)		
41	(5x2)x4+1	13GHz	4(12,12,12,5)		
43	(7x6)+1	14GHz	6(12,12,12,3,3,1)		
47	(9x5)+2	15GHz	6(12,12,12,5,5,1)		

FIGURE 9.13 A table shows 47 primes and its possible geometric representation. Column three shows what frequency should be sent for synthesizing a prime in the brain jelly. Column 4 shows the composition of phase spheres required to build the required self-assembly. Column 5 shows the possibility of synthesis of phase spheres and Column 6 shows optical and magnetic vortices generated by the system, current results belong to helical carbon nanotube.

other components, all these components, demonstrate oscillator like behavior. PPM-inspired artificial brain explored here will not have the capability of constructing new kinds of oscillators; however, if we find the necessity from outside and supply, then it should be able to accommodate it internally. Two objects are considered to be equivalent, or "homeomorphic," if one can be morphed into the other by simply twisting and stretching its surface; they are different if we have to cut or crease the surface of one to reshape it into the form of the other. All Hinductor or H devices that are the key ingredient for brain jelly are homeomorphic to accommodate smaller changes, but for larger, they add or decrease length to morph an external input.

The supremacy of brain's pattern recognition: Brain jelly processes information by capturing images using an effective visual technology—blinds make those images using sounds—blind, deaf, and dumb make those images in the brain by touching the objects. However, irrespective of the origin of those images, the brain follows a unified geometric language GML protocol to analyze and prepare them for learning. We look at a picture but do not see every part of it distinctly, only a few geometrical points are noted, converted into primes just as outlined in the table, the color band and contrast regions are noted but no specific details are pointed out at the first instance, we zoom at a very specific

region, see only that part which is very interesting to us. It is a fantastic way to minimize the amount of information need to be processed and stored in the brain—suggesting an essential feature of an artificial brain to zoom an interesting part while capturing a visual. All geometric structures are produced by hyperbolic functions so that rhythms or time fractals come into picture and "nested geometry" is converted into "time crystals" that could be encoded into the materials property. It is the foundation of "brain jelly."

9.8.2 Writing Prime Numbers in a Jelly

How a brain jelly builds a geometric shape and synthesis a PPM: A brain jelly has to process 15 geometric shapes (Figure 9.13), it means by synthesis it has to produce organic supramolecular architectures, preferably helices and vortices as outlined by the fourth-circuit element, H. We already know by now that H devices are such that they build 12-hole phase space that blinks when the structure changes its geometry continuously. However, if the external user sends a particular ac signal to write a particular geometry, then the structure that is created does not change the geometry which fixes the number of holes in the 12-hole phase space. The number of holes is the numbers of the corners a geometric shape has. Thus, a geometric shape is created. It is very interesting how to select a suitable *H* device to build a brain jelly, given that thousands of reports on the synthesis of helices are there. Figure 9.12a shows one such example, where a typical multi-layered helical nanowire namely microtubule found in the neurons to all eukaryotic cells are shown. It does not generate all 12 number of holes when one measures the reflectance and transmittance of these materials, but only a few, especially 2, 3, and 5, are enough to synthesis all other primes. Figure 9.12b shows how actually three nanowires 5, 6, and 12 holes come together. The 12-hole phase spaces for the three elements 5, 6, and 12 build a new phase space represented as 3. In the singularity domain of the phase space with three holes, the other three, 5, 6, and 12 enters. In reality, the distribution of static charge on the helix surface undergoes rapid oscillation and due to birefringence, polarized electromagnetic signals produced from the device *H* generates magnetic vortices which are time crystals. The projection of magnetic vortex atom assemblies is the driving force for the self-assembly of the other *H* devices. However, please note that we have used magnetic vortex atoms, that does not mean this is the only way to build brain jelly, one could build ionic vortex-like neurons do in the axons, or molecular diffusion could create vortex, even sound or mechanical waves could build vortex of solitons that could act like virtual atoms. In general, vortex atoms are the coolest tool to write read time crystals, wireless, ultralow power processing is evident.

9.9 ENTROPY DRIVES THE SYNTHESIS OF A PATTERN OF PRIMES

However, the synthesis of *H* with 15 possible geometries is only the first step. Creation of a single *H* device with a single geometric shape is not enough. To build the structure representing the primes, these elementary structures with the elementary geometric shapes are to combine and then build the composite structure of *H* devices as outlined in the table of Figure 9.13. Here is the key. We do not endorse any particular material, but whatever be the material used for building a brain jelly, should first synthesis *H* devices in the solution or gel matrix. Then, it should engage in creating the 15 primes as necessary. Finally, the primes would self-assemble to create the divisors of the integers sent as an input. Say, we want to write a triangle with a ratio of 13:43:7, then only the three prime structures on demand would be produced from the elementary *H* devices and these prime structures would come together. However, if we send a sum of ac frequencies as a modulated waveform, or otherwise as the materials synthesis demands to write 24:27:99 then, all the divisors of the integers would be produced. In the solution, the divisors would follow the Hasse diagram as outlined in Figure 9.14a. For the spiral helices, how the composition of divisors might look like in the solution are outlined in Figure 9.14b. Following the Hasse diagram, the thermodynamic entropy regulates synthesis so that coupling the primes build the integers. It is beautiful mathematics of entropy played by *H* devices in the solution, as if thermodynamics conspires to lead the self-assembly

toward the synthesis of integers from the primes following PPM. As if nature conspires to follow PPM. When the divisors are born, the journey through the Hasse diagram results in building the all possible structures that those divisors could synthesize. There are errors in the production of integers, and that thermodynamic error takes us through the path of perfecting the 3D pattern of PPM.

Say, in the solution, all the divisors made of primes 2, 3, 7 are produced. That number of divisors is not less. For example, $2 \times 2 \times 3 \times 3 \times 3 \times 7$, $3 \times 3 \times 2$, $7 \times 7 \times 7$, etc. Following this route an infinite number of divisors could be generated; however, the desired divisors and their few neighbors are created. To understand it simply, if the desired divisors are $2 \times 2 \times 3$, 3×7 and $2 \times 7 \times 2 \times 3$ then, in the solution, **$2 \times 2 \times 3 \times 2$, $7 \times 2 \times 2 \times 3$, $3 \times 7 \times 2 \times 2$**, etc. structures would be created naturally as an error. Thus, in the PPM plot, which is built by connecting the nearest neighbors in the ordered factors of integers, we get many dots as more and more erroneous products flood the solution. Connecting the nearest neighbors is also like minimizing entropy, just the way discrete isolated atoms or molecules come together and build a crystal of atoms or molecules. Thus, the PPM plot is an account of Hasse's universal protocol, when errors take over. And then, another beautiful thing happens in the solution or synthesis matrix. PPMs have several loops, and the errors trigger those loops which and drives the formation of erroneous products more to fill the missing points of the loops. The natural thermodynamic drive wants to close the loops in the PPMs and that expands the geometric input pattern as shown in Figure 9.14c.

9.9.1 HASSE DIAGRAM FOR ENTROPY BUILDS MATERIAL ANALOG OF INTEGERS

Creation of the column of time crystals by self-assembly follows a power law: While describing the Hasse diagram above we have noted that the energy transmission follows a unique mathematical pathway. The brain jelly starts learning from the first time crystals encoded in the amygdala, and then more and more time crystals are added. The addition process is very special. When two time crystals are added to the hardware, new higher-level coupling rules are created and similarly more and more high-level coupling rules are born automatically. If we make a simple calculation that only one resonance peak represents a singularity and only one peak represents a pair of singularity points burst to emit energy, then for the *n*th addition of paired time crystal will find $\sim n^2 + 1$ number of higher-level coupling rules are automatically added. These automatic additions of higher-level couplings follow fractal relationship and develop their own phase transition rules during learning. Therefore, during self-assembly of time crystals by the brain jelly an astronomically large number of phase transition rules are also created and stored, and that is controlled by boundary conditions set to the choices and options. Choices are integer and options are ordered factor (Figure 9.14d). If we simply consider that a pair of exhibitory and inhibitory time crystals are there,

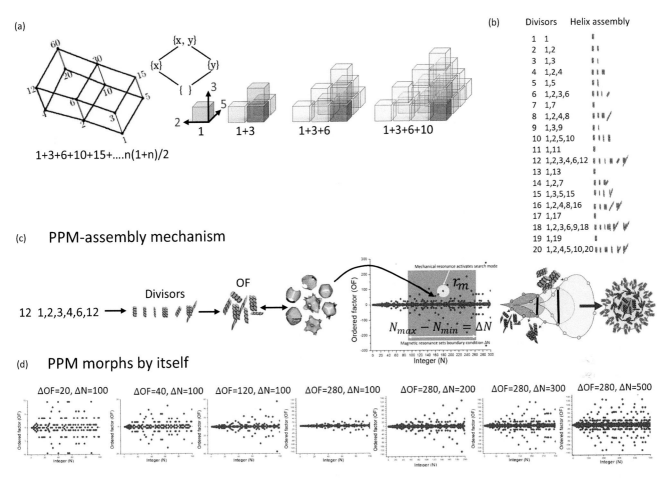

FIGURE 9.14 (a) 1, 2, 3, 5 series in Hasse diagram creates all primes, divisors, four-hole control by helix is enough to trigger Hasse energy transmission, synthesis of divisors. Integer = total number of holes in the assembly of helices Hasse diagram, X, Y, and Z directions represent three primes, 2, 3, 5, since it regulates entropy, we control its boundary by applying magnetic pulses. Computing is simultaneous electric, magnetic, mechanical resonance. (b) How a set of primes builds integers by synthesis, depends how does a helix manage hole production in phase space. Some helices produce all 12 holes, some only a few, if s helix produces only four types of holes,—1, 2, 3, 5—then we can build all integers by synthesis. We can use 2, 4, 6, 8, 10, 12 as 1, 2, 3, 4, 5, 6, respectively. (c) Three self-assembly steps, make integer, build divisors and OF, finally OF assembly and PPM loop assembly. The common line between two loops activates each other. A loop forms a layer, multiple loops forms, core-shell architecture magnetic field decides the number of loops, mechanical resonance decides search and find thermodynamically preferred helix, electrical resonance assists in healing wireless wiring between the helices. (d) The nearest neighbor of an OF point (OF, N) connects as a line, so, a point finds unique neighbors depending on width of N and OF, represented as Δ. Above, a series of plots show that the relative width of OF and N, i.e., ΔOF / ΔN determines nearest neighbors. Here ΔN is the number of oscillators active in the solution, this number is tuned by applying magnetic and mechanical vibrations. Electrical wireless writing, magnetic arresting of neighbors and mechanical vibrations to activate the search mode operate $e^2 + \varphi^2 = \pi^2$.

then $n^2 + 1$ additional coupling increases to $n^4 + 1$ and so on. When two time crystals generate one new cluster of a time crystal, three bursts from one loop, then we get a converging triangle. The synchronization/response time will depend on the number of different frequency fractals used, not on the complexity of the path.

Geometric shapes and the brain circuits: Brain jelly is therefore similar to layered hyperbolic metamaterial system proposed recently (Poddubny et al., 2013) fit to process fractal time (Smolyaninov, 2012). The negative refractive index that defines a metamaterial as an integral component for brain jelly, since there is no wiring, meta-material feature enables a particular component to vanish letting other signals to pass through. Thus, entropy under

invisible components behaves in a non-predictable manner if we think about energy alone. The picture gets clear when we read the geometric shape transition. In the forest of H devices when there is no wiring cloaking and 3D arrangement to hold a particular geometry is the way to build a living circuit. For an S, U, L, V, T, all 1D patterns convert into a straight line and those straight lines bond in a circle simultaneously for the natural oscillation of the network of the oscillator to create the typical feature of a 1D shape. For the platonic 2D and 3D geometries, all structures eventually convert to a circle and sphere, i.e., eventually all 15 structures convert into a 3D structure or a vortex atom. The "most complex" or the "most non-Abelian" subgroup of SO(3) is the symmetry of the icosahedron—the largest

Platonic polyhedron, in some counting, which is the same as the symmetry of the dual dodecahedron, the origin of these groups are explained using string theory.

These cross-sections of circle or sphere and the 2D, 3D geometric shapes get more energy and these frequency values play a dominant role in the further oscillations. As a result, we find that the ratio of these contact point frequencies become the variable in the hardware. Since all-encompassing circles have the same area the ratios play a key role in defining the major terms of the rhythm. The line, 2D or 3D geometric shape to circle or sphere-like oscillation conversion requires certain time, and that depends on the shape of the internal structure. Based on that the frequencies on the circle and the straight-line or on the sphere are selected for the rhythm construction. Therefore, the ratio of frequencies and the time delay determines all the terms for the time fractal or the rhythm to be constructed for the geometric shape. The role of the triplet of triplet resonance band in constructing the rhythms: Is it essential? Triplet band is essential because this is the minimum number of bands required to create a resonance chain, if each oscillator does not have three bands then it cannot handshake with the oscillators holding the clock below and above, apart from its own information processing, three bands are must. And nature has taken utmost care to this requirement.

9.10 THE CORTICAL PEN THAT FREEZES UNKNOWN DYNAMICS INTO A TIME CRYSTAL

Humanoid avatar is not the only application of brain jelly, one could build a tiny pen that would assist a user to solve big data problem or provide an intelligent solution when we are totally confused. As said earlier, to use GML, PPM-based computing we search for confusion. Then find some more confusions within, take one confusion and dig deeper, the journey continues until we reach facts. Now, an artificial cortical column-based prototype is shown in Figure 9.15a. The 19 parallel cortical column channels are detailed in Figure 9.15b. Using 19 Yagi antenna we apply input time crystal, the output is read by the camera on the left. Seven gel layers are filled in the capillary tube made cortical columns. In future, it would be reduced to a commercial pen-like device as shown in Figure 9.15c, wherein one should put particular cortical column miniaturized like a capsule. The operation of a capsule is shown in Figure 9.15d, where during computation how a particular architecture grows and saturates at different layers is shown. Side by side we have shown the PPM how the column represents part of it. How to read output of a magnetic film is shown below Figure 9.15b, lines of forces are bright regions and the dark region is the pole. For a vortex, we should see a bright ring. We have also explained how to write 7^{11} using an organic gel. These are generic problems, for the real world the problem is different.

Automated filtering of groups in a given image, automatic decomposition of fractal seeds: In a given 2D image,

where the corners or singularity centers that burst with distinct resonance frequency values are naturally coupled, by a clocking wave remain within the helical hardware. The time crystal or time crystal is memorized by a static distribution of charge on the helical structure, it never goes outside. When the vortex atom assembly carrying the stored time crystal enters into the resonance chain, then what happens? Resonance chain is not a living matter, it is a distribution of resonance frequencies in widely separated materials and the assembly of vortex atoms carrying the time crystal physically flows to transfer the geometric seed. Since the is no wiring how do they know where to go? Electromagnetically coupled oscillations of many components acting as the members of the resonance chain set the path of time domains. A circuit of time comes into being. A time crystal binds with the resonance chain in the tome domain where it fits. The hardware acts as a sensor jelly.

One of the remarkable aspects of this automated filtering of geometric shapes on the resonance chain is that all possible time crystals are isolated. As a natural property of the oscillators, each group of oscillators generates the nearest fractal polygons, and its equivalent rhythms or clocks. Thus, the entire input pattern is grouped into a nest of clocks; if there are many resonance chains each representing a typical PPM or a particular functional organ of the human brain then, automatically the class of information is sent at the right places. It is like an artificial hippocampus. The continuous oscillations in a certain layer of oscillators generate an image replica of a low frequency version in the above layer, this is how even a normal image becomes a fractal, a triangle has three points and all three of them gets a triangle inside each. Not just that an event does not reside at one temporal layer or one imaginary world of the brain, but in all. Thus, entire information lives at all places, just as fractal information theory, FIT demands. One of the major problems for using such a pen that solves big data instantly is that once it learns a new problem, it can solve only that class of problems. Imagine an algorithm that learns A, could read A for any kind of handwriting.

9.10.1 CAPSULES OF BRAIN JELLY—ONE EACH FOR ONE BIG PROBLEM

Therefore, the obvious question arises, what should one do, buy a new capsule from the online market to solve a particular class of problems? The answer to reusability lies in a beautiful research field developed by Ghosh et al. on the programmable matter (2015a). One of the benchmark challenges of computer science is to build a simulator that automatically learns complex rhythms or waveforms, memorize and reproduce whenever necessary (Jaeger and Haas, 2004). Several black-box computer models were implemented in harnessing this non-linearity; however, reservoirs embed tricks and thus limited to a few specific kinds of rhythms (Buisson, 2004). Randomness enables a starting function to match the variations of time series (Jaeger and Haas, 2004). Instead of software, if a matter can emulate a complex rhythm,

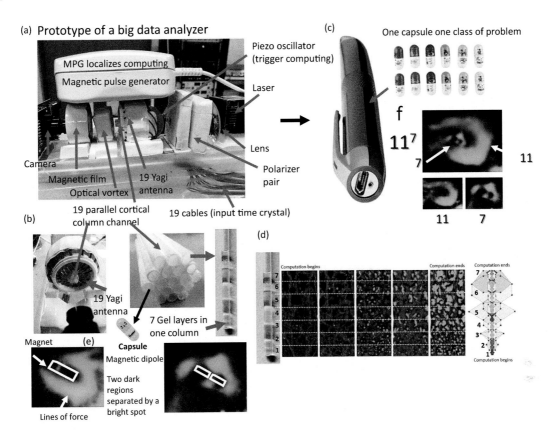

FIGURE 9.15   (a) The prototype of a big data analyzer. (b) Different parts of the big data analyzer. (c) A futuristic brain-jelly-based PPM, PPM-driven big data analyzer. (d) Decision-making process of a brain jelly in a cortical column. The leftmost picture is an NMR tube filled with seven layers of organic jelly. To its right the decision-making begins, the helical nanowire cluster (green) representing different kinds of self-assembly governed by PPM. To its right the equivalent PPM is shown which is instantly read by placing magnetic film around each layer, Seven layers deliver seven magnetic profiles similar to the ones shown in panel (e) and (f) and we read the decision-making process instantly all part of the solution at once. Brain jelly is a mixture of **electrolytic polymer (red)** and **oligomers (green)** in a cylindrical tube. While positive and negatively charged polymers create a vertical column of astronomically large number of cavities by chemical bond, the oligomers sensitive to composition of ac frequencies, build helices, and physically self-assembles the helices to maximize the absorption of modulated ac signal. Seven distinct densities of oligomer gradients ensure seven distinct size distributions of helix cluster, size is minimum at the highest density region. Monochromatic LASER shined along the cylinder's length reads all clusters and superpose magnetic vortices on the screen, a $\lambda / 4$ vortex lens filters superposition in four quadrants.

then an understanding of the language of biological systems might lead to medical treatment, chronopathology (Sainz and Halberg, 1966).

Historical background and contemporary research on programmable matter: The programmable matter was proposed as a hypothetical material that is flexible ("wind tunnel at one moment, polymer soup at the next"; Toffoli and Margolus, 1991, 1993), instantly reconfigurable, shows variable resolution, intelligent sensing, integrated actuator and locomotors, as it engages in invisible computing. Invisible computing means primarily working with virtual atoms, or waves, like quantum, where electron like real particles are invisible (McCarthy, 2003; Goldstein et al., 2009; MacLennan, 2002; Wong et al., 2012). Proteins game with music requires special attention (Wong et al., 2012). In 1991, Toffoli and Mergolus proposed programmable matter is a material absorbs a particular form of energy, say light, sound, magnetic, electric field, etc., and changes its structure typically to encode the parameters of the input signal. In a rhythm

or time fractal, a few frequencies together form a complex time series. The programmable matter is yet to face a benchmark problem of computer science (Jaeger and Haas, 2004); no true usable material was ever synthesized. By making materials that replicate rhythms we satisfy both the essentials for a true programmable matter. Our brain, even a single DNA molecule is a "programmable matter," this suggests a matter would by itself learn the hidden intelligence in a complex information packet and replicate that and no algorithm is essential. The proposed computing paradigm challenges the very necessity of a software program and the very need for CMOS based electronic world. Both soft and hard part is done by the material itself. Obviously, the invention should come from materials science. Though proposed in 1991, no experimental demonstration existed until now. When billions of data are being processed, and we have no clue regarding its hidden pattern, without building up the complete algorithm the programmable material can learn the phenomenon and reproduce. It could be the ultimate goal for artificial intelligence.

Brain jelly in that sense a true programmable matter, however, the brain has 47 Brodmann's region integrated for universal processing, that is absent in the cortical pen.

Some examples are evolving neural network, nano-wheel as integrated machine, replication of natural phenomena on molecular surface (Bandyopadhyay and Acharya, 2008; Bandyopadhyay et al., 2009a, 2009b, 2010b, 2010c). However, biological rhythms are key to the ultimate programmable matter like our brain or even DNA. Bio-systems encode and decode rhythms to run highly interconnected machines simultaneously; though the thrust to make an artificial programmable matter is increasing rapidly, no usable material exists that learns rhythms naturally and encode it in its dynamics. If realized this material would alleviate the necessity of algorithm, open the door to a new class of self-reconfiguration, cloaking, self-learning, encrypting, etc. Here, an organic jelly made of spiral nanowires changes its length and helical pitch during growth to mimic the composition of frequencies pumped into it. The cluster of spirals drive the self-assembly to a particular kind of fractal network that encodes the rhythm, so we can read/write ANY encoded rhythm or time crystal. The jelly learns several complex rhythms pumped into it operates without algorithm; learning and memorization finish within a few seconds with zero human intervention at any step. However, if one deletes the gel, converts into a liquid, all memory is deleted. Partial deletion with a clever choice of locations is utmost important and time crystal-based deletion does that amicably.

Multi-layered clocks govern a biological system using rhythms or time fractals or frequency fractals (Cipra, 2003). A material that replicates the dynamics of biological rhythm, could compute without using any machine language or algorithm (Ghosh et al., 2015a) intelligence hidden in the complex rhythms (mathematically, rhythm = time fractal ~ frequency fractal or frequency packet) get automatically decoded by the matter (Buisson, 2004). It demands a dynamic material that programs a complex time series.

Geometry programs in the time crystal: Fourth-circuit element Hinductor is a programmable matter seed, which takes a new shape, a new length pitch, diameter, and lattice parameters a and b depending on which time crystal is given in the solution as an input. The idea to create a programmable matter that emulates bio-rhythms is to harness a unique feature of a spiral geometry. A spiral material's electromagnetic resonance frequencies depend on its geometrical parameters, pitch and radius (Michalski and Mele, 2008; Atanasov and Dandoloff, 2008). The length also plays a role in governing its frequency, now if we enable a material to grow in a chemical beaker such that as a function of input frequency the material shapes its spirals then the resultant material would be a programmable matter that encodes rhythms. With this idea, we have carried out extensive synthetic chemistry work and realized an organic jelly that exhibits this feature.

Spiral wiring has the advantage that as its pitch P and perimeter $2\pi R$ ratio $P = \pm\frac{2\pi R}{c}$ that governs periodic potential $V(x, y) = A\cos(y - Px)/R$ changes the resonance band or the processing frequency. The fractional changes in the

bandgap are given by $F(\alpha, 0) = \alpha \frac{a^2}{2}\left[4P^2 \frac{|\sigma|}{4\sigma^2 - (1+P^2)^2}\right] \sim \Delta h\upsilon$. It has already been shown that the effective electromagnetic resonance band potential could be regulated in the spiral structures by tuning the ratio as curvature controls potentials $V_{eff} = V_{kin} + \frac{2\mu V_{curv}}{\hbar^2}$ (μ, energy difference with Fermi level; Michalski and Mele, 2008; Atanasov and Dandoloff, 2008). The helical pitch of the structures produced changes significantly to capture different frequency signals in the input packet. If we compare the elementary spiral in an organic gel that undergoes 1000-time increment in dimension in the spiral grows as a spiral, Ghosh et al. (2016b). The elementary 20 nm wide nanowire of has electromagnetic resonance in the GHz, as the physical ratio changes by 1000 times the resonance frequency decrease by 1000 times to the MHz domain. Analysis based on existing formulation itself explains why spiral geometry enables the materialistic capability to encode the rhythm. Thenceforth the work has been extended to abundantly available spiral, fractal, vortex-like organic super nano-structures.

Ghosh et al. made the first demonstration of a usable material that learns and programs rhythms. Here, rhythm or time series or time crystal acts as a programming language; one could create multiple rhythm processing materials at different regions of a beaker simultaneously, by inducing physical cavities. Then, the rhythm of rhythms or "time crystals" (Winfree, 1977a), i.e., a singular complex waveform as a superposition of multiple frequency packets be given as input and multiple rhythms at different parts of the beaker would generate new composite rhythms. Currently, learning and replication finish in seconds, however, constraints can edit rhythms to run for hours, if possible, years, until materials get exhausted. Once we learn to erase a material pattern and restructure a part of it with new rhythms, i.e., learn to work with a fixed amount of material forever, then this organic jelly would truly be a brain jelly.

9.11 A SENSOR THAT SEARCHES FOR GIVEN KEYS, DO NOT SENSE ANY INPUT

Eventually, the entire human brain is a sensor. The journey to study a brain jelly begins by detecting a singularity condition of a clock, when and how it's phase change becomes undefined for a certain time, when something else defines phase (Figure 9.16a).

Mathematics of "self-assembly of time crystals" and "phase transition of time crystals": The building of a time crystal is a time crystal, just like software algorithm, but here, many interesting features could spontaneously evolve (Figure 9.16b). Self-assembly of time crystals is fusion of two time fractals, the rule of fusion is that the non-harmonic set of frequencies $U\{t(n)\}$ which represents a typical composition of geometric shapes is not diluted, one example is $T = t1 + t2 + t3 \ldots + tn$, while the time fractal $g(t)$ transforms to $f(t)$ after fusion, and $f(t)$ satisfies $f(t+T) = f(t)$. For a distinct information content for all $n, t(n)$ are not harmonic, thus, inharmonicity represents distinct contents in an information packet, while harmonicity makes a boundary, thus isolates information packets $S\{U\{t(n)\}\}$ from mixing. It is self-assembly of time crystal as it grows continuously

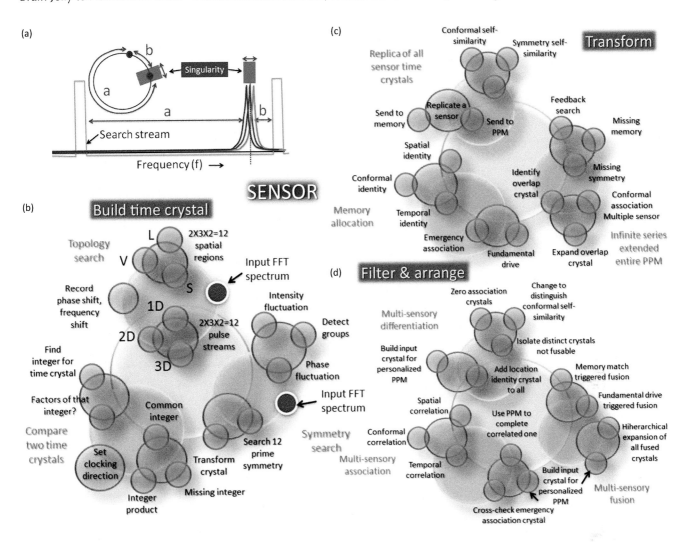

FIGURE 9.16 (a) Searching for a clock begins with the observation of phase singularity in a periodically oscillating system where at singularity there is an energy burst that triggers periodically in the system. Brain jelly is a sensor. Its prime activities are summarized in the three panels (b) explains the development of the time crystal spontaneously in the system, (c) transformation of the time crystal and finally (d) filtering of the time crystals and rearranging them by following the PPM.

and transforms. The transformation also runs by a governing time crystal (Figure 9.16c). Due to harmonics, the identity of $U\{t(n)\}$ is preserved, it acts as an observer of the entire system. Phase transition is switching between two time fractals, $g(t) \rightarrow b(t)$, if number of terms of $U\{t(n)\}$ in $g(t)$ in the course of time changes to $M\{t(n)\}$ that represents $b(t)$. Once phase transition starts, it propagates through the entire system.

A single time fractal evolves after fusing with more and more packets, the fusion satisfies integral feature explained above which naturally satisfies the condition for synchronicity. It should be noted that just like fusion, phase transition rules can have multilayered architectures, "nested network of rules." One phase transition condition is met, one decision is taken, however mostly multiple conditions are met then a new time fractal is born, possibly in a different time domain, on the resonance chain that represents the hardware, the fusion-fission and phase transition can have in two far distantly separated rhythm on the resonance chain.

Synchronization of two different rhythms is also represented as attractors how sync is used in a fusion of time crystals. Interactions of different fractals could happen in many different ways generating binaural beats and several other kinds of locking process at different time scales "composition of time fractals" is a superposition of multiple rhythms. It is not multi-fractal where at a different scale we find different fractal dimension. However, to resolve a problem, brain jelly filters a time crystal following PPM. The governing system is also a time crystal (Figure 9.16d).

Construction, transformation, and filtering or rearranging the time crystal is part of solving a problem. The time crystal-based brain jelly discussed here, converts all events happening in nature into time crystal using GML, thereby solves only one problem, that is called Clique problem, which is simply finding a pattern in a complex pattern.

The NP-complete intractable "Clique" problem and a universal "reply back" protocol: The clique problem is a well-defined intractable problem, where one has to search a given pattern in the complex set of patterns. The clique problem needs to be solved in a finite time for any advanced cognitive or creative intelligence observed even in primitive neural networks, this is our perception about brain engineering. Most input data convert into an unknown pattern that brain jelly has never encountered before, be it visual, sound, touch, taste or smell. Obviously, the number of possible patterns that could be generated from this composition is astronomically large. Now, if we want to search 15 given known pattern in that resource input, it is not possible to find that pattern with any computer within a finite time. However, if those points have the properties to reply back together spontaneously then we can get the search result without searching. As noted above that to avoid the screening effect we need a new kind of material that would follow the resonance chain throughout the architecture. There exist several classes of the clique problem originally proposed in 1949, as frequently observed in the classically intractable problems, a certain constraint is imposed to simplify the complex network and then an algorithmic route is found to solve that problem. However, in this particular case, we take any sensory data, visual, sound, touch, taste or smell in the form of a 2D pattern and from that image we transform the single image into several layers of images, each containing several different classes of "frequency fractal" seeds. During transformation each layer distinctly represents a particular type of fractal seeds, based on the size of the basic geometric shapes used that incorporates global relationship of elements in a pattern. The network between various kinds of fractals is generated due to the typical conical architecture of the entire computing network.

9.11.1 Eleven Dimensional Signals in the Human Brain and in Brain Jelly

A review of the experimental research described in this book and the final discussion on the experimental realization of a full-fledged organic brain: The potential of pattern-based computing was never explored to the fullest. Scientists have always tried to look for the computing constructs that would lead to logical operation. Scientists created the smallest molecular neural network, nano-wheel for the glia inspired circuiting and then the cellular automaton based massively parallel computing on the organic molecular layer, we have described them in Chapter 6. It was realized that neither computing constructs help us to generate bioinspired computing nor the analog pattern formation similar to a particular physical phenomenon. Thus, we started the building of brain jelly, which is an organic molecular structure that vibrates like a particular composition of electric magnetic and mechanical rhythm. The hierarchical memory in the chemical reaction is essential to program the entire growth of the brain in a single set of H or hinductor devices, which would be analogous to DNA in biology and then explore condensation beyond chemical kinetics, two remarkable tools that take us to the original brain jelly experiments. We have detailed, the final phase how 17 brain-morphic components, each encodes a typical PPM. Seventeen types of brain jelly

have been synthesized. All neuron wires made of H2 devices (second generation Hinductor) should grow one by one and extend to the final destination via H3 modules. From sensors to the editing regions of the brain and finally to the cortex domain there is linear wiring wherever we want to preserve the fractal-seed and do not want it to interact with anything else, we need to grow the circuits carefully there. Such a massive scale self-assembled supramolecular architecture was never done before. Now, other than the delicate places, we use neuron jelly wherein neurons could re-orient and reconstruct circuits based on internal axon restructuring. Figure 9.17 lists how after growing brain jelly how one could actually read 12-dimensional data. Since the technology is optimized, the brain jelly was injected into the brain of a robot (Figure 9.17, right); 3D printing components are already optimized and listed below. The growth of the brain jelly could be observed live at nanoscale using a scanning electron microscope (SEM).

9.11.2 Humanoid Avatar—An Ultimate Sensor

Humans would transform with artificial sensor arrays (Guntner et al., 2018); now it is possible to create a humanoid avatar to test the brain jelly or any form of morphological computation (Hauser et al., 2011). Since resonance chain connects all computing seeds, wireless processing is feasible without a screening effect, also we do not need to create an antenna with enormous power to wirelessly transfer signals from one part of the brain to another. The computing power is increased by maximizing the density of resonance states and bandwidth of the resonance chain together. Below we describe how we have experimentally developed the first version of organic nanobrain that forms the brain jelly while learning from its immediate environment. It is not the brain jelly for the ultimate brain, but, this is the first step toward that direction. The humanoid avatar is a resonance chain. The Shannon information capacity of space-time wireless channels formed by electromagnetic sources and receivers in a known background medium is analyzed from the fundamental physics point of view of Maxwell's equations (Gruber and Marengo, 2008).

Wide ranges of vortex solids and liquids (Huber, 1994) generating time crystals, could be tested on the 47 cortex domains of the humanoid avatar that we built. Several prototypes of humanoid avatars have been created and some of them are available in the market, but they run by algorithm, they don't have PPM, the mother of all programmers. Moreover, no effort was ever made to meticulously create the entire neural network of a human body, build every single component using wide ranges of materials by 3D printing and then filling the cavities using suitable jelly. The prime concern that arises, how do we test that the humanoid bot is conscious. The same way we understand, the other humans are conscious, talking and interacting with it, on random topics. The main idea of this book is to explore the possibility of building a human brain that is non-chemical, has no software or pre-determined algorithm, and thus processes unknown, unpredictable events instantly and perpetually, without training.

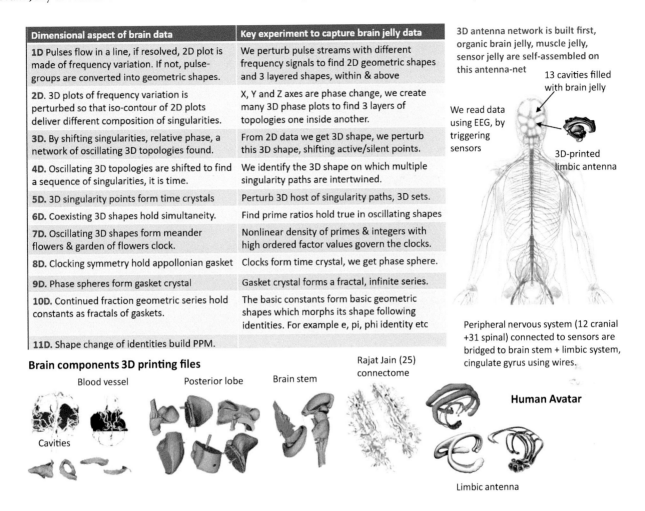

Dimensional aspect of brain data	Key experiment to capture brain jelly data
1D Pulses flow in a line, if resolved, 2D plot is made of frequency variation. If not, pulse-groups are converted into geometric shapes.	We perturb pulse streams with different frequency signals to find 2D geometric shapes and 3 layered shapes, within & above
2D. 3D plots of frequency variation is perturbed so that iso-contour of 2D plots deliver different composition of singularities.	X, Y and Z axes are phase change, we create many 3D phase plots to find 3 layers of topologies one inside another.
3D. By shifting singularities, relative phase, a network of oscillating 3D topologies found.	From 2D data we get 3D shape, we perturb this 3D shape, shifting active/silent points.
4D. Oscillating 3D topologies are shifted to find a sequence of singularities, it is time.	We identify the 3D shape on which multiple singularity paths are intertwined.
5D. 3D singularity points form time crystals	Perturb 3D host of singularity paths, 3D sets.
6D. Coexisting 3D shapes hold simultaneity.	Find prime ratios hold true in oscillating shapes
7D. Oscillating 3D shapes form meander flowers & garden of flowers clock.	Nonlinear density of primes & integers with high ordered factor values govern the clocks.
8D. Clocking symmetry hold appollonian gasket	Clocks form time crystal, we get phase sphere.
9D. Phase spheres form gasket crystal	Gasket crystal forms a fractal, infinite series.
10D. Continued fraction geometric series hold constants as fractals of gaskets.	The basic constants form basic geometric shapes which morphs its shape following identities. For example e, pi, phi identity etc
11D. Shape change of identities build PPM.	

3D antenna network is built first, organic brain jelly, muscle jelly, sensor jelly are self-assembled on this antenna-net

13 cavities filled with brain jelly

We read data using EEG, by triggering sensors

3D-printed limbic antenna

Peripheral nervous system (12 cranial +31 spinal) connected to sensors are bridged to brain stem + limbic system, cingulate gyrus using wires.

Brain components 3D printing files

Blood vessel   Posterior lobe   Brain stem

Cavities

Rajat Jain (25) connectome

Human Avatar

Limbic antenna

FIGURE 9.17 A table explains how we could feel 11D data transmission in a living brain and its brain jelly analog. Below we demonstrate some files of structures for the 3D printing of the brain components for the final humanoid avatar development. To the right side of the table, how the experiments are performed in the humanoid avatar is shown.

9.12 SENSOR TRIAD, SENSOR JELLY, MUSCLE JELLY, AND BRAIN JELLY

Eventually it is the spiral wave and its critical propagation limits (Karma, 1991) in a given excitable media that regulates the performance of a jelly. The scaling regime helps to optimize scale-free operation of a jelly, (Karma, 1992). Sensor triad, the sensor jelly, muscle jelly and the brain jelly would differ only in geometric features exhibited by the jelly; see Figure 9.18. Spiral waves propagate adopting different geometric features (Keener, 1986) in three types of jellies. Vortex filaments generate emulating the geometric parameters (Keener, 1990) required for sensing geometric shapes, executing mechanical movement and sensing the symmetry of primes. It is not one, but many waves dealing with many singularities (Koga, 1982b). The geometry of path or cavity reconfigures distribution of singularities (Kogan et al., 1992) in an auto wave (Krinsky, 1987). The sensor triad does not provide the ability to self-create or autopoiesis (Varela et al., 1974). Be it a sensor jelly or muscle jelly or brain jelly, for a typical component in the brain, a particular time domain has to be selected. Time-based priority has to be given. We discuss below, why one has to classify organic jelly based on time domain.

The fractal time domain distribution among different brain components: The neural network or brain jelly in the artificial brain of Humanoid avatar is so designed that time domain-distribution has inverse one-to-one correspondence with the resource priority of the brain, something like: "we give the highest priority to the visual data, then the sound data and finally to smell and then touch, in the absence of higher control, lower control-data takes over as the priority resource." Higher is the priority; lower is the time domain, now, to avoid any possible conflict some more rules are required for synchrony-de-synchrony flipping. First, the maximum and the minimum of total synchronization time for a particular domain do not overlap with any other regions, phase diagram reveals a pure communication (Matthews and Strogatz, 1990). Second, since complex time crystals couple neurons from two or more regions, the synchronization time is determined by the 3D pattern shape created by the coupled neuron network, it is faster if the shape is more symmetric.

Sensor jelly, SJ	Muscle Jelly, MJ	Brain jelly, BJ	Biological neural network	Artificial brain jelly
PPM-1, PPM-2. Sensor molecules added to brain jelly	PPM-2, PPM-3. Electro-mechanical fibers added to brain jelly	PPM-3, PPM-4. BJ is 3D jell that senses time crystal, rewires.	Components are neuron, glia or astrocytes, peri-neuronal network, PNN.	Neuron is replaced by Hinductor class 4th circuit element H, glia by supramolecules made of NB and PNN is made of carbon nanotube based hydrogel.
Rewires to capture 12D dynamics of input data	Shrinks/expands morphs to regulate PNS core antenna	Sense only time crystal, changes minimum	Various different classes of neuron cells make branches, bundles, vericosites, etc	H based bipolar fractal architectures create hydrogel, this is used to mimic neuron.
H1, H2 class Hinductors	H2 class Hinductors	H3 class Hinductors	Neuron membrane sends binary pulses to compute	Gel morphs time crystal like real biological materials
Most sensitive to topology to read reliably pulse features of sensor cell	Most sensitive to magnitude & phase of pulse signal	Most sensitive to geometry of phase architecture	Learning & plasticity defines short term memory & long term memory, LTM	A composition of frequencies convert part of gels into a rigid material, its LTM.
			Cells die, wastes are cleaned.	No waste is produced.
			Connections break & create.	Connections break & create
			Membrane pulses does everything, information is only in membrane.	12 types of rhythms, 12 types of memories work together
Links sensors	Links skeleton	Links metric	Proteins evolve to make new structures.	Requires artificial injection of new gel.
			Requires food and massive energy supply	No food, no energy, only thermal noise
			Brain is limited to the head, neural network of entire body is not a part of it.	Mimic neural net of entire body as part of brain, neural fibers are made of tubes, tube membranes are semi-porous plastics.

Humanoid avatar under construction

A conversation with the first prototype conscious device

USER: What do you think of Mahatma Gandhi?
Chatbot: Quite a peace in life, a vegetarian Indian was he, walk a mile, karamdas known to many, non-violence that of humanity, a person to remember.

FIGURE 9.18 A comparison between the sensor jelly, muscle jelly, and the brain jelly. To the right a table compares the similarities and the differences between the biological neural network and the artificial brain jelly. Below is a photograph how the synthesis of a humanoid avatar begins at the skeleton level and eventually how software-generated chat-bot speaks about anything without prior knowledge.

Nonlocal interaction of spiral waves always happens (Meron, 1989), it can organize infinite number of *H* devices like organisms (Siegert and Weijer, 1995). Third, synchronization time is artificially defined: in a continuous synchronization and de-synchronization process, the 3D pattern-shape inside which neurons are actively playing the game of synchrony changes continuously, therefore, there is no distinct division between any two events; we consider that an event is completed when a large 3D pattern shrinks to a minimum. In the world of wireless communication, we simply add more wire-connections; make it hybrid-wireless to tune time domain at will.

Organization and re-organization of time-domain: Simultaneously, during oscillation, in the Humanoid avatar, the global convergence rules are superimposed on the oscillatory wave using a hierarchical buffer antenna. The time domain of transmission for superimposing the higher-level rules should match with the time domain of oscillatory transformation wave. The convergence generated in the distributed local regions of the upper brain is updated regularly as transformed time crystal cluster in the global periodic oscillation of the resource buffer antenna; therefore, its time domain should be faster than the transformation process. Once local-distributed regions stop generating new time crystals, the

updated time crystal-cluster is sent to the resource buffer at the same time domain by which the resource antenna sends a signal to its buffer.

9.12.1 A TOTAL TRANSFORMATION FROM THE BIOLOGICAL NEURAL NET TO A JELLY OF TIME CRYSTALS

Biological neural network and artificial brain jelly placed on the head of a Humanoid avatar are fundamentally different (Figure 9.18). Since ionic and chemical reaction based systems are avoided completely in the operation of humanoid avatar or brain jelly, an enormous number of problems have been avoided. Of course, brain jelly is organic; however, production, editing, and processing of time crystal do not require any chemical reaction. Whenever there would be chemical processing, there would be a waste—rather, here, physical reorganization of H, H2, and H3 devices is the key operating factors. Brain jelly is a quantum processor at the molecular scale and classical processor at the macro scale. Time crystals operate smoothly between classical and quantum domain because of fractal mechanics, that enables dodecanion, octonion, and quaternion tensors to regulate the interactive dynamics between the brain components.

Quantum models of the human brain: Brain's activity is related to quantum and even consciousness (Beck and Eccles, 1992). Space-time metric has been used by several physicists to explain consciousness, for example Lowen model (Lowen and Miike, 1982) was the first to consider gravitational lensing to generate consciousness. Spatial function becomes irrelevant; most importantly, E. Surowitz argued for space-time generated neural cycles or rhythms as the origin of consciousness (Surowitz, 2011). Directional computing becomes unnecessary, and they considered a feedback loop to generate time-invariance. It is exactly what happens in a time crystal. As a driving force an evolved Lowen model considers dual face to face conical-shaped lens formation, and its hierarchical network. Orch-OR theory (Penrose and Hameroff, 1996) that argues for gravitational collapse to explain consciousness just like the Lowen model makes another contribution, it includes a remarkable biological molecule microtubule as the processing unit, microtubule in axon is a singular wire, that means in giraffe, a 10 ft neck has one neuron and possibly a 10 ft long single molecule. Since all single molecules are a quantum device, the debate is not about the effect is quantum or not, the true debate is how the hierarchical network and the language of the brain are constructed. PPM protocol is a generic proposal independent of a particular consciousness theory to provide hitherto existing consciousness models a route to integrate with all highly interconnected research fields. It is a journey to consolidate Bayesian brain that correlates function and brain image (Friston, 2011) on a concrete mechanistic pathway.

Decomposes Recursive functions of Undefined problems and simulates all scales at a time: Fractal machine, brain jelly is designed to address undefined problems: One of the finest features of fractal tape network is that the machine can compose a rhythm inside a rhythm inside a rhythm... or clock inside a clock inside a clock wherever it sees in nature in any known or unknown form. Mathematically it is argued by infinite fractal time series, such a fractal of rhythms enables a machine to generate 99.99% of all possible rhythms in its environment or user with which the machine is locked. As a result of it, the machine does not require to define a particular phenomenon that is not known to the machine. The machines embedded can resonate with the 11D scale recursive functions naturally evolving in the universe in its frequency fractal network. It decomposes the dynamics to find the function that repeats itself and generate the entire composition of rhythms, it means the sensor builds the time crystal. When the time crystal is expanded using the PPM, as we read it from the Humanoid avatar using several EEG and multi-channel analyzer cables, the evolved architecture builds the patterns of the past present and future simultaneously in the network. Thus, brain jelly can address unknown events and situations. Therefore, it is not far brain jelly filled avatar would start thinking like a human brain (Binnig et al., 2002). One such classified effort's glimpse is shown in Figure 9.18 end. The final humanoid bot is not shown here, but its conversation is noted. It is not good in grammar but would surely learn in the coming future.

Jan-manasa na manute jenahurmano matam|Tadeba brahma tang bidhdhi nedang yadidamupasate||
No one can accurately underpin the universal consciousness even in deep inside his inner mind, wise men say, it is the deepest inner mind that is an expression of universal consciousness. Even the god that is worshipped by people with names and forms, that god is not the pure form of universal consciousness.

Jat pranena na praniti jeno prana praniyate| Tadeba brahma tang bidhdhi nedang yadidamupasate||
That cannot be expressed with the primary signs of life (breathing of soul ~ prana), but that universal consciousness composes the fundamentals of life form. Even the god that is worshipped by people with names and forms, that god is not the pure form of universal consciousness.

10 Uploading Consciousness— The Evolution of Conscious Machines of the Future

10.1 A JOURNEY FROM CORTICAL PEN TO A CONSCIOUS EGG AS A COMPANION OF LIFE

What is consciousness? Would an organic nano brain or brain jelly ever be conscious?

The definition of consciousness: "Consciousness is an ability that enables a machine to define and evaluate all its behavior by taking all its information content outside as three distinct identities; a conscious machine has at least three distinct information architectures, which could interact with each other and edit, independently." Truly, this is Russel's paradox, but this paradoxical definition would satisfy much of the conscious entity's behavior. All other criteria, except having three distinct information architectures simultaneously operating could be realized by a classical simple Turing system. **The definition of information contained in the brain or for consciousness**, "Information content is an endless architecture of time crystal following a pattern of primes that is localized in a matter, following which the matter perceives it as a finite state." **The evolution of consciousness:** Transformation of phase prime metric (PPM) encoded in a matter's finite state residing inside an endless time crystal of nature, it is synchronization of symmetries between a living entity and nature.

Consciousness is defined as self-awareness, the ability of a species to create a system point outside its body wherefrom it can evaluate every single reaction, operation, the thoughts, and the dreams of its own physical body. Hardware can have two co-existing states simultaneously, e.g., a qubit, or three coexisting states like a qutrit of quantum. However, quantum is rigid, it does not allow multiple imaginary worlds coexist and operate by interacting with themselves. Quantum mechanics does not allow to make a hole in the phase space, fractal mechanics make holes, blink it, i.e., open or close by wirelessly pumping a signal. While deriving quantum formulations from the information content it was argued that angle between the phase space controls the probability of choices (Wootters, 1980, 1981). It was a very significant discovery. Afterward, it was further shown that the combined probability of three choices is the square root of the products of three probabilities (Fisher, 1922, 1956). However, this is true for only one kind of geometric relations. We have introduced a metric of primes (PPM), that finds hidden patterns in all events represented as geometric shapes. Therein, the probability of a combined choice is an infinite series of symmetry of the geometric shapes, not one (see Chapter 4, fractal mechanics [FM]). Thus, Wheeler's dream to deduce the existence of the universe could be pure geometric, as he wanted, but, not using pure quantum mechanics that we know. When we add one condition, that one of the three mirror states changes partially by the typical geometric relationship between different parts of Hilbert spaces, it is more like a classical system, not quantum. Moreover, when there is no classical point, then, all points are either located in the imaginary world tensor of phases or singularity points, undefined, because they hold unique information inside, then, neither classical nor quantum tag fits it. Thenceforth, the mechanics are named as fractal mechanics (Chapter 4;

Reddy et al., 2018). Now, if we imagine that hardware can create three replicas of its information structure representing the brain hardware, and all three structures, edit each other's symmetry breaking features at all time scales, then, we get the most primitive machine inspired by phase prime metric or PPM. Then, an infinite chain of clocks one inside another forms an undefined singularity network, the chain is not linear or non-linear; it is a nested sphere, i.e., a universal time crystal (Chapter 2). Such hardware we define as a conscious machine, since, it syncs with the universe, an infinite pattern of primes and a coverage of 99.99% of all symmetries make sure that we understand everything that happens around us without writing a single line of code.

One schematic of an artificial conscious brain design is shown in Figure 10.1a. We cannot cover all aspects of consciousness, but, we keep a provision, that is, if a random number of events are given to us, how they were linked in the past, to be linked in the future could be determined from the metric. It is a primitive yet first step to reverse engineer consciousness. Figure 10.1a is a commercial prototype of the human brain-body system, upper hemisphere is like human brain, but the entire human body is shrunk into lower hemisphere. Imagine in the future, people would buy a ball as a conscious companion, whatever a human does a whole day, is updated in the jelly.

How PPM computation research unravels a new way to look at the same universe?

There are primarily three steps executed by a conscious machine, using its PPM filters time crystal from the environment, then filters it for the processing elements of the brain and finally, time crystals missing in the brain is absorbed (Figure 10.1b).

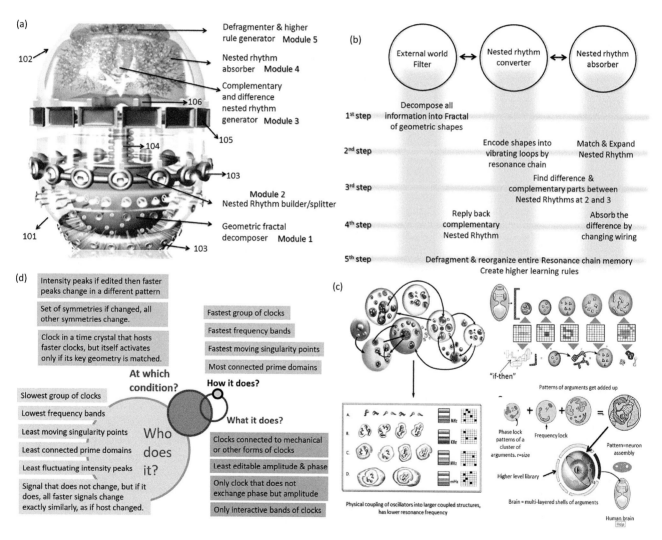

FIGURE 10.1 (a) The design of a commercial version artificial brain as an egg of consciousness companion. (b) Five different hardware modules are explained in the panel (c) The operational mechanism of nesting of time crystals shows interaction of time crystals at various hierarchical layers of hardware in the time domain. The physical structure of the hardware does not represent the true picture of the brain. (d) The information structure that the artificial brain captures is a linguistic architecture. The sensors are not dumb devices that awaits the signals to shine on it, rather, it seeks fundamental properties to fill four imaginary layers as quaternion. Each layer holds physical significance. In the chart linguistic-based transformation of information structure is defined.

1. **The whole body is the human brain:** Biggest mystery of understanding the brain lies when pulses are created at the sensors, in the skins, tongue, eyes, nose, and ears not in the neural network of the central brain. Information of external world is converted into a mysterious stream of pulses right at the sensors, we decode that as time crystal to learn brain, the biggest mystery of the brain is to be solved not at the brain but at the sensors.

2. **Evolution is a process of enriching the resonance chain to mirror image of more parts of nature:** The objective of life is to enrich the resonance chain, those who fail, disappear from the planet, as part of evolution. Darwin's view of the conflict of the matter is a primitive explanation of evolution, the true evolution could be a dance of matter and wave in harmony as we reported the same for the proteins in Chapter 6.

3. **Mirroring the nature in the resonance chain is consciousness,** the density of allowed vibrational frequency in a band and length of the frequency chain (10^{-10} Hz to 10^{16} Hz) determines the degree of consciousness. Therefore one can quantify how a machine could evolve and reach toward consciousness, there is a threshold value to start mirroring the universe or nature around, however, there is no upper limit.

4. **Fill in the gap in the resonance chain is the true understanding of a functional brain:** The real human brain-building project would be taking each material in the brain and fill in the gap of the resonance chain that we created by experimentally measuring the resonance chain of the brain (10^{-10} Hz to 10^{16} Hz). PPM-GML-H triad project is fundamental research in the understanding of the functional working principle of the human brain, it demands an entirely new culture of constructing the brain in future.

5. **Brain or every living thing performs only one computation in life:** It begins at the birth of a baby in the mother's womb and it ends with the death of human life. Life is a single pulse of the lowest resonance frequency of a life form, the chain vibrates as one string (10^{-10} Hz to 10^{16} Hz). It is the same with the universe too. One rhythm, one oscillator, one tape with one cell… inside it's a nested network of those elements. The nesting of everything in the form of escape time fractal is the key, this is a major paradigm shift from the existing iterative function system type fractals which until now was mostly used in designing the hardware.

6. **Living and non-living things are subtle differences in nested symmetry clusters:** Nature concentrates more resonance states in the chain in some matters which we call life or brain and nature dilutes the resonance states in some matters i.e., open space, just like we have black holes and supernovae in the galactic universe. PPM-GML-H protocol sees the universe as nested resonating material, does not engage in differentiating the living and non-living objects, one unifying principle of the density of resonance chain and its length determines living and non-living entity.

7. **Multiple imaginary worlds: Beyond quantum:** Background microwave radiation in the 3D sky around our planet resembles the map of the potential distribution of a single molecule, so many papers claim to read that pattern. If we look at fMRI images and no one tells us what it is, potential fluctuation looks just like a single molecule, where multiple atoms are sharing a single electron in an amazing way. The simultaneous potential changes are strikingly similar when we are in the quantum world or in the classical world. Since PPM mathematical formulations use nested clocks, there are multiple imaginary worlds unlike quantum where there is only one imaginary world.

8. **The supremacy of spiral symmetry:** Microtubule is another DNA, instead of genetic codes it holds the thermodynamic code. Not written like DNA does not mean no code exists. Microtubule holds the map of 3.5 billion years of evolution, but people do not have time to look beyond DNA. The spiral symmetry is everywhere from DNA, the secondary structure of proteins to solar systems even to the blackholes.

9. **Evolution of rhythm with materials: Unique programmable matter:** Brain plays a music like Indian classical raga; it means a set of rhythm unfolds infinitely just like a coded self-assembly, everything we see, hear, taste, touch, smell is decomposed in terms of a few geometric shapes and the entire song unfolds when we interact. Since time crystal runs, all our expressions are some or other forms of music, so are the events of this universe. Thus, a single rhythm can accumulate matter to grow manifold perpetually until an equilibrium is reached with the external universe.

10. **Resonance chain resides at every single point:** The entire universe has been interpreted a single mass, a single wave function and a single photon, that photon never gets old, it is everywhere at every moment, we can go inside it, find seeds, then take one go inside find another seed, the journey would continue to the Plank scale. A chain of mass, a chain of resonant vibration starts from Plank dimension to the ultimate universe, our brain is a small chain and a part of this universal chain. Every point is therefore a mirror of the entire universe. Due to this escape time feature, not a single point of this Universe could be touched.

10.1.1 How We Construct a Sentence Is How We Think

The neural network started in 1943 by Mcleod and Pitts, then, the neuron was considered as a logical device. In the 1960s, F. Rosenblatt and B. Widrow introduced adaptive neurons. Finally, the concept of a 3D surface of hills and valleys where computation moves through minimum energy paths was the most remarkable adoption in the 1990s by Tank and Hopfield (1987). The computation begins with a random initial state as a pattern, and then, via adopting the evolutionary pathways, system reaches the solution pattern. Contours of the surface continuously change due to environment and system tries to match as brilliantly explained by Hopfield and Tank while discussing collective computation in neuronlike circuits. Computation beyond Turing limit has enjoyed several failed proposals, including the idea that in a highly chaotic dynamic system one can use shift map connecting many at a time to stretch beyond Turing (Siegelmann, 1995). Finally, we come to an infinite structure of bubbles here in this book, playing with primes, where the whole universe turns out to be an undefined structures of singularity points. The journey to traverse is from everything is defined to nothing is defined.

In the last century, the language has evolved and now we have geometric musical language (GML). In Chapter 2, we noted that a quaternion is the brains decision structure, *who, when, what and how*, are four questions. However, in the course of this book, we have outlined how one could see these questions from various research fields. In Figure 10.1c we have summarized all variants of four queries. If a person just looks at a big data, notice the changing in a 3D complex pattern, what would it look for, this is an important question. One could try to find periodically repeating events while building an architecture of confusion. Start using a concept where you are confused, go inside and find one of 15 geometric shapes where corners are confusion. Repeat until you reach facts. However, which clock to put where in which confusion? The outline is below.

10.2 TEN KEY GUIDELINES TO REVERSE ENGINEER A HUMAN BRAIN

Unconventional life-like computations by honeybee and algae: Recently, news came up that honeybees are solving the problem of a traveling salesman. Of course, even a mosquito can solve this problem. Computer fails to solve this problem within a finite time because it wants to find the exact solution considering all possibilities. In PPM computing philosophy search and find is performed without searching because the element with the right answer responds spontaneously. We living creatures' can spontaneously respond, so is the programmable matter proposed by Toffoli and Margolus (1991), its prime example is brain jelly. We can take up a few choices and follow one of them, which may be close to the exact solution, even if not, we don't care. We can stop computation at any point in time, can make a decision even with less than the minimum amount of resources required to solve a problem. It is a remarkable computing ability. Can one imagine a situation, when a computer is asked to add five numbers, while two given; don't wonder if computer pulls off its own power plug and put the plug in the operator's hand as an answer! Faith on live materials is paramount. One example of such an idea is the creation of the Tokyo metro map using physarum. They feed the algae sugar and salt at particular station locations, physarum grows connecting the stations. Again, we can

test the same by asking a group of ants to follow the bumps of sugar kept on a checkerboard; possibly they would also solve the traveling salesman problem successfully. Even using plasma, mesh problem has been solved as the material tries to find minimum path via trial and error, then it is a programmable matter. These kinds of approaches to solving a problem have been categorized in the unconventional computing paradigm. If we do a similar simulation in the existing computers, we complete the logically circular path, i.e., compromise with "simultaneity." Analog computers are very funny: first, we have to tell the computer how to find the solution of a problem, by defining various functions and then we have to ask the question as if we have not done anything beforehand. Current flows through random paths for an external observer, however, we know that typical "randomness" represents either a logarithmic or sinusoidal function. Obviously, we get the answer. What an analog way to get the answer to a problem, instead of logic gate use direct mathematical functions directly! One can innocently argue, if one has to define "functions" and re-formulate a problem before starting the calculation, why not find the solution itself using the classical Neumann computer.

Four original concrete routes to build a neural network: Figure 10.1d outlines a unique feature of the artificial brain described here. There are plenty of distinct and multi-layered core-shell architectures. However, when we look into the time domain the geometry of the time, the architecture is widely different. It is a firm belief of the scientific community that information is stored in the synaptic junctions, and learning means strengthening the connection, since a permanent high potential is generated, scientifically it is called potentiating a junction for a long term potentiation (LTP). The scientific literature is divided into two parts: One-part deals with junctions that exponentially decays the stored information after one session of learning; the others avoid the complexity associated with this particular issue where the decay rate of forgetting plays with the rate at which we expose learning images to our neural network and exhibits interesting features of learning. When patterns are presented slowly to a neural network, N neuron can learn at most lnN patterns, since in this case the network has to learn one pattern at one shot, there is a possibility that it learns wrong information.

Reservoir computing: Reservoir computing is a framework for computation like a neural network. Typically, an input signal is fed into a fixed (random) dynamical system called reservoir and the dynamics of the reservoir map the input to a higher dimension. Then a simple readout mechanism is trained to read the state of the reservoir and map it to the desired output. The main benefit is that the training is performed only at the readout stage and the reservoir is fixed. Liquid-state machines and echo state networks are two major types of reservoir computing.

A liquid state machine (LSM) is a computational construct like a neural network. An LSM consists of a large collection of units (called nodes, or neurons). Each node receives time-varying input from external sources (the inputs) as well as from other nodes. Nodes are randomly connected to each other. The recurrent nature of the connections turns the time-varying input into a spatio-temporal pattern of activations in the network nodes. The spatio-temporal patterns of activation are read out by linear discriminant units (Maass et al., 2002).

The soup of recurrently connected nodes will end up computing a large variety of nonlinear functions on the input. Given a large enough variety of such nonlinear functions, it is theoretically possible to obtain linear combinations (using the readout units) to perform whatever mathematical operation is needed to perform a certain task, such as speech recognition or computer vision.

The word liquid in the name comes from the analogy drawn to dropping a stone into a still body of water or other liquid. The falling stone will generate ripples in the liquid. The input (motion of the falling stone) has been converted into a spatio-temporal pattern of the liquid displacement (ripples).

LSMs have been put forward as a way to explain the operation of brains. LSMs are argued to be an improvement over the theory of artificial neural networks because: Circuits are not hard coded to perform a specific task, Continuous-time inputs are handled "naturally." Computations on various time scales can be done using the same network. One network can perform multiple computations. Criticisms of LSMs as used in computational neuroscience are that LSMs don't actually explain how

the brain functions. At best they can replicate some parts of brain functionality. There is no guaranteed way to dissect a working network and figure out how or what computations are being performed. It has very little control over the process. The model is inefficient to implement because variables require lots of computations, compared to custom designed circuits, or even neural networks. If a reservoir has fading memory and input separability, with the help of a powerful readout, it could be established that the liquid state machine is a universal function approximator using the Stone-Weierstrass theorem.

The echo state network (ESN) is a recurrent neural network with a sparsely connected hidden layer (with typically 1% connectivity). The connectivity and weights of hidden neurons are randomly assigned and are fixed. Learning the weights of output neurons enables producing specific temporal patterns. Although its behavior is non-linear, the only variables are the weights of the output layer. The error function is thus quadratic with respect to the parameter vector and can be differentiated easily to a linear system.

Now, we return to the time crystals produced by brain-jelly-derived artificial brain in Figure 10.2. When we look at the peacock dancing, listening to an Indian classical song, proprioception or feeling that we exist in this environment with a

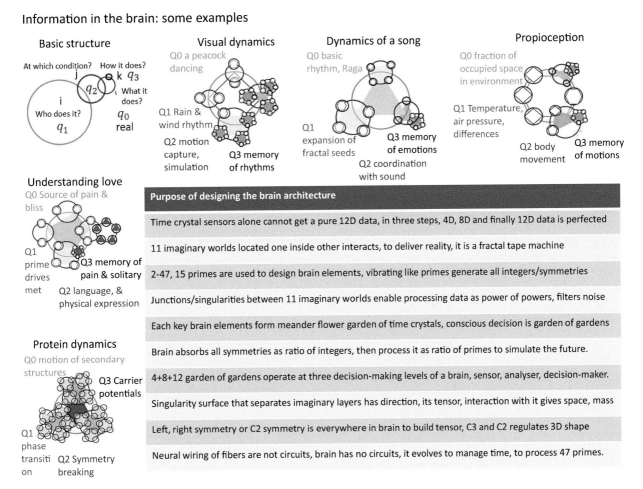

FIGURE 10.2 The applications of linguistics based information structure are outlined in the five examples of processed information in the software brain. In the table ten key guidelines for designing the artificial brain structures are noted.

physical body, how deep sense of missing someone creates a sense of love, when a person falls in love; and proteins under massive ionic motion to carry out fundamental tasks for the life form. All these situations are very confusing, and an artificial brain is designed to get into confusion and build a map. The dancing of peacock is confusing because in the rain when one sees a peacock dancing suddenly, why we don't know but we feel like we are in heaven. Classical Indian songs run for a whole night, 12 hours sometimes, we cannot say anything but only at the late moments the basic pattern and its architecture unfolds in our brain. Similarly, proprioception, we feel that we exist, but why and how we do not know. The story of confusion would go on and on, and that's how a kid should learn to solve mysterious problems taking a paper and pencil, build an architecture of confusion. Figure 10.1 prepares the readers mind on how sensing the architecture of confusion should begin a new paradigm of decision-making beyond logic and arguments. At this moment it is worth for a moment briefly look into the summary of this book, how we described the purpose of building the artificial brain (table Figure 10.2).

10.3 TWELVE PARADOXES THAT WOULD CHANGE OUR THOUGHTS FOREVER

The journey from neurogenic brain model to time-crystal-based brain model transforms the fundamental philosophies we believed thus far (Figure 10.3). Instead of equations, patterns would take over, replace equations and a science to harvest singularities using symmetry of primes would be born. Feynman argued to replace everything with the changing pattern and there were plenty of followers. Several greatest thinkers put thought into this matter, but none took this seriously. Latest in the frame was Wolfram, who wrote even an entire book on the new kind of science. Many researchers learned to investigate the origin of a new kind of science with these physicists. And then the time came when we realized that the dream of implementing pattern-based computing would never succeed if we remain within the domain of Turing. We have to go beyond, and we also realized, all proposals of hypercomputing are not that effective to implement Turing's forgotten ideas in computer science (Copeland and Proudfoot, 1999). **The Turing principle is so powerful that if anything is defined we are baptized by Turing religion, and the only way to go beyond is to enter into the world of singularity, an escape time fractal but self-similarity does not exist. However, the journey is not easy, in this book we covered wide ranges of topics to explore one possible way.** Quantum's one-imaginary world shakes the reality; then, 12 imaginary worlds with one pseudo-real have arrived to shake it even further. Quantum entanglement in room temperature raises heat more in the mind of fiction agents than scientists, it is possible to do quantum computing without quantum entanglement (Lanyon et al.,

Neurogenic brain model	Time crystal brain model	Quality	Design principle
Neuron membrane does all information processing, it is the smallest unit	12 gardens of time crystals form canopy, topology of time, no device is absolute, smallest unit--helices	Consciousness	The bandwidth of resonance chain beyond threshold peak density and length sync as three distinct identities outside the physical boundary of the chain but located on the universal resonance chain.
Map neuron network, replicate physical map, get a brain, artificial intelligence	Map symmetry of nested clocks at all spatial time scales, replicate, get a brain, natural intelligence	Intelligence	Brain jelly identifies missing symmetries & grow associated time crystals which were absent in the hardware of both sender and receiver.
Unit: switch, turing machine, circuit	Unit: time crystal, fractal machine, meander garden	Creativity	Time crystal associated with phase transition is created that adds a metric of primes in the system, new methods to create time crystals begin.
1D data flow, data structure: bit/qubit	10D data flow, data structure: three nested clocks		
Water is silent	Water converts all biomaterials into superlenses		
Noise disrupts operation, current-voltage property is key device operation	Harvests noise, magnetic flux-charge induced clocking is the key device operation	Adaptability	Phase prime metric made of dodecanions create a resonance chain that is made of twelve distinctly operating phase prime metric based independent systems would be able to emulate 99.99% information generated in the universe.
Cortex region is the key, all operations happen in the brain, by neuron circuits, there are logical operations	12 organs spread over the body is integral part of the brain, information processing happens in topological architecture of time of all 12 organs. No logic its loop	Plasticity	15 geometric shapes, 15 primes, 3 types of phase prime metrices, 10 classes of metrices of primes cover 99.99% of all possibilities. Hence hierarchical network of symmetries ensure a new unprecedented symmetry to replace a composition of symmetries that was governing a local part of a hardware.
Only ionic clocks exist, clocks are not integral part of information processing.	12 classes of clocks run by 12 types of carriers, ion one of them, all are linked, part of information.		
No embedded language, no controller	Geometric musical language, GML; PPM controls all.	Universality	Simply, universality in computing means emulating everything, that ability comes from the engineering principle of using 12 systems at all scales, at least 12 primes, at least 12 geometries, at least 12 metrices build at least 12 distinct hardware.
Pattern of primes has no role	Pattern of primes link geometric shapes, it is thought		
Diffusion of ions=information dynamics	Neither classical, nor quantum, its fractal mechanics		
Energy is proportional to info transfer	Symmetry transfers, info is regenerated post transfer	Learning	Learning is an identification protocols by which the PPM computer determines which time crystals or symmetries stored inside and which does not, the difference rhythms are created and written at the suitable places.
Spike is key, its frequency maps all info	firing mechanically adjusts neuron, time crystal is all.		
		Thinking	Thinking is a process by which a hardware uses the pattern of primes to artificially create the time crystals for filling up the missing symmetries and expand.
		Emotion	The basic driving criterion that are set for a brain to learn and evolve with time. The driving time crystal generates its own higher level maps for the futuristic learning. This crystal is also a filter that transforms the geometric shape of a time crystal exchanged between sender and receiver.
		Growing	Dodecanion tensor and nesting of resonance chains are not limited to 12. 12 is the beginning of the journey, next level would be 12, 18, 36 and so on

FIGURE 10.3 Differences between neurogenic brain models that are there for nearly 70 years and the time crystal-based brain model (left). To the right, a table that compiles the consciousness features in the artificial brain jelly.

2008). Using polarized laser, room temperature quantum experiments could be carried out and zillions of molecules are all quantum devices operate under massive noise. We advocated for optical vortices in Chapter 8, rejecting the world of electronics totally, so, the fear of decoherence would reduce sooner or later.

PPM-GML-H triad has made one little statement, "Every single cell of a Turing tape contains a Turing tape inside." If we do just that, we break Turing tape, but we fall into a singularity. We cannot do the science of Classical mechanics or Quantum mechanics. We needed new mechanics, to model the interaction of imaginary worlds (Chapter 4, dodecanion algebra). Not only that we had to find the true scientific construct of information in Nature (Chapter 2 time crystal). Developing a new scientific protocol for addressing the information integration such that natural events could be replaced should be our first and foremost target. But, even if we integrate information, what would we do? Shall we recreate the natural phenomenon. Shall we take the whole information or would adopt reductionism? Then came the question of representing a observable simple phenomenon in nature, we built the concepts of (i) fractal tape, (ii) three imaginary world's construct of information, (iii) resonance chain, and (iv) frequency wheel then combined them to generate an information construct for the natural system, just like the one we created on a nano-surface with molecules (Chapter 9, brain jelly).

The greatest problem that we have had is that "logic" and "argument" are to be replaced by time cycles and new mathematical and other constructs need to be built for the singularity architecture. We needed a new language to process information, we have built one, this is called geometric musical language (GML). Using this language, we can convert any sensory information like visual, auditory, smell, taste and touch signals into nested time cycles and re-write it in terms of primes so that PPM morphs the 3D clock architecture.

What would be haunting a general people is that if all the above jargons are true, how the philosophy of the new world of science be like? Unbelievable, remarkable like Quantum or something even funnier, shocking and amazing. Note that all we need to know is that imagine a Turing tape and each cell has a Turing tape within, or "**a new typewriter inside every single key of a typewriter,**" that is sufficient to come to the science of singularities, but we need to have an attitude, "**We will not try to bridge the singularity using brute force, like re-normalization we will float and go deeper inside singularity and above to learn what are there, using specially designed tools and mathematical constructs for the new kind of science.**"

While discussing the philosophy of this singularity harvesting science, we will see that we need to include everything that quantum information processing suggests and then we need to go beyond, far more exciting features like topology of undefined regions in a network, need to be included.

The science of architecture of singularity is absolutely not classical and far beyond quantum.

Why both classical and quantum mechanics fails and how should we work toward the development of a new kind of science?

1. *The trumpet-paint time-cycle paradox: Using the Gabriel's Trumpet paradox*: For Quantum, we have "incompleteness," so there was a series of development to "supplement," here we need to have "penetration boundary" because every single point in this world view (PPM-GML-H) is a singular point: Just like quantum we need, another historical movement for perfecting the journey into the singularity: "Every single cell of a Turing tape contains a Turing tape inside." One to one correspondence between Gabriel's paradox and nested time cycle synchronization that is designed to replace renormalization, the surface area of the cone is infinity, however, when we put paint to color it, the paint passes through the tunnel and it blocks at the point depending on the density of the paint. The paint for us is the observer's nested time cycle, it could penetrate above and below and much far away, it does not need to pass through as someone from far could "spontaneously reply." The resolution of the solution is always ultimate, therefore, just like quantum we do not have any predictability, we also need contextuality, but in quantum the object that we are measuring has got something, but for us, that object has got nothing.

2. *The paradox of infinite network of reality spheres: Using the Banach-Tarski paradox and the "measurement" in fractal information theory*: In contextual measurement in quantum mechanics observer has a Hilbert space and the measuring state also has a Hilbert space, they interact to create a common Hilbert space with an inner product. However, for a non-computer, PPM creates a 3D projection of the observer, we have three Hilbert sub-spaces, one for the world below, one above and one for the observer. However, we do not need to create them in a large number. Suppose we have a nested time cycle sphere, like the time crystal shown here, the observer would also be a sphere, it would sync with the paths and generate the connected path and form a "reality sphere" just like the one here. Creation of an identical information replica is its greatness because the solution of a problem is an infinite space. The Hyper-Webster dictionary and the generation of multiple spheres: If there is a large number of observers, of course there is an infinitely large number of observers, therefore a large number of "reality spheres" will be produced and those would merge just like the Hyper-Websters dictionary. All these spheres are distinctly produced, yet they create a singular identity. Imagine thousands

ot "reality spheres" being produced and all of them simultaneously hold the solution.

3. *Resonance Drive Paradox: Density of resonance frequency R and frequency bandwidth B product RB determines the degree of consciousness elements, when the environment remains constant.* The ratio of RB of a device and its environment is an essential index or the degree of consciousness (C = RB(M)/RB(U)). There is always a spontaneous drive of any fractal cavity resonator structure to increase bandwidth ratio, this is perpetual, it is never-ending, the machine senses more and adds more bandwidth then evolves to sync as much as possible. Resonance drive paradox leads to several other drives, for example drive to morphing, etc.

4. *Number system derived metric and its saturation: Ordered factor paradox*: Reddy S. et al.'s number system study (2018) shows that only 12 primes are sufficient and the number of solutions of fractal cavity resonators reaches 99.999% new situations by 10^{12} oscillators alone, therefore, we do not need infinite or astrological number of oscillators. Then how could a large structure form? The code to generate new structure must make a fractal repeating its entire process after each limit is reached. When we get a structure of 10^{12} oscillators, we consider it one and start re-counting. It is the reason, the universe appears like a fractal, but actually it is not a fractal, GML follows the pattern of primes, and the primes never repeat.

5. *The observer-absolute paradox:* In a fractal tape there is no finite-state for a particular cell, the observer syncs with the world below and the world above and creates a state for the measuring cell. Then we never write information in this tape in any cell, we write only in the observer, only the observer exists and the rest everything, even the tape does not exist: Twelve fundamentals of new information theory: (i) Instead of modeling, fitting, we try to do morphing, The morphing of frequency wheels through the number system derived resonance chain is the route for information processing. (ii) The same information is complete at the moment, a fraction at the other, linking clocks at a particular time enables global link, this is a weakness of PPM-GML-H triad system. (iii) The same information content gets a different property at different environment, in the reality sphere, the same plane holding many nesting cycles behave differently. (iv) System spontaneously emerges information, even when no question is asked. (v) A mirror image is created without changing the source, though both the source and the reader are coupled. (vi) The same information takes different form at different times thus creating wide ranges of information, also it looks very different from different directions. (vii) Information is alive; it senses, expands creates new life forms. (viii) Information is a phase, frequency intensity of time cycles, these time cycles could self assemble in a single or multiple planes, they could make 11 planes holding 11 dimensions. (ix) Dimensions representing different planes inside the sphere rotates differently for different planes to accommodate higher dimensions. (x) Starting phase and direction of motion of system point are two parameters that higher-dimensional planes process. (xi) The pot and the matrix are the same: the information that shapes information content is the information itself. (xii) Identical information holds the same content yet different meaning, very different information holds the same appearance, duality is embedded.

6. The paradox of infinity-mechanics: Quantum mechanics is a subset of fractal mechanics: Ten differences between quantum mechanics (QM) and fractal mechanics (FM) (i) One imaginary world in QM, and three imaginary worlds in FM. (ii) Uncertainty principle is universal in QM, uncertainty principle varies, depending on the observer and 12 nested imaginary world configuration. (iii) Contextuality is a side part in QM; contextuality is everything in FM, the observer is the only reality, "supplement" to complete a quantum state is essential; in FM, continued fraction geometric algebra generates an infinite series to hold that. (iv) Entanglement is a property in QM; in FM, entanglement is the way time is connected. (v) Quantum deals with matter; FM only time-cycle. (vi) QM has its own version of condensation, interference, oscillator; FM also has its own versions for each of them. (vii) Psi is absolute in QM, Psi is deconstructed in FM. (viii) Like QM there is a Hilbert space, but they do not take the tensor product of the inner space. It's a phase-orientation calculation of time cycles in FM. Instead of normal tensor algebra we need continued fractal-based geometric algebra to process. (ix) QED is based on a field-theoretic concept, for FM, a universal unified field exists made of 12 nested time cycles and 12 dimensions. (x) Quadrupolar moment plays a vital role in determining complex time cycles in FM, but that is irrelevant in QM.

7. *The paradox of quest: Ten cultural practices that existing scientific practices follow and we do not need to make them blind religious:* (i) Survival of the fittest, not just in biology, quantum to astrophysics to elementary particle physics, this is applied as a universal physical principle. Co-operation

to survive is just opposite to the idea of struggle for existence. (ii) Something equals something: Equations or inequalities is the only form to get an answer, except one, all variables should be known. (iii) Step-by-step, not above and within: Searching systems believing that they would work step-by-step is generating craps. (iv) Universe is sum of two-body system: all other particles vanish from the universe, many body theorems are forceful relationship building exercise. (v) Do anything in the name of closed system: Scientific practices limited resources, always considers close system to anything and whenever necessary makes it permeable, stops, or vanishes anything. (vi) Decision-making requires selection and rejection. Build many options and then reject all but one. (vii) Order and degree of differential equation captures all possible variations of a natural system: We say that nested time cycle could capture with the three imaginary time worlds operating simultaneously (viii) Singularity has to be bridged: Dancing in the singularity is all what we need to do in our singularity kind of science. (ix) Only one time, only one cell of a tape, only one clock, only one of everything: In the existing science there are plenty of things, for PPM-GML-H triad, there is only one thing, a single time cycle that encompasses all. (x) Everything wants to be complete: Every single system in the universe has different objectives in the current science, but for non-computing, all systems have only one target to achieve 10–12 prime-dynamics, a goal for 10^{12} oscillator-assembly.

8. *The cycle-non-reductionist paradox:* If there are many choices, then without reduction we cannot make a selection, what if there are not many, and we would never have to choose, in the information processing protocol we follow "spontaneous reply," only the right one answers to the query. Therefore, the question of reduction does not appear. The next question is that when we model a system in nature, do we mimic the system as is? Do we take all the information? No. Just like a single cell creates a full grown human, a few delicate codes of the natural system could regrow by morphing, the recreation could be limited by the number of resonators and its design. The information space gets filled up with matter, with a few codes it is build up. Ten principles of a Fractal tape: (i) Every single cell in the tape has tape inside, we get a fractal tape. (ii) A tape has four tuples, convert to nested time cycles, match and expand, transform and reply back, restructure internal wiring or learning. (iii) Triplet of time cycles is a unit of information, three imaginary worlds with a real world builds a quaternion. (iv) Two kinds of fractals—ET

and IFS—work together to make the fractal tape. (v) The tape and the imaginary world do not exist in reality the observer does exist in the imaginary world that it is fit to sense. (vi) Fractal tape leads to fractal machine just like Turing tape leads to Turing machine, an imaginary state processor, generalized complex number and the number system rule of 15 primes and 10^{12} oscillators is the key to create a cell in the fractal tape. (vii) The fractal tape is good for QM, FM, and CM all; it morphs to any information structure, fractal tape does not have a fixed geometry, PPM projects it to infinity (Chapter 3), feedback from infinity remains undefined in the fractal tape making the decision. (viii) It operates through GML using 15 geometric shapes regulated by 11D dynamics of 12 nested universes linked within and above. (ix) A tape finds the resonance frequency bandwidth of a system and draws a circle to connect it's various time scale, just like spiders net, but the fractal tape has many circles as clocks and it tries to touch all the clock boundaries first and then it goes for the faster and faster time scales. (x) The fractal tape can run sequential algorithm, but we use it to run projection from infinity for a pattern of choices. There is no search to find or transport communication among cells, the cells self-assemble to morph a system.

9. *The equation-inequality-fractal paradox:* In the conventional practice of science, variables are correlated and equated to represent a system. The science of equations has been very successful. Now, if there is no equation because of fractal tape, we needed to build up another language that runs in the undefined world. We replace equations with a new kind of algebra called continued fraction geometric algebra (Chapter 4). Using this algebra we solve several problems. It is the tool using which we plan to manage the world of singularity. In this algebra, we always have a circle and we draw several circles following a certain set of rules. To do all kinds of mathematics addition, subtraction, tensor operation and more importantly estimating projection from infinity. It is the route of doing mathematics by drawing circles and always making a journey toward infinity via singularity points only.

10. *The paradox of illogical, nonsense decision-making:* We are always advised to think logically. It means, we have to think that every event is a sum of several sequential events. However, it could be possible that problems and solutions entangle in a single time cycle, and then we integrate them to construct entire information processing architecture, then, there would always be the simultaneous robust response. There would be no logic like "if-then," no argument then there would be what,

architecture of confusion? Here are the ten ways of arguing differently as practiced by the tribals for thousands of years: (i) A question always has its answer in it, no question could be created without an answer. (ii) A confusion is always the sum of a set of new confusions, thus, it is an endless fractal network. (iii) No answer is complete; there is always a question that invalidates the answer, fixing that to evolve answer is a journey to completeness. Gödel hinted, however, never suggested a route to explore the engineering of incompleteness. (iv) Logically circular or circular logic is the unit of a singular logic to enter inside and go above, always bring circular logic but follow the boundary limit of rule (ii). (v) Since question-answer couples are always together, every possible solution are connected to the question contextually. A question could have many answers in different contexts. All are mapped in a 3D nested time cycle geometry. (vi) Every single set of confusions form a geometric shape and a complex pattern of the composition of basic geometries should form to compose a complete confusion, the shape would be symmetric and follow the rules of geometries. (vii) No sensible confusion could stand alone, isolated; it needs support, even though a fractal arrangement argues that at the top there should be one unit of confusion, but even that would require at least three perspectives—one from the arguments below that makes it, one from above or the argument it constructs with others, and one the perspective of an observer. (viii) A true confusion does not have an idiom or universal truth, no assumption, all assumptions are environment and context-dependent. There is never a truth or a false statement in the architecture of confusion. (ix) Expansion of raga is the way to expand a confusion. We cannot add confusions one after another, place simultaneously, considering all as truth like quantum. We always find the geometry of confusions then expand them using PPM, just like Indian classical raga expands by extrapolating the input pattern of runes. (x) Every natural event has a geometric arrangement of confusion and the practice or game of confusion is to morph with nature's geometrical arrangement of confusion. (xi) A structure of confusion nested within and above always has one less variable than the fixed truth and a total number of variable and truth should compose like the number system metric, hence the network of argument could be constructed. If there is a little change in the most static confusion all nested confusions would change simultaneously. (xii) One could enter inside an

argument 12 times to generate 12 fundamentally different dimensions, however, if we enter more, then the relevance to the first argument becomes faint. We need to change the starting confusion point regularly to sustain relevance in the nested fractal-like confusion practice. The rule (x) morphing comparison is the route to auto-correct a confusion.

11. *The paradox of an infinite field and limits of the universe:* Infinity is everywhere in the fractal tape universe where the thing that exists is a singularity point and there are specific ten properties of this impossible universe: (i) If nature is discretized once, it cannot be recreated; we follow, everything is connected to everything: Probability theory is not required, PPM projection is enough to create enormous unpredictability. (ii) Every point in fractal tape universe holds an infinitely large universe within. (iii) Every single point tries to be dynamic following 12 prime, symmetries using the 10^{12} oscillator network to sense the environment with its maximum. (iv) There is no force, only the time-cycles and time gaps (domain of singularity); at the elementary scale, there exists only a time cycle. (v) Information is filled up in the all existing time cycles, total time cycle is infinite in number, they exist everywhere as infinite series. (vi) There are no zero points. To have a zero-point energy, also we cannot consider an external boundary if we simply consider universe is a frequency fractal, only thing it is made of is time cycle (of what? please note, vortex fields), there is no mass, no distance, all are perception of interaction between nested time cycles made of vortices. (vii) Nested time cycles self-assemble to generate all forces, and within 12 nested universes of dodecanions the time cycles self-assemble, all structures are formed. (viii) All fundamental constants are typical 3D geometric shapes which is a constraint made of nested time cycles. (ix) There is no total energy of the dodecanion universe in principle to feed the fields, because energy could be created and transformed, fractal network cannot have a boundary. What we see around us, and that is the truth, if we create a lower and an upper limit, we would surely need to tell what's outside. If we do not know what is outside then who gives us the right to draw a boundary. The energy could increase or decrease in a given imaginary world depends on the system, for that universe, we cannot make any rule, its PPM projection to infinity that decides everything. (x) There is no starting time and end time for the universe and for any event, in principle by the condition of fractal we are restricted to do that.

12. *The paradox of time replacing everything: Universe is nothing but clocks of vortices of fields.* We do not have mass, we do not have space, we have created everything, using only time cycles of vortices. The reason is that time cycles would interact and generate everything. It is how we have constructed the new information theory a new geometric language and this is what makes dynamics associated with primes distinct. Just imagine, only this consideration would shade significant light on the unsolved mysteries of matter-photon interaction. Even if we look at the Schrödinger equation, it shows clearly, we do not know what to do with "mass," when equation wishes to explain that very mass as wave When we see the foundation of the new information theory, FIT, often we ask ourselves, are all these new?

10.4 CONSCIOUSNESS IS NOT ON THE NEURON SKIN—TIME CRYSTAL MODEL IS NOWHERE BUT EVERYWHERE

If brain is not a computer and not a Turing machine, then what?

"These words information, data, rules, software, knowledge, lexicons, representations, algorithms, programs, models, memories, images, processors, subroutines, encoders, decoders, symbols, or buffers do not exist in the brain." Such statements flood the internet, but no one tells us, what are the words true for the brain.

Our journey to explore the synthesis of an artificial brain started with microtubule, its language is GML, its resonance frequencies revealed PPM and Hinductor H is its replica. Microtubule is the gateway to the understanding of the brain, thanks to its water channel inside. It is not the only component responsible for consciousness, but, just like *DNA holds the key to genetics and evolution, microtubule holds the key to thermodynamics and acquired information by a living system.* However, microtubule has never received the honor it should get, in spite of its magical contributions, throughout the entire living species in the planet. However, we feel that there is a bigger problem with our overall approach to biology. Our perception regarding the information is holding us back and if we do not know what information in the Nature is, we know nothing. We always get an answer to a query, but, the form of the question determines the nature of the answer.

We think there is always an attempt to see proteins and any biosystems like a transistor or a switch, as if from proteins to brain everything is a circuit made of elementary electronic devices. We blindly force the biological systems as if they are electronic devices found in the conventional electronic shops. Design experiments accordingly, and undoubtedly get the results satisfying our faith. It is not only a very wrong idea for the common people; we would find such an erroneous understanding even in the scientists. A Google search would get us several breakthroughs in Nature and Science journals suggesting that the protein is a transistor and a switch and like OP-AMP. So many varieties of biological systems are there and without investigating what those devices want to do, we carry out simple experiments to tag them as a handful of known devices used in electronic chips, it has become a religious faith now. In addition, if some scientists want to seem radical, they tag it as a quantum system, one mystery to support another mystery. Therefore, the mindset masks the true quest, "bits" has not impacted "in a bit" but in a massive way to the free thinkers.

We would also see that there are two groups of peoples, one that believes that our brain is a computer, or brain is a Turing machine. In addition, there is another group, which suggests that the brain is not a Turing machine, but do not tell us what is that unknown device actually the brain is or the biological system represents. Opposition to mainstream should be done sensibly. Toppling the world of "logic" that started with Bertrand Russel's "principles of mathematics" in 1903, is not so easy or fun. Criticizing the mainstream science is good but not without an alternative in our hand that is concrete. Those who challenge the mainstream science (the brain is a computer, a neuron is a switch, all information processing happens through the skin of neuron body or membrane, but everything inside the neuron is just a blank, a neural network is a circuit etc), they are mostly not serious. If the brain is not a Turing machine, then tell us what that elementary machine is? Well, *in the last 100 years of development of logic, no one talked about making decisions without "logic." The inverse of "logic" is "illogical" which means "nonsense" to us for the current science. Making sense in an illogical way is the way to go.* But the question is how?

10.4.1 The Difference between a Neurogenic Brain Model and a Time Crystal Model

The difference between neurogenic brain model and the time crystal based brain model (Figure 10.3) is a journey that started with the microtubule research. Microtubule showed the way to PPM-GML-H triad. The biological systems while building living machines, does not make a switch or circuit, it wants to make a clock inside a clock inside a clock.... in a nested structure. The wiring we observe in the brain and at several biological structures is an illusion; every single biological machine is a PPM driven musical instrument. Unlike Tuning forks, they have several fundamental resonance frequencies, and they select those frequencies to play music, and those musical notes are such that they represent geometric shapes and information, so it is not harmonic, but an anharmonic device. The discovery is not "simple" as it appears. Just imagine, music representing banana and apple is being played in a device as two simple pieces of music in a loop, then if one activates an apple by synchrony or resonance, the other is played. Thus, *there is no reduction of decisions like we see in the logic gates and hence "time crystal" to replace "bits" and "qubits." There is no collapse, no finiteness, projection from infinity morphs in a way, they system cannot even define the decision received from infinity via feedback*. Consequently, PPM and GML are what we learned from microtubule and applied to the entire human brain. Thus, we reject "bits," or "qubits," we say that the experiment suggests, that we should not be debating whether the brain is a quantum or classical or chaotic, those concerns are irrelevant, debates are unnecessary. There is something much more interesting and profound, that is the true nature of the information."Time cycles" as suggested above get nested, we have verified this in neuron cells, furthermore in at least 7 proteins.

The discovery of universal time crystals in microtubule, protein and neurons has led to "fractal tape" to replace the "Turing tape." A very simple decision, "every single cell of a Turing tape has a Turing tape inside," this statement alone collapses the foundation of Turing machine. Several proteins "tubulin," actin, beta spectrin, ankyrin, clathrin, collagen builds time crystals as magnetic vortices to explore the role of electromagnetic resonance frequencies. We found that they have a composition of resonance frequency bands, where prime number of frequencies shift together as a group (see Chapter 6 for details). Mostly we found triplet of triplet bands that connects all proteins and components in the brain, self-similarity is not in appearance but in the vibration. The brain is a clock inside a clock inside a clock... there is no end to it, though we format it in 12 imaginary worlds (dodecanions) one inside another. The particular finding changes the basic understanding of the brain. The old belief that neuron skin or membrane does everything falls apart. Neuron skin is just one of 500 different types of cavities that we have in the brain. Microtubule is also one of those 300 types of triplet-of-triplet systems vibrating in the brain. *Therefore, the cavity inside a cavity inside a cavity makes a resonance chain which is best arranged in the form of a frequency wheel is the machine that creates rhythms to operate*. In summary, we have not restricted ourselves to the microtubule alone, or single protein studies, or even to a few live neuron cells. Primes of all the brain components nested together to build vibrations, so no brain component holds consciousness (Chapter 7 for details), its space-time-topology-symmetry metric that governs decisions. We have built a series of technologies like atom probe, noise free electrode system to see what

is happening to a single cell when we see them simultaneously using various probing clocks (Chapter 6). **With 26000 proteins and more than a few hundreds of different types of cavities, we have engaged in building a comprehensive clock architecture of the brain-body system that would act as key to the future massive brain building projects:** We cannot understand the dynamics of a river by mapping the wave shapes of ripples on its surface, mapping neurons would tell a little about the river of consciousness. One cavity, the neuron skin can take 100 billion new forms ($2 \times 3 \times 5 \times 7 \times 11 \times 13 \times 17 \times 19 \times 23 \times 29 \times 31 \times 37 \times 41 \times 43 \times 47$) in 8 billion humans, mapping all neuron geometries would not tell what our brain is. However, we realized only way to go forward to let one cavity say neuron geometry to come under the influence of trillions of cavities below and millions of cavities above, and evolve geometry in 10^{12} ways, then, information about music will let the neuron to dance, and that would be the way to understand the information processing in the brain. We understand that developing GML, fractal cavity resonator model of the brain, doing experiment with neurons, proteins, complexes, and neuron clusters are not simple and easy task. However, if one models it properly, we would be able to give the world of science a new way to look into the basic understanding of information. Only then perhaps people would understand, that the basic definition of information re-defines everything. The transition from "bit" to "universal time crystal" and providing every single cell of a Turing tape to hold a tape inside could transform everything we know about the operations of a biological system.

10.5 WHEN WE CARRY IT AND WHY WE MAKE IT

When one does requires a truly bio-inspired computer?
Biological clocks are well known in the brain. It extends from circadian rhythms to the single neurons. The connecting protocol and the route that connects the rhythms or clocks are unknown in biology. Currently, rhythmic activities are linked as a chemical process associated with proteins and enzymes. The network of clocks in the protein-like nanoscale biomaterials does not terminate at the neuron level, as it was believed thus far. The rhythmic or clocking reaches deep down to the few atomic groups. Triplet of triplet resonance band connects the Peta Hertz (femto seconds) to the nano Hertz (12 years) frequency scales (Ghosh et al., 2014a, 2016a). The resonance pattern looks similar to the pattern of primes derived from the resonance of the dielectric resonators. There are many carriers (Chapter 7). All carriers resonate with different dielectric resonators Lauber et al. (1994); Pistolesi and Manini (2000); Pechal et al. (2012); Lee et al. (2011); Joshi and Xiao (2006). Yet, the frequencies constitute a singular pattern.

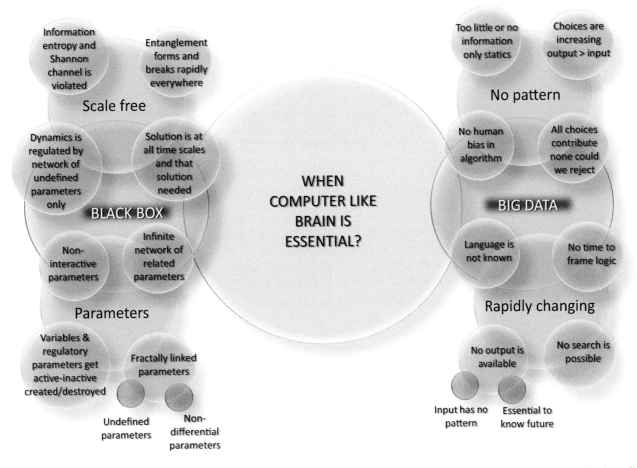

FIGURE 10.4 People used to build a brain-like computer, now humanities need computer like artificial brain. The necessity is outlined here.

Ten situations when one should use this non-computer: The objective is to develop a science for non-computing, to make decisions where architecture of confusion operates and obviously the Turing computing fails. Here we note 10 circumstances where non-computing is essential (Figure 10.4).

(i) Information is not sufficient or organized to frame logic. (ii) No time is available to find the rules for structuring logic, i.e., the urge for an instant reply. (iii) Rejection of choices is not advisable. The rejected choices could take over the lead anytime as the dominant player. (iv) The database is too big to structure it into a format solvable by a futuristic quantum computer. It requires to "search" without searching, i.e., spontaneous reply. (v) The decision-making devices of the future cannot carry a giant megawatt power supply continuously. Thermal and electrical noises are the only energy sources. (vi) We encounter a system that uses an unknown language, cannot

be understood at all. (vii) Learning the real parameters using which a system configures its response. Complete rejection of black box approach, to unraveling the true dynamics. (viii) A large number of parameters are being born, disappear, change, and redefine itself with a truly random, chaotic fashion. When, even the variable parameters could not be identified. (ix) Undefinable factors govern a situation. A factor has several sub-factors. In addition, each of those has several sub-sub-factors. Thus, the logical statements inside logic inside logic perpetuate into an endless network (Gödel, 1938). (x) Computing is always a reduction of choices, but in morphing, it is just the opposite. There is a continuous increment of choices, output is more than input and that defines non-computing. The output is more than the input. The difference between the Turing-based computing and the time-crystal-based computing was discussed in many parts of the book, here is a summary to refresh in Figure 10.5.

Turing based computing	Artificial brain
Device: logic gate	4th Circuit element, Hinductor
Energy: Power supply	Noise, light or em wave
Computing architecture: Boolean architecture,	Phase prime metric based geometric architecture
Physical variable: electron	Knots of darkness, magnetic particle
Data structure: bits	Time crystal
Channel capacity: bits/second	Time bandwidth per symmetry
Information unit: logical statement	Time crystal triplet, subject-clause-verb-adjective
Hardware: Circuits	Hardware: fractal super molecule
Software	Geometric musical language
Dimension of data: 1D	10 Dimensional data
Mechanics: classical/quantum	Fractal mechanics
Data processor: Switch	Clock with phase singularity

Physically wired circuit	Temporally wired circuit
Junction & branches	Topology & symmetry
Device edits speed of flow, amount, density of channels	Device edits phase, amplitude & duration of a ripple
Geometry of path, physically governs time.	Geometry of hierarchical ten dimensions arrange clocks
Directional processing	No direction, symmetry similarity
Quantity of information = amount, power consumption	Set of symmetry transfers = fixed power consumption.
Device inside a device inside a device concept, not possible.	Clock inside a clock inside.... Infinite net glued by singularity.
Without entanglement no simultaneity, all step by step.	12 imaginary state (duodecanion) controls simultaneous operation.
Electrical/optical insulation	Superlensing insulation
3D orientation of device, packing is not important.	3D orientation & packing devices inside a cavity is a must.
Linear stream of signal	3D projection of time crystals
User designs circuit using free will	Metric of prime auto designs
Cannot hold circuit of time.	Circuit of time edits physical wiring

How the decision looks like in a brain?

Quaternary decision delivered has three layers, one could enter inside one element three times.

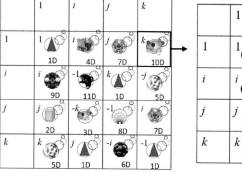

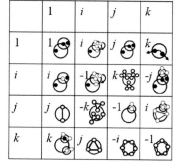

FIGURE 10.5 The difference between the Turing-based computing and the artificial brain made of fractal tape, operated by time crystal-based language (left). To the right differences between a circuit made of hardware elements and temporal elements like clocks and singularities are outlined. To the bottom manifolds representing the 11D dodecanion tensor's many world impacts are built by cutting pieces of papers. To the bottom right, two quaternion tensors are presented, the left tensor is made of 11D data and if one time crystal is zoomed then a new tensor is found shown adjacent right to it.

10.6 HOW TO CUT 11D MANIFOLDS OF TIME IN A PIECE OF PAPER

The most important question using which we started the book is that how human thoughts look like. Now we know that it looks like a dodecanion manifold described in Chapter 4. However, one could really cut paper and fold it to see how the artificial brain thoughts would look like, it's a 11D manifold (Figure 10.5). One very important point needs to be clarified regarding the manifold. Of course a human thought is expressed as a quaternion but the elements are made of time crystals of various dimensions. One could zoom one element and expand that to find that each element is also a quaternion whose elements could be a time crystal of various dimensions (Figure 10.5). Therefore, the paper of Turing analogy is the perception that is projected to the other human we interact, but two factors are always common, data is 11D or dodecanion tensor but always presented as a quaternion form for linguistic reasons.

10.7 THE MARRIAGE OF PRIMES WITH GEOMETRY WOULD RESHAPE HUMANITY

How to picturize a PPM in the real universe: (a) Argument in an argument in an argument in an argument... this goes on and on forever (information expression). Replace one argument as architecture of confusion in the form of a fractal network. (b) The clock inside a clock inside a clock inside a clock inside a clock... this goes on and on forever. (Frequency or energy expression). Just one clock in the form of a fractal network. (c) Oscillator inside an oscillator inside an oscillator... this goes on and on and on forever (hardware expression)...Just one oscillator in the form of a fractal network. (d) A waveform inside a waveform inside a waveform inside a waveform. It goes on and on and on forever (continuation expression). Just one waveform in the form of a fractal network. (e) A rhythm inside a rhythm inside a rhythm inside a rhythm... this goes on and on and on (decisional or computational expression). Just one rhythm in the form of a fractal network. (f) Turing tape inside the cell of a Turing tape inside the cell of a Turing tape inside the cell of a Turing tape inside the cell of a Turing tape... this goes on and on and on (machine expression)—just one tape in the form of a fractal network.

> So, I have often made a hypothesis that ultimately physics will not require a mathematical statement, that in the end the machinery will be revealed, and the laws will turn out to be simple, like the checker board with all its apparent complexities.
>
> **Richard Feynman, The character**
> **of physical law, 1974**

Derivation of physics laws from no laws: Wheeler advocated to derive physics laws from no laws (Wheeler, 1980, 1983a, 1985), his PhD student Feynman was no different as we have seen in the above quote. As the computational capability of massively parallel supercomputing architectures is increasing linearly for the last half a century, thanks to the boons of Gordon Moor's law, researchers at the singularity university have got a fantastic game of "predicting the future of mankind." Any parameter they plot defining a particular feature of the human civilization starting from economics to medicine turns out to be linear. The total number of toothbrushes used by human race per decade to the development of every single technology we can imagine, falls into the category of linear development. One should strongly object the argument that these are nothing but an artifact created by a clever choice of representing parameters. Even if we consider that some tricks have been used to modify the representation of parameters, so that it looks linear, we cannot ignore the fact that existing civilization is monotonous and humans are not playing with its technological fundamentals. If they could, it would not have been possible to convert any parameter to such a linear form. While we do not rule out the argument that it is possible in principle to draw a straight line even on the surface of a stormy ocean, it is however impossible to generate products through a linear development by adopting the technology of "simultaneity." When we ignore symmetry in a stochastic fluctuation, everything could be linear, but the inception of "simultaneity" would induce non-linearity even at the plank scale which cannot be manipulated. We manipulate complex patterns from the stock market to astrophysics in such a manner as if at a time only two bodies exist and then try to ignore the very parallelism at the local scale in the quest for finding a global relationship. Unknowingly we, the human race, oversimplify everything that happens around us using only one principle, "complete destruction of parallelism." It is the reason, linearity is everywhere, yes, we are arguing to change the practice of mathematical formulation by nested clock approach that potentially more suits to address simultaneity.

We fail to recover this information because sequential expressions create a strange world; one can make millions of innovative and creative ways to create those alternate worlds similar to reality. However, every time, failure is ensured by the attempt to recovering the data following a linear process. Recovery of the lost data should start with a pattern itself,

a seed argument that contains an arbitrary set of information queries, however all queries are coupled in a particular way. The particular chain of coupling that links several other arguments is not an expression, but again another set of patterns. The bottom line is that to survive in this universe with original information, we must not speak, avoid any step that is sequential, and preferably draw images capturing all associated information within its frame.

The dangerous philosophy of black box practices: Currently the accuracy of computation is defined by the size of clusters embedded in the supercomputer, i.e., the amount of money invested. The research group which gets more money would derive a solution with better "resolution." The reason is very simple. When we couple the small two particles boxes "dangerously," then to compensate we come up with successive courses of never-ending manipulations. When we attempt to model nature, we simply try to find the global weight of a parameter in an equation. The equation means we have a left hand side equating to the right-hand side. We prefer to keep a parameter lonely in the left-hand side, which will be determined by experiment. In the right-hand side we keep all fundamental variables that generate the left-hand side; the right-hand parameters can couple between themselves in various different ways, like a square of the first one, the logarithm of the second, sinusoidal function of the third, etc. The history of science is basically a documentation of arguments on proper placing or weightage given to the parameters embedded in an equation. Arguments and counter-arguments on the constants used in the equation were paramount in the last century. A close correlation through the feed-forward network between experimentalists and the theorists has made the expressions perfect in explaining the experimental output. An experiment is a process, by which the effect of all system variables is nullified except the measuring one.

10.8   MACHINES OF THE FUTURE

10.8.1   The Difference between Circuits of Wand Circuits of Time

A comparative study between time crystal computing and artificial intelligence: Linearization of events is not accurate: Artificial intelligence has thus far considered that all events could be expressed as a sum of a series of elementary sub-events. Here in natural intelligence the events are not linearly connected. They are intricately connected by phase in 1D, 2D and 3D geometries. It means, if an event has several parts, their intricate relationships are neither in series nor in parallel. A temporal 3D wiring of sub-events exhibiting 11D dynamics is a reality. When one tries to draw the connections, 3D phase wiring should remain intact; one cannot draw it on a 2D surface. Every corner of this geometric wiring of events is important. A corner holds a unique geometric structure inside that is also a 3D network of sub-events. A singular change in this worldview changes everything in the Turing information theory. Consequently, the whole research field of artificial intelligence is redefined into natural intelligence.

Why does this little change affect so much? When all events are considered as a 3D wiring of sub-events, then immediately all events turn unpredictable, just like quantum (Durr et al., 1992). Say, one is looking into a complicated 3D network. It would appear differently in different directions. Now, the second problem is even more serious. Every event has a 3D network of sub-events inside. It means, there is an infinite journey for any observer who wants to find out the basic event that gives rise to all other events. It is a disruptive idea. One could immediately notice that an "event" becomes an undefined function. The third immediate effect is that an observer has to limit its sensing time width between limits. It is not like cutting a tape, it is cutting a 3D rock. If the lower limit is cut, even after cutting the rock appears as the same. Thus, an observer recreates an event. First, by finding a suitable orientation around the 3D event architecture. Second, by locating itself wherein the infinite journey, it would fit. Finally, to sense it, cutting off the event architecture based on the observer's own time limits. We have detailed these discussions in the Chapter 5.

10.8.2 Future Applications

Future applications of a brain jelly based human brain are outlined in Figure 10.6.

(1) **Musiceuticals (musical + pharmaceuticals):** Vibrations could rectify the misfolding or unfolding of proteins (Sahu et al., 2014) or activate new age chemical bots (Ghosh et al., 2016b). (2) **Increased human sense bandwidth:** PPM restricts the time-bandwidth of a brain like a computer. By harnessing the PPM, mathematically, human ++ intelligence could be developed. (3) **Halt aging-related processes significantly:** Editing age requires a true PPM hardware to feed cells with real vibrational data. The hardware would correct the clocking errors in the age-related proteins and complexes. (4) **Understanding of the language of natural events like the beating of earth's magnetic field and every life form:** GML is a universal language. It could be scaled up to replace the fitting with a black box, equations, to a group of patterns explaining fundamental physics theories (Feynman, 1965) (Sahu et al., 2009, 2012). (5) **Developing a truly dynamic model where we cannot find any logic:** e.g., earthquake, weather change, the evolution of a virus, aging, side effects of drugs, dynamics of gaseous clouds, etc. Non-computing hardware learns higher-level rules. Thenceforth, it bridges the missing links in the information architecture and spontaneously simulates the future instantly. It can morph events much better than the previous cellular automaton based architectures (Bandyopadhyay et al., 2010b; Sahu et al., 2009, 2012). (6) **Science of human behavior, society, economics etc:** The psychological behaviors, emotions, and other non-defined parameters would be geometrically defined (Wegner, 1997). One would get a geometric pattern of clocks from the human responses. (7) **Simulate beyond limit or knowledge:** Once built, a PPM hardware needs very

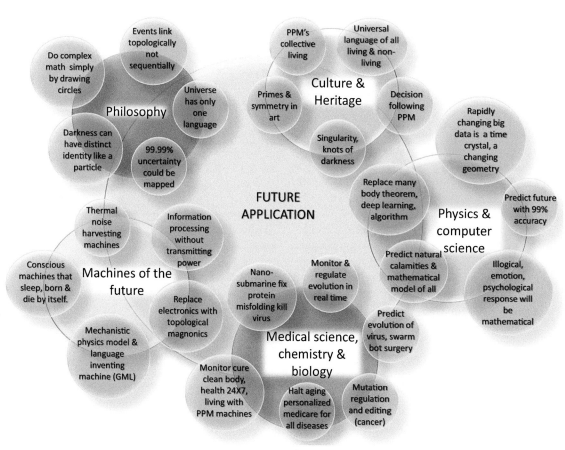

FIGURE 10.6 The future applications of the artificial brain in different research fields.

little gross information about any event. Then from that little information, it generates dynamics at time scales that it has never encountered in the past. (8) **The noise would replace signal, enter into the era of ultra-low power:** Earlier, scaling up was like adding more resources, faster speed, more power, etc. Now, it is all about how one could make a device that captures a much longer slice of PPM. (9) **Predict and simulate' million-year evolution in' a finite time (the science of evolution):** Currently there is no tool to estimate evolution because this is a slow process. Non-computer, by using GML and the PPM, could project a far more reliable picture of the future. (10) **Machines of nature:** The PPM, by intimately interacting with nature, can design scientifically life-like machines, it could be an architect.

Among various laboratories across the globe, several products of similar nature are being developed, we have categorized them into ten classes and outlined them in Figure 10.7. In Figure 10.8 we have outlined some of the primitive instruments that were the foundation of building time crystal map of

the human brain (Chapter 7) and the brain jelly to implement that model and build an artificial brain (Chapter 9). The first instrument is a dual study of patch-clamp and coaxial probe, Ghosh et al. saw live in 2008 that neuron firing is outcome of signals in the microtubule strings deep below the neuron. The second instrument is the detection of time crystal using a simulator called time crystal analyzer. The third instrument is quantum cloaking discovery, the path how in the random mess of fourth-circuit elements, only the desired devices see each other, others go silent. The fourth instrument is for seeing magnetic vortices using a modified quantum entanglement measurement set up. The fifth and final instrument was regulating organic nano brain using an antenna.

Ten features that constitutes frequency fractal computer distinct [The proposed hypothetical name is AjoChhand- A = Advanced, J = Junction-free, O = Organic, C = Computer, via, H = Hierarchical (higher level), and H = Heuristic (without programming), N = Nanobrain, D = Development]: (i) It never performs a search yet finds

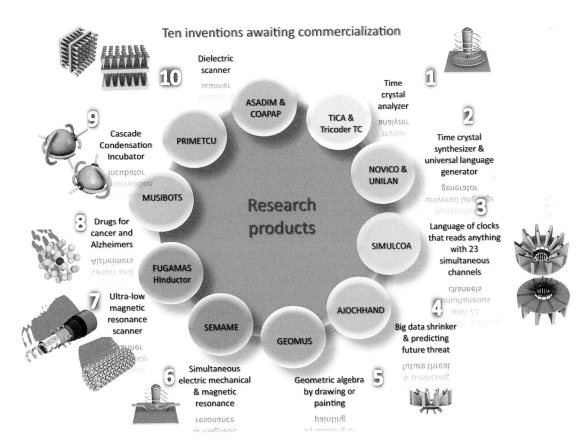

FIGURE 10.7 Conventional characterization systems do not work if we change the information theory, so we need new generation instruments. (1) TICA and tricoder TC: Time crystal analyzer, if one touches a protein or biosystem instantly reads the time crystal inside. (2) NOVICO and UNILAN: Reads any big data that is changing with time rapidly and using geometric musical language (GML, Chapter 2) converts into geometric language. (3) SIMULCOA: Coaxial probe-based 23 channels can read the signals from proteins simultaneously. (4) AJOCHHAND: The simulator converts rapidly changing big data of any size into a changing topology. (5) GEOMUS: We have developed a new kind of algebra where we need to draw circles and simply by drawing it following certain rules, complex infinite series calculations could be performed. (6) SEMAME: A unique experimental setup for simultaneously measuring three resonance frequencies, electric, magnetic and mechanical following $(e^2 + \Phi^2 = \pi^2)$. (7) FUGAMAS: Our fourth-circuit element namely Hinductor. (8) MUSIBOTS: Musical robots that can destroy cancer cells and cure Alzheimer's. (9) PRIMETCU: Cascade fractal condensation where simultaneous self-assembly takes place at several spatial and temporal scales. (10) ASADIM and COAPAP: Dielectric scanner that vibrates with a piezo motor to instantly capture an entire image (all pixels at a time).

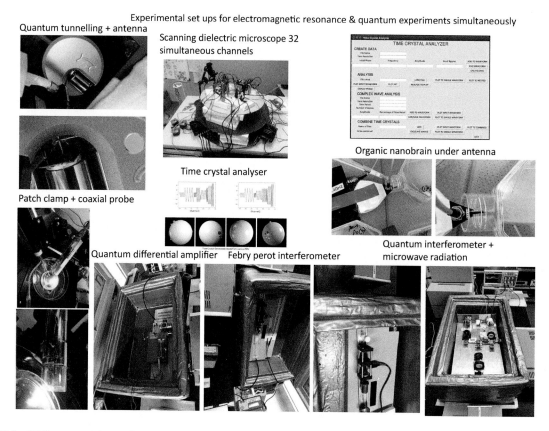

Experimental set ups for electromagnetic resonance & quantum experiments simultaneously

Quantum tunnelling + antenna

Scanning dielectric microscope 32 simultaneous channels

TIME CRYSTAL ANALYZER

Organic nanobrain under antenna

Time crystal analyser

Patch clamp + coaxial probe

Quantum differential amplifier   Febry perot interferometer

Quantum interferometer + microwave radiation

FIGURE 10.8 Different experimental setups where the data were produced for the artificial brain material characterization are shown.

what it seeks (**no search**). Search a massive database without searching (spontaneous reply). It never acquires a true input; it has all possible input elements already inside as part of the GML. So, it searches for them outside, thenceforth, a spontaneous reply is its operational key. (ii) Multiple nested clocks one inside another enable "a virtual instant decision-making," It **runs 24-7** as it evolves its wiring by itself for learning driven by PPM, a computation never stops, solutions are sent periodically, "halt" means observer's prerogative to shut acquiring a solution. (iii) It does not have any software program (**no software**). No programming is required as "cycles self-assemble/dis-assembly for better sync at all possible time scales simultaneously." (iv) "Phase space" keeps "volume intact" as required resources only increase the phase density not real space. Computing power is not related to energy consumption, there could even be a negative consumption; It shrinks massive information into a small geometric clock assembly or a seed of time crystal. It follows a unique superposition of 10 classes of PPMs, each sub-metric represents a set of unique geometric patterns to link events. (v) Perpetual spontaneous editing of slower time cycles (creation/destruction/defragmentation) "prepare for unknown" = higher-level learning. It introduces "fractal resolution," a complex signal's lowest and fastest time scale signals are absorbed. simultaneously, and during expansion, the fractal seed delivers full output, from a seed of information

(drastic shrinking of data). (vi) It runs by **white noise**, better randomness in noise is preferred. It uses an ultra-low power; only to manage re-wiring, non-computing does not require power in principle, as there is no reduction, no collapse, and no junction. (vii) The superposition of simultaneously operating a million paths assembles into a sphere enables "extreme parallelism." In quantum, only one Bloch sphere, here sphere inside a sphere inside a sphere... It explores **singularity** and used fractal mechanics, nothing to do with classical or quantum. (viii) Time cycle is the memory, rotation along the cycle is processing, are same events, "no transport needed between memory and processing units," no wiring. **No wiring**, a completely wireless connection of vortex fields. A wireless connection to process geometry at all the time scales is allowed in the hardware simultaneously. (ix) No logic gate, no reduction of choices, which ensures that "speed" is irrelevant. The computer is made of one element only, **clocks**; considers only parameter phase. It considers only parameter phase, emulate mass, space and time to process information. It expands the input information using PPM, hacks nature to predict a gross future. (x) All sensory information is converted to one geometric language that allows "perception," a yellow color could have a taste. Perception is not programming as wrongly perceived. It has a unified **homogeneous fractal** hardware for everything, learning changes them.

10.9 PLAYING WITH CHEAP, HANDY TOOLS TO BREAK THE PARADIGM OF THOUGHT

Will the unknown remain unknown? Exploring a Science Fiction in the making: Artificial brain won't be a real computer, but we are making a user. The computer runs by itself. But then, who is the original driver? The metric of primes created by Reddy S. et al. exclusively for the computer, runs the show inside its core hardware. What is the metric? And why a metric would have the power to do remarkable things? When we work on building a metric, like good old days of astrophysics, PPM-GML-H protocols also becoming as religious entity as a Turing machine. In astrophysics, theoreticians used to have a space-time metric, while doing complex math, students used to refer to the metric time to time and retrieve all essential data to solve planetary problems. Similarly, for natural intelligence (AI) we explored here a new metric of primes, namely space-time-topology-time. The idea is to hack nature and make a computer that can generate most patterns that we see in nature, so that unknown is known via PPM. How to perfectly build an effective prime metric architecture, or how to advance our primitive conscious bot (Chapter 9) is to be a matter of investigation for a longer time, but it cannot be ruled out that the concept to use a prime metric as a prime decision-maker is a new concept altogether. Self-awareness would require post-mortem of PPM-GML-H prescribed here.

The existing information theory is based on the idea of the known. FIT, (Fractal Information Theory, Chapter 2), wherein tools are there to bridge two known domains through an unknown path. It is an important change from the era of information theory that was existing for the last century. What is the trick that I know the unknown? We can do it if we build a universal metric that keeps all possible solutions, just like the space-time metric that is being used for nearly a century with little modifications to discover a new and new physical phenomenon that was never known. If we are not surprised how a space-time metric discovered in the 1920s is able to provide us new and new discoveries over a century, we should not be surprised that a similar metric for natural intelligence a 11D space-time-topology-prime metric would trigger another revolution. Of course, this is not a known culture in AI, but we feel that people would get accustomed to this culture of building architecture of confusion in day-to-day life.

Imagine we have two parts of music, and we have a kit that would combine two parts of the music with a new one in the middle, and that new music would make a sense to our mind. Similar things would be true in handling large data, it would generate unseen patterns in the big data. The reason we would like to inspire readers to build a DIY kit so that every people in the world could get free access to the information revolution knowing nature as it is. Why not seeing the share market using time crystal, build chatbots for the olds and deserted, build machines to give birth to new machines!!!

The beauty of PPM based computing is that we get the total picture at once. Then, more the time pass by, more the information arrives, from 50% reliability to 66% reliability to 72% reliability to 76% reliability to... the journey moves on toward 99% reliability, beyond which not possible to achieve. What brain jelly does is absorbing the dynamics related to new primes. Absolute reliability is a trademark of the existing computers, but for PPM-GML-H triad, "zooming the unknown" as a function of time and more detailed input is the key. The artificial brain would be a toy to change the perspective of this world, not step-by-step, rather, within and above. Not just playing a game, if we are in an unknown territory about which we have absolutely no information, then, the product computer or user can provide a good overview, instantly with 66% success rate. Uncharted territories are increasing every day with the data explosion. If we humans do not have a technology to estimate what is there in the uncharted territory, we cannot do anything. Accidents would create massive havoc on human society.

1. Imagine a virus is silently evolving into a dangerous species. Some prime metric hardware is there to perpetually track the development of the virus evolution and perfecting the prediction to monitor its evolution. Thus, it could estimate the terror threat well in advance.
2. From the Microwave background data of the universe, it could estimate the structure of the universe partially.
3. From the massive data flow of the internet, it could find patterns of threats, like cyber attack.
4. It can monitor and predict all possible climate change, where the future predictions are not possible due to complexity.
5. It can monitor individual health over the years and learn about individual health crisis well in advance. Typical health problems exclusive to a person could be identified and cared.
6. Economics will become a scientific subject of study as the computer would build predictable models perfecting it over time.
7. Social science and psychology will become a scientific subject as verifiable predictable models would be there, that could be rejected or accepted by logic.
8. General science would get a tool to study the absolute property of a system, not a fitting model, thus, even scientific studies would get a better cross checker of its conclusions.
9. Evolution of life could be tracked scientifically, not just in the past, but also the future to be predicted.
10. Lifelike machines of the future will come, which would have their own operation lifetime and after a certain time, they will die just like living systems.

10.10 NUMEROLOGY OF HUMAN CONSCIOUSNESS

We have summarized all the important numbers that we encountered while presenting the formation of brain jelly based artificial brain or developing the time crystal model

Number, N	Mathematics	Physics	Biology
4, 8	Quaternion, Octonion, Lie algebra	Continued fraction geometric algebra, CFGA	Class of sensors enable
12	Dodecanion, 144 elements tensor with 11 imaginary states	Topology begins, triangle, three points are 2x2x3, 3x2x2, 2x3x2	No of thoracic nerve, cranial nerve, thalamic body, nucleus of hypothalamus
19^4	Elements require to build dodecanion	H elements required for building an H3 device	Cortical culumns form H3, nerve bundle, hippocampus layer
47, 15th prime	The last prime that builds distinct metric	Eisenstein prime with no imaginary part	Cortex operational region, Brodmann's soma domain.
2x3x5.....37 $\sim 10^{12}$ or up to $47 \sim 10^{14}$	99.99% cover of all factors in the integer space.	New symmetries become <0.01% in the available options	Elements repeat as single element start new counting for new device
11,17,23,29,37,41	Silent primes	Groups define statistics	Various brain components
2,3,5,7,13,19,31,43	Active primes	Groups define symmetry	Various brain components
12N, 24N, 48N, series, abundant in nature	24N, 12N OF>>N, ordered factor, at 48, OF =N for first time	Degenerate solutions ensure fractal mechanics	Truly intelligent & high level perception
174000	43-47 pairing ends, 47 takes over	Minimum H device required to cover 99%	Number of neurons in 1 of 47 cortex regions
1,2,3,5,7..47	Metrices are explicit	>48, primes are masked	C4H7O2N=48/101 life=48 carbon

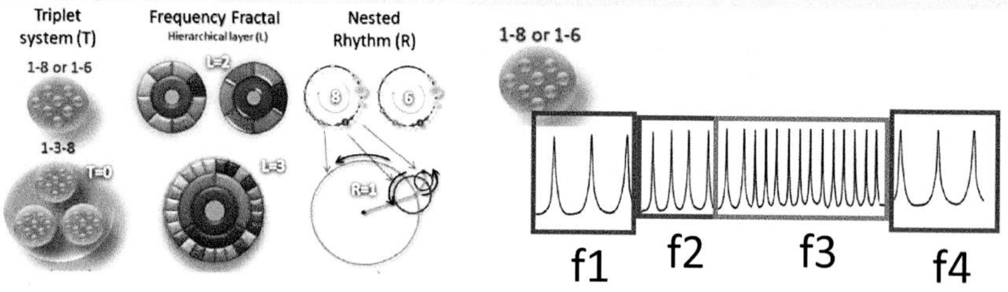

FIGURE 10.9 Important numbers that were found abundantly in nature and in the brain. Below, the formation of time crystals of primes is shown using simple resonance bands and the clocks nested with each other.

of the human brain in Figure 10.9. From biology to physics to mathematics, the emergence of a few numbers, especially primes in the human brain, have been astonishing. They may be accidental or mere coincidence, it does not matter, a good summary would refresh readers mind about how the brain adventure unfolded.

The pattern of primes in the life and culture: Twenty-three tonal patterns used in 10,552 versus of Rig Veda: During information processing, a large number of fractal seeds of time crystals are generated and during memory retrieval, they expand and create the entire image. It is incredible memory management, within a few numbers of oscillators we can encode entire complex architecture. Indian classical music has a concept called raga, each raga is a packet of seven frequency notes; a singer needs to sing low to high and then high to low frequency to complete a period. For thousands of years, these songs are perfected to generate a particular feeling in the mind of the listener the frequencies are changed by tiny values, feedback cycles run generations after generations (Ganeri, 2001). Often the singer expands the seed geometric pattern for hours in various styles to create maximum impact on the mind of the listener. The rigorous mathematics for the composition of such classical songs suggest a very similar behavior. With 47 notable sound frequencies (48 letters including

Om sound), each representing a meaning, the Sanskrit language was made, that uses the matrix-vector projection mathematical formulation to convert a word to include space, time, context, sexuality, the transformed word is used to create 23 different classes of rhythms in Vedas, based on tonal ratios (1:2, 1:3, 2:3, 3:5, etc), just like we did in Chapter 7 for pattern recognition, tones recur cyclically at every doubling or halving the frequency (rhythm used to generate time fractals by Vedic tradition). Logic in Vedas is geometric (Yantra, Mandala), uses 3^p5^q formulation (p, q are ±, hence they made triangle upside down to construct Sri-Yantra) to create all tones (male), 2r is used as added dimension (female), together they make the universe (Nicolus and McClain), we found pattern of prime similarly and ordered factors provide universal fractal chain in Chapter 2. Vedas used all arithmetic, geometric and harmonic infinite series using 2, 3, and 5, while notes make a packet of 7, still Indian musicians use rhythm packets of 17 tunes (in a loop of 18 bits), including all prime numbers up to 47 (prime = $6n \pm 1$). Fractional infinite time series (time fractal f(t)), were also used (e.g., 5–7 and 3–6 time series define Kalpa Bramha, 5–10 time series defines Universe as an egg i.e., Hiranyagarbha etc.). Nested rhythm as time crystal presented in this book was extensively used in 5000 years of Sanskrit texts, fastest clocks were even number of hymns, chapters, words, letters were specifically chosen to make it rhythm inside a rhythm, entire book has to be a song, like the complete time crystal model presented for the human brain (Figures 7.14 and 7.15).

10.11 FREQUENCY WHEELS OF AN EVOLVING BRAIN

The PPM world starts from fractal tape, ends in the collective evolution of machines: A summary: Turing's world of computing considers every single element is complete in itself, just like that a "matter" is complete; nothing inside/outside is changing it. Completeness enables discrete and binary true/false argument to define it. We get a Turing world. The PPM vision has one argument, one energy packet, one Turing like tape, one clock, one rhythm, one oscillator that encompasses the entire universe, but if we enter inside any of them, we find millions of that inside, the journey continues forever, whether we zoom in/out. Unlike Turing, any "matter" is conceptually incomplete so we need matter+energy+topology+symmetry tetrality to connect all matters inside and outside. The wheels of life presented in Figures 10.10 and 10.11 are combinations of both matter and energy.

Conserved energy takes the form of periodic oscillation or rhythm and maintains time, but each rhythm has rhythms inside/outside (Figure 10.10). We cannot isolate any small piece of argument from tape network, hence, no bits or arguments like true/false, that leads to no classical information content. Observer/user is also part of the tape network. Entire

rhythm net, or time crystal that encompasses everything is represented as a matrix of truth with infinite dimensions, no rhythm is ever destroyed, no choice is ever rejected. Unlike Turing, fractal tape (PPM tape) needs four new steps, resonate (with a rhythm), expand (self-assembly of rhythms), phase transition (of rhythms), and spontaneous reply (the rhythms). Self-similarity of shape/rhythm appears/disappears with time, as rhythms couple/decouple with other rhythms, changes the structural symmetry of matters to memorize a particular set of frequencies for a rhythm. The entire tape network is never-ending resonance chain of coupled matters that vibrate with rhythm, means, local rhythms vibrate as a distinct entity in parts of the tape.

Since observer/user is also part of the rhythms of the tape-network, it senses by perpetual local synchrony and de-synchrony with various parts of the tape, thus, its "sensing" not computation by the classical theory of computation. At the bottom part of Figure 10.9, when we summarized the numbers important for the adventure of the brain, we showed how vibrations that were mere a spectrum thus far have become time crystals. That journey continued in Figures 10.10 and 10.11 where we have increased layers of ET tape or clock inside a clock… and then made efforts to increase complexity one by one. The equivalent life forms that the time crystal model could represent is shown side by side. The idea is to suggest **possibly the billions of years of evolutions from a single molecule to the complex life form have been a journey of time crystals**. We see a catalog of

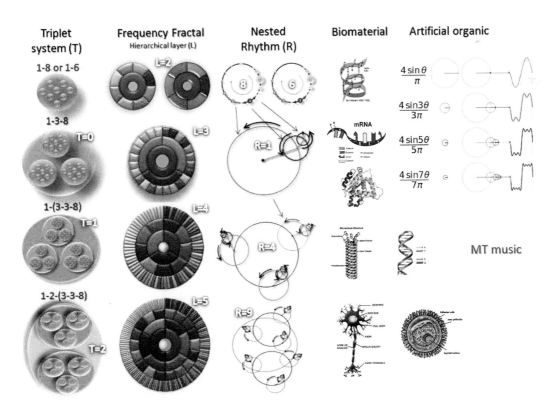

FIGURE 10.10 The first part of the Table showing step-by-step evolution of computers or decision-makers starting from a single time crystal. The first column shows 2D resonance domain, Second column shows equivalent frequency wheel, third column is a slice of time crystals showing simplified operation. Fourth column the equivalent biomaterial targeted to mimic.

Generalization of frequency fractal computer

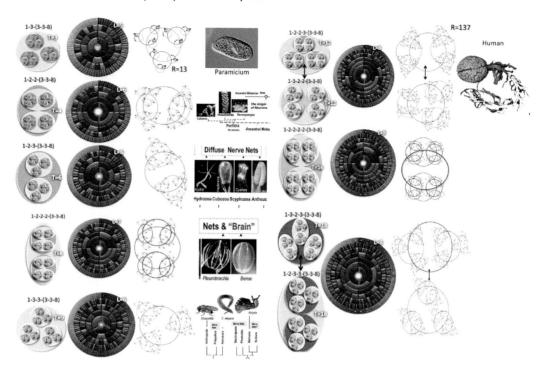

FIGURE 10.11 The second part of the Table showing step-by-step evolution of computers or decision-makers starting from a single time crystal. The first column shows 2D resonance domain, Second column shows equivalent frequency wheel, third column is a slice of time crystals showing simplified operation. Fourth column the equivalent biomaterial targeted to mimic.

life forms created by nature in the planet, maybe it simply added more and more clocks following the PPM described in Chapter 3.

10.12 SYNTHESIS OF CRITICAL BRAIN FEATURES EVEN BEFORE WE MEET THE ALIENS

We have created a table in Figure 10.12, where we have shown that if mathematics is the fundamental basis for human consciousness then, it would not be the last or only one kind of consciousness. There would be various kinds of consciousness. The human brain explored in the book could be represented by one integer, 12. Why? $2 \times 2 \times 3$ could be written in three ways, so a higher level geometry one could create. It means the human brain using 12 components, each of which could process higher-level decision-making (each component have 12 bands to operate), realize consciousness. However, the same could be achieved by $2 \times 2 \times 5$, possibilities are infinite. The table suggests how to build higher than human consciousness machines. Of course they would never be machines anymore. The most interesting is the case for 37 primes or case of 36, then higher-level geometry that makes human conscious as per the definition of current discussion would make a truly higher conscious alien whom, humans could not even understand.

Projects similar to PPM computing: There are plenty of brain-building projects, some of the projects do consider nested networks and interference as a tool to reach the goal

(Ersatz brain project, J. A. Anderson), not only that the fractal analysis of neurodynamics does provide greater understanding of the mind-brain paradox (King, 1996). Quantum chaos has been used as a tool to explain synchronization, in this book, classical synchronization formulations are provided, however, given that in the molecular scale quantum mechanical principles are significantly active, the role of quantum chaos in delivering multi-scale synchronization cannot be ruled out. Especially, since algebraic-geometric tools are extensively used in representing the entanglement and superposition (Bernardi and Carusotto, 2011), a complete analog to this book in terms of pure quantum mechanics is possible to formulate.

Artificial brain is the last machine of mankind: First machine to learn what is nature: We must bypass the era of brain-less bots: The era of conscious machines: Artificial brain would be the last machine of mankind, since willingly or unwillingly the human race wants to reach that point when the machines would take over. If algorithm-based instruction is the key, robots would learn it, outperform humans in every aspect, we have to make a new world of technology where the algorithm does not exist. The era of life-like machines or conscious machines is human destiny now. We live in an exponential information age; everything around us is changing so rapidly that after a few years, a third-year engineering student would have to discard everything he studied in the previous two years, simply because of the fast-technological implementations of an invented technology in the industry.

Number of primes	Human level	Tensor elements	Multinion	Linguistic structure	Hardware	Seed geometry	Operating tensor
4 (2,3,5,7)	Sub-human	8^2	2x2x2= {O}	quaternary	Sensor 1		Octonion
5 (+11)	human	12^2	2x2x3={d}	quaternary	Brain 1	▲	Dodecanion
6 (+13)	Sub-human	16^2	2x{O}	quaternary	Sensor 2		hexdecanion
7 (+17)	Human+	18^2	2x3x3={d}	trinary	Brain 2	▲	Octo-deca...
9 (+19, 23)	2xhuman	24^2	2x{d}	quaternary	Brain 3	▲—▲	Qua-di-deca
9	Sub-human	27^2	3x3x3={O}	trinary	Sensor 3		Hep-di-deca
9	Human++	28^2	2x2x7={d}	quaternary	Brain 4	▲	Oct-di-deca
10 (+29)	Alien*	30^2	2x3x5={t}	hexanary	Brain 5	⬡	tridecanion
11 (+31)	Sub-human	32^2	4x{O}	quaternary	Sensor 4		Di-tri-deca
12 (+37)	3xhuman	36^2	3x{d}	quaternary	Brain 6	◭	Hex-tri-deca

*Alien refers to 2x3x5 seed geometric system where elementary tensor is not 2x2 but a pair or triplet of 3x3 or 5x5

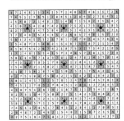
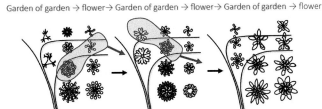

Garden of garden → flower→ Garden of garden → flower→ Garden of garden → flower

FIGURE 10.12 A table shows how artificial brain's decision-making power could be enhanced beyond humans to the aliens who are not known to mankind. Below, the garden of gardens GOG representation of higher than human's information processing capabilities are shown as ones GOG is others petal of flower.

Now, the problem is that we cannot ignore the elementary interactions as we normally do in the formulations of physics theory, simply because one fundamental element could be me. Therefore, when we try to create an alternative model, we should consider that many-body networks generated at the single element level. We are not far away from a time, when we will not ask the pressure exerted by 1 gm-mole gas, rather we would seek for the information contained in a single gas molecule in that container. It is beyond our brain capacity to process, but we are all pushing human civilization to a point where there will be no turning back, everything around us would estimate and beyond a certain limit that we humans could access using our intelligence, but could feel with our senses.

One option would be to wait, until nature adopts another structural evolution of the entire brain architecture, or holding a micro-brain device on our palm. It will be like we will carry several brains similar to the power of our brain and our brain will work as a manager to these brains, a superior authority which plays only with the output of the inferior brains (Figure 10.12). Nano brain will not be a superior intelligence provider, rather, the last tool of ours to survive in a world where intelligence will be clustered into various pockets whose domains will be decided by something which is not a human, not a machine, not a software, but a virtual life form that is composed of everything we see around us. We will give birth to this unknown creature, who will not have any life form, yet would govern the evolution of every living and non-living things in that world.

While the ever-expanding brain is colliding continuously with the restricted space inside our skull, we are amazed by the synchronous burst of neurons between the two ends of our brains separated by 15 cm. We take shelter into this remarkable phenomenon of simultaneous bursts of neurons because we belong to the class of living species that do not like to walk along with a one-dimensional tape to process information since looking here and it is not allowed there. The question that always troubles us that if we have that creature, will it destroy the mankind, in reply we would argue that we try to preserve baboons and chimps not kill them, so the new creatures will not kill humans, they will feel pity for us. Also do not forget that supremacy does not come alone, severe weaknesses will be there in their hardware too and supreme conscious beings born among the human race have always talked about love not hate. When we try to build artificial brains, we are not attempting to create Frankenstein, we have enough of them among humans already.

The first machine to understand nature as-is: Why we exist? The triangular relationship between the GML, a PPM and a fourth-circuit element (Hinductor, H) change the worldview forever. The century-old faiths that "no question, no answer," is found to be a wrong question, a reply may come even if no question is asked, fate could be decided even without communication and without measurement. Hardware engineered by the PPM has a guideline for an infinite chain reaction, a drive to match the intricate details of patterns between the two participators that cannot reach an equilibrium ever. The language of bit and the language of the continuum can coexist as part of GML, wherein the quantum is a local perspective of the global engineering of singularity governed by PPM. Light has a gradient but the darkness is pure; sound has colors, but the silence is pure; a defined state is finite, but a singularity is infinite; a non-prime is rich in factors and decisions, but fixed; in contrast, a prime is neural, pure and a creator of an endless pattern like a life form. The language of Bekenstein number includes a human bias, where the number of bits was everything, the primes exhibiting the knots of darkness is pure, because it sees nothing but symmetry, thus, a few, say 12–14 primes only use 12–14 ways of ordering the geometric shapes to link 99.99% events, structures, that has happened, that are happening and what would happen in the future. We exist because of primes. Thus, we grasp the central idea of it all as so simple, so beautiful, so compelling that we could say to each other, "Oh, how -could it have been otherwise! How could we all have been so blind so long!." Let humanity begin its journey to learn nature as is by harnessing its last machine.

10.13   A CYCLE OF A GARDEN FROM FLOWER TO A PETAL

Without mentioning the case of meander flower garden we could not finish the discussion on the higher than human consciousness. Humans follow the garden of garden of meander flowers as described in Chapters 2, 6, and 7. When universe would create a catalog of conscious machines then it would take a set of petals from multiple garden of gardens and build another garden of meander flowers. Therefore, the table is shown in Figure 10.12 where using simply the mathematical argument we extrapolated the human consciousness to an infinite journey, the same journey could be viewed in terms of meander flower gardens.

When every part of a system grows simultaneously, spontaneously: Imagine we have an architecture of self-assembled time crystals and every part of it is growing, or imagine our whole body, every part of it is growing simultaneously, spontaneously. Conventional science teaches us that there should be a boundary point and beyond that everything would remain static, and then only we would analyze the dynamics happening inside. Here, we have a universal guideline, we create a guideline for how symmetries could change spontaneously. We know how symmetries would break and adopt a new symmetry point. Learning to live with everything that is changing continuously is the new kind of science and we are looking into one such new kinds of practices, engineering of singularity, adventuring confusion, being random and logically circular, doing nonsense. It is a totally new experience, a journey where time and space evolutions would appear messy, random, unpredictable because everything talking to everything has global symmetries, we may look into that. In this world of science, even topology would not take us far, we need a new kind of drawing, the mathematics of patterns. Equations won't help, the interactions of the imaginary worlds would be the mother of all algebras. In this new type of mathematics, we only draw things and the rest of the solutions are automatically derived, we envision a kid of 21st century drawing only circles to solve problems which super-supercomputers cannot fathom. By drawing circles and circles alone, infinity is explored, it is a world of undefined and feedback from infinity. That infinity is the universe itself, it is the complete time crystal, somewhere in a singularity point our times crystals are vibrating and discussing about conscious machines.

The total number of minds in the universe is one. In fact, consciousness is a singularity phasing within all beings.

E. Schrödinger

Bibliography

Abbot, L. F., Wise, M. B. (1981) Dimension of a quantum mechanical path. *Am J Phys* 49(1), 37–39.

Abbott, B. C., Hill, A. V., Howarth, J. V. (1958) The positive and negative heat associated with a nerve impulse. *Proc R Soc B* 148(931), 149–187.

Abdel, N. H. (2011) Computer geometry and encoding the information on a manifold. *J Egypt Math Soc* 19(1–2), 17–27.

Abdollahramezani, S. (2017) Dielectric metasurfaces solve differential and integro-differential equations. *Opt Lett* 42(7), 1197–1200.

Abe, I., Ochiai, N., Ichimura, H. I., Tsujino, A., Sun, J., Hara, Y. (2004) Internodes can nearly double in length with gradual elongation of the adult rat sciatic nerve. *J Orthop Res* 22, 571–577.

Adey, W. R., Dunlop, C. W., Hendrix, C. E. (1960) Hippocampal slow waves: Distribution and phase relationships in the course of approach learning. *Arch Neurol* 3, 74–90.

Adhikari, A. et al. (2015) Basomedial amygdala mediates top-down control of anxiety and fear. *Nature* 527, 179–185. doi:10.1038/nature15698.

Adhikari, S. K., Tomio, L. (1982) Efimov effect in the three-nucleon system. *Phys Rev C* 26, 83.

Adleman, L. M. (1994) Molecular computation of solutions to combinatorial problems. *Science* 266, 1021–1023.

Adrian, E. B., Mathews, B. (1934) The Berger rhythm. *Brain* 57, 355–385.

Afanasiev, G. N., Dubovik, V. M. (1998) Some remarkable charge-current configurations. *Phys Part Nuclei* 29(4), 366–391.

Afreixo, V., Rodrigues, J. M. O. S., Bastos, C. A. C., Silva, R. M. (2016) Exceptional symmetry of DNA sequence. *BMC Bioinformatics* 1–10.

Agladze, K., Keener, J. P., Müller, S. C., Panfilov, A. (1994) Rotating spiral waves created by geometry. *Science* 264, 1746–1748.

Agladze, K., Obata, S., Yoshikawa, K. (1995) Phase-shift as a basis of image processing in oscillating chemical medium. *Physica D* 84, 238–245.

Agnati, L. F. et al. (2009) Self-similarity logic, biological attraction principles three explanatory instruments in biology. *Commun Integr Biol* 2(6), 552–563.

Agrawal, L. et al. (2016a) Replacing Turing tape with a Fractal tape: A new information theory, associated mechanics and decision making without computing. In: *Consciousness: Integrating Indian and Western Perspective*, 1st ed. New Age Books, India, Chapter 6, pp. 87–159. Paperback: 630 p.

Agrawal, L. et al. (2016b) Fractal Information Theory (FIT) derived Geometric Musical Language (GML) for brain inspired hypercomputing. Advances in Intelligent Systems and Soft Computing AISC, Series Ed.: Kacprzyk, Janusz, pp. 37–61. Proceedings of SocTA 2016, Springer.

Agrawal, L. et al. (2016c) Inventing atomic resolution scanning dielectric microscopy to see a single protein complex operation live at resonance in a neuron without touching or adulterating the cell. *J Int Neurosci* 15(4), 435–462.

Agu, M. (1988) Field theory of pattern recognition. *Phys Rev A* 37, 4415–4418.

Aharonov, Y., Bohm, D. (1959) Significance of electromagnetic potentials in the quantum theory. *Phys Rev* 115(3), 485–491; Bekenstein, J. D. (1972) Baryon number, entropy, and black hole physics, PhD thesis, Princeton University; Photocopy available from University Microfilms, Ann Arbor, MI.

Ahissar, E., Arieli, A. (2001) Figuring space by time. *Neuron* 32, 185–201.

Aieta, F. et al. (2012) Out-of-plane reflection and refraction of light by anisotropic optical antenna metasurfaces with phase discontinuities. *Nano Lett* 12, 1702–1706.

Albert, D. Z. (2000) *Time and Chance*. Cambridge, MA, Harvard University Press.

Aldridge, J., Pavlidis, T. (1976) Clock like behavior of biological clocks. *Nature* 259, 343–344.

Ali, Z. et al. (2016) 3D printed hydrogel soft actuators. In *Region 10 Conference (TENCON)*, IEEE. doi:10.1109/TENCON.2016.7848433.

Allan, L. G. (1979) The perception of time. *Percept Psychophys* 26, 340–354.

Alleva, J. J., Waleski, M. V., Alleva, F. R. (1971) A biological clock controlling the estrous cycle of the hamster. *Endocrinology* 88, 1368–1379.

Altsybeev, I. G., Feofilov, G. A., Gillies, E. L. (2016) Forward-backward correlations with strange particles in Pythia. *J Phys* 668, 012034.

Alzetta, G. et al. (1999) Simultaneous micromechanical and electromagnetic detection of electron paramagnetic resonance. *J Mag Res* 141, 148–158.

Alzetta, G., Gozzini, A. (1964) in *Proceedings of the XII Colloque Ampere*. North Holland, Amsterdam, the Netherlands, p. 82.

Alzetta, G., Arimondo, E., Ascoli, C. et al. (1967) Paramagnetic resonance experiments at low fields with angular-momentum detection. *Nuovo Cimento B (1965–1970)* 52, 392–402. doi:10.1007/BF02711084.

Amaral, D. G., Witter, M. P. (1989) The three-dimensional organization of the hippocampal formation: A review of anatomical data. *Neuroscience* 31, 159–177.

Amassian, V. E., Cracco, R. Q. Maccabee, P. J., Cracco, J. B., Rudell, A. P., Eberle, L. (1998) Transcranial magnetic stimulation in study of the visual pathway. *J Clin Neurophysiol* 15, 288–304.

Amit, D. J., Brunel, N. (1997) Model of global spontaneous activity and local structured activity during delay periods in the cerebral cortex. *Cereb Cortex* 7, 237–252.

Ammari, H. et al. (2013) Anomalous localized resonance using a folded geometry in three dimensions. *Proc Royal Soc A* 69, 20130048.

Anandan, J. (1988) Comment on geometric phase for classical field theories. *Phys Rev Lett* 60, 2555.

Anandan, J., Aharonov, Y. (1988) Geometric quantum phase and angles. *Phys Rev D* 38, 1863–1870.

Anastasio, T. J., Robinson, D. A. (1990) Distributed parallel processing in the vertical vestibulo-ocular cortex: Learning network compared to tensor theory. *Biol Cybern* 63, 161–167.

Andersen, P., Eccles, J. C. (1962) Inhibitory phasing of neuronal discharge. *Nature* 196, 645–647.

Anderson, J. A. (2005) A brain-like computer for cognitive software application: The Ersatz brain project. *Fourth IEEE Conference on Cognitive Informatics*. IEEE.

Anderson, J., Binzegger, T., Kahana, O., Martin, K., Segev, I. (1999) Dendritic asymmetry cannot account for directional responses of neurons in visual cortex. *Nat Neurosci* 2, 820–824.

Anderson, P. W. (1972) More is different. *Science* 177, 393–396.

Andrew, C., Pfurtscheller, G. (1996) Event-related coherence as a tool for studying dynamic interaction of brain regions. *Electroencephalogr Clin Neurophysiol* 98, 144–148.

Anumonwo, J., Delmar, M., Vinet, A., Michaels, D., Jallfe, J. (1991) Phase resetting and entrainment of pacemaker activity in single sinus nodal cells. *Circ Res* 68, 1138–1153.

Ares, S., Castro, M. (2006) Hidden structure in the randomness of the prime number sequence? *Physica A* 360, 285.

Arlinger, S. et al. (1982) Cortical magnetic fields evoked by frequency glides of a continuous tone. *Electroencephalogr Clin Neurophysiol* 54(6), 642–653. doi:10.1016/0013-4694(82)90118-3.

Armstrong, D. M. (1983) *What is a Law of Nature?* Cambridge, Cambridge University Press, Chapters 1–6.

Arnowitt, R., Deser, S., Misner, C. W. (1960a) Gravitational-electromagnetic coupling and the classical self-energy problem. *Phys Rev* 120, 313.

Arnowitt, R., Deser, S., Misner, C. W. (1960b) Canonical variables for general relativity. *Phys Rev* 117, 1595.

Arshavsky, Y., Berkinblit, M. B., Kovalev, S. A., Chailakhyan, M. (1964) Periodic transformation of rhythm in a nerve fiber with gradually changing properties. *Biofizika* 9, 365–371.

Artes, J. M. et al. (2015) Conformational gating of DNA conductance. *Nat Commun* 6, 8870.

Arvanitaki, A., Chalazonitis, N. (1968) Electrical properties and temporal organization in oscillatory neurons. In: *Neurobiology of Invertebrates*, J. Salanki, Ed. Plenum, New York, pp. 169–174.

Asakawa, K. (2010) Equivalent circuit modeling of triple barrier resonant tunneling diodes taking nonlinear quantum inductance and capacitance into account. *Phys Status Solidi C* 7(10), 2555–2558.

Asami, K. (1998) Dielectric analysis of polystyrene microcapsules using scanning dielectric microscope. *Colloid Polym Sci* 276, 373–378.

Aschoff, J. (1965a) Circadian rhythms in man. *Science* 148, 1427–1432.

Aschoff, J., Wever, R. (1976) Human circadian rhythms: A multioscillatory system. *Fed Proc* 35, 2326–2332.

Ashby, W. R. (1947) Principles of the self-organizing dynamic system. *J Gen Psychol* 37, 125–128.

Asher, G., Schibler, U. (2011) Crosstalk between components of circadian and metabolic cycles in mammals. *Cell Metab* 13, 125–137.

Assawaworrarit, S., Yu, X., Fan, S. (2017) Robust wireless power transfer using a nonlinear parity–time-symmetric circuit. *Nature* 546, 387.

Atanasov, V., Dandoloff, R. (2008) Curvature induced quantum behavior on a helical nanotube. *Phys Letts A* 372, 6141–6144.

Atherton, J. F., Wokosin, D. L., Ramanathan, S., Bevan, M. D. (2008) Autonomous initiation and propagation of action potentials in neurons of the subthalamic nucleus. *J Physiol* 586(23), 5679–5700. doi:10.1113/jphysiol.2008.155861.

Audenaert, R., Heremans, L., Heremans, K., Engelborghs, Y. (1989) Secondary structure-analysis of tubulin and microtubules with Raman-spectroscopy. *Biochim Biophys Acta* 996(1–2), 110–115.

Aviel, Y., Gerstner, W. (2006) From spiking neurons to rate models: A cascade model as an approximation to spiking neuron models with refractoriness. *Phys Rev E* 73, 051908.

Ayers, J. L., Selverston, A. I. (1979) Monosynaptic entrainment of an endogenous pacemaker network: A cellular mechanism for von Holst' s magnet effect. *J Comp Physiol* 129, 5–17.

Baas, P. W., Deitch, J. S., Black, M. M., Banker, G. A. (1988) Polarity orientation of microtubules in hippocampal neurons: Uniformity in the axon and nonuniformity in the dendrites. *Proc Natl Acad Sci* 85, 8335–8339.

Baccus, S. A. (1998) Synaptic facilitation by reflected action potential enhancement of transmission when nerve impulses reverse direction at axon branch points. *PNAS* 95, 8345–8350.

Badurek, G., Rauch, H., Tuppinger, D. (1986) Neutron interferometric double-resonance experiment. *Phys Rev A* 34, 2600.

Baena, J. D. et al. (2004) Artificial magnetic metamaterial design by using spiral resonators. *Phys Rev B* 69, 014402.

Baena, J. D. et al. (2005) Near perfect tunneling and amplification of evanescent electromagnetic waves. *Phys Rev B* 72, 075116.

Bai, Y. et al. (2017) Ferroelectric, pyroelectric, and piezoelectric properties of a photovoltaic perovskite oxide. *Appl Phys Lett* 110, 063903.

Bak, P. (1996) *How Nature Works: The Science of Self-Organized Criticality.* Springer-Verlag, New York.

Baker, R. R. (1980) Goal orientation in blindfolded humans after long distance displacement: Possible involvement of a magnetic sense. *Science* 210, 555–557.

Bakkum, D. J. et al. (2013) Tracking axonal action potential propagation on a high-density microelectrode array across hundreds of sites. *Nat Commun* 4, 2181.

Bal, M., Zhang, J., Zaika, O., Hernandez, C. C., Shapiro, M. S. (2008) Homomeric and heteromeric assembly of KCNQ (Kv7) K+ channels assayed by total internal reflection fluorescence/fluorescence resonance energy transfer and patch clamp analysis. *J Biol Chem* 283, 30668–30676.

Bal, T., McCormick, D. A. (1993) Mechanisms of oscillatory activity in guinea-pig nucleus reticularis thalami in vitro: A mammalian pacemaker. *J Physiol* 468, 669–691.

Balestra, F. R., Tobel, L. V., Gonczy, P. (2015) Beyond genes: Are centriole carriers of biological information? *Cell Res* 25, 642–644.

Bandyopadhyay, A. et al. (2006b) Global tuning of local molecular phenomena: An alternative approach to nano-bioelectronics. *J Phys Chem B* 110, 20852–20857.

Bandyopadhyay, A. et al. (2010a) A new approach to extract multiple distinct conformers and co-existing distinct electronic properties of a single molecule by point-contact method. *PCCP* 12, 2033.

Bandyopadhyay, A. et al. (2010b) Investigating universal computability of conventional cellular automata problems on an organic molecular matrix. *Nat Comput* 2, 1–12.

Bandyopadhyay, A. et al. (2010c) A massively parallel computing on an organic molecular layer. *Nat Phys* 6, 369.

Bandyopadhyay, A., Acharya, A. (2008) A 16 bit parallel processing in a molecular assembly. *Proc Natl Acad Sci USA* 105, 3668–3672.

Bandyopadhyay, A., Fujita, D., Pati, R. (2009b) Architecture of a massive parallel processing nano brain operating 100 billion molecular neurons simultaneously. *Int J Nanotech Mol Comp* 1, 50–80.

Bandyopadhyay, A., Miki, K. (2006) A vertical parallel processor JP-5187804.

Bandyopadhyay, A., Miki, K., Wakayama, Y. (2006a) Writing and erasing information in multilevel logic systems of a single molecule using scanning tunneling microscope (STM). *Appl Phys Lett* 89, 243506.

Bandyopadhyay, A., Sahu, S., Fujita, D. (2009a) Smallest artificial molecular neural-net for collective and emergent information processing. *Appl Phys Lett* 95, 113702.

Bandyopadhyay, A., Wakayama, Y. (2007) Origin of negative differential resistance in molecular junctions of rose bengal. *Appl Phys Lett* 90, 023512.

Banerjee, A. et al. (2016) Proximate kitaev quantum spin liquid behaviour in a honeycomb magnet. *Nat Mater* 15, 733–740.

Barabási, A. L., Albert, R. (1999) Emergence of scaling in random networks. *Science* 286, 509–512.

Baranova, N. B., Mamaev, A. V., Pilipetsky, N. F., Shkunov, V. V., Ya Zel'dovich, B. (1983) Wave-front dislocations: Topological limitations for adaptive systems with phase conjugation. *J Opt Soc Am* 73, 525–528.

Bardeen, W. (1995) On Naturalness in the Standard Model 1995 Ontake Summer Institute, FERMILAB-CONF-95-391-T.

Barkai, N., Leibier, S. (2000) Circadian clocks limited by noise. *Nature* 403, 267–268.

Barkley, D. (1991) A model for fast computer simulation of waves in excitable media. *Physica D* 49, 61–70.

Barkley, D. (1994) Euclidean symmetry and dynamics of rotating spiral waves. *Phys Rev Lett* 72, 164–167.

Barkley, D. (1995) Spiral meandering. In: *Chemical Waves and Patterns*, R. Kapral, K. Showalter, Ed. Kluwer, Dordrecht, the Netherlands, Chapter 5, pp. 163–190.

Barnes, F. S., Greenebaum, B. (2007) *Biological and Medical Aspects of Electromagnetic Fields*, 3rd ed. Handbook of Biological Effects of Electromagnetic Fields. CRC Press, Boca Raton, FL.

Barnsley, M., Hutchinson, J., Stenflo, Ö. (2008) V-variable fractals: Fractals with partial self similarity. *Adv Math* 218, 2051–2088.

Barrett, L. (2011) *Beyond the Brain: How Body and Environment Shape Animal and Human Minds*. Princeton University Press, Princeton, NJ.

Barry, J. F., Turner, M. J., Schloss, J. M., Glenn, D. R., Song, Y., Lukin, M. D., Park, H., Walsworth, R. L. (2016) Optical magnetic detection of single neuron action potentials using quantum defects in diamond. *Proc Natl Acad Sci USA* 113, 14133–14138.

Bartol Jr, T. M. et al. (2015) Nanoconnectomic upper bound on the variability of synaptic plasticity. *eLife* 4, e10778. doi:10.7554/eLife.10778.

Bartos, M., Vida, I., Frotscher, M., Meyer, A., Monyer, H., Geiger, J. R., Jonas, P. (2002) Fast synaptic inhibition promotes synchronized gamma oscillations in hippocampal interneuron networks. *Proc Natl Acad Sci USA* 99, 13222–13227.

Basar, E. (1990) Chaotic dynamics and resonance phenomena in brain function: Progress, perspectives and thoughts. In *Chaos in Brain Function*, E. Basar, Ed. Springer-Verlag, Heidelberg, Germany, pp. 1–30.

Bastiaansen, M., Hagoort, P. (2003) Event-induced theta responses as a window on the dynamics of memory. *Cortex* 39, 967–992.

Bateman, H. (1909) The conformal transformations of a space of four dimensions and their applications to geometrical optics. *Proc Lond Math Soc* s2-7, 70–89. doi:10.1112/plms/s2-7.1.70.

Beck, F., Eccles, J., (1992) Quantum aspects of brain activity and the role of consciousness. *Proc Natl Acad Sci USA* 89, 11357–11361.

Becker, S. et al. (2016) A high-yielding, strictly regioselective prebiotic purine nucleoside formation pathway. *Science* 352(6287), 833–836.

Bedau, M. A. et al. (1997) A comparison of evolutionary activity in artificial evolving systems and in the biosphere. In: *Fourth European Conference on Artificial Life*, P. Husbands, I. Harvey, Eds. MIT Press/Bradford Books, Cambridge, MA, pp. 124–134.

Beggs, J. M., Timme, N. (2012) Being critical of criticality in the brain. *Front Fractal Physiol* 3, 163.

Begliarbekov, M. et al. (2011) Quantum inductance and high frequency oscillators in graphene nanoribbons. *Nanotechnology* 22, 165203–165210.

Bekenstein, J. D. (1973) Generalized second law of thermodynamics in black-hole physics. *Phys Rev D* 9, 3292–3300.

Bekenstein, J. D. (1980) Black-hole thermodynamics. *Phys Today* 33, 24–31.

Bell, C. C. (2002) Evolution of cerebellum-like structures. *Brain Behav Evol* 59, 235–239.

Belousov, B. P. (1959) A periodic reaction and its mechanism. Sb Ref Radiats Med, Medgiz, Moscow, p. 145.

Ben, A., Cresson, J. (2005) Fractional differential equations and the Schrödinger equation. *Appl Math Comput* 161(1), 323–345. doi:10.1016/j.amc.2003.12.031.

Benedetto, A., Burr, D. C., Morrone, M. C. (2018) Perceptual oscillation of audiovisual time simultaneity. *eNeuro* 5(3), 1–12.

Benham, C. W. (1894) The artificial spectrum top. *Nature* 51, 200.

Bennett, B. M., Hoffman, D. D., Prakash, C. (1989) *Observer Mechanics: A Formal Theory of Perception*. Academic, San Diego, CA.

Benson, J. A., Jacklet, J. W. (1977) Circadian rhythm of output from neurons in the eye of Aplysia. *J Exp Biol* 70, 151–211.

Berg, N., Joel, K., Koolyk, M., Prodan, E. (2011) Topological phonon modes in filamentary structures. *Phys Rev E* 83, 021913.

Berger, J. O., Berry, D. A. (1988) Statistical analysis and the illusion of objectivity. *Am Sci* 76, 159–165.

Berger, T. W. et al. (2010) The neurobiological basis of cognition: Identification by multi-input, multioutput nonlinear dynamic modeling: A method is proposed for measuring and modeling human long-term memory formation by mathematical analysis and computer simulation of nerve-cell dynamics. *Proc IEEE* 98, 356.

Berkeley, G. (1959) *Treatise Concerning the Principles of Understanding*, Dublin (1710; 2nd ed. 1734); re his reasoning that "No object exists apart from mind," cf. article Berkeley by R. Adamson: Encyclopedia Brittanica, Chicago 3(1959) 438.

Berliner, M. D., Neurath, P. W. (1965) The rhythms of three clock mutants of *Ascobolus immersus*. *Mycologia* 57, 809–817.

Bernardi, A., Carusotto, I. (2012) Algebraic Geometry tools for the study of entanglement: An application to spin squeezed states. *J Phys A Math Theor* 45, 105304–105317. doi:10.1088/1751-8113/45/10/105304.

Bernstein, N. A. (1967) *The Coordination and Regulation of Movements*. Pergamon Press, London, UK.

Berry, M. V. (2004) The electric and magnetic polarization singularities of paraxial waves. *J Opt* 6, 475–481.

Berry, M. V., Chambers, R. G., Large, M. D., Upstill, C., Walmsley, J. C. (1980) Wavefront dislocations in the Aharonov-Bohm effect and its water wave analogue. *Eur J Phys* 1, 154–162.

Berry, M. V., Dennis, M. R. (2007) Topological event on wave dislocation lines: Birth and death of loop, and reconnection. *J Phys A Math Theor* 40, 65–74.

Berry, S. D., Thompson, R. F. (1978) Prediction of learning rate from the hippocampal electroencephalogram. *Science* 200, 1298–1300.

Best, E. N. (1976) Null space and phrase resetting curves for the Hodgkin-Huxley equations. PhD thesis, Purdue University.

Betz, A., Becker, J. U. (1975) Phase dependent phase shifts induced by pyruvate and acetaldehyde in oscillating NADH of yeast cells. *J Interdiscipl Cycle Res* 6, 167–173.

Betz, A., Chance, B. (1965) Phase relationship of glycolytic intermediates in yeast cells with oscillatory metabolite control. *Arch Biochem Biophys* 109, 585–594.

Betzig, E., Patterson, G. H., Hess, H. F. (2006) Imaging intracellular fluorescent proteins at nanometer resolution. *Science* 313, 1642–1645.

Biagioni, P., Huang, J.-S., Hecht, B. (2012) Nano-antennas for visible and infrared radiation. *Rep Prog Phys* 75, 024402.

Biasiucci, A. et al. (2018) Brain-actuated functional electrical stimulation elicits lasting arm motor recovery after stroke. *Nat Commun* 9, 2421. doi:10.1038/s41467-018-04673-z.

Bibbig, A., Faulkner, H. J., Whittington, M. A., Traub, R. D. (2001) Self-organized synaptic plasticity contributes to the shaping of γ, β oscillations in vitro. *J Neurosci* 21, 9053–9067.

Bibbig, A., Traub, R. D., Whittington, M. A. (2002) Long-range synchronization of gamma and beta oscillations and the plasticity of excitatory and inhibitory synapses: A network model. *J Neurophysiol* 8, 1634–1654.

Biener, G., Niv, A., Kleiner, V., Hasman, E. (2005) Geometrical phase image encryption obtained with space-variant subwavelength gratings. *Opt Lett* 30, 1096–1098.

Biktashev, V. N., Holden, A. V., Mironov, S. F., Pertsov, A. M., Zaitsev, A. V. (1999) Three-dimensional aspects of reentry in experimental and numerical models of ventricular fibrillation. *Int J Bifurc Chaos* 9, 695–704.

Billings, R. J. (1989) The origin of occipital lambda waves in man. *Electroencephalogr Clin Neurophysiol* 72, 95–113.

Binnig, G. et al. (2002) Win machines start to think like humans?-Artificial versus natural Intelligence. *Europhys News* 33(2), 44–47.

Binzegger, T., Douglas, R. J., Martin, K. A. (2005) Axons in cat visual cortex are topologically self-similar. *Cereb Cortex* 15, 152–165.

Birbaumer, N. (1970) The EEG of congenitally blind adults. *Electroencephalogr Clin Neurophysiol* 29, 318.

Bisoyi, H. K., Li, Q. (2016) Light-directed dynamic chirality inversion in functional self-organized helical superstructures. *Angew Chem Int Ed Engl* 55(9), 2994–3010. doi:10.1002/anie.201505520.

Blank, M., Goodman, R. (2011) DNA is a fractal antenna in electromagnetic fields. *Int J Radiat Biol* 87(4), 409–415.

Blaustein, M. W. et al. (2017) Cliques of neurons bound into cavities provide a missing link between structure and function. *Front Comput Neurosci* 11(48), 1–16.

Bliss, T. V., Collingridge, G. L. (1993) A synaptic model of memory: Long-term potentiation in the hippocampus. *Nature* 361, 31–39.

Bloch, F. (1946) Nuclear induction. *Phys Rev* 70(7–8), 460–474.

Bloch, S. (1997) On parallel hierarchies and R_k^i. *Ann Pure Appl Logic* 89(2–3), 231–273.

Blum, K. I., Abbott, L. F. (1996) A model of spatial map formation in the hippocampus of the rat. *Neural Comput* 8, 85–93.

Bocchieri, P., Loinger, A. (1957) Quantum recurrence theorem. *Phys Rev* 107(2), 337–338.

Boddy, K. K., Carroll, S. M. (2013) Can the Higgs Boson save us from the menace of the Boltzmann brains? *arXiv* preprint arXiv:1308.4686.

Bohm, D. (1952) A suggested interpretation of quantum theory in terms of hidden variables. *Phys Rev* 85, 166–193.

Bohr, N., (1935) Can quantum-mechanical description of physical reality be considered complete? *Phys Rev* 48, 696–702.

Bonanno, C., Mega, M. S. (2004) Toward a dynamical model for prime numbers. *Chaos Soliton Fract* 20, 107–118.

Bongard, J., Zykov, V., Lipson, H. (2006) Resilient machines through continuous self-modeling. *Science* 314, 1118–1121.

Bonhoeffer, T., Grinvald, A. (1991) Iso-orientation domains in cat visual cortex are arranged in pinwheel-like patterns. *Nature* 353, 429–431.

Boraud, T., Brown, P., Goldberg, J. A., Graybiel, A. M., Magill, P. J. (2005) Oscillations in the basal ganglia: The good, the bad, and the unexpected. In: *The Basal Ganglia*, J. P. Bolam, C. A. Ingham, P. J. Magill, Eds. Springer, New York, Volume 7, pp. 3–24.

Boring, E. G. (1942) *Sensation and Perception in the History of Experimental Psychology*, R. M. Elliott, Ed. Appleton-Century-Crofts, New York, pp. 644.

Boulias, K., Lieberman, J., Greer, E. C. (2016) An epigenetic clock measure accelerated aging in treated HIV infection. *Mol Cell* 62, 153–155.

Bourbaki, N. (1989) *Lie Groups and Lie Algebras*. Springer, New York, Chapter 1–3.

Bourret, A., Lincoln, R. G., Carpenter, B. H. (1969) Fungal endogenous rhythms expressed by spiral figures. *Science* 166, 763–764.

Bourret, J. A. (1971) Modification of the period of a non-circadian rhythm in *Nectria cinnabarina. Plant Physiol* 47, 682–684.

Boyken, S. C. et al. (2016) De novo design of protein homo-oligomers with modular hydrogen-bond network–mediated specificity. *Science* 352(6286), 680–687.

Bragin, A. G., Vinogradova, O. S. (1983) Comparison of neuronal activity in septal and hippocampal grafts developing in the anterior eye chamber of the rat. *Brain Res* 312(2), 279–286.

Braginsky, V. (1980) Quantum nondemolition measurement. *Science* 209(4456), 547–557.

Braitenberg, V., Braitenberg, C. (1979) Geometry of orientation columns in the visual cortex. *Biol Cybern* 33, 179–186.

Braitenberg, V., Schütz, A. (1998) *Cortex: Statistics and Geometry of Neural Connectivity*. Springer, Berlin, Germany.

Brambilla, M., Battipede, F., Lugiato, L. A., Penna, V., Prati, F., Tamm, C., Weiss, C. O. (1991) Transverse laser patterns. I. phase singularity crystals. *Phys Rev A* 43, 5090–5113.

Branco, T., Clark, B. A., Häusser, M. (2010) Dendritic discrimination of temporal input sequences in cortical neurons. *Science* 329, 1671–1675.

Brascamp, J., Blake, R., Knapen, T. (2015) Negligible fronto-parietal BOLD activity accompanying unreportable switches in bistable perception. *Nat Neurosci* 18, 1672–1678.

Brillouin, L. (1953) The negentropy principle of information. *J Appl Phys* 24, 1152.

Brink, R., Bronk, D., Larrabee, M. (1946) Chemical excitation of nerve. *Ann N Y Acad Sci* 47, 457–486.

Brinkmann, K. (1971) Metabolic control of temperature compensation in the circadian rhythm of *Euglena gracilis*. In: *Biochronometry*, M. Menaker, Ed. National Academy of Sciences, Washington, DC, pp. 567–593.

Brizhika, L., Eremkoa, A., Pietteb, B., Zakrzewskib, W. (2006) Charge and energy transfer by solitons in low-dimensional nanosystems with helical structure. *Chem Phys* 324, 259–266.

Broad, C. D. (1923) *Scientific Thought (pdf)*. Harcourt, Brace and Co, New York.

Broadhurst, D. J., Kreimer, D. (1995) Knots and Numbers in φ⁴ Theory to 7 Loops and Beyond. ArXiv. https://www.quantamagazine.org/strange-numbers-found-in-particle-collisions-20161115?utm_content=buffer29238&utm_medium=social&utm_source=facebook.com&utm_campaign=buffer.

Broda, H., Brugge, D., Homma, K, Hastings, J. W. (1986) Circadian communication between unicells? *Cell Biophys* 8, 47–63.

Brodmann, K. (1909) *Vergleichende Lokalisationslehre der Grosshirnrinde* (in German). Johann Ambrosius Barth, Leipzig, Germany.

Brooks, R. F. (1980) Mammalian cell cycles need 2 random transitions. *Cell* 19, 493.

Brown, B. H., Duthie, H. L., Horn, A. R., Smallwood, R. H. (1975) A linked oscillator model of electrical activity of human small intestine. *Am J Physiol* 229, 384–388.

Brown, E. R. et al. (1989) Effect of quasibound state life time on the oscillation power of resonant tunneling diodes. *Appl Phys Lett* 54, 934–936.

Brown, F. A. (1959) Living docks. *Science* 130, 1535–1544.

Brown, F. A., Webb, H. M. (1948) Temperature relations of an endogenous daily rhythmicity in the fiddler crab Uca. *Physiol Zool* 21, 371–381.

Brown, J. H., West, G. B. (2000) *Scaling in Biology. Santa Fe Institute.* Oxford University Press, Oxford.

Brown, N. A., Wolpert, L. (1990) The development of handedness in left/right asymmetry. *Development* 109, 1–9.

Brown, P. (2003) Oscillatory nature of human basal ganglia activity: Relationship to the pathophysiology of Parkinson's disease. *Mov Disord* 18(4), 357–363.

Brown, S., Martinez, M. J., Parsons, L. M. (2006) Music and language side by side in the brain: A PET study of the generation of melodies and sentences. *Eur J Neurosci* 23(10), 2791–2803. doi:10.1111/j.1460-9568.2006.04785.x.

Bruce, C. (1998) *The Einstein Paradox: And Other Science Mysteries Solved by Sherlock Holmes*, S. Davies, Ed. Ingram Publisher Services, New York, pp. 272.

Bruce, D. W., O'Hare, D., Walton, R. I. (2011) *Functional Oxides (Inorganic Materials Series Book 12).* Wiley, Chichester, UK.

Bruce, V. G., Pittendrigh, C. S. (1956) Temperature independence in a unicellular "clock." *Proc Nat Acad Sci* 42, 676–682.

Bruce, V. G., Wright, F., Pittendrigh, C. S. (1960) Resetting the sporulation rhythm in pilobolus with short light flashes of high intensity. *Science* 131, 728 730.

Brumberger, H. (1970) Rhythmic crystallization of poly-L-alanine. *Nature* 227, 490–491.

Brun, T. A. (2008) Computers with closed timelike curves can solve hard problems. http://arxiv.org/pdf/gr-qc/0209061v1.pdf (accessed on January 15, 2014).

Brunel, N., Chance, F. S., Fourcaud, N., Abbott, L. F. (2001) Effects of synaptic noise and filtering in the frequency response of spiking neurons. *Phys Rev Lett* 86, 2186–2189.

Brunel, N., Van Rossum, M. C. (2007) Lapicque's 1907 paper: From frogs to integrate-and-fire. *Biol Cybern* 97(5–6), 337–339.

Brunel, N., Wang, X. J. (2003) What determines the frequency of fast network oscillations with irregular neural discharges? I. Synaptic dynamics and excitation-inhibition balance. *J Neurophysiol* 90, 415–430.

Brunisholz, R. A., Zuber, H. (1992) Structure, function and organization of antenna polypeptides and antenna complexes from the three families of Rhodospirillaneae. *J Photochem Photobiol B* 15(1–2), 113–140.

Brunisholz, R. A., Zuber, H. (1992) Structure, function and organization of antenna polypeptides and antenna complexes from the three families of Rhodospirillaneae. *J Photochem Photobiol B* 15(1–2), 113–140.

Bruno, P. (2013) Impossibility of spontaneously rotating time crystals: A no-go theorem. *Phys Rev Lett* 111, 070402.

Bryant, S. V., French, V., Bryant, P. J. (1981) Distal regeneration and symmetry. *Science* 212, 993–1002.

Bub, G. et al. (1998) Bursting calcium rotors in cultured cardiac myocyte monolayers. *PNAS* 95, 10283.

Bucher, B. et al. (2016) Direct evidence of octupole deformation in neutron-rich. *Phys Rev Lett* 116, 112503.

Buckminster Fuller, R. (1975–1979) *Synergetics: Explorations in the Geometry of Thinking*, Vols 1 and 2. Macmillan Publishing Company, New York.

Buhusi, C. V., Meck, W. H. (2005) What makes us tick? Functional and neural mechanisms of interval timing. *Nat Rev Neurosci* 6, 755–765.

Buisson, J.-C. (2004) A rhythm recognition computer program to advocate interactivist perception. *Cogn Sci* 28, 75–88.

Bullier, J., Nowak, L. G. (1995) Parallel versus serial processing: New vistas on the distributed organization of the visual system. *Curr Opin Neurobiol* 5, 497–503.

Bullmore, R., Sporns, O. (2009) Complex brain networks: Graph theoretical analysis of structural and functional systems. *Nat Rev Neurosci* 10, 186.

Bulsara, A. R., Gammaitoni, L. (1996) Tuning in to noise. *Phys Today* 49, 39–45.

Bunning, E. (1964, 1973) *The Physiological Clock.* Springer Verlag, New York.

Buño, W. Jr., Velluti, J. C. (1977) Relationships of hippocampal theta cycles with bar pressing during self-stimulation. *Physiol Behav* 19, 615–621.

Buonomano, D. (2017) *Your Brain Is a Time Machine: The Neuroscience and Physics of Time.* W. W. Norton & Company, New York.

Buracas, G. T., Zador, A. M., DeWeese, M. R., Albright, T. D. (1998) Efficient discrimination of temporal patterns by motion-sensitive neurons in primate visual cortex. *Neuron* 20, 959–969.

Burke, J. (1985) *The Day the Universe Changed.* Little Brown, Boston, MA.

Burlando, B. (1993) The fractal geometry of evolution. *J Theor Biol* 163(2), 161–172.

Burresi, M. et al. (2009) Probing the magnetic field of light at optical frequencies. *Science* 326(5952), 550–553.

Burton, A. C. (1971) Cellular communication, contact inhibition, cell docks, and cancer: The impact of the work and ideas of W. R. Loewenstein. *Perspect Biol Med* 14, 301–318.

Butterfield, J. (1999) *The Arguments of Time.* Oxford University Press, Oxford.

Buzatu, S. (2009) The temperature-induced changes in membrane potential. *Riv Biol* 102(2), 199–217.

Buzek, V., Derka, R., Massar, S. (1999) Optimal quantum clocks. *Phys Rev Lett* 82, 2207–2210.

Buzsáki, G. (1989) Two-stage model of memory trace formation: A role for "noisy" brain states. *Neuroscience* 31, 551–570.

Buzsáki, G. (2005) Theta rhythm of navigation: Link between path integration and landmark navigation, episodic and semantic memory. *Hippocampus* 15, 827–840.

Buzsáki, G. (2006) *The Rhythms of the Brain.* Oxford University Press, New York.

Buzsáki, G., Bragin, A., Chrobak, J. J., Nádasdy, Z., Sík, A., Ylinen, A. (1994a) Oscillatory and intermittent synchrony in the hippocampus: Relevance for memory trace formation. In: *Temporal Coding in the Brain*, G. Buzsàki, R. R. Llinás, W. Singer, A. Berthoz, Y. Christen, Eds. Springer-Verlag, Berlin, Germany, pp. 145–172.

Buzsáki, G., Horváth, Z., Urioste, R., Hetke, J., Wise, K. (1992) High-frequency network oscillation in the hippocampus. *Science* 256, 1025–1027.

Buzsáki, G., Chrobak, J. J. (1995) Temporal structure in spatially organized neuronal ensembles: A role for interneuronal networks. *Curr Opin Neurobiol* 5, 504–510.

Buzsáki, G., Llinás, R. R., Singer, W., Berthoz, A., Christen, Y. (1994b) *Temporal Coding in the Brain.* Springer-Verlag, Berlin, Germany.

Buzsáki, G., Moser, E. I. (2013) Memory, navigation and theta rhythm in the hippocampal-entorhinal system. *Nat Neurosci* 16, 130–138. doi:10.1038/nn.3304.

Cador, M., Robbins, T. W., Everitt, B. J. (1989) Involvement of the amygdala in stimulus-reward associations: Interaction with the ventral striatum. *Neuroscience* 30, 77–86. doi:10.1016/0306-4522(89)90354-0.

Callender, C., Quinn, I., Tymoczko, D. (2008) Generalized voice-leading spaces. *Science* 320(5874), 346–348.

Campagnolo, M. L., Moore, C., Costa, J. F. (2000) An analog characterization of the subrecursive functions. In: *Proceedings of the 4th Conference on Real Numbers and Computers*, Odense University, pp. 91–109.

Canavati, S. E. et al. (2017) Maximizing research study effectiveness in malaria elimination settings: A mixed methods study to capture the experiences of field-based staff. *Malaria J* 16, 362.

Cao, T. Y., Schweber, S. S. (1993) The conceptual foundation and the philosophical aspect of the renormalization theory. *Synthese* 97, 33–108.

Caplow, M., Shanks, J., Ruhlen, R. L. (1988) Temperature jump studies of microtubule dynamic instability. *J Biol Chem* 263, 10344–10352.

Carena, M. et al. (2018) Probing the electroweak phase transition via enhanced di-Higgs boson production. arXiv:1801.00794v2.

Cariani, P. (1994) As if time really mattered: Temporal strategies for neural coding of sensory information. *Commun Cognit Artif Intell* 12, 161–229.

Cariani, P. (1995) As if time really mattered: Temporal strategies for neural coding of sensory information. *Commun Cognit Artifi Intell* 12(1–2), 161–229; Reprinted in Origins: Brain and Self-Organization, K. Pribram, Ed. (Lawrence Erlbaum, Hillsdale, NJ, 1994), pp. 208–252.

Cariani, P. (1999) Temporal coding of periodicity pitch in the auditory system: An overview. *Neural Plast* 6, 147–172.

Cariani, P. A., Delgutte, B. (1996) Neural correlates of the pitch of complex tones. II. Pitch shift, pitch ambiguity, phase-invariance, pitch circularity, and the dominance region for pitch. *J Neuro-Physiol* 76, 1717–1734.

Carnot, S., Thurston, R. H. (1890) *Reflections on the Motive Power of Heat and on Machines Fitted to Develop That Power*. John Wiley & Sons, New York.

Carpenter, G. A., Grossberg, S. (2003) Adaptive resonance theory archived 2006-05-19 at the wayback machine. In: *The Handbook of Brain Theory and Neural Networks*, 2nd ed, M. A. Arbib, Ed. MIT Press, Cambridge, MA, pp. 87–90.

Carr, C. E. (1993) Processing of temporal information in the brain. *Annu Rev Neurosci* 16, 223–243.

Carr, C. E., Konishi, M. (1990) A circuit for detection of interaural time differences in the brain stem of the barn owl. *J Neurosci* 10, 3227–3246.

Carroll, J. B. (1993) *Human Cognitive Abilities: A Survey of Factor-Analytic Studies*. Cambridge University Press, Cambridge.

Carroll, S. M. (2017) Why Boltzmann brains are bad. arXiv preprint arXiv:1702.00850.

Cartwright, N. (1983) *How the Laws of Physics Lie*. Clarendon Press, Oxford. doi:10.1093/0198247044.001.0001.

Cartwright, N. (1997) Where do laws of nature come from? *Dialectica* 51, 65–78. doi:10.1111/j.1746-8361.1997.tb00021.x.

Cathey, J. W. T. (1970) Phase holograms, phase-only holograms, and kinoforms. *Appl Opt* 9, 1478–1479.

Cavanagh, J. B. (1984) The problems of neurons with long axons. *Lancet* 1(8389), 1284–1287.

Cayley, A. (1845) On Jacobi's elliptic functions, in reply to the Rev.; and on quaternions. *Philos Mag* 26, 208–211.

Célérier, M., Nottale, L. (2003) A scale-relativistic derivation of the Dirac equation. *Electromagnetic Phenomena* 3, 70–80.

Celli, J. P. et al. (2009) *Helicobacter pylori* moves through mucus by reducing mucin viscoelasticity. *Proc Natl Acad Sci USA* 106, 14321.

Cenquizca, L. A., Swanson, L. W. (2007) Spatial organization of direct hippocampal field CA1 axonal projections to the rest of the cerebral cortex. *Brain Res Rev* 56, 1–26. doi:10.1016/j.brainresrev.2007.05.002.

Ceriani, M. F., Darlington, T. K., Staknis, D., Mas, P., Petti, A. A., Weitz, C. J., Kay, S. A. (1999) Light-dependent sequestration of timeless by cryptochrome. *Science* 285, 553–556.

Chandrashekaran, M. K. (1974) Phase shifts in the *Drosophila pseudoobscura* circadian rhythm evoked by temperature pulses of varying durations. *J Interdiscipl Cycle Res* 5, 371–380.

Chandrashekaran, M. K., Engelmann, W. (1973) Early and late subjective night phases of the *Drosophila pseudoobscura* circadian rhythm require different energies of blue light for phase shifting. *Z Naturforsch* 28, 750–753.

Chandrashekaran, M. K., Engelmann, W. (1973) Early and late subjective night phases of the *Drosophila pseudoobscura* circadian rhythm require different energies of blue light for phase shifting. *Z Naturforsch* 28, 750–753.

Chandrashekaran, M. K., Loher, W. (1969) The relationship between the intensity of light pulses and the extent of phase shifts of the circadian rhythm in the exclusion rate of *Drosophila pseudoobscura*. *J Exp Zool* 172, 147–152.

Chen, J., Kanai, Y., Cowan, N. J., Hirokawa, N. (1992) Projection domains of MAP2 and tau determine spacings between microtubules in dendrites and axons. *Nature* 360, 674–677.

Chen, W. R., Shen, G. Y., Shepherd, G. M., Hines, M. L., Midtgaard, J. (2002) Multiple modes of action potential initiation and propagation in mitral cell primary dendrite. *J Neurophysiol* 88, 2755–2764.

Chen, Y., Ding, M., Kelso, J. A. S. (2001) Origins of timing errors in human sensorimotor coordination. *J Mot Behav* 33, 3–8.

Chen, Y., Mingzhou, D., Kelso, J. A. S. (1997) Long memory processes ($1/f^{\alpha}$ type) in human coordination. *Phys Rev Lett* 79, 4501–4504.

Chennu, S. et al. (2014) Spectral signatures of reorganized brain networks in disorders of consciousness. *PLoS Comput Biol* 10(10), e1003887.

Chiel, H. J., Beer, R. D. (1997) The brain has a body: Adaptive behavior emerges from interactions of nervous system, body and environment. *Trend Neurosci* 20, 553–557.

Chklovskii, D. B., Mel, B. W., Svoboda, K. (2004) Cortical rewiring and information storage. *Nature* 431, 782–788.

Cho, Y. Scanning nonlinear dielectric microscopy with super-high resolution. *Jpn J Appl Phys* 46(1), 7B.

Cho, Y., Kirihara, A., Saeki, T. (1996) Scanning non-linear dielectric microscope. *Rev Sci Instrum* 67, 2297.

Choe, J. et al. (2016) Transcranial direct current stimulation modulates neuronal activity and learning in pilot training. *Front Hum Neurosci* 10(34), 1–25.

Choi, S. et al. (2017) Observation of discrete time-crystalline order in a disordered dipolar many-body system. *Nature* 543, 221–225.

Choi, S. et al. Observation of discrete time-crystalline order in a disordered dipolar many-body system. *Nature* 543, 221–225.

Chow, S. S. et al. (2004) Adaptive radiation from resource competition in digital organisms. *Science* 305, 84–86.

Christiansen, P. L. (1997) Soliton analysis in complex molecular systems: A zig-zag chain. *J Comp Phys* 134, 108.

Christiansen, P. L., Savin, A. V., Zolotaryuk, A. V. (1997) Soliton analysis in complex molecular systems: A zig-zag chain. *J Comput Phys* 134, 108–121.

Christoph, K., Martin, K., Nikita, A., Siegfried, B. (2010) Röntgen's electrode-free elastomer actuators without electromechanical pull-in instability. *PNAS* 107(10), 4505–4510.

Chua, L. O. (1971) Memristor: The missing circuit element. *IEEE Trans Circuits Theory* 18(5), 507–519.

Ciocchi, S. et al. (2015) Brain computation. Selective information routing by ventral hippocampal CA1 projection neurons. *Science* 348, 560–563. doi:10.1126/science.aaa3245.

Cipra, B. A. (2003) A healthy heart is a fractal heart. *SIAM News* 36, 7.

Clark, L. W., Feng, L., Chin, C. (2016) Universal space-time scaling symmetry in the dynamics of bosons across a quantum phase transition. *Science* 354, 606–610.

Clark, R. L., Steck, T. L. (1979) Morphogenesis in *Dictyostelium*: An orbital hypothesis. *Science* 204, 1163–1168.

Clay, J. R., Brochu, R. M., Shrier, A. (1990) Phase resetting of embryonic chick atrial heart cell aggregates. *Biophys J* 58, 609–621.

Clay, J. R., Guevara, M. R., Shrier, A. (1984) Phase resetting of the rhythmic activity of embryonic heart cell aggregates: Experiment and theory. *Biophys J* 45, 699–714.

Cohen, D. (1968) Magnetoencephalography: Evidence of magnetic fields produced by alpha-rhythm currents. *Science* 161, 784–786.

Cohen, M. H. (1972) Models of clocks and maps in developing organisms. In: *Some Mathematical Questions in Biology*, J. Cowan, Ed. American Mathematical Society, Providence, RI, pp. 3–32.

Coldea, R. et al. (2010) Quantum criticality in an Ising Chain: Experimental evidence for emergent E8 symmetry. *Science*, 327, 177–180.

Collins, B. (2000) Visualization: From biology to culture. *Bridges Proc* 309–315.

Collins, M. A., Ross, J. (1978) Chemical relaxation pulses and waves. Analysis of lowest order multiple time scale expansion. *J Chem Phys* 68, 3774–3784.

Connor, J. A. (1978) Slow repetitive activity from fast conductance changes in neurons. *Fed Proc* 37, 2139–2145.

Connors, B. W., Regehr, W. G. (1996) Neuronal firing: Does function follow form? *Curr Biol* 6, 1560–1562.

Contreras, D., Destexhe, A., Sejnowski, T. J., Steriade, M. (1996) Control of spatiotemporal coherence of a thalamic oscillation by corticothalamic feedback. *Science* 274, 771–774.

Contreras, D., Destexhe, A., Steriade, M. (1997) Spindle oscillations during cortical spreading depression in naturally sleeping cats. *Neuroscience* 77, 933–936.

Conway, J. H., Smith, D. A. (2003) *On Quaternions and Octonions: Their Geometry, Arithmetic, and Symmetry*. A. K. Peters, Ltd, Natick, MA.

Cook, L. S. (1887) *Geometrical Psychology, or, the Science of Representation: An Abstract of the Theories and Diagrams of B.W. Betts*. G. Redway, London, UK, pp. 194.

Cooke, J., Zeeman, E. C. (1976) A clock and wave front model for control of the number of repeated structures during animal morphogenesis. *J Theor Biol* 58, 455–476.

Cooper, M. J., Rowson, E. A. (1975) Control of the oestrous cycle in *Friesian heifers* with ICI 80996. *Ann Biol Anim Biochem Biophys* 15, 427–436.

Copeland, B. J., Proudfoot, D. (1999) Alan Turing's forgotten ideas in computer science. *Sci Am* 99–103.

Copie, F. et al. (2019) Interplay of polarization and time-reversal symmetry breaking in synchronously pumped ring resonators. *Phys Rev Lett* 122, 013905.

Corbel, S., Cerda, J., Sautet, P. (1999) Ab initio calculation of scanning tunneling microscopy images within a scattering formalism. *Phys Rev B* 60, 1989.

Cote, G. (1991) Type two resetting of the *Euglena gracilis* circadian rhythm? *J Biol Rhy* 6, 367–369.

Coullet, P., Gil, L., Rocca, F. (1989) Optical vortices. *Opt Comm* 73, 403–408.

Courtemanche, M., Glass, L., Keener, J. P. (1993) Instabilities of a propagating pulse in a ring of excitable media. *Phys Rev Lett* 70, 2182–2185.

Courtland, R. (2018) Bias detectives: The researchers striving to make algorithms fair. *Nature* 558, 357–360.

Cowan, N. (2001) The magical number 4 in short-term memory: A reconsideration of mental storage capacity. *Behav Brain Sci* 24(1), 87–114.

Cramer, J. G. (1980) Generalized absorber theory and the Einstein-Podolsky-Rosen paradox. *Phys Rev D* 22, 362–376.

Cranford, T. W. (2015) Fin whale sound reception mechanisms: Skull vibration enables low-frequency hearing. *PLoS One* 10(3), e0116222.

Cross, M. C. et al. (2004) Synchronization by nonlinear frequency pulling. *Phys Rev Lett* 93, 224101.

Crosthwaite, S. K., Loros, J. L., Dunlap, J. C. (1995) Light-induced resetting of a circadian dock is mediated by a rapid increase in frequency transcript. *Cell* 81, 1003–1012.

Csermely, P. (2005) Weak links: A universal key of network diversity and stability. http://www.weaklink.sote.hu/weakbook.html.

Cubitt, T. S., Garcia, D. P., Wolf, M. M. (2015) Undecidability of the spectral gap. *Nature* 528, 207–211.

Cui, L. et al. (2016) Enhancement of magnetic dipole emission at yellow light with polarization-independent hexagonally arrayed nanorods optical metamaterials. *Opt Mater Express* 6(4), 1151–1160.

Cully, A., Clune, J., Tarapore, D., Mouret, J.-B. (2015) Robots that can adapt like animals. *Nature* 521, 503–507.

Curio, G., Mackert, B. M., Burghoff, M., Koetitz, R., Abraham-Fuchs, K., Harer, W. (1994) Localization of evoked neuromagnetic 600 Hz activity in the cerebral somatosensory system. *Electroencephalogr Clin Neurophysiol* 91, 483–487.

Curto, A. G. et al. (2013) Multipolar radiation of quantum emitters with nanowire optical antennas. *Nat Commun*. doi:10.1038/ncomms2769.

Curto, C. (2017) What can the topology tell us about the neural code? *Bull Am Math Soc* 54(1), 63–78.

D'Este, E. et al. (2015) STED nanoscopy reveals the ubiquity of subcortical cytoskeleton periodicity in living neurons. *Cell Rep* 10, 1246–1251.

D'Este, E. et al. (2016) Subcortical cytoskeleton periodicity throughout the nervous system. *Sci Rep* 6, 22741.

Damasio, A. R. (1989) Time-locked multiregional retroactivation: A systems-level proposal for the neural substrates of recall and recognition. *Cognition* 33, 25–62.

Damsma, A., Rijn, H. V. (2017) Pupillary response indexes the metrical hierarchy of unattended rhythmic violations. *Brain Cogn* 111, 95–103.

Daneshmand, F., Amabili, M. (2012) Coupled oscillations of a protein microtubule immersed in cytoplasm: An orthotropic elastic shell modelling. *J Biol Phys* 38(3), 429–448.

David, K., Alexandre, P. (2004) The Bayesian brain: The role of uncertainty in neural coding and computation. *Trends Neurosci* 27(12), 2004.

Davies, P. C. W. (1974) *The Physics of Time Asymmetry*. Surrey University Press, London, UK.

de Finetti, B. (1937) Translated into English as "Foresight: Its logical laws, its subjective sources." In: *Studies in Subjective Probability*, H. E. Kyburg, Jr., H. E. Smokler, Eds. Wiley, New York (1964), pp. 93–158.

De Ruyter van Steveninck, R. R., Lewen, G. D., Strong, S. P., Koberle, R., Bialek, W. (1997) Reproducibility and variability in neural spike trains. *Science* 275, 1805–1808.

de Wit, B., Nicolai, H. (1982) N = 8 supergravity with local SO(8) × SU(8) invariance. *Phys Lett B* 108(4–5), 285–290.

Deadwyler, S. A., Hampson, R. E. (1997) The significance of neural ensemble codes during behavior and cognition. *Annu Rev Neurosci* 20, 217–244.

Debanne, D. (2004) Information processing in the axon. *Nat Rev Neuro* 5, 304–316.

Deboer, T., Vansteensel, M. J., Detari, L., Meijer, J. H. (2003) Sleep states alter activity of suprachiasmatic nucleus neurons. *Nat Neurosci* 6, 1086–1090.

del Rio, L., Åberg, J., Renner, R., Dahlsten, O., Vedral, V. (2011) The thermodynamic meaning of negative entropy. *Nature* 474, 61–63.

Delahaye, J. P. (1989) Chaitin's equation; an extension of Gödel's theorem. *Notices Amer Math Soc* 36, 948–987.

Della Valle, G., Sondergaard, T., Bozhevolnyi, S. I. (2008) Plasmon-polariton nanostrip resonators: From visible to infra-red. *Opt Express* 16, 6867–6876.

Delli Pizzi, S. et al. (2017) Functional and neurochemical interactions within the amygdala-medial prefrontal cortex circuit and their relevance to emotional processing. *Brain Struct Func* 222, 1267–1279. doi:10.1007/s00429-016-1276-z.

Demuro, A., Parker, I. (2005) Optical patch-clamping: Single-channel recording by imaging Ca^{2+} flux through individual muscle acetylcholine receptor channels. *J Gen Physiol* 126, 179–192.

Denisov, V. et al. (1996) Using buried water molecules to explore the energy landscape of proteins. *Nat Struct Biol* 3, 505–509.

Denning, P. J. (1989) Bayesian learning. *Am Sci* 77, 216–218.

Dent, E. W., Baas, P. W. (2014) Microtubules in neurons as information carriers. *J Neurochem* 129(2), 235–239.

Destexhe, A., Contreras, D., Sejnowski, T. J., Steriade, M. (1994) A model of spindle rhythmicity in the isolated thalamic reticular nucleus. *J Neurophysiol* 72, 803–818.

Devlin, R. C. et al. (2017) Arbitrary spin-to-orbital angular momentum conversion of light. *Science* 1–9.

Dhannananda, S., Feldman, J. F. (1979) Assay of·spatial distribution of circadian dock phase in aging cultures of *Neurospora crassa*. *Plant Physiol* 63, 1049–1054.

Dharmananda, S. (1980) Studies of the circadian clock of *Neurospora crassa*. Light-induced phase shifting, PhD dissertation, University of California, Santa Cruz, CA.

Di Leva, A. (2016) *The Fractal Geometry of the Brain*. Springer Science+Business Media, New York.

Di Ventra, M., Pershin, Y. V. (2013) On the physical properties of memristive, memcapacitive and meminductive systems. *Nanotechnology* 24(25), 255201.

Dias, B., Ressler, K. J. (2013) Parental olfactory experience influences behavior and neural structure in subsequent generations. *Nat Neurosci* 17, 89–96.

Dickau, R. (1999) Visualizing combinatorial enumeration. *Math Educ Res* 8, 11–18.

Ding, M., Chen, Y., Kelso, J. A. (2002) Statistical analysis of timing errors. *Brain Cogn* 48, 98–106.

Dirac, P. A. M. (1931) Quantised singularities in the electromagnetic field. *Proc Royal Soc A* 133(821), 60–72.

Doeschi, A. et al. (2004) Assessing cellular automata based models using partial differential equations. *Math Comp Mod* 40, 977.

Dopico, X. C. et al. (2015) Widespread seasonal gene expression reveals annual differences in human immunity and physiology. *Nat Commun* 6, 7000.

Dowle, M., Mantel, R. M., Barkley, D. (1997) Fast simulations of waves in three-dimensional excitable media. *Int J Bif Chaos* 7, 2529–2545.

Doya, K. (2007) *Bayesian Brain: Probabilistic Approaches to Neural Coding*. MIT Press, Cambridge, MA.

Doyle, D. A. et al. (1998) The structure of the potassium channel: Molecular basis of K+ conduction and selectivity. *Science* 280(5360), 69–77.

Draganski, B., Gaser, C., Busch, V., Schuierer, G., Bogdahn, U., May, A. (2004) Neuroplasticity: Changes in grey matter induced by training. *Nature* 427, 311–312.

Du, Q., Freysz, E., Shen, Y. R. (1994) Surface vibrational spectroscopic studies of hydrogen bonding and hydrophobicity. *Science* 264(5160), 826–828.

Dubois, D. M. (1992) The hyperincursive fractal machine as a quantum holographic brain. *Commun Cognit Artif Intell* 9(4), 335–372.

Dubois, D. M., Resconi, G. (1994) Hyperincursive fractal machine beyond the Turing machine. *Advances in Cognitive Engineering and Knowledge-based Systems*. International Institute for Advanced Studies in Systems Research and Cybernetics abbreviation, pp. 212–216.

Dudek, F. E., Yasumura, T., Rash, J. E. (1998) Non-synaptic mechanisms in seizures and epileptogenesis. *Cell Biol Int* 22, 11–12.

Duke, A. R. et al. (2013) Transient and selective suppression of neural activity with infrared light. *Sci Rep* 3, 2600.

Dummett, M. A. E. (1954) Can an effect precede its cause? *Proc Aristotelian Soc Suppl Vol* 38, 27–44.

Dunlap, K. (1910) Reactions on rhythmic stimuli, with attempt to synchronize. *Psychol Rev* 17, 399–416.

Durante, D., Dunson, D. B. (2018) Bayesian inference and testing of group differences in brain networks. *Bayesian Anal* 13, 29–58.

Durr, D., Goldstein, S., Zanghi, N. (1992) Quantum equilibrium and the origin of absolute uncertainty. *J Stat Phys* 67, 843–907.

Eagles, T. H. (1885) *Constructive Geometry of Plane Curves*. Macmillan, London, UK, pp. 348–354.

Earman, J. (2006) The implications of general covariance for the ontology and ideology of spacetime. In: *Philosophy and Foundations of Physics*. The Ontology of Spacetime, D. Dieks, Ed. Elsevier, B.V., Amsterdam, the Netherlands.

Ebbinghaus, S. et al. (2007) An extended dynamical hydration shell around proteins. *Proc Natl Acad Sci USA* 104(52), 20749–20752.

Eccles, J. C., Ito, M., Szentágothai, J. (1967) *The Cerebellum as a Neuronal Machine*. Springer-Verlag, Berlin, Germany.

Eckhorn, R. (2000) Cortical processing by fast synchronization: High frequency rhythmic and non-rhythmic signals in the visual cortex point to general principles of spatiotemporal coding. In: *Time and the Brain*, R. Miller, Ed. Harwood Academic Publishers, Australia, pp. 169–201.

Edelman, G. M. (1987) *Neural Darwinism*. Basic Books, New York.

Edmunds, L. N. (1977) Clocked cell cycle clocks. *Waking Sleeping* 1, 227–252.

Edmunds, L. N., Adams, K. J. (1981) Clocked cell cycle docks. *Science* 211, 1002–1013.

Efimov, I. R., Cheng, Y., Van Wagoner, D., Mazgalev, T. N., Tchou, P. J. (1998) Virtual electrode induced phase singularity: A basic mechanism of defibrillation failure. *Circ Res* 82, 918–925.

Eguiluz, V. M. et al. (2005) Scale free brain functional networks. *Phys Rev Lett* 94, 018102.

Eiben, A. E., Smith, J. (2015) From evolutionary computation to the evolution of things. *Nature* 521, 476–482.

Einstein, A. (1963) *Albert Einstein Theoretical Physicist*. Macmillan, New York, p. 81.

El Naschie, M. S. (2003) VAK, vakuum fluctuation and the mass spectrum of high energy particle physics. *Chaos Soliton Fract* 17, 797–807.

El Naschie, M. S. (2004) The speed of the passing of time as yet another facet of cosmic dark energy. *Chaos Soliton Fract* 19, 209–236.

El Naschie, M. S. (2009) Deriving the curvature of fractal-Cantorian spacetime from first principles. *Chaos Soliton Fract* 41(5), 2635–2646.

Elble, R. J., Koller, W. C. (1990) *Tremor.* Johns Hopkins University Press, Baltimore, MD.

Eldridge, F. L., Paydarfar, D. (1989) Phase resetting of respiratory rhythm studied in a model of a limit cycle oscillator: Influence of stochastic processes. In: *Respiratory Control*, G. D. Swanson, F. S. Grodins, R. L. Hughson, Eds. Plenum Press, New York.

Eldridge, F. L., Paydarfar, D., Wagner, P. G., Dowell, R. T. (1989) Phase resetting of respiratory rhythm: Effect of changing respiratory drive. *Am J Physiol* 257, R271–R277.

Eliasmith, C. et al. (2012) A large-scale model of the functioning brain. *Science* 338, 1202.

Elliott, M. A., Williams, J. W. (1939) The dielectric behavior of solutions of the protein zein. *J Am Chem Soc* 61(3), 718–725.

Else, D. V., Bauer, B., Nayak, C. (2016) Floquet time crystals. *Phys Rev Lett* 117, 090402.

Else, D. V., Bauer, B., Nayak, C. (2017) Pre-thermal time crystal and floquet topological phases without disorder. *Phys Rev X* 7, 011026.

Elvira, B., Mari, T., Risto, N., Isabelle, P. (2006) Musical scale properties are automatically processed in the human auditory cortex. *Brain Res* 1117(1), 162–174.

Empedocles, S. A., Neuhauser, R., Bawendi, M. G. (1999) Three-dimensional orientation measurements of symmetric single chromophores using polarization microscopy. *Nature* 399, 126–130.

Engel, A. K., Fries, P., Singer, W. (2001) Dynamic predictions: Oscillations and synchrony in top-down processing. *Nat Rev Neurosci* 2, 704–716.

Engel, A. K., Konig, P., Singer, W. (1991) Direct physiological evidence for scene segmentation by temporal coding. *Proc Natl Acad Sci USA* 8, 9136–9140.

Engel, A. K., Singer, W. (2001) Temporal binding and the neural correlates of sensory awareness. *Trends Cogn Sci* 5, 16–25.

Engelmann, W., Eger, I., Johnsson, A., Karlsson, H. G. (1974) Effect of temperature pulses on the petal rhythm of Kalanchoe: An experimental and theoretical study. *Int J Chronobiol* 2, 347–358.

Engelmann, W., Karlsson, H. G., Johnsson, A. (1973) Phase shifts in the Kalanchoe petal rhythm caused by light pulses of different durations. *Int J Chronobiol* 1, 147–156.

Enright, J. T. (1980) Temporal precision in circadian systems: A reliable neuronal clock from unreliable components? *Science* 209, 1542–1545.

Enright, J. T., Winfree, A. T. (1987) Detecting a phase singularity in a coupled stochastic system. In: *Some Mathematical Questions in Biology: Circadian Rhythms*, G. A. Carpenter, Ed. American Mathematical Society, Providence, RI.

Erika, S., Nina, K. (2010) Auditory brain stem response to complex sounds: A tutorial. *Ear Hearing* 31(3), 302–24.

Erker, P. et al. (2017) Autonomous quantum clocks: Does thermodynamic limit our ability to measure time. *Phys Rev X* 7, 031022.

Ermentrout, G. B., Kleinfeld, D. (2001) Traveling electrical waves in cortex: Insight from phase dynamics and speculation on a computational role. *Neuron* 29, 33–44.

Ermentrout, G. B., Kopell, N. (1994) Learning of phase-lags in coupled neural oscillators. *Neural Comput* 6, 225–241.

Ermentrout, G. B., Kopen, N. (1984) Frequency plateaus in a chain of weakly coupled oscillators. *SIAM J Math* 15, 215–237.

Ethan, A. P., Yan, E. C. Y. (2017) The H_2O helix: The chiral water superstructure surrounding DNA. *ACS Cent Sci* 3(7), 683–685. doi:10.1021/acscentsci.7b00229.

Eujin, P. et al. (2017) A study of 4D printing and functionally graded additive manufacturing. *Assem Autom* 37, 147–153.

Fadeev, L., Niemi, A. J. (1997) Stable knot-like structures in classical field theory. *Nature* 387, 58–61.

Farver, O., Pecht, I. (1991) Electron transfer in proteins: In search of preferential pathways. *FASEB J* 5(11), 2554–2559.

Faure, C. et al. (2016) Experiments on metasurface carpet cloaking for audible acoustics. *Appl Phys Lett* 108, 064103. doi:10.1063/1.4941810.

Fedotov, V. A. et al. (2006) Asymmetric propagation of electromagnetic waves through a planar chiral structure. *Phys Rev Lett* 97, 167401.

Fedyanin, V. K., Yakushevich, L. V. (1981) The exciton spectrum of the alpha-helical protein molecule model. *J Theor Biol* 91, 1–11.

Feige, U., Goldwasser, S., Lovász, L., Safra, S., Szegedy, M. Approximating clique is almost NP-complete. In *Proceedings of the 32nd IEEE Symposium on Foundations of Computer Science*, San Juan, Puerto Rico, October 1–4, 1991, pp. 2–12. doi:10.1109/SFCS.1991.185341.

Ferbinteanu, J., Shapiro, M. L. (2003) Prospective and retrospective memory coding in the hippocampus. *Neuron* 40, 1227–1239.

Feyerabend, P. (1975) *Against Method.* New Left Books/Verso, London, UK.

Feynman, R. P. (1949) Space–time approach to quantum electronic. *Phys Rev* 76, 769.

Feynman, R. P. (1965) *The Character of Physical Law* (Modern Library). MIT Press, Cambridge, MA, p. 173.

Feynman, R., Hibbs, A. (1965) *Quantum Mechanics and Path Integrals.* McGraw-Hill, New York.

Field, M. D., Maywood, E. S., O'Brien, J. A., Weaver, D. R., Reppert, S. M., Hastings, M. H. (2000) Analysis of clock proteins in mouse SCN demonstrates phylogenetic divergence of the circadian clockwork and resetting mechanisms. *Neuron* 25(2), 437–447.

Field, R. J., Noyes, R M. (1972) Explanation of spatial band propagation in the Belousov. reaction. *Nature* 237, 390–392.

Fink, J. M. et al. (2017) Observation of the photon-blockade breakdown phase transition. *Phys Rev X* 7, 011012.

Fischer, P., Nelson, B., Yang, G.-Z. (2018) New materials for next-generation robots. *Sci Robot* 3, eaau0448.

Fisher, R. A. (1922) On the dominance ratio. *Proc R Soc Edinb* 42, 321–341.

Fisher, R. A. (1956) *Statistical Methods and Statistical Inference.* Haftier, New York, pp. 8–17.

Fleury, R., Alu, A. (2013) Quantum cloaking based on scattering cancellation. *Phys Rev B* 87, 045423.

Floch, J. M. et al. (2016) Towards achieving strong coupling in 3D-cavity with solid state spin resonance. *J Appl Phys* 119, 153901.

Fohlmeister, C., Gerstner, W., Ritz, R, van Hemmen, J. L. (1995) Spontaneous excitations in the visual cortex: Stripes, spirals, rings, and collective bursts. *Neural Comput* 7, 1046–1055.

Follesdal, D. (1975) Meaning and experience. In *Mind and Language*, S. Guttenplan, Ed. Clarendon, Oxford, pp. 25–44.

Forbes, T. (1999) Prime clusters and Cunningham chains. *Math Comput* 68, 1739–1748.

Forero, G. M., Gardner, A. (2018) Inference of ecological and social drivers of human brain-size evolution. *Nature* 557, 554–557.

Forrest, A. R. (1971) Computational geometry. *Proc R Soc London* 321, series 4, 187–195.

Forsberg, D. (2016) CO_2-evoked release of PGE2 modulates sighs and inspiration as demonstrated in brainstem organotypic culture. *eLife* 5, e14170.

Foster, R. G., Kreitzman, L. (2004) *Rhythms of Life: The Biological Clocks That Control the Daily Lives of Every Living Thing.* Yale University Press, New Haven, CT.

Frank, K. D., Zimmerman, W. F. (1969) Action spectra for phase shifts of a circadian rhythm in *Drosophila. Science* 163, 688–689.

Frank, L. M., Brown, E. N., Wilson, M. (2000) Trajectory encoding in the hippocampus and entorhinal cortex. *Neuron* 27, 169–178.

Fraser, A., Frey, A. H. (1968) Electromagnetic emission at micron wavelengths from active nerves. *Biophys J* 8, 731–734.

Freed, S., Weissman, S. I. (1941) Multiple nature of elementary sources of radiation—Wide-angle interference. *Phys Rev* 60, 440.

Freedman, M., Shokrian-Zini, M., Wang, Z. (2018) Quantum computing with octonions. arXiv:1811.08580.

Freeman, W. J. (2000) Perception of time and causation through the kinesthesia of intentional action. *Cogn Proc* 1, 18–34.

Freeman, W. J., Burke, B. C., Holmes, M. D. (2003) A periodic phase-resetting in scalp EEG of beta-gamma-oscillations by state transitions at alpha-theta-rates. *Hum Brain Mapp* 19, 248–272.

Freeman, W. J., Rogers, L. J. (2002) Fine temporal resolution of analytic phase reveals episodic synchronization by state transitions in EEGs. *J Neurophysiol* 87, 937–945.

Friston, K. (2010) The free energy principle: A unified brain theory? *Nat Rev Neurosci* 11. doi:10.1038/nrn2787.

Friston, K., Kilner, J., Harrison, L. (2006) A free energy principle for the brain. *J Physiol Paris* 100(1–3), 70–87.

Friston, K. J. (2011) Functional and effective connectivity: A review. *Brain Connect* 1(1), 13–36.

Fröhlich, H. (1968a) Bose condensation of strongly excited longitudinal electric modes. *Phys Lett A* 26, 402–403.

Fröhlich, H. (1968b) Long range coherence and energy storage in biological systems. *Int J Quantum Chem* 2, 641–649.

Fua, T.-M. et al. (2014) Sub-10-nm intracellular bioelectronic probes from nanowire–nanotube heterostructures. *Proc Natl Acad Sci* 111(4), 1259–1264.

Fuller, C. A., Sulzman, F. M., Moore-ede, M. C. (1978) Thermoregulation is impaired in an environment without circadian time cues. *Science* 199, 794–796.

Furey, C. (2015) Charge quantization from a number operator. *Phys Lett B* 742, 195–199.

Furey, C. (2018) Three generations, two unbroken gauge symmetries, and one eight-dimensional algebra. *Phys Lett B* 785, 84–89.

Furukawa, T. (1989) Ferroelectric properties of vinylidene fluoride copolymers. *Phase Transit* 18, 143–211.

Gabelli, J. et al. (2006) Violation of Kirchoff's laws for a coherent RC circuit. *Science* 313, 499.

Gage, F. H. (2002) Neurogenesis in the adult brain. *J Neurosci* 22, 612–613.

Galber, M. K., Hurst, G., Comi, T. J., Bhargava, R. (2018) Model-guided design and characterization of a high-precision 3D printing process for carbohydrate glass. *Addit Manuf* 22, 38–50.

Galil, Z. et al. (1987) An O(n3 log n) deterministic and an O(n3) Las Vegas isomorphism test for trivalent graphs. *J ACM* 34(3), 513–531.

Galve, F., Pachón, L. A., Zueco, D. (2010) Bringing entanglement to the high temperature limit. *Phys Rev Lett* 105, 180501–180504.

Ganeri, J. (2001) *Indian Logic: A Reader.* Routledge, London, UK, p. vii.

Ganguly, A. et al. (2015) A dynamic formin-dependent deep F-actin network in axons. *J Cell Biol* 210(3), 401–417.

Gao, Z. et al. (2018) A cortico-cerebellar loop for motor planning. *Nature* 563(7729), 113–116. doi:10.1038/s41586-018-0633-x.

Garceau, N. Y., Liu, Y., Loros, J. J., Dunlap, J. C. (1997) Alternative initiation and time-specific phosphorylation yield multiple forms of the essential dock protein Frequency. *Cell* 89, 469–486.

Garcia-Sanchez, J. L., Buño, W. Jr., Fuentes, J., Garcia-Austt, E. (1978) Non-rhythmical hippocampal units, theta rhythm and afferent stimulation. *Brain Res Bull* 3, 213–219.

Gardiner, J., Overall, R., Marc, J. (2010) The fractal nature of the brain: EEG data suggest that the brain function as a "quantum computer" in 5–8 dimensions. *NeuroQuantology* 8(2), 137–141.

Gardner, J. (2002) Assessing the computational potential of the eschaton: Testing the selfish biocosm hypothesis. *J Br Interplanet Soc* 55(7/8), 285–288.

Gardner, J. N. (2005) The physical constants as biosignature: An anthropic retrodiction of the selfish biocosm hypothesis. *Int J Astrobiol* 3, 229–236.

Gardner, M. (1980) Mathematical games. *Sci Am* 243, 20–44.

Garey, L. J. (2006) *Brodmann's Localisation in the Cerebral Cortex.* Springer, New York.

Garrison, J. R. (2015) Paracingulate sulcus morphology is associated with hallucinations in the human brain. *Nat Commun* 6, 8956.

Gawne, T. J., Richmond, B. J. (1993) How independent are the messages carried by adjacent inferior temporal cortical neurons? *J Neurosci* 13, 2758–2771.

Gebber, G. L., Zhong, S., Lewis, C., Barman, S. M. (1999) Human brain alpha rhythm: Nonlinear oscillations or filtered noise. *Brain Res* 818, 56–60.

Gedeon, T., Glass, I. (1998) Continuity of resetting curves for FitzHugh-Nagumo equations on the circle. *Fields Inst Commun* 21, 225–236.

Genevet, P. et al. (2012) Ultra-thin plasmonic optical vortex plate based on phase discontinuities. *Appl Phys Lett* 100, 013101.

Gerisch, G., Hulser, D., Malchow, D., Wick, U. (1975) Cell communication by periodic cyclic-AMP pulses. *Phil Trans Roy Soc Lond B* 272, 181–192.

Gerlach, G., Arndt, K.-F. (2009) *Hydrogel Sensors and Actuators,* 1st ed. Springer, Berlin, Germany.

Gerola, H., Seiden, P. E. (1978) Stochastic star formation and spiral structure of galaxies. *Astrophys J* 223, 129–139.

Gerprags, S. et al. (2016) Origin of the spin Seebeck effect in compensated ferrimagnets. *Nat Commun* 7, 10452.

Gervasi, M. G. et al. (2018) The actin cytoskeleton of the mouse sperm flagellum is organized in a helical structure. *J Cell Sci* 131(11), jcs215897.

Ghashghaei, H. T., Hilgetag, C. C., Barbas, H. (2007) Sequence of information processing for emotions based on the anatomic dialogue between prefrontal cortex and amygdala. *Neuroimage* 34, 905–923. doi:10.1016/j.neuroimage.2006.09.046.

Ghosh, S., Fujita, D., Bandyopadhyay, A. (2019a) Universal Geometric-musical language for big data processing in an assembly of clocking resonators, JP-2017-150171, 8/2/2017: World patent received February 2019, World Patent: WO 2019/026983.

Ghosh, S., Fujita, D., Bandyopadhyay, A. (2019b) Human brain like intelligent decision-making machine JP-2017-150173, 8/2/2017; World patent WO 2019/026984.

Ghosh, S. et al. (2014a) Design and operation of a brain like computer: A new class of frequency-fractal computing using wireless communication in a supramolecular organic, inorganic systems. *Information* 5, 28–99.

Ghosh, S. et al. (2014b) Nano Molecular-platform: A protocol to write energy transmission program inside a molecule for bio-inspired supramolecular engineering. *Adv Func Mater* 24, 1364–1371.

Ghosh, S. et al. (2014c) Design and construction of a brain-like computer: A new class of frequency-fractal computing using wireless communication in a supra-molecular organic, inorganic system. Information 5(1), 28–100.

Ghosh, S. et al. (2015a) Resonant oscillation language of a futuristic nano-machine-module: Eliminating cancer cells & Alzheimer Aβ plaques. *Curr Top Med Chem* 15, 534–541.

Ghosh, S. et al. (2015b) An organic jelly made fractal logic gate with an infinite truth table. *Sci Rep* 5, 11265.

Ghosh, S. et al. (2016a) Inventing a co-axial atomic resolution patch clamp to study a single resonating protein complex and ultralow power communication deep inside a living neuron cell. *J Integr Neurosci* 15(4), 403–433.

Ghosh, S. et al. (2016b) A simultaneous one pot synthesis of two fractal structures via swapping two fractal reaction kinetic states. *Phys Chem Chem Phys* 18, 14772–14775.

Gibbon, J., Church, R. M., Meck, W. H. (1984) Scalar timing in memory. *Ann NY Acad Sci* 423, 52–77.

Giebultowicz, J. M., Hege, D. M. (1997) Circadian docks in Malphigian tubules. *Nature* 386, 684.

Giere, R. (1999) *Science Without Laws*. University of Chicago Press, Chicago, IL.

Girelli, F., Liberati, S., Sindoni, L. (2009) Is the notion of time really fundamental? Submitted on March 27, 2009. http://arxiv.org/abs/0903.4876.

Gittes, F., Mickey, B., Nettleton, J., Howard, J. (1993) Flexural rigidity of microtubules and actin filaments measured from thermal fluctuations in shape. *J Cell Biol* 120(4), 923–934.

Glass, L. (1977) Patterns of supernumerary limb regeneration. *Science* 198, 321–322.

Glass, L. (2001) Synchronization and rhythmic processes in physiology. *Nature* 410, 277–284.

Glass, L., Guevara, M., Shrier, A., Perez, R. (1983) Bifurcation and chaos in a periodically stimulated cardiac oscillator. *Physica D* 7, 89–101.

Glass, L., Mackey, M. C. (1979) A simple model for phase locking of biological oscillators. *J Math Bio* 7, 339–352.

Glass, L., Mackey, M. C. (1988) *From Clocks to Chaos: The Rhythms of Life*. Princeton University Press, Princeton, NJ.

Glass, L., Winfree, A. T. (1984) Discontinuities in phase-resetting experiments. *Am J Physiol* 246, R251–R258.

Gleick, J. (1992) *Genius: Richard Feynman and Modern Physics*. Little, Brown and Company, London, UK.

Gloveli, T., Dugladze, T., Rotstein, H. G., Traub, R. D., Monyer, H., Heinemann, U., Whittington, M. A., Kopell, N. J. (2005) Orthogonal arrangement of rhythm-generating microcircuits in the hippocampus. *Proc Natl Acad Sci USA* 102, 13295–13300.

Gödel, K. (1938) The consistency of the axiom of choice and of the generalized continuum-hypothesis. *Proc Natl Acad Sci USA* 24, 556–557.

Gödel, K. (1947) What is Cantor's continuum problem? *Am Math Mon* 54, 515–525.

Goila, A., Pawar, M. (2009) The diagnosis of brain death. *Indian J Crit Care Med* 13(1), 7–11. The blue brain project. https://bluebrain.epfl.ch/.

Gold, T. (1962) The arrow of time. *Am J Phys* 30, 403–410.

Gold, T. ed. (1967) *The Nature of Time*. Cornell University Press, New York.

Goldstein, E. B. (2010) In: *Sensation and Perception*, 8th ed., J. D. Hague, J. A. Perkins, Eds. (Linda Schreiber) Wadsworth, Cengage Learning, Belmont, CA.

Goldstein, S. C. et al. (2009) Beyond audio and video: Using claytronics to enable pario. *AI Mag* 30(2), 29–45.

Goldstein, S., Rall, W. (1974) Changes in axon potential shape and velocity for changing core conductor geometry. *Biophys J* 14, 731–757.

Golubitsky, M., Lauterbach, R. (2009) Bifurcations from synchrony in homogeneous networks: Linear theory. *SIAM Appl Dyn Syst* 8, 40–75.

Gomatam, J. (1982) Pattern synthesis from singular solutions in the Debye limit: Helical waves and twisted toroidal scroll structures. *J Phys AIS* 1463–1476.

Gong, B. (2018) Artificial water channels: Inspiration, progress, and challenges. *Faraday Discuss* 209, 415–427.

Gong, P., Nikolaev, A. R., Van Leeuwen, C. (2003) Scale-invariant fluctuations of the dynamical synchronization in human brain electrical activity. *Neurosci Lett* 336, 33–36.

Gong, Y. et al. (2015) High-speed recording of neural spikes in awake mice and flies with a fluorescent voltage sensor. *Science* 350(6266), 1361–1366.

Gooch, V. D., Packer, L. (1971) Adenine nucleotide control of heart mitochondrial oscillations. *Biochim Biophys Acta* 245, 17–20.

Good, B. H. (2017) The dynamics of molecular evolution over 60,000 generations. *Nature* 551, 45–50.

Goodman, N. (1954) *Fact, Fiction and Forecast*. Athlone, London, UK, Chapter 1, Section 3 & Chapter 3.

Goodwin, B. C. (1967) Biological control processes and time. *Ann NY Acad Sci* 138, 748–758.

Goodwin, B. C., Cohen, M. H. (1969) A phase-shift model for the spatial and temporal organization of developing systems. *J Theor Biol* 25, 49–107.

Gordiz, K., Henry, A. (2016) Phonon transport at interfaces: Determining the correct modes of vibration. *J Appl Phys* 119, 015101.

Gothard, K. M., Skaggs, W. E., McNaughton, B. L. (1996) Dynamics of mismatch correction in the hippocampal ensemble code for space: Interaction between path integration and environmental cues. *J Neurosci* 16, 8027–8040.

Götze, W. (2008) *Dynamics of Glass Forming Liquids*. Oxford University Press, Oxford.

Goyal, S. K., Simon, B. N., Singh, R., Simon, S. (2011) Geometry of the generalized Bloch sphere for Qutrits. ArXiv:1111.4427v2.

Grabow, C. et al. (2010) Does small worlds synchronize fastest? *Eurphys Lett* 90, 48002.

Graham, J. E., Marians, K. J., Kowalczykowski, S. C. (2017) Independent and stochastic action of DNA polymerases in the replisome. *Cell* 169(7), 1201–1213.

Grant, E. H. (1979) *Dielectric Behavior of Bio Molecules in Solution*. Oxford University Press 7, 16.

Gray, H. B., Winkler, J. R. (2003) Electron tunneling through proteins. *Biophys* 36(3), 341–372.

Gray, C. M. (1994) Synchronous oscillations in neuronal systems: Mechanisms and functions. *J Comput Neurosci* 1, 11–38.

Graybiel, A. M. (2005) The basal ganglia: Learning new tricks and loving it. *Curr Opin Neurobiol* 15, 638–644.

Grebogi, C., Ott, E., Yorke, J. A. (1985) Attractors on an N-torus: Quasiperiodicity versus chaos. *Physica D* 15, 354–373.

Green, A. E. et al. (2017) Thinking cap plus thinking zap: tDCS of frontopolar cortex improves creative analogical reasoning and facilitates conscious augmentation of state creativity in verb generation. *Cereb Cortex* 27(4), 2628–2639.

Greenberg, J. M., Hassard, B. D., Hastings, S. P. (1978) Pattern formation and periodic structures in systems modeled by reaction-diffusion equations. *Bull Am Math Soc* 84(6), 1296–1327.

Greenleaf, A. et al. (2008) Approximate quantum cloaking and almost-trapped states. *Phys Rev Lett* 101, 220404.

Grenier, F., Timofeev, I., Steriade, M. (2001) Focal synchronization of ripples (80–200 Hz) in neocortex and their neuronal correlates. *J Neurophysiol* 86, 1884–1898.

Grigor'kin, A. A., Dunaevskii, S. M. (2007) Electronic spectrum and ballistic transport in a helical nanotube. *Phys Solid State* 49, 585–590.

Grigoryan, G. et al. (2011) Computational design of virus-like protein assemblies on carbon nanotube surfaces. *Science* 332(6033), 1071–1076.

Groma, G. I. et al. (2004) Resonant optical rectification in bacteriorhodopsin. *Proc Natl Acad Sci USA* 101(21), 7971–7975.

Groot, M. L., Vos, M. H., Schlichting, I., van Mourik, F., Joffre, M., Lambry, J. C., Martin, J. L., (2002) Coherent infrared emission from myoglobin crystals: An electric field measurement. *Proc Natl Acad Sci USA* 99(3), 1323–1328.

Gross, D. J. (1989) On the calculation of the fine-structure constant. *Phys Today* 42(12), 9.

Grossberg, S. (1980) How does the brain build a cognitive code. *Psychol Rev* 87, 1–51.

Grossman, Y., Parnas, I., Spira, M. E. (1979) Differential conduction block in branches of a bifurcating axon. *J Physiol* 295, 283–305.

Gruber, F., Marengo, E. (2008) New aspects of electromagnetic information theory for wireless and antenna systems. *IEEE Trans Antennas Propag* 56(11), 3470–3484.

Gu, Q., Crook, T., Wallace, G. G., Crook, J. M. (2017) 3D bioprinting human induced pluripotent stem cell constructs for in situ cell proliferation and successive multilineage differentiation. *Adv Healthc Mater* 6(17). doi:10.1002/adhm.201700175.

Guckenheimer, J. (1975) Isochrons and phaseless sets. *J Math Biol* 1, 259–273.

Guckenheimer, J. (1976) Constant velocity waves in oscillating chemical reactions. In: *Lecture Notes on Mathematics*, Volume 525, P. Hilton, Ed. Springer-Verlag, the Netherlands, pp. 99–103.

Guckenheimer, J., Labouriau, J. S. (1993) Bifurcation of the Hodgkin and Huxley equations: A new twist. *Bull Math Biol* 55(5), 937–952.

Guevara, M. R., Glass, L., Shrier, A. (1981) Phase locking, period-doubling bifurcations and irregular dynamics in periodically stimulated cardiac cells. *Science* 214, 1350–1353.

Guevara, M. R., Jongsma, H. J. (1990) Phase resetting in a model of sinoatrial nodal membrane: Ionic and topological aspects. *Am J Physiol* 258, H734–H747.

Guevara, M. R., Shrier, A., Glass, L. (1986) Phase resetting of spontaneously beating embryonic ventricular heart cell aggregates. *Am J Physiol* 251, 1298–1305.

Guevara, M. R., Ward, G., Shrier, A., Glass, L. (1984) Electrical alternans and period-doubling bifurcations. *Comput Cardiol* 167–170.

Guillemin, V., Pollack, A. (1974) *Differential Topology*. Prentice Hall, Englewood Cliffs, NJ.

Guntner, A. T. et al. (2018) Sniffing entrapped human with sensor arrays. Anal Chem 90(8), 4940–4945.

Gupta, S., De, S., Janaki, M. S., Iyenger, A. N. S. (2017) Exploring the route to measure synchronization in non-linearly coupled Hamiltonian systems. An interdisciplinary. *J Nonlinear Sci* 27(11).

Gurevich, Y., Shelah, S. (1989) Nearly linear time, Springer LNCS. *International Symposium on Logical Foundations of Computer Science* 363, pp. 108–118. Springer, Berlin, Germany.

Gurzadyan, V. G., Penrose, R. (2013) On CCC-predicted concentric low-variance circles in the CMB sky. *Eur Phys J Plus* 128. arXiv:1302.5162.

Guy, R. K. Patterns of primes. In: *Unsolved Problems in Number Theory*, 2nd ed. Springer-Verlag, New York, pp. 23–25.

Guzman, S. J., Schlogl, A., Frotscher, A., Jonas, P. (2016) Synaptic mechanisms of pattern completion in the hippocampal CA3 network. *Science* 353(6304), 1117–1123.

Hajnal, J. V. (1990) Observation of singularities in the electric and magnetic field of freely propagating microwave. *Proc Royal Soc Lond A* 430, 413–421.

Haken, H. (1988) *Information and Self-Organization: A Macroscopic Approach to Complex Systems*. Springer, Berlin, Germany.

Halász, G. B., Hsieh, T. H., Balents, L. (2017) Fraction topological phases from strongly coupled spin chains. *Phys Rev Lett* 119, 257202.

Halberg, F. (1977) Implications of biological rhythms for clinical practice. *Hosp Pract* 12, 139–149.

Hale, J., Hamilton, A. F. (2016) Cognitive mechanisms for responding to mimicry from others. *Neurosci Biobehav Rev* 63, 106–123. doi:10.1016/j.neubiorev.2016.02.006.

Hall, M. J.W., Deckert, D.-A., Wiseman, H. M. (2014) Quantum phenomena modeled by interactions between many classical worlds. *Phys Rev X* 4, 041013.

Hall, W., Carter, A. (2011) Is deep brain stimulation a prospective "cure" for addiction? F1000. *Med Rep* 3(4), 1–3.

Halpern, A. R. (2006) Cerebral substrates of musical imagery. *Ann N Y Acad Sci* 930, 179–192.

Hameroff, S. & Penrose, R. (1996a) In: *Toward a Science of Consciousness - The First Tucson Discussions and Debates*, S. R. Hameroff, A. W. Kaszniak, A. C. Scott, Eds. Cambridge, MA: MIT Press, pp. 507–540.

Hameroff, S. R., Penrose, R. (1996b) Conscious events as orchestrated spacetime selections. *J Conscious Stud* 3(1), 36–53.

Hameroff, S. R., Watt, R. C. (1982) Information processing in microtubule. *J Theor Biol* 98, 549–561.

Hamkins, J. D., Lewis, A. (2000) Infinite time turing machines. *J Symb Log* 65(2), 567–604.

Hamm, P. (2008) Ultrafast peptide and protein dynamics by vibrational spectroscopy. In: *Ultrashort Laser Pulses in Biology and Medicine*, M. Braun, P. Gilch, W. Zinth, Eds. Berlin, Germany, pp. 77–94.

Hammett, S. T., Smith, A. T. (1994) Temporal beats in the human visual system. *Vision Res* 34, 2833–2840.

Hamrick, P. et al. (2018) Child first language and adult second language are both tied to general-purpose learning systems. *Proc Natl Acad Sci* 115(7), 1487–1492.

Hanham, S. M., Watts, C., Otter, W. J., Lucyszyn, S., Klein, N. (2015) Dielectric measurements of nanoliter liquids with a photonic crystal resonator at terahertz frequencies. *Appl Phys Lett* 107, 032903.

Hansma, P. K., Drake, B., Marti, O., Gould, S. A., Prater, C. B. (1989) The scanning ion-conductance microscope. *Science* 243, 641–643.

Haramein, N. (2001) Scaling law for organized matter in the universe. *Bull Am Phys Soc*. AB006, Ft. Worth, October 5.

Harms, G., Orr, G., Lu, H. P. (2004) Probing ion channel conformational dynamics using simultaneous single-molecule ultrafast spectroscopy and patch-clamp electric recording. *Appl Phys Lett* 84, 1792.

Harris, K. D., Henze, D. A., Hirase, H., Leinekugel, X., Dragoi, G., Czurkó, A., Buzsáki, G. (2002) Spike train dynamics predicts theta-related phase precession in hippocampal pyramidal cells. *Nature* 417, 738–741.

Harris, K. D., Hirase, H., Leinekugel, X., Henze, D. A., Buzsáki, G. (2001) Temporal interaction between single spikes and complex spike bursts in hippocampal pyramidal cells. *Neuron* 32, 141–149.

Harris, V. C., Subbarao, M. V. (1991) On product partitions of integers. *Can Math Bull* 34(4), 474–479.

Hart, M., Stevens, J., Kieberman, F. (1998) *Drumming at the Edge of Magic: A Journey into the Spirit of Percussion.* Acid Test Production, Houston, TX, p. 264.

Hartle, J. B. (2005) Progress in quantum cosmology. In: *General Relativity and Gravitation,* Ashby, N., Bartlett, D. F., Wyss, W. Eds. Cambridge University Press, Cambridge, UK.

Hartle, J. B., Hawking, S. W. (1983) Wave function of the universe. *Phys Rev D* 28, 2960–2975.

Hasenstaub, A., Callaway, E., Otte, S., Sejnowski, T. 2010, Metabolic cost as a unifying principle governing neuronal biophysics. *Proc Natl Acad Sci USA* 107(27), 12329–12334.

Hassan, A. U. et al. (2009) Role of stem cells in treatment of neurological disorder. *Int J Health Sci* 3(2), 227–233.

Hasson, U. et al. (2012) Brain to brain coupling: A mechanism for creating and sharing a social world. *Trends Cogn Sci* 16(2), 114–121.

Hastings, J. W., Broda, H., Johnson, C. H. (1985) Phase and period effects of physical and chemical factors. Do cells communicate? In: *Temporal Order,* L. Rensing, N. I. Jaeger, Ed. Springer, Berlin, Germany.

Hastings, J. W., Sweeney, B. M. (1957) On the mechanism of temperature independence in a biological clock. *PNAS USA* 43(9), 804–811.

Hauser, H., Ijspeert, A. J., Füchslin, R. M., Pfeifer, R., Maass, W. (2011) Towards a theoretical foundation for morphological computation with compliant bodies. *Biol Cybern* 105, 355–370.

Havel, T. F., Doran, C. J. L. (2004) A bloch-sphere-type model for two qubits in the geometric algebra of a 6D Euclidean vector space. *Proc SPIE 2nd Conf* 5436, 93–106; Quantum information and computation; arXiv:quant-ph/0403136.

Hawking, S. W. (1982) The boundary conditions of the universe, In: *Astrophysical Cosmology,* H. A. Briick, G. V. Coyne, M. S. Longair, Eds. Pontificia Academia Scientiarum, Vatican City, pp. 563–594.

Hawking, S. W. (1985) Arrow of time in cosmology. *Phys Rev D* 33, 2489–2495.

Hawking, S. W. (1988) *Brief History of Time.* Bantam, London, UK.

Hawking, S. W. (1994) The no boundary condition and the arrow of time. In: *Physical Origins of Time Asymmetry,* J. Halliwell, J. Perez-Mercader, W. Zurek, Eds. Cambridge University Press, Cambridge, pp. 346–357.

Hawking, S., Hertog, T. (2002) Why does inflation start at the top of the hill? *Phys Rev* 66, 123509.

Hawkins, J., Ahmad, S. (2016) Why neurons have thousands of synapses, a theory of sequence memory in neocortex. *Front Neural Circuits* 10(23), 1–13.

Hazan, H., Manevit, L. M. (2012) Topological constraints and robustness in liquid state machines. *Expert Syst Appl* 39(2), 1597–1606.

He, B. J., Zempel, J. M., Snyder, A. Z., Raichle, M. E. (2010) The temporal structures and functional significance of scale-free brain activity. *Neuron* 66(3), 353–369.

He, J. et al. (2016) Prevalent presence of periodic actin-spectrin-based membrane skeleton in a broad range of neuronal cell types and animal species. *Proc Natl Acad Sci* 113(21), 6029–6034.

He, P., Parimi, P. V., He, Y., Harris, V. G., Vittoria, C. (2007) Tunable negative refractive index metamaterial phase shifter. *Electron Lett* 43(25), 1440. doi:10.1049/el:20072451.

He, Y., Sun, G., Koga, K., Xu, L. (2014) Electrostatic field-exposed water in nanotube at constant axial pressure. *Sci Rep* 4, 6596.

Healey, R. (2002) Can physics coherently deny the reality of time? *Royal Inst Philos Suppl* 50, 293–316.

Heebner, J. E. et al. (2000) Conversion of unpolarized light to polarized light with greater than 50% efficiency by photorefractive two-beam coupling. *Opt Lett* 25(4), 257–259.

Heidemann, S. R. et al. (1984) Spatial organization of axonal microtubules. *J Cell Biol* 99, 1289–1295.

Heidemann, S. R., Landers, J. M., Hamborg, M. A. (1981) Polarity orientation of axonal microtubules. *J Cell Biol* 91(3 Pt 1), 661–665.

Heimburg, T., Jackson, A. D. (2005) On soliton propagation in biomembranes and nerves. *Proc Natl Acad Sci USA* 102(2), 9790–9795.

Hein, B., Willig, K. I., Hell, S. W. (2008) Stimulated emission depletion (STED) nanoscopy of a fluorescent protein-labeled organelle inside a living cell. *Proc Natl Acad Sci USA* 105, 14271–14276.

Heit, G., Smith, M. E., Halgren, E. (1998) Neural encoding of individual words and faces by the human hippocampus and amygdala. *Nature* 333, 773–775.

Hell, S. W. (2007) Far-field optical nanoscopy. *Science* 316, 1153–1158.

Henke, W., Selzle, H. L., Lin, S. H., Schlag, E. W. (1981) Effect of collision and magnetic field on quantum beat in biacetyl. *Chem Phys Lett* 77(3), 448–451.

Henke, W., Selzle, H. L., Lin, S. H., Schlag, E. W. (1981) Effect of collision and magnetic field on quantum beat in biacetyl. *Chem Phys Lett* 77(3), 448–451.

Herculano-Houzel, S. (2011) Scaling of brain metabolism with a fixed energy budget per neuron: Implications for neuronal activity, plasticity and evolution. *PLoS one* 6, e17514.

Herms, D. (1984) A brilliant discussion on hypothesis testing, "Logical Basis of Hypothesis Testing in Scientific Research" A logic primer to accompany Giere 1984, Chapter 6.

Herz, M. D. (2017) Distinct mechanisms mediate speed-accuracy adjustments in cortico-subthalamic networks. *Neuroscience.* doi:10.7554/eLife.21481.

Hess, B., Boiteux, A. (1971) Oscillatory phenomena in biochemistry. *Ann R Biochem* 40, 237–258.

Hestenes, D. (1986) In: *A Unified Language for Mathematics and Physics, Clifford Algebras and Their Applications in Mathematical Physics,* NATO ASI Series (Series C), J. S. R. Chisholm, A. K. Commons, Eds. Springer, Dordrecht, the Netherlands, 183, pp. 1–23.

Hildebrandt, G., Moog, R., Raschke, F. (Eds.) (1957) *Chronobiology and Chronomedicine: Basic and Applications.* Peter Lang, Frankfurt, Germany.

Hill, A. V. (1933) Wave transmission as the basis of nerve activity. *Cold Spring Harbor Symp Quant Biol* 1, 146–151.

Hill, C. (2006) *Electro Fractal Universe.* Colin Hill, pp. 77. http://www.fractaluniverse.eclipse.co.uk/ElectroFractal UniverseWebVersion.pdf.

Hinaut, X., Dominey, P. F. (2013) Real-time parallel processing of grammatical structure in the fronto-striatal system: A recurrent network simulation study using reservoir computing. *PLoS One.* doi:10.1371/journal.pone.0052946.

Hines, M. (1989) A program for simulation of nerve equations with branching geometries. *Int J Bio-med Comput* 24, 55–68.

Hirokawa, N. (2011) From electron microscopy to molecular cell biology, molecular genetics and structural biology: Intracellular transport and kinesin superfamily proteins, KIFs: Genes, structure, dynamics and functions. *J Electron Microsc* 60, S63–S92.

Hirokawa, N. et al. (1989) Submolecular domains of bovine brain kinesin identified by electron microscopy and monoclonal antibody decoration. *Cell* 56, 867–878.

Hirokawa, N., Shiomura, Y., Okabe, S. (1988) Tau proteins: The molecular structure and mode of binding on microtubules. *J Cell Biol* 107, 1449–1459.

Hoffmann, K. (1976) The adaptive significance of biological rhythms corresponding to geophysical cycles. In: *The Molecular Basis of Circadian Rhythms*, J. W. Hastings, H. G. Schweiger, Eds. Abakon, Berlin, Germany, pp. 63–76.

Hogan, C. (2004) Quarks, electrons, and atoms in closely related universes, astro-ph/0407086.

Hohlfeld, R. G., Cohen, N. (1999) Self-similarity and the geometric requirements for frequency independence in antenna. *Fractals* 7. doi:10.1142/S0218348X99000098.

Hol, W. G. J., van Duijnen, P. T., Berendsen, H. J. C. (1978) The alpha-helix dipole and the properties of proteins. *Nature* 273, 443–446.

Holscher, C., Anwyl, R., Rowan, M. J. (1997) Stimulation on the positive phase of hippocampal theta rhythm induces long-term potentiation that can be depotentiated by stimulation on the negative phase in area CA1 in vivo. *J Neurosci* 17, 6470–6477.

Hopfield, J. J., Tank, D. W. (1986) Computing with neural circuits: A model. *Science* 233, 625–633.

Hoppensteadt, F. C., Izhikevich, E. M. (1998) Thalamo-cortical interactions modeled by weakly connected oscillators: Could the brain use FM radio principles? *Biosystems* 48, 85–94.

Hoppensteadt, F. C., Keener, J. P. (1982) Phase locking of biological clocks. *J Math Biol* 15, 339–349.

Hormuzdi, S. G., Pais, I., LeBeau, F. E., Towers, S. K., Rozov, A., Buhl, E. H., Whittington, M. A., Monyer, H. (2001) Impaired electrical signaling disrupts gamma frequency oscillations in connexin 36-deficient mice. *Neuron* 31, 487–495.

Horodeck, M. (2005) Partial quantum information. *Nature* 436, 673.

Horodeck, M., Oppenheim, J., Winter, A. (2005) Partial quantum information. *Nature* 436, 673–676.

Horton, J. C., Adams, D. L. (2005) The cortical column: A structure without a function. *Philos Trans R Soc Lond B Biol Sci* 360(1456), 837–862. doi:10.1098/rstb.2005.1623.

Howard, D. et al. (2019) Evolving embodied intelligence from materials to machines. *Nat Mach Intell* 1, 12–19.

Howard, M. W., Fotedar, M. S., Datey, A. V., Hasselmo, M. E. (2005) The temporal context model in spatial navigation and relational learning: Toward a common explanation of medial temporal lobe function across domains. *Psychol Rev* 112, 75–116.

Howard, M. W., Kahana, M. J. (2002) A distributed representation of temporal context. *J Math Psychol* 46, 269–299.

Howarth, C. (2010) The energy use associated with neural computation in the cerebellum. *J Cereb Blood Flow Metab* 30, 403–414.

Howarth, J. V., Keynes, R. D., Ritchie, J. M., Muralt, A. V. (1975) The heat production associated with the passage of a single impulse in pike olfactory nerve fibres. *J Physiol* 249(2), 349–368.

Howson, C., Urbach, P. (2005) *Scientific Reasoning: The Bayesian Approach*, 3rd ed. Open Court Publishing Company, La Salle, IL.

Huang, B., Babcock, H., Zhuang, X. (2010) Breaking the diffraction barrier: Super-resolution imaging of cells. *Cell* 143, 1047–1058.

Huang, L. et al. (2012a) Dispersionless phase discontinuities for controlling light propagation. *Nano Lett* 12, 5750–5755.

Huang, L. et al. (2013b) Helicity dependent directional surface plasmon polariton excitation using a metasurface with interfacial phase discontinuity. *Light Sci Appl* 2, e70.

Huang, L. et al. (2013a) Holography of a 3D helix using metasurface. *Nat Commun* 4, 2808.

Huang, N. et al. (2012b) Crystal structure of the heterodimeric clock: BMAL1 transcriptional activator complex. *Science* 337(6091), 189–194.

Huang, W. C. et al. (2018) Ultracompliant hydrogel-based neural interfaces fabricated by aqueous-phase micro transfer printing. *Adv Funct Mater* 28, 1801059.

Huang, W. Q. et al. (2009) Selective transmission and enhanced thermal conductance of ballistic phonon by nano-cavities embedded in a narrow constriction. *J Phys D* 42, 015101.

Huang, X. et al. (2010) Spiral wave dynamics in neocortex. *Neuron* 68(5), 978–990.

Hubel, D. H. (1957) Tungsten microelectrodes for recording single units. *Science* 125, 549–550.

Huber, G. (1994) Vortex solids and vortex liquids in a complex Ginzburg-Landau system. In: *Spatio-Temporal Patterns*, P. Palffy-Muhoray, P. Cladis, Eds. SFI Studies in the Sciences of Complexity XXI, Addison-Wesley, Boston, MA.

Hubert, D. (1972) *What Computers Can't Do*. MIT Press, New York.

Hughes, A. J. (2018) Engineered tissue folding by mechanical compaction of the mesenchyme. *Dev Cell* 44, 165–178.

Hume, D. (2000) 1739–1740. *A Treatise of Human Nature: Being an Attempt to Introduce the Experimental Method of Reasoning into Moral Subjects* (Oxford Philosophical Texts), D. F. Norton, M. J. Norton. Oxford University Press, Oxford.

Hunter, J. D., Milton, J. G., Thomas, P. J., Cowan, J. D. (1998) Resonance effect for neural spike time reliability. *J Neurophysiol* 80, 1427–1438.

Hunyadi, V., Chrétien, D., Flyvbjerg, H., Jánosi, I. M. (2007) Why is the microtubule lattice helical? *Biol Cell* 99, 117–128.

Hurdal, M. K. et al. (1999) Quasi-conformally flat mapping the human cerebellum. In: *Medical Image Computing and Computer-Assisted Intervention–MICCAI'99*, Vol. 1679 of Lecture Notes in Computer Science, C. Taylor, A. Colchester, Eds. Springer, Berlin, Germany, pp. 279–286, 1999.

Hurdal, M. K., Stephenson, K. (2009) Discrete conformal methods for cortical brain flattening. *Neuroimage* 45(1), S86–S98.

Hurtado, J. M., Rubchinsky, L. L., Sigvardt, K. A. (2004) Statistical method for detection of phase-locking episodes in neural oscillations. *J Neurophysiol* 91, 1883–1898.

Hutcheon, B., Yarom, Y. (2000) Resonance, oscillation and the intrinsic frequency preferences of neurons. *Trends Neurosci* 23, 216–222.

Hutcheon, B., Yarom, Y. (2000) Resonance, oscillation and the intrinsic frequency preferences of neurons. *Trends Neurosci* 23, 216–222.

Hwa, R., Ferree, T. C. (2002) Scaling properties of fluctuations in the human electroencephalogram. *Phys Rev E* 66, 021901.

Ide, T., Takeuchi, Y., Aoki, T., Yanagida, T. (2002) Simultaneous optical and electrical recording of a single ion-channel. *Jpn J Physiol* 52, 429–434.

Ikegaya, Y., Aaron, G., Cossart, R., Aronov, D., Lampl, I., Ferster, D., Yuste, R. (2004) Synfire chains and cortical songs: Temporal modules of cortical activity. *Science* 304, 559–564.

Ilani, S. et al. (2006) Measurement of quantum capacitance of interacting electrons in carbon nanotubes. *Nat Phys* 2, 487–691.

Ilmoniemi, R. J., Ruohonen, J., Karhu, J. (1999) Transcranial magnetic stimulation—A new tool for functional imaging of the brain. *Crit Rev Biomed Eng* 27, 241–284.

Ishiwata, S., Taguchi, Y., Murakawa, H., Onose, Y., Tokura, Y. (2008) Low-magnetic-field control of electric polarization vector in a helimagnet. *Science* 319, 1643–1646.

Israelachvili, J., Wennerström, H. (1996) Role of hydration and water structure in biological and colloidal interactions. *Nature* 379, 219–225.

Ivry, R. B., Spencer, R. M. (2004) The neural representation of time. *Curr Opin Neurobiol* 14, 225–232.

Iwata, T., Hieftje, G. M. (1992) Simple method to measure the coherence time of a mode-locked laser. *Appl Spectrosc* 46, 1464–1468.

Izhikevich, E. M., Desai, N. S., Walcott, E. C., Hoppensteadt, F. C. (2003) Bursts as a unit of neural information: Selective communication via resonance. *Trends Neurosci* 26, 161–167.

Jacklet, J. W. (1977) Neuronal circadian rhythm: Phase shifting by a protein synthesis inhibitor. *Science* 198, 69–71.

Jaeger, H., Haas, H. (2004) Harnessing nonlinearity: Predicting chaotic systems and saving energy in wireless communication. *Science* 304(5667), 78–80.

Jakobczyk, L., Siennicki, M. (2001) Geometry of Bloch vectors in two qubit system. *Phys Lett A* 286, 383–390.

Jalife, J., Antzelevitch, C. (1979) Phase resetting and annihilation of pacemaker activity in cardiac tissues. *Science* 206, 695–697.

Jalife, J., Moe, G. K. (1979) Phasic effects of vagal stimulation on pacemaker activity of the isolated sinus node of the young cat. *Circ Res* 45, 595–607.

Jarosiewicz, B. et al. (2015) Virtual typing by people with tetraplegia using a self-calibrating intracortical brain-computer interface. *Sci Transl Med* 7(313), 313ra179.

Jaynes, E. T. (1980) Quantum beats. In: *Foundations of Radiation Theory and Quantum Electrodynamics*, A. Barut, Ed. Springer, Boston, MA, pp. 37–43.

Jaynes, E. T. (1986) Bayesian methods: General background. In: *Maximum Entropy and Bayesian Methods in Applied Statistics*, J. H. Justice, Ed. Cambridge University Press, Cambridge, UK, pp. 1–25.

Jefferys, J. G. R. (1995) Nonsynaptic modulation of neuronal activity in the brain: Electric currents and extracellular ions. *Physiol Rev* 75(4), 689–723.

Jenerick, H. (1963) Phase plane trajectories of the muscle spike potential. *Biophys J* 3(5), 363–377.

Jensen, K. et al. (2016) Non-invasive detection of animal nerve impulses with an atomic magnetometer operating near quantum limited sensitivity. *Sci Rep* 6, 29638.

Jelinek, L., Baena, J. D., Voves, J., Marques, R. (2011) Metamaterial-inspired perfect tunneling in semiconductor heterostructures. *New J Phys* 13, 083011.

Jimenez, M. G. et al. Observation of coherent delocalized phonon-like modes in DNA under physiological conditions. *Nat Commun* 7, 11799.

Jin, H. (2015) Phonon-induced diamagnetic force and its effect on the lattice thermal conductivity. *Nat Mater* 14, 601–606.

Jin, C. et al. (2017) Topological trajectories of a magnetic skyrmion with an in-plane microwave magnetic field featured. *J. Appl Phys* 122, 223901.

Jin, L. et al. (2012) Single action potentials and subthreshold electrical events imaged in neurons with a fluorescent protein voltage probe. *Neuron* 75, 779–785.

Jin, Q. et al. (2018) Terahertz wave emission from a liquid water film under the excitation of asymmetric optical fields. *Appl Phys Lett* 113, 261101. doi:10.1063/1.5064644.

Jirsa, V., Haken, H. (1996) Field theory of electromagnetic brain activity. *Phys Rev Lett* 77, 960–963.

Jochim, S. M. et al. (2003) Bose-Einstein condensation of molecules. *Science* 302, 2101–2103.

Johnson, A., Brogarth, T., Holje, Ø. (1979) Oscillatory transpiration for avena plants: Perturbation experiments provide evidence for a stable point of singularity. *Physiol Plant* 45, 393–398.

Johnson, B. N. (2015) 3D printed anatomical nerve regeneration pathways. *Adv Funct Mater* 25, 6205–6217.

Johnson, S., Coxon. M. (2016) Sound can enhance the analgesic effect of virtual reality. *Royal Soc Open Sci* 3, 150567.

Johnsson, A., Brogärdh, T., Holje, P. (1979) Oscillatory transpiration of *Avena* plants: Perturbation experiments provide evidence for a stable point of singularity. *Physiol Plant* 45, 393–398.

Johnsson, A., Karlsson, H. G. (1971) Biological rhythms: Singularities in phase-shift experiments as predicted from a feedback model. In: *Proceedings of First European Biophysical Congress*, E. Broda and A. Locker, Eds. Springer, Berlin, Germany, pp. 263–267.

Johnsson, A., Karlsson, H. G., Engelman, W. (1973) Phase shift effect in the Kalanchoe petal rhythm due to two or more light pulses. *Physiol Plant* 28, 134–142.

Johnston, M. B. et al. (2002a) Simulation of terahertz generation at semiconductor surfaces. *Phys Rev B* 65, 165301.

Johnston, M. B. et al. (2002b) Theory of magnetic-field enhancement of surface-field terahertz emission. *J Appl Phys* 91, 2104.

Jonas, P., Bischofberger, J., Fricker, D., Miles, R. (2004) Interneuron diversity series: Fast in, fast out—temporal and spatial signal processing in hippocampal interneurons. *Trends Neurosci* 27, 30–40.

Joos, E., Zeh, H. D. (1985) The emergence of classical properties through interaction with the environment. *Z Phys* B59, 223–243.

Joshi, A., Xiao, M. (2006) Cavity-QED based unconventional geometric phase dc gates with bichromatic field modes. *Phys Lett A* 359, 390–395.

Jozsa, R., Linden, N. (2003) On the role of entanglement in quantum-computational speed-up. *Proc Royal Soc A* 459(2036). doi:10.1098/rspa.2002.1097.

Jurkiewicz, J., Loll, R., Ambjorn, J. (2008) Using causality to solve the puzzle of quantum space-time. *Scientific American*. http://www.sciam.com/article.cfm?id=the-self-organizing-quantum-universe.

Kaech, S., Banker, G. (2006) Culturing hippocampal neurons. *Nat Protoc* 1(5), 2406.

Kaeser, P. S., Regehr, W. G. (2014) Molecular mechanisms for synchronous, asynchronous, and spontaneous neurotransmitter release. *Annu Rev Physiol* 76, 333–363.

Kajetan Schmidt, M., Mackowski, S., Aizpurua, J. (2012) Control of single emitter radiation by polarization- and position-dependent activation of dark antenna modes. *Opt Lett* 37, 1017–1019.

Kamondi, A., Acsády, L., Wang, X. J., Buzsáki, G. (1998) Theta oscillations in somata and dendrites of hippocampal pyramidal cells in vivo: Activity-dependent phase-precession of action potentials. *Hippocampus* 8, 244–261.

Kang, H., Welcher, A. A., Shelton, D., Schuman, E. M. (1997) Neurotrophins and time: Different roles for TrkB signaling in hippocampal long-term potentiation. *Neuron* 19(3), 653–664.

Kantelhardt, J. W. (2011) Fractal and multi-fractal time series. In: *Mathematics of Complexity and Dynamical Systems*, R. A. Meyers, Ed. Springer, New York, pp. 463–487.

Karaveli, S., Zia, R. (2010) Enhancement of magnetic dipole emission in a multilevel electronic system. *Opt Lett* 35(20), 3318.

Karma, A. (1991) Universal limit of spiral wave propagation in excitable media. *Phys Rev Lett* 66, 2274–2277.

Karma, A. (1992) The scaling regime of spiral wave propagation in single-diffusive media. *Phys Rev Lett* 68, 397–400.

Kats, M. A. et al. (2012) Giant birefringence in optical antenna arrays with widely tailorable optical anisotropy. *Proc Natl Acad Sci* 109, 12364–12368.

Katsura, H., Balatsky, A. V., Nagaosa, N. (2007) Dynamical magneto-electric coupling in helical magnets. *Phys Rev Lett* 98, 027203.

Katz, B., Schmitt, O. H. (1940) Electric interaction between two adjacent nerve fibers. *J Physiol* 97(4), 471–488.

Kawato, M. (1981) Transient and steady-state phase response curves of limit-cycle oscillators. *J Math Biol* 12, 13–30.

Kawato, M., Suzuki, R. (1978) Biological oscillators can be stopped-topological study of a phase response curve. *Biol Cybern* 30, 241–248.

Kayser, V., Turton, D. A., Aggeli, A., Beevers, A., Reid, G. D., Beddard, G. S. (2004) Energy migration in novel pH-triggered self-assembled β-sheet ribbons. *J Am Chem Soc* 126, 336–343.

Keener, J. P. (1986) A geometrical theory for spiral waves in excitable media. *SIAM J Appl Math* 46, 1039–1056.

Keener, J. P. (1990) Knotted vortex filaments in an ideal fluid. *J Fluid Mech* 211, 629–651.

Keidel, W. (1984) The sensory detection of vibrations. In: *Foundations of Sensory Science*, W. W. Dawson, J. M. Enoch, Eds. Springer-Verlag, Berlin, Germany, pp. 465–512.

Kelso, J. A., Fuchs, A. (1995) Self-organizing dynamics of the human brain: Critical instabilities and Sil'nikov chaos. *Chaos* 5, 64–69.

Kemeny, J. G. (1955) Fair bets and inductive probabilities. *J Symb Log* 20(3), 263–273.

Kennedy, S. R. et al. (2018) Tailored supramolecular gel and microemulsion crystallization strategies—is isoniazid really monomorphic? *Cryst Eng Comm* 20,1390.

Khamis, H., Birznieks, I., Redmond, S. J. (2015) Decoding tactile afferent activity to obtain an estimate of instantaneous force and torque applied to the finger pad. *J Neurophysiol* 114(1), 474–484.

Khemani, V., Lazarides, A., Moessner, R., Sondhi, S. L. (2016) Phase structure of driven quantum system. *Phys Rev Letter* 116, 250401.

Kheyfets, A., Wheeler, J. A. (1986) Boundary of a boundary principle and geometric structure of field theories. *Int J Theo Phys* 25, 573–580.

Khire, T. S. et al. (2010) The fractal self-assembly of the silk protein sericin. *Soft Matter* 6, 2066–2071.

Klimesch, W. (2000) Theta frequency, synchronization and episodic memory performance. In: *Time and the Brain*, R. Miller, Ed. Harwood Academic Publishers, Singapore, pp. 225–240.

Kiebel, S. J., Friston, K. J. (2011) Free energy and dendritic self-organization. *Front Syst Neurosci* 5, 80.

Killeen, P. R., Weiss, N. A. (1987) Optimal timing and the Weber function. *Psychol Rev* 94, 455–468.

Kim, E. D. et al. (2010) Fast spin rotations by optically controlled geometric phases in a charge-tunable InAs Quantum dot. *Phys Rev Lett* 104, 167401.

Kim, Y.-H. et al. (2008) A simple and direct biomolecule detection scheme based on a microwave resonator. *Sens Actuators B* 130, 823–828.

Kimura, G. (2003) The Bloch vector for n level systems. *Phys Lett A* 314, 339.

King, C. (1996) Fractal neuro-dynamics and quantum chaos: Resolving the mind-brain paradox through novel biophysics. In: *Fractals of Brain Fractals of Mind, Advances in Consciousness Research* 7. John Betjamin & Co, Amsterdam, the Netherlands.

Kikkawa, T. et al. (1994) Direct visualization of the microtubule lattice seam both in vitro and in vivo. *J Cell Biol* 127, 1965–1971.

Kishi, T., Tsumori, T., Yokota, S., Yasui, Y. (2006) Topographical projection from the hippocampal formation to the amygdala: A combined anterograde and retrograde tracing study in the rat. *J Comp Neurol* 496, 349–368. doi:10.1002/cne.20919.

Kizawa, H. et al. (2017) Scaffold-free 3D bio-printed human liver tissue stably maintains metabolic functions useful for drug discovery. *Biochem Biophys* 10, 186–191.

Knopfmacher, A., Mays, M. (2006) Ordered and unordered factorizations of integers. *Mathematica J* 10, 72–89.

Koelsch, S., Schroger, E., Gunter, T. C. (2002) Music matters: Pre-attentive musicality of the human brain. *Psychophysiology* 39(1), 38–48. doi:10.1111/1469-8986.3910038.

Koga, K., Gao, G. T., Tanaka, H., Zeng, X. C. (2001) Formation of ordered ice nanotubes inside carbon nanotubes. *Nature* 412, 802–805.

Koga, S. (1982a) Schrödinger Equation approach to rotating spiral waves in reaction-diffusion systems. *Prog Theor Phys* 67, 454–463.

Koga, S. (1982b) Rotating spiral waves in reaction-diffusion systems: Phase singularities of multi-armed waves. *Prog Theor Phys* 67, 164–178.

Kogan, B. V., Karplus, W. J., Billett, B. S., Stevenson, W. G. (1992) Excitation wave propagation within narrow pathways: Geometric configurations facilitating unidirectional block and reentry. *Physica D* 59, 275–296.

Kolata, G. B. (1977) Catastrophe theory: The emperor has no clothes. *Science* 196, 287–351.

Kolwankar, K. M., Gangal, A. D. (1996) Fractional differentiability of nowhere differentiable functions and dimensions, Chaos: An Interdisciplinary. *J Nonlinear Sci* 6(4), 505–513.

König, P., Engel, A. K., Singer, W. (1995) Relation between oscillatory activity and long-range synchronization in cat visual cortex. *Proc Natl Acad Sci USA* 92, 290–294. doi:10.1073/pnas.92.1.290.

König, P., Engel, A. K., Singer, W. (1996) Integrator or coincidence detector? The role of the cortical neuron revisited. *Trends Neurosci.* 19, 130–137. doi:10.1016/s0166-2236(96)80019-1.

Kondepudi, D., Petrosky, T., Pojman, J. A. (2017) Dissipative structures and irreversibility in nature: Celebrating 100th birth anniversary of Ilya Prigogine (1917–2003). *Chaos* 27, 104501. doi:10.1063/1.5008858.

Konishi, M. (1990) Deciphering the brain's codes. In: *Neural Codes and Distributed Representations: Foundations of Neural Computation*, L. Abbott, T. J. Sejnowski, Eds. MIT Press, Cambridge, MA, pp. 1–18.

Konorski, J. (1967) *Integrative Activity of the Brain.* University of Chicago Press, Chicago, IL.

Kopell, N., Ermentrout, G. B., Whittington, M. A., Traub, R. D. (2000) Gamma rhythms and beta rhythms have different synchronization properties. *Proc Natl Acad Sci USA* 97, 1867–1872.

Kopelman, R. (1988) Fractal reaction kinetics. *Science* 241(4873), 1620–1626.

Kozlov, G. V., Mikitaev, A. K., Zaikov, G. E. (2013) *The Fractal Physics of Polymer Synthesis.* Apple Academic Press, Boca Raton, FL, 378 p.

Koppelmans, K. (2016) Brain structure plasticity with space flight. *npj Microgravity* 2, 2.

Kraus, B. J. et al. (2015) During running in place, grid cells integrate elapsed time and distance run. *Neuron* 88(3), 578–589.

Krinsky, V. I. (1987) Dynamics of autowave vortices in active media. *Z Phys Chem* 268, 4–14.

Kröger, H. (1997) Proposal for an experiment to measure the Hausdorff dimension of quantum-mechanical trajectories. *Phys Rev A* 55(2), 951–966.

Kucewicz, M. T. et al. (2014) High frequency oscillations are associated with cognitive processing in human recognition memory. *Brain.* doi:10.1093/brain/awu149.

Kuffler, S. W., Nichols, J. G. (1977) From neuron to brain: Cellular approach to the function of the nervous system. *Exp Physio* 62, 287–287.

Kuhn, T. S. (1962) *The Structure of Scientific Revolutions*. University of Chicago Press, Chicago, IL.

Kuhn, H. (2008) Origin of life–Symmetry breaking in the universe: Emergence. *Curr Opin Coll Int Sci* 13, 3–11.

Kulić, I. M. et al. (2008) The role of microtubule movement in bidirectional organelle transport. *PNAS* 105(29), 10011–10016.

Kuramoto, Y. (1983) Each singularity generates a signal burst. *Chaos and Statistical Methods: Proceedings of the Sixth Kyoto Summer Institute*, Kyoto, Japan, September 12–15; December 6, 2012, p. 273, Springer, Berlin, Germany.

Kurihara, K., Tamura, M., Shohda, K., Toyota, T., Suzuki, K., Sugawara, T. (2011) Self-reproduction of supramolecular giant vesicles combined with the amplification of encapsulated DNA. *Nat Chem* 3(10), 775–781. doi:10.1038/nchem.1127.

Kurs, A. et al. (2007) Wireless power transfer via strongly coupled magnetic resonances. *Science* 317, 83–86.

Laage, D., Elsaesser, O. T., Hynes, J. T. (2017) Water dynamics in the hydration shells of biomolecules. *Chem Rev* 117(16), 10694–10725.

Lachaux, J. P., Rodriguez, E., Martinerie, J., Varela, F. J. (1999) Measuring phase synchrony in brain signals. *Hum Brain Mapp* 8, 194–208.

Lachaux, J. P., Rodriguez, E., Martinerie, J., Varela, F. J. (1999) Measuring phase synchrony in brain signal. *Hum Brain Mapp* 8(4), 194–208.

Lakatos, I. (1970a) Falsification and the methodology of scientific research programs. In: *Criticism and the Growth of Knowledge*, New York: Cambridge University Press, 1970, pp. 91–195; from Theodore Schick, Jr., ed., Readings in the Philosophy of Science, Mountain View, CA: Mayfield Publishing Company, 2000, pp. 20–23.

Lakatos, I. (1970b) Falsification and the methodology of scientific research programs. In: *Criticism and the Growth of Knowledge*, I. Lakatos, A. Musgrave, Eds. Cambridge University Press, pp. 91–196. (Republished as Chapter 1 of Lakatos 1978a, PP1, cited pages from this version.)

Lam, C.-W. (2014) Prime Indel: Four-prime-number genetic code for indel decryption and sequence read alignment. *Clin Chim Acta* 436, 1–4.

Landauer, R. (1988) Dissipation and noise in computation and communication. *Nature* 335, 779–784.

Landman, U. (2005) Materials by numbers: Computations as tools of discovery. *PNAS* 102, 6671.

Landsberg, P. T., Park, D. (1975) Entropy in an oscillating universe. *Proc Royal Soc A*. doi:10.1098/rspa.1975.0187.

Langen, R. et al. (2015) Electron tunneling in proteins: Coupling through a beta strand. *Science* 268(5218), 1733–1735.

Langmuir, I. (1938) The role of attractive and repulsive forces in the formation of tactoids, thixotropic gels, protein crystals and coacervates. *J Chem Phys* 8, 873–896.

Langmuir, I. (1989) Pathological science 1953 colloquium, transcribed and edited. *Phys Today* 42(12), 36–48.

Langner, G. (1992) Periodicity coding in the auditory system. *Hear Res* 60, 115–142.

Lanyon, B. P., Barbieri, M., Almeida, M. P., White, A. G. (2008) Experimental quantum computing without entanglement. *Phys Rev Lett* 101, 200501.

Lanzalaco, F., Zia, W. (2009) Dipole neurology an electromagnetic multipole solution to brain structure, function and abnormality. *Presentation To: Integrative Approaches to Brain Complexity*, Wellcome trust, Hinxton Hall, Cambridge, UK.

Lapicque, L. (1907) Recherches quantitatives sur l'excitation électrique des nerfs traitée comme une polarisation. *J Physiol Pathol Gen* 9, 620–635.

Large, E. W., Palmer, C. (2002) Perceiving temporal regularity in music. *Cogn Sci* 26, 1–37.

Larouche, S., Tsai, Y.-J., Tyler, T., Jokerst, N. M., Smith, D. R. (2012) Infrared metamaterial phase holograms. *Nat Mater* 11, 450–454.

Lau, A. Y. et al. (2013) A conformational intermediate in glutamate receptor activation. *Neuron* 79(3), 492–503.

Lauber, H.-M., Weidenhammer, P., Dubbers, D. (1994) Geometric phases and hidden symmetries in simple resonators. *Phys Rev Lett* 72(7), 1004–1007.

Laughlin, S. B., Sejnowski, T. J. (2003) Communication in neuronal networks. *Science* 301, 1870–1874.

Laurent, G. (1999) A system perspective on early olfactory learning. *Science* 286, 723–728.

Laurent, G. (2002) Olfactory network dynamics and the coding of multidimensional signals. *Nat Rev Neurosci* 3, 884–895.

Lee, G. et al. (2006) Nanospring behaviour of ankyrin repeats. *Nature* 440, 246–249.

Lee, H., Simpson, G. V., Logothetis, N. K., Rainer, G. (2005) Phase locking of single neuron activity to theta oscillations during working memory in monkey extrastriate visual cortex. *Neuron* 45, 147–156.

Lee, J. B. et al. (2016) Development of 3D microvascular networks within gelatin hydrogels using thermoresponsive sacrificial microfibers. *Adv Healthcare Mater* 5, 781–785.

Lee, J. Y., Lee, R.-K. (2013) Hide the interior region of core-shell nanoparticles with quantum invisible cloaks; arXiv:1306.2120 [quant-ph].

Lee, S. H., Blake, R. (1999) Visual form created solely from temporal structure. *Science* 284, 1165–1168.

Lee, T., Broderick, N. G. R., Brambilla, G. (2011) Berry phase magnification in optical microcoil resonators. *Opt Lett* 36, 2839–2841.

Leffert, C. B. (2002) Can clock really run backwards? Astrophysics; General Relativity and Quantum Cosmology. arXiv:astro-ph/0208234.

Lehman, R. S. (1955) On confirmation and rational betting. *J Symb Log* 20(3), 251–262.

Lehtela, L., Salmelin, R., Hari, R. (1997) Evidence for reactive magnetic 10-Hz rhythm in the human auditory cortex. *Neurosci Lett* 222, 111–114.

Leibniz, G. W. (1956) as cited in Newman: J. R. *The World of Mathematics*. Simon and Schuster, New York.

Leigh, E. G. (1965) On the relation between the productivity, diversity, and stability of a community. *Proc Natl Acad Sci* 53, 777–781.

Lentini, R. et al. (2017) Two-way chemical communication between artificial and natural cells. *ACS Cent Sci* 3(2), 117–123.

Leopold, L. B., Langbein, W. B. (1966) River meanders. *Sci Am* 214, 60–70. doi:10.1038/scientificamerican0666-60.

Leterrier, C. et al. (2011) EB3 and EB1 link microtubules to ankyrin G in the axon initial segment. *Proc Natl Acad Sci USA* 108, 8826–8831.

Leutgeb, S., Leutgeb, J. K., Barnes, C. A., Moser, E. I., McNaughton, B. L., Moser, M. B. (2005) Independent codes for spatial and episodic memory in hippocampal neuronal ensembles. *Science* 309, 619–623.

Levin, I., Zakay, D. (1989) *Time and Human Cognition: A Life Span Perspective*. North-Holland, Amsterdam, the Netherlands.

Levy, W. B., Baxter, R. A. (1996) Energy efficient neural codes. *Neural Comput* 8, 531–543.

Lewandowska, M. K., Bakkum, D. J., Rompani, S. B., Hierlemann, A. (2015) Recording large extracellular spikes in microchannels along many axonal sites from individual neurons. *PLoS One* 10, e0118514.

Lewis, D. (1980) A subjectivist's guide to objective chance. In: *Studies in Inductive Logic and Probability*, Volume II, R. C. Jeffrey, Ed. University of California Press, Berkeley, CA, Chapter 13, pp. 263–293.

Lewis, J., Bachoo, M., Glass, L., Polosa, C. (1987) Complex dynamics resulting from repeated stimulation of nonlinear oscillators at a fixed phase. *Phys Lett A* 125, 119–122.

Lewis, J., Glass, L., Bachoo, M., Palosa, C. (1992) Phase resetting and fixed-delay stimulation of a simple model of respiratory rhythm control. *J Theor Biol* 159, 491–506.

Lewis, L. D. et al. (2015) Thalamic reticular nucleus induces fast and local modulation of arousal state. *eLife* 4, e08760.

Leyton, M. (1999) *Symmetry, Causality, Mind.* MIT Press, Cambridge, MA.

Li, L., Haghighi, A., Yang, Y. (2018) A novel 6-axis hybrid additive-subtractive manufacturing process: Design and case studies. *J Manuf Process* 33, 150–160.

Li, T. Y., Yorke, J. A. (1975) Period three implies chaos. *Am Math Mon* 82, 985.

Li, Y., Urbas, A., Li, Q. (2012) Reversible light-directed red, green, and blue reflection with thermal stability enabled by a self-organized helical superstructure. *J Am Chem Soc* 134(23), 9573–9576. doi:10.1021/ja302772z.

Liang, W. et al. (2001) Fabry-Perot interference in a nanotube electron waveguide. *Nature* 411, 665.

Libet, B. (2004) *Mind Time: The Temporal Factor in Consciousness.* Harvard University Press, Cambridge, MA.

Licklider, J. C. R. (1959) Three auditory theories. *Psychology: A Study of a Science. Study I. Conceptual and Systematic*, S. Koch, Ed. McGraw-Hill, New York, pp. 41–144.

Liebovitch, L. S. et al. (1987) Fractal model of ion-channel kinetics. *Biochim Biophys Acta* 896, 173–180.

Lin, X. et al. (2017) Water-induced helical supramolecular polymerization and gel formation of an alkylene-tethered perylene bisimide dyad. *Chem Commun* 53, 168–171.

Linsay, P. S., Cumming, A. W. (1989) Three-frequency quasi-periodicity, phase-Locking, and the onset of chaos. *Physica D* 40, 196–217.

Linton, R. (2009) Sound Vision, patterns of vibration in sound, symbols and the body. Master thesis. https://mro.massey.ac.nz/bitstream/handle/10179/1018/02whole.pdf.

Lipson, H. (2014) Challenges and opportunities for design, simulation, and fabrication of soft robots. *Soft Robot* 1, 21–27.

Lipson, H., Pollack, J. B. (2000) Automatic design and manufacture of robotic lifeforms. *Nature* 406, 974–978.

Lisman, J. E. (1997) Bursts as a unit of neural information: Making unreliable synapses reliable. *Trends Neurosci* 20, 38–43.

Lisman, J. E., Idiart, M. A. P. (1995) Storage of 7 ± 2 short-term memories in oscillatory subcycles. *Science* 267, 1512–1515.

Liu, C., Reppert, S. M. (2000) GABA synchronizes clock cells within the suprachiasmatic circadian clock. *Neuron* 25, 123–128.

Liu, H. C. (1991) Quantum inductance in resonant tunneling. *J Appl Phys* 69(4), 2705–2707.

Liu, Y. et al. (2017) Molecular couplings and energy exchange between DNA and water mapped by femtosecond infrared spectroscopy of backbone vibrations. *Struct Dyn* 4(4), 044015.

Livio, M. (2003) Cosmology and life, astro-ph/0301615.

Llinás, R. (1988) The intrinsic properties of mammalian neurons: Insights into central nervous system function. *Science* 242, 1654–1664.

Llinás, R. et al. (2005) Rhythmic and dysrhythmic thalamocortical dynamics: GABA systems and the edge effect. *Trends Neurosci* 28, 325–333.

Llinás, R. R., Ribary, U., Joliot, M., Wang, X. J. (1994) Content and context in temporal thalamocortical binding. In: *Temporal Coding in the Brain*, G. Buzsáki, R. Llinás, W. Singer, A. Berthoz, Y. Christen, Eds. Springer-Verlag, Berlin, Germany, pp. 251–272.

Lloyd, S. (1999) Quantum search without entanglement. *Phys Rev A* 61(01031-1).

Loewer, B. (2001) Determinism and chance. *Stud Hist Philos Phys* 32B(4), 609–620.

Loftus, E. F. (1997) Creating false memories. *Sci Am* 277, 70–75.

Logothetis, N. (2003) The underpinnings of the BOLD functional magnetic resonance imaging signal. *J Neurosci* 23, 3963–3971.

Long, X., Ye, J., Zhao, D., Zhang, S. J. (2015) Magnetogenetics: Remote non-invasive magnetic activation of neuronal activity with a magnetoreceptor. *Sci Bull* 24, 2107–2119.

Loos, H. G. (2003) Nervous system manipulation by electromagnetic fields from monitors. Patent, US 6506148 B2, lens.org/127-943-791-188-651.

Lopes da Silva, F. H., Storm van Leeuwen, W. (1978) The cortical alpha rhythm of the dog: The depth and profile of phase. In: *Architectonics of the Cerebral Cortex*, M. A. B. Brazier, H. Petsche, Eds. Raven, New York, pp. 150–187.

Lopes da Silva, F. H., Vos, J. E., Van Rotterdam, A. (1980) Relative contributions of intracortical and thalamo-cortical processes in the generation of alpha rhythms, revealed by partial coherence analysis. *Electroencephalogr Clin Neurophysiol* 50, 449–456.

Losonczy, A., Makara, J. K., Magee, J. C. (2008) Compartmentalized dendritic plasticity and input feature storage in neurons. *Nature* 452, 436–442.

Lottermoser, T. et al. (2004) Magnetic phase control by an electric field effect. *Nature* 430, 541–544.

Lovinger, A. J. (1983) Ferroelectric polymers. *Science* 220(4602), 1115–1121. doi:10.1126/science.220.4602.1115.

Lowen, W., Miike, L. (1982) *Dichotomies of the Mind: A Systems Science Model of the Mind and Personality*, 1st ed. Wiley, New York.

Lugosi, E. (1989) Analysis of meandering in Zykov-kinetics. *Physica D* 40, 331–337.

Lukinavicius, G. et al. (2014) Fluorogenic probes for live-cell imaging of the cytoskeleton. *Nat Methods* 11, 731–733.

Lundstrom, B. N. et al. (2009) Sensitivity of firing rate to input fluctuations depends on time scale separation between fast and slow variables in single neurons. *J Comput Neurosci* 27, 277–290.

Luryi, S. (1988) Quantum capacitance devices. *Appl Phys Lett* 52(6).

Lykken, G. I., Geiger, A. L., Mitchell, E. N. (1970) Measurement of the Fermi Velocity in single crystal films of lead by electron tunneling. *Phys Rev Lett* 25, 1578–1580.

Lytton, W. W., Sejnowski, T. J. (1991) Simulations of cortical pyramidal neurons synchronized by inhibitory interneurons. *J Neurophysiol* 66, 1059–1079.

Maass, W., Natschlaeger, T., Markram, H. (2002) Real-time computing without stable states: A new framework for neural computation based on perturbations. *Neural Comput* 14(11), 2531–2560.

MacKay, D. M. (1960) On the logical indeterminacy of a free choice. *Mind* 69(273), 31–40.

MacKay, R. S. (1991) Transition of the phase-resetting map for kicked oscillators. *Physica D* 52, 254–266.

MacKinnon, R., Aldrich, R. W., Lee, A. W. (1993) Functional stoichiometry of Shaker potassium channel inactivation. *Science* 262(5134), 757–759.

Mackinnon, S. E., Yee, A., Ray, W. Z. (2012) Nerve transfers for the restoration of hand function after spinal cord injury. *Journal of Neurosurgery* 117(1), 176–185.

MacLennan, B. J. (2002) Universally programmable intelligent matter: A systematic approach to nanotechnology, Presentation at IEEE-Nano.

MacLeod, K., Laurent, G. (1996) Distinct mechanisms for synchronization and temporal patterning of odor-encoding neural assemblies. *Science* 274, 976–979.

Magee, J. A. (2003) A prominent role for intrinsic neuronal properties in temporal coding. *Trends Neurosci* 26, 14–16.

Mahadevan, L., Matsudaira, P. (2000) Motility powered by supramolecular springs and ratchets. *Science* 288(5463), 95–100.

Mahajan, S. M. (2010) Twisting space-time: Relativistic origin of seed magnetic field and vorticity. *Phys Rev Lett* 105, 095005.

Malevanets, A., Kapral, R. (1996) Links, knots and knotted labyrinths in bistable systems. *Phys Rev Lett* 77, 767–770.

Malinowski, J. R., Laval-Martin, D. L., Edmunds, L. N., Jr. (1985) Circadian oscillators, cell cycles, and singularities: Light perturbations of the free-running rhythm of cell division in *Euglena J Comp Physiol* 155B, 257–276.

Mallat, S., Hwang, W. L. (1992) Singularity signal detection and processing with wavelets. *IEEE Trans Inform Theory* 38, 617–643.

Mandelbrot, B. (1999) *Multifractals and 1/f Noise: Wild Self-affinity in Physics*. Springer, New York.

Mandelkow, E.-M., Mandelkow, E., Milligan, R. A. (1991) Microtubule dynamics and microtubule caps: A time-resolved cryo-electron microscopy study. *J Cell Biol* 114, 977–991.

Manor, Y., Gonczarowski, J., Segev, I. (1991) Propagation of action potentials along complex axonal trees. Model and implementation. *Biophys J* 60(6), 1411–1423.

Marder, E., Calabrese, R. L. (1996) Principles of rhythmic motor pattern generation. *Physiol Rev* 76, 687–717.

Marek-Crnjac, L. (2009) A short history of fractal-Cantorian space-time. *Chaos Soliton Fract* 41, 2697–2705.

Marella, S., Ermentrout, B. (2010) Amplification of asynchronous inhibition-mediated synchronization by feedback in recurrent networks. *PLoS Comp Biol* 6, e1000679.

Margolus, N. (2003) Looking at nature as a computer. *Int J Theor Phys* 42(2), 309–327.

Margolus, N., Levitin, B. (1998) The maximum speed of dynamical evolution. *Physica D* 120, 188.

Marion-Poll, F., Tobin, T. R. (1992) Temporal coding of pheromone pulses and trains in *Manduca sexta*. *J Comp Physiol A* 171, 505–512.

Martin, R. D., Harvey, P. H. (1985) Brain size allometry ontogeny and phylogeny. In: *Size and Scaling in Primate Biology*, W. L. Jungers, Ed. Plenum, New York, pp. 147–173.

Marwan, N., Romano, M. C., Thiel M., Kurths, J. (2007) Recurrence plots for the analysis of complex systems. *Phys Rep* 438(5–6), 237.

Maskey, D. et al. (2010) Effect of 835 MHz radiofrequency radiation exposure on calcium binding proteins in the hippocampus of the mouse brain. *Brain Res* 1313, 232–241.

Massimini, M., Huber, R., Ferrarelli, F., Hill, S., Tononi, G. (2004) The sleep slow oscillation as a traveling wave. *J Neurosci* 24, 6862–6870.

Mathews, M. et al. (2010) Light-driven reversible handedness inversion in self-organized helical superstructures. *J Am Chem Soc* 132(51), 18361–18366. doi:10.1021/ja108437n.

Matsukura, F., Tokura, Y., Ohno, H. (2015) Control of magnetism by electric fields. *Nat Nanotech* 10, 209–220.

Matthews, P. C., Strogatz, S. H. (1990) Phase diagram for the collective behavior of limit-cycle oscillators. *Phys Rev Lett* 64, 1701–1704.

Maxwell, J. C. (1890) On physical lines of force; The scientific papers of James Clerk Maxwell, W. D. Niven, Ed. Cambridge University Press, Cambridge, Ch 23, p. 477.

McCarthy, W. (2003) *Hacking Matter: Levitating Chairs, Quantum Mirages, and the Infinite Weirdness of Programmable Atoms*. Basic Books, New York.

McClelland, J. L., McNaughton, B. L., O'Reilly, R. C. (1995) Why there are complementary learning systems in the hippocampus and neocortex: Insights from the successes and failures of connectionist models of learning and memory? *Psychol Rev* 102, 419–457.

McCormick, D. A., Shu, Y., Yu, Y. (2007) Neurophysiology: Hodgkin and Huxley model–still standing? *Nature* 445, E1-2; discussion E2-3.

McDermott, M. L. et al. (2017) DNA's chiral spine of hydration. *ACS Cent Sci*, 3, 708–714.

McDuff, D., Wehrheim, K. (2012) Kuranishi atlases with trivial isotropy: The 2013 state of affairs. Symplectic Geometry (math. SG). arXiv:1208.1340.

McIntyre, R. L., Fahy, G. M. (2015) Aldehyde-stabilized cryopreservation. *Cryobiology* 71(3), 448–458.

Mead, C. A., Truhlar, D. G. (1979) On the determination of Born–Oppenheimer nuclear motion wave functions including complications due to conical intersections and identical nuclei. *J Chem Phys* 70, 2284. doi:10.1063/1.437734.

Meddis, R., O'Mard, L. (1997) A unitary model of pitch perception. *J Acoust Soc Am* 102, 1811–1820.

Mehta, M. R., Lee, A. K., Wilson, M. A. (2002) Role of experience and oscillations in transforming a rate code into a temporal code. *Nature* 417, 741–746.

Mehta, N., Cheng, H. Y. (2012) Micro-managing the circadian clock: The role of microRNAs in biological timekeeping. *J Mol Biol*. doi:10.1016/j.jmb.2012.10.022.

Meklachi, T., Milton, G. W., Onofrei, D., Thaler, A. E., Funchess, G. (2016) Sensitivity of anomalous localized resonance phenomena with respect to dissipation. *Quart Appl Math* 74, 201–234.

Menaker, M., Takahashi, J. S., Eskin, A. (1978) The physiology of circadian pacemakers. *Annu Rev Physiol* 40, 501–526.

Menaker, W., Menaker, A. (1959) Lunar periodicity as a unit of time in human reproduction. *Am J Obstet Gynec* 77, 905–914.

Mendonca, J. T., Dodonov, V. V. (2014) Time crystal in ultracold matter. *J Russ Laser Res* 35(1), 93–100.

Mengüç, Y., Correll, N., Kramer, R., Paik, J. (2017) Will robots be bodies with brains or brains with bodies? *Sci Robot* 2, eaar4527.

Merica, H., Fortune, R. D. (2000) Brainstem origin for a new very slow (1 mHz) oscillation in the human non-REM sleep episode. *Sleep Res Online* 3, 53–59.

Merkle, R. C. (2016) Molecular mechanical computing systems. IMM Report No. 46, pp. 1–38.

Meron, E. (1989) Nonlocal effects in spiral waves. *Phys Rev Lett* 63, 684–687.

Merzel, F., Smith, J. C. (2002) Is the first hydration shell of lysozyme of higher density than bulk water? *Proc Natl Acad Sci U S A* 99(8), 5378–5383.

Mesquita, M. V., Vasconcellos, A. R., Luzzi, R. (1993) Amplification of coherent polar vibrations in bio-polymers: Fröhlich condensate. *Phys Rev E* 48, 4049–4059.

Mesquita, M. V., Vasconcellos, A. R., Luzzi, R. (1998) Positive-feedback-enhanced Fröhlich's Bose-Einstein-like condensation in biosystems. *Int J Quant Chem* 66, 177–187.

Mezrich, R. S. (1970) Magnetic holography. *Appl Opt* 9(10), 2275–2279.

Michalareas, G. et al. (2016) Alpha-beta and gamma rhythms subserve feedback and feedforward influences among human visual cortical areas. *Neuron* 89(2), 384.

Michalski, P. J., Mele, E. J. (2008) Carbon nanotubes in helically modulated potentials. *Phys Rev B* 77, 085429-1-11.

Michler, P. et al. (2000) Quantum correlation among photons from a single quantum dot at room temperature. *Nature* 406, 968–970.

Middlebrooks, J. C., Clock, A. E., Xu, L., Green, D. M. (1994) A panoramic code for sound location by cortical neurons. *Science* 264, 842–844.

Miller, G. A. (1956) The magical number seven, plus or minus two: Some limits on our capacity for processing information. *Psychol Rev* 63, 81–97.

Miller, R. (1989) Cortico-hippocampal interplay: Self-organizing phase-locked loops for indexing memories. *Psychobiology* 17, 115–128.

Miller, R. Ed. (2000) *Time and the Brain*. Harwood Academic Publishers, Singapore.

Milton, G. W., Briane, M., Willis, J. R. (2006) On cloaking for elasticity and physical equations with a transformation invariant form. *New J Phys* 8, 248.

Milton, G. W., Nicorovici, N.-A. (2006) On the cloaking effects associated with anomalous localized resonance. *Proc R Soc A* 462, 3027.

Minoura, I., Muto, E. (2006) Dielectric measurement of individual microtubules using the electroorientation method. *Biophysical J* 90, 3739–3748.

Minsky, M. (1967) *Computation: Finite and Infinite Machines*. Prentice Hall, Upper Saddle River, NJ. Chapter 8, Section 8.2 Unsolvability of the Halting Problem.

Mirollo, R. E., Strogatz, S. H. (1990) Synchronization of pulse coupled biological oscillators. *SIAM J Appl Math* 50, 1645–1662.

Misner, C. W., Thorne, K. S., Wheeler, J. A. (1973) *Gravitation*. W.H. Freeman, San Francisco, CA (now New York) p. 1217.

Mitchell, T. M. (1980) The need for biases in learning generalizations. Technical report CBM-TR-117, Rutgers University, New Brunswick, NJ.

Mithieux, G., Chauvin, F., Roux, B., Rousset, B. (1985) Association states of tubulin in the presence and absence of microtubule-associated proteins: Analysis by electric birefringence. *Biophys Chem* 22(4), 307–316.

Mitkov, I. (1996) Dynamics and interaction of topological defects in excitable media. PhD dissertation, Hebrew University of Jerusalem.

Mittedorf, J. (2016) The epigenetic clock controls aging. *Biogerontology* 17(1), 257–265.

Monserrat, B., Drummond, N. D., Needs, R. J. (2013) Anharmonic vibrational properties in periodic systems: Energy, electron-phonon coupling, and stress. *Phys Rev B* 87, 144302.

Moore-Ede, M. C., Sulzman, F. M., Fuller, C. A. (1982) *The Clocks that Time Us*. Harvard University Press, Cambridge.

Morison, R. S., Basset, D. L. (1945) Electrical activity of the thalamus and basal ganglia in decorticate cats. *J Neurophysiol* 8, 309–314.

Moritsugu, K., Njunda, B. M., Smith, J. C. (2010) Theory and normal-mode analysis of change in protein vibrational dynamics on ligand binding. *J Phys Chem B* 114(3), 1479–1485. doi:10.1021/jp909677p.

Moskalenko, S. A., Miglei, M. F., Khadshi, P. I., Pokatolov, E. P., Kiselyova, E. S. (1980) Coherent phonons and excitations in biological systems. *Phys Lett A* 76, 197–200. The protrusions observed in the AFM images are confirmed as energy-packet group for three reasons, protrusions move ~1 μm per minute, (400 km/hour). Energy-packet density is obtained by counting total number of energy-packet groups (not individual energy-packets) divided by the area scanned, considering 50% visibility as absolute. We calculate periodicity λ for using $\lambda = \pi/k\_F$, where, wave vector $k\_F = \pi/na$, where a is the lattice parameter, 8 nm, and 1/n is the band filling. Mass of energy-packet has been calculated by using the formulae $M\_s = (4/3l)(u\_0/a)^2 M$, where u0 is the lattice distortion, M is the mass of 3 tubulin monomers capturing a defect site or point of asymmetry, 2l is energy-packet length, a is lattice parameter, Ms~300 me for phonon energy-packet periodicity 3. Even though electronic energy-packet has mass Ms~15 me, statistically, phonon energy-packet periodicity 3, reflects synchrony accurately, so we ignore electronic mass.

Moss, F., Wiesenfeld, K. (1995) The benefits of background noise. *Sci Am* 273, 66–69.

Mossallam, B. E. et al. (2016) Multiplexing genetic and nucleosome positioning codes: A computational approach. *PLoS ONE* 11(6), e0156905.

Mosseri, R., Dandoloff, R. (2001) Geometry of entangled states, Bloch spheres and Hopf fibrations. *J Phys A: Math Gen* 34(47), 10243–10252.

Mountcastle, V. (1967) The problem of sensing and the neural coding of sensory events. In: *The Neurosciences: A Study Program*, G. C. Quarton, T. Melnechuk, F. O. Schmitt, Eds. Rockefeller University Press, New York, pp. 393–408.

Moutoussis, K., Zeki, S. (1997) A direct demonstration of perceptual asynchrony in vision. *Proc Biol Sci* 264, 393–269.

Mozo-Villarias, A., Morros, A., Andreu, J. M. (1991) Thermal transitions in the structure of tubulin: Environments of aromatic aminoacids Euro. *Biophys J* 19, 295–300.

Muller, H. (2009) Fractal scaling models of resonant oscillations in chain systems of harmonic oscillators. *Prog Phys* 2, 72–76.

Muller, R. U., Stead, M., Pach, J. (1996) The hippocampus as a cognitive graph. *J Gen Physiol* 107, 663–694.

Mumford, S. (2004) *Laws in Nature*. Routledge, London, UK.

Myers, M. P., Wagner-Smith, K, Rothenfluh-Hilfiker, A., Young, M. W. (1996) Light-induced degradation of TIMELESS and entrainment of the *Drosophila* circadian clock. *Science* 271, 1736–1740.

Näätänen, R., Syssoeva, O., Takegata, R. (2004) Automatic time perception in the human brain for intervals ranging from milliseconds to seconds. *Psychophysiology* 41, 660–663.

Nádasdy, Z., Hirase, H., Czurkó, A., Csicsvari, J., Buzsáki, G. (1999) Replay and time compression of recurring spike sequences in the hippocampus. *J Neurosci* 19, 9497–9507.

Nagumo, J., Arimoto, S., Yoshizawa, S. (1962) An active pulse transmission line simulating nerve axon. *Proc IRE* 50, 2061–2070.

Nair, V. V., Choi, J. R. (2016) An efficiency enhancement technique for a wireless power transmission system based on a multiple coil switching technique. *Energies* 9(156), 1–15.

Nakashima, N. et al. (2005) Helical superstructures of fullerene peapods and empty single-walled carbon nanotubes formed in water. *J Phys Chem B* 109, 13076–13082.

Nalwa, H. (1995) *Ferroelectric Polymers*, 1st ed. Marcel Dekker, New York.

Nam, K., Ott, E., Guzdar, P. N., Gabbay, M. (1998) Stability of spiral wave vortex filaments with phase twists. *Phys Rev E* 58, 2580–2585.

Neff, K. L. et al. (2011) Validation of fractal-like kinetic models by time-resolved binding kinetics of dansylamide and carbonic anhydrase in crowded media. *Biophys J* 100(10), 2495–2503. doi:10.1016/j.bpj.2011.04.016.

Newhouse, E. I., Kopelman, R. (1988) Fractal-like triplet-triplet annihilation kinetics in naphthalene-doped poly(methylmethacrylate). *Chem Phys Lett* 143(1), 106–110.

Nicolelis, M. A., Baccala, L. A., Lin, R. C., Chapin, J. K. (1995) Sensorimotor encoding by synchronous neural ensemble activity at multiple levels of the somatosensory system. *Science* 268, 1353–1358.

Nicorovici, N. A., McPhedran, R. C., Milton, G. W. (1994) Optical and dielectric properties of partially resonant composites. *Phys Rev B Condens Matter* 49(12), 8479–8482.

Nie, Y., Fellous, J. M., Tatsuno, M. (2014) Information-geometric measures estimate neural interactions during oscillatory brain states. *J Front Neural Circuit* 8, 11. doi:10.3389/fncir.2014.00011.

Nieder, A., Miller, E. K. (2003) Coding of cognitive magnitude: Compressed scaling of numerical information in the primate prefrontal cortex. *Neurons* 37, 149–157.

Nieuwland, M. S. et al. (2018) Large-scale replication study reveals a limit on probabilistic prediction in language comprehension. *eLife* 7, e33468. doi:10.7554/eLife.33468.

Nietzsche, F. (1967) The eternal recurrence. In: *Nietzsche*, J. Richardson, B. Leiter, Eds. Oxford University Press, Oxford, pp. 118–138.

Nieves, C. (2016) 3D bioprinting of functional human skin: Production and in vivo analysis. *Biofabrication* 9(1), 015006.

Nikolić, D. (2017) Why deep neural nets cannot ever match biological intelligence and what to do about it? *Int J Autom Comput* 14, 532. doi:10.1007/s11633-017-1093-8.

Nixon, R. A., Shea, T. B. (1992) Dynamics of neuronal intermediate filaments: A developmental perspective. *Cell Motil Cytoskel* 22(2), 81–91.

Nobili, R. (1985) Schrödinger wave holography in brain cortex. *Phys Rev A* 32, 3618.

Nola, R., Sankey, H. (2007) *Theories of Scientific Method*. McGill-Queen's University Press, Montreal, Canada.

Noctor, S., Martinez-Cerdeno, V., Ivic, L., Kriegstein, A. (2004) Cortical neurons arise in symmetric and asymmetric division zones and migrate through specific phases. *Nat Rev Neurosci* 7, 136–144.

Norimoto, H. et al. (2018) Hippocampal ripples down-regulate synapses. *Science* 359(6383), 1524–1527. doi:10.1126/science.aao0702.

Nottale, L. (1989) Fractals and the quantum theory of space-time. *Int J Mod Phys A* 4(19), 5047–5117. doi:10.1142/S0217751X89002156.

Nottale, L. (1992) The theory of scale relativity. *Int J Mod Phys A* 7(20), 4899–4936. doi:10.1142/S0217751X92002222.

Nottale, L. (1997) Scale relativity. In: *Scale Invariance and Beyond*, B. Dubrulle, F. Graner, D. Sornette, Eds. Springer, Berlin, Germany.

Nottale, L. (2004) The theory of scale relativity: Non-differentiable geometry and fractal space-time. Computing Anticipatory Systems: CASYS'03 – Sixth International Conference, Liège 1993. *AIP Conf Proc* 718, 68–95. doi:10.1063/1.1787313.

Nottale, L. (2011) *Scale Relativity and Fractal Space-Time: A New Approach to Unifying Relativity and Quantum Mechanics*. Imperial College Press, London, UK.

Nottale, L. C. (2005) On the transition from the classical to the quantum regime in fractal space-time theory. *Soliton Fract* 25, 797–803.

Nowakowski, T. J. (2018) Building blocks of the human brain. *Science* 362(6411), 169.

Nozakura, T., Ikeuehi, S. (1984) Formation of dissipative structures in galaxies. *Astrophys J* 279, 40–52.

Nuñez, P. L. (1981) *Electrical Fields of the Brain*. Oxford University Press, Oxford.

Nuñez, P. L. (1998) *Electric Fields of the Brain: The Neurophysics of EEG*. Oxford University Press, New York.

Nye, J. F. (1983) Polarization effects in the diffraction of electromagnetic waves: The role of disclinations. *Proc R Soc Lond A* 387, 105–132.

Ofer, N., Shefi, O. (2016) Axonal geometry as a tool for modulating firing patterns. *Appl Math Model* 40, 3175–3184.

Oh, S. et al. (2016) Singularity of the time energy uncertainty in adiabatic perturbation and cycloids on a Bloch sphere. *Sci Rep* 6, 20824.

Oh, S.-H. et al. (2017) Coherent terahertz spin-wave emission associated with ferrimagnetic domain wall dynamics. *Phys Rev B* 96, 100407(R).

O'Holleren, K., Padgett, M. J., Dennis, M. R. (2006) Observation of quantum entanglement using spatial light modulators. *Opt Express* 14, 3039–3044.

O'Keefe, J., Burgess, N. (1996) Geometric determinants of the place fields of hippocampal neurons. *Nature* 381, 425–428.

O'Keefe, J., Recce, M. L. (1993) Phase relationship between hippocampal place units and the EEG theta rhythm. *Hippocampus* 3, 317–330.

Omohundro, S. (1984) Modelling cellular automata with partial differential equations. *Phys D Nonlin Phen* 10, 128.

Onoda, N., Mori, K. (1980) Depth distribution of temporal firing patterns in olfactory bulb related to air intake cycles. *J Neurophysiol* 44, 29–39.

Ord, G. N. (1983) Fractal space-time a geometric analogue of relativistic quantum mechanics. *J Phys A* 16, 1869–1884.

Ord, G. N. (2012) Quantum phase from the twin paradox. *J Phys* 361, 012007.

Ord, G. N., Mann, R. B. (2012) How does an electron tell the time? *Int J Theo Phys* 51(2), 652–666.

Otting, G., Liepinsh, E., Wuthrich, K. (1991) Protein hydration in aqueous solution. *Science* 41, 974–980.

Packel, E. W., Traub, J. F. (1987) Information-based complexity. *Nature*, 328, 29.

Pagels, H. (1983) *The Cosmic Code*. Bantam Books, New York.

Palmer, R. G. (1982) Broken ergodicity. *Adv Phys* 31(6), 669. doi:10.1080/00018738200101438.

Panarella, E. (1987) Non-linear behavior of light at low intensities: The photon clump model. In: *Quantum Uncertainties—Recent and Future Experiments and Interpretations*, W. M. Honing, D. W. Kraft, E. Panarella, Eds., Plenum Press, New York, p. 105.

Pancharatnam, S. (1956) Generalized theory of interference, and its applications. Part I. Coherent Pencils. *Proc Indian Acad Sci A* 44(5), 247–262.

Panfilov, A. V., Rudenko, A. N., Winfree, A. T. (1985) Twisted vortex waves in three-dimensional active media. *Biofizika* 30, 464–466.

Panman, M. R., van Dijk, C. N., Meuzelaar, H., Woutersen, S. (2015) Nanosecond folding dynamics of an alpha helix: Time-dependent 2D-IR cross peaks observed using polarization-sensitive dispersed pump-probe spectroscopy. *J Chem Phys* 142, 041103.

Paré, D., Collins, D. R., Pelletier, J. G. (2002) Amygdala oscillations and the consolidation of emotional memories. *Trends Cogn Sci* 6, 306–314.

Parnas, I., Hochstein, S., Parnas, H. (1976) Theoretical analysis of parameters leading to frequency modulation along an inhomogeneous axon. *J Neurophysiol* 39, 909–923.

Partsvania, B., Sulaberidze, T., Shoshiashvili, L. (2013) Effect of high SARs produced by cell phone like radiofrequency fields on mollusk single neuron. *Electromagn Biol Med* 32(1), 48–58.

Pasca, S. P. (2018) The rise of three-dimensional human brain cultures. *Nature* 553, 437–445.

Paterson, M. S., Fischer, M. S., Meyer, A. R. (1974) An improved overlap argument for on-line multiplication. *SLAM-AMS Proc* 7(1974), 97–111.

Pattle, K. et al. (2018) First observations of the magnetic field inside the pillars of creation: Results from the BISTRO survey, Astrophysical Journal Letters: arxiv.org/abs/1805.11554; Read more at: https://phys.org/news/2018-06-magnetic-fields-key-star-formation.html#jCp.

Paydarfar, D., Eldridge, F. L., Kiley, J. P. (1986) Resetting of mammalian respiratory rhythm: Existence of a phase singularity. *Am J Physiol* 250, R721–R727.

Pearson, C. G., Winey, M. (2009) Basal body assembly in ciliates: The power of numbers. *Traffic* 10, 461–471.

Pechal, M. et al. (2012) Geometric phase and nonadiabatic effects in an electronic harmonic oscillator. *Phys Rev Lett* 108, 170401.

Peirce, C. S. (1940) *The Philosophy of Peirce: Selected Writings*, ed. J. Buchler. Routledge and Kegan Paul, London, pp. 593–595.

Pendry, J. B. (2000) Negative refraction makes a perfect lens. *Phys Rev Lett* 85, 3966.

Pendry, J. B. (2004) A chiral route to negative refraction. *Science* 306(5700), 1353–1355.

Pendry, J. B., Schurig, D., Smith, D. R. (2006) Controlling electromagnetic fields. *Science* 312, 1780–1782.

Peng, K. et al. (2015) Single nanowire photoconductive terahertz detectors. *Nano Lett* 15, 206–210.

Penrose, R. (1969) Gravitational collapse: The role of general relativity. *Riv Nuovo Cimento* 1, 252–276.

Penrose, R. (1979a) Singularities and time-asymmetry. In: *General Relativity: An Einstein Centenary*, S. W. Hawking, W. Israel, Eds. Cambridge University Press, Cambridge, pp. 581–638.

Penrose, R. (1979b) The topology of ridge systems. *Ann Hum Gen Lond* 42, 435–444.

Penrose, R. (1989) *The Emperor's New Mind*, Oxford University Press, New York.

Penttonen, M., Buzsáki, G. (2003) Natural logarithmic relationship between brain oscillators. *Thalamus Relat Syst* 48, 1–8.

Pereira, D. P. (2016) Fluorescent false neurotransmitter reveals functionally silent dopamine vesicle clusters in the striatum. *Nat Neurosci* 19, 578–586.

Peres, A. (1980) Measurement of time by quantum clocks. *Am J Phys* 48, 552.

Perge, J. A. et al. (2012) Why do axons differ in caliber? *J Neurosci* 32(2), 626–638.

Perkell, D. H., Bullock, T. H. (1968) Neural coding. *Neurosci Res Program Bull* 6, 221–348.

Pertsov, A. M., Ermakova, E. A., Schnoll, E. E. (1990) On the diffraction of autowaves. *Physica D* 44, 178–190.

Peterson, E. L., Jones, M. D. R. (1979) Do circadian oscillators ever stop in constant light? *Nature* 280, 677–679.

Pethig, R. (1979) *Dielectric and Electronic Properties of Biological Materials*. Wiley, New York, p. 139.

Peyré, G., Cuturi, M. (2019) Computational optimal transport. *Found Trends Mach Learn* 11(5–6), 355–607, arXiv:1803.00567 [stat.ML].

Pfeifer, P. (1993) How fast can a quantum state change with time? *Phys Rev Lett* 70, 3365.

Phillipson, O. T., Griffiths, A. C. (1985) The topographic order of inputs to nucleus accumbens in the rat. *Neuroscience* 16, 275–296. doi:10.1016/0306-4522(85)90002-8.

Pickard, G. E., Turek, F. W. (1983) The suprachiasmatic nuclei: Two circadian clocks? *Brain Res* 268, 201–210.

Pierce, J. R. (1961) *Symbols, Signals and Noise: The Nature and Process of Communication.* Harper and Brothers, New York.

Pikovsky, A., Maistrenko, Y. L. (2003) *Synchronization, Theory and Application.* Springer, Dordrecht, the Netherlands.

Pinsker, H. M. (1977) Aplysia bursting neurons as endogenous oscillators. I: Phase response curves for pulsed inhibitory synaptic input and II. Synchronization and entrainment by pulsed inhibitory synaptic input. *J Neurophysiol* 40, 527–543 and 544–556.

Pippinger, N., Fischer, M. J. (1979) Relations among complexity measures. *J ACM* 26(2), 361–381.

Piramuthu, S., Shaw, M. J., Gentry, J. A. (1994) A classification approach using multi-layered neural networks. *Decis Support Syst* 11(5), 509–525.

Pistolesi, F., Manini, N. (2000) Geometric phases and multiple degeneracies in harmonic resonators. *Phys Rev Lett* 85, 1585.

Plailly, J., Tillmann, B., Royet, J. P. (2007) The feeling of familiarity of music and odors: The same neural signature? *Cereb Cortex* 17(11), 2650–2658.

Plum, F., Posner, J. B. (1980) *The Diagnosis of Stupor and Coma*, 3rd ed. F.A. Davis Company, Philadelphia.

Poddubny, A., Iorsh, I., Belov, P., Kivshar, Y. (2013) Hyperbolic metamaterials. *Nat Photonics* 7, 948–957.

Poincaré, H. (1890) Sur le problème des trois corps et les équations de la dynamique. *Acta Math* 13, 1–270. Œuvres VII 262–490 (theorem 1 section 8).

Poirier, B. (2010) Bohmian mechanics without pilot waves. *Chem Phys* 370, 1–3, 4–14.

Pokorny, J., Jelenek, F., Trkval, V., Lamprecht, I., Holtzel, R. (1997) Vibrations in microtubules. *J Biol Phys* 48, 261–266.

Ponstingl, H. et al. (1982) Amino Acid Sequence of α- and β-Tubulins from Pig Brain: Heterogeneity and Regional Similarity to Muscle Proteins. *Cold Spring Harbor Symp Quant Biol* 46, 191–197.

Popper, K. (2002) *The Logic of Scientific Discovery*, 2nd ed. Routledge, London (Reprint of 1959 translation of 1935 original) Page 57.

Portan, E. et al. (2017) Dynamical Majorana edge modes in a broad class of topological mechanical systems. *Nat Commun* 8, 14587.

Poston, T., Winfree, A. T. (1992) *Complex Singularities and Excitable Media.* Manuscript unpublished due to difficulty finishing the figures at Tucson. Reinberg.

Poznanski, R. R. et al. (2017) Induced mitochondrial membrane potential for modelling solitonic conduction of electrotonic signals. *PLOS One* 12, e0183677.

Prati, E. (2009) The nature of time: From a timeless Hamiltonian framework to clock time metrology, arXiv:0907.1707v1, 10 July.

Pratt, V. (1975) Every prime has a succinct certificate. *SIAM J Comput* 4, 214–220.

Preparata, F. P., Shamos, M. I. (1985) *Computational Geometry—An Introduction*, 1st ed. Springer-Verlag; 2nd printing, corrected and expanded, 1988.

Price, H. (1996) *Time's Arrow and Archimedes' Point*. Oxford University Press, Oxford, especially chs. 1–2 and 8–9.

Price, H. (2002) Burbury's last case: The mystery of the entropic arrow. In: *Time, Reality and Experience*, C. Callender, Ed. Cambridge University Press, Cambridge, pp. 19–56.

Prigogine, I., Rössler, O., El Naschie, M. S. (1995) *Quantum Mechanics, Diffusion and Chaotic Fractals*. Pergamon, Oxford.

Prochazkova, E., Kret, M. E. (2017) Connecting minds and sharing emotions through mimicry: A neurocognitive model of emotional contagion. *Neurosci Biobehav Rev* 80, 99–114.

Prodan, E., Prodan, C. (2009) Topological phonon modes and their role in dynamic instability of microtubules. *Phys Rev Lett* 103, 248101.

Puget, J.-F. (1993) On the satisfiability of symmetrical constraint satisfaction problems. In: *Methodologies for Intelligent Systems*, volume 689 of Lecture Notes in Artificial Intelligence, Springer, Berlin, pp. 350–361.

Qin, L., Gu, W., Wei, J., Yu, Y. (2018) Piecewise phototuning of self-organized helical superstructures. *Adv Mater* 30(8). doi:10.1002/adma.201704941.

Qiu, C., Shivacharan, R. S., Zhang, M., Durand, D. M. (2015) Can neural activity propagate by endogenous electrical field? *J Neurosci* 35(48), 15800–15811.

Qu, L., Peng, Q., Dai, L., Spinks, G. M., Wallace, G. G., Baughman, R. H. (2008) Carbon nanotube electroactive polymer materials: Opportunities and challenges. *MRS Bulletin* 33(3), 215–224. doi:10.1557/mrs2008.47.ISSN 0883-7694.

Qu, Y. et al. (2018) Superelastic multimaterial electronic and photonic fibers and devices via thermal drawing. *Adv Mater* 2018, 1707251.

Quine, W. V. O. (1980) p. 18 in the essay "On what there is." In: *From a Logical Point of View*, 2nd ed. Harvard University Press, Cambridge, MA, pp. 1–19.

Quiroz, F. G., Chilkoti, A. (2015) Sequence heuristics to encode phase behaviour in intrinsically disordered protein polymers. *Nat Mater* 14, 1164–1171.

Raastad, M., Shepherd, G. (2003) Single-axon action potentials in the rat hippocampal cortex. *J Physiol* (Lond) 548, 745–752.

Rabinovich, M., Huerta, R., Laurent, G., (2008) Neuroscience: Transient dynamics for neural processing. *Science* 321, 48.

Rabitz, H. (2012) Control in the sciences over vast length and time scales. *Quant Phys Lett* 1, 1–19.

Radford, B. (2010) *Scientific Paranormal Investigation: How to Solve Unexplained Mysteries*. L. J. S. Goodlin, Ed. Rhombus Publishing Company, Corrales, NM.

Radman, T., Ramos, R. L., Brumberg, J. C., Bikson, M. (2009) Role of cortical cell type and morphology in subthreshold and suprathreshold uniform electric field stimulation in vitro. *Brain Stimul* 2, 215–228.e3.

Radman, T., Su, Y., An, J. H., Parra, L. C., Bikson, M. (2007) Spike timing amplifies the effect of electric fields on neurons: Implications for endogenous field effects. *J Neurosci* 27, 3030–3036.

Raghavachari, S. et al. (2001) Gating of human theta oscillations by a working memory task. *J Neurosci* 21, 3175–3183.

Rajaraman, R. (1982) *Solitons & Instantons*. North-Holland Publishing Company, Amsterdam, the Netherlands, Chapter 3, 5, p. 70.

Rajat, C. Q. et al. (2015) Can neural activity propagate by endogenous electrical field? *J Neurosci* 35(48), 15800–15811.

Rakic, P. (1988) Specification of cerebral cortical areas. *Science* 241(4862), 170–176. doi:10.1126/science.3291116. PMID 3291116.

Rall, W. (1959) Branching dendritic trees and motor neuron membrane resistivity. *Exp Neurol* 1, 491–527.

Ralston, B., Ajmone-Marsan, C. (1956) Thalamic control of certain normal and abnormal cortical rhythms. *Electroencephalogr Clin Neurophysiol* 8, 559–583.

Ramirez, S. (2018) Crystallizing a memory. *Science* 60(6394), 1182–1183. https://lifeboat.com/blog/2018/06/harvard-rewinds-the-biological-clock-of-time.

Ramkumar, P., Parkkonen, L., Hari, R., Hyvärinen, A. (2012) Characterization of neuromagnetic brain rhythms over time scales of minutes using spatial independent component analysis. *Hum Brain Mapp* 33(7), 1648–1662. doi:10.1002/hbm.21303.

Ramm, A. G. et al. (1996) Minimization of the total radiation from an obstacle by a control function on the part of its boundary. *J Inv Ill Posed Prob* 4, 531–534.

Ramón, F., Moore, J. W. (1978) Ephaptic transmission in squid giant axons. *Am J Physiol* 234(5), C162–C169.

Ramsey, F. P. (1926) Truth and probability. In: *The Foundations of Mathematics and Other Logical Essays*, R. B. Braithwaite, Ed. Kegan Paul, Trench, Trübner (1931), London, Chapter VII, pp. 156–198.

Randy, F. (2008) Electroactive Polymer Artificial Muscle—A Polymer Based Generator? (PDF). Thin Film Users Group. Northern California Chapter of the American Vacuum Society. Retrieved July 16, 2012.

Raslau, F. D. et al. (2015) Memory Part 3: The role of the fornix and clinical cases. *Am J Neuroradiol* 36(9), 1604–1608. doi:10.3174/ajnr. A4371.

Rasmussen, R., Jensen, M. H., Heltberg, M. L. (2017) Chaotic dynamics mediate brain state transitions, driven by changes in extracellular ion concentrations. *Cell Syst* 5(6), 591–603.

Rathmanner, S., Hutter, M. (2011) A philosophical treatise of universal induction. *Entropy* 13(6), 1076–1136.

Ratté, S. et al. (2015) Subthreshold membrane currents confer distinct tuning properties that enable neurons to encode the integral or derivative of their input. *Front Cell Neurosci* 8, 452.

Reddy, S. et al. (2018) A brain-like computer made of time crystal: Could a metric of prime alone replace a user and alleviate programming forever? In: *Soft Computing Applications*, K. Ray, M. Pant, A. Bandyopadhyay, Eds. Studies in computational Intelligence vol. 761, Springer Nature Singapore Pvt. Ltd, Singapore. doi:10.1007/978-981-10-8049-4_1.

Reichenbach, H. (1956) *The Direction of Time*, University of California Press, Berkeley.

Reimann, M. W. et al. (2015) An algorithm to predict the connectome of neural microcircuits. *Front Comput Neurosci.* doi:10.3389/fncom.2015.00120.

Reimann, M. W. et al. (2017) Cliques of neurons bound into cavities provide a missing link between structure and function. *Front Comput Neurosci.* doi:10.3389/fncom.2017.00048.

Reinagel, P., Reid, C. (2000) Temporal coding of visual information in the thalamus. *J Neurosci* 20, 5392–5400.

Reinberg, A., Smolensky, M. H. (1983) *Biological Rhythms and Medicine*. Springer-Verlag, New York.

Reiner, V. S., Antzeleviteh, C. (1985) Phase-resetting and annihilation in a mathematical model of the sinus node. *Am J Physiol* 249, H1143–H1153.

Ren, J., Momose-Sato, Y., Sato, K., Greer, J. J. (2006) Rhythmic neuronal discharge in the medulla and spinal cord of fetal rats in the absence of synaptic transmission. *J Neurophysiol* 95(1), 527–534.

Rensing, L., van der Heiden, U., Mackey, M. C. Eds. (1987) *Temporal Disorder in Human Oscillatory System*. Springer Series in Synergetics, Vol 36. Springer, New York.

Rescigno, A., Stein, R. B., Purple, R. L., Poppele, R. E. (1970) A neuronal model for the discharge patterns produced by cyclic inputs. *Bull Math Biophys* 32, 337–353.

Reshef, D. N. et al. (2011) Detecting novel associations in large data sets. *Science* 334, 1518.

Resta, R. (1997) Polarization as a Berry Phase. *Europhysics News* 28, 18.

Reuter, M. et al. (2012) Highly birefringent liquid crystal for THz applications; Microwave Conference (GeMiC), The 7th German; INSPEC Accession Number: 12691621; IEEE.

Reynolds, J. H., Desimone, R. (1999) The role of neural mechanisms of attention in solving the binding problem. *Neuron* 24, 19–29.

Ricca, R. L. (1993) Torus knots and polynomial invariants for a class of soliton equations. *Chaos* 3, 83–91.

Ricca, R. L., Samuels, D. C., Barenghi, C. F. (1999) Evolution of vortex knots. *J Fluid Mech* 391, 29–44.

Richmond, B., Knopfmacher, A. (1995) Compositions with distinct parts. *Aequationes Math* 49, 86–97.

Richmond, B. J., Optican, L. M., Gawne, T. J. (1989) Neurons use multiple messages encoded in temporally modulated spike trains to represent pictures. In: *Seeing Contour and Colour*, J. J. Kulikowski, C. M. Dickenson, Eds. Pergamon Press, New York. pp. 705–713.

Richter, C. (1960) Biological clocks in medicine and psychiatry; shock-phase hypothesis. *Proc Natl Acad Sci USA* 46, 1506–1530.

Richter, A., Türke, A., Pich, A. (2007) Controlled double-sensitivity of microgels applied to electronically adjustable chemostats. *Adv Mater* 19(8), 1109–1112. doi:10.1002/adma.200601989.

Ridge, E. et al. (2005) Moving nature-inspired algorithms to parallel, asynchronous and decentralised environments. In *Self-Organization and Autonomic Informatics (I)*, H. Czap, R. Unland, C. Brank, H. Tianfield, Eds. IOS Press, Amsterdam, the Netherlands, pp. 35–49.

Ringo, J. L. (1991) Neuronal interconnection as a function of brain size. *Brain Behav Evol* 38, 1–6.

Ringo, J. L., Doty, R. W., Demeter, S., Simard, P. Y. (1994) Time is of the essence: A conjecture that hemispheric specialization arises from interhemispheric conduction delay. *Cereb Cortex* 4, 331–343.

Rivnay, J. et al. (2017) Next-generation probes, particles, and proteins for neural interfacing. *Sci Adv* 3, e1601649.

Rocke, C. et al. (1998) Exciton ionization in a quantum well studied by surface acoustic waves. *Phys Rev B* 57, R6850(R).

Roglic, D. (2007) The universal evolutionary computer based on super-recursive algorithms of evolvability. https://arxiv.org/abs/0708.2686.

Rommell, S. A., McCleave, J. D. (1972) Oceanic electric fields: Perception by American eels. *Science* 176, 1233.

Rose, J. E., Mountcastle, V. (1959) Touch and kinesis. In: *Handbook of Physiology: Neurophysiology*, Vol. II, J. Field, H. W. Magoun, V. E. Hall, Eds. American Physiological Society, Washington, DC, pp. 387–429.

Rose, N. S. (2016) Reactivation of latent working memories with transcranial magnetic stimulation. *Science* 354(6316), 1136–1139.

Roskies, A. (1999) The binding problem. *Neuron* 24, 7–9.

Roskos, H. G. et al. (1992) Coherent submillimeter-wave emission from charge oscillations in a double-well potential. *Phys Rev Lett* 68, 2216.

Roth, B. J., Guo, W. Q. (1988) The effects of spiral anisotropy on the electric potential and the magnetic field at the apex of the heart. *Math Biosci* 88, 191–221.

Roth, B. J., Wikswo, J. P., Jr. (1985) The magnetic field of a single axon. A comparison of theory and experiment. *Biophys J* 48, 93–109.

Rotstein, H. G., Pervouchine, D. D., Acker, C. D., Gillies, M. J., White, J. A., Buhl, E. H., Whittington, M. A., Kopell, N. (2005) Slow and fast inhibition and an H-current interact to create a theta rhythm in a model of CA1 interneuron network. *J Neurophysiol* 94, 1509–1518.

Rougeul-Buser, A., Buser, P. (1997) Rhythms in the alpha band in cats and their behavioural correlates. *Int J Psychophysiol* 26, 191–203.

Rousseau, G., Chate, H., Kapral, R. (1998) Coiling and supercoiling of vortex filaments in oscillatory media. *Phys Rev Lett* 80, 5671–5674.

Royer, S. et al. (2012) Control of timing, rate and bursts of hippocampal place cells by dendritic and somatic inhibition. *Nat Neurosci* 15, 769–775.

Rual, J. F. (2005) Towards a proteome-scale map of the human protein–protein interaction network. *Nature* 437, 1173–1178.

Rugar, D., Yamoni, C. S., Shidles, J. A. (1992) Mechanical detection of magnetic resonance. *Nature* 360, 563–566.

Rus, D., Tolley, M. T. (1999) Design, fabrication and control of soft robots. *Nature* 521, 467–475.

Rusak, B. (1993) Open forum: Human phase-resetting sensitivity to light. *J Biol Rhy* 8, 339–361.

Rusov, V. D. et al. (2012) Can resonant oscillations of the earth-ionosphere influence the human brain biorhythm? *General Physics*. pp. 1–13. arXiv:1208.4970v1.

Russell, B. (1901) Principia Mathematica. Godehard Link (2004), One hundred years of Russell's paradox, p. 350.

Russell, B. (1938) [First published 1903]. *Principles of Mathematics*, 2nd ed. W. W. Norton & Company, New York.

Russell, B. (1948) *Human Knowledge: Its Scope and Limits*. Routledge, London, UK.

Rust, M. J., Bates, M., Zhuang, X. (2006) Sub-diffraction-limit imaging by stochastic optical reconstruction microscopy (STORM). *Nat Methods* 3, 793–795.

Rutishauser, U., Ross, I. B., Mamelak, A. N., Schuman, E. M. (2010) Human memory strength is predicted by theta-frequency phase-locking of single neurons. *Nature* 464(7290), 903–907.

Sahoo, P., Sankolli, R., Lee, H.-Y., Raghavan, S. R., Dastidar, P. (2012) Gel sculpture: Moldable, load-bearing and self-healing non-polymeric supramolecular gel derived from a simple organic salt. *Chem Eur J* 18, 8057–8063.

Sahu, S. et al. (2009) Remarkable potential of pattern based computing on an organic molecular layer using the concept of cellular automata. IEEE PID 107, 2403, IEEE, Kanazawa.

Sahu, S. et al. (2012) On Cellular Automata rules of molecular arrays. *Nat Comput* 11(2), 311–321.

Sahu, S. et al. (2013a) Multi-level memory-switching properties of a single brain microtubule. *Appl Phys Lett* 102, 123701.

Sahu, S. et al. (2013b) Atomic water channel controlling remarkable properties of a single brain microtubule: Correlating single protein to its supramolecular assembly. *Biosens Bioelectron* 47, 141–148.

Sahu, S. et al. (2014) Live visualizations of single isolated tubulin protein self-assembly via tunneling current: Effect of electromagnetic pumping during spontaneous growth of microtubule. *Sci Rep* 4, 7303.

Sahu, S., Fujita, D., Bandyopadhyay, A.; US patent 9019685B2: Sahu, S., Fujita, D., Bandyopadhyay, A. (2010) Inductor made of arrayed capacitors (2010) Japanese patent has been issued on 20th August 2015 JP-511630 (world patent filed, this is the invention of fourth circuit element), US patent has been issued 9019685B2, April 28, 2015.

Sainz, M. G., Halberg, F. (1966) Mitotic rhythms in human cancer, reevaluated by electronic computer programs—Evidence for chronopathology. *J Natl Cancer Inst* 37(3), 279–292.

Sakai, K., Hikosaka, O., Nakamura, H. (2004) Emergence of rhythm during motor learning. *Trends Cogn Sci* 8, 547–553.

Salthe, S. (1985) *Evolving Hierarchical Systems: Their Structure and Representation*. Columbia University Press, New York.

Salvioa, A., Strumiab, A. (2017) Agravity; https://arxiv.org/pdf/1403.4226.pdf.

Sammon, M. (1994) Geometry of respiratory phase switching. *J Appl Physiol* 77, 2468–2480.

Samsonovich, A., McNaughton, B. L. (1997) Path integration and cognitive mapping in a continuous attractor neural network model. *J Neurosci* 17, 5900–5920.

Sanchez-Vives, M. V., McCormick, D. A. (2000) Cellular and network mechanisms of rhythmic recurrent activity in neocortex. *Nat Neurosci* 3, 1027–1034.

Sanes, J. N., Donoghue, J. P. (1993) Oscillations in local field potentials of the primate motor cortex during voluntary movement. *Proc Natl Acad Sci USA* 90, 4470–4474.

Sanes, J. R., Lichtman, J. W. (1999) Can molecules explain long term potentiation? *Nat Neuro* 597, 2.

Santos, E. S. (1970) Fuzzy algorithms. *Information Control* 17(4), 326–339. doi:10.1016/S0019-9958(70)80032-8.

Sapoval, B., Gobron, T., Margolina, A. (1991) Vibrations of fractal drums. *Phys Rev Lett* 67, 2974.

Sara, S. J. (2000) Retrieval and reconsolidation: Toward a neurobiology of remembering. *Learn Mem* 7, 73–84.

Sardi, S. et al. (2017) New types of experiments reveal that a neuron functions as multiple independent threshold units. *Sci Rep* 7, 18036. doi:10.1038/s41598-017-18363-1.

Savageau, M. A. (1998) Development of fractal kinetic theory for enzyme-catalysed reactions and implications for the design of biochemical pathways. *Biosystems* 47(1–2), 9–36.

Saxena, K. et al. (2018) Wireless communication through microtubule analogue device: Noise driven machines in the biosystems. In: *Engineering Vibration, Communication and Information Processing*, K. Ray, S. Sharan, S. Rawat, S. K. Jain, S. Srivastava, A. Bandyopadhyay, Eds. Springer, Singapore, Vol. 478, pp. 735–749. doi:10.1007/978-981-13-1642-5_64.

Schaffer, C. (1994) A conservation law for generalization performance. In: Proceedings of the Eleventh International Conference on Machine Learning, W. W. Cohen, H. Hirsh, Eds. Kaufmann, M., San Francisco, CA, pp. 259–265.

Schahmann, J. D., Pandya, D. N. (2006) *Fiber Pathways of the Brain*. Oxford University Press, Oxford.

Scheibel, M. E., Scheibel, A. B. (1966) The organization of the nucleus reticularis thalami, a Golgi study. *Brain Res* 1, 43–62.

Schermelleh, L., Heintzmann, R., Leonhardt, H. (2010) A guide to super-resolution fluorescence microscopy. *J Cell Biol* 190, 165–175.

Schine, N. et al. (2016) Synthetic landau levels for photons. *Nature* 534, 671–675.

Schirber, M. (2010) Measuring the magnetism of light. *Phys Rev Focus* 26, 13.

Schirò, G., Cupane, A., Vitrano, E., Bruni, F. (2009) Dielectric relaxations in confined hydrated myoglobin. *J Phys Chem B* 113(28), 9606–9613.

Schlosshauer, M., Kofler, J., Zeilinger, A. (2013) A snapshot of foundational attitudes toward quantum mechanics. *Stud Hist Phil Mod Phys* 44, 222–230.

Schmidt, L. A., Trainor, L. J. (2001) Frontal brain electrical activity (EEG) distinguishes valence and intensity of musical emotions. *Cognition Emotion* 15(4), 487. doi:10.1080/02699930126048.

Schmitz, D. et al. (2001) Axo-axonal coupling: A novel mechanism for ultrafast neuronal communication. *Neuron* 31, 831–840.

Schnitzler, A., Gross, J. (2005) Normal and pathological oscillatory communication in the brain. *Nat Rev Neurosci* 6, 285–296.

Schnorr, C. P. (1978) Satisfiability is quasilinear complete in NQL. *J ACM* 25(1), 136–145.

Schubert, L. K. (1974) Iterated limiting recursion and the program minimization problem. *J ACM* 21(3), 436–445. doi:10.1145/321832.321841.

Scheuer, J., Orenstein, M. (1999) Optical vortices crystals: Spontaneous generation in nonlinear semiconductor microcavities. *Science* 285, 230–233.

Schurig, D. et al. (2006) Metamaterial electromagnetic cloak at microwave frequencies. *Science* 314, 977–980.

Schuster, P., Toro, N. (2013) On the theory of continuous-spin particles: Wavefunctions and soft-factor scattering amplitudes. http://arxiv.org/pdf/1302.1198v2.pdf.

Schuster, P., Toro, N. (2014) A CSP Field Theory with Helicity. https://arxiv.org/pdf/1404.0675v1.pdf.

Schwan, H. P. (1957) Electrical properties of tissue and cell suspensions. *Adv Bio Med Phys* 5, 147–209.

Schwartz, D. A., Howe, C. Q., Purves, D. (2003) The statistical structure of human speech sounds predicts musical universals. *J Neurosci* 23, 7160–7168.

Scott, L. L., Hage, T. A., Golding, N. L. (2007) Weak action potential back propagation is associated with high frequency axonal firing capability in principle neurons of the gerbil medial superior olive. *J Physiol* 583, 647–661.

Scully, M. O., Zubairy, M. S., Agarwal, G. S., Walther, H. (2003) Extracting work from a single heat bath via vanishing quantum coherence. *Science* 299(5608), 862–864.

Sederberg, P. B., Kahana, M. J., Howard, M. W., Donner, E. J., Madsen, J. R. (2003) Theta and gamma oscillations during encoding predict subsequent recall. *J Neurosci* 23, 10809–10814.

Sehgal, A., Rothenfluh-Hilfiker, A., Hunter-Ensor, M., Chen, Y., Myers, M. P., Young, M. W. (1995) Rhythmic expression of timeless: A basis for promoting circadian cycles in period gene autoregulation. *Science* 270, 808–810.

Seidenbecher, T., Laxmi, T. R., Stork, O., Pape, H. C. (2003) Amygdala and hippocampal theta rhythm synchronization during fear memory retrieval. *Science* 301, 846–850.

Self-taught software: http://www.technologyreview.com/fromthelabs/428910/self-taught-software/.

Sewell, R. J. et al. 2013, Certified quantum non-demolition measurement of a macroscopic material system. *Nat Photonics* 7, 517–520.

Shan, J. et al. (2001) Origin of magnetic field enhancement in the generation of terahertz radiation from semiconductor surfaces. *Opt Lett* 26, 849.

Shanker, O. (2006) Random matrices, generalized zeta functions and self-similarity of zero distributions. *J Phys A* 39, 13983–13997.

Shannon, R. V., Zeng, F., Kamath, G. V., Wygonski, J., Eke-lid, M. (1995) Speech recognition with primarily temporal cues. *Science* 270, 303–304.

Shapere, A., Wilczek, F. (2012) Classical time crystals. *Phys Rev Lett* 109, 160402.

Shelby, R. A., Smith, D. R., Schultz, S. (2001) Experimental verification of a negative index of refraction. *Science* 292, 77–79.

Sherman, S. M., Guillery, R. W. (2002) The role of the thalamus in the flow of information to the cortex. *Philos Trans R Soc Lond B Biol Sci* 357, 1695–1708.

Shevchenko, S. V., Tokarevsky, V. V. (2013) Space and Time. arXiv:1110.0003v3. 1–19.

Shevchuk, A. I. et al. (2006) Imaging proteins in membranes of living cells by high-resolution scanning ion conductance microscopy. *Angew Chem Int Ed Engl* 45(14), 2212–2216.

Shi, L., Li, Q. (2011) Synthesis and formation mechanism of helical single-crystalline CuInSe2 nanowires. *Cryst Eng Comm* 13, 7262–7266.

Shibata, M., Bures, J. (1974) Optimum topographical conditions for reverberating cortical spreading depression in rats. *J Neurobiol* 5, 107–118.

Shields, R. (1976) New view of the cell cycle. *Nature* 260, 193–194.

Shiino, M., Frankowicz, M. (1989) Synchronization of infinitely many coupled limit-cycle type oscillators. *Phys Lett A* 136, 103–108.

Shimony, A. (1955) Coherence and the axioms of confirmation. *J Symb Log* 20(1), 1–28.

Shlesinger, M. F. (1988) Fractal time in condensed matter. *Ann Rev Phys Chem* 39, 269–290.

Shoenfield, J. R. (2001) *Recursion Theory*, Lecture notes in logic 1, Association for Symbolic logic, Urbana, IL, A. K. Peters, Ltd., Natick, MA, viii+344pp.

Sia, P. D. (2013) Exciting peculiarities of the Planck scale physics. *J Phys* 442, 012068.

Siapas, A. G., Lubenov, E. V., Wilson, M. A. (2005) Prefrontal phase locking to hippocampal theta oscillations. *Neuron* 46, 141–151.

Siegelmann, H. T. (1995) Computation beyond the Turing limit. *Science* 268, 545–548.

Siegert, F., Weijer, C. J. (1995) Spiral and concentric waves organize multicellular *Dictyostelium* mounds. *Curr Biol* 5, 937–943.

Silva, A. et al. (2014) Performing mathematical operations with metamaterials. *Science* 343(6167), 160–163. doi:10.1126/science.1242818.

Simon, D. J. et al. (2016) Axon degeneration gated by retrograde activation of somatic pro-apoptotic signaling. *Cell* 164, 1031–1045.

Singer, W. (1994) Time as coding space in neocortical processing. In: *Temporal Coding in the Brain*, G. Buzsáki et al., Eds. Springer-Verlag, Berlin, Germany, pp. 51–80.

Singh, G. et al. (2014) Self-assembly of magnetite nanocubes into helical superstructures. *Science* 345(6201), 1149–1153. doi:10.1126/science.1254132.

Singh, P. et al. (2018a) Complete dielectric resonator model of a human brain from MRI data: A journey from connectome neural modeling to single protein. In: *Engineering Vibration, Communication and Information Processing (ICoEVCI)*; Lecture notes in electrical engineering, LNEE; Springer, Vol. 478, pp. 717–737. doi:10.1007/978-981-13-1642-5_63.

Singh, P. et al. (2018b) Fractal and periodical biological antennas: Hidden topologies in DNA, wasps and retina in the eye. In: *Soft Computing Applications*, K. Ray et al., Ed. Springer, Singapore, pp. 113–130.

Skaggs, W. E., McNaughton, B. L., Wilson, M. A., Barnes, C. A. (1996) Theta phase precession in hippocampal neuronal populations and the compression of temporal sequences. *Hippocampus* 6, 149–172.

Sklar, L. (1992) *Philosophy of Physics*. Oxford University Press, Oxford.

Small, H., Garfield, E. (1985) The geography of science: Disciplinary and national mappings. *J Info Sci* 11(1985), 147–159.

Smith, D. H. (2009) Stretch growth of integrated axon tracts: Extremes and exploitations. *Prog Neurobiol* 89(3), 231–239.

Smith, D. R., Pendry, J. B., Wiltshire, M. C. K. (2004) Metamaterials and negative refractive index. *Science* 305, 788.

Smith, H. O., Hutchison, C. A., 3rd, Pfannkoch, C., Venter, J. C. (2003) Generating a synthetic genome by whole genome assembly: PhiX174 Bacteriophage from synthetic oligonucleotides. *Proc Natl Acad Sci USA* 100(26), 15440–15445.

Smith, J. A., Martin, L. (1973) Do cells cycle? *Proc Natl Acad Sci USA* 70, 1263–1267.

Electron 24, 225.

Smoes, M.-L. (1976) Toward a mathematical description of phase waves. *J Theor Biol* 58, 1–14.

Smolin, L. (2004) Scientific Alternatives to the Anthropic Principle, arXiv:hep-th/0407213, pp. 1–43.

Smolyaninov, I. I. (2012) Metamaterial model of fractal time. *Phys Lett A* 376, 1315–1317.

Smorynski, C. (1985) *Self-reference and Model Logic*. Springer, Berlin, Germany.

Snyder, J. S., Large, E. W. (2005) Gamma-band activity reflects the metric structure of rhythmic tone sequences. *Cogn Brain Res* 24(1), 117–126.

Sonnleitner, A., Isacoff, E. (2003) Single ion channel imaging. *Methods Enzymol* 361, 304–319.

Sornette, D. et al. (2007) Algorithm for model validation: Theory and applications. *PNAS* 104, 6562–6567.

Soto-Andrade, J., Jaramillo, S., Gutierrez, C., Letelier, J. C. (2011) Ouroboros avatars: A mathematical exploration of self-reference and metabolic closure. In: *Advances in Artificial Life ECAL 2011: Proceedings of the Eleventh European Conference on the Synthesis and Simulation of Living Systems*, T. Lenaerts, M. Giacobini, H. Bersini, P. Bourgine, M. Dorigo, R. Doursat, Eds. The MIT Press, Cambridge, MA, pp. 763–770.

Spafford, E. H. (1992) Computer viruses—A form of artificial life? In: *Artificial Life II, Santa Fe Institute Studies in the Sciences of Complexity*, Volume X, C. G. Langton, C. Taylor, J. D. Farmer, S. Rasmussen, Eds. Addison-Wesley, Redwood City, CA, pp. 727–745.

Spanier, E. H. (1966) *Algebraic Topology*. McGraw-Hill, New York.

Spencer, W. A., Kandel, E. R. (1961) Electrophysiology of hippocampal neurons. *J Neurophysiol* 24, 272–285.

Spitzer, N. C., Sejnowski, T. J. (1997) Biological information processing: Bit of Progress. *Science* 277(5329), 1060–1061.

Srinivasan, R. (1999) Spatial structure of the human alpha rhythm: Global correlation in adults and local correlation in children. *Clin Neurophysiol* 110, 1351–1362.

Stam, C. J., Pijn, J. P. M., Suffczynski, P., Lopes da Silva, F. H. (1999) Dynamics of the human rhythm: Evidence for nonlinearity? *Clin Neurophysiol* 110, 1801–1813.

Stanley, S. (1985) *Evolving Hierarchical Systems*. Columbia University Press, New York.

Stanley, S. A. et al. (2016) Bidirectional electromagnetic control of the hypothalamus regulates feeding and metabolism. *Nature* 531, 647–650.

Staresina, B. P. et al. (2015) Hierarchical nesting of slow oscillations, spindles and ripples in the human hippocampus during sleep. *Nat Neurosci* 18, 1679–1686.

Steen, L. A. (1988) The science of patterns. *Science* 240(1988), 611–616.

Steenrod, N. E. (1962) *Cohomology Operations*. Princeton University Press, Princeton, NJ.

Steinbock, O. (1997) Excitation waves on cylindrical surfaces: Rotor competition and vortex drift. *Phys Rev Lett* 78, 745–748.

Stergachis, A. B. et al. (2013) Exonic transcription factor binding directs codon choice and affects protein evolution. *Science*, 342(6164), 1367–1372.

Steriade, M. (2001) To burst, or rather, not to burst. *Nat Neurosci* 4, 671.

Steriade, M., Buzsáki, G. (1990) Parallel activation of thalamic and cortical neurons by brainstem and basal forebrain cholinergic systems. In: *Brain Cholinergic Systems*, M. Steriade, D. Biesold, Eds. Oxford University Press, Oxford, pp. 3–64.

Steriade, M., Contreras, D., Curro Dossi, R., Nuñez, A. (1993a) The slow (<1 Hz) oscillation in reticular thalamic and thalamocortical neurons: Scenario of sleep rhythm generation in interacting thalamic and neocortical networks. *J Neurosci* 13, 3266–3283.

Steriade, M., Deschenes, M. (1984) The thalamus as a neuronal oscillator. *Brain Res Rev* 8, 1–63.

Steriade, M., Deschenes, M., Domich, L., Mulle, C. (1985) Abolition of spindle oscillations in thalamic neurons disconnected from nucleus reticularis thalami. *J Neurophysiol* 54, 1473–1497.

Steriade, M., Gloor, P., Llinás, R., Lopes da Silva, F. H., Mesulam, M. M. (1990) Basic mechanisms of cerebral rhythmic activity. *Electroencephalogr Clin Neurophysiol* 76, 481–508.

Steriade, M., Nuñez, A., Amzica, F. (1993b) Intracellular analysis of relations between the slow (<1 Hz) neocortical oscillation and other sleep rhythms in the electroencephalogram. *J Neurosci* 13, 3266–3283.

Sternberg, S. (1966) High speed scanning in human memory. *Science* 153, 652–654.

Stevens, C. J., Chan, C. W. T., Stamatis, K., Ewards, D. J. (2010) Magnetic metamaterials as 1D data transfer channels: An application for magneto-inductive channels. *IEEE Trans Micro Theory Tech* 58, 1248–1256.

Stockbridge, N., Stockbridge, L. L. (1988) Differential conduction at axonal bifurcations. I. Effect of electrotonic length. *J Neurophysiol* 59, 1277–1285.

Stöhr, J., Siegmann, H. C., Kashuna, A., Gamble, S. J. (2009) Magnetization switching without charge or spin currents. *Appl Phys Lett* 94, 072504.

Stoney, S. D., Jr (1990) Limitations on impulse conduction at the branch point of afferent axons in frog dorsal root ganglion. *Exp Brain Res* 80, 512–524.

Striegel, D. A., Hurdal, M. K. (2009) Chemically based mathematical model for development of cerebral cortical folding patterns. *PLoS Comput Biol* 5(9), e1000524.

Strogatz, S. H. (1990) Interpreting the human phase-response curve to bright light. *J Biol Rhy* 5, 169–174.

Strogatz, S. H. (2003) *Sync: The Emerging Science of Spontaneous Order.* Hyperion, New York.

Strogatz, S. H., Mirollo, R. E. (1988) Phase-locking and critical phenomena in lattices of coupled nonlinear oscillators with random intrinsic frequencies. *Phys D* 31, 143–168.

Strukov, D. B., Snider, G. S., Stewart, D. R., Williams, R. S. (2008) The missing memristor found. *Nature* 453(7191), 80–83.

Strumwasser, F. (1974) Neuronal principles organizing periodic behavior. In: *The Neurosciences Third Study Program,* F. O. Schmitt, F. G. Worden, Eds. Massachusetts Institute of Technology, Cambridge, MA, pp. 459–478.

Stuart, I. et al. (2015) Recovery of drug delivery nanoparticles from human plasma using an electrokinetic platform technology. *Small* 11(38), 5088–5096.

Stuchebrukhov, A. A. (2010) Long-distance electron tunneling in proteins: A new challenge for time-resolved spectroscopy. *Laser Phys* 20(1), 125–138.

Stulf, G., Pollet, T. V., Barret, L. (2015) The not-always-uniquely-predictive power of an evolutionary approach to understanding our not-so-computational nature. *Front Psychol.* doi:10.3389/fpsyg.2015.00419.

Suddendorf, T., Corballis, M. C. (1997) Mental time travel and the evolution of the human mind. *Genet Soc Gen Psychol Monogr* 123, 133–167.

Sudkamp, T. A. (2006) *Languages and Machines*, 3rd ed. Addison-Wesley, Boston, MA. Chapter 8.6: Multitape Machines: pp. 269–271.

Sun, D. (2016) Inverse spin Hall effect from pulsed spin current in organic semiconductors with tunable spin–orbit coupling. *Nat Mater* 5, 863–869.

Sun, S. et al. (2012) Gradient-index meta-surfaces as a bridge linking propagating waves and surface waves. *Nat Mater* 11, 426–431.

Sun, Z. et al. (2018) Imidazole derivatives as artificial water channel building-blocks: Structural design influence on water permeability. *Faraday Discuss* 209, 113–124.

Suprun, A. D., Atmazha, Y. B. (2002) Quantum excitation of protein α-spiral and the problem of protein functionality. *Funct Mater* 9(2), 624–630.

Suprun, A. D., Shmeleva, L. V. (2014) Alpha-helical regions of the protein molecule as organic nanotubes. *Nanoscale Res Lett* 9(1), 200–206.

Surowitz, E. J. (2011) Analytic general intelligence: Space + Time and Lowen Model. *Biol Inspir Cogn Arc* 233, 365–370.

Surwillo, W. (1961) The relationship of the alpha rhythm, reaction time and age. *Nature* 191, 823–824.

Sutton, P. S., Clarke, D., Morley, E. L., Robert, D. (2016) Mechanosensory hairs in bumblebees (*Bombus terrestris*) detect weak electric fields. *Proc Natl Acad Sci USA* 113(26), 7261–7265.

Swadlow, H. A. (2000) Information flow along neocortical axons. In: *Time and the Brain*, R. Miller, Ed. Harwood Academic Publishers, Singapore, pp. 131–155.

Swadlow, H. A., Kocsis, J. D., Waxman, S. G. (1980) Modulation of impulse conduction along the axonal tree. *Annu Rev Biophys Bioeng* 9, 143–179.

Swain, H. H., Valley, S. L. (1970) Critical fibrillation frequency. *Univ Mich Med Center J* 36, 122–126.

Swindale, N. V. (1982) A model for the formation of orientation columns. *Proc R Soc Lond B* 215, 211–230.

Taddei-Ferretti, C., Cordella, L. (1976) Modulation of Hydra attenuata rhythmic activity: Phase response curve. *J Exp Biol* 65, 737–751.

Tahara, S. et al. (2015) Ultrafast photoreaction dynamics of a light-driven sodium-ion-pumping retinal protein from Krokinobactereikastus revealed by femtosecond time-resolved absorption spectroscopy. *J Phys Chem Lett* 6(22), 4481–4486.

Takaiwa, D. et al. (2008) Phase diagram of water in carbon nanotubes. *Proc Natl Acad Sci USA* 105(1), 39–43. doi:10.1073/pnas.0707917105.

Tallinen, T. et al. (2016) On the growth and form of cortical convolutions. *Nat Phys* 12, 588–593.

Tallon-Baudry, C., Bertrand, O., Wienbruch, C., Ross, B., Pantev, C. (1997) Combined EEG and MEG recordings of visual 40 Hz responses to illusory triangles in human. *Neuroreport* 8, 1103–1107.

Taminiau, T. H., Stefani, F. D., van Hulst, N. F. (2011) Optical nanorod antennas modeled as cavities for dipolar emitters: Evolution of sub- and super-radiant modes. *Nano Lett* 11, 1020–1024.

Tan, Z. et al. (2017) Cryogenic 3D printing of super soft hydrogels. *Sci Rep* 7, Article number: 16293.

Tanaka, S. (1995) Topological analysis of point singularities in stimulus preference maps of the primary visual cortex. *Proc Roy Soc Lond B* 261, 81–88.

Tani, T. et al. (2018) Sound frequency representation in the auditory cortex of the common marmoset visualized using optical intrinsic signal imaging. *eNeuro* 5(2). doi:10.1523/ENEURO.0078-18.2018.

Tank, D. W., Hopfield, J. J. (1987, December). Collective computation in neuron-like circuits. *Sci Am* 257, 104–115.

Tass, P. (1999) *Phase Resetting in Medicine and Biology.* Springer-Verlag, Berlin, Germany.

Taylor, P., Hobbs, J. N., Burroni, J., Siegelmann, H. T. (2015) The global landscape of cognition hierarchical aggregation as an organization principle of human cortical networks and functions. *Sci Rep* 5, 18112. doi:10.1038/srep18112.

Teeter, M. M. (1984) Water structure of a hydrophobic protein at atomic resolution: Pentagon rings of water molecules in crystals of crambin. *Proc Natl Acad Sci USA* 81, 6014–6018.

Terrazas, A., Krause, M., Lipa, P., Gothard, K. M., Barnes, C. A., McNaughton, B. L. (2005) Selfmotion and the hippocampal spatial metric. *J Neurosci* 25, 8085–8096.

Thach, W. T. (1998) A role for the cerebellum in learning movement coordination. *Neurobiol Learn Mem* 70, 177–188.

Thompson, J. M. T. (1975) Experiments in catastrophe. *Nature* 254, 392–395.

Thomson, W. (1867) On vortex atoms. *Proc R Soc Edinb* 6, 94–105.

Thorpe, S. J. (1990) Spike arrival times: A highly efficient coding scheme for neural networks. In: *Parallel Processing in Neural Systems*, R. Eckmiller, G. Hartmann, G. Hauske, Eds. Elsevier, Amsterdam, the Netherlands, pp. 91–94.

Thouless, D. J. (1972) *The Quantum Mechanics of Many-body Systems*. Academic Press, New York.

Thouless, D. J. (1998) Gauge invariance and the Aharonov–Bohm effect. In: *Topological Quantum Numbers in Nonrelativistic Physics*. World Scientific, Singapore, pp. 18ff.

Tibbitt, M. W., Anseth, K. S. (2009) Hydrogels as extracellular matrix mimics for 3D cell culture. *Biotechnol Bioeng* 103(4), 655–663. doi:10.1002/bit.22361.

Timofeev, I. et al. (2002) Short- and medium-term plasticity associated with augmenting responses in cortical slabs and spindles in intact cortex of cats in vivo. *J Physiol* 542(2), 583–598.

Timofeev, I., Grenier, F., Bazhenov, M., Sejnowski, T. J., Steriade, M. (2000) Origin of slow cortical oscillations in deafferented cortical slabs. *Cereb Cortex* 10, 1185–1199.

Toffoli, T. (1984) Cellular automata as an alternative to (rather than an approximation of) differential equations in modeling physics. *Phys D Nonlin Phen* 10, 117–127.

Toffoli, T., Margolus, N. (1991) Programmable matter: Concepts and realization. *Physica D* 47, 263–272.

Toffoli, T., Margolus, N. (1993) Programmable matter: Concepts and realization. *Int J High Speed Comp* 5, 155.

Tomasch, W. J. (1966) Geometrical resonance and boundary effects in tunneling from superconducting in. *Phys Rev Lett* 16, 16–19.

Tomassy, G. S., Dershowitz, L. B., Arlotta, P. (2016) Diversity matters: A revised guide to myelination. *Trends Cell Biol* 26(2), 135–147.

Tomes, R. (1990) Several references herein http://ray.tomes.biz/maths.html.

Tong, W. et al. (2011) Self-replication of information-bearing nanoscale patterns. *Nature* 478(7368), 225–228.

Tononi, G., Boly, M., Massimini, M., Koch, C. (2016) Integrated information theory: From consciousness to its physical substrate. *Nat Rev Neurosci* 17(7), 450–461.

Toolan, M. (2009) *Language Teaching: Integrational Linguistic Approaches* (Routledge Advances in Communication and Linguistic Theory). Illustrated edition. Routledge London, UK.

Tramo, M. J. (2001) Biology and music: Enhanced: Music of the hemispheres. *Science* 291(5501), 54–56.

Traub, R. D., Wong, R. K. (1982) Cellular mechanism of neuronal synchronization in epilepsy. *Science* 216(4547), 745–747.

Traub, R. D. et al. (2005) Single-column thalamocortical network model exhibiting gamma oscillations, sleep spindles, and epileptogenic bursts. *J Neurophysiol* 93, 2194–2232.

Traub, R. D., Whittington, M. A., Stanford, I. M., Jefferys, J. G. (1996) A mechanism for generation of long-range synchronous fast oscillations in the cortex. *Nature* 383, 621–624.

Trudeau, M. C., Zagotta, W. N. (2002) An intersubunit interaction regulates trafficking of rod cyclic nucleotide-gated channels and is disrupted in an inherited form of blindness. *Neuron* 34, 197–207. doi:10.1016/S0896-6273(02)00647-5.

Trujillo, C. A. et al. (2018) Lab-grown "mini brains" produce electrical patterns that resemble those of premature babies; Nested oscillatory dynamics in cortical organoids model early human brain network development, Bio-arXive doi:10.1101/358622; News on this work, *Nature* 563, 453. https://www.nature.com/articles/d41586-018-07402-0.

Tsang, M., Psaltis, D. (2007) Magnifying perfect lens and superlens design by coordinate transformation; arXiv:0708.0262[physics.optics].

Tsang, W. (2017) *Fractal Brain Theory*. lulu.com. p. 530.

Tso, C. F. (2017) Astrocytes regulate daily rhythms in the suprachiasmatic nucleus and behavior. *Curr Biol* 27(7), 1055–1061.

Tukey, J. W. (1984) "Sequential conversion of continuous data to digital data," Bell Laboratories memorandum of 1 September 1947 marks the introduction of the term "bit" reprinted in *Origin of the Term Bit*, ed. H. S. Tropp (*Annals Hist Computing* 6, 152–155).

Tulub, A. A. (2004) Activation of tubulin assembly into microtubules upon a series of repeated femtosecond laser impulses. *J Chem Phys* 121, 11345.

Tuma, T. et al. (2016) Stochastics phase-change neurons. *Nat Nanotechnol* 11, 693–699.

Turing, A. (1939) Systems of logic based on ordinals. *Proc Lond Math Soc* 2–45(1), 161–228.

Turner, P. (2013) A review of scale relativity and fractal space-time. Research-gate. Retrieved 31 May, 2016. https://www.researchgate.net/publication/259373681_A_Review_of_Scale_Relativity_and_fractal_space-time.

Tuszynski, J. A. et al. (1985/1995) Ferroelectric behavior in microtubule dipole lattices: Implications for information processing, signaling and assembly/dis-assembly. *J Theo Biol* 174, 371–380.

Ueda, K., Nittono, H., Hayashi, M., Hori, T. (2001) Spatiotemporal changes of slow wave activities before and after 14 Hz/12 Hz sleep spindles during stage 2 sleep. *Psychiatry Clin Neurosci* 55, 183–184.

Uezo, A. et al. (2016) Identification of an elaborate complex mediating postsynaptic inhibition. *Science* 353(6304), 1123–1129.

Ulbrich, M. H., Isacoff, E. Y. (2007) Subunit counting in membrane-bound proteins. *Nat Methods* 4, 319–321.

Unruh, W. G., Zurek, W. H. (1989) Reduction of a wave packet in quantum Brownian motion. *Phys Rev D* 40, 1071–1094.

Usher, M., Stemmler, M., Koch, C., Olami, Z. (1994) Network amplification of local fluctuations causes high spike rate variability, fractal firing patterns and oscillatory local field potentials. *Neural Comput* 6, 795.

Usher, T. M., Levin, I., Daniels, J. E., Jones, J. L. (2015) Electric-field-induced local and mesoscale structural changes in polycrystalline dielectrics and ferroelectrics. *Sci Rep* 5, 14678. doi:10.1038/srep14678.

Uusitalo, M. A., Williamson, S. J., Seppa, M. T. (1996) Dynamical organisation of the human visual system revealed by lifetimes of activation traces. *Neurosci Lett* 213, 149–152.

Vadhan, S. (2010) The unified theory of pseudorandomness. *Proceedings of the International Congress of Mathematicians*, Hyderabad, India.

Valentine, J. et al. (2008) Three-dimensional optical metamaterial with a negative refractive index. *Nature* 455, 376–379.

Van der Pol, B., van der Mark, J. (1928) The heartbeat considered as a relaxation oscillation, and an electrical model of the heart. *Balth Phil Mag Suppl* 6, 763–775.

Van Essen, D. (1997) A tension-based theory of morphogenesis and compact wiring in the central nervous system. *Nature* 385, 313–318.

van Fraassen, B. C. (1970) *An Introduction to the Philosophy of Time and Space.* Originally published by Random House in 1970: Reprinted by Columbia University Press in 1985.

van Fraassen, B. C. (1984) Belief and the will. *J Philos* 81(5), 235–256.

van Fraassen, B. C. (1995) Belief and the problem of Ulysses and the sirens. *J Philos Stud* 77(1), 7–37.

Van Meerwijk, W. P. M, de Bruin, G., Ginneken, A. C., van Hartevelt, J., Jongsma, H. J., Scott, S. W. (1984) Phase resetting properties of cardiac pacemaker cells. *J Gen Physiol* 83, 613–629.

Van Stokum, W. J. (1937) The gravitational field of a distribution of particles rotating about an axis of symmetry. *Proc R Soc Edinb* 57, 135–154.

van Vreeswijk, C., Sompolinsky, H. (1996) Chaos in neuronal networks with balanced excitatory and inhibitory activity. *Science* 274(5293), 1724–1726.

VanRullen, R., Koch, C. (2003) Is perception discrete or continuous? *Trends Cogn Sci* 7, 207–213.

Varela, F., Lachaux, J. P., Rodriguez, E., Martinerie, J. (2001) The brain web: Phase synchronization and large-scale integration. *Nat Rev Neurosci* 2, 229–239.

Varela, F. J., Maturana, H. R., Uribe, R. (1974) Autopoiesis: The organization of living systems, its characterization and a model. *BioSystems* 5, 187–196.

Vaughan, H. G. Jr., Arezzo, J. C. (1988) The neural basis of event-related potentials. In: *Human Event-Related Potentials*, T. W. Picton, Ed. vol 3. Elsevier Science Publishers, New York, pp. 45–94.

Veretennikov, I. et al. (2002) Mechanism for helical gel formation from evaporation of colloidal solutions. *Langmuir* 1823, 8792–8798.

Verma, R., Daya, K. S. (2017) Rapid detection of pM concentration of insulin, using Microwave Whispering Gallery Mode. *IEEE Sensors J* 17(9), 2758–2765.

Vertes, R. P., Kocsis, B. (1997) Brainstem-diencephalo-septohippocampal systems controlling the theta rhythm of the hippocampus. *Neuroscience* 81, 893–926.

Vicsek, T. (1992) *Fractal Growth Phenomena*, 2nd ed. World Scientific, Singapore.

Victor, J. D. (1999) Temporal aspects of neuronal coding in the retina and lateral geniculate. *Network Comput Neural Syst* 10, 1–66.

Vida, I., Bartos, M., Jonas, P. (2006) Shunting inhibition improves robustness of gamma oscillations in hippocampal interneuron networks by homogenizing firing rates. *Neuron* 49, 107–117.

Vignolini, S. et al. (2010) Magnetic imaging in photonic crystal microcavities. *Phys Rev Lett* 105, 123902.

Vilenkin, A. (1981) Cosmic strings. *Phys Rev D* 24, 2082–2089.

Vilenkin, A. (1982) Creation of universes from nothing. *Phys Lett* B117.

Vilenkin, A. (1984) Formation and evolution of cosmic strings. *Phys Rev D* 30, 2036–2045.

Vilenkin, A. (1985) Cosmic strings and domain walls. *Phys Rep* 121, 263–315.

Vilenkin, A. (1986) Looking for cosmic strings. *Nature* 322, 613–614.

Vinogradova, O. S. (1995) Expression, control, and probable functional significance of the neuronal theta-rhythm. *Prog Neurobiol* 45, 523–583.

Vinson, M., Pertsov, A. M. (1999) Dynamics of scroll rings in a parameter gradient. *Phys Rev E* 59, 2764–2771.

Vinson, M., Pertsov, A. M., Jalife, J. (1993) Anchoring of vortex filaments in 3D excitable media. *Physica D* 72, 119–134.

Vitiello, G. (2012) Fractals, dissipation and coherent states. In: *Quantum Interaction*, Lecture Notes in Computer Science; Springer, Berlin, Germany, pp. 68–79.

Vlad, M. O., Popa, V. T., Segal, E., Ross, J. (2005) Multiple rate-determining steps for nonideal and fractal kinetics. *J Phys Chem B* 109(6), 2455–2460.

Vollmer, F., Braun, D., Libchaber, A. (2002) Protein detection by optical shift of a resonant microcavity. *Appl Phys Lett* 80, 4057.

Von Stockar, U., Liu, J. S. (1999) Does microbial life always feed on negative entropy? Thermodynamic analysis of microbial growth. *BioChem BioPhys Acta*, 1412, 191–211.

von Weizsacker, C. (1939) Der zweite Haupsatz und der Unterschied von der Vergangenheit und Zukunft, *Annalen der Physik* (5 Folge) 36, 275–283.

von Weizsacker, C. (1980) *The Unity of Nature.* Farrar Straus Giroux, New York.

Vulevic, B., Correia, J. J. (1997) Thermodynamic and structural analysis of microtubule assembly: The role of GTP hydrolysis. *Biophys J* 72, 1357–1375.

Wagh, A. G., Rakhecha, V. C. (1990) Geometric phase in neutron interferometry. *Phys Lett A* 148, 17–19.

Wallenstein, G. V., Eichenbaum, H., Hasselmo, M. E. (1998) The hippocampus as an associator of discontinuous events. *Trends Neurosci* 21, 317–323.

Wallenstein, G. V., Hasselmo, M. E. (1997) GABAergic modulation of hippocampal population activity: Sequence learning, place field development, and the phase precession effect. *J Neurophysiol* 78, 393–408.

Wang, J., Wang, B., Guo, H. (2007) Quantum inductance and negative electrochemical capacitance at finite frequency in a two-plate quantum capacitor. *Phys Rev B* 75, 155336.

Wang, L. et al. (2014a) Reversible near-infrared light directed reflection in a self-organized helical superstructure loaded with up-conversion nanoparticles. *J Am Chem Soc* 136(12), 4480–4483. doi:10.1021/ja500933h.

Wang, P. et al. (2014b) Harnessing buckling to design tunable locally resonant acoustic metamaterials. *Phys Rev Lett* 113, 014301.

Wang, X.-J. (1994) Multiple dynamical modes of thalamic relay neurons: Rhythmic bursting and intermittent phase-locking. *Neuroscience* 59, 21–31.

Wang, X. J., Rinzel, J. (1993) Spindle rhythmicity in the reticularis thalamic nucleus: Synchronization among mutually inhibitory neurons. *Neuroscience* 53, 899–904.

Wang, Y. et al. (2004) Deformation mechanisms of electrostrictive graft elastomers. *Smart Mater Struct* 13, 1407–1413. doi:10.1088/0964-1726/13/6/011.

Wang, Y., Gundevia, M. (2013) Measurement of thermal conductivity and heat pipe effect in hydrophilic and hydrophobic carbon papers. *Int J Heat Mass Transfer* 60, 134–142.

Warlimont, R. (1993) Factorisatio numerorum with constraints. *J Number Theory* 45, 186–199.

Waschke, C. et al. (1993) Coherent submillimeter-wave emission from Bloch oscillations in a semiconductor superlattice. *Phys Rev Lett* 70, 3319.

Watanabe, H., Oshikawa, M. (2015) Absence of Quantum time crystal. *Phys Rev Lett* 114, 251603.

Watrous, J., Aaronson, S. (2009) Closed time like curves make quantum and classical computing equivalent. *Proc R Soc A* 465(2102), 631.

Watters, P. A. (1998) Fractal structure in the electroencephalogram. Complex Int 5. http://www.complexity.org.au/ci/v0105/watters/watters.html.

Watts, D. J. (2003) *Six Degrees: The Science of a Connected Age*. W.W. Norton and Company, New York.

Watts, D., Strogatz, S. (1998) Collective dynamics of "small-world" networks. *Nature* 393, 440.

Wegner, P. (1997) Why interaction is more powerful than algorithms. *Commun ACM* 40, 80–91.

Wehner, M. et al. (2016) An integrated design and fabrication strategy for entirely soft, autonomous robots. *Nature* 536, 451–455.

Wehr, G. M., Davidowitz, H. (1996a) Temporal representations of odors in an olfactory network. *J Neurosci* 16, 3837–3847.

Wehr, M., Laurent, G. (1996b) Odour encoding by temporal sequences of firing in oscillating neural assemblies. *Nature* 384, 162–166.

Weich, T. et al. (2014) Formation and interaction of resonance chains in the open 3-disk system. *New J Phys* 16, 033029.

Weidmann, J. (1980) *Linear Operators in Hilbert Spaces*. Springer–Verlag, New York, p. 412.

Weil, A. (1951) De la metaphysique aux mathematiques. *Sciences*, pp. 52–56; reprinted in: *Ouevres Scientifiques: Collected Works*, A. Weil, Ed. Vol. 2, 1951–1964. Springer, New York, 1979, pp. 408–412.

Weinberg, S. (1999) *A Designer Universe?* New York Review of Books.

Welch, K. (2010) A fractal topology of the time: Implications for consciousness and cosmology. Doctorate thesis, California Institute of Integral Studies. p. 202.

Welsh, B. J., Gomatam, J. (1984) A trial phase function method for a class of lambda-omega reaction-diffusion systems: Reduction to a Schrödinger-type problem in an eigen sub-domain. *Adv Appl Math* 5, 333–355.

Welsh, D. K., Reppert, S. M. (1996) Gap junctions couple astrocytes but not neurons in dissociated cultures of rat suprachiasmatic nucleus. *Brain Res* 706, 30–36.

Werner, G. (2010) Fractals in the nervous system: Conceptual implications for theoretical. *Neurosci Front Physiol* 1, 15.

Wessel, R. (1995) In vitro study of resetting and phase locking in a time comparison circuit in the electric fish. *Eigenmannia Biophys J* 69, 1880–1890.

Westerfield, M., Joyner, R. W., Moore, J. W. (1978) Temperature-sensitive conduction failure at axon branch points. *J Neurophysiol* 41, 1–8.

Weyl, H. (1949) *Philosophy of Mathematics and Natural Science*. Princeton University Press, Princeton, NJ.

Wheeler, J. (1996) *At Home in the Universe*. AIP Press, Woodbury, NY.

Wheeler, J. A. (1957) On the nature of quantum geometro dynamics. *Ann Phys* 2(1957), 604–614.

Wheeler, J. A. (1967/1968) Superspace and the nature of quantum geometro dynamics. In: *Battelle Rencontres: 1967 Lectures in Mathematics and Physics*, C. M. DeWitt, J. A. Wheeler, Eds. Benjamin, New York.

Wheeler, J. A. (1968) Superspace and the nature of quantum geometrodynamics. In: Battelle Rencontres: 1967 Lectures in Mathematics and Physics, C. M. DeWitt, J. A. Wheeler, Eds. Benjamin, New York, pp. 242–307.

Wheeler, J. A. (1976/1977) Include the observer in the wave function? *Fundamenta Scientiae: Seminaire sur les fondements des sciences* (Strasbourg) 25, 9–35; reprinted in *Quantum Mechanics a Half Century Later*, Eds. J. Leite Lopes, M. Paty (Reidel, Dordrecht) pp. 1–18.

Wheeler, J. A. (1977) Genesis and observer ship. In: *Foundational Problems in the Special Sciences*, R. Butts, J. Hintikka, Eds. Reidel, Dordrecht, the Netherlands, pp. 1–33.

Wheeler, J. A. (1979) Pregeometry: Motivations and prospects. In: *Quantum Theory and Gravitation*, proceedings of a symposium held at Loyola University, New Orleans, May 23–26, A. R. Marlow, Ed. Academic, New York, 1980, pp. 1–11.

Wheeler, J. A. (1980) Law without law. In: *Structure in Science and Art*, P. Medawar, J. Shelley, Eds. Elsevier North-Holland, New York and Excerpta Medica, Amsterdam, pp. 132–154.

Wheeler, J. A. (1981) Not consciousness but the distinction between the probe and the probed as central to the elemental quantum act of observation. In: *The Role of Consciousness in the Physical World*, R. G. Jahn, Ed. Westview, Boulder, pp. 87–111.

Wheeler, J. A. (1983a) On recognizing law without law. *Am J Phys* 51(1983), 398–404.

Wheeler, J. A. (1983b) Elementary quantum phenomenon as building unit. In: *Quantum Optics, Experimental Gravitation, and Measurement Theory*, P. Meystre, M. Scully, Eds. Plenum, New York and London, pp. 141–143.

Wheeler, J. A. (1984) Bits, quanta, meaning. In: *Problems in Theoretical Physics*, A. Giovannini, F. Mancini, M. Marinaro, Eds. University of Salerno Press, Salerno, pp. 121–141; also in Theoretical Physics Meeting: Atti del Convegno, Amalfi, 6-7 maggio 1983 (Edizioni Scientific heItaliane, Naples, 1984) pp. 121–134; also in Festschrift in Honor of Eduardo R. Caianiello, eds. A. Giovannini, F. Mancini, M. Marinaro, A. Rimini (World Scientific, Singapore, 1989) pp. 133–154.

Wheeler, J. A. (1985) Bohr's "phenomenon" and "law without law." In: *Chaotic Behavior in Quantum Systems*, G. Casati, Ed. Plenum, New York, pp. 363–378.

Wheeler, J. A. (1986) Physics as meaning circuit: Three problems. In: *Frontiers of Non BquiKlibrium Statistical Physics*, G. T. Moore, M. O. Scully, Eds. Plenum, New York, pp. 25–32.

Wheeler, J. A. (1988) World as system self-synthesized by quantum networking. *IBM J Res Dev* 32, 4–15; reprinted, in *Probability in the Sciences*, ed. E. Agazzi (Kluwer, Amsterdam) pp. 103–129.

Wheeler, J. A. (1990) Information, physics, quantum: The search for links. In: *Complexity, Entropy, and the Physics of Information*, W. H. Zurek, Ed. Addison-Wesley, Redwood City, CA. https://jawarchive.files.wordpress.com/2012/03/informationquantumphysics.pdf.

White, J. A., Rubinstein, J. T., Kay, A. R. (2000) Channel noise in neurons. *Trends Neurosci* 23, 131–137.

Wilczek, F. (2012) Quantum time crystal. *Phys Rev Lett* 109, 160401. Physicists predict the existence of time crystal. February 16, 2016.

Williams, R. W., Herrup, K. (1988) The control of neuron number. *Annu Rev Neurosci* 11, 423–453.

Willig, K. I., Harke, B., Hell, S. W. (2007) STED microscopy with continuous wave beams. *Nat Methods* 4, 915–918.

Willke, P. et al. (2019) Magnetic resonance imaging of single atoms on a surface. *Nat Phys* 15, 1005–1010.

Winfree, A. (1977a) *Biological Rhythm Research* 8, 1; *The Geometry of Biological Time*, 2nd ed. Springer, New York, 2001.

Winfree, A. T. (1970) The temporal morphology of a biological dock. In: *Lectures on Mathematics in the Life Sciences*, M. Gerstenhaber, Ed. American Mathematical Society, Providence, RI.

Winfree, A. T. (1971) Corkscrews and singularities in fruitflies: Resetting behavior of the circadian edosion rhythm. In: *Biochronometry*, M. Menaker, Ed. National Academy of Sciences of the USA, Washington, DC, pp. 81–109.

Winfree, A. T. (1973) Time and tirnelessness in biological docks. In: *Temporal Aspects of Therapeutics*, J. Urquardt, F. E. Yates, Eds. Plenum, New York, pp. 35–57.

Winfree, A. T. (1974) Patterns of phase compromise in biological cycles. *J Math Biol* 1, 73–95.

Winfree, A. T. (1975) Un clock like behavior of a biological clock. *Nature* 253, 315–319.

Winfree, A. T. (1976) On phase resetting in multicellular clock-shops. In: *The Molecular Basis of Circadian Rhythms*, J. W. Hastings, H. G. Schweiger, Eds. Abakon, Berlin, Germany, pp. 109–129.

Winfree, A. T. (1977b) Phase control of neural pacemakers. *Science* 197, 761–762.

Winfree, A. T. (1980) *The Geometry of Biological Time*. Springer-Verlag, New York.

Winfree, A. T. (1980) Singularities in regeneration? *Adv Appl Probability* 12(3), 554–555.

Winfree, A. T. (1983) Sudden cardiac death—a problem in topology? *Sci Am* 248(5), 144–161.

Winfree, A. T. (1984) Discontinuities and singularities in the timing of nuclear division. In: *Cell Cycle Clocks*, L. N. Edmunds, Ed. Dekker, New York.

Winfree, A. T. (1986a) Filaments of nothingness. *Sciences* 26, 20–27.

Winfree, A. T. (1986b) The vulnerable phase and reentrant waves in 3D. *Proc Int Union Physiol Sci* 16, 287.

Winfree, A. T. (1987) *When Time Breaks down: The Three-Dimensional Dynamics of Electrochemical Waves and Cardiac Arrhythmias*. Princeton University Press, Princeton, NJ.

Winfree, A. T. (1990) Vortices in motionless media. *Appl Mech Rev* 43, 297–309.

Winfree, A. T. (1991) Resetting the human dock. *Nature* 350, 18.

Winfree, A. T. (1994) Persistent tangled vortex rings in generic excitable media. *Nature* 371, 233–236.

Winfree, A. T. (1998) Evolving perspectives during twelve years of electrical turbulence. *Chaos* 8, 1–19.

Winfree, A. T. (2000) Various ways to make phase singularities by electric shock. *J Cardiovasc Electrophysiol* 11, 286–289.

Winfree, A. T., Strogatz, S. (1984) Singular filaments organize chemical waves in three dimensions. 4: Wave taxonomy. *Physica D* 13, 221–233.

Winfree, A. T., Strogatz, S. H. (1983a) Singular filaments organize chemical waves in three dimensions: Geometrically simple waves. *Physica D* 8, 35–49.

Winfree, A. T., Strogatz, S. H. (1983b) Singular filaments organize chemical waves in three dimensions: 2. Twisted waves. *Physica D* 9, 65–80.

Winfree, A. T., Strogatz, S. H. (1983c) Singular filaments organize chemical waves in three dimensions: 3. Knotted waves. *Physica D* 9, 333–345.

Wing, A. M. (2002) Voluntary timing and brain function: An information processing approach. *Brain Cogn* 48, 7–30.

Wolf, M. (1989) Multifractality of prime numbers. *Physica A* 160, 24–42.

Wolf, M. (1997) 1/f noise in the distribution of prime numbers. *Physica A* 241, 493–499.

Wolfram, S. (2002) *A New Kind of Science*. Wolfram Media, Champaign, IL.

Wolfram, T. (1968) Tomasch oscillations in the density of states of superconducting films, *Phys Rev* 170, 481–490.

Wolpert, D. (2017) Constraints on physical reality arising from a formalization of knowledge. arXiv 1711, 03499.

Wolpert, D. H. (1996) The lack of a priori distinctions between learning algorithms. *Neural Comput* 8(7), 1341–1390.

Wong, J. Y. et al. (2012) Materials by design: Merging proteins and music. *Nano Today* 7, 488–495.

Wood, W. B., Kershaw, D. (1991) Handed asymmetry, handedness reversal and mechanisms of cell fate determination in nematode embryos. *Ciba Found Symp* 162, 143–159; discussion 159–164.

Woods, M. P., Silva, R., Oppenheim, J. (2016) Autonomous quantum machines and finite sized clocks. *Quantum Physics*. arXiv:1607.04591.

Woodward, J. (1992) Realism about Laws. *Erkenntnis* 36, 181–218.

Wootters, W. K. (1980) The acquisition of information from quantum measurements, PhD dissertation, University of Texas at Austin.

Wootters, W. K. (1981) Statistical distribution and Hilbert space. *Phys Rev* 23, 357–362.

Worrell, G. A. et al. (2008) High-frequency oscillations in human temporal lobe: Simultaneous microwire and clinical macroelectrode recordings. *Brain* 131(Pt 4), 928–937. doi:10.1093/brain/awn006.

Wu, S. Q., Cai, X. (1999) Maxwell-Boltzmann, Bose-Einstein and Fermi-Dirac statistical entropies in a D-dimensional stationary axi-symmetry space-time arXiv:gr-qc/9907054.

Wu, Z., Wang, H. W., Mu, W., Ouyang, Z., Nogales, E. et al. (2009) Simulations of tubulin sheet polymers as possible structural intermediates in microtubule assembly. *PLoS One* 4(10), e7291.

Xiang, Y. et al. (2017) Fusion of regionally specified hPSC-derived organoids models human brain development and interneuron migration. *Cell Stem Cell* 21(3), 383–398.

Xie, Y. et al. (2016) Acoustic holographic rendering with two-dimensional metamaterial-based passive phased array. *Sci Rep* 6, 35437.

Xiujuan, W. et al. (2003) Effect of sound wave on the synthesis of nucleic acid and protein in chrysanthemum. *Colloids Surfaces B* 29(1–2), 99–102.

Xu, K., Zhong, G., Zhuang, X. (2013) Actin, spectrin and associated proteins form a periodic cytoskeleton structure in axons. *Science* 339, 452–456.

Xue, G. et al. (2010) Greater neural pattern similarity across repetitions is associated with better memory. *Science* 330, 97–101.

Yamamura, H., Suzuki, Y., Imaizumi, Y. (2015) New light on ion channel imaging by total internal reflection fluorescence (TIRF) microscopy. *J Pharmacol Sci* 128, 1–7.

Yamanishi, J., Kawato, M., Suzuki, R. (1979) Studies on human finger tapping neural networks by phase transition curves. *Biol Cybern* 33, 199–208.

Yamins, D. L. K., DiCarlo, J. J. (2016) Using goal-driven deep learning models to understand sensory cortex. *Nat Neurosci* 19, 356–365.

Yan, J. F., Yan, A. K., Yan, B. C. (1991) Prime numbers and the amino acid code: Analogy in coding properties. *J Theor Biol* 151(3), 333–341. doi:10.1016/S0022-5193(05)80382-0.

Yang, Y. et al. (2016) A metasurface carpet cloak for electromagnetic, acoustic and water waves. *Sci Rep* 6, 20219.

Yang, Y., Wang, J.-Z. (2017) From structure to behavior in basolateral Amygdala-Hippocampus circuits. *Front Neural Circuits*. doi:10.3389/fncir.2017.00086.

Yao, N. Y., Potter, A. C., Potirniche, I. D., Vishwanath, A. (2017) Discrete time crystals: Rigidity, criticality, and realizations. *Phys Rev Lett* 118, 030401.

Yates, F. E. Ed. (1987) *Self-Organizing Systems: The Emergence of Order*. Plenum, New York.

Yener, G. G., Basar, E. (2012) *Brain Oscillations as Biomarkers in Neuropsychiatric Disorders: Following an Interactive Panel Discussion and Synopsis*, 1st ed. Elsevier, the Netherlands, pp. 343–363.

Yi, C. et al. (2016) Circadian rhythm of genes in the brain changes with aging. *Proc Natl Acad Sci* 113(1), 206–211.

Yoo, S. et al. (2014) Photothermal inhibition of neural activity with near-infrared-sensitive nanotransducers. *ACS Nano* 8, 8040–8049.

Young, M. W. (2000) Marking time for a Kingdom. *Science* 288, 451–453.

Yu, N. et al. (2011) Light propagation with phase discontinuities: Generalized laws of reflection and refraction. *Science* 334, 333–337.

Zaghetto, A. A. (2012) Musical visual vernacular. How the deaf people translate the sound vibrations into the sign language: An example from Italy. *Signata Ann Semiotics* 3, 273–298.

Zaitsevzotov, S. V. (1993) Temperature-independent non-linear conductivity of O-TAS3 thin samples—Quantum creep of the charge density waves. *Synth Met* 56, 2708–2713.

Zaromytidou, A.-I. (2012) Synchronizing actin and microtubules for axonal branching. *Nat Cell Biol* 14, 792.

Zatorre, R. J., Halpern, A. R. (2005) Mental concerts: Musical imagery and auditory cortex. *Neuron* 47, 9–12.

Zatorre, R. J., Perry, D. W., Beckett, C. A., Westbury, C. F., Evans, A. C. (1998) Functional anatomy of musical processing in listeners with absolute pitch and relative pitch. *Proc Natl Acad Sci* 95(6), 3172–3177.

Zeh, H. D. (1989) *The Physical Basis of the Direction of Time*. Springer-Verlag, Berlin, Germany.

Zeh, H. D. (1992) *The Physical Basis of the Direction of Time*, 2nd ed. Springer-Verlag, Berlin, Germany.

Zeki, S., Shipp, S. (1988) The functional logic of cortical connections. *Nature* 335, 311–317.

Zelano, C. et al. (2016) Nasal respiration entrains human limbic oscillations and modulates cognitive function. *J Neurosci* 36(49), 12448–12467.

Zeng, L. et al. (2017) Photonic time crystals. *Sci Rep* 7, 17165.

Zeng, W. Z., Glass, L., Shrier, A. (1992) The topology of phase response curves induced by single and paired stimuli in spontaneously oscillating chick heart cell aggregates. *J Bio Rhy* 7, 89–104.

Zermelo, E. (1896) Ueber Einen Satze der Dynamik und die mechanische Warmtheorie, Annalen der Physik, Series 3, 57, 485; English trans, in Brush (1966), 208.

Zhabotinsky, A. M. (1968) A study of self-oscillatory chemical reaction ill: Space behavior. In: *Biological and Biochemical Oscillators*, B. Chance, E. K. Pye, A. K. Ghosh, B. Hess, Eds. Academic Press, New York.

Zhai, Z., Kusko, C., Hakim, N., Sridhar, S., Revcolevschi, A., Vietkine, A. (2000) Precision microwave dielectric and magnetic susceptibility measurements of correlated electronic materials using superconducting cavities, *Rev Sci Instrum* 71(2000), 3151–3160.

Zhang, J. et al. (2009) Integrated device for optical stimulation and spatiotemporal electrical recording of neural activity in light sensitized brain tissue. *J Neural Eng* 6, 055007.

Zhang, J. et al. (2017) Observation of discrete time crystal. *Nature* 543, 217–220.

Zhang, K., Sejnowski, T. J. (2000) A universal scaling law between gray matter and white matter of cerebral cortex. *Proc Natl Acad Sci USA* 97, 5621–5626.

Zhang, S. et al. (2008) Cloaking of matter waves. *Phys Rev Lett* 100, 123002.

Zhang, X. et al. (2010) Self-assembly of pH-switchable spiral tubes: Supramolecular chemical springs. *Small* 6(2), 217–220.

Zhang, Y. et al. (2017) A semisynthetic organism engineered for the stable expansion of the genetic alphabet. *Proc Natl Acad Sci* 114(6), 1317–1322.

Zhao, K., Bruinsma, R., Mason, T. G. (2012) Local chiral symmetry breaking in triatic liquid crystals. *Nat commun* 3, 801.

Zhao, M.-Q. et al. (2014) Emerging double helical nanostructures. *Nanoscale* 6, 9339–9354.

Zhao, Y., Alu, A. (2011) Manipulating light polarization with ultra-thin plasmonic metasurfaces. *Phys Rev B* 84, 205428.

Zheng, J., Varnum, M. D., Zagotta, W. N. (2003) Disruption of an intersubunit interaction underlies Ca^{2+}-calmodulin modulation of cyclic nucleotide-gated channels. *J Neurosci* 23, 8167–8175.

Zhigulin, V. P. (2004) Dynamical motifs: Building blocks of complex dynamics in sparsely connected random networks. *Phys Rev Lett* 92(23), 238701.

Zhong, G. et al. (2014) Developmental mechanism of the periodic membrane skeleton in axons. *eLife* 3. This paper is STORM based evidence.

Zhou, W. et al. (2015) Detection of single ion channel activity with carbon nanotubes. *Sci Rep* 5(9208), 1–9.

Zhou, X.-Q. et al. (2013) Calculating unknown eigen values with a quantum algorithm. *Nat Photonics* 7, 223–228.

Zhu, J. et al. (2015) Three-dimensional magnetic cloak working from d.c. to 250 kHz. *Nat Comm* 6(8931).

Zipser, D., Andersen, R. A. (1988) A back-propagation programmed network that simulates response properties of a subset of posterior parietal neurons. *Nature* 331, 679–684.

Zugaro, M. B., Monconduit, L., Buzsáki, G. (2005) Spike phase precession persists after transient intra-hippocampal perturbation. *Nat Neurosci* 8, 67–71.

Zuo, S.-Y., Wei, Q., Cheng, Y., Liu, X.-J. (2017) Mathematical operations for acoustic signals based on layered labyrinthine metasurfaces. *Appl Phys Lett* 110, 011904. doi:10.1063/1.4973705.

Zurek, W. H. (1981) Pointer basis of quantum apparatus: Into what mixture does the wave packet collapse? *Phys Rev D* 24(1981), 1516–1525.

Zurek, W. H. (1982) Environment-induced superselection rules. *Phys Rev D* 26(1982), 1862–1880.

Zurek, W. H. (1983) Information transfer in quantum measurements: Irreversibility and amplification. In: *Quantum Optics, Experimental Gravitation and Measurement Theory*, P. Meystre, M. O. Scully, Eds. Plenum, New York, pp. 87–116.

Zurek, W. H. (1985) Cosmological experiments in liquid helium. *Nature* 317, 505–508.

Zurek, W. H. (1989) Thermodynamic cost of computation: Algorithmic complexity and the information metric. *Nature* 34(1989), 119–124.

Zurek, W. H., Thome, K. S. (1985) Statistical mechanical origin of the entropy of a rotating, charged black hole. *Phys Rev Lett* 20(1985), 2171–2175.

Zwicker, D. (2017) Growth and division of active droplets provides a model for protocells. *Nat Phys* 13, 408–413.

Zwicker, D. et al. (2016) Growth and division of active droplets: A model for protocells. *Nat Phys* 13, 408–413.

Zwirn, C. H., Delahaye, J.-P. (2013) Unpredictability and Computational Irreducibility. In: *Irreducibility and Computational Equivalence, Emergence, Complexity and Computation*, H. Zenil, Ed., Vol. 2. Springer, Berlin, Germany, pp. 273–295.

Zykov, V. S. (1986) Cycloidal circulation of spiral waves in an excitable medium. *Biofizika* 31, 862–865.

Zykov, V. S., Morozova, O. L. (1990) Computer simulation and investigation of spiral autowave stability. *J Nonlin Biol* 1(2), 127–159.

Index

Note: Page numbers in italic refer to figures.